Lecture Notes in Mathematics

Edited by A. Dold and B. Eckmann

975

Radical Banach Algebras and Automatic Continuity

Proceedings of a Conference Held at California State University, Long Beach, July 17 – 31, 1981

Edited by J. M. Bachar, W. G. Bade, P. C. Curtis Jr., H. G. Dales, and M. P. Thomas

Springer-Verlag
Berlin Heidelberg New York 1983

Editors

John M. Bachar
Department of Mathematics
California State University, Long Beach
Long Beach, CA 90840, USA

William G. Bade
Department of Mathematics,
University of California, Berkeley
Berkeley, CA 94720, USA

Philip C. Curtis
Department of Mathematics,
University of California, Los Angeles
Los Angeles, CA 90024, USA

H. Garth Dales
School of Mathematics, University of Leeds
Leeds LS2 9JT, England

Marc P. Thomas
Department of Mathematics
California State College, Bakersfield
Bakersfield, CA 93309, USA

AMS Subject Classifications (1980): 03 E 50, 04 A 30, 13-02, 13 A 10, 13 G 05, 13 J 05, 46-02, 46-06, 46 H 05, 46 H 10, 46 H 15, 46 H 20, 46 J 05, 46 J 10, 46 J 15, 46 J 20, 46 J 25, 46 J 30, 46 J 35

ISBN 3-540-11985-X Springer-Verlag Berlin Heidelberg New York
ISBN 0-387-11985-X Springer-Verlag New York Heidelberg Berlin

Printing and binding: Beltz Offsetdruck, Hemsbach/Bergstr.
2146/3140-543210

DEDICATION

This volume is dedicated to Charles Rickart, one of the first to consider questions of automatic continuity. His contributions to the theory of Banach algebras underlie many of the problems considered here.

PREFACE

This volume contains the contributions to the Conference on Radical Banach Algebras and Automatic Continuity, held at the California State University, Long Beach, July 13 - 17, 1981, and the following study period from July 18 - 31. The purpose of the conference was to present recent developments in these two areas and to explore the connections between them.

The articles given here represent expanded versions of conference talks, together with solutions of various problems that were presented and discussed. The papers contain, in varying degrees, historical background, syntheses and expository accounts, and the development of new ideas and results. Further details of the papers are given in the Introduction. The volume concludes with a list of unsolved problems.

The editors, who also served as the organizing committee, wish to thank the adminstration of the California State University, Long Beach, and particularly President Stephen Horn, for generous financial support, and for the excellent working conditions that were provided for the Conference. We are also grateful for additional financial support from the National Science Foundation, and for the travel grant from the North Atlantic Treaty Organization which enabled E. Albrecht, G. R. Allan, H. G. Dales, and M. Neumann to come to California to discuss their work.

Finally we wish to thank Elaine Barth, who typed the entire manuscript of this volume with its many corrections. Through her patience and outstanding skill she has made a major contribution to this endeavor.

The editors

John Bachar, Long Beach
William Bade, Berkeley
Philip Curtis, Los Angeles
Garth Dales, Leeds
Marc Thomas, Bakersfield

TABLE OF CONTENTS

Table of Contents - continued

PARTICIPANTS

E. Albrecht (University of Saarland, Saarbrücken, West Germany)

G. R. Allan (Cambridge University, Cambridge, England)

R. Arens (UCLA, Los Angeles, CA)

B. Aupetit (University of Laval, Quebec, Canada)

J. M. Bachar, Jr. (California State University Long Beach, Long Beach, CA)

G. F. Bachelis (Wayne State University, Detroit, MI)

W. G. Bade (UC Berkeley, Berkeley, CA)

T. G. Cho (Sogang University, Seoul, Korea)

P. C. Curtis, Jr. (UCLA, Los Angeles, CA)

H. G. Dales (University of Leeds, Leeds, England)

Y. Domar (University of Uppsala, Uppsala, Sweden)

J. Esterle (University of Bordeaux, Talence, France)

T. Gamelin (UCLA, Los Angeles, CA)

S. Grabiner (Pomona College, Claremont, CA)

N. Grønbaek (University of Copenhagen, Copenhagen, Denmark)

M. J. Hoffman (California State University, Los Angeles, Los Angeles, CA)

D. L. Johnson (Hughes Aircraft, Culver City, CA)

H. Kamowitz (University of Massachusetts, Boston, MA)

J. Koliha (University of Melbourne, Melbourne, Australia)

K. B. Laursen (University of Copenhagen, Copenhagen, Denmark)

J. A. Lindberg, Jr. (Syracuse University, Syracuse, NY)

R. J. Loy (Australian National University, Canberra, Australia)

G. H. Meisters (University of Nebraska-Lincoln, Lincoln, NE)

M. Neumann (University of Essen, Essen, West Germany)

B. Rentzsch (TRW, Redondo Beach, CA)

C. Rickart (Yale University, New Haven, CT)

M. P. Thomas (California State College Bakersfield, Bakersfield, CA)

J. C. Tripp (Southeast Missouri State University, Cape Girardeau, MO)

N. N. P. Viet (UC Berkeley, Berkeley, CA)

S. Walsh (UC Berkeley, Berkeley, CA)

G. A. Willis (University of New South Wales, Kensington, New South Wales, Australia)

W. Zame (SUNY, Buffalo, NY)

INTRODUCTION

The basic problem of automatic continuity theory is to give algebraic conditions which ensure that a linear operator between, say, two Banach spaces is necessarily continuous. This problem is of particular interest in the case of a homomorphism between two Banach algebras. Other automatic continuity questions arise in the study of derivations from Banach algebras to suitable modules and in the study of translation invariant functionals on function spaces.

There is a fundamental connection between questions of automatic continuity and the structure of radical Banach algebras. For example, the recent construction of a discontinuous homomorphism from $C(X)$, the algebra of all continuous, complex-valued functions on an infinite compact space X, depends on important structural properties of certain commutative radical Banach algebras.

The 30 papers in this volume present the latest developments in these two theories and explore the connections between them. Section I is devoted to the general theory of commutative radical Banach algebras. In [E.1], Jean Esterle gives a comprehensive classification of commutative radical Banach algebras based on the types of semigroups which these algebras contain. Esterle shows in this paper precisely which commutative radical Banach algebras R with unit adjoined can serve as the range of a discontinuous homomorphism from $C(X)$. Assuming the continuum hypothesis, such a discontinuous homomorphism ν from $C(X)$ into $R \oplus \mathbb{C}e$ exists if and only if R contains a rational semigroup over $\mathbb{Q}^+$. An equivalent condition is that R contains a non-nilpotent element of finite closed descent. This paper also contains some new short proofs of earlier theorems of the author, e.g., that each epimorphism from $C(X)$ onto a Banach algebra is automatically continuous.

In [E.2] Esterle investigates the question of whether or not a commutative radical Banach algebra must contain a non-trivial closed ideal. Substantial partial results are obtained on this fundamental open problem, which in turn is related to the invariant subspace problem for Banach spaces. Improvements of some recent results on invariant subspaces are given as well.

Section II is concerned with particular examples of radical Banach algebras. In [Da.1], H. G. Dales gives a survey of radical convolution algebras on the line and half line. The algebra $L^1(\omega)$ on $\mathbb{R}^+$, where ω is a rapidly decreasing weight function, has been much studied in recent years. Particular interest has centered on the problem of determining for which radical weights, ω, every closed ideal of $L^1(\omega)$ is a standard ideal, that is, an ideal consisting of those functions with support in an interval $[\alpha,\infty)$. In [Do.1], Y. Domar gives the first results on this problem, showing that for a wide class of radical weights ω on $\mathbb{R}^+$ each closed ideal is indeed standard. In [Do.2], Domar gives the final details of a solution to the problem of when spectral analysis holds for the analogous Beurling algebras on $\mathbb{R}$.

If ω is a rapidly decreasing weight sequence on $\mathbb{Z}^+$, then $\ell^1(\omega)$ is a radical Banach algebra of power series. At present there are no known examples of weight sequences for which $\ell^1(\omega)$ contains non-standard ideals. A construction by N. K. Nikolskii around 1970 of a weight sequence with this property has been shown to be incorrect. The problem of characterizing those weight sequences for which each ideal is standard motivates several of the papers in Section II. The multiplier algebras for the power series algebras $\ell^1(\omega)$ are discussed in [Bad] and [La]. Certain inequalities which may be relevant to the closed ideal problem are discussed in [Al.1] and [W.1]. In [Th], M. Thomas describes a class of weights for which each closed ideal in $\ell^1(\omega)$ is standard; this is the first new result of this type in about 10 years.

The papers in Section III concern the automatic continuity of homomorphisms and derivations. In [Au], B. Aupetit gives a new proof of the well known theorem of B. Johnson that a semi-simple Banach algebra necessarily has a unique Banach algebra topology. This proof involves techniques from the theory of subharmonic functions, and the ideas should be applicable to other automatic continuity problems. Uniqueness of norm for nonsemisimple algebras is considered by R. J. Loy in [Lo.1]. Automatic continuity for local algebras is surveyed by E. Albrecht and M. Neumann in [A-N 1,2].

The problem of whether or not a homomorphism between C^* algebras must be continuous has resisted solution for some time. Partial results on this question and related problems are given by Albrecht and Dales in [A-Da]. Automatic continuity questions for derivations on group algebras are discussed in [W.2] by G. Willis. The more difficult problems here involve nonamenable groups, particularly the free group on two generators.

The automatic continuity of certain linear functionals on Banach algebras is discussed in Section IV. Translation invariant functionals are surveyed by G. Meisters in [M]. The most intractable problem which remains in this area is whether or not each translation invariant linear functional on the space of continuous functions on $\mathbb{R}$ with compact support is necessarily continuous. This question is discussed by R. J. Loy in [Lo.2]. The theory of positive linear functionals and traces is considered in [Da.2].

The volume concludes with a list of open problems, some well known and others posed and discussed at the conference.

I. GENERAL THEORY OF RADICAL BANACH ALGEBRAS

ELEMENTS FOR A CLASSIFICATION OF COMMUTATIVE RADICAL BANACH ALGEBRAS

J. Esterle

1. Introduction

The purpose of this paper is the investigation of the relationships between some natural algebraic and topological properties for commutative radical Banach algebras. We shall see that a lot of apparently unrelated properties are in fact equivalent, or stronger or weaker. This enables us to give a classification of infinite-dimensional, commutative radical Banach algebras into nine classes. The first one, the class of such algebras $\mathcal{R}$ for which the set of principal ideals of $\mathcal{R} \oplus \mathbb{C}e$ is not linearly ordered by inclusion, is the biggest. In fact it is so big that it contains all infinite-dimensional, commutative radical Banach algebras. Then the classes get smaller and smaller, and the ninth one is the class of commutative radical Banach algebras which possess a nonzero analytic semigroup $(a^t)_{\mathrm{Re}\, t>0}$ over the right-hand half-plane such that $\sup\limits_{\substack{0<|t|\leq 1\\ \mathrm{Re}\, t>0}} \|a^t\| < +\infty$. This ninth class does not contain all commutative radical Banach algebras with bounded approximate identities (it is shown in Section 7 that $L^1(\mathbb{R}^+, e^{-t^2})$ does not belong to the ninth class). This suggests that we introduce a tenth one, the class of commutative radical Banach algebras which contain a nonzero analytic semigroup $(a^t)_{\mathrm{Re}\, t>0}$ over the right-hand half-plane such that $\sup\limits_{\mathrm{Re}\, t>0} \|a^t\| < +\infty$, but this tenth class is unfortunately empty (see Section 7). This fact is an easy consequence of the Ahlfors-Heins theorem for bounded analytic functions over the half-plane.

The three main themes of our investigations are (i) partial approximate identities, (ii) semigroups, and (iii) factorization or division properties. By a partial approximate identity (p.a.i.) in a Banach algebra $\mathcal{R}$ we mean a sequence (e_n) of elements of $\mathcal{R}$ such that $b = \lim\limits_{n\to\infty} e_n b$ for some nonzero $b \in \mathcal{R}$. Many radical Banach algebras do not have any p.a.i.; for example, radical Banach algebras of power series, or trivial radical Banach algebras in which the product of any two elements equals zero. But the non-existence of p.a.i.'s in a radical Banach algebra $\mathcal{R}$ implies a nice algebraic property. We show in Section 3 that every nonzero element of such an algebra $\mathcal{R}$ is equal to a finite product of irreducible elements of $\mathcal{R}$. This fact is of course obvious in the examples mentioned above, but is not immediate in the case of the algebras $\mathcal{R}_4$, $\mathcal{R}_5$ and $\mathcal{R}_6$ of Section 7.

There are several natural ways to try to get a p.a.i. The nicest ones would be to find a nonzero continuous semigroup $(a^t)_{t>0}$, or a nonzero continuous semigroup $(a^t)_{t>0}$ bounded at the origin, or a nonzero analytic semigroup $(a^t)_{\mathrm{Re}\,t>0}$ over the right-hand half-plane, etc. Another natural way would be to look for a nonzero element a and a sequence (λ_n) of nonzero complex numbers which converges to zero such that $a(a+\lambda_n e)^{-1} z \xrightarrow[n\to\infty]{} z$ for some nonzero element z. (If a were invertible, which is a nonsense in a radical algebra, then $a(a+\lambda_n e)^{-1}$ would converge to the unit element which radical algebras fail to have.)

We show in Section 3 that it is possible to get such a sequence (λ_n) and such a z if and only if there exists in the algebra an a-divisible subspace, a useful condition in automatic continuity theory. (See for example Thomas' work [38].)

A special class of p.a.i.s is given by p.a.i.s of the form (ae_n). Such a p.a.i. exists if and only if $z \in [z\,a\mathcal{R}]^-$ for some nonzero $z \in \mathcal{R}$, and we show in Section 3 that this condition holds also if and only if there exists in $\mathcal{R}$ an a-divisible subspace.

Allan's theory of elements of finite closed descent [2] is also obviously related to p.a.i.s. A nonnilpotent element b of $\mathcal{R}$ is said to be of finite closed descent if and only if $b^n \in [b^{n+1}\mathcal{R}]^-$ for some $n \in \mathbb{N}$. In other terms such an element exists if and only if $a \in [a^2\mathcal{R}]^-$ for some nonzero $a \in \mathcal{R}$, the same as above with $z = a$ (here just take $a = b^n$). It was shown by the author in [23] that if this condition is satisfied then $\mathcal{R}$ does possess nonzero rational and real semigroups. We prove in Section 5 a converse. If $\mathcal{R}$ possesses a nonzero rational semigroup then $\mathcal{R}$ possesses nonnilpotent elements of finite closed descent (Theorem 5.3). The class of these algebras (Class V in the classification given in Section 9) is exactly the class of commutative radical Banach algebras $\mathcal{R}$ such that a discontinuous homomorphism from $C(K)$ into $\mathcal{R} \oplus \mathbb{C}e$ can be constructed.

The details of the classification are given in Section 9. We now give a brief outline of the paper. In Section 2 we give definitions, notations and elementary properties of modules over Banach algebras. For example, if $\mathfrak{m}$ is a Banach module over a Banach algebra A and (a_n) a sequence of elements of A, we define the set $\varprojlim a_n \dots a_1 A$ (which is a subset of $\bigcap_{n\in\mathbb{N}} a_n \dots a_1 A$). Also we recall some basic facts concerning the weighted algebras $\ell^1(S,\omega)$ where S is a monoid and ω a weight over S.

The only really new result in this section is the construction of a bounded approximate identity of the form $u(u+\lambda e)^{-1}$ (where $\lambda \to 0^+$) in commutative, separable Banach algebras with b.a.i.

In Section 3 we investigate p.a.i.s for a Banach module $\mathfrak{m}$ over a radical Banach algebra $\mathcal{R}$. This leads to two results, Corollary 3.5 and Theorem 3.7, which give respectively algebraic conditions equivalent to the existence of a p.a.i. or to the existence of a p.a.i. of the form ae_n. Theorem 3.7 has a consequence concerning the hyperinvariant subspace problem. If a quasinilpotent operator u acting on a Banach space E has no nontrivial closed hyperinvariant subspace then for every subset T of $\mathbb{C} - \{0\}$ such that 0 is a limit point of T there exists a dense subspace F_T of E such that $\inf_{\lambda \in T} \|u(u + \lambda e)^{-1}x - x\| = 0$ for every $x \in F_T$. The key result of Section 3 is the fact that, if a sequence (x_n) of elements of a Banach module $\mathfrak{M}$ over a radical Banach algebra $\mathcal{R}$ satisfies $x_n \in \mathcal{R}x_{n+1}$ for every $n \in \mathbb{N}$, then there exists $x \in (\bigcup_n \mathcal{R}x_n)^-$ such that $[\mathcal{R}x]^- = [\bigcup_n \mathcal{R}x_n]^-$. The proof of this result, which is based upon the category theorem, uses the trivial fact that every element of $\mathcal{R}$ is a limit of a sequence of invertible elements of $\mathcal{R} \oplus \mathbb{C}e$.

In Section 4 we show that, if $(a^t)_{t>0}$ is any real semigroup in a separable Banach algebra A, then there exists a sequence (t_n) of positive reals such that $t_n \xrightarrow[n\to\infty]{} 0$ and $a^{t+t_n} \xrightarrow[n\to\infty]{} a^t$ for every $t > 0$. As far as the author is aware, all results of Sections 3 and 4 are new.

In Section 5 we study commutative radical Banach algebras which possess non-nilpotent elements of finite closed descent. The lengthy computations of [23] show that these algebras possess nonzero rational, and even real, semigroups. In fact, if the continuum hypothesis is assumed, then, for every fully ordered abelian group G having the power of the continuum, there exists a one-to-one homomorphism from G^+ into the multiplicative set of nonzero elements of the algebra (this fact leads to the existence of discontinuous homomorphisms from $C(K)$ into $\mathcal{R} \oplus \mathbb{C}e$). We give in Section 5 a very short proof of the existence of nonzero rational semigroups $(a^t)_{t\in\mathbb{Q}^+}$, where $[a^t\mathcal{R}]^-$ does not depend on t, in such an algebra $\mathcal{R}$. The proof is based upon the Mittag-Leffler theorem on inverse limits (see Section 2). Zouakia has recently obtained in his thesis [41] a much shorter proof of the author's construction of [23]. This proof is analogous to Zouakia's formulation, given in this Volume [40], of the author's theory of Cohen elements, and it also uses the Mittag-Leffler theorem.

As mentioned above, we prove conversely in Section 5 that if a commutative radical Banach algebra possesses a nonzero rational semigroup or, more generally, a sequence (a_n) with $a_1 \neq 0$, $a_n \in a_{n+1}^2 \mathcal{R}$ for every $n \in \mathbb{N}$, then $\mathcal{R}$ possesses elements of finite closed descent. This gives a complete characterization of commutative radical Banach algebras $\mathcal{R}$ such that discontinuous homomorphisms from $C(K)$ into $\mathcal{R} \oplus \mathbb{C}e$ can be constructed. These results apply in particular to the weighted algebra $\ell^1(\mathbb{Q}^+, e^{-t^2})$. It is perhaps surprising to observe that from some points of view the situation concerning radical Banach algebras is better than the situation concerning semisimple ones. The semisimple algebra $\ell^1(\mathbb{Q}^+)$ possesses, of course, a nonzero rational semigroup but it does not possess either any nonzero real semigroup or any nonzero element of finite closed descent. (This is proved at the end of Section 6 by using Bohr's theory of almost periodic functions [8].)

In Section 6 we show that, if a Banach algebra A possesses a nonzero continuous semigroup $(a^t)_{t>0}$ over the positive reals, then A possesses another nonzero semigroup $(b^t)_{t>0}$ over the positive reals which is infinitely differentiable over $(0,\infty)$ and which satisfies $[b^tA]^- = [b^{t'}A]^- = [\bigcup_{s>0} a^sA]^-$ for every $t,t'>0$. Getting an infinitely differentiable semigroup from a continuous one is easy, but it is more difficult to obtain the condition $[b^tA]^- = [b^{t'}A]^-$ for every $t,t' \neq 0$ when the given semigroup $(a^t)_{t>0}$ is unbounded at the origin. (The condition on $[b^tA]^-$ implies in particular that b^t is of finite closed descent.) This result is, in fact, true for each Banach algebra A, radical or not. Roughly speaking, the problem consists in finding a function $f \in L^1(\mathbb{R}^+)$ decreasing to zero extremely fast (without vanishing over $[0,\alpha]$ a.e. for any $\alpha > 0$) such that the Laplace transform of f is an outer function on the right-hand half-plane. This is done using some lemmas about convex functions which seem to be related to Mandelbrojt's theory of the regularization of functions (but I was not able to find any reference). All the work of Section 6 was obtained in collaboration with Paul Koosis during the Spring of 1979 and has never been published.

Section 7 is devoted to counterexamples. Ten commutative radical Banach algebras are constructed to show that most of the results of Sections 3, 5 and 6 giving equivalent conditions for commutative radical Banach algebras are essentially best possible. For example, if $\mathcal{R}$ is an integral domain and if some nonzero element of $\mathcal{R}$ possesses roots of all orders, then $\mathcal{R}$ possesses a nonzero rational semigroup (see Section 2), hence a nonzero real semigroup, hence elements of finite closed descent, etc. In Section 7 we produce a commutative radical Banach algebra (which is not an integral domain) having a nonzero element with roots of all orders, but which does not even possess any nontrivial partial approximate identity.

Also we construct in Section 7 a continuum $(\mathcal{R}_\eta)_{0<\eta\leq\pi/2}$ of commutative radical Banach algebras such that $\mathcal{R}_\eta$ possesses a nonzero analytic semigroup $(a^t)_{t\in U_\eta}$ in the open angle $U_\eta = \{z \in \mathbb{C} - \{0\} \mid 0 < |\text{Arg } z| < \eta\}$, but does not possess any nonzero analytic semigroup in a larger open angle. The proof uses computations of inverse Laplace transforms and results of [21].

In Section 8 we give very short proofs of results about rates of decrease of sequences of powers previously proved by the author in [22] and [25] using unpleasant computations of infinite products. The result about prime ideals in radical Banach algebras with p.a.i. [22] (which was the key of the proof of continuity of epimorphisms from $C(K)$) is proved in Section 8 in a few lines using a very nice trick of Bouloussa [10]. Also we extend the result of [22], which asserts that $\mathbb{C}$ is the only Banach algebra which is a valuation ring. Using results of Section 3 we show that if the set of principal ideals of a commutative Banach algebra A with unit forms a chain, then A is isomorphic with $\mathbb{C}[X]/X^p\mathbb{C}[X]$ for some positive integer p (so that A is finite-dimensional).

All these results lead to the classification given in Section 9. We conclude Section 9 by a list of open questions. In particular I was not able to produce a commutative radical Banach algebra with p.a.i. which does not possess any p.a.i. of the form (ae_n). If A has a bounded p.a.i., such an a must exist. This follows from Grønbaek's factorization theorem [28].

The main open question is the following. Does the existence of a nonzero rational semigroup in a commutative radical Banach algebra $\mathfrak{R}$ imply the existence of a nonzero continuous semigroup $(a^t)_{t>0}$ over the positive reals in $\mathfrak{R}$? I was not even able to decide whether or not the algebra $\ell^1(\mathbb{Q}^+, e^{-t^2})$ possesses such a continuous real semigroup. A positive answer to that question is suggested by the fact that radical Banach algebras having nonzero rational semigroups possess in fact nonzero real semigroups $(b^t)_{t>0}$ where $b^t = \lim_{n\to\infty} b^{t+t_n}$ for every $t > 0$, (t_n) being a suitable sequence of positive reals which converges to zero.

I must thank B. Chevreau, John Bachar, P. C. Curtis, H. G. Dales, S. Grabiner, N. Grønbaek, Paul Koosis and A. M. Sinclair for several valuable discussions (the discussions with Paul Koosis were so valuable that the credit for the results of Section 6 belongs mostly to him). H. G. Dales read this paper in detail, and his careful criticism has enabled me to improve the redaction. I am grateful to him for this and for other suggestions.

2. Definitions, preliminaries

We consider only algebras over the field of complex numbers. Recall that, if A is a complex algebra, the Jacobson radical Rad(A) of A is the intersection of the kernels of all strictly irreducible representations of A. If A is a Banach algebra then Rad(A) is the union of all left ideals of A all elements of which are quasinilpotent. In the commutative case Rad(A) is just the set of all quasinilpotent elements of A.

A left A-module over a complex algebra A is a pair $(\varphi, \mathfrak{M})$, where $\mathfrak{M}$ is a complex linear space and φ is an algebra homomorphism from A into $L(\mathfrak{M})$, the algebra of all linear maps from $\mathfrak{M}$ into itself. We will usually write $a \cdot x$ instead of $\varphi(a)[x]$ if $a \in A$, $x \in \mathfrak{M}$. If A is a Banach algebra, a left A-module $\mathfrak{M}$ is called a Banach A-module if $\mathfrak{M}$ is a Banach space and if φ is a continuous map from A into $\mathcal{L}(\mathfrak{M})$, the algebra of all bounded linear maps from A into itself.

If A is a nonunital complex algebra we denote by $A^{\#}$ the algebra obtained by adding a unit e to A, and if A is unital we just put $A^{\#} = A$. If A is a nonunital Banach algebra, then, if we put $\|a + \lambda e\| = \|a\| + |\lambda|$ $(x \in A, \lambda \in \mathbb{C})$, we obtain a Banach algebra norm on $A^{\#}$. If $\mathfrak{M}$ is a Banach A-module put $(a + \lambda e)\cdot x = a\cdot x + \lambda x$ $(a \in A, x \in \mathfrak{M}, \lambda \in \mathbb{C})$. There exists a constant $K \geq 1$ such that $\|a \cdot x\| \leq K\|a\|\|x\|$ $(a \in A,\ x \in \mathfrak{m})$ and we obtain $\|(a+\lambda e)\cdot x\| \leq K[\|a\| + |\lambda|]\|x\|$. So $\mathfrak{m}$ becomes a Banach $A^{\#}$-module.

The following well-known argument shows that we may always renorm $\mathfrak{m}$ to get $K = 1$ in the above inequalities.

Put $p(x) = \sup\limits_{\substack{b \in A^{\#} \\ \|b\| \leq 1}} \|b \cdot x\|$.

Then $\|x\| \leq p(x) \leq K\|x\|$, so that p is equivalent to the given norm. Also, if $\|b'\| \leq 1$, then

$$p(b'x) = \sup_{\|b\| \leq 1} p(b'bx) \leq p(x) \qquad (x \in \mathfrak{m}),$$

and $p(bx) \leq \|b\|p(x)$ $(x \in \mathfrak{m},\ b \in A)$. We will always assume that $K = 1$. In other terms we will always suppose that the map $\varphi : A \to \mathcal{L}(\mathfrak{m})$ is a contraction.

DEFINITION 2.1. Let A be a complex algebra, let $\mathfrak{m}$ be an A-module and let $(\alpha_n)_{n\in\mathbb{N}}$ be a sequence of elements of A. We denote by $\underleftarrow{\mathrm{Lim}}\ \alpha_1 \dots \alpha_n \mathfrak{m}$ the set of all elements $X = (x_n)_{n\in\mathbb{N}}$ of $\mathfrak{m}^{\mathbb{N}}$ such that $x_n = \alpha_n x_{n+1}$ for every $n \in \mathbb{N}$. We denote by $\underleftarrow{\lim}\ \alpha_1 \dots \alpha_n \mathfrak{m}$ the set of all elements x of $\mathfrak{m}$ such that $x = x_1$ for some element $(x_n)_{n\in\mathbb{N}}$ of $\underleftarrow{\mathrm{Lim}}\ \alpha_1 \dots \alpha_n \mathfrak{m}$.

In other terms $\underleftarrow{\mathrm{Lim}}\ \alpha_1 \dots \alpha_n \mathfrak{m}$ is the inverse limit of the projective system (E_n, θ_n), where $E_n = \mathfrak{m}$ and $\theta_n : E_{n+1} \to E_n$ is the map $x \mapsto \alpha_n \cdot x$ for every $n \in \mathbb{N}$. Also $\underleftarrow{\lim}\ \alpha_1 \dots \alpha_n \mathfrak{m}$ is the image of $\underleftarrow{\mathrm{Lim}}\ \alpha_1 \dots \alpha_n \mathfrak{m}$ by the first projection.

REMARK 2.2. Assume that $a \cdot x \neq 0$ for every nonzero $a \in A$ and every nonzero $x \in \mathfrak{m}$. Then $\bigcap\limits_{n\in\mathbb{N}} \alpha_1 \dots \alpha_n \mathfrak{m} = \underleftarrow{\lim}\ \alpha_1 \dots \alpha_n \mathfrak{m}$.

<u>Proof</u>. The set $\underleftarrow{\lim}\ \alpha_1 \dots \alpha_n \mathfrak{m}$ is always contained in $\bigcap\limits_{n\in\mathbb{N}} \alpha_1 \dots \alpha_n \mathfrak{m}$. Now consider $x \in \bigcap\limits_{n\in\mathbb{N}} \alpha_1 \dots \alpha_n \mathfrak{m}$. There exists a sequence $(x_n)_{n\geq 2}$ of elements of $\mathfrak{m}$ such that $x = \alpha_1 \dots \alpha_n x_{n+1}$ $(n \in \mathbb{N})$. Then $\alpha_1 \dots \alpha_{n-1}(x_n - \alpha_n x_{n+1}) = 0$. If $\alpha_1 \dots \alpha_{n-1} = 0$ for some $n \geq 2$, then trivially $\bigcap\limits_{n\in\mathbb{N}} \alpha_1 \dots \alpha_n \mathfrak{m} = \{0\} = \underleftarrow{\lim}\ \alpha_1 \dots \alpha_n \mathfrak{m}$. If not, we obtain $x_n = \alpha_n x_{n+1}$ $(n \geq 2)$ and $x = \alpha_1 x_2$, so that $(x, x_2, \dots, x_n, \dots) \in \underleftarrow{\mathrm{Lim}}\ \alpha_1 \dots \alpha_n \mathfrak{m}$ and $x \in \underleftarrow{\lim}\ \alpha_1 \dots \alpha_n \mathfrak{m}$, the desired result.

REMARK 2.3. There exists a sequence (α_n) of elements of A such that $\underleftarrow{\lim}\ \alpha_1 \dots \alpha_n \mathfrak{m} \neq \{0\}$ if and only if there exists a sequence (x_n) of elements of

A such that $x_1 \neq 0$, and $x_n \in Ax_{n+1}$ for every $n \in \mathbb{N}$.

Proof. Let $(x_n) \in \mathfrak{m}^{\mathbb{N}}$ and $(\alpha_n) \in A^{\mathbb{N}}$. Then $(x_n) \in \underleftarrow{\mathrm{Lim}}\, \alpha_1 \ldots \alpha_n \mathfrak{m}$ if and only if $x_n = \alpha_n x_{n+1}$ for every $n \in \mathbb{N}$. So if x is given there exists a sequence (x_n) satisfying $x_1 = x$, $x_n \in Ax_{n+1}$ $(n \in \mathbb{N})$ if and only if $x \in \underleftarrow{\lim}\, \alpha_1 \ldots \alpha_n \mathfrak{m}$ for some sequence (α_n) of elements of A.

DEFINITION 2.4. Let A be a complex algebra, and let $\mathfrak{m}$ be a left-A-module which is also an algebra. We define a product on $A \times \mathfrak{m}$ by the formula $(a,x)(b,y) = (ab, ay + bx + xy)$ $(a,b \in A, x,y \in \mathfrak{m})$. We denote by $A \oplus \mathfrak{m}$ the linear space $A \times \mathfrak{m}$ equipped with this product, and we identify $(a,0)$ with a $(a \in A)$ and $(0,x)$ with x $(x \in \mathfrak{m})$.

PROPOSITION 2.5. If A and $\mathfrak{m}$ are commutative, $A \oplus \mathfrak{m}$ is a complex commutative algebra. If A and $\mathfrak{m}$ are commutative Banach algebras and if $\mathfrak{m}$ is a Banach A-module, then $A \oplus \mathfrak{m}$ is a Banach algebra with respect to the norm $\|(a,x)\| = \|a\| + \|x\|$, and $A \oplus \mathfrak{m}$ is radical if both A and $\mathfrak{m}$ are radical.

Proof. A routine computation shows that the product is associative if A and $\mathfrak{m}$ are commutative (this may fail if commutativity of A and $\mathfrak{m}$ is not assumed). The other assertions are clear.

DEFINITION 2.6. Let A be a Banach algebra, let $\mathfrak{m}$ be a Banach A-module, and let x be a nonzero element of $\mathfrak{m}$. We will say that A possesses partial approximate identities for x (x - p.a.i.) if and only if $x \in [Ax]^-$. If A is a Banach algebra we will say that A has a nontrivial partial approximate identity (p.a.i.) if and only if $a \in [Aa]^-$ for some nonzero $a \in A$.

One of the main themes of this paper is the study of nice p.a.i. in radical Banach algebras. The author thinks that the nicest possible p.a.i. is given by an analytic semigroup $(a^t)_{\mathrm{Re}\, t > 0}$ such that $\sup\limits_{\substack{\mathrm{Re}\, t > 0 \\ |t| \leq 1}} \|a^t\| < +\infty$. Interesting p.a.i. are given by continuous bounded semigroups over the positive reals, or continuous semigroups over the positive reals unbounded at the origin, etc. The following result gives another direction in the research of p.a.i.

THEOREM 2.7. Let A be a Banach algebra. If A possesses a two-sided bounded approximate identity, then for every $x \in A$ there exists $b \in A$ such that $\mathrm{Sp}(b) \cap (-\infty, 0) = \emptyset$, $\|b(b + \lambda e)^{-1}\| < K$ for every $\lambda > 0$ and $\lim\limits_{\lambda \to 0^+} \|b(b + \lambda e)^{-1}x - x\| = 0$, where K is the bound of the approximate identity of A.

Proof. Recall that A possesses a two-sided bounded approximate identity bounded by K means that, for every finite family $a_1,\ldots,a_n$ of elements of A and every $\varepsilon > 0$, there exists $f \in A$ with $\|f\| < K$, $\|a_i f - a_i\| < \varepsilon$, and $\|fa_i - a_i\| < \varepsilon$ $(i = 1,\ldots,n)$.

In fact, as shown by Altman [4], this condition is satisfied when it holds in the case $n = 1$. A discussion of weaker equivalent conditions can be found in [9], Chapter I, §11. A theorem of Sinclair [36] shows that there exists in A a continuous semigroup $(a^t)_{t>0}$ over the positive reals such that $x \in a^t A$ $(t > 0)$, $\sup_{t>0} \|a^t\| \leq K$ and $\lim_{t\to 0^+} a^t \cdot x = x$, where x is any given element in A.

If $f \in L^1(\mathbb{R}^+)$, let us consider the Bochner integral $\int_0^\infty f(t)a^t dt$. The map $\varphi : f \mapsto \int_0^\infty f(t)a^t dt$ is, as observed in [36], an algebra homomorphism from $L^1(\mathbb{R}^+)$ into A. Also $\|\varphi(n\psi_{[\tau,\tau+1/n]}) - a^\tau\| = \|\int_\tau^{\tau+1/n} n(a^t - a^\tau)dt\| \leq \sup_{\tau\leq t\leq\tau+1/n} \|a^t - a^\tau\|$, and so $a^\tau \in [\varphi(L^1(\mathbb{R}^+))]^-$ for $\tau \in \mathbb{R}^+$.

Now put $b = \varphi(u)$, where u is the function $t \mapsto e^{-t}$. Then $\mathfrak{L}(u)(z) = 1/(z+1)$, $Sp(u) \subset \{z \in \mathbb{C} \mid \operatorname{Re} z \geq 0\}$ and

$$\mathfrak{L}(u(u+\lambda e)^{-1})(z) = \frac{1}{1+\lambda+\lambda z} = \mathfrak{L}(u_\lambda)(z),$$

where u_λ is the function $t \mapsto (1/\lambda)e^{-(1+1/\lambda)t}$ and $\mathfrak{L}$ is the Laplace transform. An easy verification shows that $\|f * u_\lambda - f\| \xrightarrow[\lambda\to 0^+]{} 0$ and that $\|u_\lambda\| \leq 1$ for every $\lambda > 0$.

Also $\|y - b(b+\lambda e)^{-1}y\| \xrightarrow[\lambda\to 0^+]{} 0$ for every $y \in [\varphi(L^1(\mathbb{R}^+))]^- \cdot A$. In particular, this occurs for every $y \in [\bigcup_{t>0} a^t A] \ni x$. Since $\|b(b+\lambda e)^{-1}\| \leq K$ for every $\lambda > 0$, this proves the theorem.

This suggests that we should look for a p.a.i. of the form $a(a+\lambda_n e)^{-1}$ if $\mathfrak{M}$ is a module over a radical Banach algebra $\mathfrak{R}$. We will see in the following section that such p.a.i.s exist for some nonzero $x \in \mathfrak{M}$ if a has a divisible subspace, in the following classical sense.

DEFINITION 2.8. ([38]) Let E be a Banach space and let $u \in \mathfrak{L}(E)$. A linear subspace F of E is said to be <u>u-divisible</u> if $(u+\lambda e)F = F$ for every $\lambda \in \mathbb{C}$. Similarly, if A is a Banach algebra, if $\mathfrak{M}$ is a Banach A-module and if $b \in A$, then a linear subspace F of $\mathfrak{M}$ is said to be <u>b-divisible</u> if $(b+\lambda e)F = F$ for every $\lambda \in \mathbb{C}$.

PROPOSITION 2.9. Let u be a quasinilpotent operator acting on a Banach space E. Then $\varprojlim u^n(E)$ is the largest u-divisible subspace of E.

<u>Proof</u>. Let F be a u-divisible subspace of E and let $x \in F$. There exists a sequence $(x_n)_{n\geq 1}$ of elements of F such that $x_1 = x$ and $x_n = u(x_{n+1})$

for every $n \in \mathbb{N}$. So $x \in \varprojlim u^n(E)$, and hence $F \subset \varprojlim u^n(E)$.

Now let $x \in \varprojlim u^n(E)$. There exists $(x_n)_{n\in\mathbb{N}} \in \varprojlim{}' u^n(E)$ such that $x = x_1$. As $x_n = u(x_{n+1})$ for every $n \in \mathbb{N}$, $(x_{n+1})_{n\in\mathbb{N}} \in \varprojlim{}' u^n(E)$. In particular, $x_2 \in \varprojlim u^n(E)$, and $\varprojlim u^n(E) \subset u(\varprojlim u^n(E))$.

Take $v \in \mathcal{L}(E)$ such that $uv = vu$. If $(x_n)_{n\in\mathbb{N}} \in \varprojlim{}' u^n(E)$, we have $x_n = u(x_{n+1})$, so $v(x_n) = u[v(x_{n+1})]$ $(n \in \mathbb{N})$ and $(v(x_n))_{n\in\mathbb{N}} \in \varprojlim{}' u^n(E)$. This shows that $v(\varprojlim u^n(E)) \subset \varprojlim u^n(E)$. In particular, $(u + \lambda e)(\varprojlim u^n(E)) \subset \varprojlim u^n(E)$ $(\lambda \in \mathbb{C})$, and $(u + \lambda e)^{-1}(\varprojlim u^n(E)) \subset \varprojlim u^n(E)$ $(\lambda \in \mathbb{C},\ \lambda \neq 0)$. This achieves the proof.

Now let A be a Banach algebra and let $a \in A$, with $a \neq 0$. The map $x \mapsto x\, a/\|a\|$ is a surjective map from A onto Aa and as a linear space Aa is isomorphic with A/H_a where $H_a = \{b \in A \mid ba = 0\}$. Let $|||\cdot|||$ be the linear norm induced on Aa by the norm of A/H_a via this isomorphism. The following fact is well known.

REMARK 2.10. We have $|||cb||| \leq \|c\|\ |||b|||$ $(b \in Aa,\ c \in A)$ and $|||b||| \geq \|b\|$ $(b \in Aa)$. In particular, $(Aa, |||\cdot|||)$ is a Banach algebra, and if $\mathfrak{M}$ is a Banach A-module then $\mathfrak{M}$ is a Banach Aa-module.

<u>Proof</u>. We may assume $\|a\| = 1$. Let $b \in Aa$. If $da = b$ we have $cda = cb$, $|||cb||| \leq \|cd\| \leq \|c\|\ \|d\|$ so $|||cb||| \leq (\inf_{da=b} \|d\|)\ \|c\| = \|c\|\ |||b|||$ for every $c \in A$. Also, $|||b||| = \inf_{da=b} \|d\| \geq \inf_{da=b} \|d\|\ \|a\| \geq \|b\|$. These inequalities imply that $(A, |||\cdot|||)$ is a Banach algebra. If $\mathfrak{M}$ is a Banach A-module then $\|bx\| \leq \|b\|\ \|x\| \leq |||b|||\ \|x\|$ $(b \in Aa, x \in \mathfrak{M})$ and $\mathfrak{M}$ is a Banach Aa-module.

A <u>rational semigroup</u> in a complex algebra A is a map $t \mapsto a^t$ from the set $\mathbb{Q}^+$ of strictly positive rationals into A such that $a^{t+t'} = a^t a^{t'}$ $(t,t' \in \mathbb{Q}^+)$. If $a = a^1$ for some rational semigroup $(a^t)_{t\in\mathbb{Q}^+}$ then a possesses roots of all orders. The converse does not hold in general, but it is true if A has no divisors of zero.

REMARK 2.11. Let A be a complex algebra with no divisors of zero, and let $a \in A$. If a possesses roots of all orders, there exists a rational semigroup $(a^t)_{t>0}$ in A such that $a = a^1$.

<u>Proof</u>. Let $x, y \in A$ such that $x^n = y^n$. We have $(x - y)(x - e^{2i\pi/n}y) \ldots (x - e^{2(n-1)i\pi/n}y) = 0$ so that $x = e^{2ki\pi/n}y$ for some $k \in \{0,\ldots,n-1\}$. Now assume that an element a of A has roots of all orders. Then, if $b^{n!} = a$ and

$c^{(n+1)!} = a$ we have $be^{2ik\pi/n!} = c^{n+1}$ for some $k \leq n! - 1$, so that $(e^{-2ik\pi/(n+1)!}c)^{n+1} = b$.

Hence we can construct a sequence (a_n) of elements of A such that $a_1 = a$, $a_n = a_{n+1}^{n+1}$ for every $n \in \mathbb{N}$. Now if $i,j \in \mathbb{N}$ put $a^{i/j} = a_j^{i(j-1)!}$. Routine verifications that we omit show that $a^{i'/j'} = a^{i/j}$ if $i'/j' = i/j$ and that the map $r \mapsto a^r$ $(r \in \mathbb{Q}^+)$ defines a rational semigroup in A with $a^1 = a$.

The above remark will have some interest in Section 5, where we discuss radical algebras which possess nonzero rational semigroups. The following certainly well-known construction will be useful in Section 7 in the construction of counterexamples. Let S be an associative monoid (this just means that an associative law is defined on S). We will say that S is abelian when its law is abelian, and we will say that S is cancellative if $s \circ t = s \circ t'$ and $t \circ s = t' \circ s$ both imply that $t = t'$ for each element s of S. As we will need only abelian monoids here we will use additive notation.

DEFINITION 2.12. A weight on a monoid S is a map ω from S into the strictly positive reals such that $\omega(t+t') \leq \omega(t)\omega(t')$ for every $t,t' \in S$. A weight ω over S is said to be radical when $\lim_{n\to\infty} [\omega(nt)]^{1/n} = 0$ for every $t \in S$.

If S is a monoid and if ω is a weight over S we denote by $\ell^1(S,\omega)$ the complex linear space of all families $(\lambda_t)_{t\in S}$ of complex numbers such that $(\lambda_t\omega(t))_{t\in S}$ is summable.

If $u = (\lambda_t)_{t\in S} \in \ell^1(S,\omega)$ put $\operatorname{Supp} u = \{t \in S \mid \lambda_t \neq 0\}$. If $(\lambda_t)_{t\in S} \in \ell^1(S,\omega)$ put $\|(\lambda_t)_{t\in S}\|_\omega = \sum_{t\in S} |\lambda_t|\omega(t)$. Then $\ell^1(S,\omega)$ equipped with that norm is a linear space, and $(\lambda_t)_{t\in S} = \sum_{t\in S} \lambda_t X^t$, where X^t is the family which equals 1 at t and vanishes elsewhere.

Note that, if $u = \sum_{t\in S} \lambda_t X^t$ and $v = \sum_{t\in S} \mu_t X^t$ belong to $\ell^1(S,\omega)$, then we have, for every $t \in S$,

$$\sum_{r+s=t} |\lambda_r||\mu_s| \leq \sum_{r+s=t} |\lambda_r||\mu_s| \frac{\omega(r)\omega(s)}{\omega(t)} \leq \frac{1}{\omega(t)} \|u\|_\omega \|v\|_\omega ,$$

so the family $(\lambda_r\mu_s)_{r+s=t}$ is summable for every $t \in S$. Moreover

$$\sum_{t\in S} \left| \sum_{r+s=t} \lambda_r\mu_s \right| \omega(t) \leq \sum_{\substack{r\in S\\ s\in S}} |\lambda_r| \, |\mu_s| \, \omega(r)\omega(s) = \|u\|_\omega\|v\|_\omega .$$

A routine verification shows that the product $(u,v) \mapsto u \cdot v = \sum_{t\in S} \left(\sum_{r+s=t} \lambda_r\mu_s\right)X^t$

is associative.

Now if the weight ω is radical, $\|X^{nt}\|^{1/n} = (\omega(nt))^{1/n} \underset{n\to\infty}{\longrightarrow} 0$ for every $t \in S$, and the quasinilpotents are dense in $\ell^1(S,\omega)$. Hence, $\ell^1(S,\omega)$ is a radical algebra. We have thus proved the following remark.

REMARK 2.13. For every monoid S and for every weight ω on S, $\ell^1(S,\omega)$ is a Banach algebra which is radical if the weight ω is radical. Note that if S has a radical weight ω then S cannot have any unit element.

We conclude this section with the Mittag-Leffler theorem for inverse limits. Recall that, if $(E_n)_{n\in\mathbb{N}}$ is any countable family of sets and if a map $\theta_n : E_{n+1} \to E_n$ is given for every $n \in \mathbb{N}$, the family $(E_n,\theta_n)_{n\in\mathbb{N}}$ is called a <u>projective system</u>. The inverse limit $\underleftarrow{\text{Lim}}\,(E_n,\theta_n)$ is the set of all families $(y_n)_{n\in\mathbb{N}}$ belonging to the Cartesian product $E_1 \times E_2 \times \cdots \times E_n \times \cdots$ which satisfy $y_n = \theta_n(y_{n+1})$ for every $n \in \mathbb{N}$. We will denote by $\underleftarrow{\lim}\,(E_n,\theta_n)$ the set of all $y \in E_1$ such that there exists $(y_n)_{n\in\mathbb{N}} \in \underleftarrow{\text{Lim}}\,(E_n,\theta_n)$ satisfying $y = y_1$.

The following result differs slightly from the version of the Mittag-Leffler theorem given by Bourbaki [11]. (Bourbaki has a weaker hypothesis concerning the density of ranges of θ_n, but he assumes that all the θ_i are uniformly continuous.)

THEOREM 2.14. ([11], Chapitre II, §3, Théorème 1) Let (E_n,θ_n) be a projective system of complete metric spaces. Assume that θ_n is continuous and that $\theta_n(E_{n+1})$ is dense in E_n for every $n \in \mathbb{N}$. Then $\underleftarrow{\lim}\,(E_n,\theta_n)$ is dense in E_1.

<u>Proof</u>. For every $n \in \mathbb{N}$ denote by d_n the metric of E_n. Let $y \in E_1$ and take $\varepsilon > 0$. We define by induction a family $(x_n)_{n\geq 2}$ belonging to the Cartesian product $E_2 \times E_3 \times \cdots \times E_n \times \cdots$ satisfying the following conditions:

$$d_1(y,\theta_1(x_2)) < \frac{\varepsilon}{2} ;$$

$$d_n(x_n,\theta_n(x_{n+1})) < \frac{\varepsilon}{2^n} \quad (n \geq 2);$$

$$d_k(\theta_k \circ \cdots \circ \theta_{n-1}(x_n),\ \theta_k \circ \cdots \circ \theta_n(x_{n+1})) < \frac{\varepsilon}{2^n} \quad (n \geq 2,\ k \leq n-1).$$

Since $\theta_1(E_2)$ is dense in E_1 it is possible to find $x_2 \in E_2$ satisfying the first condition. Now assume that $x_2,\dots,x_n$ have been chosen. There exists a sequence $(z_p)_{p\in\mathbb{N}}$ of elements of E_{n+1} such that $x_n = \lim_{p\to\infty} \theta_n(z_p)$. So $\theta_k \circ \cdots \circ \theta_{n-1}(x_n) = \lim_{p\to\infty} \theta_k \circ \cdots \circ \theta_n(z_p)$ for every $k \leq n-1$, and, taking $x_{n+1} = z_p$ with p large enough we can arrange that the desired conditions are satisfied. We thus see that we can construct the sequence (x_n) by induction.

If $k \in \mathbb{N}$, the sequence $((\theta_k \circ \cdots \circ \theta_{n-1})(x_n))_{n\geq k}$ is a Cauchy sequence. Denote by y_k its limit. We have, for every $n \geq k+1$, $(\theta_k \circ \cdots \circ \theta_{n-1})(x_n) =$

$\theta_k(\theta_{k+1}\circ\cdots\circ\theta_{n-1}(x_n))$. This shows that $y_k = \theta_k(y_{k+1})$ for every $n \in \mathbb{N}$, and that $y_1 \in \varprojlim (E_n,\theta_n)$. We have

$$\begin{aligned} d_1(y,y_1) &= \lim_{n\to\infty} d_1(y,\theta_1\circ\cdots\circ\theta_{n-1}(x_n)) \\ &\le d_1(y,\theta_1(x_2)) + \sum_{n=2}^{\infty} d_n(\theta_1\circ\cdots\circ\theta_{n-1}(x_n),\theta_1\circ\cdots\circ\theta_{n-1}\circ\theta_n(x_{n+1})) \\ &< \frac{\varepsilon}{2} + \sum_{n=2}^{\infty}\frac{\varepsilon}{2^n} = \varepsilon . \end{aligned}$$

This achieves the proof of the theorem.

Note that the Baire category theorem can be deduced from the Mittag-Leffler theorem. If E is a complete metric space and if $(U_n)_{n\in\mathbb{N}}$ is a countable family of dense open subsets of E, we have $\bigcap_{n\in\mathbb{N}} U_n = \bigcap_{n\in\mathbb{N}} V_n$, where $V_n = U_1\cap\cdots\cap U_n$. The sequence (V_n) is decreasing, and V_n is a dense open subset of E for every $n \in \mathbb{N}$. Now every open subset of a complete metric space is homeomorphic with a complete metric space [32], and we can apply Theorem 2.14 to the projective system (V_n,i_n) where i_n is the natural injection from V_{n+1} into V_n. We thus see that $\varprojlim (V_n,i_n)$ is dense in E. But this just means that $\bigcap_{n\in\mathbb{N}} V_n$ is dense in E. The Mittag-Leffler theorem will be used in Section 5 (and implicitly in Proposition 3.1 of Section 3), and the Baire category theorem will be used in Sections 3, 5 and 8. The use of these theorems will completely avoid computations of infinite products used in the author's original proofs of several previously published ([22], [23], [25]) or unpublished results given in this paper.

3. Partial approximate identities and factorization properties for Banach modules over radical algebras

The following easy proposition shows that the existence of a partial approximate identity for a module $\mathfrak{M}$ over a Banach algebra A always implies some purely algebraic properties.

PROPOSITION 3.1. Let A be a Banach algebra, let $\mathfrak{M}$ be a Banach A-module, and let $x \in \mathfrak{M}$. If $x \in [Ax]^{-}$, then for every $\varepsilon > 0$ there exists a sequence (e_n) in A and $y \in \varprojlim e_1 \ldots e_n \mathfrak{M}$ such that $\|x - y\| < \varepsilon$.

Proof. We define by induction a sequence (e_n) of elements of A such that

$$\|e_1x - x\| < \frac{\varepsilon}{2}, \quad \|e_nx - x\| < \frac{\varepsilon}{2^n}\,(1+\|e_1\|)^{-1}\ldots(1+\|e_{n-1}\|)^{-1}$$

for every $n \ge 2$.

The existence of such a sequence is clear.

Put, for $k \geq 1$, $n \geq k$,

$$\gamma_{k,n} = e_k \cdots e_n x .$$

Then $\|\gamma_{k,n} - \gamma_{k,n+1}\| \leq \|e_k \cdots e_n\| \, \|x - xe_{n+1}\| < \varepsilon/2^{n+1}$ $(k \in \mathbb{N}, n \geq k)$. So the sequence $(\gamma_{k,n})_{n \geq k}$ is Cauchy, and hence convergent. Denote its limit by y_k.

Since $\gamma_{k,n} = e_k \gamma_{k+1,n}$ $(n \geq k+1)$, we obtain $y_k = e_k y_{k+1}$ for every $k \in \mathbb{N}$. So $y_1 \in \varprojlim e_1 \cdots e_n \mathfrak{M}$. Also, $\|y_1 - x\| \leq \|xe_1 - x\| + \sum_{n=1}^{\infty} \|\gamma_{1,n+1} - \gamma_{1,n}\| < \varepsilon$, and hence the proposition is proved.

The proposition could be formally deduced from the Mittag-Leffler theorem on inverse limits, but we will not do this here because that proof would be much longer than the direct one given above.

COROLLARY 3.2. Let A be a Banach algebra, let $\mathfrak{M}$ be a Banach A-module and let $a \in A$, $x \in \mathfrak{M}$. If $x \in [Aax]^-$, and if $ab = ba$ for every $b \in A$, then $x \in \varprojlim [a^n \mathfrak{M}]^-$. In particular if $x \neq 0$ then $\lim a^n \mathfrak{M} \neq \{0\}$.

<u>Proof</u>. Put $A' = Aa$. Then A' is a Banach algebra and $\mathfrak{M}$ is a A'-Banach module by Remark 2.10. Also $x \in [A'x]^-$, so, for every $\varepsilon > 0$ there exists a sequence (e_n) of elements of A' and a sequence $(y_n) \in \varprojlim e_1 \cdots e_n \mathfrak{M}$ such that $\|x - y_1\| < \varepsilon$. For every $n \in \mathbb{N}$ we can write $e_n = b_n a$, where $b_n \in A$.

Put $z_1 = y_1$, $z_2 = b_1 y_2, \ldots, z_n = b_1 \cdots b_{n-1} y_n$ $(n \geq 2)$. Then $z_n = b_1 \cdots b_{n-1} y_n = b_1 \cdots b_{n-1} b_n a y_{n+1} = a b_1 \cdots b_n y_{n+1} = a z_{n+1}$ $(n \geq 2)$, and similarly $z_1 = y_1 = b_1 a y_2 = a b_1 y_2 = a z_2$. So $y_1 = z_1 \in \varprojlim a^n \mathfrak{M}$. This proves the corollary.

Proposition 3.1 shows that, if A has a nontrivial p.a.i. for $\mathfrak{M}$, then $\varprojlim e_n \cdots e_1 \mathfrak{M} \neq \{0\}$ for some sequence (e_n) in A. The converse of this result in the case of a radical algebra is given by the following theorem.

THEOREM 3.3. Let $\mathfrak{R}$ be a radical Banach algebra, and let $\mathfrak{M}$ be a Banach $\mathfrak{R}$-module. Then, for every sequence (x_n) of elements of $\mathfrak{M}$ such that $x_n \in \mathfrak{R} x_{n+1}$ for every $n \in \mathbb{N}$, there exists $x \in [\bigcup_{n \in \mathbb{N}} \mathfrak{R} x_n]^-$ such that $[\mathfrak{R} x]^- = [\bigcup_{n \in \mathbb{N}} \mathfrak{R} x_n]^-$. Moreover, the set of all such x is dense in $[\bigcup_{n \in \mathbb{N}} \mathfrak{R} x_n]^-$.

<u>Proof</u>. Put $G = \operatorname{Inv} \mathfrak{R}^{\#}$, $I = [\bigcup_{n \in \mathbb{N}} \mathfrak{R} x_n]^-$. Also for $n, p \in \mathbb{N}$ put

$$\Omega_{n,p} = \left\{ x \in I \mid \inf_{a \in \mathfrak{R}} \|x_n - ax\| < \frac{1}{p} \right\}.$$

We see that $\Omega_{n,p}$ is an open subset of I, and that $x_m \in \bigcap_{p \in \mathbb{N}} \Omega_{n,p}$ for every $m \geq n+1$. Now, if $x \in \Omega_{n,p}$ and $u \in G$ there exists $a \in \mathfrak{R}$ such that $\|x_n - ax\| < 1/p$, so that $\|x_n - (au^{-1})ux\| < 1/p$, and hence $ux \in \Omega_{n,p}$. Now if $v \in \mathfrak{R}$ then

$vx = \lim_{\lambda\to 0} (\lambda e + v)\cdot x \in \Omega_{n,p}$, so that $\mathfrak{R}\Omega_{n,p} \subset \Omega_{n,p}$ and $\bigcup_{m>n} \mathfrak{R}x_m \subset \Omega_{n,p}$. As $x_k \in \mathfrak{R}x_{n+1}$ for every $k \leq n$ we obtain $I = \overline{\Omega_{n,p}}$, so that $\Omega_{n,p}$ is dense in I. It follows from the Baire category theorem that $\Omega = \bigcap_{n,p\in\mathbb{N}} \Omega_{n,p}$ is a dense G_δ subset of I. Let $x \in \Omega$. We have $x_n \in \mathfrak{R}x$ for every $n \in \mathbb{N}$, so that $I = [\mathfrak{R}x]^-$, which achieves the proof of the theorem.

This theorem is also true for more general radical topological algebras; see Bouloussa [10].

COROLLARY 3.4. Let $\mathfrak{R}$ be a radical Banach algebra and let $\mathfrak{M}$ be a Banach $\mathfrak{R}$-module. The following conditions imply each other.

(1) $\mathfrak{R}$ possesses x-partial approximate identities for some nonzero $x \in \mathfrak{M}$.

(2) There exists a sequence (α_n) of elements of $\mathfrak{R}$ such that $\varprojlim \alpha_1 \ldots \alpha_n \mathfrak{M} \neq \{0\}$.

(3) There exists a sequence (x_n) of elements of $\mathfrak{M}$ such that $x_1 \neq 0$ and $x_n \in \mathfrak{R}x_{n+1}$ for every $n \in \mathbb{N}$.

<u>Proof</u>. The fact that (2) and (3) imply each other was noticed in Remark 2.3, and the other assertions follow from Proposition 3.1 and Theorem 3.3.

COROLLARY 3.5. Let $\mathfrak{R}$ be a radical Banach algebra. The following conditions imply each other.

(1) $a \in [\mathfrak{R}a]^-$ for some nonzero $a \in \mathfrak{R}$.

(2) $\varprojlim a_1 \ldots a_n \mathfrak{R} \neq \{0\}$ for some sequence (a_n) of elements of $\mathfrak{R}$.

(3) There exists in $\mathfrak{R}^\#$ a strictly increasing sequence of left principal ideals.

<u>Proof</u>. The equivalence between (1) and (2) is given by Corollary 3.4, and (2) implies that $x_n \in \mathfrak{R}x_{n+1}$ $(n \in \mathbb{N})$ for some sequence (x_n) of elements of $\mathfrak{R}$ such that $x_1 \neq 0$. We have for every $n \in \mathbb{N}$, $\mathfrak{R}^\# x_n \subset \mathfrak{R}^\# x_{n+1}$. If $x_{n+1} \in \mathfrak{R}^\# x_n$ for some $n \in \mathbb{N}$ we have $x_n = bx_{n+1}$, $x_{n+1} = (\lambda e + c)x_n$ where $\lambda \in \mathbb{C}$, $b,c \in \mathfrak{R}$. So $x_n = b(\lambda e + c)x_n$. Since $e - b(\lambda e + c)$ is invertible in $\mathfrak{R}^\#$ this implies that $x_n = 0$, and $x_1 = 0$, a contradiction. So the sequence $(\mathfrak{R}^\# x_n)$ is strictly increasing, and (3) holds.

Conversely assume that $(\mathfrak{R}^\# x_n)$ is a strictly increasing sequence of left principal ideals in $\mathfrak{R}^\#$. Then $x_n \in \mathfrak{R}$ for every n (for otherwise $x_n \in \mathrm{Inv}(\mathfrak{R}^\#)$, $\mathfrak{R}^\# x_n = \mathfrak{R}^\#$), and we may assume that $x_1 \neq 0$. We obtain $x_n = (\lambda_n e + b_n)x_{n+1}$ for every $n \in \mathbb{N}$, where $\lambda_n \in \mathbb{C}$, $b_n \in \mathfrak{R}$. In fact $\lambda_n = 0$, because otherwise $x_{n+1} \in \mathfrak{R}^\# x_n$. So $x_n \in \mathfrak{R}x_{n+1}$ for every $n \in \mathbb{N}$ and (2) is satisfied as observed in Remark 2.3 and Corollary 3.4.

COROLLARY 3.6. Let $\mathfrak{R}$ be a commutative radical Banach algebra. If $\mathfrak{R}$ has no nontrivial partial approximate identities, then every nonzero element of $\mathfrak{R}$ is the product of a finite number of irreducible elements of $\mathfrak{R}$.

Proof. (The following argument can be found in any algebra course for French freshmen.) Denote by S the set of nonzero elements of $\mathcal{R}$ which cannot be written as a finite product of irreducible elements of $\mathcal{R}$. If $S \neq \emptyset$ let $a \in S$. Then $a = bb'$ with $b, b' \in \mathcal{R}$. At least one of the elements b and b', say b', belongs to S (the complement of S is stable under products). Using this observation we can construct by induction a sequence $(a_n)_{n \geq 1}$ of elements of S and a sequence $(b_n)_{n \geq 1}$ of elements of $\mathcal{R}$ such that $a_1 = a$, $a_n = b_n a_{n+1}$ for every $n \geq 1$. As $a \neq 0$ this implies that $\mathcal{R}$ has a nontrivial p.a.i. (Corollary 3.4), and the result follows.

Corollary 3.2 shows that if $x \in [Aax]^-$ for some nonzero element x of a Banach A-module $\mathfrak{m}$ where a commutes with all elements of A, then $\varprojlim a^n \mathfrak{m} \neq \{0\}$. The following theorem shows in particular that the converse is true when a is quasinilpotent.

THEOREM 3.7. Let A be a Banach algebra, let $\mathfrak{m}$ be a Banach A-module and let a be a quasinilpotent element of A such that $ab = ba$ for every $b \in A$. The following conditions imply each other.

(1) There exists a nonzero element x of $\mathfrak{m}$ such that $x \in [Aax]^-$.

(2) $\varprojlim a^n \mathfrak{m} \neq \{0\}$.

(3) There exists a nontrivial a-divisible subspace of $\mathfrak{m}$.

(4) There exists a nonzero element y of $\mathfrak{m}$ and a sequence (P_n) of polynomials with complex coefficients such that $y = \lim_{n\to\infty} aP_n(a)y$.

(5) For every sequence (λ_n) of nonzero complex numbers such that $\lim_{n\to\infty} \lambda_n = 0$ there exists some nonzero $z \in \mathfrak{m}$ such that $z = \lim_{i\to\infty} (a + \lambda_{n_i} e)^{-1} az$ for some subsequence $(\lambda_{n_i})_{i \in \mathbb{N}}$ of (λ_n).

Proof. Corollary 3.2 shows that (1) implies (2), and the fact that (2) and (3) are equivalent is given by Proposition 2.9. Applying Theorem 3.3 to the closed subalgebra of A generated by a we see that (2) implies (4).

If (4) holds for a sequence (P_n) of polynomials and a nonzero $y \in \mathfrak{m}$ we can construct by induction an increasing sequence (m_n) of integers such that $\|aP_n(a)aP_{m_n}(a)y - aP_n(a)y\| < \frac{1}{n}$ $(n \in \mathbb{N})$. So $y = \lim_{n\to\infty} P_n(a)P_{m_n}(a)a^2y$, and (1) is satisfied.

Since $(a + \lambda e)^{-1}$ is a limit of polynomials in a for every $\lambda \neq 0$, we see that (5) implies (4).

Now assume that (2) holds. Let (λ_n) be a sequence of nonzero complex numbers such that $\lim_{n\to\infty} \lambda_n = 0$, and let (x_n) be a sequence of elements of $\mathfrak{m}$ such that $x_1 \neq 0$, $x_n = ax_{n+1}$ for every $n \in \mathbb{N}$. Put $I = [\bigcup_{n \in \mathbb{N}} Aax_n]^-$, and put, for $p \in \mathbb{N}$,

$$\Omega_p = \{x \in I \mid \inf_{n\in\mathbb{N}} \|x - a(a+\lambda_n e)^{-1}x\| < \frac{1}{p}\}.$$

The set Ω_p is clearly open for every $p \in \mathbb{N}$. Now let $y \in I$ and take $\varepsilon > 0$. There exists $u \in A$ and $m \in \mathbb{N}$ such that $\|y - ux_m\| < \varepsilon/2$.

Since $x_m = ax_{m+1}$, there exists $n \in \mathbb{N}$ such that $\|u(a+\lambda_n e)x_{m+1} - ux_m\| < \min\{1/p, \varepsilon/2\}$. We have $a(a+\lambda_n e)^{-1} u(a+\lambda_n e)x_{m+1} = aux_{m+1} = uax_{m+1} = ux_m$, so that $u(a+\lambda_n e)x_{m+1} \in \Omega_p$. As $\|y - u(a+\lambda_n e)x_{m+1}\| < \varepsilon$, this implies that Ω_p is dense in I. The Baire category theorem shows that $\Omega = \bigcap_{p\in\mathbb{N}} \Omega_p$ is dense in I, and clearly each nonzero element z of Ω satisfies (5). Hence (5) follows from (2).

To construct the subsequence (λ_{n_i}), just note that $x \neq a(a+\lambda_n e)^{-1}x$ for every $x \neq 0$ since a is quasinilpotent. This proves the theorem.

REMARK 3.8. Note that in Theorem 3.7 if we put $\mathfrak{m}_1 = \{x \in \mathfrak{m} \mid x \in [Aax]^-\}$, $\mathfrak{m}_2 = \{x \in \mathfrak{m} \mid \inf_{P\in\mathbb{C}[X]} \|x - aP(a)x\| = 0\}$, $\mathfrak{m}_{(\lambda_n)} = \{x \in \mathfrak{m} \mid \inf_{n\in\mathbb{N}} \|x - a(\lambda_n e + a)^{-1}x\| = 0\}$ where (λ_n) is a sequence of nonzero complex numbers which converges to zero, then $\mathfrak{m}_1 \supset \mathfrak{m}_2 \supset \mathfrak{m}_{(\lambda_n)}$, and $\overline{\mathfrak{m}}_1 = \overline{\mathfrak{m}}_2 = \overline{\mathfrak{m}}_{(\lambda_n)} = [\underleftarrow{\lim}\, a^n\mathfrak{m}]^-$. This follows directly from the proofs. Similar remarks hold for Corollary 3.4 and Corollary 3.5.

We will say that a commutative radical Banach algebra $\mathfrak{R}$ satisfies the <u>Condition 3.7</u> when some element a of $\mathfrak{R}$ satisfies the conditions of Theorem 3.7, where $\mathfrak{m}$ is the algebra $\mathfrak{R}$ taken as a module over itself, the module law being the product in $\mathfrak{R}$. Note that, if $\mathfrak{R}$ is an integral domain, condition (2) of Theorem 3.7 means simply that $\bigcap_{n\in\mathbb{N}} a^n\mathfrak{R} \neq \{0\}$. Also the sets $\mathfrak{R}_1$, $\mathfrak{R}_2$, $\mathfrak{R}_{(\lambda_n)}$ defined as in Remark 3.8 are ideals of $\mathfrak{R}$ and so is $\bigcap_{n\in\mathbb{N}} a^n\mathfrak{R}$. So, if $\mathfrak{R}$ is an integral domain and if $a \in \mathfrak{R}$ satisfies the conditions of Theorem 3.7 there exists $x \in \bigcap_{n\in\mathbb{N}} a^n\mathfrak{R}$ with $x \neq 0$ satisfying all the Conditions (1), (4), (5) (of course, x depends on the sequence (λ_n)).

Theorem 3.7 has also a consequence for the hyperinvariant subspace problem.

COROLLARY 3.9. Let u be a nonzero quasinilpotent operator on a Banach space E. If u has no nontrivial hyperinvariant subspace, then, for every sequence (λ_n) of nonzero complex numbers which converges to zero, there exists a dense subspace F of E and a subsequence (λ_{n_i}) of (λ_n) such that $\lim_{i\to\infty} u(u+\lambda_{n_i} e)^{-1}y = y$ for every $y \in F$.

<u>Proof</u>. Put $A = \{v \in \mathcal{L}(E) \mid uv = vu\}$. Since u has no nontrivial hyperinvariant subspace, $[Aux]^- = E \ni x$ for every nonzero $x \in E$, because $\operatorname{Ker} u = \{0\}$. The corollary then follows from Theorem 3.7 and Remark 3.8.

REMARK 3.10. Let $\mathfrak{R}$ be a commutative radical Banach algebra and let $\mathfrak{m}$ be a Banach $\mathfrak{R}$-module. If $\mathfrak{R}$ possesses a nontrivial <u>bounded</u> p.a.i. for $\mathfrak{m}$, then an

unpublished result of the author shows that some $a \in \mathfrak{R}$ satisfies the conditions of Theorem 3.7. We will not reproduce this proof here because the result is a consequence of a better one due to Niels Grønback [28]. Assume that $\mathfrak{R}$ has a nontrivial bounded p.a.i. There exists $x \in \mathfrak{m}$ and a bounded sequence (e_n) of elements of $\mathfrak{R}$ such that $x = \lim_{n\to\infty} e_n x$. Put $\mathfrak{m}_1 = \{y \in \mathfrak{m} \mid y = \lim_{n\to\infty} e_n y\}$. Then $\mathfrak{m}_1$ is closed because (e_n) is bounded. Grønbaek's theorem shows then that for every $y \in \mathfrak{m}_1$ there exists $a \in \mathfrak{R}$ such that $y \in \bigcap_{n\in\mathbb{N}} a^n \mathfrak{m}_1$, but his proof shows in fact that $y \in \varprojlim a^n \mathfrak{m}_1$. Of course, this implies that $\mathfrak{R}$ and $\mathfrak{m}$ satisfy the conditions of Theorem 3.7, but it shows more generally that $\mathfrak{m}_1 = \bigcup_{a\in\mathfrak{R}} \varprojlim a^n \mathfrak{m}_1$.

4. An automatic continuity result for real semigroups in separable Banach algebras

Let $(s_\tau)_{\tau\in S}$ be a Hamel basis of the reals over the rationals such that $s_\tau > 0$ for every $\tau \in S$ (here S is a suitable set of indexes). Let f be any map from S into $\mathbb{R}$ such that $f(\tau) > 0$ for every $\tau \in S$. If $\tau_1,\dots,\tau_k \in S$, $r_1,\dots,r_k \in \mathbb{Q}$ put $\varphi(r_1 s_{\tau_1} + \dots + r_k s_{\tau_k}) = [f(\tau_1)]^{r_1} \dots [f(\tau_k)]^{r_k}$. Then φ is well defined on $\mathbb{R}$, and $\varphi(t+t') = \varphi(t)\varphi(t')$ $(t,t' \in \mathbb{R})$. Now let $(a^t)_{t>0}$ be a continuous semigroup in a Banach algebra A with $a^t \neq 0$ for at least some $t > 0$. For $t > 0$, set $b^t = \varphi(t)a^t$.

Then $(b^t)_{t>0}$ is a real semigroup in A, and $\|b^t\| = |\varphi(t)|\ \|a^t\|$ for every $t > 0$. Since f was an arbitrary positive-valued function on S, we see that the semigroup (b^t) may be very discontinuous.

The following elementary result shows nevertheless that, if the algebra is separable, then all the real semigroups contained in it must have some partial continuity properties.

THEOREM 4.1. Let A be a separable Banach algebra and let $(a^t)_{t>0}$ be a semigroup in A over the positive reals. Then there exist two sequences (r_n) and (s_n) of positive reals such that $\lim_{n\to\infty} r_n = \lim_{n\to\infty} s_n = 0$ and such that $\lim_{n\to\infty} \|a^{t+r_n} - a^t\| = \lim_{n\to\infty} \|a^{t-s_n} - a^t\| = 0$ for every $t > 0$.

Proof. Take $t_0 > 0$ and suppose that there exists $\alpha > 0$ such that $\inf_{0<h\leq\alpha} \|a^{t_0} - a^{t_0+h}\| > 0$. We may assume that $\alpha < t_0$. Put, for $t \leq t_0$,

$$\delta(t) = \inf_{0<h\leq\alpha} \|a^t - a^{t+h}\| .$$

Since $\|a^{t_0} - a^{t_0+h}\| \leq \|a^{t_0-t}\|\ \|a^t - a^{t+h}\|$ $(t < t_0,\ h > 0)$ we have $\delta(t_0) \leq \|a^{t_0-t}\|\ \delta(t)$ for every $t < t_0$.

If $m \in \mathbb{N}$ put $\Omega_m = \{t \in [t_0-\alpha, t_0) \mid \|a^{t_0-t}\| \leq m\}$. Then $[t_0-\alpha, t_0) = \bigcup_{n\in\mathbb{N}} \Omega_n$, so there exists $n \in \mathbb{N}$ such that Ω_n is uncountable.

We have, for every $t \in \Omega_n$,

$$\delta(t) \geq \frac{\delta(t_0)}{\|a^{t_0-t}\|} \geq \frac{\delta(t_0)}{n}.$$

Put $\varepsilon = \delta(t_0)/2n$ and put, for $t > 0$, $B_t = \{x \in A \mid \|x - a^t\| < \varepsilon\}$.

If $t, t' \in \Omega_n$ with $t < t'$, we have $0 < t' - t < \alpha$, and $\|a^t - a^{t'}\| \geq \delta(t) > 2\varepsilon$. So $B_t \cap B_{t'} = \emptyset$, and the family $(B_t)_{t\in\Omega_n}$ is an uncountable family of pairwise disjoint open balls. This contradicts the separability of A, hence $\inf_{0<h\leq\alpha} \|a^{t+h} - a^t\| = 0$ for every $t > 0$ and every $\alpha > 0$. Now for every $n \in \mathbb{N}$ we can choose $r_n \in (0,1/n)$ such that

$$\|a^{(1/n)+r_n} - a^{1/n}\| < \frac{1}{n} \inf_{m<n} \{(1 + \|a^{1/m} - a^{1/n}\|)^{-1}\}.$$

We obtain $\|a^{(1/m)+r_n} - a^{1/m}\| < 1/n$ for every $m \in \mathbb{N}$ and every $n > m$, and hence $\lim_{n\to\infty} \|a^{(1/m)+r_n} - a^{1/m}\| = 0$ $(m \in \mathbb{N})$.

If $t > 0$ there exists $m \in \mathbb{N}$ such that $1/m < t$ and $\limsup_{n\to\infty} \|a^{t+r_n} - a^t\| \leq \|a^{t-(1/m)}\| \limsup_{n\to\infty} \|a^{(1/m)+r_n} - a^{1/m}\| = 0$.

Using similar arguments, it is easy to construct a sequence (s_n) of positive reals such that $s_n \xrightarrow[n\to\infty]{} 0$, and such that $\|a^{t-s_n} - a^t\| \xrightarrow[n\to\infty]{} 0$ for every $t > 0$. This proves the theorem.

COROLLARY 4.2. Let A be a commutative, separable Banach algebra. There exists a nonzero closed ideal I of A which possesses bounded approximate identities if and only if A possesses a nonzero bounded semigroup $(a^t)_{t>0}$ over the positive reals.

Proof. If some nonzero closed ideal I of A possesses b.a.i., then I is separable and it follows from a theorem of Sinclair [36] that there exists an analytic semigroup $(a^t)_{\mathrm{Re}\, t>0}$ in I such that $\sup_{t>0} \|a^t\| < +\infty$ (and $(a^t I)^- = I$ for every $t > 0$).

Now assume that A possesses a nonzero bounded semigroup $(a^t)_{t>0}$. Put $M = \sup_{t>0} \|a^t\|$, $I = [\bigcup_{t>0} a^t A]^-$. Let (t_n) be the sequence given by Theorem 4.1. Then $\lim_{n\to\infty} xa^{t_n} = x$ for every x in $\bigcup_{t>0} a^t A$. Since $\|a^{t_n}\| \leq M$ for every $n \in \mathbb{N}$, $\lim_{n\to\infty} xa^{t_n} = x$ for every $x \in I$ and so I possesses bounded approximate identities as required.

REMARK 4.3. Theorem 4.1 does not extend to nonseparable Banach algebras.

consider the Banach algebra $\ell^1(\mathbb{R}^+)$ defined as in Remark 2.13. Then $(X^t)_{t>0}$ is a real semigroup in $\ell^1(\mathbb{R}^+)$ and $\|X^t - X^{t'}\| = 2$ if $t \neq t'$. Corollary 4.2 does not extend either to that algebra. To see this, denote by L_t the linear form $\sum_{t>0} \lambda_t X^t \mapsto \lambda_t$. The family $(|L_t|)_{t>0}$ defines a locally convex topology, and from any bounded sequence (f_n) of $\ell^1(\mathbb{R}^+)$ it is possible to find a subsequence (f_{n_i}) of (f_n) and an element f of $\ell^1(\mathbb{R}^+)$ such that $\lim_{i\to\infty} L_t(f_{n_i} - f) = 0$ for every $t > 0$. In fact, there exists a countable set $S \subset \mathbb{R}^+$ such that $L_t(f_n) = 0$ for every $n \in \mathbb{N}$ and every $t \notin S$, and the weak topology on $\ell^1(S)$ considered as the dual of $c_0(S)$ equals on bounded sets the topology defined by the family of seminorms $(|L_t|)_{t\in S}$.

It can be easily checked that $L_t(gf_{n_i} - gf) \xrightarrow[i\to\infty]{} 0$ for every $t > 0$ and every $g \in \ell^1(\mathbb{R}^+)$.

Now take $g \in \ell^1(\mathbb{R}^+)$ and suppose that there exists a bounded sequence (f_n) of elements of $\ell^1(\mathbb{R}^+)$ such that $g = \lim_{n\to\infty} g f_n$. We may assume without loss of generality that there exists $f \in \ell^1(\mathbb{R}^+)$ such that $\lim_{n\to\infty} L_t(f - f_n) = 0$ for every $t > 0$.

We obtain $L_t(fg) = \lim_{n\to\infty} L_t(f_n g) = L_t(g)$ for every $t > 0$, so $fg = g$. The map $\sum_{t>0} \lambda_t X^t \mapsto \sum_{t>0} \lambda_t \delta_t$, where we denote by δ_t the Dirac measure at t, is an isometry from $\ell^1(\mathbb{R}^+)$ into $M(\mathbb{R}^+)$, the convolution algebra of measures with support in $[0,+\infty)$. It follows from the Titchmarsh convolution theorem for measures (see [15], p. 156 or [14]) that $M(\mathbb{R}^+)$ is an integral domain. If $g \neq 0$, the equality $fg = g$ would imply that f is a unit for $\ell^1(\mathbb{R}^+)$. But $\ell^1(\mathbb{R}^+)$ can be identified algebraically with a subalgebra of the radical algebra $\ell^1(\mathbb{R}^+, e^{-t^2})$, and so it cannot possess any idempotent. Hence $g = 0$, which proves our assertion. Note that similar arguments can be applied to the algebra $\ell^1(\mathbb{R}^+, e^{-t^2})$ itself, and even in the radical case we see that Theorem 4.1 and Corollary 4.2 may fail completely for nonseparable Banach algebras.

5. Semigroups and elements of finite closed descent in commutative radical Banach algebras

G. R. Allan introduced in [2] the important notion of elements of finite closed descent (f.d. elements) in a Banach algebra A. An element a of A is said to be of finite closed descent if $a^n \in [Aa^{n+1}]^-$ for some $n \in \mathbb{N}$. An easy induction given in [2] shows that $a^n \in [Aa^m]^-$ for every $m > n$ and, in particular, that $a^n \in [Aa^{2n}]^-$ with $a^n \neq 0$. So the existence of nonnilpotent f.d. elements is equivalent to the existence of nonzero elements b of A such that $b \in [b^2A]^-$. It was proved by the author in [23] that, if $\mathfrak{R}$ is commutative and radical, and if such elements exist in $\mathfrak{R}$, then $\mathfrak{R}$ possesses nonzero rational semigroups and even nonzero real semigroups (not necessarily continuous). Moreover, if the continuum hypothesis be assumed there exists a discontinuous homomorphism from $C(K)$ into $\mathfrak{R} \oplus \mathbb{C}e$ for each infinite compact space K. Also, $\mathfrak{R}$ contains a copy of every nonunital complex algebra which is an integral domain of cardinality $2^{\aleph_0}$.

Before showing that the existence of rational semigroups implies the existence of elements of finite closed descent, we will give a very short proof of one of the results of [23]. The proof is based upon the Mittag-Leffler theorem on inverse limits (Theorem 2.14).

THEOREM 5.1. ([23]) Let $\mathfrak{R}$ be a commutative radical Banach algebra, and let $a \in \mathfrak{R}$. If $a^m \in [a^{m+1}\mathfrak{R}]^-$ for some $m \in \mathbb{N}$, there exists a semigroup $(c^t)_{t\in\mathbb{Q}^+}$ over the positive rationals such that $c^t \in \bigcap_{n\in\mathbb{N}} a^n\mathfrak{R}$, and such that $[c^t\mathfrak{R}]^- = [a^m\mathfrak{R}]^-$ for every $t \in \mathbb{Q}^+$.

Proof. Put $\mathfrak{R}_1 = a^m\mathfrak{R}$, and equip $\mathfrak{R}_1$ with the norm $|||\cdot|||$ defined in Section 2. Then $\mathfrak{R}_1$ is radical (we have $|||c^n||| \le \|c^{n-1}\|\ |||c|||$ for every $n \in \mathbb{N}$ and every $c \in \mathfrak{R}_1$). There exists a sequence (u_n) of elements of $\mathfrak{R}$ such that $a^m = \lim_{n\to\infty} a^{4m}\cdot u_n$. Then $|||a^{2m} - a^{5m}\cdot u_n||| \xrightarrow[n\to\infty]{} 0$. Now put $\mathfrak{R}_2 = [a^{2m}\mathfrak{R}_1]^-$ (the closure is taken in $\mathfrak{R}_1$ with respect to the norm $|||\cdot|||$). Then $a^{2m} \in \mathfrak{R}_2$, and $[a^{2m}\mathfrak{R}_2]^- = ((a^{2m}\mathfrak{R}_1)^-\cdot a^{2m})^- \supset [a^{4m}\mathfrak{R}_1]^-$. As $a^{2m} \in [a^{4m}\mathfrak{R}_1]^-$, $[a^{4m}\mathfrak{R}_1]^- \supset [a^{2m}\mathfrak{R}_1]^- = \mathfrak{R}_2$. So $[a^{2m}\mathfrak{R}_2]^- = \mathfrak{R}_2$.

Put $b = a^{2m}$, and put $\Omega = \{x \in \mathfrak{R}_2 \mid [x\mathfrak{R}_2]^- = \mathfrak{R}_2\}$. Then $\Omega = \{x \in \mathfrak{R}_2 \mid b \in [x\mathfrak{R}_2]^-\} = \bigcap_{p\in\mathbb{N}} \{x \in \mathfrak{R}_2 \mid \inf_{y\in\mathfrak{R}_2} |||b - xy||| < 1/p\}$. We thus see that Ω is a G_δ in $\mathfrak{R}_2$ and that Ω is homeomorphic to a complete metric space. Also, Ω is stable under products. Note that $bG \subset \Omega$, where $G = \mathrm{Inv}\,\mathfrak{R}_2^\#$, so that $b\mathfrak{R}_2 \subset [bG]^- \subset \overline{\Omega}$, and hence Ω is dense in $b\mathfrak{R}_2$ (here we use again the obvious fact that G is dense in $\mathfrak{R}_2^\#$ because $\mathfrak{R}_2$ is radical).

For $n \in \mathbb{N}$, denote by $\theta_n : \Omega \to \Omega$ the map $x \mapsto x^{n+1}$. Then θ_n is continuous. Also, $b^nG \subset \Omega$, so that $b^n\mathfrak{R}_2 \subset [b^nG]^- \subset \overline{\Omega}$ and b^nG is dense in $\mathfrak{R}_2$ and a fortiori in Ω for every $n \in \mathbb{N}$. As $\mathfrak{R}_2$ is radical, we have $G = \exp \mathfrak{R}_2^\#$.

(To see this well known fact note that $e + x = \exp[\mathrm{Log}(e+x)]$ for every $x \in \mathfrak{R}_2$, where $\mathrm{Log}(e+x) = \sum_{n\ge 1} (-1)^{n+1}x^n/n$, and that every element u of G can be written $u = (\exp\alpha)(e+x)$ where $\alpha \in \mathbb{C}$ and $x \in \mathfrak{R}_2$.)

Now let $u \in b^{n+1}G$. We have $u = b^{n+1}\exp v = (b\exp(v/n+1))^{n+1}$ where v is some element of $\mathfrak{R}_2^\#$. So $b^{n+1}G \subset \theta_n(\Omega)$ and $\theta_n(\Omega)$ is dense in Ω.

It now follows from the Mittag-Leffler theorem on inverse limits (Theorem 2.14) that for every element x of Ω and every $\varepsilon > 0$ there exists a sequence (x_n) of elements of Ω such that $x_n = x_{n+1}^{n+1}$ for every $n \in \mathbb{N}$ and such that

$\|x - x_1\| < \varepsilon$. In particular

$$x_p = x_q^{q!/p!} \quad (p,q \in \mathbb{N}, \quad q \geq p).$$

Put $c^{i/j} = x_j^{i(j-1)!}$ $(i,j \in \mathbb{N})$.

Then $c^{i/j} = x_{nj}^{(i/j)(nj)!}$ for every $n \in \mathbb{N}$, and routine verifications show that, if $i/j = i'/j'$, then $c^{i/j} = c^{i'/j'}$ and that $c^{r+r'} = c^r \cdot c^{r'}$ for every $r, r' \in \mathbb{Q}^+$.

Since $c^r \in \Omega$ for every $r \in \mathbb{Q}^+$, we have $[c^r \mathcal{R}_2]^- = \mathcal{R}_2$ so a fortiori $c^r \mathcal{R}$ is dense in $a^{2m}\mathcal{R}_1 = a^{3m}\mathcal{R}$ in the sense of the norm of $\mathcal{R}$, for every $r > 0$. Since $[a^{3m}\mathcal{R}]^- = [a^m\mathcal{R}]^-$, we obtain $[c^r\mathcal{R}]^- = [a^m\mathcal{R}]^-$ for every $r \in$

Also $c^{r/p} \subset a^{2m}\mathcal{R}$, and so $c^r \in a^{2mp}\mathcal{R}$ $(r \in \mathbb{Q}^+,\ p \in \mathbb{N})$ and $c^r \in \bigcap_{n \in \mathbb{N}} a^n\mathcal{R}$ $(r \in \mathbb{Q}^+)$, which achieves the proof of the theorem.

REMARK 5.2. (1) The set Δ of all elements c of $\bigcap_{n\in\mathbb{N}} a^n\mathcal{R}$ such that $c = c^1$ for some rational semigroup $(c^t)_{t>0}$ satisfying the conditions of the theorem is dense in $[a^m\mathcal{R}]^-$. This can be deduced from the proof of the theorem, but follows also immediately from the fact that, if $c \in \Delta$, then $cG \subset \Delta$ (use the fact that $G = \exp \mathcal{R}^{\#}$) so that $[a^m\mathcal{R}]^- = [c\mathcal{R}]^- = \overline{\Delta}$.

G. R. Allan proved in [1, Lemma 1] that, if a is of finite closed descent, then the map $x \mapsto ax$ is injective on $\bigcap_{n\in\mathbb{N}} Aa^n$. This shows that $\bigcap_{n\in\mathbb{N}} Aa^n$ is equal to the set $\varprojlim a^nA$ (defined in Section 2) in that case.

We now give a complete characterization of commutative radical Banach algebras $\mathcal{R}$ such that there exist discontinuous homomorphisms from $C(K)$ into $\mathcal{R} \oplus \mathbb{C}e$.

THEOREM 5.3. Let $\mathcal{R}$ be a commutative radical Banach algebra. The following conditions imply each other.

(1) $a \in [a^2\mathcal{R}]^-$ for some nonzero $a \in \mathcal{R}$.

(2) There exists a sequence (a_n) of elements of $\mathcal{R}$ such that $a_1 \neq 0$, $a_n \in a_{n+1}^2\mathcal{R}^{\#}$ for every $n \in \mathbb{N}$.

(3) There exists a nonzero rational semigroup $(a^t)_{t\in\mathbb{Q}^+}$ in $\mathcal{R}$.

(4) There exists a nonzero real semigroup $(b^t)_{t>0}$ in $\mathcal{R}$, and a sequence (t_n) of positive reals such that $t_n \to 0$ and $b^{t+t_n} \to b^t$ as $n \to \infty$, and $[b^t\mathcal{R}]^- = [b^{t'}\mathcal{R}]^-$ for every $t, t' > 0$.

(5) (If the continuum hypothesis is assumed.) There exists a discontinuous homomorphism from $C(K)$ into $\mathcal{R}^{\#}$ for each infinite compact space K.

(6) (If the continuum hypothesis is assumed.) For every complex commutative algebra A without unit and without divisors of zero such that $\operatorname{card} A = 2^{\aleph_0}$,

there exists a one-to-one algebra homomorphism from A into $\mathfrak{R}$.

Proof. The fact (independent of the continuum hypothesis) that (5) implies (1) was proved by G. R. Allan [2], and the equivalence between (5) and (6) is given in [20]. Clearly (4), implies (3) and (3) implies (2). The fact that (1) implies (5) is proved in [23].

It follows also from [23] that if (1) holds there exists in $\mathfrak{R}$ a real semigroup $(b^t)_{t>0}$ such that $[b^t\mathfrak{R}]^- = [a^2\mathfrak{R}]^-$ for every $t > 0$. But if (1) holds, there exists a sequence (e_n) of elements of $\mathfrak{R}$ such that $a = \lim_{n\to\infty} a^2 e_n$. Let $\mathfrak{R}_1$ be the closed subalgebra of $\mathfrak{R}$ generated by $\{a\} \cup \{e_n\}_{n\in\mathbb{N}}$. Then $\mathfrak{R}_1$ is separable, $a \in [a^2\mathfrak{R}_1]^-$ and the real semigroup $(b^t)_{t>0}$ obtained above can be constructed in $\mathfrak{R}_1$. Since $\mathfrak{R}_1$ is separable the existence of the sequence (t_n) follows from Theorem 4.1 hence (1) implies (4).

Now assume that (2) holds. It follows from Theorem 3.3 that there exists $a \in [\bigcup_{n\in\mathbb{N}} a_n\mathfrak{R}]^-$ such that $[\bigcup_{n\in\mathbb{N}} a_n\mathfrak{R}]^- = [a\mathfrak{R}]^-$. In particular $a_{n+1} \in [a\mathfrak{R}]^-$, $a_{n+1}^2 \in [a^2\mathfrak{R}]^-$ and $a_n \in a_{n+1}^2\mathfrak{R}^{\#} \subset [a^2\mathfrak{R}]^-$ for every n. So $[\bigcup a_n\mathfrak{R}]^- = [a^2\mathfrak{R}]^-$, and hence $a \in [a^2\mathfrak{R}]^-$. Since $a_1 \neq 0$, we see that $a \neq 0$ and (1) holds. This achieves the proof of the theorem.

COROLLARY 5.4. Let ω be a weight over the positive rationals such that $\lim_{n\to\infty} [\omega(nt)]^{1/n} = 0$ for every positive rational t. The algebra $\ell^1(\mathbb{Q}^+, \omega)$ satisfies the conditions of Theorem 5.3. In particular, $\ell^1(\mathbb{Q}^+, e^{-t^2})$ possesses dense principal ideals, nonzero real semigroups and there exists a discontinuous homomorphism from $C(K)$ into $\ell^1(\mathbb{Q}^+, e^{-t^2}) \oplus \mathbb{C}e$ (if the continuum hypothesis is assumed).

Proof. $(X^t)_{t\in\mathbb{Q}^+}$ is a nonzero rational semigroup in $\ell^1(\mathbb{Q}^+,\omega)$, and $\ell^1(\mathbb{Q}^+,\omega)$ is a radical algebra (Remark 2.13). Note also that the sequence $(X^{1/2^n})$ satisfies condition (2) of the theorem, and Theorem 3.3 gives in fact an element a of $\ell^1(\mathbb{Q}^+,\omega)$ such that $[\bigcup_{n\in\mathbb{N}} X^{1/2^n} \ell^1(\mathbb{Q}^+,\omega)]^- = [a \cdot \ell^1(\mathbb{Q}^+,\omega)]^-$ so $[a \cdot \ell^1(\mathbb{Q}^+,\omega)]^- = \ell^1(\mathbb{Q}^+,\omega)$.

REMARK 5.5. We noticed above that, if (a_n) is a sequence of elements of $\mathfrak{R}$ satisfying condition (2) of Theorem 5.3, then there exists $a \in \mathfrak{R}$ satisfying (1) such that $[\bigcup a_n\mathfrak{R}]^- = [a\mathfrak{R}]^-$. Also we noticed that the semigroup $(b^t)_{t>0}$ given in (4) can be arranged to satisfy $[b^t\mathfrak{R}]^- = [a\mathfrak{R}]^-$ where a is a nonzero element of $\mathfrak{R}$ such that $a \in [a^2\mathfrak{R}]^-$.

Now denote by $\mathfrak{m}_1$ the set of elements a of $\mathfrak{R}$ satisfying (1), by $\mathfrak{m}_2$ the set of elements a of $\mathfrak{R}$ such that $a_1 = a$ for some sequence (a_n) satisfying (2), by $\mathfrak{m}_3$ the set of elements a of $\mathfrak{R}$ such that $c^1 = a$ for some rational semigroup $(c^t)_{t>0}$, and by $\mathfrak{m}_4$ the set of elements a of $\mathfrak{R}$ such that $a = b^1$ for some real semigroup (b^t) satisfying (4). Note that $aG \subset \mathfrak{m}_i$ for every $a \in \mathfrak{m}_i$ $(i = 1,2,3,4)$, where $G = \operatorname{Inv} \mathfrak{R}^{\#} = \exp \mathfrak{R}^{\#}$. One deduces easily from this fact and from the above remarks that $\overline{\mathfrak{m}}_1 = \overline{\mathfrak{m}}_2 = \overline{\mathfrak{m}}_3 = \overline{\mathfrak{m}}_4$.

There is some analogy between the long computations of [19] and [23]. The proof of Theorem 5.1 (which is the first step of the construction of [23]) given here suggests that these long computations could be avoided by a suitable use of the Mittag-Leffler theorem. Zouakia actually obtained in his thesis [41] such a construction. His version of the "theory of Cohen elements" [19] appears in this Volume [40].

6. Continuous semigroups and elements of finite closed descent in Banach algebras (in collaboration with Paul Koosis)

The results of Section 5 show that, if $\mathfrak{R}$ is radical and if $\mathfrak{R}$ possesses a rational semigroup, then $\mathfrak{R}$ possesses elements of finite closed descent. We will show at the end of this section that this result may fail for general (non-radical) commutative Banach algebras. The Banach algebra $\ell^1(\mathbb{Q}^+)$ has a rational semigroup, but $f \notin [f^2\ell^1(\mathbb{Q}^+)]^-$ for every nonzero element f of $\ell^1(\mathbb{Q}^+)$. Similarly, the Banach algebra $\ell^1(\mathbb{R}^+)$ possesses a nonzero real semigroup, but it does not possess any element of finite closed descent.

On the other hand, the existence of a nonzero real <u>continuous</u> semigroup in a Banach algebra A implies that A possesses elements of finite closed descent. More precisely, if $(a^t)_{t>0}$ is any real continuous semigroup in A, then there exists an infinitely-differentiable semigroup $(b^t)_{t>0}$ in A such that $[b^sA]^- = [\bigcup_{t>0} a^tA]^-$ for every $s > 0$. The key to this result is the fact that there exist functions in $L^1(\mathbb{R}^+)$ decreasing to zero arbitrarily quickly at the origin (without vanishing a.e. over $[0,\alpha]$ for any $\alpha > 0$) whose Laplace transform is an outer function on the right-hand half-plane. The main ideas of the construction seem related to some work of Mandelbrojt, but I was not able to find a reference. All this section is an unpublished joint work by Paul Koosis and the author.

In the following we consider continuous weights over $\mathbb{R}^+$, that is, we consider continuous positive functions satisfying $\omega(t+t') \le \omega(t)\omega(t')$ $(t,t' > 0)$. The elements of $L^1(\mathbb{R}^+,\omega)$ are the measurable functions f such that $\int_0^\infty |f(t)|\omega(t)dt < +\infty$, where we equate functions which agree almost everywhere. The product is the usual convolution product $(f * g)(s) = \int_0^s f(s-t)g(t)dt$, and the norm $\|\cdot\|_\omega$ of $L^1(\mathbb{R}^+,\omega)$ is given by the formula $\|f\|_\omega = \int_0^\infty |f(t)|\omega(t)dt$.

The following proposition is well known.

PROPOSITION 6.1. Let $(a^t)_{t>0}$ be a continuous semigroup in a Banach algebra A, and let ω be a weight such that $\omega(t) \ge \|a^t\|$ for every $t > 0$. The map

$\varphi : f \mapsto \int_0^\infty f(t)a^t dt$ is a continuous algebra homomorphism from $L^1(\mathbb{R}^+,\omega)$ into A, and $a^t \in [\varphi(L^1(\mathbb{R}^+,\omega))]^-$ for every $t > 0$.

Proof. The above integral is a Bochner integral ([29], Chapter 1, Sec. 3.7) computed in the closed, separable subalgebra of A generated by the (a^t). We have

$$\|\varphi(f)\| \le \int_0^\infty |f(t)|\|a^t\|dt \le \int_0^\infty |f(t)|\omega(t)dt = \|f\|_\omega ,$$

so that φ is continuous. Now let ℓ be a continuous linear form on A, and let $f,g \in L^1(\mathbb{R}^+,\omega)$. Then

$$\ell[\varphi(f*g)] = \int_0^\infty (f*g)(t)\ell(a^t)dt = \int_0^\infty \left[\int_0^t f(t-u)g(u)du\right]\ell(a^t)dt$$

$$= \int_0^\infty g(u)\left[\int_u^\infty f(t-u)\ell(a^t)dt\right]du = \int_0^\infty g(u)\left[\int_0^\infty f(t)\ell(a^{t+u})dt\right]du$$

$$= \ell\left[\iint_{\mathbb{R}^+\times\mathbb{R}^+} g(u)f(t)a^{t+u}dtdu\right] = \ell[\varphi(f)\varphi(g)] ,$$

so that $\varphi(f*g) = \varphi(f)\varphi(g)$ and φ is an algebra homomorphism.

Fix $t > 0$ and denote by u_n the characteristic function of $[t,t+1/n]$. We have

$$\|\varphi(nu_n) - a^t\| = \left\|\int_0^\infty nu_n(s)a^s ds - a^t\right\| = \left\|\int_0^\infty nu_n(s)(a^s - a^t)ds\right\|$$

$$= \left\|\int_t^{t+1/n} nu_n(s)(a^s - a^t)ds\right\| \le \sup_{t\le s\le t+1/n} \|a^s - a^t\| .$$

So $a^t = \lim_{n\to\infty} \varphi(nu_n)$, which achieves the proof of the proposition.

We now wish to exhibit in all the algebras $L^1(\mathbb{R}^+,\omega)$ some continuous semigroups $(f^t)_{t>0}$ such that f^t generates a dense principal ideal of $L^1(\mathbb{R}^+,\omega)$ for every $t > 0$. The proof will be based upon a succession of lemmas.

LEMMA 6.2. Let ω be a decreasing weight over $\mathbb{R}^+$. Then $L^1(\mathbb{R}^+,\omega)$ is a Banach $L^1(\mathbb{R}^+)$-module (the module law being given by the usual convolution of functions). Also, if $f \in L^1(\mathbb{R}^+) \cap L^1(\mathbb{R}^+,\omega)$ and if $f*L^1(\mathbb{R}^+)$ is dense in $L^1(\mathbb{R}^+)$, then $f*L^1(\mathbb{R}^+,\omega)$ is dense in $L^1(\mathbb{R}^+,\omega)$ (with respect to the norm $\|\cdot\|_\omega$).

Proof. Let $t > 0$ and let $g \in L^1(\mathbb{R}^+,\omega)$. Since $\omega(t) > 0$ for every $t > 0$, and since ω is decreasing, we have

$$\int_0^t |g(s)|ds \le \frac{1}{\omega(t)} \int_0^\infty |g(s)|\omega(s)ds < \infty .$$

So $g \in L^1_{loc}$, and $f*g$ is defined for every $f \in L^1(\mathbb{R}^+)$. We have

$$\int_0^\infty |(f*g)(t)|\omega(t)dt \le \int_0^\infty \omega(t)\left[\int_0^t |f(t-s)g(s)|ds\right] dt$$

$$\le \int_0^\infty \left[\int_0^t |f(t-s)||g(s)|\omega(s)ds\right] dt$$

$$= \int_0^\infty \left[\int_s^\infty |f(t-s)|dt\right] |g(s)|\omega(s)ds = \|f\|\|g\|_\omega .$$

This shows that $f*g \in L^1(\mathbb{R}^+,\omega)$ and that $\|f*g\|_\omega \le \|f\| \|g\|_\omega$. So $L^1(\mathbb{R}^+,\omega)$ is a Banach $L^1(\mathbb{R}^+)$-module.

The continuous functions with compact support in $(0,\infty)$ are dense in $L^1(\mathbb{R}^+,\omega)$, and they belong to $L^1(\mathbb{R}^+,\omega) \cap L^1(\mathbb{R}^+)$. Such a function f can be written in the form $f = g*h$, where $g,h \in L^1(\mathbb{R}^+)$, by Cohen's factorization theorem ([12] or [9], p. 61). It follows from Titchmarsh's convolution theorem that g or h, say h, vanishes almost everywhere over $[0,\delta]$ for some $\delta > 0$ and thus belongs to $L^1(\mathbb{R}^+,\omega)$. So $L^1(\mathbb{R}^+) * L^1(\mathbb{R}^+,\omega)$ is dense in $L^1(\mathbb{R}^+,\omega)$. As $L^1(\mathbb{R}^+)$ has bounded approximate identities, $L^1(\mathbb{R}^+) * L^1(\mathbb{R}^+,\omega)$ is closed in $L^1(\mathbb{R}^+,\omega)$ ([36], Theorem 1). So $L^1(\mathbb{R}^+) * L^1(\mathbb{R}^+,\omega) = L^1(\mathbb{R}^+,\omega)$.

Now let $f \in L^1(\mathbb{R}^+) \cap L^1(\mathbb{R}^+,\omega)$ and assume that $[f*L^1(\mathbb{R}^+)]^- = L^1(\mathbb{R}^+)$. Let $h \in L^1(\mathbb{R}^+,\omega)$. We have $h = g*k$ where $g \in L^1(\mathbb{R}^+)$ and $k \in L^1(\mathbb{R}^+,\omega)$, and then $\|g - f*f_n\| \xrightarrow[n\to\infty]{} 0$ for some sequence (f_n) in $L^1(\mathbb{R}^+)$.

We obtain

$$\limsup_{n\to\infty} \|h - f*f_n*k\|_\omega \le \|k\|_\omega \lim_{n\to\infty} \|g - f*f_n\| = 0 .$$

Since $f_n*k \in L^1(\mathbb{R}^+,\omega)$ for every $n \in \mathbb{N}$, this achieves the proof of the lemma.

We state as a lemma a classical result of Nyman. (An accessible proof of this result is given by Dales [14] in this Volume.)

LEMMA 6.3. (Nyman, [34]). Let f be an element of $L^1(\mathbb{R}^+)$. Then the principal ideal generated by f is dense in $L^1(\mathbb{R}^+)$ if and only if f satisfies the following two conditions.

(1) $\mathcal{L}(f)(z) \ne 0$ $(z \in \mathbb{C},\ \mathrm{Re}\, z \ge 0)$.

(2) f does not vanish a.e. on any segment $[0,\alpha]$ with $\alpha > 0$.

The following lemma is elementary, and certainly well known.

LEMMA 6.4. Let ω be a continuous function over $(0,\infty)$ such that $\limsup_{t\to\infty} \omega(t) < +\infty$ and $\lim_{t\to 0^+} \omega(t) = +\infty$. Then there exists a convex decreasing positive function m over $(0,+\infty)$ which is twice continuously differentiable and satisfies the condition that $\limsup_{t\to 0^+} \omega(t)\exp(-\alpha m(t/\alpha)) < +\infty$ for every $\alpha > 0$.

We need the two following remarks.

(1) If f is a continuous, positive function over $(0,\infty)$ such that $\limsup_{t\to\infty} f(t) < +\infty$, there exists a continuous decreasing function g over $(0,\infty)$ such that $g(t) \geq f(t)$ for every $t > 0$.

(2) Let g be a continuous decreasing function over $(0,\infty)$. There exists a continuously differentiable positive function h over $(0,\infty)$ such that $h(t) \geq g(t)$ for every $t > 0$.

To prove (1), put $g(t) = \sup_{s \geq t} f(s)$. Then $g(t) \geq f(t)$ $(t > 0)$. Also, if $t,h > 0$, then $g(t) \geq g(t+h)$ and $g(t) = \max\{g(t+h), \sup_{t\leq s\leq t+h} f(s)\}$. The continuity of g then follows from the uniform continuity of f on closed bounded intervals of $(0,\infty)$.

To prove (2), put $h(t) = (2/t)\int_{t/2}^{t} g(s)ds$. Then $h(t) \geq \left(\frac{2}{t}\right)\left(\frac{t}{2}\, g(t)\right) = g(t)$ $(t > 0)$, and also h is continuously differentiable.

Now let ω be a function satisfying the conditions of Lemma 6.4. Then there exists a decreasing, continuous function f_0 such that $f_0(t) \geq \omega(t)$ $(t > 0)$.

The functions $t \mapsto [f_0(t/n)]^n$ form a countable set and so there exists a continuous function f_1 such that $\liminf_{t\to 0^+} f_1(t)/[f_0(t/n)]^n = +\infty$ $(n \in \mathbb{N})$. We may assume that $f_1(t) \geq 2$ for every $t > 0$.

There exists a decreasing, continuous function f_2 such that $f_2(t) \geq f_1(t)$ $(t > 0)$, and there exists a continuously differentiable function f_3 such that $f_3(t) \geq \operatorname{Log} f_2(t)$ $(t > 0)$. If f_3 is not a decreasing function there exists $t_0 > 0$ such that $f_3'(t_0) = 0$, since $\liminf_{t\to 0^+} f_3(t) = +\infty$. Put

$$f_4(t) = f_3(t_0) \quad \text{if} \quad t > t_0\,,$$

$$f_4(t) = \int_t^{t_0} |f_3'(s)|ds - f_3(t_0) \quad \text{if} \quad t \leq t_0\,.$$

Then $f_4'(t) \leq 0$ for every $t > 0$, and $f_4(t) \geq f_3(t)$ for every $t > 0$. Also, $f_4(t) \geq \operatorname{Log} 2$ for every $t > 0$. Now set

$$m(t) = \int_0^{1/t} f_4(1/(s+1))ds\,.$$

Then m is twice continuously differentiable and decreasing over $(0,\infty)$.

Also,

$$m(t) \geq \int_{(1/t)-1}^{1/t} f_4(1/(s+1))ds \geq f_4(t) \quad (t \in (0,1)).$$

Since $s \to f_4(1/(s+1))$ is increasing, the function $\varphi : u \to \int_0^u f_4(1/(s+1))ds$ is convex. The function $\rho : t \mapsto 1/t$ is convex. So $m = \varphi \circ \rho$ is convex (φ is increasing). We have $m(t) \geq \operatorname{Log} f_1(t)$ for every $t \leq 1$, so that $e^{m(t)} \geq f_1(t)$ $(t \leq 1)$. Now fix $\alpha > 0$. There exists $n \in \mathbb{N}$ such that $1/n < \alpha$, and so

$$\alpha m\left(\frac{t}{\alpha}\right) \geq \frac{m(nt)}{n} \geq \operatorname{Log}[f_1(nt)]^{1/n} \qquad \left(t \leq \frac{1}{n}\right).$$

Since $[f_1(nt)]^{1/n} \geq f_0(t) \geq \omega(t)$ if t is small enough, we obtain

$$\liminf_{t \to 0^+} \omega(t) \exp[-\alpha m(t/\alpha)] \leq 1$$

for every $\alpha > 0$, and the lemma is proved.

The following two lemmas might well have been proved by Mandelbrojt in the thirties, but I was not able to find a reference.

LEMMA 6.5. Let m be as in Lemma 6.4. Then there exists a positive, concave, continuously differentiable increasing function φ over $(0,\infty)$ satisfying the following conditions.

(1) $\dfrac{\varphi(\sigma)}{\sigma} \xrightarrow[\sigma\to\infty]{} 0$ and $\dfrac{\varphi(\sigma)}{\sigma}$ is decreasing over $(0,\infty)$.

(2) $\varphi(\sigma) \xrightarrow[\sigma\to\infty]{} \infty$.

(3) $m(t) = \sup\limits_{\sigma > 0} (\varphi(\sigma) - \sigma t)$ for every $t > 0$.

<u>Proof</u>. For $\sigma \geq 0$, set

$$\varphi(\sigma) = \inf_{t>0} (m(t) + \sigma t).$$

The infimum is attained when $m'(t) = -\sigma$, so $\varphi(\sigma) = \sigma\rho(\sigma) + m(\rho(\sigma))$, where ρ is the function $\sigma \mapsto (m')^{-1}(-\sigma)$ and where $(m')^{-1}$ is the inverse of m'. As m is convex, positive and decreasing, m' increases from $-\infty$ to zero when t runs over $(0,\infty)$, so that ρ is well defined.

If $\sigma, \sigma' > 0$, we have

$$\varphi\left(\frac{\sigma+\sigma'}{2}\right) = \inf_{t>0} \left\{\frac{\sigma}{2}t + \frac{\sigma'}{2}t + m(t)\right\} \geq \frac{\varphi(\sigma)+\varphi(\sigma')}{2}.$$

Since ρ is continuous, φ is continuous and the above inequality shows that φ is concave. We have $\varphi(-m'(t)) = m(t) - tm'(t)$, and so $\dfrac{\varphi(-m'(t))}{-m'(t)} = \dfrac{-m(t)}{m'(t)} + t$.

The derivative of the above function is $\frac{mm''}{m'^2}$ so that $\frac{\varphi(-m'(t))}{-m'(t)}$ is increasing over $(0,\infty)$ and $\frac{\varphi(\sigma)}{\sigma}$ is decreasing over $(0,\infty)$. Also, $\varphi(\sigma) - t \leq m(t)$ for every $t > 0$, $\sigma > 0$ and $\varphi(-m'(t)) + tm'(t) = m(t)$. So $m(t) = \sup_{\sigma>0} (\varphi(\sigma) - \sigma t)$.

If $\varphi(\sigma) \geq \alpha\sigma$ for some $\alpha > 0$ and every $\sigma > 0$, then $m(\alpha/2) \geq \sup_{\sigma>0} \frac{\alpha\sigma}{2} = +\infty$, a contradiction.

This shows that $\frac{\varphi(\sigma)}{\sigma} \longrightarrow 0$ as $\sigma \to \infty$.

Since $\rho : \sigma \mapsto (m')^{-1}(-\sigma)$ is continuously differentiable, φ is continuously differentiable over $(0,\infty)$.

Put $\psi(t) = \varphi(-m'(t))$ $(t > 0)$. Then $\psi'(t) = -tm''(t)$, so that $\psi(t)$ is decreasing and $\varphi(\sigma)$ is increasing over $(0,\infty)$.

Also $\varphi(-m'(t)) = m(t) - tm'(t) \geq m(t)$, so $\liminf_{t\to0^+} \varphi(-m'(t)) = +\infty$ and $\liminf_{\sigma\to\infty} \varphi(\sigma) = +\infty$.

This achieves the proof of the lemma.

Note that φ can be extended by continuity at 0.

LEMMA 6.6. Let φ be as in Lemma 6.5. There exists a positive harmonic function F over $\Pi = \{z \in \mathbb{C} \mid \operatorname{Re} z > 0\}$ such that $F(\sigma + iy) \geq \varphi(\sigma)$ $(\sigma > 0,\ y \in \mathbb{R})$, and $\frac{F(\sigma)}{\sigma} \to 0$ as $\sigma \to \infty$.

Proof. Put $\rho(\sigma) = \varphi(\sigma) - \sigma\varphi'(\sigma)$ $(\sigma > 0)$.

Then $-\rho(\sigma)/\sigma^2$ is the derivative of $\varphi(\sigma)/\sigma$, so that $\rho(\sigma) \geq 0$ for every $\sigma > 0$. Let $\sigma, \sigma' > 0$ with $\sigma < \sigma'$. We have $\rho(\sigma') - \rho(\sigma) = \varphi(\sigma') - \varphi(\sigma) + \sigma\varphi'(\sigma) - \sigma'\varphi'(\sigma') \geq \varphi(\sigma') - \varphi(\sigma) + \sigma\varphi'(\sigma') - \sigma'\varphi'(\sigma') = [\varphi'(\sigma'') - \varphi'(\sigma')](\sigma' - \sigma)$ for some $\sigma'' \in (\sigma, \sigma')$. Since φ is concave, we obtain $\rho(\sigma') \geq \rho(\sigma)$ and so ρ is increasing. In particular, $\lim_{\sigma\to0^+} \rho(\sigma)$ exists.

We have, for $b > a > 0$,

$$\int_a^b \frac{\rho(\sigma)}{\sigma^2}\, d\sigma = \frac{\varphi(a)}{a} - \frac{\varphi(b)}{b} \quad \text{so} \quad \frac{\varphi(a)}{a} = \int_a^\infty \frac{\rho(\sigma)}{\sigma^2}\, d\sigma \quad (a > 0)\,.$$

In particular $\int_{-\infty}^{\infty} \frac{\rho(|\sigma|)d\sigma}{1+\sigma^2} < +\infty$. Put $U(x+iy) = \frac{1}{\pi}\int_{-\infty}^{\infty} \frac{x\rho(|\sigma|)}{x^2 + (y-\sigma)^2}\, d\sigma$ $(x > 0,\ y \in \mathbb{R})$. As is well known, U is a positive harmonic function ([30], Chapter 8). Also,

$$\frac{U(x+iy)}{x} = \frac{1}{\pi}\int_{-\infty}^{\infty} \frac{\rho(|\sigma|)d\sigma}{x^2 + (y-\sigma)^2} = \frac{1}{\pi}\int_{-\infty}^{\infty} \frac{\rho(|\sigma+y|)}{x^2+\sigma^2}\, d\sigma$$

$$\geq \frac{1}{\pi}\int_x^\infty \frac{\rho(|\sigma+y|)}{x^2+\sigma^2}\, d\sigma \geq \frac{1}{2\pi}\int_x^\infty \frac{\rho(|\sigma+y|)}{\sigma^2}\, d\sigma\,.$$

If $y \geq 0$, we obtain $\frac{U(x+iy)}{x} \geq \frac{1}{2\pi}\int_x^\infty \frac{\rho(\sigma)}{\sigma^2}\,d\sigma = \frac{\varphi(x)}{2\pi x}$. Since $U(x+iy) = U(x-iy)$, the same inequality holds for $y \leq 0$, and $U(x+iy) \geq \varphi(x)/2\pi$ $(x > 0, y \in \mathbb{R})$. Also, $\frac{\pi U(x)}{x} = \int_{-\infty}^{\infty} \frac{\rho(|\sigma|)}{x^2+\sigma^2}\,d\sigma$. Let (x_n) be a sequence of positive reals such that $\lim_{n\to\infty} x_n = +\infty$, and $x_n \geq 1$ for every $n \in \mathbb{N}$. It follows from the dominated convergence theorem that $\pi U(x_n)/x_n \xrightarrow[n\to\infty]{} 0$. So $U(x)/x \xrightarrow[x\to\infty]{} 0$. Taking $F = 2\pi U$, we obtain a harmonic function which satisfies the conditions of the lemma.

THEOREM 6.7. Let ω be a continuous, decreasing weight over $(0,\infty)$. Then $L^1(\mathbb{R}^+,\omega)$ contains a continuous semigroup $(a^t)_{t>0}$ such that $(a^t \cdot L^1(\mathbb{R}^+,\omega))^- = L^1(\mathbb{R}^+,\omega)$ for every $t > 0$.

Proof. Suppose first that $\liminf_{t\to 0^+} \omega(t) = +\infty$. Let m be the convex function associated with ω as in Lemma 6.4, let φ be the concave function associated with m as in Lemma 6.5, and let F be the positive harmonic function associated with φ as in Lemma 6.6. Then there exists an analytic function G over Π such that $F = \operatorname{Re} G$. Put, for $t > 0$, $\operatorname{Re} z > -1$,

$$f^t(z) = \exp[-tG(z+1) - t(z+1)^{1/2}].$$

Here we denote by $(z+1)^{1/2}$ the complex number $\exp[\frac{1}{2}\operatorname{Log}(z+1)]$ if $\operatorname{Re} z > -1$, where $\operatorname{Log} u$ is the determination of the logarithm in Π which takes real values over $\mathbb{R}^+$. Clearly $f^{t+t'}(z) = f^t(z)\cdot f^{t'}(z)$.

Also $|f^t(z)| \leq |\exp[-t(z+1)^{1/2}]| \leq \exp[-t\sqrt{|z+1|}/\sqrt{2}]$ $(t > 0, \operatorname{Re} z > -1)$.

So the inverse Laplace transform $\mathcal{L}^{-1}(f^t)$ exists and

$$|\mathcal{L}^{-1}(f^t)(x)| = \frac{1}{2\pi}\left|\int_{-\infty}^{\infty} e^{-x/2}\, e^{ixy}\, f^t(-\tfrac{1}{2}+iy)\,dy\right| \leq K(t)e^{-x/2},$$

where $K(t)$ depends only on t. Hence $a^t = \mathcal{L}^{-1}(f^t) \in L^1(\mathbb{R}^+)$ for every $t > 0$.

Now if $0 < \alpha < \beta$, $|f^t(z)|$ converges uniformly to zero over $[\alpha,\beta]$ when $|z| \to \infty$ with $\operatorname{Re} z \geq -\frac{1}{2}$.

It follows from this observation that $\sup_{\operatorname{Re} z \geq -\frac{1}{2}} |f^{t+h}(z) - f^t(z)| \xrightarrow[h\to 0]{} 0$ for every $t > 0$.

We have

$$|a^t(x) - a^{t+h}(x)| \leq e^{-x/2} \sup_{\operatorname{Re} z \geq -\frac{1}{2}} |f^{(t/2)+h}(z) - f^{t/2}(z)| \left(\frac{1}{2\pi}\int_{-\infty}^{\infty} |f^{t/2}(-\tfrac{1}{2}+iy)|\,dy\right).$$

So

$$\|a^t - a^{t+h}\| \leq \left[\frac{1}{\pi} \int_{-\infty}^{\infty} |f^{t/2}(-\tfrac{1}{2} + iy)| dy \right] \sup_{\text{Re } z \geq -\frac{1}{2}} |f^{(t/2)+h}(z) - f^{t/2}(z)| \|e^{-x/2}\|,$$

and hence $(a^t)_{t>0}$ is a continuous semigroup in $L^1(\mathbb{R}^+)$.

If $x > 0$, $\sigma > -1$ and $t > 0$ we have

$$\begin{aligned} |a^t(x)| &\leq \left| \frac{1}{2\pi} \int_{-\infty}^{\infty} f^t(\sigma + iy)\, e^{\sigma x + ixy} dt \right| \\ &\leq \frac{1}{2\pi} \left[\int_{-\infty}^{\infty} |e^{-t(1+\sigma+iy)^{1/2}}| dy \right] \sup_{y \in \mathbb{R}} e^{-tF(\sigma+1+iy)+\sigma x} \\ &\leq \frac{1}{2\pi} \left[\int_{-\infty}^{\infty} |e^{-t(1+\sigma+iy)^{1/2}}| dy \right] e^{\sigma x - t\varphi(\sigma+1)} \end{aligned}$$

Note that $\text{Re}(\sigma + iy)^{1/2} = |\sigma + iy|^{1/2} \cos \frac{\text{Arg}(\sigma + iy)}{2} \geq \frac{|y|^{1/2}}{\sqrt{2}}$ if $\sigma > 0$. So there exists $M(t) > 0$ such that

$$|a^t(x)| \leq M(t)e^{-x} \cdot \inf_{\sigma > 0} e^{\sigma x - t\varphi(\sigma)} = M(t)e^{-x}\, e^{-tm(x/t)} \leq M(t)e^{-tm(x/t)}.$$

We obtain $\limsup_{x \to 0^+} |a^t(x)\omega(x)| \leq M(t) \limsup_{x \to 0^+} \omega(x)e^{-tm(x/t)} < +\infty$ for every $t > 0$.

Since ω is decreasing and since $a^t \in L^1(\mathbb{R}^+)$, this implies that $a^t \in L^1(\mathbb{R}^+,\omega)$ for every $t > 0$.

We have $\|a^{t+h} - a^t\|_\omega \leq \|a^{t/2}\|_\omega \, \|a^{(t/2)+h} - a^{t/2}\|$, so that $(a^t)_{t>0}$ is a continuous semigroup in $L^1(\mathbb{R}^+,\omega)$. Let $t > 0$ and assume that $a^t(x) = 0$ a.e. over $[0,\alpha]$ for some $\alpha \geq 0$. Then $f^t(z) = \mathcal{L}(a^t)(z) = \int_\alpha^\infty a^t(x)e^{-xz}dx$, so that $|f^t(z)| \leq e^{-\alpha \text{Re } z}\|a^t\|$ $(\text{Re } z \geq 0)$. Hence $e^{-t\sqrt{x+1}} \cdot e^{-tF(x+1)} \leq Ke^{-\alpha x}$ (where K does not depend on x) for every $x > 0$. Since $\frac{F(x)}{x} \xrightarrow[x \to \infty]{} 0$ we obtain eventually $e^{\alpha x} \leq Ke^{t\sqrt{x+1}} \cdot e^{(\alpha/2)(x+1)}$ which shows that $\alpha = 0$. Clearly, $\mathcal{L}(a^t)$ does not have any zero z with $\text{Re } z \geq 0$. It follows then from Nyman's theorem that $[a^t \cdot L^1(\mathbb{R}^+)]^- = L^1(\mathbb{R}^+)$ for every $t > 0$. Using Lemma 6.2 we see that $[a^t \cdot L^1(\mathbb{R}^+,\omega]^- = L^1(\mathbb{R}^+,\omega)$ and the theorem is now proved in the case that ω is decreasing and $\liminf_{t \to 0^+} \omega(t) = +\infty$.

If ω is decreasing and bounded, the natural injection from $L^1(\mathbb{R}^+)$ into $L^1(\mathbb{R}^+,\omega)$ is continuous and has dense range, and we can take for example $a^t(x) = e^{-x}t^{x-1}/\Gamma(t)$, which gives, as is well known, a continuous (and even analytic) semigroup in $L^1(\mathbb{R}^+)$ satisfying the desired conditions in $L^1(\mathbb{R}^+)$, and hence in $L^1(\mathbb{R}^+,\omega)$. This concludes the proof of the theorem.

THEOREM 6.8. Let A be a Banach algebra, and let $(a^t)_{t>0}$ be a continuous semigroup in A over the positive reals. There exists an infinitely differentiable semigroup $(b^t)_{t>0}$ in A such that $[b^tA]^- = [\bigcup_{s>0} a^sA]^-$ for every $t > 0$.

Proof. Denote by ν the spectral radius of $a = a^1$. Then $\|a^t\|^{1/t} \xrightarrow[t\to\infty]{} \nu$ so that $(1+\nu)^{-t} a^t \xrightarrow[t\to\infty]{} 0$ and so we may assume without loss of generality that $\|a^t\| \xrightarrow[t\to\infty]{} 0$. Put $\varphi(t) = \sup_{s\geq 0} \|a^{t+s}\|$. As in the proof of Lemma 6.4 we see that φ is continuous over $(0,\infty)$ and decreasing. Also,

$$\varphi(t+t') = \sup_{s\geq 0} \|a^{t+t'+s}\| \leq \|a^t\| \sup_{s\geq 0} \|a^{t'+s}\| = \|a^t\|\varphi(t') \leq \varphi(t)\varphi(t'),$$

since $\|a^t\| \leq \varphi(t)$ for every $t > 0$. If φ is bounded over $(0,\infty)$, put $\omega(t) = \sup_{s>0} \varphi(s)$. Then ω is constant and $\omega(t) \geq 1$ (necessarily $\varphi(t/n) \geq [\varphi(t)]^{1/n}$ for every $t > 0$ and every $n \in \mathbb{N}$). If φ is unbounded over $(0,\infty)$ there exists $t_0 > 0$ such that $\varphi(t_0) = 1$ because $\varphi(t) \xrightarrow[t\to\infty]{} 0$. Then put $\omega(t) = \varphi(t)$ if $t \leq t_0$, $\omega(t) = 1$ if $t \geq t_0$. We clearly obtain a continuous, positive, decreasing weight, even if a^t is a nilpotent semigroup.

Let θ be the map from $L^1(\mathbb{R}^+,\omega)$ into A defined in Proposition 6.1, and let $(v^t)_{t>0}$ be a continuous semigroup in $L^1(\mathbb{R}^+,\omega)$ such that v^t generates a dense principal ideal in $L^1(\mathbb{R}^+,\omega)$ for every $t > 0$. Put $c^t = \theta(v^t)$ $(t > 0)$. Then $(c^t)_{t>0}$ is a continuous semigroup in A.

Also, $a^s \in [\theta(L^1(\mathbb{R}^+,\omega))]^- \subset [c^tA]^-$ for every $s,t > 0$ so $[c^tA]^- \supset [\bigcup_{s>0} a^sA]^-$. Since $c^t \in \theta(L^1(\mathbb{R}^+,\omega))$, we have $c^t \in (\mathrm{lin}\{a^s\}_{s>0})^-$, and finally $[c^tA]^- = [\bigcup_{s>0} a^sA]^-$ for every $t > 0$.

Now let $(d^t)_{t>0}$ be the infinitely-differentiable semigroup in $L^1(\mathbb{R}^+)$ constructed in [21], Lemma 3.4.

We have $d^t(x) = 0$ if $x \leq t$, $d^t(x) = \dfrac{t}{\sqrt{\pi}(x-t)^{3/2}} \exp\left(-\dfrac{t^2}{x-t}\right)$ if $x > t$. It is verified in [21] that $(d^t)_{t>0}$ is in fact an infinitely-differentiable semigroup in $L^1(\mathbb{R}^+,\omega)$ for every continuous weight ω such that $\sup_{s>0} \omega(1+s) < +\infty$.

By multiplying $(c^t)_{t>0}$ by $e^{-\lambda t}$ for some suitable $\lambda \in \mathbb{R}$ if necessary, we may assume that $\sup_{t\geq 1} \|c^t\| < +\infty$. Put $b^t = \int_0^\infty d^t(s)c^s ds$. Then $(b^t)_{t>0}$ is the image of $(d^t)_{t>0}$ by the continuous homomorphism $\psi : g \mapsto \int_0^\infty g(s)c^s ds$ from $L^1(\mathbb{R}^+, \|c^t\|)$ into A. So $(b^t)_{t>0}$ is infinitely differentiable.

Also, as noticed in [21], we have $d^t = \delta_t * \mathcal{L}^{-1}(e^{-2tz^{1/2}})$ for every $t > 0$, where δ_t is the Dirac measure at t.

It follows from Nyman's theorem that the principal ideal generated by $\mathcal{L}^{-1}(e^{-2tz^{1/2}})$ is dense in $L^1(\mathbb{R}^+)$ for every $t > 0$. Let χ_n be the characteristic function of $[0,1/n]$. For every $n \in \mathbb{N}$ there exists by the above observation a sequence (u_m) in $L^1(\mathbb{R}^+)$ such that $\delta_t * \chi_n = \lim_{m\to\infty} d^t * u_m$. Since

$L^1(\mathbb{R}^+, \|c^t\|) \cap L^1(\mathbb{R}^+)$ is dense in $L^1(\mathbb{R}^+)$, we may assume that $u_m \in L^1(\mathbb{R}^+, \|c^t\|)$ for every $m \in \mathbb{N}$, and since $\delta_t * \chi_n$ and $d^t * u_m$ vanish over $[0,t)$, we have $\|\delta_t * \chi_n - d^t * u_m\|_{\omega_1} \xrightarrow[m\to\infty]{} 0$, where $\omega_1(t) = \|c^t\|$ $(t > 0)$.

So $\psi(\delta_t * \chi_n) \in [b^t A]^-$ for every $t > 0$. As $\psi(n\delta_t * \chi_n) \xrightarrow[n\to\infty]{} c^t$ we obtain $c^t \in [b^t A]^-$. Since $b^t \in [\bigcup_{s>0} c^s A]^- = [c^t A]^-$, we have $[b^t A]^- = [c^t A]^- = [\bigcup_{s>0} a^s A]^-$ for every $t > 0$, and the theorem is proved.

REMARK 6.9. (1) Let ω be any continuous weight over $(0,\infty)$. Multiplying the above semigroup d^t by $e^{\lambda t}$ with a suitable $\lambda \in \mathbb{R}$ if necessary, we obtain a semigroup d'^t such that $[\bigcup_{t>0} d'^t L^1(\mathbb{R}^+,\omega)]^- = L^1(\mathbb{R}^+,\omega)$. Theorem 6.8 then shows that Theorem 6.7 extends to any continuous weight ω over $(0,\infty)$.

(2) The proof of Theorem 6.7 shows that there exist functions $f \in L^1(\mathbb{R})^+$ which decrease arbitrarily fast at zero without vanishing over $[0,\alpha]$ for any $\alpha > 0$, and whose Laplace transform is an outer function on the half-plane. This result might be known, but I was not able to find any reference for it.

(3) We used Nyman's theorem [14], [34], in the proof of Theorem 6.7. The proof of this theorem is rather difficult, but we apply it to functions whose Laplace transform is analytic in the half-plane $\operatorname{Re} z > -1$. In that case much easier proofs are available (see [37], Appendix, for example), but we will not enter into this here.

We mentioned before the fact that the existence of nonzero rational semigroups, or even nonzero real semigroups in a semisimple Banach algebra A does not imply the existence of elements of finite closed descent. Examples are given in the following theorem.

THEOREM 6.10. (i) The commutative semisimple Banach algebra $\ell^1(\mathbb{R}^+)$ possesses a nonzero real semigroup, but $\bigcap_{n\in\mathbb{N}} f^n \ell^1(\mathbb{R}^+) = \{0\}$ for every $f \in \ell^1(\mathbb{R}^+)$. In particular $\ell^1(\mathbb{R}^+)$ does not possess any nonzero element of finite closed descent.
(ii) The commutative, semisimple, separable Banach algebra $\ell^1(\mathbb{Q}^+)$ possesses a nonzero rational semigroup, but does not possess any nonzero real semigroup.

Proof. If $t > 0$, denote by δ_t the Dirac measure at t. The map $\sum_{r>0} \lambda_r X^r \mapsto \sum_{r>0} \lambda_r \delta_r$ is an isometry from $\ell^1(\mathbb{R}^+)$ into $M(\mathbb{R}^+)$, the usual convolution algebra of bounded measures with support in $(0,\infty)$. (Recall that if $\mu \in M(\mathbb{R}^+)$, $\|\mu\| = \int_0^\infty d|\mu|(t)$ where $|\mu|$ is the total variation of μ.)

If $f \in \ell^1(\mathbb{R}^+)$, say $f = \sum_{r>0} \lambda_r X^r$, put, for $\operatorname{Re} z \geq 0$, $\mathcal{L}(f)(z) = \sum_{r>0} \lambda_r e^{-rz}$. Denote by A^∞ the algebra of all continuous, bounded functions over the closed right-hand half-plane $\overline{\Pi} = \{z \in \mathbb{C} \mid \operatorname{Re} z \geq 0\}$ which are analytic for $\operatorname{Re} z > 0$. The uniqueness theorem for Laplace transforms of measures shows that $\mathcal{L}(f) \neq 0$ if $f \neq 0$, and clearly $\mathcal{L}(f) \in A^\infty$ if $f \in \ell^1(\mathbb{R}^+)$. More precisely, $\mathcal{L}(f)$ is an almost periodic function with absolutely convergent Fourier coefficients in the sense of Bohr [8]. The notations being as above, put $\delta(f) = \inf\{t > 0 \mid \lambda_t \neq 0\}$.

It follows from a classical result of Bohr (see [33], Chapter 6, Section 2, Theorem 2) that, if $\delta(f) = 0$, or if $\lambda_{\delta(f)} = 0$, then for every $n \in \mathbb{N}$ there exists $z_n \in \mathbb{C}$ such that $\mathcal{L}(f)(z_n) = 0$ and $\mathrm{Re}\, z_n > n$. In particular $\mathcal{L}(f)$ possesses a zero z_0 such that $\mathrm{Re}\, z_0 > 0$. Now, if $g \in \bigcap_{n\in\mathbb{N}} f^n \ell^1(\mathbb{R}^+)$, then $\mathcal{L}(g)$ has a zero of infinite order at z_0, so $\mathcal{L}(g)(z) = 0$ for every z with $\mathrm{Re}\, z \geq 0$, and $g = 0$ by the uniqueness theorem for Laplace transforms of measures.

Now assume that $\delta(f) > 0$. Then $\delta(f^n) \geq n\delta(f)$ (in fact equality holds by Titchmarsh's convolution theorem for measures) and, if $g \in \bigcap_{n\in\mathbb{N}} f^n \ell^1(\mathbb{R}^+)$, then $\delta(g) = +\infty$, that is $g = 0$. So $\bigcap_{n\in\mathbb{N}} f^n \ell^1(\mathbb{R}^+) = \{0\}$ for every $f \in \ell^1(\mathbb{R}^+)$. It follows then from a result by G. R. Allan [1] that $\ell^1(\mathbb{R}^+)$ does not possess any nonzero element of finite closed descent (this follows also from Corollary 3.2 or could easily be proved directly in our case using Cauchy's inequalities).

The above results extend, of course, to $\ell^1(\mathbb{Q}^+)$, for it may be considered as a closed subalgebra of $\ell^1(\mathbb{R}^+)$. Let $f \in \ell^1(\mathbb{Q}^+)$ with $f \neq 0$. If $\delta(f) = 0$, or if $\delta(f) > 0$ and $\lambda_{\delta(f)} = 0$, then $\mathcal{L}(f)$ vanishes at some $z_0 \in \mathbb{C}$ with $\mathrm{Re}\, z_0 > 0$. Thus $\mathcal{L}(f)$ cannot have any root in A^∞ of order greater than ρ where ρ is the order of the zero at z_0. So there exists no rational semigroup $(a^t)_{t\in\mathbb{Q}^+}$ in $\ell^1(\mathbb{Q}^+)$ such that $f = a^1$.

Now assume that $\ell^1(\mathbb{Q}^+)$ possesses a nonzero real semigroup $(a^t)_{t>0}$. As $\ell^1(\mathbb{Q}^+)$ is an integral domain (this follows from the fact that the map $f \mapsto \mathcal{L}(f)$ is one-to-one), we have $a^t \neq 0$ for every $t > 0$. Put $\rho_t = \delta(a^t)$ $(t > 0)$. Since a^t belongs to a rational semigroup of $\ell^1(\mathbb{Q}^+)$ we have $\rho_t > 0$ and ρ_t belongs to the support of a^t. So $\rho_t \in \mathbb{Q}^+$. Also it follows from the definition of the product of $\ell^1(\mathbb{Q}^+)$ that $\rho_{t+t'} = \rho_t + \rho_{t'}$ $(t,t' > 0)$.

This equality shows that the map $t \mapsto \rho_t$ has to be both one-to-one and countably valued over $(0,\infty)$, a contradiction.

So $\ell^1(\mathbb{Q}^+)$ does not possess any nonzero real semigroup, which achieves the proof of the theorem.

It follows from the results of Section 4 that, if a separable Banach algebra A possesses a nonzero real semigroup, then A has nontrivial p.a.i.. I do not know whether A necessarily possesses nonzero elements of finite closed descent (this is of course the case when A is radical).

7. Examples and counterexamples

The aim of this section is to give concrete examples of radical Banach algebras satisfying some properties investigated before in this paper without satisfying some other ones. We will first give examples related to Sections 3 and 5, and then show that radical Banach algebras may possess analytic semigroups $(a^t)_{|\mathrm{Arg}\, t| < \eta}$ in some angles without having any analytic semigroups in larger angles. We will show too that analytic semigroups $(a^t)_{\mathrm{Re}\, t > 0}$ bounded in the right-hand half-unit disc do exist in some radical Banach algebras without divisors of zero, but do not exist in all radical Banach algebras with b.a.i.

We first give an example of a radical Banach satisfying the conditions of Corollary 3.5, (the existence of a nontrivial p.a.i. in this algebra is nontrivial). I claimed in various places, including at the Conference, that such an algebra did not satisfy the Condition 3.7 of Section 3, but it appears that I cannot prove it.

Let S_1 be the set of all sequences $(t_n)_{n \geq 0}$ of integers which vanish eventually and which satisfy the conditions that $t_0 \geq 0$, $\inf_{n \geq 1} t_n \geq -t_0$, $\sup_{n \geq 0} t_n > 0$.

We equip S_1 with coordinatewise addition (it is easy to see that S_1 is stable under sums). Now let $(r_n)_{n \geq 0}$ be a sequence of positive reals such that the set $\{r_n\}_{n \geq 0}$ is linearly independent over $\mathbb{Q}$ and such that $r_n < 2^{-n}$ $(n \in \mathbb{N})$ and $r_0 = 1$. Then $\sum_{n=1}^{\infty} r_n < 1$. If $\sigma = (t_n)_{n \geq 0} \in S_1$, put $\varphi(\sigma) = \sum_{n \geq 0} t_n r_n$. Then φ is one-to-one.

Also $\varphi(\sigma + \sigma') = \varphi(\sigma) + \varphi(\sigma')$ $(\sigma, \sigma' \in S_1)$.

If $t_0 = 0$, then $t_n \geq 0$ for every $n \geq 1$, so $\varphi(\sigma) > 0$. If $t_0 > 0$ then $\varphi(\sigma) \geq t_0 - t_0(\sum_{n \geq 1} r_n) > 0$.

Now put $\omega(\sigma) = e^{-(\varphi(\sigma))^2}$ $(\sigma \in S_1)$. We clearly obtain a radical weight over S_1.

PROPOSITION 7.1. The commutative radical Banach algebra $\mathcal{R}_1 = \ell^1(S_1, \omega)$ is an integral domain, and it satisfies the conditions of Corollary 3.5.

<u>Proof</u>. If $u = \sum_{\sigma \in S_1} \lambda_\sigma X^\sigma$ is an element of $\ell^1(S_1, \omega)$, put $\psi(u) = \sum_{t \in \varphi(S_1)} \lambda_{\varphi^{-1}(t)} X^t$. The map ψ is well defined, and ψ is an algebra homomorphism from $\ell^1(S_1, \omega)$ into $\ell^1(\mathbb{R}^+, e^{-t^2})$. As $\sum_{t \in \varphi(S_1)} |\lambda_{\varphi^{-1}(t)}| e^{-t^2} = \sum_{\sigma \in S_1} |\lambda_\sigma| \omega(\sigma)$, ψ is an isometry. We may identify $\ell^1(\mathbb{R}^+, e^{-t^2})$ with a subalgebra of the convolution measure algebra $M(\mathbb{R}^+, e^{-t^2})$ as in Section 4, so $\ell^1(\mathbb{R}^+, e^{-t^2})$ is an integral domain (the fact that $M(\mathbb{R}^+, e^{-t^2})$ is an integral domain follows from Titchmarsh's convolution theorem for measures).

For $m \geq 0$ put $\sigma_m = (\delta_{m,n})_{n \geq 0}$, where $\delta_{m,n}$ is the usual Kronecker symbol. Then $\sigma_0 - \sigma_1 - \cdots - \sigma_m \in S_1$ for every $m \in \mathbb{N}$, $X^{\sigma_0} \neq 0$ and $X^{\sigma_0 - \sigma_1 - \cdots - \sigma_m} = X^{\sigma_{m+1}} \cdot X^{\sigma_0 - \sigma_1 - \cdots - \sigma_{m+1}}$ for every m. So $\mathcal{R}_1$ satisfies condition (2) of Corollary

3.5, since $X^{\sigma_0} \in \underleftarrow{\lim}\, X^{\sigma_m} \dots X^{\sigma_1} \mathfrak{R}_1$.

I thought that $a \notin [a^2\mathfrak{R}_1]^-$ for every nonzero $a \in \mathfrak{R}_1$, but I was unable to prove this.

Now denote by S_2 the set of all elements (t_0,t_1) of the additive group $\mathbb{Z} \times \mathbb{Z}$ such that $t_0 \geq 0$ and such that $t_1 > 0$ if $t_0 = 0$.

Then S_2 is stable under addition, so that S_2 is an abelian monoid. Let $f \in \ell^1(\mathbb{Q}^+, e^{-t^2})$ such that $f\ell^1(\mathbb{Q}^+, e^{-t^2})$ is dense in $\ell^1(\mathbb{Q}^+, e^{-t^2})$.

It follows from a result of Allan [1] that $I = \bigcap_{n\in\mathbb{N}} f^n \ell^1(\mathbb{Q}^+, e^{-t^2})$ is dense in $\ell^1(\mathbb{Q}^+, e^{-t^2})$, and hence not reduced to $\{0\}$. Let g be a nonzero element of I. If $p/q > 0$ denote by g^p/f^q the unique element h of $\ell^1(\mathbb{Q}^+, e^{-t^2})$ such that $hf^q = g^p$.

Put $\omega[t_0,t_1)] = \|g^{t_0} f^{t_1}\|$ if $t_0, t_1 \geq 0$ and $\omega[(t_0,t_1)] = \|g^{t_0}/f^{t_1}\|$ if $t_0 > 0$, $t_1 < 0$. Then ω is clearly a weight over S_2 and $[\omega(n\sigma)]^{1/n} \xrightarrow[n\to\infty]{} 0$ for every $\sigma \in S_2$, since $\ell^1(\mathbb{Q}^+, e^{-t^2})$ is radical.

THEOREM 7.2. The commutative radical Banach algebra $\mathfrak{R}_2 = \ell^1(S_2,\omega)$ satisfies the conditions of Theorem 3.7, but does not satisfy the conditions of Theorem 5.3.

<u>Proof</u>. We first show that $\mathfrak{R}_2$ does not satisfy the conditions of Theorem 5.3. Let $u = \sum_{t\in S_2} \lambda_t X^t$ be a nonzero element of $\mathfrak{R}_2$. If $t_0 = 0$ for some $(t_0,t_1) \in$ Supp u, denote by p the smallest integer such that $(0,p) \in$ Supp u. Then $p > 0$. If $v \in \ell^1(S_2,\omega)$ and if $(0,q) \in \text{Supp}\, u^2v$, then certainly $q = p_1 + p_2 + p_3$ where $(0,p_1) \in$ Supp u, $(0,p_2) \in$ Supp u, $(0,p_3) \in$ Supp v. So $q > 2p$ and $(0,p) \notin \text{Supp}\, u^2v$.

This shows that $\inf\|u - u^2v\|_\omega \geq |\lambda_\sigma|\omega(\sigma) > 0$, where $\sigma = (0,p)$.

If $t_0 > 0$ for every $(t_0,t_1) \in$ Supp u, let p be the smallest positive integer such that $(p,t) \in$ Supp u for some $t \geq 0$. Choose such an integer t and put $\sigma = (p,t)$. A similar argument shows that if $v \in \ell^1(S_2,\omega)$ and if $(t_0,t_1) \in \text{Supp}\, u^2v$ then $t_0 \geq 2p$. So $\sigma \notin \text{Supp}\, u^2v$ and hence $\inf\|u - u^2v\|_\omega \geq |\lambda_\sigma|\omega(\sigma) > 0$. This shows that $\ell^1(S_2,\omega)$ does not satisfy the conditions of Theorem 5.3.

Now put $a = X^{(0,1)}$. Then $X^{(1,0)} \in \underleftarrow{\lim}\, a^n\mathfrak{R}_2$, so a satisfies the conditions of Theorem 3.7.

I was not able to prove that $\mathfrak{R}_2$ is an integral domain (and it might be wrong). So we give another example.

It follows from [6], Theorem 3.2, that, if ω is a weight over $\mathbb{R}^+$, then for every sequence (t_n) of positive reals such that $\liminf_{n\to\infty} t_n^{1/n}/[\omega(n\delta)]^{1/n} = +\infty$ for every $\delta > 0$, there exists $f \in L^1(\mathbb{R}^+,\omega)$ such that $\|f^{*n}\| < t_n$ for every $n \in \mathbb{N}$ and such that f does not vanish a.e. over $[0,\alpha]$ for any $\alpha > 0$. Put $\omega(t) = e^{-t^3}$.

The sequence $(t_n) = e^{-n^2}$ satisfies the desired condition. Let $f \in L^1(\mathbb{R}^+, e^{-t^3})$ be as above. By Domar's theorem [16] the principal ideal generated by f in $L^1(\mathbb{R}^+, e^{-t^3})$ is dense in $L^1(\mathbb{R}^+, e^{-t^3})$, so that $\bigcap_{n\in\mathbb{N}} f^n L^1(\mathbb{R}^+, e^{-t^3}) \neq \{0\}$.

Denote by $\mathfrak{m}$ the set of all elements g of $L^1(\mathbb{R}^+, e^{-t^3})$ which vanish a.e. over $[0,1]$. The map $\varphi : \sum_{n\geq 1} \lambda_n X^n \mapsto \sum_{n\geq 1} \lambda_n f^n$ is an algebra homomorphism from $\ell^1(\mathbb{N}, e^{-n^2})$ into $L^1(\mathbb{R}^+, e^{-t^3})$ and it defines an $\ell^1(\mathbb{N}, e^{-n^2})$-module structure over $\mathfrak{m}$. Denote by $\mathfrak{R}_3$ the Banach algebra $\ell^1(\mathbb{N}, e^{-n^2}) \oplus \mathfrak{m}$ (Definition 2.4). We clearly obtain a commutative radical Banach algebra.

THEOREM 7.3. The commutative radical Banach algebra $\mathfrak{R}_3$ satisfies Condition 3.7 and it is an integral domain, but $\mathfrak{R}_3$ does not satisfy the conditions of Theorem 5.3.

Proof. First note that $\operatorname{Ker}\varphi$ is a closed ideal of $\ell^1(\mathbb{N}, e^{-n^2})$ so if $\operatorname{Ker}\varphi \neq \{0\}$ then $\operatorname{Ker}\varphi = [X^p \ell^1(\mathbb{N}, e^{-n^2})]^-$ for some $p > 0$ by a well known result of S. Grabiner [27]. As $\varphi(X^p) = f^p \neq 0$ for every $p \in \mathbb{N}$, we see that φ is in fact one-to-one. Also, $\ell^1(\mathbb{N}, e^{-n^2})$ is an integral domain.

Now if $(a+x)(b+y) = 0$ where $a,b \in \ell^1(\mathbb{N}, e^{-n^2})$ and $x,y \in \mathfrak{m}$, we have $ab = 0$, $xy + bx + ay = 0$. So a or b, say b, equals 0. We get $ay + xy = 0$. Put $\alpha(u) = \sup\{\alpha \geq 0 \mid u = 0 \text{ a.e. over } [0,\alpha]\}$. We have $\alpha(ay) = \alpha(y) + \alpha(a) = \alpha(y) < \alpha(xy)$ if $y \neq 0$ (here we use Titchmarsh's convolution theorem), so $y = 0$ and $b + y = 0$. This shows that $\mathfrak{R}_3$ is an integral domain.

Let x be a nonzero element of $\bigcap_{n\in\mathbb{N}} f^n L^1(\mathbb{R}^+, e^{-t^3})$. Then $xy \in (\bigcap_{n\in\mathbb{N}} f^n L^1(\mathbb{R}^+, e^{-t^3})) \cap \mathfrak{m}$, and $xy \neq 0$ for every nonzero $y \in \mathfrak{m}$. Also $y(x/f^n) \in \mathfrak{m}$ for every $n \geq 1$. This shows that $\bigcap_{n\in\mathbb{N}} X^n \mathfrak{R}_3 = \varprojlim X^n \mathfrak{R}_3$ is not reduced to zero, and hence $\mathfrak{R}_3$ satisfies the conditions of Theorem 3.7.

Now let $a + x$ ($a \in \ell^1(\mathbb{N}, e^{-n^2})$, $x \in \mathfrak{m}$) be a nonzero element of $\mathfrak{R}_3$. If $a \neq 0$ we have $a = \sum_{n\geq p} \lambda_n X^n$ with $\lambda_p \neq 0$ and $\|(a+x) - (a+x)^2(b+y)\| \geq |\lambda_p| e^{-p^2} > 0$ ($b \in \ell^1(\mathbb{N}, e^{-n^2})$, $y \in \mathfrak{m}$). If $a = 0$ then $\|x - x^2(b+y)\| \geq \int_{\alpha(x)}^{2\alpha(x)} e^{-t^3} |x(t)|\,dt > 0$ ($b \in \ell^1(\mathbb{N}, e^{-n^2})$, $y \in \mathfrak{m}$). This achieves the proof.

If $\mathfrak{R}$ is a radical Banach algebra, denote by $\mathfrak{R}^{[n]}$ the set of all products of n elements of $\mathfrak{R}$. If $\mathfrak{R}$ satisfies the conditions of Corollary 3.5, then obviously $\bigcap_{n\in\mathbb{N}} \mathfrak{R}^{[n]} \neq \{0\}$. The following example shows that the condition $\bigcap_{n\in\mathbb{N}} \mathfrak{R}^{[n]} \neq \{0\}$ is strictly weaker than the conditions of Corollary 3.5, even if we limit our attention to the class of commutative radical Banach algebras without divisors of zero.

Denote by S_3 the set of positive rationals r which can be written in the form $r = \frac{m_1}{p_1} + \cdots + \frac{m_k}{p_k}$, where $k \geq 1$, $m_1,\ldots,m_k \geq 1$ and where $p_1,\ldots,p_k$ are prime numbers.

For $r \in S_3$, put $\omega(r) = e^{-r^2}$. The Banach algebra $\mathfrak{R}_4 = \ell^1(S_3, e^{-r^2})$ is commutative and radical and it can be identified with a subalgebra of $\ell^1(\mathbb{Q}^+, e^{-t^2})$, so that $\ell^1(S_3, e^{-r^2})$ is an integral domain.

THEOREM 7.4. The commutative radical Banach algebra $\mathfrak{R}_4$ is an integral domain, and it possesses a nonzero element with roots of all prime orders, but $\mathfrak{R}_4$ does not satisfy the conditions of Corollary 3.5.

Proof. If p is prime then $1/p \in S_3$, so $X = (X^{1/p})^p$ has roots of all prime orders.

Let u be a nonzero element of $\mathfrak{R}_4$. If the support of u has a smallest element t, then $t \notin \mathrm{Supp}\, uv$ for each $v \in \mathfrak{R}_4$. Hence, $\inf_{v\in\mathfrak{R}_4} \|u - uv\| \geq |\lambda_t|\, e^{-t^2} > 0$.

Now suppose that the support of u has no smallest element. If $r \in S_3$ and $r < 1$, then, using elementary arguments of number theory, we see that the decomposition $r = \frac{m_1}{p_1} + \cdots + \frac{m_k}{p_k}$ with $p_1,\ldots,p_k$ prime and $m_1,\ldots,m_k$ positive is unique. In this case, put $\varphi(r) = m_1 + \cdots + m_k$. Note that, if $r, r' \in (0,1)$ and $r + r' \in (0,1)$, then $\varphi(r + r') = \varphi(r) + \varphi(r')$.

Now write $u = \sum_{t\in S_3} \lambda_t X^t$ and put $\alpha = \inf\{t \in S_3 \mid \lambda_t \neq 0\}$. We may choose $\eta > 0$ such that $(\alpha, \alpha + \eta)$ does not contain any integer. Let n be the greatest integer verifying $n \leq \alpha$. Then $(\alpha, \alpha + \eta) \cap \mathrm{Supp}\, u \neq \emptyset$. If $r \in (\alpha, \alpha + \eta)$ put $s(r) = r - n$ so that $s(r) \in (0,1)$.

Now choose $t \in (\alpha, \alpha + \eta) \cap \mathrm{Supp}\, u$ such that $\varphi(s(t)) \leq \varphi(s(r))$ for every $r \in (\alpha, \alpha + \eta) \cap \mathrm{Supp}\, u$. Let $v \in \mathfrak{R}_4$. If $\tau \in \mathrm{Supp}\, uv$, $\alpha < \tau < \alpha + \eta$ we have $\tau = \tau_1 + \tau_2$ where $\tau_1 \in \mathrm{Supp}\, u$ and $\tau_2 \in \mathrm{Supp}\, v$. So $\alpha < \tau_1 < \tau$ and $\tau_2 \in (0,1)$. We obtain $s(\tau) = s(\tau_1) + \tau_2$ so that $\varphi(s(\tau)) = \varphi(s(\tau_1)) + \varphi(\tau_2) > \varphi(s(\tau_1)) \geq \varphi(s(t))$.

So $t \neq \tau$, $t \notin \mathrm{Supp}\, uv$ and $\inf_{v\in\mathfrak{R}_4} \|u - uv\| \geq |\lambda_t|\, e^{-t^2} > 0$. This shows that $\mathfrak{R}_4$ has no nontrivial p.a.i., and the theorem is proved.

We noticed in Section 2 that, if $\bigcap_{n\in\mathbb{N}} a^n\mathfrak{R} \neq \{0\}$ for some element a of a complex algebra $\mathfrak{R}$ then $\varprojlim_{n\in\mathbb{N}} a^n\mathfrak{R} = \bigcap_{n\in\mathbb{N}} a^n\mathfrak{R} \neq \{0\}$ if $\mathfrak{R}$ is also assumed to be an integral domain. So, if $\bigcap_{n\in\mathbb{N}} a^n\mathfrak{R} \neq \{0\}$, and if $\mathfrak{R}$ is a commutative radical Banach algebra without divisors of zero, then a satisfies the conditions of

Theorem 3.7. An example of Bade, Curtis and Laursen [5] (based upon an unpublished idea of Thomas) shows that this may no longer be the case if the algebra $\mathcal{R}$ is not an integral domain. We will now show that there even exists a commutative radical Banach algebra $\mathcal{R}$ which does not satisfy the conditions of Corollary 3.5, but possesses an element a such that $\bigcap_{n\in\mathbb{N}} a^n\mathcal{R} \neq \{0\}$.

Consider the linear space $F = \prod_{p\in\mathbb{N}} \mathbb{C}^p$. If $x = (x_p)_{p\in\mathbb{N}} \in F$, we write $x_p = (x_{n,p})_{n\leq p}$ and we put $\|x_p\|_p = \sup_{n\leq p}|x_{n,p}|e^{(n+p)^2}$. Here, $x_{1,p},\dots,x_{p,p}$ belong to $\mathbb{C}$.

Put $E = \{(x_p)_{p\in\mathbb{N}} \in F \mid \sup_{p\in\mathbb{N}}\|x_p\|_p < +\infty\}$. Then E is a Banach space with respect to the norm $\|(x_p)_{p\in\mathbb{N}}\| = \sup_{p\in\mathbb{N}}\|x_p\|_p$. If $x = (x_p)_{p\in\mathbb{N}} \in E$, put $u(x) = (y_p)_{p\in\mathbb{N}}$ where $y_p = (y_{1,p},\dots,y_{p,p})$ $(p \in \mathbb{N})$ with $y_{1,1} = 1/k\ (\sum_{p=2}^{\infty} x_{1,p})$, $k = e^2(\sum_{p\geq 2} e^{-p^2})$ and $y_{n,p} = x_{n+1,p}$ if $p \geq 2$, $n \leq p-1$, $y_{p,p} = 0$ for every $p \geq 2$. Note that $\sup_{p\in\mathbb{N}}|x_{1,p}|e^{(p+1)^2} \leq \|x\|$ so that $|x_{1,p}| \leq \|x\|e^{-(p+1)^2}$ and

$$|y_{1,1}|e^2 \leq \frac{1}{\sum_{p\geq 2} e^{-p^2}}\ \|x\| \left(\sum_{p\geq 2} e^{-(p+1)^2}\right) \leq \|x\|e^{-1}.$$

Also,

$$\sup_{n\leq p}|y_{n,p}|e^{(n+p)^2} = \sup_{n\leq p-1}|x_{n+1,p}|e^{(n+p)^2}$$

$$\leq \sup_{n\leq p-1}|x_{n+1,p}|\ e^{(n+1+p)^2}\ e^{-1} \leq \|x\|e^{-1}.$$

So $\|y_p\|_p \leq e^{-1}\ \|x\|$ for every $p \in \mathbb{N}$. We obtain, for every $x \in E$, $\|u(x)\| \leq e^{-1}\|x\|$. So u is well defined, and $u \in \mathcal{L}(E)$, $\|u\| \leq e^{-1}$. If $m \geq 2$, the notations being as above, we obtain $u^m(x) = (z_p)_{p\in\mathbb{N}}$ where $z_p = (z_{n,p})_{n\leq p}$ for every $p \in \mathbb{N}$ with $z_{1,1} = (1/k)\sum_{p\geq m} x_{m,p}$, $z_{n,p} = 0$ if $p \geq 2$, $n \geq p-m+1$, and $z_{n,p} = x_{n+m,p}$ if $p \geq 2$, $n \leq p-m$. We have

$$|z_{1,1}| \leq \frac{1}{k}\sum_{p\geq m}|x_{m,p}| \leq \frac{\|x\|}{k}\left(\sum_{p\geq m} e^{-(p+m)^2}\right)$$

$$\leq \frac{e^{-m^2}}{k}\left(\sum_{p\geq m} e^{-p^2}\right)\|x\| \leq e^{-m^2-2}\ \|x\|.$$

Also, if $n \geq p-m+1$ and if $p \geq 2$, we have

$$|z_{n,p}|e^{(n+p)^2} = |x_{n+m,p}|e^{(n+p)^2} \leq |x_{n+m,p}|e^{(n+m+p)^2} \cdot e^{(n+p)^2-(n+m+p)^2}$$

$$\leq e^{-m^2}\,|x_{n+m+p}|e^{(n+m+p)^2} \leq e^{-m^2}\|x\|.$$

So $\|z_p\|_p \leq e^{-m^2}\|x\|$ $(p \in \mathbb{N})$ and $\|u^m(x)\| \leq e^{-m^2}\|x\|$ for every $x \in E$. This shows that $\|u^m\| \leq e^{-m^2}$ $(m \in \mathbb{N})$. If $\sum_{n\geq 1} \lambda_n X^n \in \ell^1(\mathbb{N}, e^{-n^2})$ then the series $\sum_{n\geq 1} \lambda_n e^{-n^2} u^n$ is absolutely convergent in $\mathcal{L}(E)$. This defines a module action of $\ell^1(\mathbb{N}, e^{-n^2})$ over E. Equip E with the trivial product $(x,y) \mapsto 0$ $(x \in E, y \in E)$ and put $\mathfrak{R}_5 = \ell^1(\mathbb{N}, e^{-n^2}) \oplus E$ (see Definition 2.4). The algebra $\mathfrak{R}_5$ is clearly commutative and radical, and we have the following theorem.

THEOREM 7.5. The commutative radical Banach algebra $\mathfrak{R}_5$ does not satisfy the conditions of Corollary 3.5, but $\bigcap_{n\in\mathbb{N}} a^n \mathfrak{R}_5 \neq \{0\}$ for some element a of $\mathfrak{R}_5$.

<u>Proof</u>. Let y be the sequence $(y_p)_{p\in\mathbb{N}}$, where $y_1 = 1$ and $y_p = 0$ for every $p \geq 2$. If $m \geq 1$ denote by y_m the sequence $(y_p^{(m)})_{p\in\mathbb{N}}$ defined by $y_p^{(m)} = 0$ if $p \neq m$, $y_{n,m}^{(m)} = 0$ if $n < m$, $y_{m,m}^{(m)} = k$. We have $y = u^m(y_m)$ for every $m \geq 1$ so that $y \in \bigcap_{n\in\mathbb{N}} a^n \mathfrak{R}_5$, where $a = X$.

Now if P is any complex polynomial without constant coefficient, if $x = (x_p)_{p\in\mathbb{N}}$ is any element of E and if $y = (y_p)_{p\in\mathbb{N}} = P(u)(x)$, we have $y_{n(p),p} = 0$ for every $p \geq 2$, where $n(p)$ is the largest integer less than or equal to p such that $x_{n(p),p} \neq 0$, (if no such integer exists we have, of course, $y_{n,p} = 0$ for every $n \leq p$). Since the polynomials in X are dense in $\ell^1(\mathbb{N}, e^{-n^2})$, the same property holds for bx where b is any element of $\ell^1(\mathbb{N}, e^{-n^2})$.

Now let $\alpha = a + x$ $(a \in \ell^1(\mathbb{N}, e^{-n^2}),\ x \in E)$ be any nonzero element of $\mathfrak{R}_5$, and let $\beta = b + y$ $(b \in \ell^1(\mathbb{N}, e^{-n^2}),\ y \in E)$ be an element of $\mathfrak{R}_5$. If $a = \sum_{n\geq 1} \lambda_n X^n \neq 0$ we have $(a+x)(b+y) = \sum_{n\geq 1} \mu_n X^n + z$, with $z \in E$ and $\mu_s = 0$, where s is the smallest integer such that $\lambda_s \neq 0$. So $\|a+x-(a+x)(b+y)\| \geq |\lambda_s|e^{-s^2}$. If $a = 0$ then $x = (x_p)_{p\in\mathbb{N}} \neq 0$. If $x_p = 0$ for every $p \geq 2$ then $bx = 0$ and $\|x(b+y) - x\| = \|x\| > 0$. If $x_p \neq 0$ for some $p \geq 2$, then $\|x(b+y) - x\| \geq |x_{n(p),p}|e^{-(p+n(p))^2}$. We thus see that $\inf_{\beta\in\mathfrak{R}_5} \|\alpha\beta - \beta\| > 0$ for every nonzero $\alpha \in \mathfrak{R}_5$. So $\mathfrak{R}_5$ does not satisfy the conditions of Corollary 3.5.

This achieves the proof.

We saw in Section 2 that, if an element a of a complex algebra A without divisors of zero has roots of all orders, then there exists a rational semigroup $(a^t)_{t\in\mathbb{Q}^+}$ such that $a^1 = a$. So, if such a nonzero element exists in a commutative, radical Banach algebra $\mathfrak{R}$, and if $\mathfrak{R}$ does not have any divisor of zero, then $\mathfrak{R}$ satisfies the conditions of Theorem 5.3.

We will now construct an example which shows that this may no longer be the case if $\mathfrak{R}$ has divisors of zero.

Let G be the abelian group with a countable family $(\delta_n)_{n\in\mathbb{N}}$ of generators and with the relations $n\delta_n = \delta_1$ $(n \in \mathbb{N})$. Note that every $s \in G$ can be written uniquely in the form $s = \sum_{n\geq 1} s(n)\delta_n$ where $0 \leq s(n) < n$ for every $n \geq 2$. We call this decomposition the standard decomposition of s. Put, for $s \in G$, the notations being as above, $\varphi(s) = \sum_{n\geq 1} s(n)/n$.

If $s = \sum_{n\geq 1} \lambda_n\delta_n$ is any decomposition of s as a finite sum of positive or negative multiples of the δ_i's, we have in fact $\varphi(s) = \sum_{n\geq 1} \lambda_n/n$ and $\varphi(s+s') = \varphi(s) + \varphi(s')$ $(s,s' \in G)$.

Denote by S_4 the set of all elements $s = \sum_{n\geq 1} s(n)\delta_n$ of G (the decomposition being the standard decomposition) such that $s(n) \geq 0$ for every n and such that $\sup_{n\in\mathbb{N}} s(n) > 0$.

Put, for $s \in S_4$,

$$\omega(s) = e^{-[\varphi(s)]^2}.$$

The set S_4 is clearly stable under sums, and ω is a radical weight over S_4. So $\ell^1(S_4,\omega)$ is a commutative radical Banach algebra.

THEOREM 7.6. The commutative radical Banach algebra $\mathfrak{R}_6 = \ell^1(S_4,\omega)$ possesses a nonzero element a having roots of all orders, but $\mathfrak{R}_6$ does not satisfy the conditions of Corollary 3.5.

Proof. Put $a = X^{\delta_1}$. Then $a = X^{n\delta_n} = (X^{\delta_n})^n$ for every $n \in \mathbb{N}$, so a has roots of all orders.

If $s \in S$ and if $s = \sum_{n\geq 1} s(n)\delta_n$ is the standard decomposition of s, put $\psi(s) = \sum_{n\geq 1} s(n)$, $\theta(s) = s(1)$. Now let x be a nonzero element of $\mathfrak{R}_6$. Put $p = \inf_{s\in \operatorname{Supp} x} \theta(s)$ and among the elements s of $\operatorname{Supp} x$ such that $\theta(s) = p$ choose s_0 such that $\psi(s_0)$ is minimal.

Now let y be any element of $\mathfrak{R}_6$ and let $s \in \operatorname{Supp} xy$. Then $s = r + t$, where $r \in \operatorname{Supp} x$, $t \in \operatorname{Supp} y$. Let $r = \sum_{n\geq 1} r(n)\delta_n$, $s = \sum_{n\geq 1} s(n)\delta_n$, and $t = \sum_{n\geq 1} t(n)\delta_n$ be the standard decompositions of r, s and t. If $r(n) + t(n) \leq n-1$ for every $n \geq 2$ then $\psi(r) + \psi(t) = \psi(s)$. If $s(1) = p$, then

$r(1) = p$, $\psi(r) \geq \psi(s_0)$ and $\psi(s) > \psi(r) \geq \psi(s_0)$, so that $s \neq s_0$. If $s(1) \neq p$, then obviously $s \neq s_0$.

Now, if $r(n) + t(n) \geq n$ for some $n \geq 2$, we have $s(1) > r(1) + t(1) \geq r(1) \geq p = s_0(1)$, so $s \neq s_0$. We thus see that $s_0 \notin \mathrm{Supp}\, xy$ for every $y \in \mathfrak{R}_6$. So $\inf_{y \in \mathfrak{R}_6} \|x - xy\| \geq |\lambda_{s_0}|\omega(s_0) > 0$ and hence $\mathfrak{R}_6$ does not satisfy the conditions of Corollary 3.5.

It was shown in [22] that, if $b \in [b\mathfrak{R}]^-$ for some nonnilpotent element of a commutative radical Banach algebra $\mathfrak{R}$, then the set of prime ideals of $\mathfrak{R} \oplus \mathbb{C}e$ does not form a chain (a short proof of this result is given in the next section). We now show that the set of primes of $\mathfrak{R} \oplus \mathbb{C}e$ may form a chain even if $b \in [b\mathfrak{R}]^-$ for some nonzero element b of $\mathfrak{R}$.

Using the same notations as in Theorem 7.3, denote by $\mathfrak{M}_1$ the Banach algebra obtained by equipping $\mathfrak{M} = \{g \in L^1(\mathbb{R}^+, e^{-t^3}) \mid g \equiv 0$ a.e. over $[0,1]\}$ with the trivial product $uv = 0$ for every $u,v \in \mathfrak{M}$. Put $\mathfrak{R}_7 = \ell^1(\mathbb{N}, e^{-n^2}) \oplus \mathfrak{M}_1$ (this direct sum is defined in Definition 2.4), where the module action of $\ell^1(\mathbb{N}, e^{-n^2})$ is the same as in Theorem 7.3.

PROPOSITION 7.7. The commutative radical Banach algebra $\mathfrak{R}_7$ possesses an element satisfying the condition of Theorem 3.7, but the only prime ideals of $\mathfrak{R}_7$ are $\mathfrak{R}_7$ and $\mathfrak{M}_1$.

Proof. The module action being the same as in Theorem 7.3, there exists $u \in \mathfrak{M}_1$ such that $u \in \varprojlim X^n \mathfrak{M}_1$, and $u \neq 0$. So $\varprojlim X^n \mathfrak{R}_7 \neq \{0\}$ and $\mathfrak{R}_7$ satisfies Condition 3.7.

All elements of $\mathfrak{M}_1$ are nilpotent, so every prime ideal of $\mathfrak{R}_7$ contains $\mathfrak{M}_1$. Let $\alpha = a + x$ and $\beta = b + y$ be two elements of $\mathfrak{R}_7$ such that $\alpha\beta \in \mathfrak{M}_1$ (where $a,b \in \ell^1(\mathbb{N}, e^{-n^2})$, $x,y \in \mathfrak{M}_1$). Then $ab + xa + yb \in \mathfrak{M}_1$, so that $ab = 0$. Since $\ell^1(\mathbb{N}, e^{-n^2})$ is an integral domain, one of the two elements a or b, say a, equals zero, and $\alpha \in \mathfrak{M}_1$. So $\mathfrak{M}_1$ is a prime ideal of $\mathfrak{R}_7$. Let J be a prime ideal of $\mathfrak{R}_7$. The projection $\rho : a + x \mapsto a$ from $\mathfrak{R}_7$ onto $\ell^1(\mathbb{N}, e^{-n^2})$ is an algebra homomorphism, so that $\rho(J)$ is a prime ideal of $\ell^1(\mathbb{N}, e^{-n^2})$. It follows from a result of Grabiner [27] that either $\rho(J) = \{0\}$ or $\rho(J) = \ell^1(\mathbb{N}, e^{-n^2})$ (Grabiner's proof is reproduced in [22]). Since $\mathfrak{M}_1 \subset J$ we obtain $J = \mathfrak{M}_1$ in the first case and $J = \mathfrak{R}_7$ in the second case.

REMARK 7.8. The fact that $b \in [b\mathfrak{R}]^-$ for some nonnilpotent $b \in \mathfrak{R}$ is not necessary for the set of primes of $\mathfrak{R} \oplus \mathbb{C}e$ not to form a chain, even if $\mathfrak{R}$ is an integral domain. An easy example is given by weighted algebras of power series in two variables X_1 and X_2 (certainly X_1 does not divide any power of X_2, and X_2 does not divide any power of X_1).

We now turn to examples related to analytic semigroups. A. M. Sinclair showed in [36] that if a commutative Banach algebra A possesses a bounded approximate identity, then for every $x \in A$ there exists a semigroup $(a^t)_{\operatorname{Re} t > 0}$ analytic over the right-hand half-plane such that $x \in a^t A$ for every t, and such that $\sup_{t>0} \|a^t\| < +\infty$. The computations of [36] show, in fact, that the semigroup $(a^t)_{\operatorname{Re} t>0}$ is bounded on the set $\{t \in \mathbb{C} \mid 0 < |t| \leq 1, |\operatorname{Arg} t| < \alpha\}$ for every $\alpha \in (0,\pi/2)$ (and even on the sector $\{t \in \mathbb{C} \setminus \{0\} \mid |\operatorname{Arg} t| < \alpha\}$ in the radical case). Using the Ahlfors-Heins theorem (see [7], Section 7, [14], or [37], Appendix 1) as in [24], it is possible to show that, if $(a^t)_{\operatorname{Re} t>0}$ is a nonzero analytic semigroup in a radical Banach algebra, then

$$\int_{-\infty}^{\infty} \frac{\operatorname{Log}^+\|a^{1+iy}\|}{1+y^2}\, dy = +\infty$$

(see [37, Chapter 5]). In particular, radical Banach algebras cannot possess nonzero analytic semigroups which are bounded on the open right-hand half-plane. So the only reasonable condition in the radical case is $\sup\{\|a^t\| \mid \operatorname{Re} t > 0, 0 < |t| \leq 1\} < +\infty$. It is not always the case that a radical Banach algebra with bounded approximate identity possesses a nonzero analytic semigroup which satisfies this condition, and so Sinclair's estimates are essentially best-possible. To see this, we need a lemma which is a weak version of [37, Theorem 5.14].

LEMMA 7.9. Let A be a nonzero commutative radical Banach algebra and let $M(A)$ be the multiplier algebra of A. If A possesses an analytic semigroup $(a^t)_{\operatorname{Re} t>0}$ such that $[a^1A]^- = A$, and $\sup_{0<|t|\leq 1} \|a^t\| < +\infty$, then the group of invertible elements of $M(A)$ possesses $2^{\aleph_0}$ different connected components.

<u>Proof</u>. An elementary computation given in [29], Theorem 17.9.1 shows that $\lim_{\substack{t\to iy \\ \operatorname{Re} t>0}} a^t u$ exists for every $y > 0$ and every $u \in [a^1 \cdot A]^- = A$. The algebra A has a b.a.i. given by the bounded sequence $(a^{1/n})_{n\in\mathbb{N}}$. We may assume, by renorming A if necessary as in [36], that A is isometrically embedded in $M(A)$. The strong limit $\lim_{\substack{t\to iy \\ \operatorname{Re} t>0}} a^t$ defines an element a^{iy} of $\mathcal{L}(A)$ which belongs in fact to $M(A)$, so we can write $a^{iy}u$ instead of $a^{iy}(u)$. Note that $(a^t)_{\operatorname{Re} t\geq 0}$ is a strongly continuous semigroup in $M(A)$.

Put $G = \operatorname{Inv} M(A)$, and assume that a^{iy} and $a^{iy'}$ belong to the same compontent of G.

Then $a^{i(y-y')}$ belongs to the principal component of G (i.e., the component of the unit element e of $M(A)$). So by [9], Corollary 1.8.8 we have $a^{i(y-y')} \in \exp M(A)$. We may assume that $t_0 = y - y' > 0$ and that $a^{it_0} = \exp(it_0 u)$ for some $u \in M(A)$. Put $b^t = a^t \exp(-tu)$ $(\mathrm{Re}\, t \geq 0)$.

The semigroup $(b^t)_{\mathrm{Re}\, t \geq 0}$ is strongly continuous in $M(A)$, and $(b^t)_{\mathrm{Re}\, t > 0}$ is analytic in A. Since $b^{it_0} = e$, $\sup_{y \in \mathbb{R}} \|b^{iy} v\| = \sup_{|y| \leq t_0} \|b^{iy} v\| < +\infty$ for every $v \in A$, so that $\sup_{y \in \mathbb{R}} \|b^{iy}\| < +\infty$.

Since A is radical, we obtain $\sup_{\mathrm{Re}\, t > 0} \|b^t\| < +\infty$. As mentioned before, the Ahlfors-Heins theorem implies that $(b^t)_{\mathrm{Re}\, t > 0}$ is identically zero. Since $a^t = b^t \exp(tu)$ for every t, we get $a^1 = 0$, $A = [a^1 A]^- = \{0\}$, a contradiction. This proves our lemma.

We can now prove the following theorem.

THEOREM 7.10. The commutative radical Banach algebra $L^1(\mathbb{R}^+, e^{-t^2})$ (which has a b.a.i.) does not possess any nonzero analytic semigroup $(a^t)_{\mathrm{Re}\, t > 0}$ bounded in the half-disc $\Delta = \{t \in \mathbb{C} \mid 0 < |t| \leq 1,\ \mathrm{Re}\, t > 0\}$.

Proof. It is well known that the multiplier algebra of $L^1(\mathbb{R}^+, e^{-t^2})$ is the convolution measure algebra $M(\mathbb{R}^+, e^{-t^2})$, and that the unique character of $M(\mathbb{R}^+, e^{-t^2})$ is the point mass at zero ([29], §4.4).

Now if $(a^t)_{\mathrm{Re}\, t > 0}$ is an analytic semigroup in $L^1(\mathbb{R}^+, e^{-t^2})$, let ℓ be a continuous linear form over $L^1(\mathbb{R}^+, e^{-t^2})$ vanishing over $a^s L^1(\mathbb{R}^+, e^{-t^2})$, where $s > 0$. The function $t \mapsto \ell(a^t)$ is analytic over the right-hand half-plane and it vanishes if $\mathrm{Re}\, t > s$, so it vanishes identically. This shows that $a^s \in [a^t L^1(\mathbb{R}^+, e^{-t^2})]^-$ for every $t, s > 0$. If the semigroup is not identically zero, then $a^t \neq 0$ for every $t > 0$, and in particular $a^1 \neq 0$. Also, $a^1 \in [a^2 L^1(\mathbb{R}^+, e^{-t^2})]^-$, so a^1 cannot vanish a.e. over $[0,\alpha]$ for any $\alpha > 0$. By Domar's theorem [16], we see that $a^1 L^1(\mathbb{R}^+, e^{-t^2})$ is dense in $L^1(\mathbb{R}^+, e^{-t^2})$. As $M(\mathbb{R}^+, e^{-t^2})$ has only one character, its group of invertible elements is connected, and the fact that $\sup_{t \in \Delta} \|a^t\| < +\infty$ follows from the lemma.

It is easy to produce a radical Banach algebra such that $\sup_{t \in \Delta} \|a^t\| < +\infty$ for some nonzero analytic semigroup $(a^t)_{\mathrm{Re}\, t > 0}$. Denote by $\mathcal{C}_0$ the algebra of all continuous functions F over the closed right-hand half-plane $\{z \in \mathbb{C} \mid \mathrm{Re}\, z \geq 0\}$ which are analytic for $\mathrm{Re}\, z > 0$ and which satisfy $\lim_{\substack{|z| \to \infty \\ \mathrm{Re}\, z > 0}} |F(z)| = 0$. Put, for $\mathrm{Re}\, t > 0$, $\mathrm{Re}\, z \geq 0$, $a^t(z) = 1/(z+1)^t$ where $(z+1)^t = \exp[t\, \mathrm{Log}\,(z+1)]$ and the logarithm takes real values on the positive reals. The map $t \mapsto a^t$ defines an analytic semigroup $(a^t)_{\mathrm{Re}\, t>0}$ in $\mathfrak{A}_0$. Put $\|F\|_\infty = \sup_{\mathrm{Re}\, z\geq 0} |F(z)|$ $(F \in \mathfrak{A}_0)$.

A routine computation shows that $\sup_{0<|t|\leq 1} \|a^t\| \leq \exp(\pi/2)$. Denote by α the function $z \mapsto e^{-z}$. Then $\alpha\mathfrak{A}_0$ is a closed ideal, $\mathfrak{A}_0/\alpha\mathfrak{A}_0$ is radical, and $a^t \notin \alpha\mathfrak{A}_0$ for each t with $\mathrm{Re}\, t > 0$, so that the image of $(a^t)_{\mathrm{Re}\, t>0}$ in the quotient algebra $\mathfrak{A}_0/\alpha\mathfrak{A}_0$ gives the desired example.

We give now another example which is an integral domain. Put, if $\mathrm{Re}\, t > 0$ and $x > 0 : a^t(x) = x^{t-1}/\Gamma(t)\, e^{-x}$. It is well known that $(a^t)_{\mathrm{Re}\, t>0}$ is an analytic semigroup in $L^1(\mathbb{R}^+)$, and hence in $L^1(\mathbb{R}^+, e^{-t^2})$ (one way to see this is to notice that it is the inverse Laplace transform of the semigroup used below in $\mathfrak{A}_0$).

We have

$$\|a^t\| = \frac{1}{|\Gamma(t)|}\left(\int_0^\infty x^{(\mathrm{Re}\, t)-1} e^{-x} dx\right) = \frac{\Gamma(\mathrm{Re}\, t)}{|\Gamma(t)|}.$$

Using standard estimates for the gamma function, we obtain $\|a^{1+t}\| \leq ke^{(\pi/2)|\mathrm{Im}\, t|} \leq ke^{(\pi/2)|t|}$ for $\mathrm{Re}\, t > 0$, where k is some constant.

Put $I = \{u \in L^1(\mathbb{R}^+, e^{-t^2}) \mid \sup_{\mathrm{Re}\, t>0} \|ua^t\| e^{-(\pi/2)|t|} < +\infty\}$. Then I is a dense ideal of $L^1(\mathbb{R}^+, e^{-t^2})$ because $a \in I$. Put $p(u) = \sup_{\mathrm{Re}\, t>0} \|ua^t\| e^{-(\pi/2)|t|}$ for every $u \in I$ and put, for every $v \in L^1(\mathbb{R}^+, e^{-t^2})$,

$$|||v||| = \sup_{\substack{u\in I\\ u\neq 0}} \frac{p(uv)}{p(u)}.$$

Routine verifications given in [26] show that $|||v||| \leq \|v\|$ for every $v \in L^1(\mathbb{R}^+, e^{-t^2})$, and that the completion $\mathfrak{R}_8$ of $(L^1(\mathbb{R}^+, e^{-t^2}), |||\cdot|||)$ is a Banach algebra and an integral domain. Also, $[a^1\mathfrak{R}_8]^- = \mathfrak{R}_8$ because the injection from $(L^1(\mathbb{R}^+, e^{-t^2}), \|\cdot\|)$ into $(\mathfrak{R}_8, |||\cdot|||)$ is continuous and has dense range. Also, if $0 < |t| \leq 1$ and $u \in I$, we have $p(ua^t) = \sup_{\mathrm{Re}\, s>0} \|ua^t a^s\| e^{-(\pi/2)|s|} \leq e^{(\pi/2)|t|} \sup_{\mathrm{Re}\, s>0} \|ua^t a^s\| e^{-(\pi/2)|s+t|} \leq e^{\pi/2} p(u)$. So $\sup_{0<|t|\leq 1} |||a^t||| \leq e^{\pi/2}$. The semigroup $(a^t)_{\mathrm{Re}\, t>0}$ is clearly analytic in $\mathfrak{R}_8$. We have proved the following result.

THEOREM 7.11. The commutative radical Banach algebra $\mathfrak{R}_8$ is an integral domain, and it possesses an analytic semigroup $(a^t)_{\mathrm{Re}\, t>0}$ such that $[a^t\mathfrak{R}_8] = \mathfrak{R}_8$ $(\mathrm{Re}\, t > 0)$ which is bounded over the half-disc $\Delta = \{t \in \mathbb{C}\ 0 < |t| \leq 1\}$.

Note that $x = \lim_{\substack{t\to 0\\ \mathrm{Re}\, t>0}} xa^t$ for every $x \in \mathfrak{R}_8$, which gives in some sense the best possible bounded approximate identity for a radical Banach algebra.

The following example shows that a radical Banach algebra may possess an analytic semigroup $(a^t)_{\mathrm{Re}\,t>0}$ without having any closed ideal which possesses a b.a.i. Let $(a^t)_{\mathrm{Re}\,t>0}$ be the analytic semigroup in $L^1(\mathbb{R}^+, e^{-t^2})$ introduced above. By the Johnson-Varopoulos extension of Cohen's factorization theorem [31], [39], there exists an element f of $L^1(\mathbb{R}^+, e^{-t^2})$ such that $a^{1/n} \in fL^1(\mathbb{R}^+, e^{-t^2})$ for every $n \in \mathbb{N}$. Put $\mathcal{R}_9 = fL^1(\mathbb{R}^+, e^{-t^2})$ and equip $\mathcal{R}_9$ with the norm $|||\cdot|||$ introduced in Section 2. Then $\mathcal{R}_9$ is a radical Banach algebra (use Remark 2.10) and we obtain the following.

PROPOSITION 7.12. The commutative radical Banach algebra $\mathcal{R}_9$ is an integral domain which does not possess any nonzero closed ideal with bounded approximate identity, and $\mathcal{R}_9$ possesses an analytic semigroup $(a^t)_{\mathrm{Re}\,t>0}$ such that $[a^t\mathcal{R}_9]^- = \mathcal{R}_9$ (Re $t > 0$).

Proof. As $a^t = a^{1/n} \cdot a^{t-(1/n)}$ if $1/n < \mathrm{Re}\, t$, we have $(a^t)_{\mathrm{Re}\,t>0} \subset \mathcal{R}_9$. If $\mathrm{Re}\, s > 0$, let $b(s)$ be the derivative of the function $t \mapsto a^t$ at s.

Using the semigroup law, we obtain $b(s+s') = a^s b(s')$ ($\mathrm{Re}\, s > 0$, $\mathrm{Re}\, s' > 0$) and $b(s) \in \mathcal{R}_9$ for every s with $\mathrm{Re}\, s > 0$. Now $\overline{\lim_{h\to 0}} \left|\left|\left| \frac{a^{s+h} - a^s}{h} - b(s) \right|\right|\right| \le |||a^{s/2}||| \ \overline{\lim_{h\to 0}} \left\| \frac{a^{(s/2)+h} - a^{s/2}}{h} - b(s/2) \right\| = 0$, so that $(a^t)_{\mathrm{Re}\,t>0}$ is analytic in $\mathcal{R}_9$. If $g \in \mathcal{R}_9$ then $g = fh$ where $h \in L^1(\mathbb{R}^+, e^{-t^2})$. Fix $t > 0$. There exists a sequence (u_n) in $L^1(\mathbb{R}^+, e^{-t^2})$ such that $\|h - u_n a^t\| \xrightarrow[n\to\infty]{} 0$.

So $|||fu_n a^t - g||| \le \|f\| \, \|u_n a^t - h\| \xrightarrow[n\to\infty]{} 0$. (Here we use the definition of the norm $|||\cdot|||$ given in Section 2.) This shows that $[a^t\mathcal{R}_9]^- = \mathcal{R}_9$. Also $|||b^n|||^{1/n} \le |||b|||^{1/n} \|b^{n-1}\|^{1/n} \le K(b) \, |||b|||^{1/n} \|f^{n-1}\|^{1/n}$ (where $K(b)$ is a constant depending on b) for every $b \in \mathcal{R}_9$.

Since $\lim_{n\to\infty} \|f^{n-1}\|^{1/n} = 0$, it follows from a result of Allan and Sinclair [3] that $\mathcal{R}_9$ does not possess any nonzero closed subalgebra with b.a.i., and hence it does not possess any nonzero closed ideal with b.a.i.

REMARK 7.13. We showed above that, if $\mathcal{R}$ is a Banach algebra, if (a^t) is an analytic semigroup in some angle starting at the origin, and if $f \in \mathcal{R}$ is a divisor of $a^{1/n}$ for each $n \in \mathbb{N}$ then (a^t) is an analytic semigroup in $(f\mathcal{R}, |||\cdot|||)$.

We now exhibit examples of commutative radical Banach algebras with nonzero analytic semigroups in some open angles which do not possess nonzero analytic semigroups in larger angles.

Denote by $\operatorname{Log} z$ the determination of the logarithm for $\operatorname{Re} z \geq 0$, $z \neq 0$, which takes positive values over the positive reals. Put $z^\alpha = \exp(\alpha \operatorname{Log} z)$ for $\operatorname{Re} z \geq 0$, $z \neq 0$, and set $u_\alpha^t(z) = e^{-t^2}$ $|\operatorname{Arg} t| < (1-\alpha)\pi/2$, $t \neq 0$, $\alpha \in (0,1)$.

We denote by $\mathfrak{A}_0$ the algebra of all continuous complex-valued functions F over the half-plane $U = \{z \in \mathbb{C} \mid \operatorname{Re} z \geq 0\}$ which are analytic for $\operatorname{Re} z > 0$ and which satisfy $\lim_{\substack{|z|\to\infty \\ \operatorname{Re} z \geq 0}} |F(z)| = 0$. We equip $\mathfrak{A}_0$ with the norm $\|F\| = \sup_{\operatorname{Re} z \geq 0} |F(z)|$ $(F \in \mathfrak{A}_0)$. Routine verifications that we omit show that $u_\alpha^t \in \mathfrak{A}_0$ if $|\operatorname{Arg} t| < (1-\alpha)\pi/2$ and that the map $t \mapsto u_\alpha^t$ defines an analytic semigroup in $\mathfrak{A}_0$ over the sector $\Delta_\alpha = \{t \in \mathbb{C} - \{0\} \mid |\operatorname{Arg} t| < (1-\alpha)\pi/2\}$.

If $t \in \Delta_\alpha$, then the function $y \mapsto u_\alpha^t(iy)$ belongs to $L^1(\mathbb{R})$. The inverse Laplace transform $v_\alpha^t = \mathcal{L}^{-1}(u_\alpha^t)$ is well defined and belongs to $C_0(\mathbb{R}^+) \subset L^1(\mathbb{R}^+, e^{-t^2})$. So $(v_\alpha^t)_{t\in\Delta_\alpha}$ is a semigroup in $L^1(\mathbb{R}^+, e^{-t^2})$.

Denote by ω_α^t the derivative of u_α^t with respect to t. Then ω_α^t is the function $z \mapsto -z^\alpha e^{-tz^\alpha}$ and the function $y \mapsto \omega_\alpha^t(iy)$ belongs to $L^1(\mathbb{R})$ for $t \in \Delta_\alpha$.

Denote by $\|\cdot\|_\infty$ the supremum norm over $L^\infty(\mathbb{R}^+)$. We have, for $x \geq 0$, $t \in \Delta_\alpha$, $t/2 + h \in \Delta_\alpha$:

$$\left| \frac{v_\alpha^{t+h}(x) - v_\alpha^t(x)}{h} - \mathcal{L}^{-1}(\omega_\alpha^t)(x) \right| \leq \frac{1}{2\pi} \left[\int_{-\infty}^{\infty} |u_\alpha^{t/2}(iy)|\,dy \right] \times \sup_{y\in\mathbb{R}} \left| \frac{u_\alpha^{(t/2)+h}(iy) - u_\alpha^{t/2}(iy)}{h} - \omega_\alpha^{t/2}(iy) \right| .$$

So

$$\left\| \frac{v_\alpha^{t+h} - v_\alpha^t}{h} - \mathcal{L}^{-1}(\omega_\alpha^t) \right\|_\infty \leq \frac{1}{2\pi} \left| \int_{-\infty}^{\infty} |u_\alpha^{t/2}(iy)|\,dy \right| \cdot \left\| \frac{u_\alpha^{(t/2)+h} - u_\alpha^t}{h} - \omega_\alpha^{t/2} \right\|$$

and the function $t \mapsto v_\alpha^t$ is an analytic map from Δ_α into $L^\infty(\mathbb{R}^+)$. This shows that $(v_\alpha^t)_{t\in\Delta_\alpha}$ is an analytic semigroup in $L^1(\mathbb{R}^+, e^{-t^2})$ for every $\alpha \in (0,1)$.

We have, for $x \geq 0$, $n \in \mathbb{N}$, $\sigma \geq 0$,

$$|v_\alpha^{n+1}(x)| = \frac{1}{2\pi} \left| \int_{-\infty}^{\infty} e^{\sigma x} e^{-(n+1)(\sigma+iy)^\alpha} \cdot e^{ixy}\,dy \right|$$

$$\leq \frac{1}{2\pi} \left[\int_{-\infty}^{\infty} e^{-\operatorname{Re}(\sigma+iy)^\alpha}\,dy \right] \sup_{y\in\mathbb{R}} \{e^{\sigma x - n \operatorname{Re}(\sigma+iy)^\alpha}\} .$$

Since $|\mathrm{Arg}(\sigma + iy)^{\alpha}| \leq \pi\alpha/2$, $\mathrm{Re}(\sigma + iy)^{\alpha} \geq |\sigma^2 + y^2|^{\alpha/2} \cos \pi\alpha/2$, and so we obtain

$$|v_{\alpha}^{n+1}(x)| \leq \frac{1}{2\pi}\left[\int_{-\infty}^{\infty} e^{-|y|^{\alpha}\cos(\pi\alpha/2)}dy\right] e^{\sigma x - n\sigma^{\alpha}\cos\pi\alpha/2} \qquad (x \geq 0,\ n \in \mathbb{N},\ \sigma \geq 0).$$

If $x > 0$, an elementary computation gives $\inf_{\sigma>0} (\sigma x - n\sigma^{\alpha} \cos \pi\alpha/2) =$ $x^{-\alpha/(1-\alpha)} (n \cos \alpha\pi/2)^{1/(1-\alpha)} \alpha^{\alpha/(1-\alpha)} (\alpha - 1)$.

We obtain $|v_{\alpha}^{n+1}(x)| \leq \lambda_{\alpha} \exp[-\mu_{\alpha} x^{-\alpha/(1-\alpha)} n^{1/(1-\alpha)}]$ $(x > 0,\ n \in \mathbb{N})$, where λ_{α} and μ_{α} are positive constants.

If $\alpha > 0$, consider the function $\theta_{\alpha} : z \mapsto \exp(-z^{\alpha}/\mathrm{Log}(z+2))$.

An easy verification shows that $\theta_{\alpha} \in \mathcal{G}_0$, and that the function $y \to \exp(-(iy)^{\alpha}/\mathrm{Log}(iy+2))$ belongs to $L^1(\mathbb{R})$. So $\mathcal{L}^{-1}(\theta_{\alpha})$ is well defined and belongs to $L^{\infty}(\mathbb{R}^+) \subset L^1(\mathbb{R}^+, e^{-t^2})$.

Also, $u_{\alpha}^t \in \theta_{\alpha}\mathcal{G}_0$, and $\mathcal{L}^{-1}(u_{\alpha}^t/\theta_{\alpha}) \in L^{\infty}(\mathbb{R}^+) \subset L^1(\mathbb{R}^+, e^{-t^2})$ for every $t \in \Delta_{\alpha}$. Similarly, $\theta_{\alpha} \in u_{\beta}^t \mathcal{G}_0$ and $\mathcal{L}^{-1}(\theta_{\alpha}/u_{\beta}^t) \in L^1(\mathbb{R}^+, e^{-t^2})$ for every $t \in \Delta_{\beta}$ if $\beta < \alpha$.

If $\eta \in (0,\pi/2)$ put $\alpha = 1 - 2\eta/\pi$, $f_{\eta} = \mathcal{L}^{-1}(\theta_{\alpha})$, $\mathcal{R}_{\eta} = f_{\eta}L^0(\mathbb{R}^+, e^{-t^2})$.

THEOREM 7.14. The commutative radical Banach algebra $\mathcal{R}_{\eta}$ is an integral domain and it possesses an analytic semigroup $(a^t)_{t\in U_{\eta}}$ defined in the open sector $U_{\eta} = \{t \in \mathbb{C} - \{0\} \mid |\mathrm{Arg}\ t| < \eta\}$ and satisfying $(a^t\mathcal{R}_{\eta})^- = \mathcal{R}_{\eta}$ for every $t \in U_{\eta}$. However, if $\eta' > \eta$ then $\mathcal{R}_{\eta}$ does not possess any nonzero analytic semigroup over $U_{\eta'}$.

Proof. As $\alpha = 1 - 2\eta/\pi$, $\eta = (1-\alpha)\pi/2$ and so U_{η} is exactly the angle Δ_{α} introduced above. Since $\mathcal{L}^{-1}(u_{\alpha}^t/\theta_{\alpha}) \in L^1(\mathbb{R}^+, e^{-t^2})$ we have $v_{\alpha}^t \in f_{\eta}L^1(\mathbb{R}^+, e^{-t^2}) = \mathcal{R}_{\eta}$ for every $t \in \Delta_{\alpha} = U_{\eta}$. According to Remark 7.13 we see that $(v_{\alpha}^t)_{t\in U_{\eta}}$ is an analytic semigroup in $\mathcal{R}_{\eta}$. Since the function $z \mapsto \exp(-tz^{\alpha} + \varepsilon z)$ is unbounded over the positive reals for every $\varepsilon > 0$, the function v_{α}^t cannot vanish a.e. over $[0,\varepsilon]$ for any $\varepsilon > 0$. By Domar's theorem $v_{\alpha}^t L^1(\mathbb{R}^+, e^{-t^2})$ is dense in $L^1(\mathbb{R}^+, e^{-t^2})$ for every $t \in U_{\eta}$. As in the proof of Proposition 7.11, this implies that $[v_{\alpha}^t\mathcal{R}_{\eta}]^- = \mathcal{R}_{\eta}$ $(t \in U_{\eta})$.

Take $\eta' > \eta$ and let $(b^t)_{t\in U_{\eta'}}$ be an analytic semigroup over $U_{\eta'}$ in $\mathcal{R}_{\eta}$. Denote by c^t the restriction of the function b^t to $[0,1]$. Then $(c^t)_{t\in U_{\eta'}}$ is an analytic semigroup in the Volterra algebra $L^1_*(0,1)$. Let g be the restriction of f_{η} to $[0,1]$, and, for $\beta < \alpha$ let d_{β}^t be the restriction of the function

v_β^t to $[0,1]$. Then $g \in d_\beta L_*^1(0,1)$ for every $\beta < \alpha$ (where $d_\beta = d_\beta^1$).

There exist constants λ_β and μ_β such that $|v_\beta^{n+1}(x)| \leq \lambda_\beta \exp(-\mu_\beta x^{-\beta/(1-\beta)} n^{1/(1-\beta)})$. So, if $0 \leq x \leq 1$, $|d_\beta^{n+1}(x)| \leq \lambda_\beta \exp(-\mu_\beta n^{1/(1-\beta)})$ and $\|d_\beta^{n+1}\|_1 \leq \lambda_\beta \exp(-\mu_\beta n^{1/(1-\beta)})$, where we denote by $\|\cdot\|_1$ the L^1-norm over $[0,1]$.

Since $(c^t)_{t \in U_{\eta'}}$ is divisible by g in $L_*^1(0,1)$, and hence by d_β, we obtain, for every $\beta < \alpha$ and every $n \in \mathbb{N}$, $\|c^n\|_1 \leq \gamma_\beta \exp(-\delta_\beta n^{1/(1-\beta)})$ where γ_β and δ_β do not depend on n. In fact, if $\beta < \alpha$, we have, taking for example $\beta' = (\alpha+\beta)/2$,

$$\limsup_{n\to\infty} \|c^n\|_1^{1/n} \exp(n^{\beta/(1-\beta)}) \leq \limsup_{n\to\infty} \gamma_{\beta'}^{1/n} \exp[n^{\beta/(1-\beta)} - \delta_{\beta'} n^{\beta'/(1-\beta')}] = 0 .$$

Now if $(\rho^t)_{t\in U_{\eta'}}$ is any nonzero analytic semigroup over $U_{\eta'}$ in a Banach algebra $\mathfrak{R}$, it follows from [21], Corollary 3.2, that $\liminf_{n\to\infty} \exp(n^{s-1}) \|\rho^n\|^{1/n} = +\infty$ for every $s > \pi/2\eta'$. We have $1-(2\eta'/\pi) < \alpha$. Take $\beta \in (1-(2\eta'/\pi), \alpha)$. Then $1/(1-\beta) > \pi/2\eta'$.

Since $(1/(1-\beta)) - 1 = \beta/(1-\beta)$, the above estimates show that necessarily $c^1 = 0$, so $b^1 = 0$ a.e. over $[0,1]$. As $b^t \in [b^n L^1(\mathbb{R}^+, e^{-t^2})]^-$ for every $n \in \mathbb{N}$ and every $t \in U_\eta$, we have $b^t \equiv 0$ a.e. over $[0,n]$ for every $n \in \mathbb{N}$, so that $b^t = 0$ $(t \in U_{\eta'})$ and $(b^t)_{t\in U_{\eta'}}$ is the zero semigroup. This achieves the proof of the theorem.

We wish now to produce an example of a commutative radical Banach algebra which possesses some nonzero continuous semigroup $(a^t)_{t>0}$ over the positive reals, but does not possess any nonzero continuous semigroup $(b^t)_{t>0}$ analytic at some point $t_0 > 0$. A lower estimate for the rate of decrease of a semigroup analytic over $(0,\infty)$ is given in [21], Theorem 3.3 and would certainly suffice for our purpose, but this estimate is so crude that we will first improve it.

If $(b^t)_{t>0}$ is a semigroup in a Banach algebra $\mathfrak{R}$ which is analytic at t_0, then there exists a sequence $(\beta_n)_{n\geq 0}$ of elements of $\mathfrak{R}$ and $\eta > 0$ such that $b^{t_0+h} = \sum_{n=0}^\infty h^n \beta_n$ for every $h \in (-\eta,\eta)$ (of course $\beta_0 = b^{t_0}$).

If $t > t_0$, then by multiplying both members of the above equality by b^{t-t_0}, we obtain

$$b^{t+h} = \sum_{n=0}^\infty h^n \beta_n b^{t-t_0} \qquad (h \in (-\eta,\eta)) .$$

So the semigroup $(b^t)_{t>0}$ is in fact analytic at t for every $t > t_0$. It follows then from [29], Theorem 17.2.2 that there exists a domain Δ of the complex plane, which is the interior of a "spinal semimodule" and contains $(t_0,+\infty)$, and a function $W:\Delta\to\mathfrak{R}$ such that the following properties hold:

(1) $t+t'\in\Delta$ for every $t,t'\in\Delta$;

(2) $W(t+t')=W(t)W(t')$ for every $t,t'\in\Delta$;

(3) W is analytic over Δ;

(4) $W(t)=b^t$ for every $t>t_0$.

Also it follows from [29], Theorem 8.7.9, that Δ contains an angle $\{t\in\mathbb{C}\mid \operatorname{Re} t\geq\alpha, |\operatorname{Arg}(t-\alpha)|\leq\eta\}$ for some $\alpha>0$ and some $\eta\in(0,\pi/2)$.

This shows that semigroups analytic at some point are very close to semigroups analytic in angles starting at the origin. The estimate given in [21], Theorem 3.1, for semigroups analytic in angles starting at the origin holds in fact for semigroups analytic in angles starting at some point of the positive real axis. (The proof only uses the analyticity of the semigroup in some angle starting at 1.) We thus obtain the following result.

THEOREM 7.15. Let $(a^t)_{t>0}$ be a continuous semigroup in a Banach algebra such that $a^t\neq 0$ for every $t>0$. If the semigroup is analytic at t_0 for some $t_0>0$, then there exists $\gamma>0$ such that $\liminf_{n\to\infty}\exp(n^\gamma)\,\|a^n\|^{1/n}=+\infty$.

Routine computations show that if $t>0$ the function $u^t:z\mapsto \exp[-t(z+2)/\operatorname{Log}(z+2)]$ belongs to the algebra $\mathfrak{A}_0$ introduced above. If $0<\alpha<\beta$, then $u^t(z)$ converges uniformly to zero for $t\in[\alpha,\beta]$ when $|z|\to\infty$, $\operatorname{Re} z>0$. It follows easily from this fact that $(u^t)_{t>0}$ is a continuous semigroup in $\mathfrak{A}_0$. The function $\theta:z\mapsto\exp(-(z+2)/[\operatorname{Log}(z+2)]^2)$ also belongs to $\mathfrak{A}_0$.

Put $\omega^t=u^t_{1/2}u^t$, $\rho=\theta u_{1/3}$. (Here we use again the functions u_α introduced in the proof of Theorem 7.14.) Then $\omega^t\in\rho\mathfrak{A}_0$ for every $t>0$, (one checks easily that $u^t\in\theta\mathfrak{A}_0$ for every $t>0$) and $\omega^t/\rho\in(u^t_{1/2}/u_{1/3})\,\mathfrak{A}_0$. Thus, the function $y\mapsto\omega^t(iy)/\rho(iy)$ belongs to $L^1(\mathbb{R})$ for every $t>0$. Also $\theta\in u_\alpha\mathfrak{A}_0$ for every $\alpha\in(0,1)$, so that $\rho\in u_\alpha\mathfrak{A}_0$ and $y\mapsto\rho(iy)/u_\alpha(iy)$ belongs to $L^1(\mathbb{R})$ for every $\alpha\in(0,1)$. Put $f=\mathfrak{L}^{-1}(\rho)$. Then $f\in L^\infty(\mathbb{R}^+)\subset L^1(\mathbb{R}^+,e^{-t^2})$. Denote by $\mathfrak{R}_{10}$ the Banach algebra $fL^1(\mathbb{R}^+,e^{-t^2})$.

THEOREM 7.16. The commutative radical Banach algebra $\mathfrak{R}_{10}$ possesses a nonzero continuous semigroup $(a^t)_{t>0}$ over the positive reals but does not possess any nonzero continuous semigroup $(b_t)_{t>0}$ analytic at some $t_0>0$ (and $\mathfrak{R}_{10}$ is an integral domain).

Proof. Put $a^t=\mathfrak{L}^{-1}(\omega^t)$ $(t>0)$. Then $a^t\in L^1(\mathbb{R}^+,e^{-t^2})$ for every $t>0$

and so does $\mathfrak{L}^{-1}(\omega^t/\rho)$. Hence $a^t \in \mathfrak{R}_{10}$ for every $t > 0$.

We have

$$\|a^t - a^{t+h}\|_\infty \leq \frac{1}{2\pi}\left[\int_{-\infty}^{\infty} |\omega^{t/2}(iy)|\,dy\right] \|\omega^{t/2} - \omega^{(t/2)+h}\| .$$

Since the injection from $L^\infty(\mathbb{R}^+)$ into $L^1(\mathbb{R}^+, e^{-t^2})$ is continuous, we see that $(a^t)_{t>0}$ is a continuous semigroup in $L^1(\mathbb{R}^+, e^{-t^2})$. We have $|||a^{t+h} - a^t||| \leq |||a^{t/2}|||\ \|a^{(t/2)+h} - a^{t/2}\|$ $(t > 0,\ |h| < t/2)$, so $(a^t)_{t>0}$ is in fact a continuous semigroup in $\mathfrak{R}_{10}$. Certainly $a^t \neq 0$ for every $t > 0$.

Now denote by g the restriction of f to $[0,1]$. Then $\mathfrak{L}^{-1}(\rho/u_\alpha) \in L^1(\mathbb{R}^+, e^{-t^2})$ for every $\alpha \in (0,1)$ so $f \in v_\alpha L^1(\mathbb{R}^+, e^{-t^2})$ and $g \in d_\alpha L^1_*(0,1)$ where v_α and d_α have the same meaning as in the proof of Theorem 7.14.

As in the proof of Theorem 7.14, we obtain $\limsup_{n\to\infty} \|g^n\|_1 \exp(n^{\alpha/(1-\alpha)}) = 0$ for every $\alpha \in (0,1)$ so that $\limsup_{n\to\infty} \|g^n\|_1 \exp(n^\gamma) = 0$ for every $\gamma > 0$. Using Theorem 7.15, we see that, if $(c^t)_{t>0}$ is any continuous semigroup in $\mathfrak{R}_{10}$ analytic at some $t_0 > 0$ and if $(d^t)_{t>0}$ is the canonical image of $(c^t)_{t>0}$ in $L^1_*(0,1)$, then $d^t = 0$ for every $t > 0$. As c^t is analytic over (t_0,∞) any continuous linear functional ℓ such that $t \mapsto \ell(c^t)$ vanishes over $[2t_0,\infty)$ satisfies $\ell(c^{t_0}) = 0$. So $c^{t_0} \in [c^{2t_0} L^1(\mathbb{R}^+, e^{-t^2})]^-$ and since $c^{t_0} = 0$ a.e. over $[0,1]$ we have in fact $c^{t_0} = 0$. So $c^{t_0/n} = 0$ for every $n \in \mathbb{N}$ (noting that $\mathfrak{R}_{10}$ is an integral domain). Finally $c^t = 0$ for every $t > 0$, and the theorem is proved.

8. Some quick proofs, and a new theorem of Gelfand-Mazur type

We give here very short proofs of some results of [22] and [25] concerning, first, rates of decrease of sequences of powers and, second, prime ideals in radical Banach algebras. We show also that, if in a Banach algebra A with unit the principal ideals form a chain, then $A \simeq \mathbb{C}[X]/X^p\mathbb{C}[X]$ for some $p \geq 1$ and in particular A is finite-dimensional. This strengthens a result of [22] which shows that $\mathbb{C}$ is the only Banach algebra which is a valuation ring.

The proof given here of the result about primes is due to Bouloussa (who proved it in the more general context of B_0-algebras). The other proofs are new.

THEOREM 8.1. ([10] and [22]) Let $\mathfrak{R}$ be a commutative radical Banach algebra. If $b \in [b\mathfrak{R}]^-$ for some nonnilpotent $b \in \mathfrak{R}$, the set of prime ideals of $\mathfrak{R}^\#$ does not form a chain.

<u>Proof</u>. It follows from [22], Lemma 4.4 that there exists a sequence (u_m) of invertible elements of $\mathcal{R}^{\#}$ such that $\lim_{m\to\infty} bu_m = b$ and $\liminf_{m\to\infty} \|b^p u_m^{-1}\| = +\infty$ for every $p \in \mathbb{N}$. Put $I = [b\mathcal{R}]^-$ and put, for $n,p \in \mathbb{N}$,

$$\Omega_{n,p} = \left\{ (c,d) \in I \times I \mid \inf_{u\in\mathcal{R}^{\#}} \left(\|cu - c\| + \frac{1}{\|d^p u\|} \right) < \frac{1}{n} \right\}$$

The set $\Omega_{n,p}$ is clearly open. Now let $x,y \in I$ and $\varepsilon > 0$. There exists $\alpha,\beta \in \mathcal{R}$ such that $\|x - b\alpha\| < \varepsilon/2\|y - b\beta\| < \varepsilon$.

Since $\beta = \lim_{\lambda\to 0} \lambda e + \beta$ there exists $\gamma \in \mathrm{Inv}(\mathcal{R}^{\#})$ such that $\|y - b\gamma\| < \varepsilon$.

We have

$$\lim_{m\to\infty} b\alpha u_m = b\alpha,\ \liminf_{m\to\infty} \|b^p\gamma^p u_m^{-1}\| \geq \liminf_{m\to\infty} \frac{\|b^p u_m^{-1}\|}{\|\gamma^{-p}\|} = +\infty .$$

Choose $m \in \mathbb{N}$ such that $\|b\alpha u_m - b\alpha\| < \inf(\varepsilon/2,\ 1/2n)$ and $\|b^p\gamma^p u_m^{-1}\| > 2n$.

Then $\|(b\alpha u_m)u_m^{-1} - b\alpha u_m\| + 1/\|b^p\gamma^p u_m^{-1}\| < 1/n$, so that $(b\alpha u_m, b\gamma) \in \Omega_{n,p}$. As $\|x - b\alpha u_m\| < \varepsilon$ and $\|y - b\gamma\| < \varepsilon$, we see that $\Omega_{n,p}$ is dense in $I \times I$. So $\Omega = \bigcap_{n,p\in\mathbb{N}} \Omega_{n,p}$ is a dense G_δ of $I \times I$.

Put $\Omega' = \{(x,y) \in I \times I \mid (y,x) \in \Omega\}$. The map $(x,y) \mapsto (y,x)$ is a homeomorphism from $I \times I$ onto itself, so Ω' is a dense G_δ of $I \times I$ too. Using again the category theorem, we see that $\Omega \cap \Omega'$ is a dense G_δ of $I \times I$.

Let $(c,d) \in \Omega \cap \Omega'$ and $p \in \mathbb{N}$. There exist two sequences (u_n), (v_n) of elements of $\mathcal{R}^{\#}$ such that $\|cu_n - c\| \xrightarrow[n\to\infty]{} 0$, $\|d^p u_n\| \xrightarrow[n\to\infty]{} \infty$, $\|dv_n - d\| \xrightarrow[n\to\infty]{} 0$ and $\|c^p v_n\| \xrightarrow[n\to\infty]{} \infty$.

This shows that $d^p \notin c\mathcal{R}^{\#}$, $c^p \notin d\mathcal{R}^{\#}$ for every $p \in \mathbb{N}$ and by a standard result of commutative algebra there exist two prime ideals J and L of $\mathcal{R}^{\#}$ such that $c \notin J$, $d \in J$ and $c \in L$, $d \notin L$. This achieves the proof of the theorem.

The following theorem was proved in [25] by a lengthy computation of infinite products.

THEOREM 8.2 [25]. Let $\mathcal{R}$ be a commutative separable radical Banach algebra with b.a.i. If the nilpotents are dense in $\mathcal{R}$, then, for every pair of sequences (λ_n) and (μ_n) of positive reals which converge to zero there exists $x \in \mathcal{R}$ such that

$$\limsup_{n\to\infty} \frac{\|x^n\|^{1/n}}{\lambda_n} = +\infty,\ \liminf_{n\to\infty} \frac{\|x^n\|^{1/n}}{\mu_n} = 0 \text{ and } [x\mathcal{R}]^- = \mathcal{R} .$$

Proof. It follows from a theorem of Allan and Sinclair [3] that there exists $b \in \mathfrak{R}$ such that $\|b^n\|^{1/n}/\lambda_n \xrightarrow[n\to\infty]{} \infty$ and considering a common divisor of $\{b\} \cup \{f_n\}_{n\in\mathbb{N}}$ where (f_n) is a dense sequence in $\mathfrak{R}$, we see that we may assume without loss of generality that $[b\mathfrak{R}]^- = \mathfrak{R}$.

Put $\Omega = \{x \in \mathfrak{R} \mid [x\mathfrak{R}]^- = \mathfrak{R}\}$. As in Section 5 we see that Ω is a dense G_δ of $\mathfrak{R}$.

Put, for $p \in \mathbb{N}$,

$$U_p = \left\{ x \in \mathfrak{R} \mid \sup_{n\in\mathbb{N}} \frac{\|x^n\|}{p^n\lambda_n^n} > 1 \right\},$$

$$V_p = \left\{ x \in \mathfrak{R} \mid \inf_{n\in\mathbb{N}} \frac{p^n\|x^n\|}{\mu_n^n} < 1 \right\}.$$

Then V_p contains the nilpotents of $\mathfrak{R}$, so V_p is dense in $\mathfrak{R}$ for every $p \in \mathbb{N}$.

Since $\|b^n\|^{1/n}/\lambda_n \xrightarrow[n\to\infty]{} \infty$ and $\|u^{-n}\|^{1/n} \xrightarrow[n\to\infty]{} \nu(u^{-1})$, the spectral radius of u^{-1}, for every invertible element u of $\mathfrak{R}^{\#}$, we obtain for such elements

$$\frac{\|b^n u^n\|^{1/n}}{\lambda_n} \geq \frac{\|b^n\|^{1/n}}{\lambda_n \|u^{-n}\|^{1/n}} \xrightarrow[n\to\infty]{} \infty .$$

So, if we put $G = \operatorname{Inv} \mathfrak{R}^{\#}$, we have $bG \subset U_p$, $b\mathfrak{R} \subset \overline{bG} = \overline{U_p}$, and hence U_p is dense in $\mathfrak{R}$ for every $p \in \mathbb{N}$.

Thus $(\bigcap_{p\in\mathbb{N}} U_p) \cap (\bigcap_{p\in\mathbb{N}} V_p) \cap \Omega$ is dense in $\mathfrak{R}$, and all elements x of that set satisfy the conditions of the theorem.

The following theorem was proved in [22], and the proof we give here also avoids infinite products.

THEOREM 8.3. [22] Let $\mathfrak{R}$ be a commutative radical Banach algebra, and let b be a nonnilpotent element of $\mathfrak{R}$. If $b \in [b\mathfrak{R}]^-$, there exists $c \in \mathfrak{R}$ such that

$$\liminf_{n\to\infty} \frac{\|c^n\|^{1/n}}{\|b^n\|^{1/n}} = 0, \ \limsup_{n\to\infty} \frac{\|c^n\|^{1/n}}{\|b^n\|^{1/n}} = +\infty, \ [c\mathfrak{R}]^- = [b\mathfrak{R}]^- .$$

Proof. Put $I = [b\mathfrak{R}]^-$. An argument used before in this paper shows that $\Omega = \{x \in I \mid [x\mathfrak{R}]^- = I\}$ is a dense G_δ of I.

Put, for $p \in \mathbb{N}$,

$$U_p = \left\{ x \in I \mid \sup_{n\in\mathbb{N}} \frac{\|x^n\|}{p^n\|b^n\|} > 1 \right\},$$

$$V_p = \left\{ x \in I \mid \inf_{n\in\mathbb{N}} \frac{p^n\|x^n\|}{\|b^n\|} < 1 \right\}.$$

The sets U_p and V_p are clearly open in I. Since $b\mathfrak{R} \subset \bigcap_{p\in\mathbb{N}} V_p$, V_p is dense in I for every $p \in \mathbb{N}$. We have, for every $f \in \mathfrak{R}$

$$\limsup_{n\to\infty} \frac{\|b^n(2pe+f)^n\|^{1/n}}{\|b^n\|^{1/n}} \geq \limsup_{n\to\infty} \frac{1}{\|(2pe+f)^{-n}\|^{1/n}} = 2p .$$

So $b(2pe+f) \in U_p$ for every $f \in \mathfrak{R}$.

Let $u \in \mathfrak{R}$, and let (y_m) be a sequence of elements of $\mathfrak{R}$ such that $b = \lim_{m\to\infty} by_m$.

Then $\|bu - b(2pe - 2py_m + u)\| \xrightarrow[m\to\infty]{} 0$. This shows that $b\mathfrak{R} \subset \overline{U_p}$ and that U_p is dense in I for every $p \in \mathbb{N}$. It follows from the category theorem that $(\bigcap_{p\in\mathbb{N}} U_p) \cap (\bigcap_{p\in\mathbb{N}} V_p) \cap \Omega$ is a dense G_δ of I, and all elements of that set satisfy the conditions of the theorem.

It was shown in [22] that $\mathbb{C}$ is the only commutative Banach algebra which is a ring of valuation. The following theorem extends this result.

THEOREM 8.4. Let A be a commutative unital Banach algebra. If the set of principal ideals of A forms a chain, then A is isomorphic to $\mathbb{C}[X]/X^p\mathbb{C}[X]$ for some integer $p \geq 0$.

<u>Proof</u>. If A possesses two different characters χ_1 and χ_2, then there exists $a,b \in A$ with $\chi_1(a) = \chi_2(b) = 0$, $\chi_1(b) = \chi_2(a) = 1$. Clearly $a \notin bA$ and $b \notin aA$ so that $aA \not\subset bA$, $bA \not\subset aA$. Hence A is in fact a local ring. Denote by $\mathfrak{R}$ the kernel of the unique character χ of A. Then $\mathfrak{R}$ is a radical algebra.

If $b \in [b\mathfrak{R}]^-$ for some nonnilpotent $b \in \mathfrak{R}$, the element c constructed in Theorem 8.3 certainly satisfies $cA \not\subset bA$, $bA \not\subset cA$, as observed in [22], but we have also to deal here with the case where b is nilpotent. So consider a nonzero element b of $\mathfrak{R}$ such that $b \in [b\mathfrak{R}]^-$. Since $nb \in [b\mathfrak{R}]^-$ for every $n \in \mathbb{N}$, there exists a sequence (x_n) of elements of $\mathfrak{R}$ such that $b(ne+x_n) \xrightarrow[n\to\infty]{} 0$.

Put $u_n = (n+1)e + x_n$. Then $bu_n \xrightarrow[n\to\infty]{} b$.

Denote by χ the unique character of A and put, for $x \in A$,

$$|||x||| = \sup_{\substack{y\in[b\mathfrak{R}]^- \\ \|y\|\leq 1}} \|xy\| .$$

We obtain an algebra seminorm, $|||e||| = 1$ as $b \neq 0$, and $|||x||| = \|\tilde{x}\|$, where $\tilde{x}$ is the element of $\mathcal{L}(I)$ defined by the formula $\tilde{x}(y) = xy$ and where $I = [b\mathfrak{R}]^-$.

The character χ is the only character of $(A, |||\cdot|||)$ so it is necessarily continuous with respect to $|||\cdot|||$, and more precisely $|\chi(x)| \leq |||x|||$ for every $x \in A$. So $|||u_n||| \xrightarrow[n\to\infty]{} \infty$. It follows from the Banach-Steinhaus theorem that $\|\alpha u_n\| \xrightarrow[n\to\infty]{} \infty$ for some $\alpha \in I$.

Put, for $p \in \mathbb{N}$, $\Omega_p = \{(x,y) \in I \times I \mid \inf_{u\in A} (\|x - ux\| + 1/\|yu\|) < 1/p\}$. Then $(b,\alpha) \in \Omega_p$ for every $p \in \mathbb{N}$. Fix $\varepsilon > 0$ and $(x,y) \in I \times I$. There exists $v \in \mathfrak{R}$ such that $\|x - bv\| < \varepsilon$. Also $\|y - (y + (\varepsilon\alpha/2\|\alpha\|))\| < \varepsilon$.

Clearly $bvu_n \xrightarrow[n\to\infty]{} bv$. If $\limsup_{n\to\infty} \|yu_n\| = +\infty$, then $(bv,y) \in \Omega_p$. If not, then $\|(y + (\varepsilon\alpha/2\|\alpha\|))u_n\| \to +\infty$, so that $(bv, y + (\varepsilon\alpha/2\|\alpha\|)) \in \Omega_p$. This shows that Ω_p is dense in $I \times I$. So $\Omega = \bigcap_{p\in\mathbb{N}} \Omega_p$ is a dense G_δ of $I \times I$ and $\Omega' = \{(x,y) \in I \times I \mid (y,x) \in \Omega\}$ is a dense G_δ of $I \times I$ too. Now if $(x,y) \in \Omega \cap \Omega'$ there exist two sequences (v_p) and (v'_p) of elements of A such that $\|xv_p - x\| \xrightarrow[p\to\infty]{} 0$, $\|yv_p\| \xrightarrow[p\to\infty]{} \infty$, $\|yv'_p - y\| \xrightarrow[p\to\infty]{} \infty$ and hence $x \notin yA$, $y \notin xA$.

Thus, if A satisfies the condition of the theorem, we have $b \notin [b\mathfrak{R}]^-$ for every nonzero element b of $\mathfrak{R}$. If $b\mathfrak{R}$ is not closed in A, then, taking $c \in [b\mathfrak{R}]^-$, $c \notin b\mathfrak{R}$, we have $b \notin cA$ because $[b\mathfrak{R}]^-$ is a closed ideal of A. Thus $c \notin b \operatorname{Inv} A$. As $c \notin b\mathfrak{R}$ by hypothesis, $c \notin b[\operatorname{Inv} A \cup \mathfrak{R}] = bA$. So, if A satisfies the condition of the theorem, $b\mathfrak{R}$ is closed for every $b \in \mathfrak{R}$ (and $b \notin [b\mathfrak{R}]^- = b\mathfrak{R}$ if $b \neq 0$, but this is automatically true as $\mathfrak{R}$ is radical).

It follows then from Corollary 3.6 that every nonzero element of $\mathfrak{R}$ is a finite product of irreducible elements of $\mathfrak{R}$. Assume $\mathfrak{R} \neq \{0\}$.

Let x,y be two irreducible elements of $\mathfrak{R}$. Then either $x \in yA$ or $y \in xA$, say $y \in xA$. As y is irreducible, $y \notin x\mathfrak{R}$, so $y \in x \operatorname{Inv} A$ and $x \in y \operatorname{Inv} A$.

Thus, if we fix an irreducible element a of $\mathfrak{R}$, we have $b = a^n u$ for every nonzero $b \in \mathfrak{R}$, where $u \in \operatorname{Inv} A$, $n \in \mathbb{N}$.

If a were not nilpotent, this would imply that $\mathfrak{R}$ is an integral domain. But if $\mathfrak{R}$ is an integral domain and if $b\mathfrak{R}$ is closed with $b \neq 0$, a standard application of the closed graph theorem shows that b is not a topological divisor of zero in $\mathfrak{R}$. Since

$$\liminf_{n\to\infty} \left\| b \frac{(b^n)}{\|b^n\|} \right\| = \liminf_{n\to\infty} \left(\frac{\|b^{n+1}\|}{\|b^n\|} \right) = 0 ,$$

we get a contradiction. So a is nilpotent. Let p be the largest integer such that $a^p \neq 0$. (The case $p = 0$ gives the case where $\mathfrak{R} = \{0\}$). Every element u

of A can be written in the form $u = \lambda_0 e + u_1$ where $u_1 \in \mathcal{R}$. Now $u_1 = av_1 = a(\lambda_1 e + u_2)$ where $u_2 \in \mathcal{R}$, so $u = \lambda_0 e + \lambda_1 a + au_2$.

By an easy finite induction we can find a sequence $\lambda_0, \ldots, \lambda_p$ of complex numbers and a sequence $u_1, \ldots, u_{p+1}$ of elements of $\mathcal{R}$ such that $u = \lambda_0 e + \lambda_1 a + \cdots + \lambda_p a^p + a^p u_{p+1}$.

Since $u_{p+1} \in \mathcal{R}$, $u_{p+1} = a^k v$ where $k > 0$, and $v \in A$, so that $a^p u_{p+1} = a^{p+k} v = 0$, and $u = \lambda_0 e + \lambda_1 a + \cdots + \lambda_p a^p$. Since $a^p \neq 0$ and $a^{p+1} = 0$, the family $\{e, \ldots, a^p\}$ is linearly independent, and the above decomposition is unique. The map $\lambda_0 + \lambda_1 X + \cdots + \lambda_k X^k \mapsto \lambda_0 + \lambda_1 a + \cdots + \lambda_k a^k$ is an algebra homomorphism from $\mathbb{C}[X]$ onto A whose kernel is $X^{p+1}\mathbb{C}[X]$. Hence $A \simeq \mathbb{C}[X]/X^{p+1}\,\mathbb{C}[X]$ and the theorem is proved.

9. A classification of commutative radical Banach algebras

We summarize here the results of this paper. A set of conditions concerning infinite-dimensional, commutative, radical Banach algebras is obtained which enables us to distinguish nine classes of such algebras, these classes getting smaller and smaller. The first one is so big that it contains all these algebras. The condition which defines the ninth class suggest a tenth class, but this tenth class is so small that it is in fact empty.

We will write CRBA for "commutative, radical, Banach algebra."

Class I. Infinite-dimensional CRBA $\mathcal{R}$ such that the set of principal ideals of $\mathcal{R} \oplus \mathbb{C}e$ is not totally ordered by inclusion.

Theorem 8.4 shows that this class contains all infinite dimensional CRBA.

Class II. Infinite-dimensional CRBA $\mathcal{R}$ such that $\bigcap_{n\in\mathbb{N}} \mathcal{R}^{[n]} \neq \{0\}$.

Here, we denote by $\mathcal{R}^{[n]}$ the set of all products of n elements of $\mathcal{R}$. Banach algebras of power series give examples of integral domains which do not belong to Class II. Of course, the "trivial" algebras such that $\mathcal{R}^{[2]} = \{0\}$ do not belong to Class II.

Class III. Infinite-dimensional CRBA $\mathcal{R}$ satisfying one of the following equivalent conditions:

(3.1) $a \in [a\mathcal{R}]^-$ for some nonzero $a \in \mathcal{R}$ (i.e., $\mathcal{R}$ possesses a nontrivial p.a.i.);

(3.2) $\underleftarrow{\lim}\, a_n \cdots a_1 \mathcal{R} \neq \{0\}$ for some sequence (a_n) of elements of $\mathcal{R}$;

(3.3) $\mathcal{R}^{\#}$ possesses a strictly increasing sequence of principal ideals.

The equivalence between these conditions was given by Corollary 3.5. Of course, if $\mathcal{R}$ is an integral domain, these conditions are equivalent to

(3.4) $\bigcap_{n\in\mathbb{N}} a_n \cdots a_1 \mathfrak{R} \neq \{0\}$ for some sequence (a_n) of elements of $\mathfrak{R}$.

Condition 3.4 always implies that $\mathfrak{R}$ belongs to Class II, but the converse is false. The CRBA $\mathfrak{R}_4$ of Section 7 is an integral domain which belongs to Class II, but it does not belong to Class III, and hence does not satisfy 3.4. Also the CRBA $\mathfrak{R}_5$ of Section 7 satisfies $\bigcap_{n\in\mathbb{N}} a^n \mathfrak{R}_5 \neq \{0\}$ for some $a \in \mathfrak{R}_5$, so $\mathfrak{R}_5$ belongs to Class II, but $\mathfrak{R}_5$ does not belong to Class III (Theorem 7.5).

Class IV. Infinite-dimensional CRBA $\mathfrak{R}$ which possess an element a satisfying one of the following equivalent conditions:

(4.1) $x \in [xa\mathfrak{R}]^-$ for some nonzero $x \in \mathfrak{R}$;

(4.2) $\varprojlim a^n \mathfrak{R} \neq \{0\}$;

(4.3) there exists in $\mathfrak{R}$ a nonzero a-divisible subspace;

(4.4) $y = \lim_{n\to\infty} a P_n(a)y$ for some nonzero $y \in \mathfrak{R}$ and some sequence (P_n) of complex polynomials;

(4.5) For every subset S of $\mathbb{C} - \{0\}$ such that $0 \in \overline{S}$, there exists a nonzero $z \in \mathfrak{R}$ such that $\inf_{\lambda\in S} \|z - a(a + \lambda e)^{-1} z\| = 0$.

The equivalence between these five conditions is given by Theorem 3.7. If $\mathfrak{R}$ is an integral domain, these conditions are equivalent to:

(4.6) $$\bigcap_{n\in\mathbb{N}} a^n \mathfrak{R} \neq \{0\}.$$

Theorem 7.5 shows that an infinite dimensional CRBA with divisors of zero may satisfy 4.6 without even belonging to Class III.

Class III obviously contains Class IV, but I was not able to find any CRBA which belongs to Class IV without belonging to Class III. I am convinced that such an algebra does exist, and the algebra $\mathfrak{R}_1$ of Section 7 seems a reasonable candidate. As mentioned at the end of Section 3, if $\mathfrak{R}$ is a CRBA and if $x = \lim_{n\to\infty} xe_n$ for some nonzero $x \in \mathfrak{R}$ with $\sup_{n\in\mathbb{N}} \|e_n\| < +\infty$ then $\mathfrak{R}$ belongs to Class IV by Grønbaek's factorization theorem [28].

Class V. Infinite-dimensional CRBA $\mathfrak{R}$ satisfying the following equivalent conditions:

(5.1) $a \in [a^2\mathfrak{R}]^-$ for some nonzero $a \in \mathfrak{R}$ (i.e., $\mathfrak{R}$ possesses a nonnilpotent element of finite closed descent);

(5.2) $a_n \in a_{n+1}^2(\mathfrak{R} \oplus \mathbb{C}e)$ for some sequence (a_n) of elements of $\mathfrak{R}$ with $a_1 \neq 0$;

(5.3) there exists a nonzero rational semigroup $(a^t)_{t\in\mathbb{Q}^+}$ in $\mathfrak{R}$;

(5.4) there exists a nonzero real semigroup $(b^t)_{t>0}$ in $\mathfrak{R}$ and a sequence (t_n)

of positive reals such that $t_n \xrightarrow[n\to\infty]{} 0$, $b^t = \lim_{n\to\infty} b^{t+t_n}$ for every $t > 0$, and $[b^t \mathfrak{R}]^- = [b^{t'} \mathfrak{R}]^-$ for every $t, t' > 0$.

If the continuum hypothesis is assumed, these conditions are equivalent to the following ones:

(5.5) there exists a discontinuous homomorphism from $C(K)$ into $\mathfrak{R} \oplus ¢e$, where K is any infinite compact space;

(5.6) for every commutative complex algebra A without unit which is an integral domain and which has the power of the continuum, there exists a one-to-one algebra homomorphism from A into $\mathfrak{R}$.

The equivalence between the above conditions was given by Theorem 5.3. Of course, Condition 5.1 is stronger than Condition 4.1, so Class IV contains Class V. Theorems 7.2 and 7.3 give examples of infinite-dimensional CRBA which belong to Class IV without belonging to Class V (the example of Theorem 7.3 is an integral domain).

If $\mathfrak{R}$ is an integral domain, then Condition 5.3 is satisfied if and only if some nonzero element of $\mathfrak{R}$ possesses roots of all orders, as observed in Section 2. But the CRBA $\mathfrak{R}_6$ of Section 7, which is not an integral domain, possesses such an element and does not even belong to Class III (of course, it belongs to Class II).

We mentioned above the fact that, if $\mathfrak{R}$ possesses a nontrivial bounded p.a.i., then $\mathfrak{R}$ belongs to Class IV. Now identify the algebra $\mathfrak{R}_4$ of Section 7 (which is a subalgebra of $\ell^1(\mathbb{Q}^+, e^{-t^2}))$ with its image in the measure algebra $M(\mathbb{R}^+, e^{-t^2})$, put $\mathfrak{m} = \{f \in L^1(\mathbb{R}^+, e^{-t^2}) \mid f \equiv 0 \text{ a.e. over } [0,1]\}$. Then $\mathfrak{R}_4 \oplus \mathfrak{m}$ is a closed subalgebra of $M(\mathbb{R}^+, e^{-t^2})$. (The sum is direct as $M(\mathbb{R}^+, e^{-t^2})$ is the topological direct sum of the subspaces of singular and absolutely continuous measures.) Denote by (p_n) the sequence of all prime numbers written in increasing order. Since $\delta_{1/p_n} * f \xrightarrow[n\to\infty]{} f$ for every $f \in \mathfrak{m}$, $\mathfrak{R}_4 \oplus \mathfrak{m}$ has a nontrivial bounded p.a.i., but it is easy to see that $\inf_{z\in\mathfrak{R}_4\oplus\mathfrak{m}} \|x - x^2 z\| > 0$ for every nonzero $x \in \mathfrak{R}_4 \oplus \mathfrak{m}$. So $\mathfrak{R}_4 \oplus \mathfrak{m}$ does not belong to Class V and the existence of a bounded nontrivial p.a.i. in a CRBA $\mathfrak{R}$ only implies that $\mathfrak{R}$ belongs to Class IV.

<u>Class VI</u>. Infinite-dimensional CRBA $\mathfrak{R}$ which satisfy one of the following two equivalent conditions:

(6.1) $\mathfrak{R}$ possesses a nonzero continuous semigroup $(a^t)_{t>0}$ over the positive reals;

(6.2) $\mathfrak{R}$ possesses a nonzero infinitely differentiable semigroup $(b^t)_{t>0}$ over the positive reals such that $[b^t \mathfrak{R}]^- = [b^{t'} \mathfrak{R}]^-$ for every $t, t' > 0$.

The equivalence between (6.1) and (6.2) was proved in Section 6 (and does not involve the radicality of $\mathfrak{R}$). The existence of a CRBA belonging to Class V without belonging to Class VI is the main open problem mentioned at the end of this paper.

Class VII,η ($0 < \eta \leq \pi/2$). Infinite-dimensional CRBA which possess a nonzero analytic semigroup $(a^t)_{t \in U_\eta}$ over the open sector $U_\eta = \{t \in \mathbb{C} - \{0\} \mid |\text{Arg } t| < \eta\}$.

Theorem 7.14 shows that, for every $\eta \in (0,\pi/2)$, there exists an infinite dimensional CRBA $\mathfrak{R}$ which belongs to the Class VII,η without belonging to the Class VII,η' for any $\eta' > \eta$.

Theorem 7.16 shows that there exists an infinite-dimensional CRBA which belongs to Class VI without belonging to Class VII,η for any $\eta > 0$, and Proposition 7.12 shows that there exists an infinite dimensional CRBA which belongs to the Class VII,$\pi/2$ without belonging to the following class.

Class VIII. Infinite-dimensional CRBA $\mathfrak{R}$ which satisfy one of the following equivalent conditions:

(8.1) some nonzero closed ideal I of $\mathfrak{R}$ possesses a bounded approximate identity;

(8.2) there exists in $\mathfrak{R}$ a nonzero analytic semigroup $(a^t)_{\text{Re } t > 0}$ over the right-hand half-plane such that $\sup_{t>0} \|a^t\| < +\infty$.

The equivalence between (8.1) and (8.2) follows from a work of A. M. Sinclair [36]. We showed in Section 4 that, if $\mathfrak{R}$ is separable, conditions (8.1) and (8.2) are equivalent to the following condition:

(8.3) $\mathfrak{R}$ possesses a nonzero bounded semigroup $(b^t)_{t>0}$ over the positive reals.

Theorem 7.9 shows that $L^1(\mathbb{R}^+, e^{-t^2})$ belongs to Class VIII without belonging to the ninth one that we now define.

Class IX. Infinite-dimensional CRBA which contain a nonzero analytic semigroup $(a^t)_{\text{Re } t>0}$ over the right-hand half-plane such that $\sup_{0<|t|\leq 1} \|a^t\| < +\infty$.

Theorem 7.11 gives an example of an infinite dimensional CRBA which is an integral domain and which belongs to Class IX.

The natural tenth class would obviously be the class of CRBA's which contain nonzero, bounded, analytic semigroups $(a^t)_{\text{Re } t>0}$ over the right-hand half-plane. Unfortunately, as observed in Section 7 just before Lemma 7.9, this class is empty. (Note that such semigroups exist in abundance in $C_0(L)$, the algebra of continuous functions vanishing at infinity on a locally compact space L.) Of course, some other reasonable conditions about CRBA may be considered. For example, if we denote by $\mathfrak{R}^n$ the linear space spanned by $\mathfrak{R}^{[n]}$ (which is an ideal of $\mathfrak{R}$), we may introduce the condition $\bigcap_{n\in\mathbb{N}} \mathfrak{R}^n \neq \{0\}$ which is weaker than the condition $\bigcap_{n\in\mathbb{N}} \mathfrak{R}^{[n]} \neq \{0\}$. Other natural classes are the following

Class P. Infinite-dimensional CRBA in which the prime ideals are not totally ordered by inclusion.

Class F. Infinite-dimensional CRBA $\mathfrak{R}$ which possess some nonzero element which is not the product of any finite family of irreducible elements of $\mathfrak{R}$.

Theorem 8.1 shows that every CRBA of Class III which is an integral domain belongs to Class P, and Remark 7.8 gives obvious examples of CRBA which are integral domains and which belong to Class P without even belonging to Class II. Also Theorem 8.1 shows that every CRBA of Class V belongs to Class P (if $a \in [a^2\mathfrak{R}]^-$, then $a \in [a^n\mathfrak{R}]^-$ for every $n \in \mathbb{N}$ so a is not nilpotent if $a \neq 0$). On the other hand Proposition 7.7 gives an example of a CRBA $\mathfrak{R}_7$ which belongs to Class IV without belonging to Class P.

Also, Corollary 3.6 shows that Class F is contained in Class III. I do not know any example of an algebra of Class III which does not belong to Class F.

To conclude we give a list of questions which arise naturally from the investigations of this paper.

Question 1. Does there exist an infinite-dimensional CRBA of Class III in which every nonzero element can be written as a finite product of irreducible elements?

Question 2. Does there exist an infinite-dimensional CRBA of Class II which is an integral domain in which the set of primes is totally ordered?

Question 3. Does there exist an infinite-dimensional CRBA $\mathfrak{R}$ such that $\bigcap_{n\in\mathbb{N}} \mathfrak{R}^n \neq \{0\}$, but $\bigcap_{n\in\mathbb{N}} \mathfrak{R}^{[n]} = \{0\}$.

Question 4. Does every infinite-dimensional CRBA which possesses a nonzero continuous semigroup $(a^t)_{t>0}$ analytic at some $t_0 > 0$ possess a nonzero analytic semigroup $(b^t)_{t>0}$? (This question is suggested by the remarks made just before Theorem 7.15.)

Question 5. Does there exist an infinite-dimensional CRBA which belongs to Class III without belonging to Class IV?

Question 6. Does there exist an infinite dimensional CRBA which belongs to Class V without belong to Class VI?

Question 6bis. In particular, does the weighted algebra $\ell^1(\mathbb{Q}^+, e^{-t^2})$ possess a nonzero continuous semigroup $(a^t)_{t>0}$ over the positive reals?

The algebra $\mathfrak{R}_1$ of Section 7 seems to be a good candidate to give a positive answer to Question 5. Question 6 is, of course, the main one. A negative answer would imply that Class V and VI are equal and strengthen the results of [26]: the closed ideal problem for infinite-dimensional CRBA would be reduced to a problem concerning CRBA with bounded approximate identities!

I suspect, unfortunately, that the anser to Question 6bis is negative, but I was not able to prove it. It is easy to see, using the natural weak* topology on $\ell^1(\mathbb{Q}^+, e^{-t^2})$, that $\ell^1(\mathbb{Q}^+, e^{-t^2})$ cannot possess any nonzero continuous semigroup $(a^t)_{t>0}$ bounded over the positive reals, but the argument fails for continuous semigroups $(a^t)_{t>0}$ unbounded at the origin.

Reference

[1] G. R. Allan, Embedding the algebra of all formal power series in a Banach algebra, Proc. London Math. Soc., (3) 25 (1972), 329-340.

[2] ________, Elements of finite closed descent in a Banach algebra, J. London Math. Soc., 7 (1973), 462-466.

[3] G. R. Allan and A. M. Sinclair, Power factorization in Banach algebras with a bounded approximate identity, Studia Math., 56 (1976), 31-38.

[4] M. Altman, Contracteurs dans les algèbres de Banach, C. R. Acad. Sci. Paris, Ser. AB, 272 (1971), 1388-1389.

[5] W. G. Bade, P. C. Curtis, Jr. and K. B. Laursen, Divisible subspaces and problems in automatic continuity, Studia Math., 48 (1980), 159-186.

[6] W. G. Bade and H. G. Dales, Norms and ideals in radical convolution algebras, J. Functional Analysis, 41 (1981), 77-109.

[7] R. P. Boas, Entire Functions, Academic Press, New York, 1954.

[8] H. Bohr, Almost Periodic Functions, Chelsea Publishing Company, New York (1947).

[9] F. F. Bonsall and J. Duncan, Complete Normed Algebras, Springer-Verlag, Berlin, Heidelberg, New York, 1973.

[10] S. H. Bouloussa, Caracterisation des algèbres de Fréchet qui sont des anneaux de valuation, J. London Math. Soc., (2) 25 (1982), 355-364.

[11] N. Bourbaki, Topologie Générale, Chapitre II, Hermann, Paris, 1960.

[12] P. J. Cohen, Factorization in group algebras, Duke Math. J., 26 (1959), 199-206.

[13] H. G. Dales, A discontinuous homomorphism from C(X), Amer. J. Math., 101 (1979), 647-734.

[14] ________, Convolution algebras on the real line, this Volume.

[15] ________, Automatic continuity: a survey, Bull. London Math. Soc., 10 (1978), 129-183.

[16] Y. Domar, A solution of the translation-invariant subspace problem for weighted L^p on $\mathbb{R}$, $\mathbb{R}^+$ or $\mathbb{Z}$, this Volume.

[17] J. Esterle, Solution d'un problème d'Erdös, Gillman et Henriksen et application à l'étude des homomorphismes de $C(K)$, Acta Math., (Hungarica), 30 (1977), 113-127.

[18] ________, Sur l'existence d'un homomorphisme discontinu de $C(K)$, Proc. London Math. Soc., (3) 36 (1978), 46-58.

[19] ________, Injection de semigroupes divisibles dans des algèbres de convolution et construction d'homomorphismes discontinus de $C(K)$, Proc. London Math. Soc., (3) 36 (1978), 59-85.

[20] ________, Homomorphismes discontinus des algèbres de Banach commutatives separables, Studia Math., 66 (1979), 119-141.

[21] ________, Rates of decrease of sequences of powers in commutative radical Banach algebras, Pacific J. Math., 94 (1981), 61-82.

[22] ________, Theorems of Gelfand-Mazur type and continuity of epimorphisms from $C(K)$, J. Functional Analysis, 36 (1980), 273-286.

[23] ________, Universal properties of some commutative radical Banach algebras, J. für die Reine und ang. Math., 321 (1981), 1-24.

[24] ________, A complex variable proof of the Wiener Tauberian theorem, Ann. Inst. Fourier, (2) 30 (1980), 91-96.

[25] ________, Irregularity of the rate of decrease of sequences of powers in the Volterra algebra, Canad. J. Math., 33 (1981), 320-324.

[26] ________, Quasimultipliers, representations of H^∞, and the closed ideal problem for commutative Banach algebras, this Volume.

[27] S. Grabiner, Derivations and automorphisms of Banach algebras of power series, Memoirs Amer. Math. Soc., 146 (1975).

[28] N. Grønbaek, Power factorization in Banach modules over commutative radical Banach algebras, Math. Scand., 50 (1982), 123-134.

[29] E. Hille and R. S. Phillips, Functional Analysis and Semi-groups, Colloquium Publication Series, Vol. 31, Amer. Math. Soc., Providence, Rhode Island, 1957.

[30] K. Hoffman, Banach Spaces of Analytic Functions, Prentice-Hall Inc., Englewood Cliffs, N.J., 1962.

[31] B. E. Johnson, Continuity of centralizers on Banach algebras, J. London Math. Soc., 41 (1966), 639-640.

[32] K. Kuratowski, Topology, Vol. I, Academic Press, New York and London, 1966.

[33] B. J. Levin, Distribution of Zeros of Entire Functions, Translations of Math. Monographs, Vol. 5, revised edition, Amer. Math. Soc., Providence, Rhode Island, 1964.

[34] B. Nyman, On the one dimensional translation group on certain function spaces, Thesis, Uppsala (1950).

[35] C. E. Rickart, General Theory of Banach Algebras, Van Nostrand, New Jersey, 1960.

[36] A. M. Sinclair, Bounded approximate identities, factorization, and a convolution algebra, J. Functional Analysis, 29 (1978), 308-318.

[37] __________, Continuous Semigroups in Banach Algebras, London Math. Soc. Lecture Notes, 63, Cambridge University Press, 1982.

[38] M. P. Thomas, Algebra homomorphisms and the functional calculus, Pacific J. Math., 79 (1978), 251-269.

[39] N. T. Varopoulos, Continuité des formes linéaires positives sur une algèbre de Banach avec involution, C. R. Acad. Sci. Paris, Ser. AB 258 (1964), 1121-1124.

[40] F. Zouakia, The theory of Cohen elements, this Volume.

[41] __________, Semigroupes réels dans certaines algèbres de Banach commutatives radicales, Thèse de 3e cycle, Bordeaux, Juin 1980.

U. E. R. de Mathematiques et Informatique
Universite de Bordeaux I
351 Cours de la Libération
33405 Talence, France

QUASIMULTIPLIERS, REPRESENTATIONS OF H^∞, AND THE CLOSED IDEAL PROBLEM FOR COMMUTATIVE BANACH ALGEBRAS

J. Esterle

1. Introduction

The motivation for this paper is the following unsolved problem: Does every infinite-dimensional, commutative Banach algebra possess a proper closed ideal (that is, a closed ideal $I \neq \{0\}$ which does not equal the whole algebra)? The answer is of course yes if $\mathfrak{A}$ is not radical (take the kernel of a character), or if $\mathfrak{A}$ possesses divisors of zero. So this so-called "closed ideal problem" concerns commutative radical Banach algebras which are integral domains. A commutative Banach algebra which does not possess any proper closed ideal will be called topologically simple.

We are still far from giving an answer to the closed ideal problem, but we obtain some partial results in Section 8. We may summarize these results as follows.

THEOREM. Assume that there exists an infinite-dimensional, topologically simple Banach algebra $\mathfrak{A}$ which possesses a nonzero continuous semigroup $(a^t)_{t>0}$ over the positive reals. Then the following properties hold.

(1) There exists a topologically simple Banach algebra with bounded approximate identity.

(2) There exists a Banach space E and a linear contraction T acting on E satisfying:

- (a) T does not possess any nontrivial closed hyperinvariant subspace;
- (b) there exists an isometric isomorphism $\varphi : H^\infty \to \mathcal{L}(E)$ such that T is the image of the position function $\alpha : z \mapsto z$;
- (c) the spectrum (in fact the left essential spectrum) of T equals the closed unit disc.

(3) If the topologically simple Banach algebra $\mathfrak{A}$ is spanned by a nonzero continuous semigroup $(a^t)_{t>0}$ over the positive reals, there exists a Banach space F and a linear contraction S acting on F satisfying:

- (a) S has no nontrivial closed invariant subspace;
- (b) $Sp(S) = \{1\}$;
- (c) there exists a norm-decreasing one-to-one homomorphism $\psi : H^\infty \to \mathcal{L}(E)$ such that S is the image of the position function $\alpha : z \mapsto z$.

The Brown-Chevreau-Pearcy theorem [15] shows that every linear contraction

T acting on a Hilbert space whose spectrum is "rich" (which is certainly the case if the spectrum of T equals the closed unit disc) possesses a nontrivial closed invariant subspace. It is not known whether or not such a contraction possesses a proper hyperinvariant subspace. It is also not known whether or not the Brown-Chevreau-Pearcy theorem remains true for contractions on a Hilbert space whose spectrum contains the unit circle. (Foias, Pearcy and Sz.-Nagy obtain in [30] some partial results in this direction.) We thus see that the results of Section 8 lead to problems which remain unsolved even in the case of Hilbert spaces.

The Brown-Chevreau-Pearcy theorem uses the representations of H^∞ associated to completely non-unitary contractions on a Hilbert space by the Nagy-Foias functional calculus ([29], Chapter III). Apostol [6] extended the Brown-Chevreau-Pearcy theorem to representations of H^∞ (where the spectrum of the image of the position function is "rich") on some Banach spaces. So there is some hope to get at least a proper invariant subspace for the operator T introduced in the second assertion of the theorem (but the Banach space E we get is highly non-reflexive). We did not investigate that point.

Note that the results of Section 8 concern the closed ideal problem for commutative Banach algebras which possess nonzero continuous real semigroups $(a^t)_{t>0}$. On the other hand, the author showed in [24] that every commutative radical Banach algebra $\mathfrak{R}$ such that $x \in [x^2\mathfrak{R}]^-$ for some nonzero $x \in \mathfrak{R}$ possesses nonzero real semigroups $(a^t)_{t>0}$, and it follows from [25], Section 4, that it is possible to get semigroups $(a^t)_{t>0}$ satisfying $\|a^{t+t_n} - a^t\| \to 0$ as $n \to \infty$ for every $t > 0$, where (t_n) is a sequence of positive reals which converges to zero (the sequence depending on the semigroup). The author's original construction given in [24] has been nicely improved recently by Zouakia [51] in his thesis. (This construction is strongly related to his version of the theory of Cohen elements given in these Proceedings, [50].) It remains unknown whether the existence of a nonzero real semigroup in a radical Banach algebra $\mathfrak{R}$ implies the existence of a nonzero real continuous semigroup in $\mathfrak{R}$. This question is the main problem raised in [25], and a positive answer (which seems unfortunately unlikely) would give a significant progress for the closed ideal problem.

The closed ideal problem is a special case of the hyperinvariant subspace problem. Recall that a closed subspace F of a Banach space E is said to be <u>invariant (respectively, hyperinvariant)</u> for an operator $T \in \mathcal{L}(E)$ if $T(F) \subset F$ (respectively, $S(F) \subset F$ for every $S \in \mathcal{L}(E)$ such that $ST = TS$). An example of a Banach space E and an operator $T \in \mathcal{L}(E)$ without nontrivial invariant subspace has been constructed by Enflo in [23], but some computations remain to be checked in it and it seems unlikely that this construction could be changed to give a construction of a topologically simple Banach algebra.

A survey of recent results concerning invariant or hyperinvariant subspaces is given by Beauzamy in his talk at the Bourbaki seminar [10]. Lomonosov's theorem shows that every operator which commutes with a nonzero compact operator has a proper hyperinvariant subspace (see [39] and [41], Chapter 7). Since elements of $L^1(\mathbb{R}^+, e^{-t^2})$ act compactly ([8]), this shows that every closed subalgebra of $L^1(\mathbb{R}^+, e^{-t^2})$ possesses a proper closed ideal, but Lomonosov's method does not seem to give a key for the closed ideal problem in the general case.

Many invariant or hyperinvariant subspace theorems, like Wermer's theorem [49], consist in fact in showing that the closed subalgebra of $\mathcal{L}(E)$ generated by T, or generated by T and T^{-1}, is not an integral domain. We discuss in the appendix a paper by Beauzamy [9] which shows that a contraction T such that (T^n) does not strongly converge to 0, and such that a sequence (x_n) with $x_1 \neq 0$, $x_n = T(x_{n+1})$ for every n does not grow too fast in norm, has a proper hyperinvariant subspace. The first condition shows that T acts as an isometry on some Banach space containing E and the second one shows that T acts as a "Wermer operator" on some other Banach space continuously contained in E, which leads to a functional calculus $\varphi : \sum_{n\in\mathbb{Z}} \lambda_n X^n \mapsto \sum_{n\in\mathbb{Z}} \lambda_n T^{-n}$ if T is one-to-one, where $\sum_{n\in\mathbb{Z}} \lambda_n T^{-n}$ is a possibly unbounded closed operator (T is not assumed to be invertible). Here, the sequence $(\lambda_n)_{n\in\mathbb{Z}}$ satisfies some growth condition which ensures that there exists $f = \sum_{n\in\mathbb{Z}} \lambda_n X^n$, $g = \sum_{n\in\mathbb{Z}} \mu_n X^n$ with $\varphi(f)\varphi(g) = 0$, $\varphi(f) \neq 0$, $\varphi(g) \neq 0$. This point of view is probably simpler than the point of view of [9], but if T is a multiplier on a Banach algebra A satisfying these conditions, then A is not an integral domain. The only method which could give some results for the closed ideal problem seems to be related to Atzmon's paper [7]. Atzmon shows in [7] that every operator T which satisfies $\|T^n\| = \underline{O}(n^k)$ and $\|T^{-n}\| = \underline{O}(\exp cn^{1/2})$ as $n \to \infty$ for some $c > 0$ has a proper hyperinvariant subspace. Stated in this form, the theorem shows in fact that the algebra generated by T and T^{-1} is not an integral domain, but if $Sp(T) = \{1\}$ it is possible to take the hypothesis $\|\ell \circ T^{-n}\| = \underline{O}(\exp cn^{1/2})$, where ℓ is a nonzero continuous linear functional, instead of the hypothesis $\|T^{-n}\| = \underline{O}(\exp cn^{1/2})$. Standard computations about Laguerre polynomials show that $\|u(\delta - u)^n\|_1 \leq 1$ for every $n \in \mathbb{N}$ where $u : t \mapsto e^{-t}$ belongs to $L^1(\mathbb{R}^+)$ (here, powers are convolution powers). So $\|(\delta - u)^n\| = \underline{O}(n)$, and, using Sinclair's map [45], we see that elements x such that $\|(e - x)^n\| = \underline{O}(n)$ exist in abundance in commutative Banach algebras with bounded approximate identities. Applying the above improvement of Atzmon's theorem to u it is possible to obtain all the standard closed ideals of $L^1(\mathbb{R}^+, e^{-t^2})$ (hence all the closed ideals of $L^1(\mathbb{R}^+, e^{-t^2})$ by Domar's theorem [22]). In fact, we give in the appendix improvements of Atzmon's theorems based upon the Paley-Wiener theorem. If $\|T^n\| = \underline{O}(n^k)$, there exists a continuous map

$t \mapsto u_t$ from $[0,\infty)$ into $\mathcal{L}(E)$ such that $(u_t)_{t\geq 0}$ is "nearly a semigroup". A resolvent formula

$$\int_0^\infty e^{-\lambda t}\, u_t dt = (T-I)^{2k+2}(\lambda(T-I)-T-I)^{-1}$$

holds for $\operatorname{Re}\lambda > 0$, and, applying the Paley-Wiener theorem to the function $\lambda \mapsto \langle \int_0^\infty e^{-\lambda} u_t(x)dt, \ell\rangle$, it is possible to show that T has a proper invariant subspace if $|\langle (T-I)^n x, \ell\rangle|^{1/n} = \underline{O}(1/n)$ as $n \to \infty$ for some nonzero $x \in E$ and some nonzero $\ell \in E^*$. (In fact, T has in general an uncountable chain of invariant subspaces.) Similarly, T has a proper hyperinvariant subspace if $\|(T-I)^n \circ \ell\|^{1/n} = \underline{O}(1/n)$ as $n \to \infty$. The bridge between conditions concerning $\langle (T-I)^n x, \ell\rangle$ and conditions of Atzmon type concerning $\langle T^{-n}x, \ell\rangle$ is given by a theorem of Cartwright ([12], Theorem 10.2.1), and a classical extension of the Phragmén-Lindelöf principle to functions of zero exponential type when $\operatorname{Sp}T = \{1\}$. This method, developed in detail in the appendix, is similar to the method used by the author in his unpublished 1979 UCLA postgraduate course to study nilpotent continuous semigroups. (These results have fortunately been recently published, with some improvements, by A. M. Sinclair in Chapter 6 of his lecture notes [47].)

There is in this paper much more material than is strictly necessary to obtain the results of Section 8. It seems that some of the results and methods used here have their own interest. We obtain spectral mapping theorems, in particular for representations of H^∞, results about Banach algebras with bounded approximate identities (b.a.i.), inequalities in some Banach algebras, and a new spectral theory for nonunital, commutative Banach algebras with dense principal ideals which works in particular for radical Banach algebras with b.a.i.

Usually, continuous unital homomorphisms between Banach algebras reduce spectra (and non-unital homomorphisms can just add $\{0\}$ when the image of the unit element of the domain is not the unit element of the range). In particular, if $\mathfrak{A}$ is a Banach algebra and $\varphi : \mathfrak{A} \to \mathcal{L}(E)$ is a representation of $\mathfrak{A}$ on a Banach space E, then $\operatorname{Sp}[\varphi(x)] \subset \operatorname{Sp}(x) \cup \{0\}$ for every $x \in \mathfrak{A}$. Spectral mapping theorems ensure that under some conditions $\varphi(\lambda) \in \operatorname{Sp}(\varphi(x))$ if $\lambda \in \operatorname{Sp}(x)$. For example, a theorem of Foias-Mlak [27] shows that, if $\varphi : f \mapsto f(T)$ is the representation of H^∞ associated to a completely non-unitary contraction T on a Hilbert space by the Nagy-Foias calculus such that $\lambda \in \operatorname{Sp}(T)$, then $h(\lambda) \in \operatorname{Sp}(h(T))$ if $|\lambda| = 1$ and if $h(\lambda) = \lim\{h(z) : z \to \lambda, |z| < 1\}$ does exist. We give at the end of the paper a very short proof of a similar result for representations of H^∞ on arbitrary Banach spaces. (The proof uses only the fact that the ideal $\mathfrak{m}_\lambda = \{f \in \mathfrak{A}(D) | f(\lambda) = 0\}$ possesses a nice bounded approximate identity given by fractional powers of $z - \lambda$.) The Foias-Mlak theorem is used, among other tools, by Foias, Pearcy and Sz.-Nagy in [30] to produce elements h of H^∞ of norm 1 such that the spectrum of $h(T)$ equals the closed unit disc for every representation $f \mapsto f(T)$ of H^∞ on a Hilbert space such that $Sp(T)$ is connected and contains 1. We extend this result in Section 6, obtaining elements of H^∞ which possess the above property for representations of H^∞ on an arbitrary Banach space (assuming that $Im(I - T)$ is not closed and that $Sp(T)$ is connected). The method we use seems to be both simple and new. If $\mathfrak{A}$ is unital and if (y_n) is a sequence of elements of $\mathfrak{A}$ of norm 1 such that $\lim\inf_{n\to\infty} \|(a - \lambda e)y_n\| = 0$, where $a \in \mathfrak{A}$, $\lambda \in \mathbb{C}$, then $\lambda \in Sp(a)$ and $\lim\inf_{n\to\infty} \|(\varphi(a) - \lambda e_B)\varphi(y_n)\| = 0$ for every homomorphism φ from $\mathfrak{A}$ into a Banach algebra $\mathfrak{B}$ with unit element e_B. Now if for some reason $\lim\inf_{n\to\infty} \|\varphi(y_n)\| > 0$, then $\lambda \in Sp(\varphi(a))$. An inequality proved in Section 4 shows that $\lim\inf_{n\to\infty} \|e_n^2 - e_n\| \geq \frac{1}{4}$ if (e_n) is a sequential bounded approximate identity in a nonzero Banach algebra which does not possess any nonzero idempotent. The ideal $\mathfrak{m} = \{f \in \mathfrak{A}(D) \mid f(1) = 0\}$ possesses a sequential b.a.i. (e_n), and if $\varphi : f \mapsto f(T)$ is any representation of H^∞ such that the spectrum of the image T of the position function $z \mapsto z$ is connected and contains 1, then $\varphi(\mathfrak{m})$ does not contain any nonzero idempotent if $Im(I - T)$ is not closed. In that case $\lambda \in Sp(h(T))$ if $\lim\inf_{n\to\infty} \|(h - \lambda 1)(e_n^2 - e_n)\| = 0$. In fact, there exist elements h of H^∞ of norm 1 such that $\lim\inf_{n\to\infty} \|(h - \lambda 1)(e_n^2 - e_n)\| = 0$ for every λ with $|\lambda| \leq 1$, and for these elements $Sp(h(T))$ equals the closed unit disc for every representation of H^∞ satisfying the above conditions. The elements h we construct belong to a smaller algebra, the algebra $\mathfrak{A}_1^\infty$ of all elements of H^∞ which have a continuous extension to $\overline{D}\setminus\{1\}$. These elements are obtained in Section 5 by using a special bounded approximate identity which happens to exist in $\mathfrak{m}$. If we consider representations $\varphi : f \mapsto f(T)$ of H^∞ such that $Im(I - T)$ is not closed and such that $Sp(T)$ is disconnected and contains 1, then it is possible to find an element h_φ of H^∞ of norm 1 such that $Sp(h_\varphi(T))$ equals the closed unit disc, but here h_φ depends on the representation φ.

Now let $\varphi : f \mapsto f(T)$ be a representation of H^∞ on a Banach space E such that $Sp(T) = \{1\}$, $T \neq I$, let $x \in E$ such that $T^n(x) \to 0$ as $n \to \infty$ and let Ω be a countable subset of the closed unit disc. There exists an element h of H^∞ of norm 1, which depends on x, Ω, and φ such that $\lim\inf_{n\to\infty} \|(h(T) - \lambda I)^n x\|^{1/n} = 0$ for every $\lambda \in \Omega$. This implies that $\Omega \subset Sp(h(T))$. This element h is also constructed in Section 5 by using a special approximate identity in $\mathfrak{m}$. I do not know whether this "lower x-spectral radius condition" can be of any help in the research of invariant subspaces for T. This condition means heuristically that even if the spectrum of $h(T)$ equals the closed unit disc (which is the case if Ω is dense in the unit disc), $h(T)$ carries some quasinilpotency induced by the quasinilpotency of $T - I$. Note that, if $Sp(T) = \{1\}$ and if $T \neq I$, there always exists a nonzero $x \in E$ such that $T^n(x) \to 0$ as $n \to \infty$. This follows from the fact that $\|(I - T)T^n\| \to 0$, as $n \to \infty$, which is proved in the appendix by using a version of the Phragmén-Lindelöf theorem for functions of zero exponential type.

We mentioned the fact that $\liminf_{n\to\infty} \|e_n^2 - e_n\| \geq \frac{1}{4}$ for any sequential bounded approximate identity (e_n) in a commutative Banach algebra $\mathfrak{A}$ which does not possess any nonzero idempotent. In fact, if $\mathfrak{A}$ does not possess any nonzero idempotent we have $\inf\{\|x^2 - x\| : \|x\| \geq \frac{1}{2}\} \geq \frac{1}{4}$, as shown in Section 4. Some investigations about this inequality were made by Berkani [11] who showed in particular that there exists $x \in \mathfrak{A}$ with $\|x\| = \frac{1}{2}$, $\|x^2 - x\| = \frac{1}{4}$ if and only if there exists a nonzero $u \in \mathfrak{A}$ whose norm equals its spectral radius. A more general inequality is proved in the appendix, using a classical theorem of Caratheodory which shows that, if f is analytic over the open unit disc D and if $f(0) = 0$, $f(z) \neq 0$ for every $z \in D - \{0\}$, $f'(0) = 1$, then $f(D)$ contains an open disc of radius $1/16$ centered at the origin (some elliptic function makes the constant $1/16$ best possible). This inequality, which is given in Theorem 9.4, shows in particular that, if a is any element in a Banach algebra such that $Sp(a) = \{1\}$ and $a \neq e$, then $\liminf_{n\to\infty} n\|a^n - a^{n+1}\| \geq 1/96$ (we must have $\lim_{n\to\infty} \|a^n - a^{n+1}\| = 0$ by Theorem 9.1). The constant $1/96$ is certainly not best possible, and this circle of ideas could merit more investigations.

Sections 4 and 5 directly concern Banach algebras with b.a.i. In Section 4, the inequality $\liminf_{n\to\infty} \|e_n^2 - e_n\| \geq \frac{1}{4}$ answers in some sense the heuristic question: how far is a bounded approximate identity from being an identity? This distance between e_n and e_n^2 can be observed in any Banach algebra $\mathfrak{A}$ with b.a.i. which does not possess a nonzero idempotent. In fact such an inequality is not that surprising, at least if $pq \neq p$ if $p,q \in \mathfrak{A}$, $p \neq 0$. In that case the function $x \mapsto x^2 - x$ is one-to-one, and the inverse function is defined and continuous on a neighborhood of the origin, so that, if $x_n^2 - x_n \to 0$ as $n \to \infty$, then $x_n \to 0$ as $n \to \infty$. The value $\frac{1}{4}$ is sharp, and the proof of the inequality needs a more precise argument given in Section 4. Lower estimates for $\liminf_{n\to\infty} \|e_n(e - e_n)^k\|$ can be deduced from Theorem 9.4 in the appendix in the case where $pq \neq p$ if $p,q \in \mathfrak{A}$, $p \neq 0$.

In Section 5, we are concerned with a special class of Banach algebras with b.a.i. that we call Banach algebras with <u>regular sequential b.a.i</u>. A regular sequential b.a.i. is a sequence (e_n) such that $xe_n \to x$ as $n \to \infty$ for every $x \in \mathfrak{A}$ and such that $\|(1 - \lambda)e_n + \lambda e\| = 1$ for every $n \in \mathbb{N}$, where $|\lambda| = 1$ and where $(\arg \lambda)/\pi$ is irrational. For every $\mu \in \mathbb{C}$ with $|\mu| = 1$ it is then possible to find a sequential b.a.i. $(e_n(\mu))$ depending on μ, satisfying $\|(1 - \mu)e_n(\mu) + \mu e\| = 1$ for every $n \in \mathbb{N}$. If $\mathfrak{A}$ has no nonzero hermitian element, then some numerical range argument shows that $\sup_{|\mu|=1} \|(1 - \mu)x + \mu e\| > 1$ for every nonzero $x \in \mathfrak{A}$, and the sequence $e_n(\mu)$ has to depend on μ.

Now consider the algebra $\mathfrak{M}(\mathfrak{A})$ of all multipliers on $\mathfrak{A}$, and denote by $\Omega(\mathfrak{A})$ the closure of the set $\{T \in \mathfrak{M}(\mathfrak{A}) \mid \|T\| = 1, T - e \in \mathfrak{A}\}$ with respect to the strong topology. This set is always convex and stable under products, but might reduce to $\{e\}$. If $\mathfrak{A}$ has a regular b.a.i. then $\Omega(\mathfrak{A})$ becomes stable by products by complex numbers of modulus 1. Using the Baire category theorem it is then possible to show that $\Omega(\mathfrak{A})$ contains multipliers S on $\mathfrak{A}$ such that $\liminf_{n\to\infty} \|(S - \lambda e)y_n\| = 0$ for every element λ of the closed unit disc, where (y_n) is a given sequence of elements of $\mathfrak{A}$ such that $xy_n \to 0$ as $n \to \infty$ for every $x \in \mathfrak{A}$. Taking $y_n = e_n - e_n^2$, where (e_n) is a sequential approximate identity in $\mathfrak{A}$, we obtain contractions S in $\mathfrak{M}(\mathfrak{A})$ such that $Sp(S)$ equals the closed unit disc $\overline{D}$ for every homomorphism φ of $\mathfrak{M}(\mathfrak{A})$ into a Banach algebra such that $[\varphi(\mathfrak{A})]^-$ has no nontrivial idempotent (this condition reduces to $\varphi(\mathfrak{A}) \neq \{0\}$ if $\mathfrak{A}$ is radical). If $\mathfrak{A}$ is a radical algebra, it is also possible, given any nonzero $x \in \mathfrak{A}$ and any countable subset Δ of $\overline{D}$, to find $T \in \mathfrak{M}(\mathfrak{A})$ with $\|T\| = 1$ and $\liminf_{n\to\infty} \|(T - \lambda e)^n x\|^{1/n} = 0$ for every $\lambda \in \Delta$. If Δ is dense in $\overline{D}$, we must have $\limsup_{n\to\infty} \|(T - \mu e)^n x\|^{1/n} = 1 + |\mu|$ for every $\mu \in \overline{D}$, but I was not able to find any interesting consequence of this strange gap between "lower x-spectral radius" and "upper x-spectral radius" for these $T - \lambda e$. Since any quotient of a uniform algebra with b.a.i. has a regular b.a.i., this theory applies to the maximal ideal $\mathfrak{m}$ of $\mathfrak{A}(D)$ discussed above, and leads to the results about representations of H^∞ obtained in Section 6.

Note that the set $\Omega(\mathfrak{A})$ is the strong closure in $\mathfrak{M}(\mathfrak{A})$ of the set of elements of the form $e - x$ where $x \in \mathfrak{A}$, $\|e - x\| = 1$. These elements have norm 1 and spectral radius 1, and if there exists $x \in \mathfrak{A}$ such that $\|e - x\| = 1$ and $\|\ell \circ x^n\| = \underline{O}(1/n)$ for some nonzero $\ell \in \mathfrak{A}^*$, the improvements of Atzmon's paper given in the appendix show that $\mathfrak{A}$ has a proper closed ideal.

We now turn to a description of Sections 2 and 7, which lead in particular to a nontrivial spectral theory for commutative radical Banach algebras with b.a.i. The radical Banach algebras with regular b.a.i. discussed in Section 5 have a very rich multiplier algebra, but an example due to Koua [37] [38] shows that there exists a commutative separable radical Banach algebra $\mathfrak{R}$ with b.a.i. such that $\mathfrak{M}(\mathfrak{R})$ reduces to $\mathfrak{R} \oplus \mathbb{C}e$. Also, the multiplier algebra of $L^1(\mathbb{R}^+, e^{-t^2})$ may be identified with the measure algebra $\mathfrak{M}(\mathbb{R}^+, e^{-t^2})$ whose carrier space reduces to a singleton (the unique character of $\mathfrak{M}(\mathbb{R}^+, e^{-t^2})$ is the function $\mu \mapsto \mu(\{0\})$, the point mass at zero). See [31], p. 149, Theorem 4.18.4. So the Gelfand theory for $\mathfrak{M}(\mathfrak{R})$ might be trivial.

We introduce in Section 2 the quasi-multipliers for commutative Banach algebras $\mathfrak{A}$ such that $x\mathfrak{A} \neq \{0\}$ for every nonzero $x \in \mathfrak{A}$, $[a\mathfrak{A}]^- = \mathfrak{A}$ for some $a \in \mathfrak{A}$. A quasi-multiplier is a possibly unbounded closed operator whose domain is an ideal containing a dense principal ideal of $\mathfrak{A}$. More precisely, a quasi-multiplier is a couple $\{T_{a/b}, \mathfrak{D}_{a/b}\}$, where $a,b \in \mathfrak{A}$, $[b\mathfrak{A}]^- = \mathfrak{A}$, $\mathfrak{D}_{a/b}$ is the set $\{x \in \mathfrak{A} \mid ax \in b\mathfrak{A}\}$, and, if $x \in \mathfrak{D}_{a/b}$, then $T_{a/b}(x)$ is the unique $y \in \mathfrak{A}$ such that $ax = by$. The quasi-multipliers form an algebra $\mathfrak{Qm}(\mathfrak{A})$ which is isomorphic to the algebra of fractions $\mathfrak{A}/S$, where $S = \{b \in \mathfrak{A} \mid [b\mathfrak{A}]^- = \mathfrak{A}\}$. This algebra has a natural family of bounded sets, the sets U such that $\bigcap_{T\in U} \mathfrak{D}_T$ contains an element x satisfying $[x\mathfrak{A}]^- = \mathfrak{A}$ and $\sup_{T\in U} \|T(x)\| < \infty$. In fact, $\mathfrak{Qm}(\mathfrak{A})$ is a "bornological algebra" [33] in the sense that the product of two pseudo-bounded sets is pseudo-bounded (pseudo-bounded sets are sets which are bounded in the above sense). In Section 2 we show that $\mathfrak{Qm}(\mathfrak{A})$ can be identified with an inductive limit of Banach spaces and we check a lot of elementary properties of quasi-multipliers. A multiplier in the usual sense is a quasi-multiplier whose domain equals $\mathfrak{A}$.

The algebra $\mathfrak{Qm}(\mathfrak{A})$ of all quasi-multipliers is too large in some sense. It contains, for example, an inverse for b if $[b\mathfrak{A}]^- = \mathfrak{A}$. We introduce, as usual with this kind of algebra, the algebra $\mathfrak{Qm}_r(\mathfrak{A})$ of all regular elements of $\mathfrak{Qm}(\mathfrak{A})$. A regular quasi-multiplier is a quasi-multiplier T such that the set $(\lambda^n T^n)_{n\in\mathbb{N}}$ is pseudo-bounded for some $\lambda > 0$. A subset of $\mathfrak{Qm}_r(\mathfrak{A})$ is said to be multiplicatively pseudo-bounded if it is contained in a set of the form λU, where $\lambda > 0$ and where U is a pseudo-bounded subset of $\mathfrak{Qm}_r(\mathfrak{A})$ stable under products.

Equipped with the family of all its multiplicatively pseudo-bounded subsets, the algebra $\mathfrak{Qm}_r(\mathfrak{A})$ is a pseudo-Banach algebra in the sense of Allan, Dales and McClure [4]. Every maximal ideal of $\mathfrak{Qm}_r(\mathfrak{A})$ is the kernel of a character of $\mathfrak{Qm}_r(\mathfrak{A})$, and $\mathfrak{Qm}_r(\mathfrak{A})$ is an inductive limit of Banach algebras. In fact, $\mathfrak{Qm}_r(\mathfrak{A})$ is in some sense a reserve of possible multipliers in the usual sense for $\mathfrak{A}$. We define in Section 7 a notion of similarity between two commutative Banach algebras $\mathfrak{A}$ and $\mathfrak{B}$ with dense principal ideals. The algebras $\mathfrak{A}$ and $\mathfrak{B}$ are said to be similar if there exists a third Banach algebra $\mathfrak{D}$ with dense principal ideals and two homomorphisms $\varphi : \mathfrak{D} \to \mathfrak{A}$ and $\psi : \mathfrak{D} \to \mathfrak{B}$ such that $\varphi(\mathfrak{D})$ is a dense ideal of $\mathfrak{A}$ and $\psi(\mathfrak{D})$ is a dense ideal of $\mathfrak{B}$. The algebras $\mathfrak{Qm}(\mathfrak{A})$ and $\mathfrak{Qm}(\mathfrak{B})$ are isomorphic (with respect to their pseudo-boundedness structures) if and only if $\mathfrak{A}$ and $\mathfrak{B}$ are similar. Also, for every pseudo-bounded subset U of $\mathfrak{Qm}(\mathfrak{A})$ stable under products, there exists a weaker algebra norm p on $\mathfrak{A}$ such that the completion $\mathfrak{B}$ of $\mathfrak{A}$ with respect to p is similar to $\mathfrak{A}$ and such that U is contained in the unit ball of $\mathfrak{M}(\mathfrak{B})$, the algebra of all multipliers of $\mathfrak{B}$ in the usual sense. We thus see that, modulo a slight change of algebra, regular quasi-multipliers can become multipliers in the usual sense. We then obtain $\mathfrak{Qm}_r(\mathfrak{A})$ as an inductive limit $\varinjlim_{\alpha} \mathfrak{M}(\mathfrak{A}_\alpha)$ where $(\mathfrak{A}_\alpha)$ is a suitable family of Banach algebras similar to $\mathfrak{A}$.

Every Banach algebra which possesses a continuous semigroup $(a^t)_{t>0}$

satisfying $[\bigcup_{t>0} a^t \mathfrak{A}]^- = \mathfrak{A}$ is similar to a Banach algebra with sequential b.a.i., and every Banach algebra with sequential b.a.i. is similar to a Banach algebra with sequential regular b.a.i. Also, if $\mathfrak{A}$ is topologically simple and if $\mathfrak{B}$ is similar to $\mathfrak{A}$ and possesses a sequential b.a.i., then $\mathfrak{B}$ is topologically simple too (the closed ideal structure of $\mathfrak{A}$ and $\mathfrak{B}$ are in fact the same if $\mathfrak{A}$ and $\mathfrak{B}$ are similar and both have a b.a.i.). So it makes no difference to consider the closed ideal problem for Banach algebras having nonzero real continuous semigroups and the closed ideal problem for Banach algebras with b.a.i. (or Banach algebras with a b.a.i. given by an analytic semigroup $(a^t)_{\mathrm{Re}\, t>0}$ bounded in the half-disc $\{z \in \mathbb{C} \mid \mathrm{Re}\, z > 0,\ |z| \leq 1\}$, etc.).

Representations of H^∞ can enter the game at that stage. We give in Section 2 a slight improvement of Sinclair's construction of continuous semigroups in Banach algebras with b.a.i. We show that, if $\mathfrak{A}$ has a sequential b.a.i., there exists a continuous semigroup $(a^t)_{t>0}$ in $\mathfrak{A}$ such that $[\bigcup_{t>0} a^t \mathfrak{A}]^- = \mathfrak{A}$ and such that the Sinclair's map $f \mapsto \int_0^\infty f(t)a^t dt$ is a one-to-one map from $L^1(\mathbb{R}^+)$ into $\mathfrak{A}$. (Using a theorem of Allan [3], Sinclair showed in [46] that these maps are necessarily one-to-one if $\mathfrak{A}$ is radical, but they are not one-to-one in general.) Using inverse Laplace transforms we then construct in Section 2 a pseudo-bounded one-to-one homomorphism φ from H^∞ into $\mathfrak{M}(\mathfrak{A})$ such that $\varphi(1-\alpha) \in \mathfrak{A}$ where α is the function $z \mapsto z$. (The map φ is pseudo-bounded in the sense that the image of the unit ball of H^∞ is a pseudo-bounded subset of $\mathfrak{M}(\mathfrak{A})$.) Then $\varphi(H^\infty) \subset \mathfrak{M}_r(\mathfrak{A}) = \varinjlim_{\alpha} \mathcal{M}(\mathfrak{A}_\alpha)$, and there exists an index α such that φ is a continuous homomorphism from H^∞ into $\mathcal{M}(\mathfrak{A}_\alpha)$. Using the results of Section 6 we obtain an element h of H^∞ of norm 1 such that the spectrum of $\varphi(h)$ in $\mathcal{M}(\mathfrak{A}_\beta)$ equals the closed unit disc for every $\beta > \alpha$. So the spectrum of $\varphi(h)$ in $\mathfrak{M}_r(\mathfrak{A})$ actually equals the closed unit disc, and using h we obtain a new homomorphism $\psi : H^\infty \to \mathfrak{M}_r(\mathfrak{A})$ such that the spectrum of $\psi(f)$ in $\mathfrak{M}_r(\mathfrak{A})$ equals the spectrum of f in H^∞ for every $f \in H^\infty$. It is then easy to see that the carrier space $\widehat{\mathfrak{M}_r(\mathfrak{A})}$ of $\mathfrak{M}_r(\mathfrak{A})$ maps continuously onto the spectrum of H^∞. So, if $\mathfrak{A}$ possesses a sequential b.a.i., the compact set $\widehat{\mathfrak{M}_r(\mathfrak{A})}$ is very rich, even if $\mathfrak{A}$ is radical. The Gelfand transform works for pseudo-Banach algebras, and we can associate to $\mathfrak{A}$ the uniform algebra obtained by taking the closure of the image of $\mathfrak{M}_r(\mathfrak{A})$ in $C(\widehat{\mathfrak{M}_r(\mathfrak{A})})$ via the Gelfand transform. (An easy way to do that consists in taking the completion of $\mathfrak{M}_r(\mathfrak{A})/\mathrm{Rad}(\mathfrak{M}_r(\mathfrak{A}))$ with respect to the quotient norm induced by the seminorm $T \mapsto \nu_r(T)$, the spectral radius of T in $\mathfrak{M}_r(\mathfrak{A})$.) We thus associate, in a "canonical" way, a large compact set to any commutative Banach algebra with b.a.i.

The results of Section 8 follow easily from the existence of the above homomorphisms φ and ψ from H^∞ into $\mathcal{M}(\mathfrak{A}_\alpha)$ because $\mathfrak{A}_\alpha$ is topologically simple if we start the game with a topologically simple commutative Banach algebra having a nonzero continuous real semigroup $(a^t)_{t>0}$. (Elements of $\mathfrak{M}(\mathfrak{A}_\alpha)$ can be seen as bounded linear operators acting in $\mathfrak{A}_\alpha$ and, since $\mathfrak{A}_\alpha$ is topologically simple, these operators cannot have any nontrivial hyperinvariant subspace.)

In view of the length of the paper, we tried to make the different sections as independent as possible. Most of them begin by a detailed summary and are concluded by some notes and remarks.

The author wishes to thank G. R. Allan, Eric Amar, John Bachar, Bernard Chevreau, Phil Curtis, H. G. Dales, José Galé Gimeno, S. Grabiner, and A. M. Sinclair for valuable letters or discussions during the preparation of this manuscript or at the Long Beach Conference.

2. Multipliers and quasimultipliers for commutative Banach algebras without unit element

Let A be a commutative complex algebra. Throughout this section we denote by $S(A)$ the set of elements of A which are not divisors of zero. We will write S instead of $S(A)$ if no confusion is possible; we are only interested in the case where $S(A) \neq \emptyset$. An immediate verification shows that $S(A)$ is stable under products.

We denote by $K(A)$ the ring of fractions with denominators in $S(A)$ and numerators in A. Recall that, as a set, we have $K(A) = A \times S(A)/\sim$, where we define the equivalence relation $\sim$ as follows:

$$(a,b) \sim (a',b') \text{ if and only if } ab' - ba' = 0 \quad (a,a' \in A, b,b' \in S(A)).$$

The equivalence class containing (a,b) will be denoted by a/b. We equip $K(A)$ with the following rules:

$$(a/b)(a'/b') = aa'/bb' \qquad (a,a' \in A, b,b' \in S(A));$$

$$a/b + a'/b' = \frac{ab' + ba'}{bb'} \qquad (a,a' \in A, b,b' \in S(A));$$

$$\lambda(a/b) = \lambda a/b \qquad (a \in A,\ b \in S(A),\ \lambda \in \mathbb{C}).$$

Routine, well-known verifications show that these rules are well defined and that we obtain a structure of a complex algebra on $K(A)$.

Let $A \oplus \mathbb{C}e$ be the algebra obtained by formally adjoining a unit to A. The map $u \mapsto ub/b$, where b is any element of $S(A)$, is an algebra homomorphism from $A \oplus \mathbb{C}e$ into $K(A)$ which is one-to-one, so we may identify $A \oplus \mathbb{C}e$ with a subalgebra of $K(A)$.

DEFINITION 2.1. A <u>generalized map</u> from A into itself is a couple $(\mathcal{D}_T, T)$, where $\mathcal{D}_T$ is a nonempty subset of A and T is a map from $\mathcal{D}_T$ into A.

DEFINITION 2.2. Let $a/b \in K(A)$. Put

$$\mathcal{D}_{T_{a/b}} = \{x \in A \mid ax \in bA\}.$$

If $x \in \mathcal{D}_{T_{a/b}}$ denote by $T_{a/b}(x)$ the unique element y of A satisfying $ax = by$. The generalized maps of the form $(\mathcal{D}_{T_{a/b}}, T_{a/b})$ will be called <u>semimultipliers</u> of A.

REMARK 2.3. If $a/b = a'/b'$, then $(\mathcal{D}_{T_{a/b}}, T_{a/b}) = (\mathcal{D}_{T_{a'/b'}}, T_{a'/b'})$.

Proof. Routine.

Note that $b \in \mathcal{D}_{T_{a/b}}$, so that $\mathcal{D}_{T_{a/b}} \cap S(A) \neq \emptyset$.

NOTATION 2.4. The set of all semimultipliers of A will be denoted by $\mathcal{SM}(A)$. We will often denote a semimultiplier by T instead of by $(\mathcal{D}_T, T)$.

REMARK 2.5. Let $T \in \mathcal{SM}(A)$. Then the following properties hold:

(1) T is a linear map from $\mathcal{D}_T$ into A;

(2) $\mathcal{D}_T$ is an ideal of A, and $T(fg) = T(f)g$ $(f \in \mathcal{D}_T, g \in A)$;

(3) $fT(g) = T(f)g$ $(f,g \in \mathcal{D}_T)$.

Proof. (1) and (2) are obvious. Now let $f,g \in \mathcal{D}_T$. We have $T = T_{a/b}$, where $a \in A$, $b \in S(A)$, say. So $fa = T(f)b$, $ga = T(g)b$, $b(T(f)g) = gaf = b(fT(g))$ and $T(f)g = fT(g)$, since $b \in S(A)$.

PROPOSITION 2.6. Let $(\mathcal{D}_S, S)$ be a generalized map on A such that $\mathcal{D}_S \cap S(A) \neq \emptyset$. If $S(f)g = fS(g)$ for every $f,g \in \mathcal{D}_S$, then there exists a unique $T \in \mathcal{SM}(A)$ such that $\mathcal{D}_S \subset \mathcal{D}_T$ and $T \mid \mathcal{D}_S = S$.

Proof. Let $b \in \mathcal{D}_S \cap S(A)$. Put $a = S(b), T = T_{a/b}$. If $f \in \mathcal{D}_S$ we have $S(f)b = fS(b) = fa$, so $f \in \mathcal{D}_T$ and $T(f) = S(f)$. Now if $T' = T_{a'/b'}$ is another semimultiplier satisfying the same conditions, we have $b \in \mathcal{D}_{T'}$ and $T'(b) = S(b) = a$, so that $ab' = ba'$, $a/b = a'/b'$, and $T = T'$.

COROLLARY 2.7. If $T \in \mathcal{SM}(A)$ then $T = T_{T(b)/b}$, where b is any element of $\mathcal{D}_T \cap S(A)$.

Proof. Put $\mathcal{D}_S = \{b\}$, $S(b) = T(b)$ and apply the proposition to $(\mathcal{D}_S, S)$.

The following definition is standard.

DEFINITION 2.8. [35] A multiplier T on A is a map T from A into itself such that $T(f)g = fT(g)$ $(f,g \in A)$. The set of all multipliers on A is denoted by $\mathcal{M}(A)$.

PROPOSITION 2.9. If $S(A) \neq \emptyset$ then $\mathcal{M}(A) = \{T \in \mathcal{SM}(A) \mid \mathcal{D}_T = A\}$.

Proof. Apply Proposition 2.6.

PROPOSITION 2.10. The map $r \mapsto T_r$ is a one-to-one map from $K(A)$ onto $\mathcal{SM}(A)$.

Proof. The map is onto by definition of $\mathcal{SM}(A)$. Now if $T_{a/b} = T_{a'/b'}$ we have $b \in \mathcal{D}_{T_{a/b}}$, $b' \in \mathcal{D}_{T_{a'/b'}}$, $T_{a/b}(b) = a$, $T_{a'/b'}(b') = a'$ so $ab' = T_{a/b}(b)b' = bT_{a/b}(b') = ba'$, and $a/b = a'/b'$.

DEFINITION 2.11. Let $r,s \in K(A)$, and $\lambda \in \mathbb{C}$. We define the sum $T_r + T_s$, the product T_rT_s and the product λT_r by the formulas $T_r + T_s = T_{r+s}$, $T_rT_s = T_{rs}$, $\lambda T_r = T_{\lambda r}$.

It follows immediately from these rules that $\mathfrak{S}\mathfrak{m}(A)$ is a complex algebra isomorphic with $K(A)$.

PROPOSITION 2.12. Let $T_1, T_2 \in \mathfrak{S}\mathfrak{m}(A)$, $\lambda_1, \lambda_2 \in \mathbb{C}$. Then $\mathfrak{D}_{T_1}\mathfrak{D}_{T_2} \subset \mathfrak{D}_{T_1T_2}$, $\mathfrak{D}_{T_1} \cap \mathfrak{D}_{T_2} \subset \mathfrak{D}_{\lambda_1T_1+\lambda_2T_2}$, $T_2(\mathfrak{D}_{T_1T_2} \cap \mathfrak{D}_{T_2}) \subset \mathfrak{D}_{T_1}$, $\mathfrak{D}_{T_1} \cap \mathfrak{D}_{T_2} \cap S(A) \neq \emptyset$. Also, we have $T_1T_2(x) = T_1(T_2(x))$ $(x \in \mathfrak{D}_{T_1T_2} \cap \mathfrak{D}_{T_2})$, and $(\lambda_1T_1 + \lambda_2T_2)(x) = \lambda_1T_1(x) + \lambda_2T_2(x)$ $(x \in \mathfrak{D}_{T_1} \cap \mathfrak{D}_{T_2})$. Moreover, if $T_1 \in \mathfrak{m}(A)$, $\mathfrak{D}_{T_2} \subset \mathfrak{D}_{T_1T_2}$.

<u>Proof</u>. We have $T_1 = T_{a_1/b_1}$, $T_2 = T_{a_2/b_2}$, say. If $T_1 \in \mathfrak{m}(A)$ and $x \in \mathfrak{D}_{T_2}$, then $a_2x = b_2y$, with $y \in A$. As $\mathfrak{D}_{T_1} = A$ there exists $z \in A$ such that $a_1y = b_1z$ and $a_1a_2x = b_1b_2z$, so $x \in \mathfrak{D}_{T_1T_2}$. Return now to the general case. We have $\lambda_1T_1 + \lambda_2T_2 = T_{u/v}$ where $u = \lambda_1a_1b_2 + \lambda_2a_2b_1$, $v = b_1b_2$, $T_1T_2 = T_{a_1a_2/b_1b_2}$. So $b_1b_2 \in \mathfrak{D}_{T_1} \cap \mathfrak{D}_{T_2}$ and $\mathfrak{D}_{T_1} \cap \mathfrak{D}_{T_2} \cap S(A) \neq \emptyset$. Now if $x \in \mathfrak{D}_{T_1} \cap \mathfrak{D}_{T_2}$ we have $b_2a_1x = b_1b_2T_1(x)$, $b_1a_2x = b_1b_2T_2(x)$, $(\lambda_1b_2a_1 + \lambda_2b_1a_2)x = b_1b_2(\lambda_1T_1(x) + \lambda_2T_2(x))$. As

$$\frac{\lambda_1b_2a_1 + \lambda_2b_1a_2}{b_1b_2} = \lambda_1 \frac{a_1}{b_1} + \lambda_2 \frac{a_2}{b_2},$$

we obtain $x \in \mathfrak{D}_{\lambda_1T_1+\lambda_2T_2}$, $\lambda_1T_1(x) + \lambda_2T_2(x) = (\lambda_1T_1 + \lambda_2T_2)x$.

Let $y \in \mathfrak{D}_{T_1}$, $z \in \mathfrak{D}_{T_2}$. Then $a_1y = b_1T_1(y)$, $a_2z = b_2T_2(z)$, $a_1a_2yz = b_1b_2T_1(y)T_2(z)$, so that $yz \in \mathfrak{D}_{T_1T_2}$.

Now let $x \in \mathfrak{D}_{T_1T_2} \cap \mathfrak{D}_{T_2}$. We have

$$a_1a_2x = b_1b_2(T_1T_2)(x), \; a_2x = b_2T_2(x)$$

so that

$$a_1b_2T_2(x) = b_1b_2(T_1T_2)(x). \text{ Hence, } a_1T_2(x) = b_1(T_1T_2(x)),$$

so that

$$T_2(x) \in \mathfrak{D}_{T_1} \text{ and } T_1(T_2(x)) = T_1T_2(x).$$

Note that if $T_1 \in \mathfrak{m}(A)$, $T_2 \in \mathfrak{S}\mathfrak{m}(A)$ then $\mathfrak{D}_{T_2T_1} \cap \mathfrak{D}_{T_1} \supset \mathfrak{D}_{T_2}$, so that $T_1(\mathfrak{D}_{T_2}) \subset \mathfrak{D}_{T_2}$.

COROLLARY 2.13. The set $\mathfrak{M}(A)$ (equipped with its natural sums and products) is a subalgebra of $\mathfrak{S}\mathfrak{M}(A)$.

Proof. If $T_1, T_2 \in \mathfrak{M}(A)$, then $\mathfrak{D}_{T_1} = \mathfrak{D}_{T_2} = A$, so that $\mathfrak{D}_{T_1} \cap \mathfrak{D}_{T_2} = A$. Proposition 2.12 shows that $\lambda_1 T_1 + \lambda_2 T_2$ has the usual sense if $\lambda_1, \lambda_2 \in \mathbb{C}$. Now let $x \in A$ and write $T_1 = T_{a_1/b_1}$, $T_2 = T_{a_2/b_2}$. Then $a_2 x = b_2 T_2(x)$, $a_1 T_2(x) = b_1 T_1(T_2(x))$, $a_1 a_2 x = b_1 b_2 T_1(T_2(x))$ so $x \in \mathfrak{D}_{T_1 T_2}$ and $T_1 T_2(x) = T_1 \circ T_2(x)$.

We now turn to the case where A is a Banach algebra.

PROPOSITION 2.14. Let $\mathfrak{a}$ be a commutative Banach algebra. Every element of $\mathfrak{S}\mathfrak{M}(\mathfrak{a})$ is closed.

Proof. Recall that a generalized map T on $\mathfrak{a}$ is said to be closed if and only if the set $\{(x, T(x))\}_{x \in \mathfrak{D}_T}$ is closed in $\mathfrak{a} \times \mathfrak{a}$. (See [31], Chapter 2, Defn. 2.11.2.) Let T be an element of $\mathfrak{S}\mathfrak{M}(\mathfrak{a})$ and let f, g be two elements of $\mathfrak{a}$ such that $f = \lim_{n\to\infty} x_n$, $g = \lim_{n\to\infty} T(x_n)$ where (x_n) is some sequence of elements of $\mathfrak{D}_T$. Let $b \in \mathfrak{D}_T \cap S(\mathfrak{a})$, and put $a = T(b)$. Then $T = T_{a/b}$ (Corollary 2.7). We have

$$af = \lim_{n\to\infty} T(b)x_n = \lim_{n\to\infty} bT(x_n) = bg .$$

So $f \in \mathfrak{D}_T$, $g = T(f)$, which proves the proposition.

DEFINITION 2.15. Let $\mathfrak{a}$ be a commutative Banach algebra. We denote by $D(\mathfrak{a})$ the set of all elements f of $\mathfrak{a}$ such that $[f\mathfrak{a}]^- = \mathfrak{a}$.

REMARK 2.16. If $D(\mathfrak{a}) \neq \emptyset$, the two following conditions imply each other:
(1) $x\mathfrak{a} \neq \{0\}$ for every nonzero $x \in \mathfrak{a}$;
(2) $S(\mathfrak{a}) \neq \emptyset$.
Moreover, if these conditions are satisfied, then $D(\mathfrak{a}) \subset S(\mathfrak{a})$.

Proof. Condition (2) always implies condition (1).

Now if (1) holds, let $x \in \mathfrak{a}$ and $f \in D(\mathfrak{a})$ be such that $xf = 0$. We have $xf\mathfrak{a} = \{0\}$, so that $x\mathfrak{a} = 0$, and hence $x = 0$. This shows that $D(\mathfrak{a}) \subset S(\mathfrak{a})$ (and in particular that $S(\mathfrak{a}) \neq \emptyset$).

DEFINITION 2.17. Let $\mathfrak{a}$ be a commutative Banach algebra such that $D(\mathfrak{a}) \neq \emptyset$ and $S(\mathfrak{a}) \neq \emptyset$. A quasimultiplier on $\mathfrak{a}$ is a semimultiplier T on $\mathfrak{a}$ such that $\mathfrak{D}_T \cap D(\mathfrak{a}) \neq \emptyset$. The set of all quasimultipliers of $\mathfrak{a}$ is denoted by $\mathfrak{Q}\mathfrak{M}(\mathfrak{a})$.

PROPOSITION 2.18. $D(\mathfrak{a})$ is stable under products.

Proof. Let $u,v \in D(\mathfrak{a})$ and let $x \in \mathfrak{a}$. There exists a sequence (y_n) of elements of $\mathfrak{a}$ such that $x = \lim_{n\to\infty} uy_n$. For every $n \in \mathbb{N}$ there exists $z_n \in \mathfrak{a}$ such that $\|y_n - vz_n\| \leq 1/n$ and $x = \lim_{n\to\infty} uvz_n$. So $uv \in D(\mathfrak{a})$.

COROLLARY 2.19. If $D(\mathfrak{a}) \neq \emptyset$ and $S(\mathfrak{a}) \neq \emptyset$, then $\mathfrak{D}\mathfrak{M}(\mathfrak{a})$ is a subalgebra of $\mathfrak{S}\mathfrak{M}(\mathfrak{a})$.

Proof. Clear.

We now wish to introduce a notion of boundedness over $\mathfrak{D}\mathfrak{M}(\mathfrak{a})$.

DEFINITION 2.20. [32] Let E be a set, and let Δ be a family of subsets of E. We say that (E,Δ) is a bornological set if the following conditions are satisfied:

(1) if $B \subset B'$ where $B' \in \Delta$, then $B \subset \Delta$;

(2) $E = \bigcup\{B : B \in \Delta\}$;

(3) if $B_1,\ldots,B_k \in \Delta$, then $B_1 \cup B_2 \cup \cdots \cup B_k \in \Delta$.

Elements of Δ are called bounded subsets of Δ.

If E is a linear space, we say that (E,Δ) is a linear convex bornological space if the following conditions are satisfied:

(1) (E,Δ) is a bornological set;

(2) if Λ_1, Λ_2 are bounded subsets of $\mathbb{C}$ and if B_1, B_2 are bounded subsets of E, then the set $\Lambda_1 B_1 + \Lambda_2 B_2 = \{u \in E \mid u = \lambda_1 x_1 + \lambda_2 x_2, \lambda_1 \in \Lambda_1, \lambda_2 \in \Lambda_2, x_1 \in B_1, x_2 \in B_2\}$ is bounded;

(3) the convex hull of any bounded subset of E is bounded;

(4) if $x \neq 0$, $\mathbb{C}x$ is unbounded.

If E is an algebra, we say that (E,Δ) is a convex bornological algebra if the following conditions are satisfied:

(1) (E,Δ) is a convex bornological linear space;

(2) if B_1, B_2 are bounded subsets of E then $B_1 B_2 = \{u \in E \mid u = x_1 x_2, x_1 \in B_1, x_2 \in B_2\}$ is bounded.

DEFINITION 2.21. [32] Let (E_1,Δ_1) and (E_2,Δ_2) be two bornological sets. A map $\varphi : E_1 \to E_2$ is bounded if $\varphi(B) \in \Delta_2$ for every $B \in \Delta_1$.

DEFINITION 2.22. [32] Let $(E_\alpha, B_\alpha)_{\alpha \in I}$ be a family of bornological sets, where I is a partially preordered set. Assume that for every $\alpha < \beta$ a map $\varphi_{\alpha,\beta} : E_\alpha \to E_\beta$ is defined. We say that $(E_\alpha, \Delta_\alpha, \varphi_{\alpha,\beta})$ is an inductive system if the following conditions hold:

(1) for every finite family $\alpha_1,\ldots,\alpha_k$ of elements of I, there exists $\alpha \in I$ such that $\alpha > \alpha_i$ for every i;

(2) if $\alpha,\beta \in I$ with $\alpha < \beta$, the map $\varphi_{\alpha,\beta}$ is bounded;

(3) if $\alpha < \beta < \gamma$, then $\varphi_{\alpha,\gamma} = \varphi_{\beta,\gamma} \circ \varphi_{\alpha,\beta}$.

If $(E_\alpha,\Delta_\alpha,\varphi_{\alpha,\beta})_{\alpha,\beta\in I}$ is an inductive system, put $\Delta = \{B \subset \varinjlim(E_\alpha,\varphi_{\alpha,\beta}) \mid B \subset \varphi_\alpha(B')$ for some $\alpha \in I$ and some $B' \in \Delta_\alpha\}$. The couple $(\varinjlim(E_\alpha,\varphi_{\alpha,\beta}),\Delta)$ is called the <u>bornological inductive limit</u> of the system $(E_\alpha,\Delta_\alpha,\varphi_{\alpha,\beta})$.

Routine verifications show that we obtain a bornological set. If (E_α,Δ_α) is a convex bornological linear space for every $\alpha \in I$, then we obtain a convex bornological linear space.

We will write $(E_\alpha)_{\alpha\in I}$ instead of $(E_\alpha,\Delta_\alpha,\varphi_{\alpha,\beta})_{\alpha,\beta\in I}$ if no confusion is possible.

DEFINITION 2.23. [32] Let (E,Δ) be a bornological linear space. We say that (E,Δ) is a <u>complete convex bornological linear space</u> if there exists an inductive system $(E_\alpha)_{\alpha\in I}$ of Banach spaces such that $(E,\Delta) = \varinjlim E_\alpha$.

DEFINITION 2.24. [32] A <u>complete convex bornological algebra</u> is a convex bornological algebra (E,Δ) such that the underlying convex bornological linear space is complete.

DEFINITION 2.25. Let $\mathfrak{A}$ be a commutative Banach algebra such that $D(\mathfrak{A}) \neq \emptyset$ and $S(\mathfrak{A}) \neq \emptyset$. A subset $\mathfrak{U}$ of $\mathfrak{M}(\mathfrak{A})$ is <u>pseudobounded</u> if there exists $a \in (\bigcap_{T\in\mathfrak{U}} \mathfrak{D}_T) \cap D(\mathfrak{A})$ such that $\sup_{T\in\mathfrak{U}} \|T(a)\| < \infty$.

THEOREM 2.26. Let $\mathfrak{G}$ be the family of all pseudobounded subsets of $\mathfrak{M}(\mathfrak{A})$. Then $(\mathfrak{M}(\mathfrak{A}),\Delta_{\mathfrak{A}})$ is a complete convex bornological algebra.

<u>Proof</u>. Put $I = D(\mathfrak{A})$. If $\mathfrak{A}$ is unital, then $D(\mathfrak{A})$ is the set of invertible elements of $\mathfrak{A}$, $\mathfrak{M}(A)$ is canonically isomorphic with $\mathfrak{A}$ and $\mathfrak{B}$ is the family of subsets of $\mathfrak{A}$ which are bounded in the usual sense.

If $\mathfrak{A}$ has no unit, let $a,a' \in D(\mathfrak{A})$ such that $a \in a'\mathfrak{A}$ and $a' \in a\mathfrak{A}$. We have $a = a'b'$ and $a' = ab$, where $b,b' \in \mathfrak{A}$, so that $a = abb'$ and $u = ubb'$ for every $u \in \mathfrak{A}$ ($a\mathfrak{A}$ is dense in $\mathfrak{A}$), a contradiction. So if we put $I = D(\mathfrak{A})$ and say that $a > a'$ if and only if $a \in a'\mathfrak{A}$ $(a,a' \in D(\mathfrak{A}))$, we obtain a partially ordered set. If $a_1,\ldots,a_k \in I$, then $a_1\ldots a_k > a_i$ $(i = 1,\ldots,k)$).

Now put, for $a \in I$

$$E_a = \{T \in \mathfrak{M}(A) \mid a \in \mathfrak{D}_T\}.$$

If $T \in E_a$ put $\|T\|_a = \|T(a)\|$.

If $\|T\|_a = 0$ then $T(a) = 0$. We have $T = T_{a_1/b_1}$, where $a_1 \in \mathcal{C}$ and $b_1 \in D(\mathcal{C})$ so $T(b_1) = a_1$ and $0 = b_1T(a) = aT(b_1) = aa_1$. Since $a \in D(\mathcal{C})$, this shows that $a_1 = 0$ and $T = 0$. So $(E_a, \|\cdot\|_a)$ is a normed linear space.

Let (T_n) be a Cauchy sequence in E_a. Then $(T_n(a))$ is a Cauchy sequence in $\mathcal{C}$. Denote by u its limit and put $T = T_{u/a}$. Then $a \in \mathcal{D}_T$, $T(a) = u$, and $\|T - T_n\|_a = \|T(a) - T_n(a)\|$, so that T is the limit of (T_n) in the normed space E_a. This shows that $(E_a, \|\cdot\|_a)$ is a Banach space.

If $a' > a$ we have $a' = ab$ for some $b \in \mathcal{C}$. If $a \in \mathcal{D}_T$ then $a' \in \mathcal{D}_T$ as $\mathcal{D}_T$ is an ideal of $\mathcal{C}$. So $E_a \subset E_{a'}$. Moreover, we have

$$\|T\|_{a'} = \|T(ab)\| \leq \|b\|\|T(a)\| = \|b\|\|T\|_a \quad (a \in E_a),$$

so the natural injection from E_a into $E_{a'}$ is bounded, and $(E_a, \|\cdot\|_a)_{a\in I}$ is an inductive system of bornological sets. It follows then immediately from the definitions that $(\mathfrak{M}(\mathcal{C}), \Delta_{\mathcal{C}}) = \varinjlim (E_a, \|\cdot\|_a)$, and so $(\mathfrak{M}(\mathcal{C}), \Delta_{\mathcal{C}})$ is a complete convex bornological space.

Let B_1, B_2 be two pseudobounded subsets of $\mathfrak{M}(\mathcal{C})$. There exists $a_1 \in \bigcap_{T\in B_1} \mathcal{D}_T$ and $a_2 \in \bigcap_{T\in B_2} \mathcal{D}_T$ such that $\sup_{T\in B_j} \|T(a_j)\| < \infty$ $(j = 1,2)$. So $a_1a_2 \in \mathcal{D}_{T_1T_2}$ for every $T_1 \in B_1$ and every $T_2 \in B_2$. We obtain

$$\sup_{\substack{T_1\in B_1\\ T_2\in B_2}} \|T_1T_2(a_1a_2)\| = \sup_{\substack{T_1\in B_1\\ T_2\in B_2}} \|T_1[T_2(a_1a_2)]\| = \sup_{\substack{T_1\in B_1\\ T_2\in B_2}} \|T_1(a_1T_2(a_2)\|$$

$$= \sup_{\substack{T_1\in B_1\\ T_2\in B_2}} \|T_1(a_1)T_2(a_2)\| \leq (\sup_{T\in B_1} \|T(a_1)\|)(\sup_{T\in B_2} \|T(a_2)\|).$$

So B_1B_2 is pseudobounded. This achieves the proof of the theorem.

DEFINITION 2.27. An element T of $\mathfrak{M}(\mathcal{C})$ is said to be <u>regular</u> if the set $(\lambda^nT^n)_{n\in\mathbb{N}}$ is pseudobounded for some $\lambda > 0$.

This notion of regularity is classical for convex bornological algebras. It is well known that the set of regular elements is an algebra if the given algebra is commutative [1]. We thus have the following result (which could be easily proved directly in our case).

PROPOSITION 2.28. The set $\mathfrak{M}_r(\mathcal{C})$ of all regular quasimultipliers on $\mathcal{C}$ is a subalgebra of $\mathfrak{M}(\mathcal{C})$.

If $x\mathfrak{A} \neq \{0\}$ for every nonzero $x \in \mathfrak{A}$ put $\tilde{a}(x) = ax$ $(a, x \in \mathfrak{A})$.

Then $\tilde{a} \in \mathfrak{M}(\mathfrak{A})$, and the map $a \mapsto \tilde{a}$ is one-to-one. So we may identify $\mathfrak{A}$ with a subalgebra of $\mathfrak{M}(\mathfrak{A})$. The above condition implies, as is well known, that all elements of $\mathfrak{M}(\mathfrak{A})$ are linear and bounded ([35], Theorem 2.1). In fact we verified this property here when $S(\mathfrak{A}) \neq \emptyset$. So $\mathfrak{M}(\mathfrak{A})$ is a closed subalgebra of $\mathfrak{L}(\mathfrak{A})$, the Banach algebra of all bounded linear maps from $\mathfrak{A}$ into itself, and the map $a \mapsto \tilde{a}$ is a norm-decreasing homomorphism from $\mathfrak{A}$ into $\mathfrak{M}(\mathfrak{A})$. So, if $S(\mathfrak{A}) \neq \emptyset$ and $D(\mathfrak{A}) \neq \emptyset$, the following inclusion holds

$$\mathfrak{A} \subset \mathfrak{M}(\mathfrak{A}) \subset \mathfrak{D}\mathfrak{M}_r(\mathfrak{A}) \subset \mathfrak{D}\mathfrak{M}(\mathfrak{A}) \subset \mathfrak{S}\mathfrak{M}(\mathfrak{A}),$$

and the injections $\mathfrak{A} \to \mathfrak{M}(\mathfrak{A})$ and $\mathfrak{M}(\mathfrak{A}) \to \mathfrak{D}\mathfrak{M}(\mathfrak{A})$ are bounded.

We say that a map $\varphi : \mathfrak{D}\mathfrak{M}(\mathfrak{A}) \to \mathfrak{D}\mathfrak{M}(\mathfrak{B})$ is <u>pseudobounded</u> if $\varphi(B)$ is pseudobounded in $\mathfrak{D}\mathfrak{M}(\mathfrak{B})$ for every pseudobounded set $\mathfrak{B}$ of $\mathfrak{D}\mathfrak{M}(\mathfrak{A})$. We use similar conventions for maps $\varphi : \mathfrak{B} \to \mathfrak{D}\mathfrak{M}(\mathfrak{A})$ if $\mathfrak{A}$ and $\mathfrak{B}$ are Banach algebras.

PROPOSITION 2.29. Let $\mathfrak{A}$ and $\mathfrak{B}$ be two commutative Banach algebras such that $x\mathfrak{A} \neq \{0\}$ and $y\mathfrak{B} \neq \{0\}$ for every nonzero $x \in \mathfrak{A}$ and every nonzero $y \in \mathfrak{B}$. Assume that $D(\mathfrak{A}) \neq \emptyset$, and let $\varphi : \mathfrak{A} \to \mathfrak{B}$ be a continuous homomorphism. If either $[\varphi(\mathfrak{A})]^- = \mathfrak{B}$ or $[\varphi(\mathfrak{A})\mathfrak{B}]^- = \mathfrak{B}$, then $D(\mathfrak{B}) \neq \emptyset$ and φ uniquely extends to an algebra homomorphism $\hat{\varphi}$ from $\mathfrak{D}\mathfrak{M}(\mathfrak{A})$ into $\mathfrak{D}\mathfrak{M}(\mathfrak{B})$. Moreover $\hat{\varphi}$ is pseudobounded.

<u>Proof</u>. Let $b \in D(\mathfrak{A})$. Then $\varphi(b)\varphi(\mathfrak{A})$ is dense in $\varphi(\mathfrak{A})$. So, if one of the above conditions is satisfied, then $\varphi(D(\mathfrak{A})) \subset D(\mathfrak{B})$ and $D(\mathfrak{B}) \neq \emptyset$.

Let $T = T_{a/b} \in \mathfrak{D}\mathfrak{M}(\mathfrak{A})$, where $a \in \mathfrak{A}$, $b \in D(\mathfrak{A})$. Put

$$\hat{\varphi}(T) = T_{\varphi(a)/\varphi(b)} .$$

If $T = T_{a'/b'}$ where $a' \in \mathfrak{A}$, $b' \in D(\mathfrak{A})$, then $a'/b' = a/b$, $a'b = b'a$, and $\varphi(a')\varphi(b) = \varphi(b')\varphi(a)$, so that $\varphi(a')/\varphi(b') = \varphi(a)/\varphi(b)$ and $\hat{\varphi}(T)$ is well defined. If $a \in \mathfrak{A}$ then $a = T_{ab/b}$, where b is any element of $D(\mathfrak{A})$, and we obtain

$$\hat{\varphi}(a) = T_{\varphi(ab)/\varphi(b)} = T_{\varphi(a)\varphi(b)/\varphi(b)} = \varphi(a) .$$

Let $T = T_{a/b}$ and $T' = T_{a'/b'}$ be two elements of $\mathfrak{D}\mathfrak{M}(\mathfrak{A})$. We have $\hat{\varphi}(TT') = \hat{\varphi}(T_{aa'/bb'}) = T_{\varphi(aa')/\varphi(bb')} = T_{\varphi(a)\varphi(a')/\varphi(b)\varphi(b')} = T_{\varphi(a)/\varphi(b)}T_{\varphi(a')/\varphi(b')} = \hat{\varphi}(T)\hat{\varphi}(T')$. We see similarly that $\hat{\varphi}$ is a linear map, so that $\hat{\varphi}$ is an algebra homomorphism.

Let $\mathfrak{U}$ be a pseudobounded subset of $\mathfrak{D}\mathfrak{M}(\mathfrak{A})$, and let $a \in (\bigcap_{T \in \mathfrak{U}} \mathcal{D}_T) \cap D(\mathfrak{A})$

be such that $\sup_{T\in\mathfrak{A}} \|T(a)\| < \infty$. Then

$$\sup_{T\in\mathfrak{A}} \|\hat{\varphi}(T)\varphi(a)\| = \sup_{T\in\mathfrak{A}} \|\hat{\varphi}(Ta)\| = \sup_{T\in\mathfrak{A}} \|\varphi(T(a))\|$$

$$\leq \|\varphi\| \sup_{T\in\mathfrak{A}} \|T(a)\| < \infty,$$

and so $\hat{\varphi}(\mathfrak{A})$ is pseudo-bounded. This achieves the proof of the proposition.

REMARK 2.30. It follows from the definition of regular quasimultipliers that $\hat{\varphi}(\mathfrak{Q}\mathfrak{M}_r(\mathfrak{A})) \subset \mathfrak{Q}\mathfrak{M}_r(\mathfrak{B})$ if $\mathfrak{A}$ and $\mathfrak{B}$ satisfy the conditions of Proposition 2.28. In particular, $\hat{\varphi}(\mathfrak{M}(\mathfrak{A})) \subset \mathfrak{Q}\mathfrak{M}_r(\mathfrak{B})$.

There is no reason in general for $\hat{\varphi}(\mathfrak{M}(\mathfrak{A}))$ to be contained in $\mathfrak{M}(\mathfrak{B})$. This is nevertheless the case if we suppose that $\mathfrak{A}$ possesses a bounded approximate identity (b.a.i). Let k be the bound of the b.a.i of $\mathfrak{A}$. The map $(a,x) \mapsto \varphi(a)x$ defines a module action of $\mathfrak{A}$ over $\mathfrak{B}$. Cohen's factorization theorem shows that every element of $\mathfrak{A}$ is the product of two elements of $\mathfrak{A}$, so if $[\varphi(\mathfrak{A})]^- = \mathfrak{B}$ then $[\varphi(\mathfrak{A})\mathfrak{B}]^- = \mathfrak{B}$. Cohen's factorization theorem for modules (see for example [45], Theorem 1) shows that $\varphi(\mathfrak{A})\mathfrak{B}$ is closed, so we have in fact $\varphi(\mathfrak{A})\mathfrak{B} = \mathfrak{B}$.

Let $T = T_{a/b} \in \mathfrak{M}(\mathfrak{A})$. Then $\mathfrak{D}_T = \mathfrak{A}$, so that $a\mathfrak{A} \subseteq b\mathfrak{A}$. Let $x \in \mathfrak{B}$. We have $x = \varphi(c)y$, where $c \in \mathfrak{A}$ and $y \in \mathfrak{B}$. So $\varphi(a)x = \varphi(a)\varphi(c)y = \varphi(ac)y$. Since $a\mathfrak{A} \subseteq b\mathfrak{A}$, $ac = bu$, where $u \in \mathfrak{A}$, and $\|u\| = \|T(c)\| \leq \|T\|\|c\|$. So $\varphi(a)x = \varphi(b)[\varphi(u)y]$, $x \in \mathfrak{D}_{T_{\varphi(a)/\varphi(b)}} = \mathfrak{D}_{\hat{\varphi}(T)}$, and $\|\hat{\varphi}(T)(x)\| = \|\varphi(u)y\| \leq \|\varphi\|\|u\|\|y\| \leq \|\varphi\|\|T\|\|c\|\|y\|$. Cohen's factorization theorem for modules shows in fact that for every $\varepsilon > 0$ we may chose c and y such that $x = \varphi(c)y$ and $\|c\|\|y\| < k\|x\| + \varepsilon$. So $\|\hat{\varphi}(T)(x)\| \leq \|\varphi\|\|T\|k\|x\|$ for every $x \in \mathfrak{B} = \mathfrak{D}_{\hat{\varphi}(T)}$. This shows that $\hat{\varphi}$ maps $\mathfrak{M}(\mathfrak{A})$ into $\mathfrak{M}(\mathfrak{B})$ and that $\|\hat{\varphi}(T)\| \leq k\|\varphi\|\|T\|$ for every $T \in \mathfrak{M}(\mathfrak{B})$.

REMARK 2.31. The notations and hypothesis being the same as in Proposition 2.29, $\hat{\varphi}$ is one-to-one if φ is one-to-one.

Proof. If $\varphi(a) \neq 0$ and $b \in D(\mathfrak{A})$, then $\varphi(b) \in D(\mathfrak{B})$, $\varphi(a)/\varphi(b) \neq 0$ and $T_{\varphi(a)/\varphi(b)} \neq 0$.

3. Pseudo-bounded homomorphisms from H^∞ into the quasimultiplier algebras of commutative Banach algebras with s.b.a.i

We denote by H^∞ the algebra of bounded analytic functions on the open unit disc. If $f \in H^\infty$, put $\|f\| = \sup_{|z|<1} |f(z)|$. Equipped with that norm H^∞ is of course a Banach algebra.

Let $\mathfrak{A}$ be a commutative Banach algebra with sequential b.a.i. (see definition below), and let $\alpha : z \mapsto z$ be the "position function". We construct in this section a pseudo-bounded homomorphism φ from H^∞ into $\mathfrak{M}_r(\mathfrak{A})$ such that $\varphi(1-\alpha) \in \mathfrak{A}$ and $[\varphi(1-\alpha)\mathfrak{A}]^- = \mathfrak{A}$. In fact, $\varphi = \hat{\theta} \circ \psi$, where ψ is a pseudo-bounded map from H^∞ into $\mathfrak{M}_r(L^1(\mathbb{R}^+))$ and where $\hat{\theta}$ is a pseudo-bounded map from $\mathfrak{M}_r(L^1(\mathbb{R}^+))$ into $\mathfrak{M}_r(\mathfrak{A})$. The map ψ is obtained using inverse Laplace transforms, and $\hat{\theta}$ is the map related to the map $\theta : L^1(\mathbb{R}^+) \to \mathfrak{A}$, which was constructed by Sinclair in [45], by the method of Proposition 2.29.

Recall that a commutative Banach algebra $\mathfrak{A}$ has a bounded approximate identity (b.a.i) if there exists $k > 0$ such that for every finite family $x_1,\ldots,x_p$ of elements of $\mathfrak{A}$ there exists a sequence (e_n) in $\mathfrak{A}$ satisfying $\|e_n\| \le k$ $(n \in \mathbb{N})$ and $\lim_{n\to\infty} x_i e_n = x_i$ $(i = 1,\ldots,p)$.

Note that a result of Altman [5] shows that it is sufficient to have this condition when $p = 1$.

DEFINITION 3.1. Let $\mathfrak{A}$ be a commutative Banach algebra. Then $\mathfrak{A}$ has a <u>sequential bounded approximate identity (s.b.a.i.)</u> if there exists a bounded sequence (e_n) of elements of $\mathfrak{A}$ such that $x = \lim_{n\to\infty} xe_n$ for every $x \in \mathfrak{A}$.

REMARK 3.2. Let $\mathfrak{A}$ be a commutative Banach algebra with a b.a.i. The following conditions are equivalent:

(1) $[a\mathfrak{A}]^- = \mathfrak{A}$ for some $a \in \mathfrak{A}$:

(2) $\mathfrak{A}$ has a sequential b.a.i.

<u>Proof</u>. If $[a\mathfrak{A}]^- = \mathfrak{A}$, then any bounded sequence (e_n) such that $a = \lim_{n\to\infty} ae_n$ will satisfy $x = \lim_{n\to\infty} xe_n$ for every $x \in \mathfrak{A}$.

Now assume that $\mathfrak{A}$ has a sequential b.a.i. given by a sequence (e_n). The usual Johnson-Varopoulos extension of Cohen's factorization theorem [18], [34], [48] shows that there exists $a \in \mathfrak{A}$ such that $e_n \in a\mathfrak{A}$ for every $n \in \mathbb{N}$. We obtain $x = \lim_{n\to\infty} xe_n \in [ax\mathfrak{A}]^- \subset [a\mathfrak{A}]^-$ for every $x \in \mathfrak{A}$, so that $[a\mathfrak{A}]^- = \mathfrak{A}$.

A well known, routine induction shows that every commutative separable Banach algebra with b.a.i. possesses some s.b.a.i. Also, if $\mathfrak{A}$ possesses a s.b.a.i., it follows from a result of Sinclair [45] that $\mathfrak{A}$ can be renormed with an equivalent norm for which the s.b.a.i. is bounded by 1.

REMARK 3.3. If $\mathfrak{A}$ has a b.a.i., then $x\mathfrak{A} = \{0\}$ implies $x = 0$.

Proof. If $x \in \mathfrak{A}$, then $x = \lim_{n\to\infty} xe_n$ where (e_n) is some sequence of elements of $\mathfrak{A}$.

Remarks 3.2 and 3.3 show that the quasimultiplier algebra $\mathfrak{QM}(\mathfrak{A})$ of $\mathfrak{A}$ is well defined if $\mathfrak{A}$ possesses a s.b.a.i.

A. M. Sinclair showed in [45] that, if $\mathfrak{A}$ is a commutative Banach algebra with b.a.i., there exists for every $x \in \mathfrak{A}$ a continuous semigroup $(a^t)_{t>0}$ in $\mathfrak{A}$ such that $x \in a^t\mathfrak{A}$ for every $t > 0$. If $[x\mathfrak{A}]^- = \mathfrak{A}$, then of course $[a^t\mathfrak{A}]^- = \mathfrak{A}$ for every $t > 0$.

Sinclair also considers in [45] the map $\theta : L^1(\mathbb{R}^+) \to \mathfrak{A}$ defined by the formula

$$\theta(f) = \int_0^\infty f(t)a^t dt \qquad (f \in L^1(\mathbb{R}^+)),$$

where $(a^t)_{t>0}$ is a bounded, continuous semigroup in $\mathfrak{A}$. He observes that θ is an algebra homomorphism and that $[\theta(L^1(\mathbb{R}^+)]\mathfrak{A} = \mathfrak{A}$ if $[\bigcup_{t>0} a^t\mathfrak{A}]^- = \mathfrak{A}$, but he leaves open the question whether or not the map is necessarily one-to-one.

The following example shows that Sinclair's map is not necessarily one-to-one. Consider the Banach algebra $\mathfrak{A} = c_0$, and for $t > 0$ denote by a^t the sequence $(e^{-n^2t})_{n\in\mathbb{N}}$. Clearly, $(a^t)_{t>0}$ is a bounded, continuous semigroup in c_0, and $[a^tc_0]^- = c_0$ for every $t > 0$. Set

$$\varphi(f) = \int_0^\infty f(t)a^t dt \quad (f \in L^1(\mathbb{R}^+)).$$

Then $\varphi(f)$ is the sequence $(\mathfrak{L}(f)(n^2))_{n\in\mathbb{N}}$. Since

$$\sum_{n=1}^{\infty} \mathrm{Re}\left(\frac{1}{n^2+1}\right) = \sum_{n=1}^{\infty} \frac{1}{n^2+1} < \infty,$$

there exists a Blaschke product $B(z)$ over the open right-hand half-plane such that $B(n^2+1) = 0$ for every $n \in \mathbb{N}$. Set

$$g(z) = \frac{B(z+1)}{(z+1)^2} \quad (\mathrm{Re}\, z > -1).$$

Then g is analytic for $\mathrm{Re}\, z > -1$, and $|g(z)| \leq 1/(z+1)^2$, so that $\mathfrak{L}^{-1}(g)$ exists and belongs to $L^1(\mathbb{R}^+, e^{t/2}) \subset L^1(\mathbb{R}^+)$. Put $f = \mathfrak{L}^{-1}(g)$. Then $\varphi(f)$ is the sequence $(B(n^2+1)/(n^2+1)^2)_{n\in\mathbb{N}}$, so that $\varphi(f) = 0$.

We will now show that, if $(a^t)_{t>0}$ is suitably chosen, the corresponding Sinclair's map φ is one-to-one (this fact might be well known). To show this we need some lemmas.

LEMMA 3.4. Let $\mathfrak{A}$ be a commutative Banach algebra with b.a.i., and let (χ_n) be a sequence of nonzero characters of $\mathfrak{A}$. For every sequence (λ_n) of real numbers such that $\lim_{n\to\infty} \lambda_n = 0$, there exists $u \in \mathfrak{A}$ such that $|\chi_n(u)| \geq \lambda_n$ for every $n \in \mathbb{N}$.

Proof. If $a \in \mathfrak{A}$ and if a sequence $x = (x_n)_{n\geq 1}$ belongs to c_0, denote by ax the sequence $(\chi_n(a)x_n)_{n\geq 1}$. We clearly define a Banach module action of $\mathfrak{A}$ over c_0. Put $f_m = (\delta_{m,n})_{n\geq 1}$, where $\delta_{m,n}$ is the usual Kronecker symbol. There exists $u \in \mathfrak{A}$ such that $\chi_m(u) \neq 0$, and there exists a sequence (e_p) in $\mathfrak{A}$ such that $u = \lim_{p\to\infty} ue_p$. So $\chi_m(e_p) \xrightarrow[p\to\infty]{} f$, and $f_m = \lim_{p\to\infty} e_p f_m \in [\mathfrak{A}c_0]^-$ for every $m \in \mathbb{N}$. So $[\mathfrak{A}c_0]^- = c_0$. Cohen's factorization theorem for modules (the version we need is contained in [45], Theorem 1) shows that in fact $c_0 = \mathfrak{A}c_0$. So there exists $u \in \mathfrak{A}$ and $(\mu_n)_{n\geq 1} \in c_0$ such that $\lambda_n = \mu_n \chi_n(u)$ for every $n \in \mathbb{N}$. So $|\chi_n(u)| \geq \lambda_n$ eventually. By multiplying u by a suitable constant we obtain the desired condition.

Let I be a closed ideal of $L^1(\mathbb{R}^+)$. Set

$$H(I) = \{z \in \mathbb{C} \mid \operatorname{Re} z \geq 0,\ \mathcal{L}(f)(z) = 0 \text{ for every } f \in I\},$$

$$\alpha(I) = \inf\{\alpha > 0 \mid f(x) = 0 \text{ a.e. over } [0,\alpha] \text{ for every } f \in I\}.$$

We state as a lemma a classical result of Nyman [40]. For a proof see Dales [20].

LEMMA 3.5. Let I be a closed ideal of $L^1(\mathbb{R}^+)$. If $H(I) = \emptyset$ and $\alpha(I) = 0$, then $I = L^1(\mathbb{R}^+)$.

LEMMA 3.6. Let $(a^t)_{t>0}$ be a bounded, continuous semigroup in a commutative Banach algebra $\mathfrak{A}$, and let $\varphi : f \mapsto \int_0^\infty f(t)a^t dt$ be the corresponding Sinclair's map. If $[a^t\mathfrak{A}]^- = [a^{t'}\mathfrak{A}]^-$ for every $t,t' > 0$, and if $[\bigcup a^t\mathfrak{A}]^-$ has no unit element, then $H(\operatorname{Ker}\varphi)$ is an unbounded subset of the right-hand half-plane.

Proof. Let Q be the canonical map from $L^1(\mathbb{R}^+)$ onto $L^1(\mathbb{R}^+)/\operatorname{Ker}\varphi$. There exists a one-to-one continuous algebra homomorphism $\overline{\varphi}$ from $L^1(\mathbb{R}^+)/\operatorname{Ker}\varphi$ into $\mathfrak{A}$ such that $\varphi = \overline{\varphi} \circ Q$. Since $[\bigcup a^t\mathfrak{A}]^-$ has no unit element, and since $\varphi(L^1(\mathbb{R}^+))\mathfrak{A}$ equals $[\bigcup a^t\mathfrak{A}]^-$ (see Sinclair [45]), the quotient algebra $L^1(\mathbb{R}^+)/\operatorname{Ker}\varphi$ has no unit element. Assume that $H(\operatorname{Ker}\varphi)$ is bounded. Put $U = L^1(\mathbb{R}^+)/\operatorname{Ker}\varphi$, and denote by $U^\#$ the Banach algebra obtained by adjoining a unit to U. The carrier space of U is well known to be the set $\{\chi \circ Q\}$, where $\{\chi\}$ is the set of characters of $L^1(\mathbb{R}^+)$ which vanish over I. So, if

$f \in L^1(\mathbb{R}^+)$, the spectrum of $Q(f)$ in $U^\#$ is the set $\{\mathcal{L}(f)(z)\}_{z \in H(\ker \varphi)} \cup \{0\}$. The function $u : x \mapsto e^{-x}$ belongs to $L^1(\mathbb{R}^+)$, and $\mathcal{L}(u)$ is the function $z \to (z+1)^{-1}$, so that if $H(\operatorname{Ker} \varphi)$ is bounded, then 0 is an isolated point of the spectrum of $Q(u)$ in $U^\#$. It follows then from an easy version of Shilov's idempotent theorem that there exists an idempotent p of $U^\#$ such that $\chi(p) = 1$ for every character χ of $U^\#$ such that $\chi[Q(u)] \neq 0$, and such that $\chi(p) = 0$ for every character χ of $U^\#$ vanishing at $Q(u)$. The unique character of $U^\#$ such that $\chi(u) = 0$ is the character whose kernel is U, so $p \in U$ and $p = Q(q)$ for some $q \in L^1(\mathbb{R}^+)$. Put $J = \{f \in L^1(\mathbb{R}^+) \mid \varphi(f) = \varphi(q)\varphi(f)\}$. Then J is clearly a closed ideal of $L^1(\mathbb{R}^+)$ and $I \subset J$, so $H(J) \subset H(I)$. Also, it follows from the definition of q that $\mathcal{L}(q)(z) = 1$ for every $z \in H(I)$. We have

$$\varphi(qu) = \overline{\varphi}(Q(qu)) = \overline{\varphi}(pQ(u)) = \overline{\varphi}(p^2Q(u)) = \varphi(q^2)\varphi(u) = \varphi(q)\varphi(qu),$$

so that $qu \in J$. If $z \in H(I)$ we obtain

$$\mathcal{L}(qu)(z) = \mathcal{L}(q)(z)\mathcal{L}(u)(z) = \frac{1}{z+1} \neq 0 \quad (\operatorname{Re} z \geq 0).$$

So $H(J) = \emptyset$ (and $J \neq \{0\}$). Put $\alpha = \alpha(J)$ and put $J' = \{f \in L^1(\mathbb{R}^+) \mid f * \delta_\alpha \in J\}$, where δ_α is the Dirac measure at α. [The function $f * \delta_\alpha$ is the function $x \mapsto f(x+\alpha)$.] It follows from the definition of α that $\alpha(J') = 0$. Also, if $f \in J'$, then $\mathcal{L}(f * \delta_\alpha)(z) = \mathcal{L}(f)(z)e^{-\alpha z}$ $(\operatorname{Re} z \geq 0)$ so that $H(J') = \emptyset$. By Nyman's theorem, $J' = L^1(\mathbb{R}^+)$. Denote by u_n the characteristic function of the interval $[\alpha, \alpha + 1/n]$. A routine computation shows that $\varphi(nu_n) \xrightarrow[n\to\infty]{} a^\alpha$. Since $u_n \in J$, we obtain $\varphi(a^\alpha)\varphi(q) = \varphi(a^\alpha)$. Since $[a^\alpha \mathfrak{A}]^- = [\bigcup_{t>0} a^t\mathfrak{A}]^-$, we obtain $\varphi(x)\varphi(q) = \varphi(x)$ for every $x \in [\bigcup_{t>0} a^t\mathfrak{A}]^-$, so that $\varphi(q)$ would be a unit for $[\bigcup_{t>0} a^t\mathfrak{A}]^-$. This contradiction proves the lemma.

We can now prove the following theorem.

THEOREM 3.7. Let $\mathfrak{A}$ be a commutative Banach algebra with s.b.a.i. If $\mathfrak{A}$ has no unit element, there exists a bounded, continuous semigroup $(a^t)_{t>0}$ in $\mathfrak{A}$ such that φ is one-to-one and satisfies $\varphi(L^1(\mathbb{R}^+))\mathfrak{A} = \mathfrak{A}$, where $\varphi : f \mapsto \int_0^\infty f(t)a^t dt$ is the corresponding Sinclair map from $L^1(\mathbb{R}^+)$ into $\mathfrak{A}$.

Proof. Let $(b^t)_{t>0}$ be a bounded, continuous semigroup in $\mathfrak{A}$ such that $[b^t\mathfrak{A}]^- = \mathfrak{A}$ for every $t > 0$, and denote by ψ the Sinclair map $f \mapsto \int_0^\infty f(t)b^t dt$. If ψ is one-to-one, there is nothing to prove (Sinclair shows in that case that $\psi(L^1(\mathbb{R}^+))\mathfrak{A} = \mathfrak{A}$; see [45].)

Now assume that $\operatorname{Ker} \psi \neq \{0\}$. It follows from Lemma 3.6 that $H(\operatorname{Ker} \psi)$ is unbounded, and so we can find a sequence (z_n) in $H(\mathrm{Ker}\,\psi)$ such that $\liminf_{n\to\infty} |z_n| = \infty$. Applying Lemma 3.4 to $L^1(\mathbb{R}^+)$ we can find $u \in L^1(\mathbb{R}^+)$ such that $|\mathcal{L}(u)(z_n)| \geq 1/n$ for every $n \in \mathbb{N}$. Now by applying a more subtle factorization theorem, also due to Sinclair [45], we can find a continuous semigroup $(c^t)_{t>0}$ in $L^1(\mathbb{R}^+)$ such that $\sup_{t>0} \|c^t\| < \infty$, $u \in c^t\, L^1(\mathbb{R}^+)$, such that

$$\mathrm{Sp}(c^t) \subset \Delta_t = \{z \in \mathbb{C} \mid 0 \leq |z| \leq 1,\ |\mathrm{Arg}\, z| < \frac{\pi t}{2}\},$$

and such that $[c^t L^1(\mathbb{R}^+)]^- = L^1(\mathbb{R}^+)$ for every $t > 0$. The map $t \mapsto \mathcal{L}(c^t)(z_n)$ is continuous, and $\mathcal{L}(c^t)(z_n) \neq 0$ $(t > 0)$ since $\mathcal{L}(u)(z_n) \neq 0$, so that there exist two sequences (α_n) and (β_n) of real numbers such that $\mathcal{L}(c^t)(z_n) = \exp(-t\alpha_n - it\beta_n)$ $(n \in \mathbb{N}, t > 0)$. Since $\mathcal{L}(c^t)(z_n) \in \Delta_t$ for every $t > 0$ we must have $|\beta_n| \leq \pi/2$. Also, $\alpha_n \geq 0$. Since $|z_n| \to \infty$ and since $u \in c^t L^1(\mathbb{R}^+)$ for every $t > 0$, we see in particular that $\exp(-\alpha_n) = |\mathcal{L}(c)(z_n)| \geq |\mathcal{L}(u)(z_n)| \geq 1/n$ eventually.

Put

$$\varphi(f) = \int_0^\infty f(t)\psi(c^t)dt \quad (f \in L^1(\mathbb{R}^+)).$$

Then φ is the Sinclair map associated to the semigroup $(\psi(c^t))_{t>0}$. Since $\psi(L^1(\mathbb{R}^+))\mathfrak{a} = \mathfrak{a}$ and since $[c^t L^1(\mathbb{R}^+)]^- = L^1(\mathbb{R}^+)$ for every $t > 0$, we see that $[\psi(c^t)\mathfrak{a}]^- = \mathfrak{a}$ $(t > 0)$ so that $\varphi(L^1(\mathbb{R}^+))\mathfrak{a} = \mathfrak{a}$.

Since Bochner integrals commute with bounded linear operators, we have $\varphi(f) = \psi[\int_0^\infty f(t)c^t dt]$ $(f \in L^1(\mathbb{R}^+))$. Now if $f \in \mathrm{Ker}\,\varphi$, then $\int_0^\infty f(t)c^t dt \in \mathrm{Ker}\,\psi$. Denote by χ_n the character $g \mapsto \mathcal{L}(g)(z_n)$ on $L^1(\mathbb{R}^+)$. Since χ_n commutes with Bochner integrals, we have

$$\chi_n\Big[\int_0^\infty f(t)c^t dt\Big] = \int_0^\infty f(t)\chi_n(c^t)dt = \int_0^\infty f(t)\exp(-t\alpha_n - it\beta_n)dt$$
$$= \mathcal{L}(f)(\alpha_n + i\beta_n).$$

So $\mathcal{L}(f)(\alpha_n + i\beta_n) = 0$ for every $f \in \mathrm{Ker}\,\varphi$. Since $\exp(-\alpha_n) \geq 1/n$ eventually, $\alpha_n \leq \log n$ eventually. Also, $|\exp(-\alpha_n)| = |\mathcal{L}(c)(z_n)| \xrightarrow[n\to\infty]{} 0$, so $\alpha_n \xrightarrow[n\to\infty]{} \infty$. We obtain

$$\mathrm{Re}\left(\frac{1}{\alpha_n + i\beta_n}\right) = \frac{\alpha_n}{|\alpha_n + i\beta_n|^2} = \frac{\alpha_n}{\alpha_n^2 + \beta_n^2} \geq \frac{\alpha_n}{\alpha_n^2 + \frac{\pi^2}{4}} \sim \frac{1}{\alpha_n}.$$

Since $\alpha_n \leq \log n$ eventually, the series $\sum_{n=p_0}^\infty 1/\alpha_n$ is divergent (we choose

p_0 such that $\alpha_n > 0$ for $n \geq p_0$). So $\sum_{n\geq 0} \mathrm{Re}(1/(\alpha_n + i\beta_n)) = \infty$. This condition implies, as is well known (see [12], Theorem 6.3.9), that every bounded holomorphic function over the right-hand half-plane vanishing at $\alpha_n + i\beta_n$ for every $n \in \mathbb{N}$ vanishes identically. So $\mathcal{L}(f) = 0$, and hence $f = 0$ for every $f \in \operatorname{Ker} \varphi$. This shows that φ is one-to-one and the theorem is proved.

We now construct a pseudo-bounded homomorphism from H^∞ into $\mathfrak{M}_r(L^1(\mathbb{R}^+))$.

LEMMA 3.8. There exists a pseudo-bounded one-to-one homomorphism ψ from H^∞ into $\mathfrak{M}_r(L^1(\mathbb{R}^+))$ which possesses the following properties:

(1) $\psi(1-\alpha) \in L^1(\mathbb{R}^+)$, where α is the position function $z \mapsto z$;

(2) $[\psi(1-\alpha)L^1(\mathbb{R}^+)]^- = L^1(\mathbb{R}^+)$.

Proof. Denote by $\mathcal{D}$ the algebra of all bounded analytic functions over the half-plane $\{z \in \mathbb{C} \mid \mathrm{Re}\, z > -1\}$. For $\mathrm{Re}\, z > -1$ put $\beta(z) = (z+1)^{-2}$.

The inverse Laplace transform of βf is well defined for every $f \in \mathcal{D}$, and $\mathcal{L}^{-1}(\beta f) \in L^1(\mathbb{R}^+, e^{\sigma t}) \subset L^1(\mathbb{R}^+)$ if $-1 < \sigma < 0$. More precisely, we have

$$\mathcal{L}^{-1}(\beta f)(x) = \frac{1}{2\pi} \int_{-\infty}^{\infty} \frac{e^{(\sigma+iy)x} f(\sigma+iy)}{(1+\sigma+iy)^2}\, dy \quad (x \geq 0)\,.$$

We obtain $|\mathcal{L}^{-1}(\beta f)(x)| \leq K_\sigma e^{\sigma x}\|f\|_\infty$ $(x \geq 0,\ \sigma > -1,\ f \in \mathcal{D})$, where $\|f\|_\infty = \sup_{\mathrm{Re}\, z > -1} |f(z)|$ and where $K_\sigma = (1/2\pi) \int_{-\infty}^{\infty} dy/[(\sigma+1)^2 + y^2]$ does not depend on f or x.

Note that $\beta = [\mathcal{L}(u)]^2$ where u is the function $u : x \mapsto e^{-x}$ $(x \geq 0)$. Since u is a polynomial generator in $L^1(\mathbb{R}^+)$ (see [20], 4.3), $u * L^1(\mathbb{R}^+)$ is certainly dense in $L^1(\mathbb{R}^+)$.

For $f \in \mathcal{D}$ denote by $\varphi(f)$ the unique quasimultiplier of $L^1(\mathbb{R}^+)$ such that $u^2 \in \mathcal{D}_{\varphi(f)}$ and $\varphi(f)(u^2) = \mathcal{L}^{-1}(\beta f)$. (By u^2 we mean $u * u$.) Since $\mathcal{L}^{-1}(\beta(\lambda f + \mu g)) = \lambda\mathcal{L}^{-1}(\beta f) + \mu\mathcal{L}^{-1}(\beta g)$ for $f,g \in \mathcal{D}$, $\lambda,\mu \in \mathbb{C}$, we see that φ is linear. Now let $f,g \in \mathcal{D}$. We have $\varphi(fg)(u^2) = \mathcal{L}^{-1}(\beta fg)$, and

$$\begin{aligned}\varphi(fg)(u^4) &= u^2 * \varphi(fg)(u^2) = \mathcal{L}^{-1}(\beta) * \mathcal{L}^{-1}(\beta fg) = \mathcal{L}^{-1}(\beta^2 fg) \\ &= \mathcal{L}^{-1}(\beta f) * \mathcal{L}^{-1}(\beta g) = \varphi(f)(u^2) * \varphi(g)(u^2) = \varphi(f)[u^2 * \varphi(g)(u^2)] \\ &= \varphi(f)[\varphi(g)(u^4)] = [\varphi(f)\varphi(g)](u^4)\end{aligned}$$

since $u^4 \in \mathcal{D}_{\varphi(f)} \cap \mathcal{D}_{\varphi(g)}$. Also, $u^4 \in D(L^1(\mathbb{R}^+))$, so it follows from Corollary 2.7 that $\varphi(fg) = \varphi(f)\varphi(g)$ and that φ is an algebra homomorphism from $\mathcal{D}$ into

$\mathfrak{M}(L^1(\mathbb{R}^+))$.

Denote by Ω the closed unit ball of $\mathfrak{D}$. We have, for every $f \in \Omega$

$$\|\varphi(f)(u^2)\| = \int_0^\infty |\mathcal{L}^{-1}(\beta f)(x)|dx \leq K_{-1/2} \int_0^\infty e^{-x/2}dx = 2K_{-1/2}.$$

This shows that ψ is pseudo-bounded. If $f \in \mathfrak{D}$ then $f/(1+\|f\|) \in \Omega$ and the set $\{\varphi(f)^n/(1+\|f\|)^n\}$ is contained in $\varphi(\Omega)$, so it is pseudo-bounded and $\varphi(f) \in \mathfrak{M}_2(L^1(\mathbb{R}^+))$.

The map $\theta : z \mapsto z/(z+2)$ conformally maps the open half-plane $\{z \in \mathbb{C} \mid \operatorname{Re} z > -1\}$ onto the open unit disc D, so the map $\rho : f \mapsto f \circ \theta$ is an isometry from $H^\infty = H^\infty(D)$ onto $\mathfrak{D}$. Denote by α the position function $z \mapsto z$. Then $\rho(1-\alpha)$ is the function $z \to 2/(z+2)$. Denote by v the function $x \mapsto 2e^{-2x}(x \geq 0)$. We have $\mathcal{L}(v) = \rho(1-\alpha)$, so that $\varphi(\rho(1-\alpha))(u^2) = \mathcal{L}^{-1}(\beta\rho(1-\alpha)) = u^2 * v$ and $\varphi(\rho(1-\alpha)) = v \in L^1(\mathbb{R}^+)$. We thus see that $\psi = \varphi \circ \rho$ possesses the desired properties. (It follows immediately from the definition of ψ that ψ is one-to-one, and an argument similar to the argument given for u shows that $[vL^1(\mathbb{R}^+)]^- = L^1(\mathbb{R}^+)$.)

Using Theorem 3.7, Lemma 3.8, Proposition 2.29 and Remarks 2.30 and 2.31, we obtain immediately the following result.

THEOREM 3.9. Let $\mathfrak{A}$ be a commutative Banach algebra with a sequential bounded approximate identity. Then there exists a one-to-one pseudo-bounded algebra homomorphism φ from H^∞ into $\mathfrak{M}(\mathfrak{A})$ such that $\varphi(H^\infty) \subset \mathfrak{M}_r(\mathfrak{A})$, $\varphi(1-\alpha) \in \mathfrak{A}$, and $[\varphi(1-\alpha)\mathfrak{A}]^- = \mathfrak{A}$, where we denote by α the position function $z \mapsto z$.

4. Distances of elements to their squares in Banach algebras without idempotents

We prove the following inequality, which will be useful in the next section.

THEOREM 4.1. Let $\mathfrak{A}$ be a Banach algebra such that 0 is the only idempotent of $\mathfrak{A}$. Then $\inf_{\|x\| \geq \frac{1}{2}} \|x - x^2\| \geq \frac{1}{4}$.

Proof. Since x commutes with x^2, we may assume without loss of generality that $\mathfrak{A}$ is commutative. Adjoin a unit e to $\mathfrak{A}$. If $\|x\| < 1$, the series $\sum_{n=1}^{\infty} (\frac{1}{2})(\frac{1}{2}-1)\dots(\frac{1}{2}-n+1)x^n/n!$ is convergent, and its sum y satisfies the equation $(e+y)^2 = e + x$.

We have

$$\begin{aligned}\|y\| &\leq \sum_{n=1}^{\infty} (-1)^{n+1} \tfrac{1}{2}(\tfrac{1}{2}-1)\dots(\tfrac{1}{2}-n+1)\|x^n\|/n! \\ &\leq -\sum_{n=1}^{\infty} \tfrac{1}{2}(\tfrac{1}{2}-1)\dots(\tfrac{1}{2}-n+1)(-\|x\|)^n/n! \\ &= 1 - \sqrt{1-\|x\|}\,.\end{aligned}$$

In particular, $\|y\| < 1$.

Now assume that there exists $x \in \mathfrak{A}$ such that $\|x\| \geq \frac{1}{2}$ and $\|x^2 - x\| < \frac{1}{4}$. We have $(e-2x)^2 = e - 4x + 4x^2$. Now $\|4x^2 - 4x\| < 1$, and so there exists $y \in \mathfrak{A}$ with $\|y\| < 1$ and $(e+y)^2 = e - 4x + 4x^2 = (e-2x)^2$. So $(y+2x)(2e+y-2x) = 0$. Put $p = y + 2x$, $q = (2x-y)/2$. Then $p, q \in \mathfrak{A}$ and $p(e-q) = 0$.

Let χ be a character of $\mathfrak{A} \oplus \mathbb{C}e$. If $\chi(p) \neq 0$, then $\chi(q) = 1$. If $\chi(p) = 0$, then $\chi(x) = -\frac{1}{2}\chi(y)$. Since $\|y\| < 1$, $|\chi(x)| < \frac{1}{2}$ and $|\chi(q)| \leq |\chi(x)| + \frac{1}{2}|\chi(y)| < 1$. Let K be the carrier space of $\mathfrak{A} \oplus \mathbb{C}e$ and put $U = \{\chi \in K \mid \chi(p) = 0\}$, $V = \{\chi \in K \mid \chi(p) \neq 0\}$. Clearly $U \cap V \neq \emptyset$, $U \cup V = K$, and V is open. But $U = \{\chi \in K \mid \chi(q) \neq 1\}$, so U is open too. As $\mathfrak{A} \oplus \mathbb{C}e$ possesses a character which vanishes over $\mathfrak{A}$, $U \neq \emptyset$. If $U = K$, then $|\chi(q)| < 1$ for every $\chi \in K$ so $e - q$ would be invertible in $\mathfrak{A} \oplus \mathbb{C}e$ and $p = 0$, $y = -2x$ which is impossible as $\|y\| < 1$, $\|x\| \geq \frac{1}{2}$. So $U \neq K$, $V \neq \emptyset$ and K is not connected. It follows then from Shilov's idempotent theorem that $\mathfrak{A} \oplus \mathbb{C}e$ possesses an idempotent f with $f \neq 0$ and $f \neq e$. Since $f(e-f) = 0$ either f or $e - f$ belongs to $\mathfrak{A}$, and so $\mathfrak{A}$ possesses a nonzero idempotent, a contradiction. The result follows.

COROLLARY 4.2. Let $\mathfrak{R}$ be a radical Banach algebra. Then $\inf_{\|x\| \geq \frac{1}{2}} \|x^2 - x\| \geq \frac{1}{4}$.

COROLLARY 4.3. Let $\mathfrak{A}$ be a commutative Banach algebra which possesses a sequential bounded approximate identity $(e_n)_{n \in \mathbb{N}}$. If $\mathfrak{A}$ does not possess any nontrivial idempotent, then $\liminf_{n\to\infty} \|e_n^2 - e_n\| \geq \frac{1}{4}$.

Proof. $\liminf_{n\to\infty} \|e_n\| \geq 1$.

REMARKS 4.4. (1) The commutative Banach algebra c_0 is not unital, and it possesses a sequential bounded approximate identity given by a sequence of idempotents, so the condition of Corollary 4.3 cannot be avoided.

(2) If there exists $x \in \mathfrak{A}$ such that $\|x\| > \frac{1}{2}$ and $\|x^2 - x\| = \frac{1}{4}$, then $\mathfrak{A}$ is not radical and $\mathfrak{A}$ is not an integral domain unless $\mathfrak{A}$ is unital. To see this note that the series $\sum_{n=1}^{\infty} |\frac{1}{2}(\frac{1}{2}-1)\ldots(\frac{1}{2}-n+1)|/n!$ is convergent (the partial sums are majorized by $\lim_{t\to 1} - \sum_{n=1}^{\infty} \frac{1}{2}(\frac{1}{2}-1)\ldots(\frac{1}{2}-n+1)(-t)^n/n! \leq \lim_{t\to 1} (1-\sqrt{1-t}\,)=1$). So in fact if $\|h\| = 1$ there exists $u \in \mathfrak{A}$ with $\|u\| \leq 1$ such that $(e+u)^2 = e+h$. The notations being as in the proof of Theorem 4.1, let $x \in \mathfrak{A}$ be such that $\|x^2 - x\| \leq \frac{1}{4}$. Then $(y+2x)(2e+y-2x) = 0$ with $\|y\| \leq 1$. If $\mathfrak{A}$ is radical, this implies that $y + 2x = 0$ so $\|x\| \leq \frac{1}{2}$. If $\mathfrak{A}$ is not unital and is an integral domain, this implies that $\mathfrak{A} \oplus \mathbb{C}e$ is an integral domain, so as $2e + y - 2x \neq 0$ we get again $y + 2x = 0$, $\|x\| \leq \frac{1}{2}$.

We see in particular that, if $f \in M_1 = \{f \in A(D) \mid f(1) = 0\}$ satisfies $\|f\| > \frac{1}{2}$, then $\|f^2 - f\| > \frac{1}{4}$. On the other hand, if $f \in C([0,1])$ is positive and satisfies $\frac{1}{2} \leq \|f\| \leq (1+\sqrt{2})/2$, one checks easily that $\|f^2 - f\| = \frac{1}{4}$.

(3) Corollary 4.3 is almost trivial in the non-radical case. In fact, if (e_n) is a s.b.a.i. for $\mathfrak{A}$ and if $\mathfrak{A}$ has no nonzero idempotent, then there exists a character χ on $\mathfrak{A}$ and $x \in \mathfrak{A}$ such that $\chi(x) \neq 0$. Since $\chi(e_n x) \xrightarrow[n\to\infty]{} \chi(x)$, we have $\lim_{n\to\infty} \chi(e_n) = 1$, and since the spectrum of e_n in $\mathfrak{A} \oplus \mathbb{C}e$ is connected and contains 0, there exists a character χ_n such that $|\chi_n(e_n)| = \frac{1}{2}$ for n large enough. So $\|e_n^2 - e_n\| \geq |\chi_n(e_n^2) - \chi_n(e_n)| \geq |\chi_n(e_n)|^2 - |\chi_n(e_n)| = \frac{1}{4}$ eventually.

It is possible to get inequalities similar to the inequality of Theorem 4.1 taking any polynomial, or even any function g analytic in a neighbourhood of the origin and having a simple zero at the origin if, say, the given algebra is radical or at least does not possess any pair (p,q) of nonzero elements satisfying $p = pq$. In that case it is possible, if $\|g(x)\|$ is small enough, to write $x = \varphi[g(x)]$ where φ is another analytic function defined in a neighbourhood of the origin. So we obtain a control over $\|x\|$ if $\|g(x)\|$ is small enough. This point of view is developed in Section 8. The interest of the inequality of Theorem 4.1 lies in the fact that the function φ corresponding to $x^2 - x$ is absolutely convergent over its circle of convergence, and allows us to obtain a sharp inequality.

5. Contractions with large spectra in the multiplier algebra of some Banach algebras with b.a.i.

In this section we introduce a new class of commutative Banach algebras with b.a.i. All Banach algebras $\mathfrak{A}$ belonging to that class enjoy the property that there exists in $\mathfrak{M}(\mathfrak{A})$ a contraction T whose spectrum equals the closed unit disc. They even possess contractions T such that the spectrum of $\varphi(T)$ equals the closed unit disc for every continuous homomorphism φ from $\mathfrak{M}(\mathfrak{A})$ into a Banach algebra $\mathfrak{B}$ such that $[\varphi(\mathfrak{A})]^-$ does not possess any nonzero idempotent. This class contains all uniform algebras with sequential b.a.i., hence all quotients of such algebras. The underlying idea to get these "permanent spectra" is fairly simple. If there exists a sequence (y_n) in $\mathfrak{A}$ such that $\inf_n \|y_n\| > 0$ and $\liminf_{n\to\infty} \|Ty_n\| = 0$ then certainly T is not invertible in $\mathfrak{M}(\mathfrak{A})$. Now, if φ is any homomorphism from $\mathfrak{M}(\mathfrak{A})$ in a Banach algebra $\mathfrak{B}$ such that $\inf_n \|\varphi(y_n)\| > 0$, then $\varphi(T)$ is not invertible either. So we get the desired result if we can find first a sequence (y_n) in $\mathfrak{A}$ such that $\inf_n \|\varphi(y_n)\| > 0$ for every homomorphism φ for which $[\varphi(\mathfrak{A})]^-$ has no nonzero idempotent, and second a contraction T in $\mathfrak{M}(\mathfrak{A})$ such that $\liminf_{n\to\infty} \|(T-\lambda e)y_n\| = 0$ for every element λ of the closed unit disc. Such a sequence (y_n) does exist. We can take $(y_n) = (e_n - e_n^2)_{n\in\mathbb{N}}$ and apply Theorem 4.1. Since $e_n - e_n^2$ strongly converges to zero in $\mathfrak{A}$, we can find the desired T under a special assumption concerning the algebra, namely the existence of a complex number μ of modulus 1 such that $(\arg \mu)/\pi$ is irrational and such that there exists a sequential b.a.i. (f_n) satisfying $\|(1-\mu)f_n + \mu e\| = 1$ for every n. We introduce the set $U(\mathfrak{A}) = \{T \in \mathfrak{M}(\mathfrak{A}) \mid \|T\| = 1,\ T - e \in \mathfrak{A}\}$ and its strong closure $\Omega(\mathfrak{A})$ in $\mathfrak{M}(\mathfrak{A})$. The set $\Omega(\mathfrak{A})$ is always stable under sums and products, and $\Omega(\mathfrak{A})$ is also convex. Under the above additional assumption we will see that $\Omega(\mathfrak{A})$ is stable under the action of complex numbers of modulus 1. These properties imply that, if (z_n) is any sequence in $\mathfrak{A}$ which strongly converges to zero, then the set of all $T \in \Omega(\mathfrak{A})$ such that $\liminf_{n\to\infty} \|(T-\lambda e)z_n\| = 0$ for every element λ of the closed unit disc is a dense G_δ in $\Omega(\mathfrak{A})$ with respect to the strong topology. These methods lead also to the construction of a contraction T in $\mathfrak{M}(\mathfrak{A})$ such that $\liminf_{n\to\infty} \|(T-\mu e)^n x\|^{1/n} = 0$ for every element μ of a given countable subset of the closed unit disc D (here, x is any fixed element of $\mathfrak{A}$), provided that $\mathfrak{A}$ is radical and satisfies the above condition. This "lower spectral radius" condition is rather surprising. In fact these contractions T satisfy in the other direction $\limsup_{n\to\infty} \|(T-\lambda e)^n y\|^{1/n} = 1 + |\lambda|$ for every $\lambda \in \mathbb{C}$ and every $y \in \mathfrak{A}$ provided that the "lower spectral radius" condition is satisfied on some dense subset of D.

These results will be applied in the next section to the study of representations of H^∞.

We first recall some basic facts about multipliers. Let $\mathfrak{A}$ be a commutative Banach algebra with b.a.i. If $T \in \mathfrak{M}(\mathfrak{A})$, put $\|T\|_{op} = \sup_{\|x\| \leq 1} \|T(x)\|$. Then $\mathfrak{M}(\mathfrak{A})$ is a Banach algebra, and $\|a\|_{op} \leq \|a\|$ for every $a \in \mathfrak{A}$. If k is a bound for the approximate identities in $\mathfrak{A}$ we have $\|a\| \leq k\|a\|_{op}$. So the injection from $\mathfrak{A}$ into $\mathfrak{M}(\mathfrak{A})$ is bicontinuous, and it is even an isometry if the b.a.i. is bounded by 1. In particular if $\mathfrak{A}$ has a sequential b.a.i., then we may renorm $\mathfrak{A}$ to get a sequential b.a.i. bounded by 1 [45], and in this situation we will write $\|T\|$ instead of $\|T\|_{op}$.

The following proposition is certainly well known. Let $\mathfrak{A}$ be a commutative Banach algebra with b.a.i. The <u>strong topology</u> on $\mathfrak{M}(\mathfrak{A})$ is the topology defined by the family $(p_x)_{x\in\mathfrak{A}}$ of seminorms, where $p_x(T) = \|T(x)\|$ $(x \in \mathfrak{A}, T \in \mathfrak{M}(\mathfrak{A}))$.

PROPOSITION 5.1. The strong topology is the restriction to $\mathfrak{M}(\mathfrak{A})$ of the usual strong topology on $\mathcal{L}(\mathfrak{A})$, $\mathfrak{M}(\mathfrak{A})$ is the closure of $\mathfrak{A}$ in $\mathcal{L}(\mathfrak{A})$ with respect to the strong topology, and the product on $\mathfrak{M}(\mathfrak{A})$ is strongly continuous. If the b.a.i. of $\mathfrak{A}$ is sequential, the closed unit ball Ω of $\mathfrak{M}(\mathfrak{A})$ is a complete metrizable space with respect to the strong topology.

<u>Proof</u>. The fact that the strong topology is the restriction to $\mathfrak{M}(\mathfrak{A})$ of the usual strong operator topology on $\mathcal{L}(\mathfrak{A})$ is clear.

Now let $T,T' \in \mathfrak{M}(\mathfrak{A})$, and let $x \in \mathfrak{A}$. By Cohen's factorization theorem we have $x = yz$ where $y \in \mathfrak{A}$, $z \in \mathfrak{A}$. So $p_x(TT') \leq p_y(T)p_z(T')$. This shows that the product on $\mathfrak{M}(\mathfrak{A})$ is a strongly continuous bilinear map from $\mathfrak{M}(\mathfrak{A}) \times \mathfrak{M}(\mathfrak{A})$ into $\mathfrak{M}(\mathfrak{A})$.

Let $T \in \mathfrak{M}(\mathfrak{A})$, and let $x_1,\ldots x_k \in \mathfrak{A}$. There exists a sequence (e_n) of elements of $\mathfrak{A}$ such that $x_i = \lim_{n\to\infty} x_i e_n$ $(i = 1,\ldots,k)$. So $p_{x_i}(T - T(e_n)) \xrightarrow[n\to\infty]{} 0$ for $i = 1,\ldots,k$. Since $T(e_n) \in \mathfrak{A}$, this shows that $\mathfrak{A}$ is strongly dense in $\mathfrak{M}(\mathfrak{A})$.

Fix $y,z \in \mathfrak{A}$. The maps $T \mapsto T(y)z$ and $T \mapsto yT(z)$ are continuous with respect to the strong topology of $\mathcal{L}(\mathfrak{A})$, and this implies that $\mathfrak{M}(\mathfrak{A})$ is strongly closed in $\mathcal{L}(\mathfrak{A})$. So $\mathfrak{M}(\mathfrak{A})$ equals the strong closure of $\mathfrak{A}$ in $\mathcal{L}(\mathfrak{A})$.

Now assume that $\mathfrak{A}$ possesses a sequential b.a.i. We may suppose that this b.a.i. is bounded by 1. Let Ω be the closed unit ball of $\mathfrak{M}(\mathfrak{A})$, and let $a \in \mathfrak{A}$ be such that $[a\mathfrak{A}]^- = \mathfrak{A}$. Put $d(T,T') = \|T(a) - T'(a)\|$. If $T \neq 0$, then $T(a) \neq 0$, so that d is a distance on $\mathfrak{M}(\mathfrak{A})$. Now let $T,T' \in \Omega$ and let $x \in \mathfrak{A}$. Then there exists $y \in \mathfrak{A}$ with $\|x - ya\| < \varepsilon/2$. We obtain

$$p_x(T - T') = \|T(x) - T'(x)\| \leq 2\|x - ay\| + \|y\|d(T,T') \leq \varepsilon + \|y\|d(T,T') .$$

This inequality shows that the restriction of p_x to Ω is uniformly continuous with respect to d, so that the restriction of the strong topology to Ω equals the topology defined by d.

Now let (T_n) be a Cauchy sequence with respect to d in Ω. As p_x is uniformly continuous with respect to d, $(T_n(x))_{n\in\mathbb{N}}$ is a Cauchy sequence in $\mathfrak{A}$ for every $x \in \Omega$. Put $T(x) = \lim_{n\to\infty} T_n(x)$ $(x \in \mathfrak{A})$. We have $\|T(x)\| = \lim_{n\to\infty} \|T_n(x)\| \leq \|x\|$. A well known, easy argument shows that T is linear, and since T is the strong limit of (T_n) in $\mathcal{L}(\mathfrak{A})$ we see that $T \in \mathfrak{M}(\mathfrak{A})$ and $\|T\| \leq 1$, so that $T \in \Omega$. Also, $d(T,T_n) = \lim_{n\to\infty} \|T(a) - T_n(a)\| = 0$, and Ω is complete with respect to d. This achieves the proof of the proposition.

The following proposition shows that under some conditions a Banach $\mathfrak{A}$-module $\mathfrak{M}$ can become a Banach $\mathfrak{M}(\mathfrak{A})$-module if $\mathfrak{A}$ possesses a b.a.i.

PROPOSITION 5.2. Let $\mathfrak{A}$ be a commutative Banach algebra with b.a.i., and let φ be a continuous homomorphism from $\mathfrak{A}$ into $\mathcal{L}(E)$, where E is a Banach space. If $\varphi(\mathfrak{A})(E)$ is dense in E, then φ uniquely extends to an homomorphism $\tilde{\varphi} : \mathfrak{M}(\mathfrak{A}) \to \mathcal{L}(E)$ which is continuous with respect to the uniform norm topology and with respect to the strong operator topology. Moreover, $\|\tilde{\varphi}\| \leq k\|\varphi\|$ where k is the bound for the b.a.i. of $\mathfrak{A}$, and $\tilde{\varphi}$ is one-to-one if φ is one-to-one.

<u>Proof</u>. Cohen's factorization theorem for modules shows that $\varphi(\mathfrak{A})E$ is a closed linear subspace of E (see for example [45], Theorem 1, where much more is proved). So $E = \varphi(\mathfrak{A})E$ in our case. Let $\tilde{\varphi}$ be any extension of φ, let $T \in \mathfrak{M}(\mathfrak{A})$, and let $x \in E$. There exists $a \in \mathfrak{A}$ and $y \in E$ with $x = \varphi(a)(y)$, and we have

$$\tilde{\varphi}(T)(x) = \tilde{\varphi}(T)(\varphi(a)y) = \tilde{\varphi}(T(a))(y) = \varphi(T(a))(y) .$$

So, if the extension $\tilde{\varphi}$ exists, it has to be unique.

Now, if $x = \varphi(a)(y) = \varphi(a')(y')$, where $x,y,y' \in E$ and $a,a' \in \mathfrak{A}$, there exists a sequence (e_n) of elements of $\mathfrak{A}$ such that $ae_n \xrightarrow[n\to\infty]{} a$ and $a'e_n \xrightarrow[n\to\infty]{} a'$. So

$$\begin{aligned}\varphi[T(a)](y) &= \lim_{n\to\infty} \varphi[T(ae_n)](y) = \lim_{n\to\infty} \varphi(T(e_n))[\varphi(a)(y)] \\ &= \lim_{n\to\infty} \varphi[T(e_n))[\varphi(a')(y')] = \lim_{n\to\infty} \varphi(T(a'e_n))(y') \\ &= \varphi(T(a'))(y') .\end{aligned}$$

Thus if we put $\tilde{\varphi}(T)(x) = \varphi(T(a))(y)$, where $x = \varphi(a)y$ is any decomposition of x, we see that $\tilde{\varphi}(T)$ is well defined. If $d \in \mathfrak{A}$, then $\tilde{\varphi}(d)(x) = \varphi(da)(y) = \varphi(d)[\varphi(a)(y)] = \varphi(d)(x)$. So $\tilde{\varphi}$ is an extension of φ. The same argument as above show that, if $\lambda, \lambda' \in \mathbb{C}$ and $x,x' \in E$, there exists a sequence (e_n) in $\mathfrak{A}$ such that $\tilde{\varphi}(T)(x) = \lim_{n\to\infty} \varphi[T(e_n)](x)$, $\tilde{\varphi}(T)(x') = \lim_{n\to\infty} \varphi[T(e_n)](x')$, and $\varphi(T)(\lambda x + \lambda' x') = \lim_{n\to\infty} \tilde{\varphi}(T(e_n))(\lambda x + \lambda' x')$. This implies that $\tilde{\varphi}(T)$ is linear over E. A routine verification shows that $\tilde{\varphi}$ is in fact an algebra homomorphism from $\mathfrak{M}(\mathfrak{A})$ into $\mathfrak{L}(E)$.

Let $x \in E$. For each $\varepsilon > 0$, we can find an element a of $\mathfrak{A}$ and an element y of E such that $x = \varphi(a)(y)$ and $\|a\|\|y\| < (k+\varepsilon)\|x\|$. So $\|\tilde{\varphi}(T)(x)\| \le \|\varphi(T(a))\|\|y\| \le (k+\varepsilon)\|\varphi\|\|T\|\|x\|$ for every $\varepsilon > 0$ and every $x \in E$, and hence $\|\tilde{\varphi}\| \le k\|\varphi\|$.

Now suppose that $\operatorname{Ker} \varphi = \{0\}$, and let $T \in \mathfrak{M}(\mathfrak{A})$. If $T \neq 0$, there exists $a \in \mathfrak{A}$ such that $T(a) \neq 0$, and there exists $x \in E$ such that $\varphi[T(a)](x) \neq 0$. So $\tilde{\varphi}[T](\varphi(a)(x)) = \varphi(T(a))(x) \neq 0$, and $\tilde{\varphi}(T) \neq 0$. Thus $\operatorname{Ker} \tilde{\varphi} = \{0\}$, and $\tilde{\varphi}$ is one-to-one.

Let $x \in E$. We have $x = \varphi(a)(y)$ where $a \in \mathfrak{A}$, $y \in E$. We obtain, for every $T \in \mathfrak{M}(\mathfrak{A})$,

$$\|\tilde{\varphi}(T)(x)\| = \|\varphi(T(a))(y)\| \le \|y\|\|\varphi\|\|T(a)\| .$$

So $\tilde{\varphi}$ is continuous with respect to the strong operator topologies. This completes the proof of the proposition.

REMARK 5.3. (1) If the b.a.i. of $\mathfrak{A}$ is bounded by one, and if φ is norm-decreasing, then $\tilde{\varphi}$ is also norm-decreasing. So if $\mathfrak{M}$ is a Banach $\mathfrak{A}$-module, where $\mathfrak{A}$ has a sequential b.a.i., we may assume that the b.a.i. is bounded by one, and if $\|ax\| \le \|a\|\|x\|$ and $[\mathfrak{A}\mathfrak{M}]^- = \mathfrak{M}$, we may extend the module action to $\mathfrak{M}(\mathfrak{A})$ and obtain $\|Tx\| \le \|T\|\|x\|$ for every $x \in \mathfrak{M}$ and every $T \in \mathfrak{M}(\mathfrak{A})$.

(2) When φ is an algebra homomorphism from $\mathfrak{A}$ into $\mathfrak{B}$, where $\mathfrak{A}$ has a sequential b.a.i. and where $[\varphi(\mathfrak{A})\mathfrak{B}]^- = \mathfrak{B}$, we may consider φ as a homomorphism from $\mathfrak{A}$ into $\mathfrak{L}(\mathfrak{B})$ and $\tilde{\varphi}(T)$ as a multiplier of $\mathfrak{B}$ for every $T \in \mathfrak{M}(\mathfrak{A})$. Then $\tilde{\varphi}(T) = \hat{\varphi}(T)$, where $\hat{\varphi}$ has the same meaning as in Proposition 2.29 and Remark 2.30. (Notice that $\hat{\varphi} \mid \mathfrak{M}(\mathfrak{A}) = \varphi$ is in fact continuous with respect to the strong operator topologies.)

DEFINITION 5.4. Let $\mathfrak{A}$ be a commutative Banach algebra. A sequence $(e_n)_{n\ge 1}$ of elements of $\mathfrak{A}$ is a <u>regular</u> bounded approximate identity (r.b.a.i.) for $\mathfrak{A}$ if it satisfies the following conditions:

(1) $a = \lim_{n\to\infty} ae_n$ for every $a \in \mathfrak{A}$;

(2) there exists a complex number λ, with $|\lambda| = 1$ and $(\arg \lambda)/\pi$ irrational, such that

$$\|(1-\lambda)e_n + \lambda e\| \le 1 \quad (n \in \mathbb{N}) .$$

THEOREM 5.5. Let $\mathfrak{A}$ be a commutative Banach algebra with r.b.a.i. For every element μ of the unit circle, there exists a sequence $(e_n(\mu))_{n\geq 1}$ of elements of $\mathfrak{A}$ such that:

(1) $a = \lim_{n\to\infty} ae_n(\mu)$ for every $a \in \mathfrak{A}$;

(2) $\|(1-\mu)e_n(\mu) + \mu e\| \leq 1$ for every $n \in \mathbb{N}$.

In particular $\mathfrak{A}$ possesses a sequential bounded approximate identity bounded by 1.

Proof. Let λ and $(e_m)_{m\in\mathbb{N}}$ be respectively the complex number and the sequence of elements of $\mathfrak{A}$ given by Definition 5.4. Since $(\arg \lambda)/\pi$ is irrational, there exists a sequence $(p_n)_{n\in\mathbb{N}}$ of positive integers such that $\mu = \lim_{n\to\infty} \lambda^{p_n}$, where $\mu \neq 1$. Note that $\|e_m\| \leq 2/|1-\lambda|$, so that the sequence (e_m) is bounded, $\mathfrak{A}$ has a s.b.a.i., and there exists $a \in \mathfrak{A}$ such that $[a\mathfrak{A}]^- = \mathfrak{A}$.

If $k \in \mathbb{N}$, we have $ae_m^k \xrightarrow[m\to\infty]{} a$, and so

$$a = \lim_{m\to\infty} [(1-\lambda)e_m + \lambda e]^p a \quad \text{for every } p \in \mathbb{N}.$$

Put

$$f_{m,p} = \frac{[(1-\lambda)e_m + \lambda e]^p - \lambda^p e}{1-\lambda^p}.$$

Then $f_{m,p} \in \mathfrak{A}$, and $\lim_{m\to\infty} af_{m,p} = a$ for every $p \in \mathbb{N}$. Also,

$$\|(1-\lambda^p)f_{m,p} + \lambda^p e\| \leq \|(1-\lambda)e_m + \lambda e\|^p \leq 1 \qquad (m,p \in \mathbb{N}).$$

If $n \in \mathbb{N}$, pick $m_n \in \mathbb{N}$ such that $\|a - af_{m_n,p_n}\| \leq 1/n$, and put $e_n(\mu) = (1-\bar{\lambda}^{p_n})/(1-\bar{\mu})\, f_{m_n,p_n}$. We have

$$\begin{aligned}\|(1-\mu)e_n(\mu) + \mu e\| &= \|(\bar{\mu}-1)e_n(\mu) + e\| \\ &= \|(\bar{\lambda}^{p_n} - 1)f_{m_n,p_n} + e\| = \|(1-\lambda^{p_n})f_{m_n,p_n} + \lambda^{p_n} e\| \leq 1\end{aligned}$$

for every $n \in \mathbb{N}$. But $\chi[(1-\mu)e_n(\mu) + \mu e] = \mu$, where χ is the character of $\mathfrak{A} \oplus \mathbb{C}e$ vanishing over $\mathfrak{A}$, so in fact $\|(1-\mu)e_n(\mu) + \mu e\| = 1$. Also,

$$ae_n(\mu) = \left(\frac{1-\bar{\lambda}^{p_n}}{1-\bar{\mu}}\right) af_{m_n,p_n} \xrightarrow[n\to\infty]{} a.$$

Since $\|e_n(\mu)\| \leq 2/|1-\mu|$ $(n \in \mathbb{N})$ and since $[a\mathfrak{A}]^- = \mathfrak{A}$, we obtain

$b = \lim_{n\to\infty} be_n(\mu)$ for every $n \in \mathbb{N}$.

We also have $\|2e_n(-1)+e\| = 1$, so that $\|e_n(-1)\| \leq 1$ for every $n \in \mathbb{N}$. This shows that $(e_n(-1))_{n\in\mathbb{N}}$ is a s.b.a.i. bounded by 1 in $\mathfrak{A}$, and hence the theorem is proved.

The following theorem gives a significant class of commutative Banach algebras which have a r.b.a.i.

THEOREM 5.6. Let $\mathfrak{A}$ be a commutative Banach algebra. If $\mathfrak{A}$ is the quotient of a uniform algebra with sequential bounded approximate identity, there exists a sequence (e_n) of elements of $\mathfrak{A}$ such that $\lim_{n\to\infty} \|ae_n - a\| = 0$ for every $a \in \mathfrak{A}$ and such that $\|(1-\lambda)e_n + \lambda e\| \leq 1$ for $n \in \mathbb{N}$, $0 \leq |\lambda| \leq 1$, and $0 \leq \arg \lambda \leq \pi$.

Proof. Denote by A_0 the algebra of all continuous complex-valued functions f over the half-plane $\Pi = \{z \in \mathbb{C} \mid \text{Re } z \geq 0\}$ which are analytic for $\text{Re } z > 0$. Then A_0 is a Banach algebra with respect to the norm $\|f\| = \sup_{\text{Re } z \geq 0} |f(z)|$. Put

$$e_n(z) = \left(1 + \frac{\sqrt{z}}{n} e^{i\pi/4}\right)^{-1} \quad (\text{Re } z \geq 0,\ n \in \mathbb{N}),$$

where $\sqrt{z} = \exp[\frac{1}{2} \log z]$ and where $\log z$ is the determination of the logarithm which takes real values over the positive reals. Note that $-\pi/4 \leq \arg \sqrt{z} \leq \pi/4$, so that $\arg \sqrt{z}\, e^{i\pi/4} \in [0,\pi/2]$ and the above function is well defined. Also, $\lim_{\substack{|z|\to\infty \\ \text{Re } z \geq 0}} |e_n(z)| = 0$, so $e_n \in \mathfrak{A}_0$ for every $n \in \mathbb{N}$.

Now put $\lambda = \rho e^{i\alpha}$, where $0 \leq \rho \leq 1$ and $0 \leq \alpha \leq \pi$. We have $(1-\lambda)e_n(z) + \lambda = (n + \lambda\sqrt{z}\, e^{i\pi/4})/(n + \sqrt{z}\, e^{i\pi/4}) = (1-\lambda)e_1(z/n^2) + \lambda$. So $\|(1-\lambda)e_n + \lambda e\| = \|(1-\lambda)e_1 + \lambda e\|$. If $\text{Re } z \geq 0$, we can write $\sqrt{z} = \delta e^{i\theta}$ where $\delta \geq 0$, and $\theta \in [-\pi/4, \pi/4]$. We obtain

$$|1 + \lambda\sqrt{z}\, e^{i\pi/4}|^2 = 1 + \rho^2\delta^2 + 2\rho\delta \cos\left(\alpha + \theta + \frac{\pi}{4}\right)$$

and

$$|1 + \sqrt{z}\, e^{i\pi/4}|^2 = 1 + \delta^2 + 2\delta \cos\left(\theta + \frac{\pi}{4}\right).$$

Since $\theta + \frac{\pi}{4} \in \left[0,\frac{\pi}{2}\right]$, $\cos\left(\theta + \frac{\pi}{4}\right) \geq \cos\left(\theta + \frac{\pi}{4} + \alpha\right)$ for $\alpha \in [0,\pi]$, so $\|(1-\lambda)e_n + \lambda\| \leq 1$ $(n \in \mathbb{N})$. Since $\lim_{|z|\to\infty} |(1-\lambda)e_n(z) + \lambda| = |\lambda| = 1$ for every $n \in \mathbb{N}$, we obtain $\|(1-\lambda)e_n + \lambda\| = 1$ if $|\lambda| = 1$. Also, $e_n(z)$ converges to 1 uniformly on bounded subsets of the closed half-plane. Since (e_n) is uniformly bounded (the above inequality gives $\|e_n\| \leq 1$ for every $n \in \mathbb{N}$), it follows easily that $\lim_{n\to\infty} \|fe_n - f\| = 0$ for every $f \in \mathfrak{A}_0$.

Now let $\mathfrak{A}$ be a uniform algebra with sequential b.a.i. It follows from Theorem 3.9 that there exists a pseudo-bounded homomorphism $\varphi : H^\infty \to \mathfrak{D}\mathfrak{m}_r(\mathfrak{A})$ with $\varphi(1-\alpha) \in \mathfrak{A}$ and $[\varphi(1-\alpha)\mathfrak{A}]^- = \mathfrak{A}$, where we denote by α the position function $z \mapsto z$.

Koua observed in his thesis [37] that, if $\mathfrak{A}$ is uniform, then $\mathfrak{m}(\mathfrak{A})$ (which is uniform too) exactly equals $\mathfrak{D}\mathfrak{m}_r(\mathfrak{A})$. So φ maps H^∞ into $\mathfrak{m}(\mathfrak{A})$ and since all homomorphisms reduce spectra of elements, φ is necessarily norm-decreasing. Put $M_1 = \{f \in A(D) \mid f(1) = 0\}$ where we denote by $A(D)$ the usual disc algebra. Then M_1 is the closure of the span of the powers of $1-\alpha$. Since $\varphi(1-\alpha)\mathfrak{A}$ is dense in $\mathfrak{A}$, we have $[\varphi(M_1)\mathfrak{A}]^- = \mathfrak{A}$. So $\varphi(M_1)\mathfrak{A} = \mathfrak{A}$ because M_1 possesses a b.a.i. But we can identify M_1 with the algebra $\mathfrak{A}_0$ defined above, using a standard conformal mapping of the open unit disc onto the right-hand half-plane. The desired result follows if $\mathfrak{A}$ is uniform. Of course, the result extends to quotients of uniform algebras with sequential b.a.i.

We introduce now a subset of $\mathfrak{m}(\mathfrak{A})$, which can reduce to $\{e\}$, but which is very rich when $\mathfrak{A}$ has a regular b.a.i.

DEFINITION 5.7. Let $\mathfrak{A}$ be a commutative non-unital Banach algebra with b.a.i. bounded by 1. We denote by $U(\mathfrak{A})$ the set of all elements T of $\mathfrak{m}(\mathfrak{A})$ such that $\|T\| \le 1$, $T - e \in \mathfrak{A}$ and we denote by $\Omega(\mathfrak{A})$ the closure of $U(\mathfrak{A})$ with respect to the strong topology.

REMARK 5.8. The sets $U(\mathfrak{A})$ and $\Omega(\mathfrak{A})$ are convex and stable under products, $\|T\| = 1$ for every $T \in U(\mathfrak{A})$ and $\|T\| \le 1$ for every $T \in \Omega(\mathfrak{A})$.

Proof. If $T \in U(\mathfrak{A})$ then $\chi(T) = 1$, where we denote by χ the character of $\mathfrak{A} \oplus \mathbb{C}e$ vanishing over $\mathfrak{A}$. So $\|T\| \ge 1$, and in fact $\|T\| = 1$. As the closed unit ball of $\mathfrak{m}(\mathfrak{A})$ is strongly closed, $\|T\| \le 1$ for every $T \in \mathfrak{m}(\mathfrak{A})$. The convexity of $U(\mathfrak{A})$ is clear, and the convexity of $\Omega(\mathfrak{A})$ follows as the strong topology is locally convex. If $T,S \in U(\mathfrak{A})$ then $T = e + x$, $S = e + y$ where $x,y \in \mathfrak{A}$. So $TS - e = x + y + xy \in \mathfrak{A}$, and $TS \in U(\mathfrak{A})$ as $\|T\| \le 1$ and $\|S\| \le 1$. Since the strong topology is an algebra topology, $\Omega(\mathfrak{A})$ is stable under products too.

LEMMA 5.9. Let $\mathfrak{A}$ be a commutative non-unital Banach algebra with r.b.a.i. Then $\lambda T \in \Omega(\mathfrak{A})$ for every $T \in \Omega(\mathfrak{A})$ and every complex number λ such that $|\lambda| \le 1$.

Proof. Let $\lambda \in \mathbb{C}$ with $|\lambda| = 1$. There exists a sequence (e_n) of elements of $\mathfrak{A}$ such that $\|e_n x - x\| \xrightarrow[n\to\infty]{} 0$ for every $x \in \mathfrak{A}$ and such that $\|(1-\bar{\lambda})e_n + \bar{\lambda}e\| = 1$ for every $n \in \mathbb{N}$. So $\|(\lambda-1)e_n + e\| = 1$ for every $n \in \mathbb{N}$ and $((\lambda-1)e_n + e)_{n\in\mathbb{N}}$

strongly converges to λe. This shows that $\lambda e \in \Omega(\mathfrak{A})$.

Since $\Omega(\mathfrak{A})$ is convex, and since the convex hull of the unit circle equals the closed unit disc, $\lambda e \in \Omega(\mathfrak{A})$ if $|\lambda| \leq 1$. Since $\Omega(\mathfrak{A})$ is stable under products, the lemma follows.

THEOREM 5.10. Let $\mathfrak{A}$ be a commutative non-unital Banach algebra with r.b.a.i. and let $(y_n)_{n\in\mathbb{N}}$ be a sequence of elements of $\mathfrak{A}$ which strongly converges to zero. There exists $T \in \mathfrak{M}(\mathfrak{A})$ such that $\|T\| = 1$ and such that $\liminf_{n\to\infty} \|(T-\lambda e)(y_n)\| = 0$ for every element λ of the closed unit disc.

<u>Proof</u>. Note that the sequence (y_n) is necessarily bounded. Put, for $m,p \in \mathbb{N}$, $|\lambda| \leq 1$,

$$V_{m,p,\lambda} = \left\{ T \in \Omega(\mathfrak{A}) \mid \inf_{n>m} \|(T-\lambda e)y_n\| < \frac{1}{p} \right\}.$$

The set $V_{m,p,\lambda}$ is clearly open with respect to the strong topology. Now let $T \in \Omega(\mathfrak{A})$ and $\lambda \in \mathbb{C}$ with $|\lambda| \leq 1$. There exists two elements μ and ν of the unit circle such that $\lambda = \frac{1}{2}(\mu+\nu)$. Since $\bar{\mu}T \in \Omega(\mathfrak{A})$ and $\bar{\nu}T \in \Omega(\mathfrak{A})$, there exist two sequences $(R_q)_{q\in\mathbb{N}}$ and $(S_q)_{q\in\mathbb{N}}$ of elements of $U(\mathfrak{A})$ such that $(R_q)_{q\in\mathbb{N}}$ strongly converges to $\bar{\mu}T$ and $(S_q)_{q\in\mathbb{N}}$ strongly converges to $\bar{\nu}T$. Thus, the sequences $(\mu R_q)_{q\in\mathbb{N}}$ and $(\nu S_q)_{q\in\mathbb{N}}$ both strongly converge to T, and so $\frac{1}{2}(\mu R_q + \nu S_q)_{q\in\mathbb{N}}$ strongly converges to T too. Since R_q and S_q belong to $U(\mathfrak{A})$, μR_q, νS_q and $\frac{1}{2}(\mu R_q + \nu S_q)$ belong to $\Omega(\mathfrak{A})$ for every $q \in \mathbb{N}$. But $R_q - e \in \mathfrak{A}$, $S_q - e \in \mathfrak{A}$ so $\frac{1}{2}(\mu R_q + \nu S_q) - \lambda e \in \mathfrak{A}$ for every $q \in \mathbb{N}$. Since $\lim_{n\to\infty} \|xy_n\| = 0$ for every $x \in \mathfrak{A}$, we see that $\frac{1}{2}(\mu R_q + \nu S_q) \in \bigcap_{p,m\in\mathbb{N}} V_{\lambda,p,m}$ for every $q \in \mathbb{N}$. So $V_{\lambda,p,m}$ is strongly dense in Ω for every $p,m \in \mathbb{N}$.

Now let K be a countable dense subset of the closed unit disc. It follows from the category theorem that the set $V = \bigcap\{V_{\lambda,p,m} : \lambda \in K, p,m \in \mathbb{N}\}$ is dense in Ω.

Let $T \in V$, and let $\mu \in \mathbb{C}$ with $|\mu| \leq 1$. There exists $N > 0$ such that $\|y_n\| \leq N$ for every $n \in \mathbb{N}$. Let p and m be two positive integers. There exists $\lambda \in K$ such that $|\mu - \lambda| < N/2p$, and there exists $n > m$ such that $\|(T-\lambda e)y_n\| < 1/2p$. Thus $\|(T-\mu e)y_n\| < 1/p$, with $n > m$. This shows that $\liminf_{n\to\infty} \|(T-\mu e)y_n\| = 0$ for every $T \in V$ and every element μ of the closed unit disc, and the theorem is proved.

COROLLARY 5.11. Let $\mathfrak{A}$ be a commutative non-unital Banach algebra with r.b.a.i. Then there exists $T \in \mathfrak{M}(\mathfrak{A})$ such that $\|T\| = 1$ and such that the spectrum of T equals the closed unit disc.

<u>Proof</u>. Let $(e_n)_{n\in\mathbb{N}}$ be a sequential bounded approximate identity for $\mathfrak{A}$.

Then for every fixed n the sequence $(e_n - e_m)_{m\in\mathbb{N}}$ strongly converges to $e_n - e$, so $1 \leq \|e_n - e\| \leq \liminf_{n\to\infty} \|e_n - e_m\|$. Hence we can construct by induction a sequence $(m_n)_{n\in\mathbb{N}}$ of positive integers such that $m_n > n$ and $\|e_n - e_{m_n}\| \geq \frac{1}{2}$ for every $n \in \mathbb{N}$. The sequence $(e_n - e_{m_n})_{n\in\mathbb{N}}$ strongly converges to zero, so there exists $T \in \Omega(\mathfrak{a})$ such that $\liminf_{n\to\infty} \|(T-\mu e)(e_n - e_{m_n})\| = 0$ for every element μ of the closed unit disc. Since $\|e_n - e_{m_n}\| \geq \frac{1}{2}$ for every $n \in \mathbb{N}$, $T - \mu e$ is not invertible, and the spectrum of T contains the closed unit disc. But T is a contraction so equality holds.

Homomorphisms usually reduce the spectra of elements. The following improvement of Corollary 5.11 is nevertheless true.

COROLLARY 5.12. Let $\mathfrak{a}$ be a commutative non-unital Banach algebra with r.b.a.i., and let φ be a continuous homomorphism from $\mathfrak{M}(\mathfrak{a})$ into a Banach algebra $\mathfrak{B}$. If $[\varphi(\mathfrak{a})]^-$ has no unit element, there exists $T \in \mathfrak{M}(\mathfrak{a})$ such that $\|T\| = 1$ and such that the spectrum of $\varphi(T)$ in $\mathfrak{B}$ equals the closed unit disc.

<u>Proof</u>. Consider again a sequential bounded approximate identity $(e_n)_{n\in\mathbb{N}}$ for $\mathfrak{a}$. Then $(\varphi(e_n))_{n\in\mathbb{N}}$ is a sequential bounded approximate identity for $[\varphi(\mathfrak{a})]^-$. Since $[\varphi(\mathfrak{a})]^-$ is not unital, we have again $\liminf_{m\to\infty} \|\varphi(e_n) - \varphi(e_m)\| \geq 1$ for every $n \in \mathbb{N}$. So we can construct a sequence (m_n) of positive integers such that $m_n > n$ and $\|\varphi(e_n) - \varphi(e_{m_n})\| \geq \frac{1}{2}$ for every $n \in \mathbb{N}$. We obtain as before an element $T \in \Omega(\mathfrak{a})$ such that $\liminf_{n\to\infty} \|(T-\mu e)(e_n - e_{m_n})\| = 0$ for every element μ of the closed unit disc. So $\liminf_{n\to\infty} \|\varphi(T)\varphi(e_n - e_{m_n}) - \mu\varphi(e_n - e_{m_n})\| = 0$ for all such μ, and since $\|\varphi(e_n - e_{m_n})\| \geq \frac{1}{2}$ for every $n \in \mathbb{N}$ the corollary follows.

The construction of the multiplier T obtained in Corollary 5.12 depends on the homomorphism φ. Using Theorem 4.1 we now avoid this restriction, assuming some stronger condition on $[\varphi(\mathfrak{a})]^-$.

COROLLARY 5.13. Let $\mathfrak{a}$ be a commutative non-unital Banach algebra with a regular bounded approximate identity. Then there exists $T \in \mathfrak{M}(\mathfrak{a})$ such that $\|T\| = 1$ and such that the spectrum of $\varphi(T)$ equals the closed unit disc for every homomorphism φ from $\mathfrak{M}(\mathfrak{a})$ into a Banach algebra satisfying the two following conditions: (1) $\varphi(\mathfrak{a}) \neq \{0\}$; (2) $[\varphi(\mathfrak{a})]^-$ does not possess any nonzero idempotent.

<u>Proof</u>. Let (e_n) be a s.b.a.i. for $\mathfrak{a}$. Then $(\varphi(e_n))$ is a s.b.a.i. for $[\varphi(\mathfrak{a})]^-$. Since $\varphi(\mathfrak{a}) \neq \{0\}$, $\liminf_{n\to\infty} \|\varphi(e_n)\| \geq 1$, and it follows from Theorem 4.1 that $\liminf_{n\to\infty} \|\varphi(e_n - e_n^2)\| \geq \frac{1}{4}$.

We can now use an argument similar to that in the proof of Corollary 5.12, using the sequence $(e_n - e_n^2)$ instead of the sequence $(e_n - e_{m_n})$. The sequence $(e_n - e_n^2)$ does not depend on φ.

Note that, if $\mathfrak{A}$ is radical, condition (2) of Corollary 5.13 is automatically satisfied. More generally, this condition is automatically satisfied if $[\varphi(\mathfrak{A})]^-$ is radical. The following easy proposition shows that this condition cannot be avoided if $\mathfrak{A}$ is semisimple.

PROPOSITION 5.14. Let $\mathfrak{A}$ be a commutative semisimple Banach algebra with sequential bounded approximate identity. If $\mathfrak{A}$ is not unital, then there exists for every T of $\mathfrak{M}(\mathfrak{A})$ a continuous homomorphism from $\mathfrak{M}(\mathfrak{A})$ into ℓ^∞ such that $[\varphi(\mathfrak{A})]^- = c_0$ and such that the spectrum of $\varphi(T)$ does not contain the unit circle.

<u>Proof</u>. Let $a \in \mathfrak{A}$ be such that $[a\mathfrak{A}]^- = \mathfrak{A}$, and assume that 0 is an isolated point of the spectrum of a in $\mathfrak{A} \oplus \mathbb{C}e$. It follows from an easy version of Shilov's idempotent theorem that there exists an idempotent f of $\mathfrak{A} \oplus \mathbb{C}e$ such that the characters of $\mathfrak{A} \oplus \mathbb{C}e$ vanishing at f are exactly the characters of $\mathfrak{A} \oplus \mathbb{C}e$ vanishing at a. So $f \in \mathfrak{A}$, and $\chi(a - af) = 0$ for every character χ of $\mathfrak{A} \oplus \mathbb{C}e$. Since $\mathfrak{A}$ is semisimple we would have $a = af$, and $\mathfrak{A}$ would be unital. So there exists a sequence (χ_n) of characters of $\mathfrak{A}$ such that $\chi_n(a) \neq 0$ for every n and such that $\lim_{n\to\infty} \chi_n(a) = 0$. All nonzero characters of $\mathfrak{A}$ can be extended to $\mathfrak{M}(\mathfrak{A})$ (just put $\chi(T) = \chi(T(a))/\chi(a)$), so we may consider the sequence (χ_n) as a sequence of characters on $\mathfrak{M}(\mathfrak{A})$.

Let $T \in \mathfrak{M}(\mathfrak{A})$. At least one of the sets $\{n \in \mathbb{N} \mid \operatorname{Re} \chi_n(T) \geq 0\}$ and $\{n \in \mathbb{N} \mid \operatorname{Re} \chi_n(T) \leq 0\}$ is infinite. So, changing if necessary the sequence (χ_n) into a subsequence, we may assume, say, that $\operatorname{Re} \chi_n(T) \geq 0$ for every $n \in \mathbb{N}$ and that the sequence $(|\chi_n(a)|)_{n\in\mathbb{N}}$ is decreasing. Let $\varphi(R) = (\chi_n(R))_{n\in\mathbb{N}}$ $(R \in \mathfrak{M}(\mathfrak{A}))$. Then φ is clearly an algebra homomorphism from $\mathfrak{M}(\mathfrak{A})$ into ℓ^∞, and $\operatorname{Sp} \varphi(T)$ is contained in the closed right-hand half-plane. Since $\chi_n(a) \xrightarrow[n\to\infty]{} 0$, we have $\chi_n(x) \xrightarrow[n\to\infty]{} 0$ for every $x \in \mathfrak{A}$, and so $[\varphi(\mathfrak{A})]^- \subset c_0$. Since the sequence $(|\chi_n(a)|)_{n\in\mathbb{N}}$ is strictly decreasing, it is an easy exercise to show that the closed span of the set $\{\varphi(a^n)\}_{n\in\mathbb{N}}$ equals c_0.

This completes the proof of the proposition.

We now focus our attention on homomorphisms φ from $\mathfrak{M}(\mathfrak{A})$ such that $[\varphi(\mathfrak{A})]^-$ is radical. In this situation there is another way to obtain contractions T in $\mathfrak{M}(\mathfrak{A})$ such that the spectrum of $\varphi(T)$ equals the closed unit disc.

THEOREM 5.15. Let $\mathfrak{A}$ be a commutative Banach algebra with a regular bounded approximate identity, let K be a countable subset of the closed unit disc, and let φ be a homomorphism from $\mathfrak{M}(\mathfrak{A})$ into a Banach algebra such that $[\varphi(\mathfrak{A})]^-$ is radical. Then there exists for every $x \in \mathfrak{A}$ an element T of $\mathfrak{M}(\mathfrak{A})$, of norm 1, such that $\liminf_{n\to\infty} \|(\varphi(T) - \mu e)^n \varphi(x)\|^{1/n} = 0$ for every $\mu \in K$.

Proof. Since $\|\varphi(Ry)\| \le \|\varphi\|\|Ry\|$ for every $R \in \mathfrak{M}(\mathfrak{A})$ and every $y \in \mathfrak{A}$, the seminorm $R \mapsto \|\varphi(Ry)\|$ is continuous with respect to the strong topology for every $y \in \mathfrak{A}$ and every homomorphism φ from $\mathfrak{M}(\mathfrak{A})$ into a Banach algebra. So the map $R \mapsto \|\varphi[(R - \mu e)^n y]\|$ is strongly continuous for every $n \in \mathbb{N}$ and every $\mu \in \mathbb{C}$.

If $x \in \mathfrak{A}$, $\mu \in \mathbb{C}$, and $m,p \in \mathbb{N}$, put

$$V_{\mu,x,m,p} = \left\{ T \in \Omega(\mathfrak{A}) \mid \inf_{n \ge m} p^{-n} \|[\varphi(T) - \mu e]^n \varphi(x)\| < 1 \right\}.$$

Then $V_{\mu,x,m,p}$ is strongly open in $\Omega(\mathfrak{A})$. Also, if $T \in \Omega(\mathfrak{A})$ and if $T - \mu e \in \mathfrak{A}$, then $\lim_{n\to\infty} \|(\varphi(T) - \mu e)^n\|^{1/n} = 0$, so $T \in \bigcap\{V_{\mu,x,m,p} : x \in \mathfrak{A}, m,p \in \mathbb{N}\}$. But, as in the proof of Theorem 5.10, we see that the set $\{T \in \Omega(\mathfrak{A}) \mid T - \mu e \in \mathfrak{A}\}$ is strongly dense in $\Omega(\mathfrak{A})$ provided that $|\mu| \le 1$, so the result follows from the category theorem.

COROLLARY 5.16. Let $\mathfrak{A}$ be a commutative radical Banach algebra with a regular bounded approximate identity, let $x \in \mathfrak{A}$, and let K be a countable subset of the closed unit disc. Then there exists $T \in \mathfrak{M}(\mathfrak{A})$ such that $\|T\| = 1$ and $\liminf_{n\to\infty} \|(T - \mu e)^n x\|^{1/n} = 0$ for every $\mu \in K$.

REMARK 5.17. If T is an invertible operator acting on a Banach space E and if x is any nonzero element of E, then $\|T^n(x)\|^{1/n} \|T^{-n}\|^{1/n} \ge \|x\|$ and $\liminf_{n\to\infty} \|T^n(x)\|^{1/n} \ge (\nu(T^{-1}))^{-1}$. Thus, if we choose an element x of $\mathfrak{A}$ such that $\varphi(x) \ne 0$ and a countable dense subset K of the closed unit disc in Theorem 5.15, we obtain a contraction $T \in \mathfrak{M}(\mathfrak{A})$ such that the spectrum of $\varphi(T)$ equals the closed unit disc. Moreover, if $[x\mathfrak{A}]^- = \mathfrak{A}$, then $\varphi(x) \ne 0$ for every homormorphism φ from $\mathfrak{A}$ into a Banach algebra such that $\varphi(\mathfrak{A}) \ne \{0\}$. We thus see that Corollary 5.11 and Corollary 5.13 follow immediately from Corollary 5.16 when the given algebra $\mathfrak{A}$ is radical.

If T is any contraction in $\mathfrak{M}(\mathfrak{A})$ and φ is any homomorphism from $\mathfrak{M}(\mathfrak{A})$ into a Banach algebra, the spectrum of $\varphi(T)$ is contained in the closed unit disc and $\limsup_{n\to\infty} \|(\varphi(T) - \mu e)^n \varphi(y)\|^{1/n} \le \limsup_{n\to\infty} \|(\varphi(T) - \mu e)^n\|^{1/n} \le 1 + |\mu|$ for every $\mu \in \mathbb{C}$ and every $y \in \mathfrak{A}$. One could ask whether or not it is possible to get $\limsup_{n\to\infty} \|(\varphi(T) - \mu e)^n (y)\|^{1/n} = 0$ for at least some element μ of the closed unit disc and some $y \in \mathfrak{A}$ with $\varphi(y) \ne 0$ in Theorem 5.15. The following result (applied to $[\varphi(\mathfrak{A})]^-$) shows that this is indeed impossible if $[x\mathfrak{A}]^- = \mathfrak{A}$ and if

K possesses more than one element.

THEOREM 5.18. Let $\mathcal{Q}$ be a commutative Banach algebra with b.a.i., let $T \in \mathfrak{M}(\mathcal{Q})$, and let $\alpha \in \mathbb{C}$ be such that $\alpha \in Sp(\varphi(T))$ for every homomorphism φ from $\mathfrak{M}(\mathcal{Q})$ into a Banach algebra satisfying $\varphi(\mathcal{Q}) \neq \{0\}$. Then $\limsup_{n\to\infty} \|(T - \mu e)^n y\|^{1/n} \geq |\mu - \alpha|$ for every $\mu \in \mathbb{C}$ and every nonzero $y \in \mathcal{Q}$.

Proof. We will use an argument similar to an argument of the next section.

Let y be a nonzero element of $\mathcal{Q}$ and let $m > 0$ be such that $m > \limsup_{n\to\infty} \|(T - \mu e)^n y\|^{1/n}$. Put $S = T - \mu e$ and $I = \{x \in \mathcal{Q} \mid \sup_{n \geq 0} \|S^n(x)\| m^{-n} < \infty\}$. Then $y \in I$, so that $I \neq \{0\}$.

Set

$$p(x) = \sup_{n \geq 0} \frac{\|S^n x\|}{m^n} \quad (x \in I).$$

Then p is a linear norm over I, $p(x) \geq \|x\|$ for every $x \in I$, and $p(R(x)) \leq \|R\| p(x)$ for every $x \in I$ and every $R \in \mathfrak{M}(\mathcal{Q})$.

Now set

$$q(R) = \sup_{\substack{x \in I \\ x \neq 0}} \frac{p[R(x)]}{p(x)} \quad (R \in \mathfrak{M}(\mathcal{Q})).$$

We obtain an algebra seminorm over $\mathfrak{M}(\mathcal{Q})$, and $q(R) \leq \|R\|$ for every $R \in \mathfrak{M}(\mathcal{Q})$. The seminorm q induces a quotient norm $\bar{q}$ over the quotient algebra $\mathfrak{M}(\mathcal{Q})/\text{Ker } q$. Denote by $\bar{R}$ the image of elements R of $\mathfrak{M}(\mathcal{Q})$ in this quotient, and denote by $\mathcal{B}$ the completion of $\mathfrak{M}(\mathcal{Q})/\text{Ker } q$ with respect to $\bar{q}$. Then $\bar{q}(\bar{R}) = q(R)$ for every $R \in \mathfrak{M}(\mathcal{Q})$. Also, $p(Sx) = \sup_{n\in\mathbb{N}} \|S^{n+1}x\| m^{-n} \leq m p(x)$ for every $x \in I$, so $\bar{q}(S) = q(S) \leq m$. Since $\mathcal{Q}$ has an approximate identity, there exists $a \in \mathcal{Q}$ such that $ay \neq 0$ and $q(a) \neq 0$, so $\bar{a} \neq 0$.

The map $R \mapsto \bar{R}$ is a continuous homomorphism ψ from $\mathfrak{M}(\mathcal{Q})$ into $\mathcal{B}$, and $\psi(\mathcal{Q}) \neq \{0\}$ since $\bar{a} \neq 0$. Since $\alpha \in Sp(\bar{T})$, $\alpha - \mu \in Sp(\bar{S})$, so certainly $|\alpha - \mu| \leq \bar{q}(S) \leq m$. This shows that $\limsup_{n\to\infty} \|(T - \mu e)^n y\|^{1/n} \geq |\alpha - \mu|$ and the theorem is proved.

COROLLARY 5.19. The notations and hypothesis being the same as in Corollary 5.16, assume that $[x\mathcal{Q}]^- = \mathcal{Q}$ and that the closure of K contains the unit circle. Then $\limsup_{n\to\infty} \|(T - \mu)^n y\|^{1/n} = 1 + |\mu|$ for every nonzero element y of $\mathcal{Q}$ and every $\mu \in \mathbb{C}$.

Proof. Since T is a contraction, $\|T - \mu e\| \leq 1 + |\mu|$ and

$\limsup_{n\to\infty} \|T - \mu e)^n y\|^{1/n} \leq 1 + |\mu|$ for every nonzero $y \in \mathfrak{A}$ and every $\mu \in \mathbb{C}$.

Since $[x\mathfrak{A}]^- = \mathfrak{A}$ for any homomorphism φ from $\mathfrak{M}(\mathfrak{A})$ into a Banach algebra such that $\varphi(\mathfrak{A}) \neq 0$, $\varphi(x) \neq 0$, and it follows from Remark 5.16 that $\lambda \in Sp(\varphi(T))$ for every $\lambda \in K$, and hence for every λ with $|\lambda| = 1$. So $\limsup_{n\to\infty} \|(T - \mu e)^n y\|^{1/n} \geq \sup_{|\lambda|=1} |\mu + \lambda| = 1 + |\mu|$ for every $\mu \in \mathbb{C}$ and every nonzero element y of $\mathfrak{A}$.

6. Representations of H^∞

We now apply the results of Section 5 to study representations $\varphi : H^\infty \to \mathcal{L}(E)$ of H^∞ in the algebra of all bounded operators acting on an arbitrary Banach space E. We will frequently use the notation $f \mapsto f(T)$ to denote such representations, where T is the image of the position function $\alpha : z \mapsto z$, and we are interested in the case where the spectrum of T contains 1.

Such representations have been extensively studied in the case of a Hilbert space. The usual Nagy-Foias functional calculus shows that, if T is any contraction acting on a Hilbert space H, there exists a closed subalgebra H_T of H^∞ containing the disc algebra, $A(D)$, which equals H^∞ if T is completely non-unitary, and a homomorphism $\varphi : f \mapsto f(T)$ from H_T into $\mathcal{L}(H)$ such that $\varphi(\alpha) = T$ (see [29], Chapter 3, Section 2). Foias, Pearcy and Sz. Nagy showed recently in [30] that there exists an element h of H^∞ of norm 1 such that the spectrum of $h(T)$ equals the closed unit disc for every completely non-unitary contraction $T \in \mathcal{L}(H)$ whose spectrum is connected and contains 1.

We extend here this result to representations $f \mapsto f(T)$ of H^∞ over an arbitrary Banach space such that the range of $I - T$ is not closed and such that the spectrum of T is connected. Our method looks completely different from the method used in [30], and was obtained independently. For representations $\varphi : f \mapsto f(T)$ of H^∞ such that the range of $I - T$ is not closed and such that the spectrum of T is disconnected we obtain an element h_φ of H^∞ of norm 1 for which the spectrum of $h_\varphi(T)$ equals the closed unit disc, but in this case h_φ depends on φ. Also, if $\varphi : f \mapsto f(T)$ is a representation of H^∞ for which the spectrum of T equals $\{1\}$ we obtain for every countable subset K of the unit circle and every x satisfying $\lim_{n\to\infty} \|T^n(x)\| = 0$ an element $h_{\varphi,x,K}$ of H^∞ such that $\liminf_{n\to\infty} \|(h_{\varphi,x,K}(T) - \lambda I)^n(x)\|^{1/n} = 0$ for every $\lambda \in K$.

The main result obtained by Foias, Pearcy and Sz. Nagy in [30] is actually not the one mentioned above. They show also that there exists an element h of H^∞ of norm 1 such that the polynomials in h are weak-star dense in H^∞ in the sense of [15] and such that the spectrum of $h^2(T)$ is rich in the sense of [15] for every completely non-unitary contraction $T \in \mathcal{L}(H)$ for which the spectrum of T is connected and contains 1. The methods of Section 5 do not give any control about weak-star density, but the author thinks that using some suitable subset of $\Omega(M_1)$ it should be possible to extend the above result to representations of H^∞ over arbitrary Banach spaces.

In fact, we do not use H^∞ in this section but the smaller algebra G_1^∞ of all elements of H^∞ which can be continuously extended to $\overline{D} - \{1\}$. (We denote by D the open unit disc, and by $\overline{D}$ its closure.) We denote by $A(D)$ the usual disc algebra and by M_1 the maximal ideal $\{f \in A(D) \mid f(1) = 0\}$. Routine verifications show that $\mathfrak{M}(M_1)$ can be identified with $\mathfrak{a}_1^\infty$ and that the strong topology over $\mathfrak{a}_1^\infty$ is the topology of uniform convergence over compact subsets of $\overline{D} - \{1\}$.

THEOREM 6.1. There exists an element h of $\mathfrak{a}_1^\infty$ of norm 1 such that the spectrum of $h(T)$ equals the closed unit disc for every representation $\varphi : f \mapsto f(T)$ of $\mathfrak{a}_1^\infty$ over a Banach space E satisfying the following conditions:

(1) the spectrum of T is connected;

(2) the range of $I - T$ is not closed.

Proof. Recall that $T = \varphi(\alpha)$ is the image of the position function $\alpha : z \mapsto z$ in this representation. Put $J = \varphi(1)$. Then J is an idempotent of $\mathfrak{L}(E)$ (J does not necessarily equal I). Note that, for any representation of $\mathfrak{a}_1^\infty$, the range of $T - I$ is closed if the range of $T - J$ is closed. To see this, consider a sequence (x_n) of elements of E such that $(T(x_n) - x_n)_{n\in\mathbb{N}}$ has a limit y in E. We have $J(y) = \lim_{n\to\infty} (T(x_n) - J(x_n))$, so, if $\operatorname{Im}(T - J)$ is closed, there exists $x \in E$ such that $J(y) = T(x) - J(x)$. So $T(-y + J(x) + J(y)) = T(J(x)) = T(x) = J(y) + J(x)$, $T(-y + J(x) + J(y)) - (-y + J(x) + J(y)) = y$, and $y \in \operatorname{Im}(T - I)$. This proves our claim.

Now assume that the range of $I - T$ is not closed. The set $\{(1-\alpha)^n\}_{n\in\mathbb{N}}$ spans M_1, so the principal ideal generated by $J - T$ is dense in $\varphi(M_1)$. If $[\varphi(M_1)]^-$ had a unit element F, $J - T$ would be invertible in $[\varphi(M_1)]^-$ and we would have $\operatorname{Im}(J - T) = \operatorname{Im} F$, which is impossible. So condition (2) ensures that $[\varphi(M_1)]^-$ is not unital, and in particular that $J \notin [\varphi(M_1)]^-$.

Set $\mathfrak{B} = \{U \in \mathfrak{L}(H) \mid UJ = JU = U\}$. Then $\mathfrak{B}$ contains $\varphi(\mathfrak{a}_1^\infty)$, and $\mathfrak{B}$ is exactly the set $\{JVJ\}_{V\in\mathfrak{L}(E)}$. If an element U of $\mathfrak{B}$ is invertible in $\mathfrak{L}(E)$, then $U(JU^{-1}J) = UU^{-1}J = IJ = J$, and similarly $(JU^{-1}J)U = J$, so that U is invertible in $\mathfrak{B}$. This shows that the spectrum of an element U of $\mathfrak{B}$ in $\mathfrak{B}$ is contained in the spectrum of U in $\mathfrak{L}(E)$.

Let $U \in \mathfrak{B}$ and let $\lambda \in \mathbb{C}$ be such that $\lambda \neq 0$ and such that $U - \lambda J$ is invertible in $\mathfrak{B}$ with inverse V. We have $(U - \lambda I)(V + \lambda^{-1}(J - I)) = (V + \lambda^{-1}(J - I))(U - \lambda I) = I$. So $U - \lambda I$ is invertible in $\mathfrak{B}$. This shows that the spectra of U in $\mathfrak{B}$ and $\mathfrak{L}(E)$ have the same set of nonzero elements, and the only possible difference between the two spectra lies at 0, which might belong to the spectrum of U in $\mathfrak{L}(E)$ without belonging to the spectrum of U in $\mathfrak{B}$. But condition (1) ensures that 0 cannot be an isolated point of the spectrum of T in $\mathfrak{L}(E)$, so in our case the two spectra agree, and the spectrum of T in $\mathfrak{B}$ is connected.

Let $\mathfrak{D}$ be the closure in $\mathfrak{B}$ of the set of all rational fractions in T with poles outside $\operatorname{Sp}_{\mathfrak{B}}(T)$. Then $\operatorname{Sp}_{\mathfrak{D}}(T) = \operatorname{Sp}_{\mathfrak{B}}(T)$, $\mathfrak{D}$ is commutative, and the map $\chi \mapsto \chi(T)$ is, as is well known, a homeomorphism from the carrier space of $\mathfrak{D}$

onto $Sp_{\mathfrak{D}}(T) = Sp_{\mathfrak{B}}(T)$. So condition (1) ensures that 0 and J are the only idempotents of $\mathfrak{D}$. If conditions (1) and (2) are satisfied, then $\varphi(M_1)^-$ has no nonzero idempotents, and Theorem 5.19 is a consequence of Corollary 5.13, since M_1 has a r.b.a.i.

If condition (1) is not assumed, a weaker result remains true.

THEOREM 6.2. Let $\varphi : f \mapsto f(T)$ be a representation of $\mathcal{C}_1^\infty$ over a Banach space E such that the range of $I - T$ is not closed. There exists an element h_φ of norm 1 of $\mathcal{C}_1^\infty$ (depending on the representation) such that the spectrum of $h_\varphi(T)$ equals the closed unit disc.

Proof. We saw in the proof of Theorem 5.20 that the fact that the range of $I - T$ is not closed implies that $(\varphi(M_1))^-$ is not unital. So the theorem is an application of Corollary 5.12.

THEOREM 6.3. Let $\varphi : f \mapsto f(T)$ be a representation of $\mathcal{C}_1^\infty$ over a Banach space E such that $Sp(T) = \{1\}$, and let K be a countable subset of the closed unit disc. If $T^n(x) \xrightarrow[n\to\infty]{} 0$, there exists an element $h_{\varphi,x,K}$ of $\mathcal{C}_1^\infty$ of norm 1 such that $\liminf_{n\to\infty} \|(h_{\varphi,x,K}(T) - \mu I)^n x\|^{1/n} = 0$ for every $\mu \in K$.

Proof. Since $Sp(T) = \{1\}$, T is invertible, and since $\varphi(1)T = T\varphi(1) = T$, $\varphi(1) = I$. So $\varphi(\alpha - 1)$ is quasi-nilpotent and $\varphi(M_1)^-$ is a radical algebra.

Now, if $T^n(x) \xrightarrow[n\to\infty]{} 0$, we have $x = \lim(I - T^n)(x) = \lim_{n\to\infty} (I - T)(T + T^2 + \cdots + T^{n-1})(x) \in [\varphi(M_1)(E)]^-$. Since M_1 possesses a b.a.i., it follows from the factorization theorem for modules that $\varphi(M_1)(E)$ is closed. Thus, there exists $f \in M_1$ and $y \in E$ such that $x = f(T)(y)$. Applying Theorem 5.15, we see that there exists $h \in \mathcal{C}_1^\infty$ such that $\liminf_{n\to\infty} \|(h(T) - \mu I)^n f(T)\|^{1/n} = 0$ for every $\mu \in K$.

The result follows.

We observed above that, if $T^n(x) \xrightarrow[n\to\infty]{} 0$, then $x \in \varphi(M_1)(E)$. In fact, if $Sp(T) = \{1\}$, then $\|T^n(I - T)\| \xrightarrow[n\to\infty]{} 0$, so that $\varphi(M_1)(E) = \{\, x \in E \mid T^n(x) \xrightarrow[n\to\infty]{} 0\}$. The fact that $\|T^n(I - T)\| \xrightarrow[n\to\infty]{} 0$ will be proved in the Appendix. Of course, if $\varphi : f \mapsto f(T)$ is any representation of H^∞ over a Banach space such that the spectral radius of T equals 1, then, using a suitable rotation of the unit disc, we obtain a new representation $f \mapsto f(T')$ such that the spectrum of T' contains 1. So the three theorems of this section give results for representations of H^∞ for which the spectral radius of T equals 1 and for which the operator $T^* : \ell \to T \circ \ell$ acting on the dual of E has no eigenvalues. We leave these statements to the reader.

REMARK 6.4. (1) If φ is any representation of M_1 on a Banach space E such that $\varphi(M_1)(E)$ is dense in E, then φ has a continuous extension to $\mathfrak{A}_1^\infty$. This follows from Proposition 2.1. In fact, Apostol shows in [6] that some suitable hypotheses on E and φ imply that φ extends to the whole algebra H^∞.

(2) In Theorems 6.2 and 6.3 we may choose h_φ or $h_{\varphi,K,x}$ in the set of elements h of H^∞ satisfying the conditions of Theorem 6.1. To see this, just remember that these elements are to be taken in some strongly dense G_δ set of $\Omega(M_1)$. Similarly, we can take a countable subset of E instead of a single element x in Theorem 6.3. Note also that the condition of Theorem 6.1 is satisfied by any element h of $\mathfrak{A}_1^\infty$ such that $\liminf_{n\to\infty} \|(T-\lambda e)(e_n^2 - e_n)\| = 0$ for every $\lambda \in \overline{D}$, where (e_n) is any sequential b.a.i. for M_1. We can take for example $e_n(z) = (1-z)^{1/n}$, and also get the above condition for any countable family of sequential b.a.i. of M_1.

(3) The set $\Omega(M_1)$ equals the whole closed unit ball of $\mathfrak{A}_1^\infty$. Of course, the position function $\alpha : z \mapsto z$ belongs to $U(M_1) = \{h \in \mathfrak{A}_1^\infty \mid \|h\| = 1,\ h - 1 \in M_1\}$ introduced in Definition 5.7. So $\alpha^n \in U(M_1)$ for every $n \in \mathbb{N}$, and in fact any unimodular element of $\mathfrak{A}(D)$ belongs to $\Omega(M_1)$. This follows from the fact that, if $|f(1)| = 1$, then $f \in f(1)\, U(M_1) \subset \Omega(M_1)$. But it follows from a result of Fisher [27] that the closed convex hull of unimodular functions of $A(D)$ equals the unit ball of $A(D)$, which is strongly dense in the unit ball of $\mathfrak{A}_1^\infty$. This proves our claim.

(4) If $\mathfrak{A}$ is a commutative Banach algebra with s.b.a.i. such that $U(\mathfrak{A}) \neq \{e\}$, it is not true in general that $\mathfrak{m}(\mathfrak{A})$ possesses elements with large spectra. Denote by $\mathfrak{U}$ the algebra of all elements $f : z \mapsto \sum_{n\geq 0} a_n z^n$ of $A(D)$ such that $\sum_{n=0}^\infty |a_n| < \infty$ and put $\|f\| = \sum_{n=0}^\infty |a_n|$ for $f \in \mathfrak{U}$. Then $\mathfrak{U}$ is a Banach algebra and $\mathfrak{M} = \{f \in \mathfrak{U} \mid f(1) = 0\}$ is a Banach algebra with b.a.i. The set $U(\mathfrak{M})$ obviously contains the position function α and its powers. On the other hand, Koua [37] showed in his thesis that $\mathfrak{m}(\mathfrak{M}/J) = \mathfrak{M}/J \oplus \mathbb{C}e$ for some closed ideal J of $\mathfrak{M}$, so the spectrum of every element of $\mathfrak{m}(\mathfrak{M}/J)$ is a singleton.

(5) Let $\mathfrak{A}$ be a commutative Banach algebra such that there exists $x \neq 0$ satisfying $\|e - x\| = 1$ for some suitable extension of the norm of $\mathfrak{A}$ to $\mathfrak{A} \oplus \mathbb{C}e$. Then the closure of the principal ideal generated by x possesses a b.a.i. To see this, put $\varphi(f) = \sum_{n\geq 0} a_n(e-x)^n$ for every element $f : z \mapsto \sum_{n\geq 0} a_n z^n$ of the algebra $\mathfrak{U}$ introduced above. This homomorphism φ is continuous from $\mathfrak{U}$ into $\mathfrak{A} \oplus \mathbb{C}e$, $\varphi(\mathfrak{M}) \subset \mathfrak{A}$ and $x = e - (e-x) = \varphi(1-\alpha) \in \varphi(\mathfrak{M})$, so $[x\mathfrak{A}]^-$ possesses a b.a.i.

(6) It would be nice to find in M_1, and more generally in any Banach algebra $\mathfrak{A}$ with regular b.a.i., a sequence (e_n) such that $\|e_n x - x\| \xrightarrow[n\to\infty]{} 0$ and $\|(1-\lambda)e_n + \lambda e\| = 1$ for every $n \in \mathbb{N}$ and every element λ of the unit circle Γ instead of getting sequences depending on λ. But this is impossible

in M_1 and in any radical Banach algebra. To see this, note that if ℓ is a continuous linear form over $\mathfrak{a}$ of norm 1 such that $\ell(e) = 1$, and if $\|(1-\lambda)a + \lambda e\| = 1$ for every $\lambda \in \Gamma$ then $|\ell(a) - \lambda/(\lambda-1)| \leq 1$ for every $\lambda \in \Gamma$. This implies that $\ell(a) \in [0,1]$ for every such ℓ, and the numerical range ([14], Chapter 2, Definition 1) of a is real. So a is hermitian in the sense of ([14], Chapter 5, Definition 1). If $\mathfrak{a}$ is radical, this implies that $a = 0$ ([14], Chapter 5, Lemma 7), and this is also the case in M_1, $L^1(\mathbb{R})$, etc. (In fact, M_1 does not even possess any nonzero elements with real spectrum, of course).

7. Similarity between Banach algebras and the pseudo Banach algebra $\mathfrak{M}_r(\mathfrak{A})$

In this section we study the structure of the algebra $\mathfrak{M}_r(\mathfrak{A})$. (Recall that $\mathfrak{M}_r(\mathfrak{A})$ is the set of all quasi-multipliers T on $\mathfrak{A}$ such that $(\lambda^n T^n)_{n\in\mathbb{N}}$ is pseudo-bounded for some $\lambda > 0$.) The family $\mathfrak{F}_{\mathfrak{A}}$ of all subsets V of $\mathfrak{M}_r(\mathfrak{A})$ such that λV is contained in a pseudo-bounded set stable under products for some $\lambda > 0$ is a natural family of "bounded sets" that we call the multiplicatively bounded sets. The pair $(\mathfrak{M}_r(\mathfrak{A}),\mathfrak{F}_{\mathfrak{A}})$ is a pseudo-Banach algebra in the sense of Allan, Dales, and McClure [4]. A pseudo-Banach algebra B is always the inductive limit $\varinjlim_{\alpha} B_\alpha$ of an inductive family of Banach algebras, in the sense that a subset V of B is bounded if and only if it is contained in the image of a bounded set of some algebra B_α. In the case of $\mathfrak{M}_r(\mathfrak{A})$ we obtain a very "concrete" family of Banach algebras whose inductive limit equals $\mathfrak{M}_r(\mathfrak{A})$. We have $\mathfrak{M}_r(\mathfrak{A}) = \varinjlim_{\alpha} M(\mathfrak{A}_\alpha)$ where $(\mathfrak{A}_\alpha)_{\alpha\in\Lambda}$ is a suitable family of Banach algebras contained in $\mathfrak{M}_r(\mathfrak{A})$ as algebras. In other terms, the regular quasi-multipliers are the quasi-multipliers which can become multipliers in the usual sense after some suitable change of the algebra. We make this "suitable change" precise by defining the notion of simularity between commutative Banach algebras $\mathfrak{A}$ and $\mathfrak{B}$ which possess some dense principal ideal. In fact $\mathfrak{A}$ and $\mathfrak{B}$ are said to be similar if there exists a Banach algebra $\mathfrak{D}$ which possesses some dense principal ideal, and two one-to-one homomorphisms $\varphi : \mathfrak{D} \to \mathfrak{A}$ and $\psi : \mathfrak{D} \to \mathfrak{B}$ such that $\varphi(\mathfrak{D})$ is a dense ideal of $\mathfrak{A}$ and $\psi(\mathfrak{D})$ is a dense ideal of $\mathfrak{B}$. If $a\mathfrak{A} \neq \{0\}$ and $b\mathfrak{B} \neq \{0\}$ for every nonzero $a \in \mathfrak{A}$ and every nonzero $b \in \mathfrak{B}$, there exists a natural isomorphism between $\mathfrak{M}(\mathfrak{A})$ and $\mathfrak{M}(\mathfrak{B})$ (hence between $\mathfrak{M}_r(\mathfrak{A})$ and $\mathfrak{M}_r(\mathfrak{B})$) which is also an isomorphism with respect to the bound structures of these algebras. Conversely, if such an isomorphism does exist the algebras $\mathfrak{A}$ and $\mathfrak{B}$ are necessarily similar. If $\mathfrak{A}$ and $\mathfrak{B}$ possess a b.a.i., similarity between $\mathfrak{A}$ and $\mathfrak{B}$ leads to a bijection between the set of closed ideals of $\mathfrak{A}$ and $\mathfrak{B}$, and there is always some relation between the sets of closed ideals of two similar Banach algebras.

The Gelfand theory extends to pseudo-Banach algebras [4]. In particular, the set $\hat{\hat{\mathfrak{A}}}$ of all characters of $\mathfrak{M}_r(\mathfrak{A})$ is compact, and the Gelfand transform is a "bounded" homomorphism from $\mathfrak{M}_r(\mathfrak{A})$ into $C(\hat{\hat{\mathfrak{A}}})$. We can thus associate to every commutative Banach algebra a compact set $\hat{\hat{\mathfrak{A}}}$ and a normed algebra whose completion is uniform, the image of $\mathfrak{M}_r(\mathfrak{A})$ in $C(\hat{\hat{\mathfrak{A}}})$ via the Gelfand transform. Using the results of Sections 3 and 6, we show in particular that, if $\mathfrak{A}$ is a commutative, radical Banach algebra with sequential b.a.i., then the corresponding normed algebra contains an isometric, spectrum-preserving copy of H^∞, and $\hat{\hat{\mathfrak{A}}}$ can be mapped continuously onto the spectrum of H^∞. This result extends to Banach algebras $\mathfrak{A}$ which possess a continuous semigroup $(a^t)_{t>0}$ such that $\bigcup_{t>0} a^t \mathfrak{A}$ is dense in $\mathfrak{A}$, because such algebras are similar to Banach algebras with b.a.i. We thus obtain a spectral theory for a large class of radical Banach algebras.

The results of this section have some applications to the closed ideal problem, which will be given in the next section.

I have to say that the notion of similarity between Banach algebras was suggested to me by Feller's work about the generation of unbounded semigroups [26].

We first collect some elementary facts.

LEMMA 7.1. Let $(\mathcal{B},p)$ be a Banach space, let $(\mathcal{A},\|.\|)$ be a commutative Banach algebra, and let $\varphi:\mathcal{B}\to\mathcal{A}$ be a continuous one-to-one linear map. If $\varphi(\mathcal{B})$ is an ideal of $\mathcal{A}$, there exists $K>0$ such that $p[\varphi^{-1}[\varphi(x)y)]] \le Kp(x)\|y\|$ for every $x\in\mathcal{B}$ and for every $y\in\mathcal{A}$.

Proof. Denote by ψ the bilinear map $(x,y)\mapsto\varphi^{-1}(\varphi(x)y)$ $(x\in\mathcal{B},\ y\in\mathcal{A})$. Fix $y\in\mathcal{A}$, and let $a,b\in\mathcal{B}$ such that $a=\lim_{n\to\infty}x_n$ and $b=\lim_{n\to\infty}\psi(x_n,y)$ for some sequence (x_n) of elements of $\mathcal{B}$. We have $\varphi(b)=\lim_{n\to\infty}\varphi(x_n)y=\varphi(a)y$, so that $b=\psi(a,y)$. The closed graph theorem shows that the map $x\mapsto\psi(xy)$ is a continuous map from $\mathcal{B}$ into itself. A similar application of the closed graph theorem shows that the map $y\mapsto\psi(x,y)$ is a continuous map from $\mathcal{B}$ into $\mathcal{A}$ for every fixed $x\in\mathcal{B}$. So ψ is a separately continuous bilinear map from $\mathcal{B}\times\mathcal{A}$ into $\mathcal{B}$, and the Banach-Steinhaus theorem shows that ψ is jointly continuous. This gives the desired constant.

PROPOSITION 7.2. Let $\mathcal{B}$, $\mathcal{A}$ and φ be as in Lemma 7.1. Then the map $(x,y)\mapsto\varphi^{-1}(\varphi(x)\varphi(y))$ defines an associative product over $\mathcal{B}$, and there exists over $\mathcal{B}$ an equivalent norm p_1 such that $p_1[\varphi^{-1}(\varphi(x)y)]\le p_1(x)\|y\|$ $(x\in\mathcal{B},\ y\in\mathcal{A})$, $\|\varphi(x)\|\le p_1(x)$ $(x\in\mathcal{B})$, and $p_1[\varphi^{-1}(\varphi(x)\varphi(y))]\le p_1(x)p_1(y)$ $(x,y\in\mathcal{B})$.

Proof. The fact that the above product is associative follows immediately from the associativity of the product in $\mathcal{A}$. Also, the map $(y,x)\mapsto\varphi^{-1}(\varphi(x)y)$ defines a module action of $\mathcal{A}$ over $\mathcal{B}$. A well-known argument (see for example [25], Section 2) gives an equivalent norm p' over $\mathcal{B}$ satisfying the first inequality and such that $p'(x)\ge p(x)$ $(x\in\mathcal{B})$. The norm $p_1 : x\to\|\varphi\|p'(x)$ gives the desired result.

PROPOSITION 7.3. Let $\mathcal{A}$ and $\mathcal{B}$ be two commutative Banach algebras, and let $\varphi:\mathcal{B}\to\mathcal{A}$ be a continuous one-to-one homomorphism such that $\varphi(\mathcal{B})$ is a dense ideal of $\mathcal{A}$.

(1) $I\mathcal{A}\subset[\varphi(\varphi^{-1}(I))]^-\subset I$ for every closed ideal I of $\mathcal{A}$.

(2) $I=[\varphi(\varphi^{-1}(I))]^-$ for every closed ideal I of $\mathcal{A}$ if $\mathcal{A}$ possesses a b.a.i.

(3) $\mathcal{B}\varphi^{-1}(\overline{\varphi(J)})\subset J\subset\varphi^{-1}(\overline{\varphi(J)})$ for every closed ideal J of $\mathcal{B}$.

(4) If $[x\mathcal{B}]^-=\mathcal{B}$ for some $x\in\mathcal{B}$, and if $\mathcal{A}$ possesses a b.a.i., then $J=\varphi^{-1}(\overline{\varphi(J)})$ for every closed ideal J of $\mathcal{B}$.

(5) If $x \in \mathfrak{D}$ satisfies $[\varphi(x)G]^- = G$, then $[x^2 \mathfrak{L}]^- = \mathfrak{L}$, and $\varphi(\mathfrak{L})$ is a dense ideal of G, where $\mathfrak{L} = [x\mathfrak{D}]^-$.

(6) The set $\mathfrak{D}^\perp = \{y \in G \mid \varphi(\mathfrak{D})y = \{0\}\}$ reduces to $\{0\}$ if G possesses a b.a.i.

(7) If $[x\mathfrak{D}]^- = \mathfrak{D}$ for some $x \in \mathfrak{D}$, then $(\theta \circ \varphi)(\mathfrak{D})\alpha \neq \{0\}$ for every non-zero $\alpha \in G/\mathfrak{D}^\perp$, where θ is the surjection $G \to G/\mathfrak{D}^\perp$.

(8) If $[x\mathfrak{D}]^- = \mathfrak{D}$ for some $x \in \mathfrak{D}$, and if $\mathfrak{D}^\perp = \{0\}$, the homomorphism $\hat{\varphi}$ from $\mathfrak{M}(\mathfrak{D})$ into $\mathfrak{M}(G)$ associated to φ by Proposition 2.29 is a pseudo-bounded isomorphism from $\mathfrak{M}(\mathfrak{D})$ onto $\mathfrak{M}(G)$, and $\hat{\varphi}^{-1}$ is pseudo-bounded.

Proof. (1) $\varphi(\varphi^{-1}(I)) \subseteq I$ for every subset I of G, so $[\varphi(\varphi^{-1}(I))]^- \subset I$ for every closed subset I of G. Now let I be a closed ideal of G, let $y \in G$, and let $z \in I$. There exists a sequence (y_n) of elements of $\mathfrak{D}$ such that $y = \lim_{n \to \infty} \varphi(y_n)$. Then $\varphi(y_n)z \in \varphi(\mathfrak{D}) \cap I$ for every $n \in \mathbb{N}$, so $\varphi^{-1}(\varphi(y_n)z)$ exists and belongs to $\varphi^{-1}(I)$ for every $n \in \mathbb{N}$. Since $yz = \lim_{n \to \infty} \varphi(y_n)z$, $yz \in \varphi(\varphi^{-1}(I))^-$, which gives the desired conclusion.

(2) If G possesses a b.a.i., then $y \in [yG]^-$ for every $y \in G$. So $I = [IG]^-$ for every closed ideal I of G.

(3) Let J be a closed ideal of $\mathfrak{D}$. We may assume that the norm p of $\mathfrak{D}$ and the norm $\|\cdot\|$ of G satisfy the inequalities of Proposition 6.2. If $x \in J$, then $x \in \varphi^{-1}(\varphi(J))$, so $J \subset \varphi^{-1}(\overline{\varphi(J)})$. Now let x be an element of $\mathfrak{D}\varphi^{-1}(\overline{\varphi(J)})$. There exists $y \in \mathfrak{D}$ and $z \in \overline{\varphi(J)}$ such that $x = y\varphi^{-1}(z)$. So there exists a sequence (z_n) of elements of J such that $\|\varphi(z_n) - z\| \xrightarrow[n \to \infty]{} 0$. We have

$$p(yz_n - x) = p(yz_n - y\varphi^{-1}(z)) = p[\varphi^{-1}[\varphi(y)[\varphi(z_n) - z]]]$$
$$\leq p(y)\|\varphi(z_n) - z\| \quad \text{for every } n \in \mathbb{N}.$$

So $p(yz_n - x) \xrightarrow[n \to \infty]{} 0$, and $x \in J$ because J is closed.

(4) Assume that G possesses a b.a.i. and that $[x\mathfrak{D}]^- = \mathfrak{D}$ for some $x \in \mathfrak{D}$. Then $[\varphi(x)G]^- = G$, so the b.a.i. is sequential, and, since $[\varphi(\mathfrak{D})]^- = G$, there exists in fact a sequence (e_n) of elements of $\mathfrak{D}$ such that $\|y - y\varphi(e_n)\| \xrightarrow[n \to \infty]{} 0$ for every $y \in G$ and such that $\sup_{n \in \mathbb{N}} \|\varphi(e_n)\| < \infty$. Then $p(uv - uve_n) \xrightarrow[n \to \infty]{} 0$ for every $u, v \in \mathfrak{D}$. Let $u, v \in \mathfrak{D}$. We obtain

$$\limsup_{n \to \infty} p(u - ue_n) \leq p(u - xv) + \limsup_{n \to \infty} p(xv - xve_n) + \limsup_{n \to \infty} p(xve_n - ue_n)$$

$$\leq p(u - xv)[1 + \limsup_{n \to \infty} \|\varphi(e_n)\|].$$

Since $[x\mathfrak{D}]^- = \mathfrak{D}$, we see that $p(u - ue_n) \xrightarrow[n \to \infty]{} 0$ for every $u \in \mathfrak{D}$. In particular

$L = [\mathfrak{A}J]^-$ for every closed ideal J of $\mathfrak{A}$. Assertion (4) follows.

(5) Let $x \in \mathfrak{A}$ such that $[\varphi(x)\mathcal{C}]^- = \mathcal{C}$. The map $(a,y) \mapsto \varphi^{-1}(a\varphi(y))$ defines a module action of $\mathcal{C}$ over $\mathfrak{A}$, and Lemma 6.1 shows that $\mathfrak{A}$ is in fact a Banach $\mathcal{C}$-module. The set $x\mathfrak{A}$ is an $\mathcal{C}$-submodule of $\mathfrak{A}$, so its closure $\mathfrak{L}$ is also an $\mathcal{C}$-submodule. This means that $\varphi(\mathfrak{L})$ is an ideal of $\mathcal{C}$. Clearly $x^2 \in \mathfrak{L}$. We have $[\varphi(x\mathfrak{A})]^- = [\varphi(x)\varphi(\mathfrak{A})]^- = [\varphi(x)\mathcal{C}]^- = \mathcal{C}$, so $\varphi(\mathfrak{L})$ is dense in $\mathcal{C}$.

Now let $u \in \mathfrak{A}$. Since $[\varphi(x^2)\mathfrak{A}]^- = ([\varphi(x)]^2\mathcal{C})^- = [\varphi(x)\mathcal{C}]^- = \mathcal{C}$, there exists a sequence (v_n) of elements of $\mathfrak{A}$ such that $\|u - x^2v_n\| \xrightarrow[n\to\infty]{} 0$. It follows from Lemma 6.1 that $p(ux - x^3v_n) \xrightarrow[n\to\infty]{} 0$. As $x^3v_n \in x^2\mathfrak{L}$, we see that $[x^2\mathfrak{L}]^- = \mathfrak{L}$.

(6) Assume that $\mathcal{C}$ possesses a b.a.i. Then $y \in [y\mathcal{C}]^- = [y\varphi(\mathfrak{A})]^-$ for every $y \in \mathcal{C}$, so $\varphi(\mathfrak{A})y = \{0\}$ implies that $y = 0$.

(7) Assume that $[x\mathfrak{A}]^- = \mathfrak{A}$ for some $x \in \mathfrak{A}$, and let $y \in \mathcal{C}$ such that $\theta(y)(\theta[\varphi(\mathfrak{A})]) = \{0\}$. Then $\theta(y)\theta(\mathcal{C}) = \{0\}$, $\theta(y\mathcal{C}) = \{0\}$, and $y\mathcal{C} \subset \mathfrak{A}^\perp$. This means that $ya\varphi(u) = 0$ for every $a \in \mathcal{C}$ and every $u \in \mathfrak{A}$. In particular $y\varphi(\mathfrak{A}) \subset [y\varphi(x\mathfrak{A})]^- = \{0\}$, so $y \in \mathfrak{A}^\perp$.

(8) Assume that $[x\mathfrak{A}]^- = \mathfrak{A}$ for some $x \in \mathfrak{A}$ and that $\mathfrak{A}^\perp = \{0\}$. Since φ is one-to-one, this implies that $y\mathfrak{A} \neq \{0\}$ for every nonzero $y \in \mathfrak{A}$, and since $[\varphi(\mathfrak{A})]^- = \mathcal{C}$ we have $[\varphi(x)\mathcal{C}]^- = \mathcal{C}$. We thus see that $\mathfrak{M}(\mathfrak{A})$ and $\mathfrak{M}(\mathcal{C})$ are well defined.

The homomorphism $\hat{\varphi}: \mathfrak{M}(\mathfrak{A}) \to \mathfrak{M}(\mathcal{C})$ associated to φ by Proposition 2.29 is pseudo-bounded. Recall that $\hat{\varphi}$ is defined by the formula $\hat{\varphi}(T_{a/b}) = T_{\varphi(a)/\varphi(b)}$ for every $a \in \mathfrak{A}$ and every $b \in \mathfrak{A}$ such that $[b\mathfrak{A}]^- = \mathfrak{A}$.

Now let $T = T_{u/v} \in \mathfrak{M}(\mathcal{C})$, where $[v\mathcal{C}]^- = \mathcal{C}$. Then $[v\varphi(\mathfrak{A})]^- = \mathcal{C}$, and there exists for every $z \in \mathfrak{A}$ a sequence (z_n) of elements of $\mathfrak{A}$ such that $\|\varphi(z) - v\varphi(z_n)\| \xrightarrow[n\to\infty]{} 0$. So $p(xz - \varphi^{-1}(\varphi(x)v)z_n) = p(xz - x\varphi^{-1}(v\varphi(z_n))) \xrightarrow[n\to\infty]{} 0$. So $\varphi^{-1}(\varphi(x)v)$ generates in $\mathfrak{A}$ a principal ideal whose closure equals $[x\mathfrak{A}]^- = \mathfrak{A}$.

Put $\alpha = \varphi^{-1}(\varphi(x)u)$, $\beta = \varphi^{-1}(\varphi(x)v)$. Then $[\beta\mathfrak{A}]^- = \mathfrak{A}$, and $\hat{\varphi}(T_{\alpha/\beta}) = T_{\varphi(x)u/\varphi(x)v} = T_{u/v}$, and $\hat{\varphi}$ is onto. It follows from Remark 2.31 that $\hat{\varphi}$ is one-to-one, and we obtain an isomorphism from $\mathfrak{M}(\mathfrak{A})$ onto $\mathfrak{M}(\mathcal{C})$. Also, $\hat{\varphi}$ is pseudo-bounded as shown in Proposition 2.29. Now let $\mathfrak{U}$ be a pseudo-bounded subset of $\mathfrak{M}(\mathcal{C})$. There exists $v \in \bigcap_{T\in\mathfrak{U}} \mathfrak{A}_T$ such that $[v\mathcal{C}]^- = \mathcal{C}$ and $\sup_{T\in\mathfrak{U}} \|T(v)\| < \infty$. We see as above that $\beta = \varphi^{-1}(\varphi(x)v)$ generates a dense ideal of $\mathfrak{A}$. We can write $T = T_{T(v)/v}$ for every $T \in \mathfrak{U}$, so $T = \hat{\varphi}(T_{\alpha/\beta})$, where $\alpha = \varphi^{-1}(\varphi(x)T(v))$. Hence $\beta \in \bigcap_{T\in\mathfrak{U}} \mathfrak{A}_{\hat{\varphi}^{-1}(T)}$ and $\hat{\varphi}^{-1}(T)(\beta) = \alpha = \varphi^{-1}(\varphi(x)T(v))$ for every $T \in \mathfrak{U}$. We obtain

$$\sup_{T\in\mathfrak{A}} p[\hat{\varphi}^{-1}(T)(\beta)] = \sup_{T\in\mathfrak{A}} p[\varphi^{-1}(\varphi(x)T(v))] \leq p(x) \sup_{T\in\mathfrak{A}} \|T(v)\| < \infty .$$

So $\hat{\varphi}^{-1}(\mathfrak{A})$ is pseudo-bounded, and $\hat{\varphi}^{-1}$ is a pseudo-bounded map from $\mathfrak{M}(\mathcal{A})$ onto $\mathfrak{M}(\mathfrak{D})$. This achieves the proof of the proposition.

DEFINITION 7.4. Let $\mathcal{A}$ and $\mathcal{B}$ be two commutative Banach algebras. We say that $\mathcal{A}$ and $\mathcal{B}$ are <u>similar</u> if there exists a Banach algebra $\mathfrak{D}$ such that $[x\mathfrak{D}]^- = \mathfrak{D}$ for some $x \in \mathfrak{D}$, and if there exist two one-to-one continuous homomorphisms $\varphi : \mathfrak{D} \to \mathcal{A}$ and $\psi : \mathfrak{D} \to \mathcal{B}$ such that $\varphi(\mathfrak{D})$ is a dense ideal of $\mathcal{A}$ and $\psi(\mathfrak{D})$ is a dense ideal of $\mathcal{B}$.

Note that this condition implies that $\varphi(x)\mathcal{A}$ is dense in $\mathcal{A}$ and that $\psi(x)\mathcal{B}$ is dense in $\mathcal{B}$.

The transitivity of this relation follows from the following.

PROPOSITION 7.5. Let $\mathcal{A}$ be a commutative Banach algebra and let $\mathcal{A}_1,\ldots,\mathcal{A}_k$ be a finite family of commutative Banach algebras similar to $\mathcal{A}$. There exists a Banach algebra $\mathfrak{D}$ such that $[x\mathfrak{D}]^- = \mathfrak{D}$ for some $x \in \mathfrak{D}$, a one-to-one homomorphism $\psi : \mathfrak{D} \to \mathcal{A}$ such that $\psi(\mathfrak{D})$ is a dense ideal of $\mathcal{A}$ and for $i = 1,\ldots,k$, a one-to-one homomorphism $\psi_i : \mathfrak{D} \to \mathcal{A}_i$ such that $\psi_i(\mathfrak{D})$ is a dense ideal of $\mathcal{A}_i$. Moreover, we may define the norm of $\mathfrak{D}$ so that φ and all the homomorphisms φ_i satisfy the inequalities of Proposition 6.2.

<u>Proof</u>. We may assume that there exists a family $\mathfrak{D}_i$ $(i=1,\ldots,k)$ of dense ideals of $\mathcal{A}$ such that each $\mathfrak{D}_i$ is complete with respect to a norm p_i stronger than the norm of $\mathcal{A}$, for which $x_i\mathfrak{D}_i$ is dense in $\mathfrak{D}_i$ for some $x_i \in \mathfrak{D}_i$, and that there exists for every i a continuous homomorphism $\varphi_i : \mathfrak{D}_i \to \mathcal{A}_i$ such that $\varphi_i(\mathfrak{D}_i)$ is a dense ideal of $\mathcal{A}_i$, φ_i being one-to-one. Also, we may assume that the norms p_i satisfy the inequalities of Proposition 6.2 with respect to the canonical injection from $\mathfrak{D}_i$ into $\mathcal{A}$.

Denote by L the subspace of $\mathcal{A}$ spanned by the set S of all products $y_1,\ldots y_k$, where $y_i \in \mathfrak{D}_i$ for every $i \leq k$, and set $q(u) = \inf \sum_{m\leq r} p_1(z_{m,1}) \cdots p_k(z_{m,k})$ $(u \in L)$, the infimum being taken over all decompositions of u as a finite sum $u = \sum_{m\leq r} z_{m,1} \cdots z_{m,k}$, where $z_{m,i} \in \mathfrak{D}_i$ for every $m \leq r$ and every $i \leq k$. We clearly obtain a linear norm over L, and it follows from the definition of L that L is an ideal of $\mathcal{A}$ contained in $\mathfrak{D}_1 \cap \mathfrak{D}_2 \cap \cdots \cap \mathfrak{D}_k$.

For $i \leq k$ denote by $\|\cdot\|_i$ the norm of $\mathcal{A}_i$. Consider $(z_1,\ldots,z_k) \in \mathfrak{D}_1 \times \mathfrak{D}_2 \times \cdots \times \mathfrak{D}_k$, and for $i \leq k$ denote by Z_i the product $\Pi_{j\neq i}\, z_j$.

Note that, if $a \in \mathcal{A}$, $a_i \in \mathcal{A}_i$, and $z \in \mathfrak{D}_i$, we have $\varphi_i(az)a_i = \varphi_i(a\varphi_i^{-1}[\varphi_i(z)a_i])$. To see this, note that if $\alpha \in \mathfrak{D}_i$ we have

$\varphi_i(\alpha z)a_i = \varphi_i[\alpha\varphi_i^{-1}(\varphi_i(z)a_i)]$. Now if $a \in \mathfrak{A}$ there exists a sequence (α_n) of elements of $\mathfrak{D}_i$ such that $\|a - \alpha_n\| \xrightarrow[n\to\infty]{} 0$. So $p_i(az - \alpha_n z) \xrightarrow[n\to\infty]{} 0$ and $p_i(a\varphi_i^{-1}(\varphi_i(z)a_i) - \alpha_n\varphi_i^{-1}(\varphi_i(z)a_i)) \xrightarrow[n\to\infty]{} 0$. The equality follows then from the continuity of φ_i. So we have $\varphi_i(Z_i z_i)a_i = \varphi_i(Z_i\varphi_i^{-1}(\varphi_i(z_i)a_i))$ for every $a_i \in \mathfrak{A}_i$. Also $p_i[\varphi_i^{-1}(\varphi_i(z_i)a_i)] \leq K_i p_i(z_i)\|a_i\|_i$. So $[\Pi_{j\neq i}\, p_j(z_j)]\; p_i[\varphi_i^{-1}(\varphi_i(z_i)a_i)]] \leq K_i\|a_i\|_i p_1(z_1)\ldots p_k(z_k)$. Here, K_i is the constant associated to φ_i by Lemma 6.1.

We deduce from the above computations that $\varphi_i(L)$ is an ideal of $\mathfrak{A}_i$ for every $i \leq k$ and that $q[\varphi_i^{-1}(\varphi_i(u)a_i)] \leq K_i q(u)\|a_i\|_i$ $(a_i \in \mathfrak{A}_i, u \in L)$. The notations being as above, we get

$$\|\varphi_i(z_1\ldots z_k)\|_i \leq K_i p_i(z_1\ldots z_k) \leq K_i p_i(z_i) \prod_{j\neq i} \|z_j\|$$

$$\leq K_i p_1(z_1)\ldots p_k(z_k).$$

So $\|\varphi_i(u)\|_i \leq K_i q(u)$ $(u \in L,\ i \leq k)$.

It follows also immediately from the definitions of L and q that $q(u) \geq \|u\|$ for every $u \in L$, and that $q(au) \leq \|a\|q(u)$ for every $u \in L$ and every $a \in \mathfrak{A}$.

Denote by $\mathfrak{L}$ the completion of L with respect to q. The restrictions of φ_i to L can be continuously extended to $\mathfrak{L}$ for every $i \leq k$, and the injection from L into $\mathfrak{A}$ extends to an algebra homomorphism θ from $\mathfrak{L}$ into $\mathfrak{A}$. Similarly, the injection from L into $\mathfrak{D}_i$ extends for every i to an homomorphism θ_i from $\mathfrak{L}$ into $\mathfrak{A}$. Clearly $\theta_i(u) = \theta(u)$ for every $i \leq k$ and for every $u \in \mathfrak{L}$, so $\tilde{\varphi}_i = \varphi_i \circ \theta$, where we denote by $\tilde{\varphi}_i$ the extension to $\mathfrak{L}$ of the restriction of φ_i to L. Fix $i \leq k$, let $u \in \tilde{\varphi}_i(\mathfrak{L})$, and let $a \in \mathfrak{A}_i$. There exists a sequence (u_n) of elements of L such that $q(u - u_n) \xrightarrow[n\to\infty]{} 0$. Since $q[\varphi_i^{-1}(\varphi_i(y)a)] \leq K_i\|a\|_i q(y)$ for every $y \in L$, the sequence $(\varphi_i^{-1}(\varphi_i(u_n)a))_{n\in\mathbb{N}}$ is a Cauchy sequence in (L,q), and so it possesses a limit, say v, in $\mathfrak{L}$. We have $\varphi_i(v) = \lim_{n\to\infty} \varphi_i(u_n)\, a = \tilde{\varphi}_i(u)a$. So $\tilde{\varphi}_i(u)\, a \in \operatorname{Im}\tilde{\varphi}_i$ for every $u \in \mathfrak{L}$, and $\tilde{\varphi}_i(\mathfrak{L})$ is an ideal of $\mathfrak{A}_i$ for every i. A similar argument shows that $\theta(\mathfrak{L})$ is an ideal of $\mathfrak{A}$. We have $[x_1\ldots x_k\mathfrak{A}]^- = \mathfrak{A}$, so L is dense in $\mathfrak{A}$ and, a fortiori, $\theta(\mathfrak{L})$ is dense in $\mathfrak{A}$. Fix again $i \leq k$. Then $x_i\mathfrak{D}_i$ is dense in $\mathfrak{D}_i$ with respect to p_i. For every $\alpha \in \mathfrak{D}_i$ there exists a sequence (α_n) of elements of $\mathfrak{D}_i$ such that $p_i(\alpha - x_i\alpha_n) \xrightarrow[n\to\infty]{} 0$. Also, for every $n \in \mathbb{N}$ there exists $\beta_n \in \mathfrak{A}$ such that $\|(\Pi_{j\neq i}\, x_j)\beta_n - \alpha_n\| < 1/n$. So $p_i(x_i\alpha_n - x_1\ldots x_k\beta_n) \xrightarrow[n\to\infty]{} 0$, and, in fact, $x_1\ldots x_k\mathfrak{D}_i$ is dense in $\mathfrak{D}_i$ with respect to p_i. So $\varphi_i(x_1 \dots x_k \mathfrak{D}_i)$ is dense in $\mathfrak{A}_i$. Since $\varphi_i(x_1 \dots x_k \mathfrak{D}_i) \subset \varphi_i(L) \subset \tilde{\varphi}_i(\mathfrak{L})$, we see in fact that $\tilde{\varphi}_i(\mathfrak{L})$ is a dense ideal of $\mathfrak{A}_i$ for every $i \leq k$.

Now let $(z_1, \dots, z_k) \in \mathfrak{D}_1 \times \mathfrak{D}_2 \times \dots \times \mathfrak{D}_k$. There exists for every $i \leq k$ a sequence $(\alpha_{n,i})$ of elements of $\mathfrak{D}_i$ such that $p_i(z_i - x_i \alpha_{n,i}) \xrightarrow[n\to\infty]{} 0$. We have, for every $n \in \mathbb{N}$.

$$\begin{aligned} &z_1 \dots z_k - x_1 \dots x_k \alpha_{n,1} \dots \alpha_{n,k} \\ &\quad = \sum_{i=1}^{k} z_1 \dots z_{i-1}(z_i - x_i \alpha_{n,i}) x_{i+1} \dots x_k \alpha_{n,i+1} \dots \alpha_{n,k} . \end{aligned}$$

It follows from this formula that

$$q(z_1 \dots z_k - x_1 \dots x_k \alpha_{n,1} \dots \alpha_{n,k}) \xrightarrow[n\to\infty]{} 0 .$$

So $x_1 \dots x_k L$ is dense in L with respect to q, and hence $x_1 \dots x_k \mathfrak{L}$ is dense in $\mathfrak{L}$ with respect to q.

It is unfortunately not clear whether or not θ and the homomorphisms $\tilde{\varphi}_i$ are one-to-one. So put $I = \operatorname{Ker} \theta$, $\mathfrak{D} = \mathfrak{L}/I$, and denote by $\bar{q}$ the quotient norm induced by q over $\mathfrak{D}$. Since $\tilde{\varphi}_i = \varphi_i \circ \theta$, $\operatorname{Ker} \tilde{\varphi}_i = \operatorname{Ker} \theta$ for every i. So there exists a homomorphism $\psi_i : \mathfrak{D} \to \mathfrak{A}_i$ which is one-to-one and satisfies $\psi_i \circ \pi = \tilde{\varphi}_i$ for every $i \leq k$, where π is the natural surjection from $\mathfrak{L}$ onto $\mathfrak{D}$. Also, there exists a one-to-one homomorphism $\psi : \mathfrak{D} \to \mathfrak{A}$ such that $\psi \circ \pi = \theta$. The ideal $\pi(x_1 \dots x_k)\mathfrak{D}$ is dense in $\mathfrak{D}$, $\psi(\mathfrak{D}) = \theta(L)$, and $\psi_i(\mathfrak{D}) = \tilde{\varphi}_i(L)$ for every $i \leq k$. So $\psi(\mathfrak{D})$ is a dense ideal of $\mathfrak{A}$, and $\psi_i(\mathfrak{D})$ is a dense ideal of $\mathfrak{A}_i$ for every $i \leq k$.

If $a_i \in \mathfrak{A}_i$ and $u \in \mathfrak{D}$, put $a_i u = \psi_i^{-1}(\psi_i(u) a_i$. Then $\mathfrak{D}$ becomes an $\mathfrak{A}_i$-module for every $i \leq k$. We saw above that, if $a \in \mathfrak{A}$ and $a_i \in \mathfrak{A}_i$, we have $\varphi_i^{-1}(\varphi_i(az)a_i) = a\varphi_i^{-1}(\varphi_i(z)a_i)$ for every $z \in \mathfrak{D}_i$. The same arguments show that $a_i(a_j u) = a_j(a_i u)$ for every $u \in \mathfrak{D}$ and for every $a_i \in \mathfrak{D}_i$, $a_j \in \mathfrak{D}_j$. Denote by S the unit ball of $\mathfrak{A}$ and by S_i the unit ball of $\mathfrak{A}_i$ for $i \leq k$.

Put, for every $u \in \mathfrak{L}$,

$$p(u) = [\sup_{\substack{a \in S \cup \{e\} \\ a_i \in S_i \cup \{e\}}} q(aa_1 \dots a_k u)](1 + \|\psi_1\|) \dots (1 + \|\psi_k\|) .$$

Then p is equivalent to q, and $q(u) \leq p(u)$ for every $u \in \mathfrak{L}$. Renorming $\mathfrak{D}$ with p, we see that ψ and all the homomorphisms ψ_i satisfy the inequalities of Proposition 6.2. This achieves the proof of the proposition.

We now turn to a complete characterization of similarity in the case of algebras $\mathfrak{A}$ such that $a\mathfrak{A} \neq \{0\}$ for every nonzero element a of $\mathfrak{A}$. We first need a lemma.

LEMMA 7.6. Let $\mathfrak{A}$ be a commutative Banach algebra such that $a\mathfrak{A} \neq \{0\}$ for every nonzero element a of $\mathfrak{A}$ and such that $[a\mathfrak{A}]^{-} = \mathfrak{A}$ for some $a \in \mathfrak{A}$, and let $\mathfrak{U}$ be a pseudo-bounded subset of $\mathfrak{M}(\mathfrak{A})$ stable under products.

Put $I = \{x \in \bigcap_{T\in\mathfrak{U}} \mathcal{D}_T \mid \sup_{T\in\mathfrak{U}} \|Tx\| < \infty\}$, and put $p(x) = \sup\{\|Tx\|: T \in \mathfrak{U} \cup \{e\}\}$ for every $x \in I$. Then I is a dense ideal of $\mathfrak{A}$, I is complete with respect to p, $[a\mathfrak{A}]^{-} = \mathfrak{A}$ for some $a \in I$, and $Tx \in I$ for every $x \in I$ and every $T \in \mathfrak{U} \cup \mathfrak{m}(\mathfrak{A})$. Moreover the following inequalities hold:

(1) $p(x) \geq \|x\|$ $(x \in I)$;

(2) $p(yx) \leq \|y\|p(x)$ $(x \in I,\ y \in \mathfrak{A})$;

(3) $p(Tx) \leq p(x)$ $(x \in I,\ T \in \mathfrak{U})$;

(4) $p(Sx) \leq p(x) \sup_{\substack{y\in\mathfrak{A}\\ y\neq 0}} \|Sy\|/\|y\|$ $(x \in I,\ S \in \mathfrak{m}(\mathfrak{A}))$.

Proof. Since $\mathfrak{U}$ is pseudo-bounded, I contains some element a of $\mathfrak{A}$ such that $[a\mathfrak{A}]^{-} = \mathfrak{A}$. The fact that I is an ideal of $\mathfrak{A}$ and that inequalities (1) and (2) hold is clear. Also, I is dense in $\mathfrak{A}$ because $a\mathfrak{A} \subset I$.

Let T, T' be two elements of $\mathfrak{U}$, and let $x \in I$. Since $x \in \mathcal{D}_{T'T} \cap \mathcal{D}_T$, we have $Tx \in \mathcal{D}_{T'}$ and $T'(Tx) = (T'T)x$ (by Proposition 2.12). So $Tx \in \bigcap_{T'\in\mathfrak{U}} \mathcal{D}_{T'}$ for every $T \in \mathfrak{U}$ and every $x \in I$. We obtain

$$\sup_{T'\in\mathfrak{U}\cup\{e\}} \|T'(Tx)\| \leq \sup_{T'\in\mathfrak{U}} \|T'x\| \leq p(x) < \infty .$$

This shows that $Tx \in I$ for every $x \in I$ and every $T \in \mathfrak{U}$ and that inequality (3) holds.

Now let $S \in \mathfrak{m}(\mathfrak{A})$, $T \in \mathfrak{U}$. It follows from Proposition 2.12 that $\mathcal{D}_T \subset \mathcal{D}_{ST} = \mathcal{D}_{TS} = \mathcal{D}_{TS} \cap \mathcal{D}_S$, so $S(\mathcal{D}_T) \subset \mathcal{D}_T$. So if $x \in I$ then $Sx \in \bigcap_{T\in\mathfrak{U}} \mathcal{D}_T$ for every $S \in \mathfrak{m}(\mathfrak{A})$ and

$$p(Sx) = \sup_{T\in\mathfrak{U}\cup\{e\}} \|T(Sx)\| = \sup_{T\in\mathfrak{U}\cup\{e\}} \|S(Tx)\| \leq (\sup_{\|y\|\leq 1} \|S(y)\|)p(x) .$$

Let (x_n) be a Cauchy sequence in (I,p). Inequality (1) shows that (x_n) is a Cauchy sequence in $\mathfrak{A}$, so it possesses a limit x in $\mathfrak{A}$. Inequality (3) shows that (Tx_n) is a Cauchy sequence in $\mathfrak{A}$ for every $T \in \mathfrak{U}$. Put $y = \lim_{n\to\infty} Tx_n$. As T is a closed generalized operator from $\mathfrak{A}$ into itself, $x \in \mathcal{D}_T$ and $y = Tx$. In particular, $x \in \bigcap_{T\in\mathfrak{U}} \mathcal{D}_T$ and

$$\sup_{T\in\mathfrak{A}\cup\{e\}} \|Tx\| = \sup_{T\in\mathfrak{A}\cup\{e\}} [\lim_{n\to\infty} \|Tx_n\|] \leq \limsup_{n\to\infty} p(x_n) < \infty .$$

So $x \in I$.

Fix $\varepsilon > 0$. There exists $k \in \mathbb{N}$ such that $p(x_n - x_m) < \varepsilon$ for every $n > k$ and every $m > k$. For every $n > k$, we obtain

$$p(x - x_n) = \sup_{T\in\mathfrak{A}\cup\{e\}} [\lim_{n\to\infty} \|Tx_m - Tx_n\|] \leq \varepsilon .$$

This shows that x is the limit of the sequence (x_n) in I with respect to p, and so I is complete with respect to p.

This achieves the proof of the lemma.

THEOREM 7.7. Let $\mathfrak{a}_1$ and $\mathfrak{a}_2$ be two commutative Banach algebras such that $x\mathfrak{a}_1 \neq \{0\}$ for every nonzero $x \in \mathfrak{a}_1$ and $y\mathfrak{a}_2 \neq \{0\}$ for every nonzero $y \in \mathfrak{a}_2$, and such that $[x\mathfrak{a}_1]^- = \mathfrak{a}_1$, $[y\mathfrak{a}_2]^- = \mathfrak{a}_2$ for some $x \in \mathfrak{a}_1$ and some $y \in \mathfrak{a}_2$. Then $\mathfrak{a}_1$ and $\mathfrak{a}_2$ are similar if and only if there exists a pseudo-bounded isomorphism φ from $\mathfrak{M}(\mathfrak{a}_1)$ onto $\mathfrak{M}(\mathfrak{a}_2)$ such that φ^{-1} is pseudo-bounded.

Proof. Assume that $\mathfrak{a}_1$ and $\mathfrak{a}_2$ are similar. There exists a commutative Banach algebra $\mathfrak{D}$ and two homomorphisms $\varphi_1 : \mathfrak{D} \to \mathfrak{a}_1$ and $\varphi_2 : \mathfrak{D} \to \mathfrak{a}_2$ satisfying the conditions of Definition 7.4. It follows from the last assertion of Proposition 6.3 that $\hat{\varphi}_2$ is a pseudo-bounded isomorphism from $\mathfrak{M}(\mathfrak{D})$ onto $\mathfrak{M}(\mathfrak{a}_2)$ and that $\hat{\varphi}_1^{-1}$ is a pseudo-bounded isomorphism from $\mathfrak{M}(\mathfrak{a}_1)$ onto $\mathfrak{M}(\mathfrak{D})$. Also, $\hat{\varphi}_1$ and $\hat{\varphi}_2^{-1}$ are pseudo-bounded so $\hat{\varphi}_2 \circ \hat{\varphi}_1^{-1}$ gives the desired isomorphism from $\mathfrak{M}(\mathfrak{a}_1)$ into $\mathfrak{M}(\mathfrak{a}_2)$.

Now assume that there exists a pseudo-bounded isomorphism φ from $\mathfrak{M}(\mathfrak{a}_1)$ onto $\mathfrak{M}(\mathfrak{a}_2)$ such that φ^{-1} is pseudo-bounded. Denote by S_1 the closed unit ball of $\mathfrak{a}_1$ and by S_2 the closed unit ball of $\mathfrak{a}_2$. Then $\varphi(S_1)$ is a pseudo-bounded subset of $\mathfrak{M}(\mathfrak{a}_2)$ stable under products, and $\varphi^{-1}(S_2)$ is a pseudo-bounded subset of $\mathfrak{M}(\mathfrak{a}_1)$ stable under products. Put

$$I = \{x \in \bigcap_{a\in S_1} \mathfrak{D}_{\varphi(a)} \mid \sup_{a\in S_1} \|\varphi(a)x\| < \infty\} ,$$

$$J = \{y \in \bigcap_{b\in S_2} \mathfrak{D}_{\varphi^{-1}(b)} \mid \sup_{b\in S_2} \|\varphi^{-1}(b)y\| < \infty\} .$$

Also put

$$p(x) = \sup_{a\in S_1\cup\{e\}} \|\varphi(a)x\| \quad (x \in I) \quad \text{and} \quad q(y) = \sup_{b\in S_2\cup\{e\}} \|\varphi^{-1}(b)y\| \quad (y \in J) .$$

It follows from the lemma that I is an ideal of $\mathfrak{A}_2$ which is complete with respect to p, and that J is an ideal of $\mathfrak{A}_1$ which is complete with respect to q. Also $\varphi(a)x \in I$ for every $a \in S_1$, hence for every $a \in \mathfrak{A}_1$, and $\varphi^{-1}(b)y \in J$ for every $b \in S_2$, hence for every $b \in \mathfrak{A}_2$. Note that $[u\mathfrak{A}_2]^- = \mathfrak{A}_2$, $[v\mathfrak{A}_1]^- = \mathfrak{A}_1$ for some $u \in I$ and some $v \in J$, and that $p(x) \geq \|x\|$ $(x \in I)$, $q(y) \geq \|y\|$ $(y \in J)$, $p(xz) \leq p(x)\|z\|$ $(x \in I,\ z \in \mathfrak{A}_2)$, $q(yz) \leq q(y)\|z\|$ $(y \in J,\ z \in \mathfrak{A}_1)$, $p(\varphi(a)x) \leq \|a\|p(x)$ $(x \in I,\ a \in \mathfrak{A}_1)$, $q(\varphi^{-1}(b)y) \leq \|b\|q(y)$ $(y \in J,\ b \in \mathfrak{A}_2)$.

Put $\mathfrak{D} = J \cap \varphi^{-1}(I)$. We define a norm over $\mathfrak{D}$ by the formula:

$$r(v) = \max\{q(v), p[\varphi(v)]\} \qquad (v \in \mathfrak{D}).$$

Let (v_n) be a Cauchy sequence in $\mathfrak{D}$. Then (v_n) is a Cauchy sequence in J and $(\varphi(v_n))$ is a Cauchy sequence in I. Denote by v the limit of (v_n) in J and denote by u the limit of $(\varphi(v_n))$ in I. Since $q(v_n - v) \xrightarrow[n\to\infty]{} 0$, there exists a sequence (λ_n) of positive reals such that $\liminf_{n\to\infty} \lambda_n = \infty$ and $\sup_{n\in\mathbb{N}} \lambda_n q(v_n - v) < \infty$. Thus, $\sup_{n\in\mathbb{N}} \|\lambda_n(v_n - v)\| < \infty$, and the sequence $(\lambda_n(v_n - v))$ is pseudo-bounded. Hence $(\lambda_n\varphi(v_n) - \lambda_n\varphi(v))_{n\in\mathbb{N}}$ is pseudo-bounded, and there exists $z \in \mathfrak{A}_2$ such that $[z\mathfrak{A}_2]^- = \mathfrak{A}_2$, $z \in \bigcap_{n\in\mathbb{N}} \mathfrak{D}_{\varphi(v_n)-\varphi(v)}$, $\sup_{n\in\mathbb{N}} \lambda_n\|[\varphi(v_n) - \varphi(v)]z\| < \infty$. Since $\liminf_{n\to\infty} \lambda_n = \infty$, $\|[\varphi(v_n) - \varphi(v)]z\| \xrightarrow[n\to\infty]{} 0$. Let $x \in I$ such that $[x\mathfrak{A}_2]^- = \mathfrak{A}_2$. Then $xz \in I$, $[xz\mathfrak{A}_2]^- = \mathfrak{A}_2$ and $\|\varphi(v_n)xz - \varphi(v)xz\| \xrightarrow[n\to\infty]{} 0$. Since $q(u - \varphi(v_n)) \xrightarrow[n\to\infty]{} 0$, we have $\|\varphi(v_n)xz - uxz\| \xrightarrow[n\to\infty]{} 0$ and $\varphi(v)xz = uxz$. It follows from Corollary 2.7 that $u = \varphi(v)$. So $v \in \mathfrak{D}$, $r(v - v_n) \xrightarrow[n\to\infty]{} 0$ and $(\mathfrak{D}, r)$ is a Banach algebra.

We know that J is an ideal of $\mathfrak{A}_1$. Now if $v \in \mathfrak{D}$ and $a \in \mathfrak{A}_1$, then $av \in J$ and $\varphi(v) \in I$. So $\varphi(av) = \varphi(a)\varphi(v) \in I$, and $av \in J \cap \varphi^{-1}(I) = \mathfrak{D}$. This shows that $\mathfrak{D}$ is an ideal of $\mathfrak{A}_1$. As $r(v) \geq q(v) \geq \|v\|$ for every $v \in \mathfrak{D}$, the injection from $\mathfrak{D}$ into $\mathfrak{A}_1$ is continuous. Since $r(v) \geq p[\varphi(v)] \geq \|\varphi(v)\|$ for every $v \in \mathfrak{D}$, φ is a continuous homomorphism from $\mathfrak{D}$ into $\mathfrak{A}_2$. Note that $\varphi(\mathfrak{D}) = I \cap \varphi(J)$, so the relation between $\varphi(\mathfrak{D})$ and $\mathfrak{A}_2$ is the same as the relation between $\mathfrak{D}$ and $\mathfrak{A}_1$, and $\varphi(\mathfrak{D})$ is an ideal of $\mathfrak{A}_2$.

Let x be an element of I such that $[x\mathfrak{A}_2]^- = \mathfrak{A}_2$, and let y be an element of J such that $[y\mathfrak{A}_1]^- = \mathfrak{A}_1$. Then $[x^2\mathfrak{A}_2]^- = \mathfrak{A}_2$, and there exists a sequence (α_n) of elements of $\mathfrak{A}_2$ such that $\|x(e - x\alpha_n)\| \xrightarrow[n\to\infty]{} 0$. So there exists a sequence (λ_n) of positive reals such that $\liminf_{n\to\infty} \lambda_n = \infty$, $\sup_{n\in\mathbb{N}} \lambda_n x(e - x\alpha_n) < \infty$, and the sequence $(\lambda_n e - \lambda_n x\alpha_n)$ is pseudo-bounded. So $\varphi^{-1}(\lambda_n e - \lambda_n x\alpha_n)$ is pseudo-bounded, and there exists $z \in \mathfrak{A}_1$, with $[z\mathfrak{A}_1]^- = \mathfrak{A}_1$, such that $\limsup_{n\to\infty} \|\lambda_n z - \lambda_n\varphi^{-1}(x\alpha_n)z\| < \infty$. We may assume that $z \in J$, and we obtain

$\|yz - [\varphi^{-1}(x)y][\varphi^{-1}(\alpha_n)z]\| \underset{n\to\infty}{\longrightarrow} 0$. Thus, $yz \in [\varphi^{-1}(x)y\mathfrak{A}_1]^-$. Note that $\varphi^{-1}(x)y \in J$, as $y \in J$, $x \in \mathfrak{A}_2$, and that $\varphi(\varphi^{-1}(x)y) = \varphi(y)x \in I$, as $x \in I$, $y \in \mathfrak{A}_1$. So $\varphi^{-1}(x)y \in \mathfrak{D}$. Since $[yz\mathfrak{A}_1]^- = \mathfrak{A}_1$, we see that $\mathfrak{D}$ is dense in $\mathfrak{A}_1$. Since $\varphi(\mathfrak{D})$ plays with respect to $\mathfrak{A}_2$ the same role as $\mathfrak{D}$ with respect to $\mathfrak{A}_1$, the ideal $\varphi(\mathfrak{D})$ is dense in $\mathfrak{A}_2$, and $[x\varphi(y)\mathfrak{A}_2]^- = \mathfrak{A}_2$.

Denote by $\mathfrak{L}$ the closure of $\varphi^{-1}(y)x\mathfrak{D}$ in $\mathfrak{D}$ with respect to r, and put $\beta = \varphi^{-1}(y)x$. It follows from the fifth assertion of Proposition 7.3 that $\mathfrak{L}$ is a dense ideal of $\mathfrak{A}_1$, that $\varphi(\mathfrak{L})$ is a dense ideal of $\mathfrak{A}_2$, and that the closure of $\beta^2\mathfrak{L}$ with respect to r equals $\mathfrak{L}$. Since $\beta^2 \in \mathfrak{L}$, $\mathfrak{A}_1$ and $\mathfrak{A}_2$ are similar and the theorem is proved.

The following consequence of Proposition 7.3 gives some relationship between the sets of closed ideals for similar Banach algebras.

THEOREM 7.8. Let $\mathfrak{A}$ and $\mathfrak{B}$ be two similar commutative Banach algebras.

(1) If $\mathfrak{A}$ is topologically simple, and if $b\mathfrak{B} \neq \{0\}$ for every nonzero $b \in \mathfrak{B}$, then $\mathfrak{B}$ is topologically simple.

(2) If $\mathfrak{A}$ and $\mathfrak{B}$ possess bounded approximate identities, there exists a bijection between the set of all closed ideals of $\mathfrak{A}$ and the set of all closed ideals of $\mathfrak{B}$.

Proof. (1) Assume that $\mathfrak{A}$ is topologically simple, and that $b\mathfrak{B} \neq \{0\}$ for every nonzero element b of $\mathfrak{B}$. Let y be any nonzero element of $\mathfrak{B}$, and let $x \in \mathfrak{D}$ be such that $[x\mathfrak{D}]^- = \mathfrak{D}$, where $\mathfrak{D}$, $\varphi : \mathfrak{D} \to \mathfrak{A}$ and $\psi : \mathfrak{D} \to \mathfrak{B}$ are the Banach algebras and the continuous homomorphisms given by Definition 7.4. Then $\psi(x\mathfrak{D})$ is dense in $\mathfrak{B}$, so $\psi(x)y \neq 0$. Put $z = \psi^{-1}(\psi(x)y)$. Then $z \neq 0$ and $\varphi(z) \neq 0$, so $[\varphi(z)\varphi(\mathfrak{D})]^- = [\varphi(z)\mathfrak{A}]^- = \mathfrak{A}$. In particular, $\varphi(x) \in [\varphi(z)\varphi(\mathfrak{D})]^-$. Using Lemma 6.1, we see that $x^2 \in [xz\mathfrak{D}]^- \subset [z\mathfrak{D}]^-$. So $[z\mathfrak{D}]^- = [x^2\mathfrak{D}]^- = \mathfrak{D}$, and the closure of $\psi(z)\mathfrak{B}$ in $\mathfrak{B}$ contains $[\psi(\mathfrak{D})]^- = \mathfrak{B}$. So $[\psi(x)y\mathfrak{B}]^- = \mathfrak{B}$ and $[y\mathfrak{B}]^- = \mathfrak{B}$. This shows that $\mathfrak{B}$ is topologically simple.

(2) Assume now that $\mathfrak{A}$ and $\mathfrak{B}$ possess bounded approximate identities, and let $\mathfrak{D}$, φ and ψ be as above. Denote by $\mathcal{J}(\mathfrak{A})$, $\mathcal{J}(\mathfrak{B})$ and $\mathcal{J}(\mathfrak{D})$ the sets of all closed ideals of $\mathfrak{A}$, $\mathfrak{B}$, $\mathfrak{D}$. Put $\theta(I) = \varphi^{-1}(I)$ for every $I \in \mathcal{J}(\mathfrak{A})$, $\rho(J) = \overline{\varphi(J)}$ for every $J \in \mathcal{J}(\mathfrak{D})$. Using assertions (2) and (4) of Proposition 7.3 we see that $(\rho \circ \theta)(I) = I$ for every $I \in \mathcal{J}(\mathfrak{A})$ and that $(\theta \circ \rho)(J) = J$ for every $J \in \mathcal{J}(\mathfrak{D})$. So θ is a bijection from $\mathcal{J}(\mathfrak{A})$ onto $\mathcal{J}(\mathfrak{D})$. There exists similarly a bijection from $\mathcal{J}(\mathfrak{B})$ onto $\mathcal{J}(\mathfrak{D})$, and the theorem is proved.

We now introduce the following definition.

DEFINITION 7.9. Let $\mathfrak{A}$ and $\mathfrak{B}$ be two similar commutative Banach algebras, and let θ be a continuous homomorphism from $\mathfrak{A}$ into $\mathfrak{B}$. We will say that θ is an <u>s-homomorphism</u> if there exists a Banach algebra $\mathfrak{D}$ and two continuous homomorphisms $\varphi : \mathfrak{D} \to \mathfrak{A}$ and $\psi : \mathfrak{D} \to \mathfrak{B}$ satisfying the conditions of Definition 7.4 such that the diagram

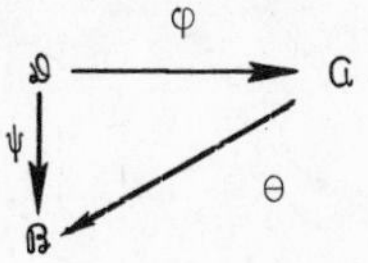

is commutative.

REMARK 7.10. Let $\mathfrak{A}$ and $\mathfrak{B}$ be two similar commutative Banach algebras, and let θ be a s-homomorphism from $\mathfrak{A}$ into $\mathfrak{B}$.

(1) $\theta(\mathfrak{A})$ is dense in $\mathfrak{B}$. Also, θ is one-to-one if $a\mathfrak{A} = \{0\}$ implies $a = 0$.

(2) If $a\mathfrak{A} = \{0\}$ implies $a = 0$ and if $b\mathfrak{B} = \{0\}$ implies $b = 0$, then $\hat{\theta}$ is a pseudo-bounded isomorphism from $\mathfrak{M}(\mathfrak{A})$ onto $\mathfrak{M}(\mathfrak{B})$, and $\hat{\theta}^{-1}$ is pseudo-bounded.

(3) If $\mathfrak{A}$ and $\mathfrak{B}$ possess bounded approximate identities, the map $I \mapsto [\theta(I)]^-$ is a bijection from the set of all closed ideals of $\mathfrak{A}$ onto the set of all closed ideals of $\mathfrak{B}$.

<u>Proof</u>. (1) $\theta[\varphi(\mathfrak{A})] = \psi(\mathfrak{D})$, and $\theta(\varphi(\mathfrak{D})) \subset \theta(\mathfrak{A})$. Now let $x \in \ker\theta$. Then $\theta(x\varphi(\mathfrak{D})) = \{0\}$, hence $\psi[\varphi^{-1}(x\varphi(\mathfrak{D}))] = \{0\}$, and $x\varphi(\mathfrak{D}) = \{0\}$. Since $\overline{\varphi(\mathfrak{D})} = \mathfrak{A}$, $x\mathfrak{A} = \{0\}$, and so $x = 0$, as required.

(2) If $T = T_{a/b} \in \mathfrak{M}(\mathfrak{D})$, then $(\hat{\theta} \circ \hat{\varphi})(T) = T_{u/v}$, where $u = \hat{\theta}(\hat{\varphi}(a)) = \psi(a)$ and $v = \theta(\varphi(b)) = \psi(b)$. So $\hat{\theta} \circ \hat{\varphi} = \hat{\psi}$ and $\hat{\theta} = \hat{\psi} \circ \hat{\varphi}^{-1}$. Since $\hat{\psi}$, $\hat{\varphi}$, $\hat{\varphi}^{-1}$, $\hat{\psi}^{-1}$ are all pseudo-bounded, $\hat{\theta}$ is pseudo-bounded. Also $\hat{\theta}$ is an isomorphism, and $\hat{\theta}^{-1} = \hat{\varphi} \circ \hat{\psi}^{-1}$ is pseudo-bounded too.

(3) Denote by $\mathfrak{J}(\mathfrak{A})$, $\mathfrak{J}(\mathfrak{B})$, $\mathfrak{J}(\mathfrak{D})$ the set of all closed ideals of $\mathfrak{A}$, $\mathfrak{B}$, $\mathfrak{D}$. Also, put $\tilde{\varphi}(I) = \overline{\varphi(I)}$, $\tilde{\psi}(I) = \overline{\psi(I)}$ for every $I \in \mathfrak{J}(\mathfrak{D})$, $\tilde{\theta}(J) = \overline{\theta(J)}$ for every $J \in \mathfrak{J}(\mathfrak{D})$. We saw in the proof of Theorem 7.8 that $\tilde{\varphi}$ is a bijection from $\mathfrak{J}(\mathfrak{D})$ onto $\mathfrak{J}(\mathfrak{A})$, and that $\tilde{\psi}$ is a bijection from $\mathfrak{J}(\mathfrak{D})$ onto $\underline{\mathfrak{J}(\mathfrak{B})}$. If Δ is any subset of $\mathfrak{A}$, $\theta(\overline{\Delta}) \subset \overline{\theta(\Delta)}$, so that $\theta(\overline{\Delta})$ actually equals $\theta(\overline{\Delta})$. Hence $\tilde{\psi}(I) = [\theta(\varphi(I))]^- = [\theta(\varphi(I)^-)]^- = \tilde{\theta}[\tilde{\varphi}(I)]$ for every $I \in \mathfrak{J}(\mathfrak{D})$, $\tilde{\theta} \circ \tilde{\varphi} = \tilde{\psi}$, $\tilde{\theta} = \tilde{\psi} \circ \tilde{\varphi}^{-1}$, and $\tilde{\theta}$ is a bijection.

The following theorem shows in particular that regular quasi-multipliers can become multipliers after some suitable change of the algebra.

THEOREM 7.11. Let $\mathfrak{A}$ be a commutative Banach algebra such that $[a\mathfrak{A}]^- = \mathfrak{A}$ for some $a \in \mathfrak{A}$ and such that $x\mathfrak{A} \neq \{0\}$ for every nonzero $x \in \mathfrak{A}$, and let $\mathfrak{U}$ be a pseudo-bounded subset of $\mathfrak{M}(\mathfrak{A})$ which is stable under products. Then there exists a commutative Banach algebra $\mathfrak{B}$ similar to $\mathfrak{A}$ such that $y\mathfrak{B} \neq \{0\}$ for every nonzero $y \in \mathfrak{B}$, and a norm-decreasing s-homomorphism θ from $\mathfrak{A}$ into $\mathfrak{B}$ satisfying the following properties:

(1) $\hat{\theta}(\mathfrak{U})$ is contained in the unit ball of $\mathfrak{M}(\mathfrak{B})$;

(2) the restriction of $\hat{\theta}$ to $\mathfrak{M}(\mathfrak{A})$ is a norm-decreasing homomorphism from $\mathfrak{M}(\mathfrak{A})$ into $\mathfrak{M}(\mathfrak{B})$;

(3) if $T \in \mathfrak{U} \cup \{S \in \mathfrak{M}(\mathfrak{A}) \mid \|S\|_{op} \leq 1\}$, and if $\hat{\theta}(T) \in \mathfrak{B}$, then $\|\hat{\theta}(T)\| \leq 1$.

Proof. Note that we do not assume that $\mathfrak{A}$ possesses a b.a.i., so that $\mathfrak{M}(\mathfrak{A})$ is equipped with the norm $\|S\|_{op} = \sup\{\|S(x)\|/\|x\| : x \in \mathfrak{A}\setminus\{0\}\}$ which may be weaker than the given norm over $\mathfrak{A}$.

Put $\mathfrak{L} = \{x \in \bigcap_{T\in\mathfrak{U}} \mathfrak{D}_T \mid \sup_{x\in\mathfrak{U}} \|Tx\| < \infty\}$. Also put $p(x) = \sup_{T\in\mathfrak{U}\cup\{e\}} \|Tx\|$ for every $x \in I$.

It follows from Lemma 7.6 that $\mathfrak{L}$ is a dense ideal of $\mathfrak{A}$ which is complete with respect to p. Also $T(\mathfrak{L}) \subset \mathfrak{L}$ for every $T \in \mathfrak{U}$. We have $\|x\| \leq p(x)$ for every $x \in \mathfrak{L}$, $p(xy) \leq p(x)\|y\|$ for every $x \in \mathfrak{L}$ and every $y \in \mathfrak{A}$, $p(Tx) \leq p(x)$ for every $x \in \mathfrak{L}$ and every $T \in \mathfrak{U}$. Note that $[a\mathfrak{A}]^- = \mathfrak{A}$ for some $a \in \mathfrak{L}$.

Set

$$q(y) = \sup\left\{\frac{p(xy)}{p(x)} : x \in \mathfrak{L}\setminus\{0\}\right\} \quad (y \in \mathfrak{A}) .$$

Then q is an algebra norm over $\mathfrak{A}$ since $ay = 0$ implies that $\mathfrak{A}y = \{0\}$, and $q(y) \leq \|y\|$ for every $y \in \mathfrak{A}$. Denote by $\mathfrak{B}$ the completion of $\mathfrak{A}$ with respect to q. Let $x \in \mathfrak{L}$ and $b \in \mathfrak{B}$. Then there exists a sequence (u_n) of elements of $\mathfrak{A}$ such that $q(b - u_n) \xrightarrow[n\to\infty]{} 0$. It follows from the definition of q that $p(xy) \leq p(x)q(y)$ for every $y \in \mathfrak{A}$, so that the sequence $(xu_n)_{n\in\mathbb{N}}$ is a Cauchy sequence in $\mathfrak{L}$ with respect to p. Denote by z its limit. We have $\limsup_{n\to\infty} q(bx - u_n x) = 0$ and $\limsup_{n\to\infty} q(z - u_n x) \leq \limsup_{n\to\infty} p(z - u_n x) = 0$. So $z = bx$, and we see that $\mathfrak{L}$ is an ideal of $\mathfrak{B}$.

Let $b \in \mathfrak{A}$ such that $[b\mathfrak{A}]^- = \mathfrak{A}$. Then $b\mathfrak{A}$ is a fortiori dense in $\mathfrak{A}$ with respect to q, so that $b\mathfrak{B}$ is dense in $\mathfrak{B}$ with respect to q.

Put $c = b^2$ and denote by $\mathfrak{D}$ the closure of $b\mathfrak{L}$ in $\mathfrak{L}$ with respect to p. It follows from the fifth assertion of Proposition 7.3 that $\mathfrak{D}$ is an ideal of $\mathfrak{A}$ which is dense in $\mathfrak{A}$ with respect to the norm $\|.\|$ and that $\mathfrak{D}$ is also an ideal of $\mathfrak{B}$ which is dense in $\mathfrak{B}$ with respect to q.

Note also that $c \in \mathfrak{D}$ and that $c\mathfrak{D}$ is dense in $\mathfrak{D}$ with respect to p. This shows that $\mathfrak{A}$ and $\mathfrak{B}$ are similar, and the natural injection θ from $\mathfrak{A}$ to $\mathfrak{B}$ is clearly a norm-decreasing s-homormorphism.

We have $q(y) = \sup_{\substack{p(x)\leq 1 \\ x\in\mathfrak{L}}} p(yx)$ for every $y \in \mathfrak{B}$. So, if $y\mathfrak{B} = \{0\}$, then $y\mathfrak{L} = \{0\}$ and $q(y) = 0$, whence $y = 0$.

Let $T \in \mathfrak{M}(\mathfrak{A})$. Assume either that $T \in \mathfrak{U}$ or that T belongs to the closed unit ball of $\mathfrak{M}(\mathfrak{A})$. Put $\hat{T} = \hat{\theta}(T)$. The definitions of $\mathfrak{D}_T$ and $\mathfrak{D}_{\hat{T}}$ are purely algebraic, so $\mathfrak{L} \subset \mathfrak{D}_{\hat{T}}$ and $\hat{T}x = Tx$ for every $x \in \mathfrak{L}$. So

$$q(\hat{T}x) = q(Tx) = \sup_{\substack{y\in\mathfrak{L} \\ y\neq 0}} \frac{p[Txy]}{p(y)} \leq \sup_{\substack{y\in\mathfrak{L} \\ y\neq 0}} \frac{p(xy)}{p(y)} = q(x) \quad (x \in \mathfrak{L}) .$$

Let $y \in \mathfrak{B}$. There exists a sequence (x_n) of elements of $\mathfrak{L}$ such that $q(y - x_n) \xrightarrow[n\to\infty]{} 0$. The above inequality shows that $(\hat{T}x_n)$ is a Cauchy sequence in $\mathfrak{B}$. Since $\hat{T}$ is a closed generalized map over $\mathfrak{B}$, $y \in \mathfrak{D}_{\hat{T}}$ and $\hat{T} \in \mathfrak{M}(\mathfrak{B})$. So $q(\hat{T}x) \leq q(x)$ for every $x \in \mathfrak{B}$ and $\|\hat{T}\|_{op} \leq 1$.

Now assume either that $T \in \mathfrak{U}$ or that $T \in \{S \in \mathfrak{M}(\mathfrak{A}) \mid \|S\|_{op} \leq 1\}$ and that $\hat{T} \in \mathfrak{B}$. Since $\mathfrak{B}$ is the completion of $\mathfrak{A}$ with respect to q, we have $q(\hat{T}) = \sup_{x\in\mathfrak{L}} p(\hat{T}x)/p(x) \leq 1$. This achieves the proof of the theorem.

COROLLARY 7.12. Let $\mathfrak{A}$ be a commutative Banach algebra. If $\mathfrak{A}$ possesses a nonzero continuous semigroup $(a^t)_{t>0}$ over the positive reals, and if $x\mathfrak{A} \neq \{0\}$ for every $x \in \mathfrak{A}$, then $I = [\bigcup a^t\mathfrak{A}]^-$ is similar to a commutative Banach algebra with b.a.i. More precisely, there exists a positive real α and a norm-decreasing s-homomorphism θ from I into a commutative Banach algebra $\mathfrak{B}$ with b.a.i. similar to I such that $\hat{\theta} \mid \mathfrak{M}(I)$ is a norm-decreasing homomorphism from $\mathfrak{M}(I)$ into $\mathfrak{M}(\mathfrak{B})$ and such that $\sup_{t>0} \|\exp(-\alpha t)\theta(a^t)\| \leq 1$.

<u>Proof</u>. It follows from [31], Theorem 7.6.1, that $\limsup_{n\to\infty} \|a^t\|^{1/t} < \infty$. Let $\alpha > \limsup_{n\to\infty} \|a^t\|^{1/t}$, and put $b^t = \exp(-\alpha t)a^t$ $(t>0)$. Then $(b^t)_{t>0}$ is a continuous semigroup in $\mathfrak{A}$, $I = [\bigcup_{t>0} b^t\mathfrak{A}]^-$, and $\sup_{t\geq s} \|b^t\| < \infty$ for every $s > 0$.

Put $\mathfrak{L} = \{x \in \mathfrak{A} \mid \sup_{t>0} \|xb^t\| < \infty\}$. Then $\mathfrak{L}$ contain b^t for every $t > 0$.

Put $p(x) = \max(\|x\|, \sup_{t>0}\|xb^t\|)$ $(x \in \mathfrak{L})$. Exactly as in the proof of Lemma 7.6, we see that $\mathfrak{L}$ is complete with respect to p and that $\mathfrak{L}$ is an ideal of $\mathfrak{A}$. Also, the norm of $\mathfrak{L}$ satisfies $p(xy) \leq p(x)\|y\|$ $(x \in \mathfrak{L}, y \in \mathfrak{A})$. So $\|b^t - b^{t+h}\| \leq p(b^{t/2})\|b^{t/2} - b^{(t/2)+h}\|$ if $t>0$, $h > -t/2$, and $(b^t)_{t>0}$ is a continuous semigroup in $(\mathfrak{L},p)$.

Put $\mathfrak{D} = [\bigcup_{t>0} b^t\mathfrak{L}]^-$, the closure being taken with respect to p. Then $b^t \in \mathfrak{D}$, and in fact $\mathfrak{D} = [\bigcup_{t>0} b^t\mathfrak{L}]^- \subset [\bigcup_{t>0} b^{t/2}\mathfrak{D}]^- = [\bigcup_{t>0} b^t\mathfrak{D}]^-$. It follows from [25], Section 6, that there exists in $\mathfrak{D}$ another continuous semigroup

$(c^t)_{t>0}$ such that $[c^s\mathfrak{D}]^- = [\bigcup_{t>0} b^t\mathfrak{D}]^- = \mathfrak{D}$ for every $s > 0$. Since $b^{t/2} \in \mathfrak{D}$ $b^t\mathfrak{A} \subset \mathfrak{D}$ for every $t > 0$, so $I \subset \overline{\mathfrak{D}}$. As $\mathfrak{D} \subset I$, we see that $\mathfrak{D}$ is dense in I with respect to the norm of $\mathfrak{A}$. But $c^s\mathfrak{D}$ is dense in $\mathfrak{D}$ with respect to the norm of $\mathfrak{A}$, so that $[c^sI]^- = I$ for every $s > 0$. Since $c^s \in \mathfrak{L}$, $\sup_{t>0} \|c^s b^t\| < \infty$, and we see that $(b^t)_{t>0}$ is pseudo-bounded subset of I stable under products.

It follows from the theorem that there exists a commutative Banach algebra $\mathfrak{B}$ similar to I and a norm-decreasing s-homomorphism θ from I into $\mathfrak{B}$ such that $\hat{\theta}\,|\,\mathfrak{M}(I)$ is a norm-decreasing homomorphism from $\mathfrak{M}(I)$ into $\mathfrak{M}(\mathfrak{B})$ and such that $\|\theta(b^t)\| \leq 1$ for every $t > 0$. Since θ is a s-homomorphism, $\theta(I)$ is dense in $\mathfrak{B}$. Also $\|\theta(b^t)\theta(b^{1/n}) - \theta(b^t)\| \xrightarrow[n\to\infty]{} 0$ for every $t > 0$. Since $\|\theta(b^{1/n})| \leq 1$ for every n, $\|x\theta(b^{1/n}) - x\| \xrightarrow[n\to\infty]{} 0$ for every $x \in [\bigcup_{t>0} \theta(b^t)\mathfrak{B}]^- = \mathfrak{B}$, and $\mathfrak{B}$ possesses a b.a.i.

Note that $\|\theta(b^{1/n})\| \geq \|\theta(b^1)\|^{1/n}$ for every n, so $\sup_{n\in\mathbb{N}} \|\theta(b^{1/n})\| \geq 1$, and in fact $\sup_{t>0} \|\theta(b^t)\| = 1$. This achieves the proof of the corollary.

We want now to describe the mathematical object obtained when equipping $\mathfrak{D}\mathfrak{M}_r(\mathfrak{A})$ with the following family of "bounded" sets.

DEFINITION 7.13. A subset U of $\mathfrak{D}\mathfrak{M}_r(\mathfrak{A})$ is said to be <u>multiplicatively pseudo-bounded (m-pseudo-bounded)</u> if there exists a pseudo-bounded set V of $\mathfrak{D}\mathfrak{M}_r(\mathfrak{A})$ stable under products and a real $\lambda > 0$ such that $U \subset \lambda V$. The family of all m-pseudo-bounded subsets of $\mathfrak{D}\mathfrak{M}_r(\mathfrak{A})$ is denoted by $\mathfrak{F}_{\mathfrak{A}}$.

A convex bornological algebra $\mathfrak{A}$ (see Section 2) is said to be <u>multiplicatively convex</u> if $(\mathfrak{A},\Delta) = \varinjlim(E_\alpha, \|.\|_\alpha)$, where $(E_\alpha, \|.\|_\alpha)$ is an inductive family of normed algebras (the "bornological" inductive limit is to be taken in the sense of Section 2, and here Δ denotes the family of all bounded subsets of $\mathfrak{A}$). Such an algebra is said to be <u>complete</u> if $(\mathfrak{A},\Delta) = \varinjlim(E_\alpha, \|.\|_\alpha)$, where $(E_\alpha, \|.\|_\alpha)$ is an inductive family of Banach algebras. The notion of multiplicatively convex complete bornological algebra is exactly equivalent to the notion of pseudo-Banach algebra introduced by Allan, Dales, and McClure in [4], and we prefer to use their terminology here. An abstract result of the theory of convex, complete bornological algebras states that, if the algebra is commutative, then the subalgebra of regular elements (i.e. elements x such that $\{\lambda^n x^n\}_{n\in\mathbb{N}}$ is bounded for some $\lambda > 0$) is a pseudo-Banach algebra with respect to the "bornology" constructed from the given one as in Definition 7.13. (See for example Akkar's thesis [1]). We will not need this abstract result here because we will get a "concrete" description of $(\mathfrak{D}\mathfrak{M}_r(\mathfrak{A}),\mathfrak{F}_{\mathfrak{A}})$ as an inductive limit of Banach algebras.

Let $(\mathfrak{A}_\alpha, \|.\|_\alpha)_{\alpha\in\Lambda}$ be the family of all commutative Banach algebras which possess the following properties (here Λ is a suitable set).

(1) $\mathfrak{A}_\alpha$ is contained in $\mathfrak{M}_r(\mathfrak{A})$ as an algebra.

(2) There exists a Banach algebra $(\mathfrak{D}_\alpha, p_\alpha)$ satisfying

(i) $\mathfrak{D}_\alpha \subset \mathfrak{A} \cap \mathfrak{A}_\alpha$,

(ii) the injections $\mathfrak{D}_\alpha \to \mathfrak{A}$, $\mathfrak{D}_\alpha \to \mathfrak{A}_\alpha$ are continuous,

(iii) $\mathfrak{D}_\alpha$ is both an ideal of $\mathfrak{A}$ and an ideal of $\mathfrak{A}_\alpha$,

(iv) there exists $x \in \mathfrak{D}_\alpha$ such that $x\mathfrak{D}_\alpha$ is dense in $\mathfrak{D}_\alpha$ with respect to p_α.

These conditions imply of course that $\mathfrak{A}_\alpha$ is similar to $\mathfrak{A}$.

It is easy to see that, if $\mathfrak{B}$ is any Banach algebra similar to $\mathfrak{A}$ such that $b\mathfrak{B} \neq \{0\}$ for every nonzero $b \in \mathfrak{B}$, then $\theta^{-1}(\mathfrak{B})$ belongs to the above class, where $\theta : \mathfrak{M}_r(\mathfrak{A}) \to \mathfrak{M}_r(\mathfrak{B})$ is the pseudo-bounded isomorphism associated to the similarity relation between $\mathfrak{A}$ and $\mathfrak{B}$.

REMARK 7.14. (1) $a\mathfrak{A}_\alpha \neq \{0\}$ for every $\alpha \in \Lambda$ and every $a \in \mathfrak{A}_\alpha$.

(2) If $\alpha, \beta \in \Lambda$ and if $\mathfrak{A}_\alpha$ contains $\mathfrak{A}_\beta$, then the injection from $\mathfrak{A}_\beta$ into $\mathfrak{A}_\alpha$ is a continuous s-homomorphism from $\mathfrak{A}_\beta$ into $\mathfrak{A}_\alpha$.

Proof. (1) Let $a \in \mathfrak{A}_\alpha$ such that $a\mathfrak{A}_\alpha = 0$ and let $\mathfrak{D}_\alpha$ be as above. Then $ax = 0$ for every $x \in \mathfrak{D}_\alpha$, and we may choose x such that $x\mathfrak{D}_\alpha$ is dense in $\mathfrak{D}_\alpha$. Hence $x\mathfrak{A}$ is dense in $\mathfrak{A}$. So $a = T_{ax/x} = 0$.

(2) Assume that $\mathfrak{A}_\alpha \supset \mathfrak{A}_\beta$, and let $x \in \mathfrak{A}_\beta$ and $y \in \mathfrak{A}_\alpha$ be such that $\|x - x_n\|_\beta \xrightarrow[n\to\infty]{} 0$ and $\|y - x_n\|_\alpha \xrightarrow[n\to\infty]{} 0$ for some sequence (x_n) of elements of $\mathfrak{A}_\beta$. Let $\mathfrak{D}_\alpha$, $\mathfrak{D}_\beta$ be as above, and let $u \in \mathfrak{D}_\alpha$, $v \in \mathfrak{D}_\beta$ be such that $[u\mathfrak{D}_\alpha]^- = \mathfrak{D}_\alpha$ and $[v\mathfrak{D}_\beta]^- = \mathfrak{D}_\beta$. Then $p_\alpha(ux - ux_n) \xrightarrow[n\to\infty]{} 0$ and $p_\beta(vy - vx_n) \xrightarrow[n\to\infty]{} 0$, and hence $\|uvx - uvx_n\| \xrightarrow[n\to\infty]{} 0$, $\|uvy - uvx_n\| \xrightarrow[n\to\infty]{} 0$ and $uvx = uvy$. Since $[uv\mathfrak{A}]^- = \mathfrak{A}$, it follows from Corollary 2.7 that $x = y$. The continuity of the injection $\mathfrak{A}_\beta \to \mathfrak{A}_\alpha$ follows then from the closed graph theorem.

Now consider the following diagram, where the arrows represent natural injections.

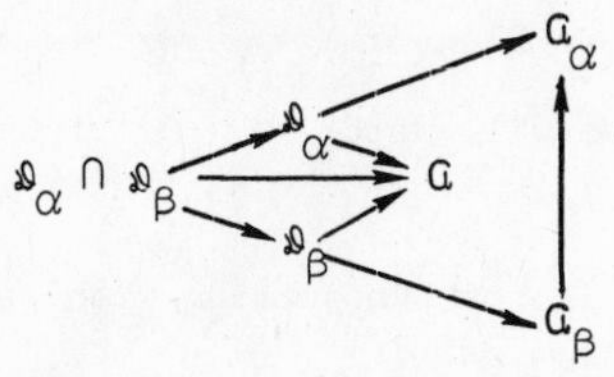

The construction made in the proof of Proposition 7.5 gives an ideal $\mathfrak{D}_{\alpha,\beta}$ of $\mathfrak{A}$ contained in $\mathfrak{D}_\alpha \cap \mathfrak{D}_\beta$ such that $\mathfrak{D}_{\alpha,\beta}$ is both a dense ideal of $(\mathfrak{A}_\alpha, \|.\|_\alpha)$ and a dense ideal of $(\mathfrak{A}_\beta, \|.\|_\beta)$. Also, $\mathfrak{D}_{\alpha,\beta}$ is complete with respect to a norm $p_{\alpha,\beta}$, and possesses principal ideals which are dense with respect to that norm. The diagram

$$
\begin{array}{ccc}
 & & \mathfrak{A}_\alpha \\
\mathfrak{B}_{\alpha,\beta} \to \mathfrak{B}_\alpha \cap \mathfrak{B}_\beta & \nearrow & \uparrow \\
 & \searrow & \mathfrak{A}_\beta
\end{array}
$$

(where arrows represent again the natural injections), shows that the injection from $\mathfrak{A}_\beta$ into $\mathfrak{A}_\alpha$ is an s-homomorphism.

We wish now to identify $Q\mathfrak{M}_r(\mathfrak{A}_\alpha)$ with $Q\mathfrak{M}_r(\mathfrak{A})$. To do this in a canonical way we need the following proposition.

PROPOSITION 7.15. Let $\alpha \in \Lambda$ and let $\mathfrak{B}_\alpha$ be as above. Denote by $\varphi_\alpha : \mathfrak{B}_\alpha \to \mathfrak{A}_\alpha$ and by $\psi_\alpha : \mathfrak{B}_\alpha \to \mathfrak{A}$ the natural injections. Then, if $T \in Q\mathfrak{M}_r(\mathfrak{A}_\alpha)$, $(\hat{\psi}_\alpha \circ \hat{\varphi}_\alpha^{-1})(T)$ does not depend on the choice of $\mathfrak{B}_\alpha$.

Proof. Let $(\mathfrak{B}'_\alpha, p'_\alpha)$ be another Banach algebra contained in $\mathfrak{A} \cap \mathfrak{A}_\alpha$ which possesses the same properties as $(\mathfrak{B}_\alpha, p_\alpha)$, and let $\psi'_\alpha : \mathfrak{B}'_\alpha \to \mathfrak{A}$ and $\varphi'_\alpha : \mathfrak{B}'_\alpha \to \mathfrak{A}_\alpha$ be the natural injections. Then there exists $x \in \mathfrak{B}_\alpha$ and $x' \in \mathfrak{B}'_\alpha$ such that $[x\mathfrak{B}_\alpha]^- = \mathfrak{B}_\alpha$ and $[x\mathfrak{B}'_\alpha]^- = \mathfrak{B}'_\alpha$, the closures being taken respectively with respect to p_α and p'_α. Consider $T \in Q\mathfrak{M}_r(\mathfrak{A}_\alpha)$. We have $T = T_{u/v}$ where $[v\mathfrak{A}_\alpha]^- = \mathfrak{A}_\alpha$. As has been seen several times before, $vx\mathfrak{B}_\alpha$ is dense in $\mathfrak{B}_\alpha$ and $vx'\mathfrak{B}'_\alpha$ is dense in $\mathfrak{B}'_\alpha$ (with respect to p_α and p'_α), so that $[vx\mathfrak{A}]^- = [vx'\mathfrak{A}]^- = \mathfrak{A}$, $(\hat{\psi}_\alpha \circ \hat{\varphi}_\alpha^{-1})(T)(vx) = ux$, and $(\hat{\psi}'_\alpha \circ \hat{\varphi}'^{-1}_\alpha)(T)vx' = ux'$. Thus, the two quasimultipliers have the same action on vxx', and hence are equal by Corollary 2.7.

We now identify $Q\mathfrak{M}_r(\mathfrak{A}_\alpha)$ with $Q\mathfrak{M}_r(\mathfrak{A})$, using the map $T \mapsto \hat{\psi}_\alpha \circ \hat{\varphi}_\alpha^{-1}(T)$. Of course, this identification is an isomorphism with respect to the "bornological" structures of $Q\mathfrak{M}_r(\mathfrak{A})$ and $Q\mathfrak{M}_r(\mathfrak{A}_\alpha)$. In particular, $\mathfrak{M}(\mathfrak{A}_\alpha)$ appears as a subalgebra of $Q\mathfrak{M}_r(\mathfrak{A})$. Note that every bounded subset of $\mathfrak{M}(\mathfrak{A}_\alpha)$ is pseudo-bounded in $Q\mathfrak{M}_r(\mathfrak{A})$.

DEFINITION 7.16. Let $\alpha, \beta \in \Lambda$. We will write $\alpha \geq \beta$ if the two following conditions are satisfied.

(1) $\mathfrak{A}_\beta \subset \mathfrak{A}_\alpha$, and this injection is norm-decreasing with respect to the norms $\|.\|_\alpha$ and $\|.\|_\beta$.

(2) $\mathfrak{M}(\mathfrak{A}_\beta) \subset \mathfrak{M}(\mathfrak{A}_\alpha)$, and this injection is norm-decreasing with respect to the operator norms on $\mathfrak{M}(\mathfrak{A}_\beta)$ and $\mathfrak{M}(\mathfrak{A}_\alpha)$.

The following structure theorem is a consequence of Theorem 7.11.

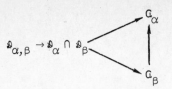

THEOREM 7.17. The system $(\mathfrak{M}(\mathfrak{A}_\alpha))_{\alpha\in\Lambda}$ is inductive, and $\mathfrak{DM}_r(\mathfrak{A})$ is a pseudo-Banach algebra isomorphic with $\varinjlim_{\alpha\in\Lambda} \mathfrak{M}(\mathfrak{A}_\alpha)$.

<u>Proof</u>. Here we use of course, the injections $\theta_{\beta,\alpha} : \mathfrak{M}(\mathfrak{A}_\beta) \to \mathfrak{M}(\mathfrak{A}_\alpha)$ $(\beta,\alpha \in \Lambda,\ \beta \leq \alpha)$ to define the inductive limit. The inductive limit $\varinjlim_{\alpha\in\Lambda} \mathfrak{M}(\mathfrak{A}_\alpha)$ may be identified as a set with $\bigcup_{\alpha\in\Lambda} \mathfrak{M}(\mathfrak{A}_\alpha)$. (Recall that we identified above $\mathfrak{DM}_r(\mathfrak{A}_\alpha)$ with $\mathfrak{DM}_r(\mathfrak{A})$ for every $\alpha \in \Lambda$.)

Theorem 7.11 shows that for every pseudo-bounded subset U of $\mathfrak{DM}_r(\mathfrak{A})$ stable under products there exists a norm-decreasing s-homomorphism θ from $\mathfrak{A}$ into a Banach algebra $\mathfrak{B}$ similar to $\mathfrak{A}$ and satisfying $b\mathfrak{B} \neq \{0\}$ for every $b \in \mathfrak{B}$ such that $\hat{\theta}(U)$ is contained in the unit ball of $\mathfrak{M}(\mathfrak{B})$. We have the following commutative diagram:

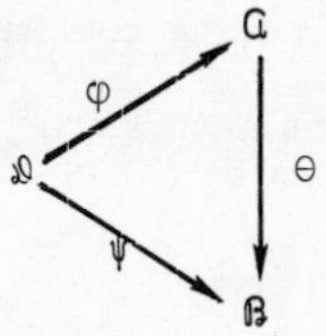

Here, $\varphi(\mathfrak{D})$ is a dense ideal of $\mathfrak{A}$, and $\psi(\mathfrak{B})$ is a dense ideal of $\mathfrak{B}$. We may identify $\mathfrak{D}$ with an ideal of $\mathfrak{A}$ ($\mathfrak{D}$ is actually constructed as an ideal of $\mathfrak{A}$ in the proof of Theorem 7.11.) Now put $\mathfrak{C} = \hat{\theta}^{-1}(\mathfrak{B})$. Then $\mathfrak{C}$ is a subalgebra of $\mathfrak{DM}_r(\mathfrak{A})$ containing $\mathfrak{A}$, and hence it contains $\mathfrak{D}$. Put $r(x) = q(\hat{\theta}(x))$ for $x \in \mathfrak{C}$. (We denote by p the norm of $\mathfrak{D}$ and by q the norm of $\mathfrak{B}$, as in the proof of Theorem 7.11). Then $(\mathfrak{C},r)$ is of course a Banach algebra, and $\mathfrak{D}$ is an ideal of $\mathfrak{C}$ which is dense in $\mathfrak{C}$ with respect to r.

Denote by i the injection from $\mathfrak{A}$ into $\mathfrak{C}$. Then $\hat{i}(U) \subset \mathfrak{M}(\mathfrak{C})$. Identifying $\mathfrak{DM}_r(\mathfrak{C})$ as before to its image in $\mathfrak{DM}_r(\mathfrak{A})$ via $\hat{i}^{-1}$, we obtain $U \subset \mathfrak{M}(\mathfrak{C})$ and $\mathfrak{C}$ belongs to the family $(\mathfrak{A}_\alpha)_{\alpha\in\Lambda}$. Also, $r_{op}(T) \leq 1$ for every $T \in U$.

Now, if V is any multiplicatively pseudo-bounded subset of $\mathfrak{DM}_r(\mathfrak{A})$, there exists a pseudo-bounded subset U of $\mathfrak{DM}_r(\mathfrak{A})$ stable under products and there exists $\lambda > 0$ such that $V \subset \lambda U$. We thus see that $V \subset \mathfrak{M}(\mathfrak{A}_\alpha)$ and that V is bounded in $\mathfrak{M}(\mathfrak{A}_\alpha)$ with respect to the operator norm on $\mathfrak{M}(\mathfrak{A}_\alpha)$ associated to $\|.\|_\alpha$. In particular, singletons are multiplicatively pseudo-bounded, so $\bigcup_{\alpha\in\Lambda} \mathfrak{M}(\mathfrak{A}_\alpha) = \mathfrak{DM}_r(\mathfrak{A})$, and it follows from the above discussion that $\mathfrak{F}_\mathfrak{A} = \bigcup_{\alpha\in\Lambda} \{V \in \mathfrak{M}(\mathfrak{A}_\alpha) \mid \sup_{T\in V} \|T\|_{\alpha,op}\} < \infty$.

It remains to show that the system $(\mathfrak{M}(\mathfrak{A}_\alpha))_{\alpha\in\Lambda}$ is inductive. If $\beta \leq \gamma \leq \alpha$, then $\mathfrak{A}_\beta \subset \mathfrak{A}_\gamma \subset \mathfrak{A}_\alpha$ and $\mathfrak{M}(\mathfrak{A}_\beta) \subset \mathfrak{M}(\mathfrak{A}_\gamma) \subset \mathfrak{M}(\mathfrak{A}_\alpha)$, the injections being norm-decreasing with respect to the norms $\|.\|_\beta$, $\|.\|_\gamma$, $\|.\|_\alpha$ and $\|.\|_{\beta,op}$, $\|.\|_{\gamma,op}$, $\|.\|_{\alpha,op}$. So the relation $\leq$ is a preorder over Λ, and of course the injections

$\theta_{\beta,\gamma} : \mathfrak{M}(\mathfrak{A}_\beta) \to \mathfrak{M}(\mathfrak{A}_\gamma)$, $\theta_{\gamma,\alpha} : \mathfrak{M}(\mathfrak{A}_\gamma) \to \mathfrak{M}(\mathfrak{A}_\alpha)$, and $\theta_{\beta,\alpha} : \mathfrak{M}(\mathfrak{A}_\beta) \to \mathfrak{M}(\mathfrak{A}_\alpha)$ satisfy $\theta_{\beta,\alpha}$, $\theta_{\beta,\gamma} \circ \theta_{\gamma,\alpha}$ if $\beta \geq \gamma \geq \alpha$.

Now let $\alpha_1, \ldots, \alpha_k$ be any finite family of elements of Λ. We did not use above the full power of Theorem 7.11. The last assertion of Theorem 7.11 shows in fact that for every pseudo-bounded subset U of $\mathfrak{DM}_r(\mathfrak{A})$ stable under products there exists $\alpha \in \Lambda$ such that U is contained in the unit ball of $\mathfrak{M}(\mathfrak{A}_\alpha)$ and such that $U \cap \mathfrak{A}_\alpha$ is contained in the unit ball of $\mathfrak{A}_\alpha$. Also, $\mathfrak{A} \subset \mathfrak{A}_\alpha$, $\mathfrak{M}(\mathfrak{A}) \subset \mathfrak{M}(\mathfrak{A}_\alpha)$, the injections being norm-decreasing.

For $i \leq k$, let S_i be the unit ball of $\mathfrak{M}(\mathfrak{A}_{\alpha_i})$, put $U = S_1 \ldots S_k$, and let α be as above. Fix $i \leq k$, and let $(\mathfrak{B}_i, p_i)$ be a Banach algebra contained in $\mathfrak{A} \cap \mathfrak{A}_{\alpha_i}$ which is both a dense ideal of $\mathfrak{A}_{\alpha_i}$ and a dense ideal of $\mathfrak{A}$ (the injections of $\mathfrak{B}_i$ onto $\mathfrak{A}$ and $\mathfrak{A}_{\alpha_i}$ being continuous). In the following commutative diagram, the arrows represent natural injections and are continuous.

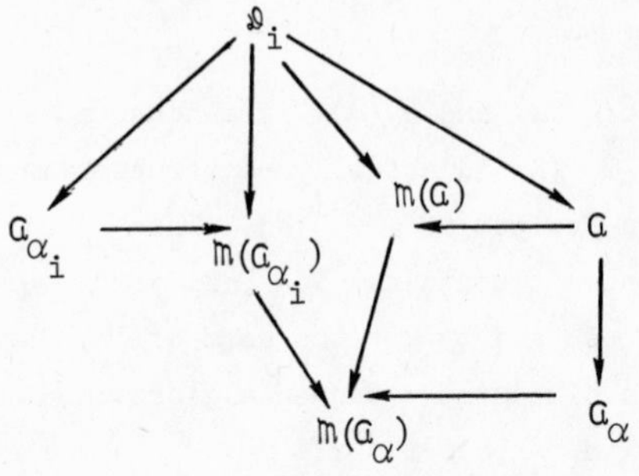

Let $u \in \mathfrak{B}_i$ such that $\|u\|_{\alpha_i} \leq 1$. Then $u \in \mathfrak{A}_\alpha \cap U$, so $\|u\|_\alpha \leq 1$, and hence $\|x\|_\alpha \leq \|x\|_{\alpha_i}$ for every $x \in \mathfrak{B}_i$. Now let $v \in \mathfrak{A}_{\alpha_i}$. There exists a sequence (u_n) of elements of $\mathfrak{B}_i$ such that $\|u_n - v\|_{\alpha_i} \xrightarrow[n\to\infty]{} 0$. The above inequality shows that (u_n) has a limit u in $\mathfrak{A}_\alpha$, and $\|u - u_n\|_{\alpha,op} \xrightarrow[n\to\infty]{} 0$. Since $\|u_n - v\|_{\alpha,op} \xrightarrow[n\to\infty]{} 0$, we obtain $v = u$, so that $\mathfrak{A}_{\alpha_i} \subset \mathfrak{A}_\alpha$. Moreover, $\|u\|_\alpha = \lim_{n\to\infty} \|u_n\|_\alpha \leq \lim_{n\to\infty} \|u_n\|_{\alpha_i} = \|u\|_{\alpha_i}$, and the inclusion from $\mathfrak{A}_{\alpha_i}$ into $\mathfrak{A}_\alpha$ is norm-decreasing. So $\alpha \geq \alpha_i$ for every i, which achieves the proof of the theorem. (Remark 7.14 shows that the injection from $\mathfrak{A}_{\alpha_i}$ into $\mathfrak{A}_\alpha$ is an s-homomorphism.)

The natural subalgebra of $\mathfrak{DM}_r(\mathfrak{A})$ associated to $\mathfrak{A}$ is of course $\bigcup_{\alpha\in\Lambda} \mathfrak{A}_\alpha$. The norm on $\mathfrak{A}_\alpha$ is not in general the restriction to $\mathfrak{A}_\alpha$ of the norm of $\mathfrak{M}(\mathfrak{A}_\alpha)$. Nevertheless, we have the following.

REMARK 7.18. Let $\alpha \in \Lambda$, and let $U \subset \mathcal{A}_\alpha$. If U is multiplicatively pseudo-bounded in $\mathfrak{M}_r(\mathcal{A})$, then there exists $\beta \geq \alpha$ such that U is bounded in $\mathcal{A}_\beta$.

<u>Proof</u>. There exists β such that U is a bounded subset of $\mathcal{M}(\mathcal{A}_\beta)$, and we may suppose that $\beta \geq \alpha$. Then $U \subset \mathcal{A}_\beta$, and U is bounded with respect to the norm $\|\cdot\|_{\beta,op}$. Denote by $\mathcal{C}$ the closure of $\mathcal{A}_\beta$ in $\mathcal{M}(\mathcal{A}_\beta)$. Then $\mathcal{A}_\beta$ is an ideal of $\mathcal{C}$. (It is easy to see and is well known that $\mathcal{A}_\beta$ is always an ideal of $\mathcal{M}(\mathcal{A}_\beta)$.)

We have the following diagram

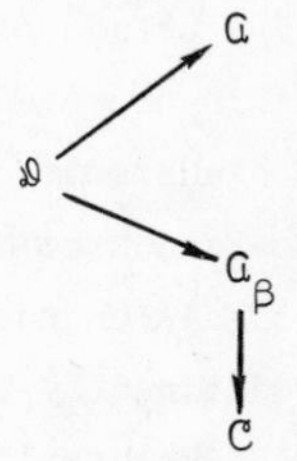

Using the construction of Proposition 7.5 as in Remark 7.14, we see that there exists a Banach algebra $\mathcal{D}' \subset \mathcal{A}_\beta \cap \mathcal{D}$ continuously contained in $\mathcal{A}$ and $\mathcal{C}$ such that $[x\mathcal{D}']^- = \mathcal{D}'$ for some $x \in \mathcal{D}'$, and such that $\mathcal{D}'$ is both a dense ideal of $\mathcal{A}$ and a dense ideal of $\mathcal{C}$. So $\mathcal{C} = \mathcal{A}_\gamma$ for some $\gamma \in \Lambda$ and U is bounded in $\mathcal{A}_\beta$.

Remark 7.18 shows that the bound structures induced by $\mathcal{A}_r = \varinjlim_{\alpha\in\Lambda} \mathcal{A}_\alpha$ and $\mathfrak{M}_r(\mathcal{A}) = \varinjlim_{\alpha\in\Lambda} \mathcal{A}_\alpha$ over each $\mathcal{A}_\beta$ are the same.

We now introduce the following notion, which is standard in "bornology."

DEFINITION 7.19. Let (x_n) be a sequence in $\mathfrak{M}_r(\mathcal{A})$. The sequence (x_n) is said to be <u>Mackey-convergent</u> to an element x of $\mathfrak{M}_r(\mathcal{A})$ if there exists a sequence (λ_n) of positive reals satisfying $\liminf_{n\to\infty} \lambda_n = \infty$ such that the sequence $(\lambda_n(x_n - x))$ is multiplicatively pseudo-bounded. The element x is then called the <u>Mackey-limit</u> of the sequence (x_n). One defines in a similar way <u>Mackey-closed</u> sets and the <u>Mackey-closure</u> of a set.

Note that the fact that (x_n) is Mackey convergent to x just means that there exists some $\alpha \in \Lambda$ such that $x \in \mathcal{M}(\mathcal{A}_\alpha)$, $x_n \in \mathcal{M}(\mathcal{A}_\alpha)$ for every n, and $\|x - x_n\|_{\alpha,op} \xrightarrow[n\to\infty]{} 0$. We now obtain the following.

PROPOSITION 7.20. The Mackey-closure of $\mathcal{A}$ in $\mathfrak{M}_r(\mathcal{A})$ is the set $\bigcup_{\alpha\in\Lambda} \mathcal{A}_\alpha$.

<u>Proof</u>. If $x \in \mathfrak{A}_\alpha$ for some α, then $x \in \mathfrak{A}_\beta$ for some β such that $\mathfrak{A}_\beta$ contains $\mathfrak{A}$ (we may write $\mathfrak{A} = \mathfrak{A}_{\alpha_0}$, with $\alpha_0 \in \Lambda$ and take $\beta \geq \alpha$, $\beta \geq \alpha_0$). The injection from $\mathfrak{A}$ onto $\mathfrak{A}_\beta$ is an s-homomorphism, so $\mathfrak{A}$ is dense in $\mathfrak{A}_\beta$, and some sequence (x_n) of elements of $\mathfrak{A}$ converges to x in $\mathfrak{A}_\beta$. Hence, it Mackey-converges to x in $\mathfrak{M}(\mathfrak{A})$. Now let $x \in \mathfrak{M}_r(\mathfrak{A})$ be such that x is the Mackey-limit in $\mathfrak{M}_r(\mathfrak{A})$ of some sequence (x_n) of elements of $\mathfrak{A}$. Writing as before $\mathfrak{A} = \mathfrak{A}_{\alpha_0}$ we may find $\beta \geq \alpha_0$ such that $x_n \in \mathfrak{A}_\beta$ for every n, $x \in \mathfrak{m}(\mathfrak{A}_\beta)$, and $\|x_n - x\|_{\beta,op} \xrightarrow[n\to\infty]{} 0$. The closure $\mathfrak{C}$ of $\mathfrak{A}_\beta$ in $\mathfrak{m}(\mathfrak{A}_\beta)$ may be written $\mathfrak{C} = \mathfrak{A}_\gamma$ for some $\gamma \in \Lambda$, so $x \in \mathfrak{C} \subset \bigcup_{\alpha\in\Lambda} \mathfrak{A}_\alpha$. This proves the proposition.

Standard results about commutative pseudo-Banach algebras (see [4] or [33]) show that the spectra of each element in these algebras is compact. Moreover, every maximal ideal is the kernel of a character, and the Gelfand theory applies to such algebras. If $\mathfrak{U}$ is a pseudo-Banach commutative unital algebra, the carrier space $\hat{\mathfrak{U}}$ of $\mathfrak{U}$ is the set of all characters on $\mathfrak{U}$. The space $\mathfrak{U}$ is compact with respect to the topology defined by the distances $d_x : (\chi,\chi') \mapsto |\chi(x) - \chi'(x)|$ $(x \in \mathfrak{U})$. If $\mathfrak{U}$ is an inductive limit $\mathfrak{U} = \varinjlim_{\zeta\in\Delta} \mathfrak{U}_\zeta$, where $(\mathfrak{U}_\zeta)_{\zeta\in\Delta}$ is a family of Banach algebras, then $\hat{\mathfrak{U}} = \varprojlim_{\zeta\in\Delta} \hat{\mathfrak{U}}_\zeta$, the inverse limit being taken in the topological sense. The Gelfand transform $\mathfrak{H} : \mathfrak{U} \to C(\hat{\mathfrak{U}})$ is defined in the usual way.

We thus see that the function $\nu : x \to \nu(x) = \|\mathfrak{H}(x)\| = \sup\{|\lambda| : \lambda \in \mathrm{Spec}(x)\}$ is a seminorm on every commutative pseudo-Banach algebra $\mathfrak{U}$. If $\mathfrak{U} = \varinjlim_{\zeta\in\Delta} \mathfrak{U}_\zeta$ then of course $\mathrm{Spec}_{\mathfrak{U}}(x) = \bigcap_{\zeta\in\Delta} \mathrm{Spec}_{\mathfrak{U}_\zeta}(x)$.

DEFINITION 7.21. Let $\mathfrak{A}$ be a commutative Banach algebra such that $[a\mathfrak{A}]^- = \mathfrak{A}$ for some $a \in \mathfrak{A}$ and such that $x\mathfrak{A} \neq \{0\}$ for every $x \in \mathfrak{A}$. The <u>regular spectral radius</u> $\nu_r(x)$ of an element x of $\mathfrak{M}_r(\mathfrak{A})$ is the spectral radius of x in $\mathfrak{M}_r(\mathfrak{A})$.

PROPOSITION 7.22. (1) If $x \in \mathfrak{A}_\alpha$, $\mathrm{Spec}_{\mathfrak{A}_\alpha}(x) = \mathrm{Spec}_{\mathfrak{M}_r(\mathfrak{A})}(x)$.
(2) $\nu_r(x) = \inf_{x\in\mathfrak{M}_r(\mathfrak{A}_\alpha)} \|x\|_{\alpha,op}$ for every $x \in \mathfrak{M}_r(\mathfrak{A})$, and $\nu_r(x) = \inf_{x\in\mathfrak{A}_\alpha} \|x\|_\alpha$ for every $x \in \mathfrak{A}_r$.

<u>Proof</u>. (1) As observed by Koua [37] in his thesis, every character of $\mathfrak{A}$ extends to $\mathfrak{M}_r(\mathfrak{A})$, so that the first assertion is a consequence of the isomorphism between $\mathfrak{M}_r(\mathfrak{A})$ and each of the $\mathfrak{M}_r(\mathfrak{A}_\alpha)$. (The extension formula is $\chi(T_{a/b}) = \chi(a)/\chi(b)$, since $\chi(b) \neq 0$ for every nonzero character χ of $\mathfrak{A}$ if $[b\mathfrak{A}]^- = \mathfrak{A}$.)

(2) Let $x \in \mathfrak{DM}_r(\mathfrak{A})$, and let Ω be the spectrum of x in $\mathfrak{DM}_r(\mathfrak{A})$. Then $\lambda \in \mathrm{Spec}_{\mathfrak{M}(\mathfrak{A}_\alpha)}(x)$ for every $\lambda \in \Omega$ and every α such that $x \in \mathfrak{M}(\mathfrak{A}_\alpha)$. So $\|x\|_{\alpha,op} \geq \max_{\lambda\in\Omega} |\lambda| = \nu_r(x)$.

Now take $\mu > \nu_r(x)$. Standard considerations about intersections of compact sets show that there exists $\alpha \in \Lambda$ such that $x \in \mathfrak{M}(\mathfrak{A}_\alpha)$ and such that the spectral radius of x in $\mathfrak{M}(\mathfrak{A}_\alpha)$ is strictly smaller than μ. So the sequence (x^n/μ^n) is bounded in $\mathfrak{M}(\mathfrak{A}_\alpha)$, hence it is multiplicatively pseudo-bounded in $\mathfrak{DM}_r(\mathfrak{A})$, and there exists $\beta \in \Lambda$ such that $x \in \mathfrak{M}(\mathfrak{A}_\beta)$, $\|x/\mu\|_{\beta,op} \leq 1$. So $\|x\|_{\beta,op} \leq \mu$, and the desired equality follows. The argument given in Remark 7.18 shows that $\inf_{x\in\mathfrak{A}_\alpha} \|x\|_\alpha = \inf_{x\in\mathfrak{M}(\mathfrak{A}_\alpha)} \|x\|_{\alpha,op}$ for every $x \in \mathfrak{A}_r$, which achieves the proof.

We now apply the results of this chapter to commutative Banach algebras with sequential b.a.i. Theorem 3.9, Theorem 7.11, and Corollary 5.12 lead to the following.

THEOREM 7.23. Let $\mathfrak{A}$ be a commutative non-unital Banach algebra with sequential b.a.i. There exists a commutative Banach algebra $\mathfrak{B}$ with b.a.i. similar to $\mathfrak{A}$ which possesses the following properties.

(1) There exists a one-to-one norm-decreasing homomorphism $\varphi : H^\infty \to \mathfrak{B}$ such that $\varphi(M_1) \subset \mathfrak{B}$, $[\varphi(M_1)\mathfrak{B}]^- = \mathfrak{B}$ where $M_1 = \{f \in A(D) \mid f(1) = 0\}$.

(2) There exists an isometry ψ from H^∞ into $\mathfrak{M}(\mathfrak{B})$ such that the left essential spectrum of $\psi(\alpha)$ equals the closed unit disc, where α is the position function $z \mapsto z$.

<u>Proof</u>. Theorem 3.9 gives a pseudo-bounded one-to-one homomorphism θ from H^∞ into $\mathfrak{DM}(\mathfrak{A})$ such that $\theta(1-\alpha) \in \mathfrak{A}$ and $[\theta(1-\alpha)\mathfrak{A}]^- = \mathfrak{A}$. Since the unit ball S of H^∞ is stable under products, we may apply Theorem 7.11 to $\theta(S)$ and obtain a norm-decreasing s-homomorphism ω from $\mathfrak{A}$ into a Banach algebra $\mathfrak{B}$ similar to $\mathfrak{A}$ such that $(\hat{\omega}\circ\theta)(S)$ is contained in the unit ball of $\mathfrak{M}(\mathfrak{B})$. So $\varphi = \hat{\omega}\circ\theta$ is a norm-decreasing homomorphism from H^∞ into $\mathfrak{M}(\mathfrak{B})$.

Since $\mathfrak{A}$ possesses a b.a.i., and since $\omega(\mathfrak{A})$ is dense in $\mathfrak{B}$ (Remark 7.10), $\mathfrak{B}$ possesses a b.a.i. Thus, $\mathfrak{B}$ is a closed ideal of $\mathfrak{M}(\mathfrak{B})$. Since $[(1-\alpha)M_1]^- = M_1$, we obtain $\varphi(M_1) \subset [\varphi(1-\alpha)\varphi(M_1)]^- \subset \mathfrak{B}$. But since $[\theta(1-\alpha)\mathfrak{A}]^- = \mathfrak{A}$, $[\varphi(1-\alpha)\omega(\mathfrak{A})]^- = [\omega(\mathfrak{A})]^- = \mathfrak{B}$, which proves the first assertion.

Note that $\mathfrak{B}$ is not unital, because of the similarity between $\mathfrak{A}$ and $\mathfrak{B}$. It follows from Corollary 5.12 that there exists $h \in \mathfrak{A}_1^\infty \subset H^\infty$ such that $\|h\| = 1$ and such that the spectrum of $\varphi(h)$ in $\mathfrak{M}(\mathfrak{B})$ equals the closed unit disc. More precisely, there exists a sequential approximate identity (e_n) of $\mathfrak{B}$ and a sequence (m_n) of integers with $m_n > n$ for every n such that $\|e_{m_n} - e_n\| \geq \frac{1}{2}$ for every n and such that $\liminf_{n\to\infty} \|(\varphi(h) - \lambda e)(e_{m_n} - e_n)\| = 0$ for every element λ of the closed unit disc. Put $y_n = e_{m_n} - e_n$. Then $xy_n \xrightarrow[n\to\infty]{} 0$ for every $x \in \mathfrak{B}$. If some subsequence of the sequence (y_n) had a limit α, we would have $\alpha x = 0$ for every $x \in \mathfrak{B}$, and hence $\alpha = 0$ ($\mathfrak{B}$ possesses a b.a.i.), which is impossible as $\|y_n\| \geq \frac{1}{2}$ for every n. This means that the left essential spectrum of $\varphi(h)$ contains the closed unit disc D. Hence both sets are equal, since $\varphi(h)$ is a contraction.

If $f \in H^\infty$, denote by $\rho(f)$ the function $z \mapsto f(h(z))$. Then ρ is a norm-decreasing homomorphism from H^∞ into itself, so that $\psi = \varphi \circ \rho$ is a norm-decreasing homomorphism from H^∞ into $\mathfrak{M}(\mathfrak{B})$. Since $\rho(\alpha) = h$, the left essential spectrum of $\psi(\alpha)$ equals the closed unit disc. Now, if $f \in H^\infty$ and $\lambda = f(t)$ with $|t| < 1$, we can write $f - \lambda = (\alpha - t)g$ where $g \in H^\infty$. Thus, $\psi(f) - \lambda e = (\psi(\alpha) - te)\psi(g)$, and so $\psi(f) - \lambda e$ is not invertible. Hence, $\operatorname{Spec} \psi(f) = \psi(\operatorname{Spec} f)$ for every $f \in H^\infty$, $\|\psi(f)\| \geq \sup_{|z|<1} |f(z)| = \|f\|$, and in fact $\|\psi(f)\| = \|f\|$ as ψ is norm-decreasing. So ψ is an isometry, and the theorem is proved.

We do not resist here the temptation to use the full power of the results of Section 5 to construct a homomorphism from H^∞ into $\mathfrak{M}_r(\mathfrak{A})$ which is an "isometry" with respect to the pseudo-Banach structure of $\mathfrak{M}_r(\mathfrak{A})$. We use the notation $f \mapsto f(T)$ to denote a homomorphism of H^∞, where T is the image of the position function α.

THEOREM 7.24. Let $\mathfrak{A}$ be a commutative Banach algebra with sequential b.a.i. Assume either that $\mathfrak{A}$ is radical or that $\mathfrak{A}$ is an integral domain. Then there exists a one-to-one pseudo-bounded homomorphism $\psi : h \mapsto h(T)$ from H^∞ into $\mathfrak{M}_r(\mathfrak{A})$ which possesses the following properties.

(1) The spectrum of $f(T)$ in $\mathfrak{M}_r(\mathfrak{A})$ equals the spectrum of f in H^∞ for every $f \in H^\infty$.

(2) If $U \subset H^\infty$, then $\psi(U)$ is pseudo-bounded in $\mathfrak{M}_r(\mathfrak{A})$ if and only if U is bounded in H^∞.

Proof. We write as before $\mathfrak{M}_r(\mathfrak{A}) = \varinjlim_{\alpha\in\Lambda} \mathfrak{M}(\mathfrak{A}_\alpha)$. If $\mathfrak{A}$ is an integral domain, then $\mathfrak{M}_r(\mathfrak{A})$ is an integral domain, so $\mathfrak{A}_\alpha$ is an integral domain for every α. Also, any Banach algebra similar to a radical Banach algebra is radical (the proof is easy) so, if $\mathfrak{A}$ is radical, each $\mathfrak{A}_\alpha$ is radical. In any case, the algebras $\mathfrak{A}_\alpha$ do not possess any nonzero idempotent. Corollary 5.13 applied to M_1 gives an element h of H^∞ such that $\|h\| = 1$ and such that the spectrum of $\theta(h)$ equals the closed unit disc for every homomorphism θ from H^∞ into a Banach algebra such that $\theta(M_1) \neq \{0\}$ and such that $[\theta(M_1)]^-$ has no nonzero idempotent.

Now let φ be the first homomorphism from H^∞ into a Banach algebra similar to $\mathfrak{A}$ constructed in Theorem 7.23, and put $\psi = \varphi \circ \omega$, where h is as above and where $\omega(f)(z) = f[h(z)]$ $(|z| \leq 1,\ f \in H^\infty)$. We may consider ψ and φ as pseudo-bounded homomorphisms from H^∞ into $\mathfrak{M}_r(\mathfrak{A})$. In particular, we may assume that φ is a norm-decreasing homomorphism from H^∞ into $\mathfrak{m}(\mathfrak{A}_\alpha)$ and that $\varphi(M_1) \subset \mathfrak{A}_\alpha$, where α is some element of Λ. Let $\beta \in \Lambda$ be such that $\varphi(h) \in \mathfrak{m}(\mathfrak{A}_\beta)$, and choose γ with $\gamma \geq \alpha$, $\gamma \geq \beta$. Then we may consider φ as a norm-decreasing homomorphism from H^∞ into $\mathfrak{m}(\mathfrak{A}_\gamma)$, and $\varphi(M_1) \subset \mathfrak{A}_\alpha \subset \mathfrak{A}_\gamma$. Also $Sp_{\mathfrak{m}(\mathfrak{A}_\gamma)}\, \varphi(h) \subset Sp_{\mathfrak{m}(\mathfrak{A}_\beta)}\, \varphi(h)$. Since $\mathfrak{A}_\gamma$ does not possess any nonzero idempotent, $Sp_{\mathfrak{m}(\mathfrak{A}_\gamma)}\, \varphi(h)$ equals the closed unit disc D. So

$$D \subset \bigcap \{Sp_{\mathfrak{m}(\mathfrak{A}_\beta)}\, \varphi(h) : \varphi(h) \in \mathfrak{m}(\mathfrak{A}_\beta)\} = Sp_{\mathfrak{M}_r(\mathfrak{A})}\, \varphi(h)\,.$$

Since φ reduces spectra, we see that the spectrum of $\varphi(h)$ in $\mathfrak{M}_r(\mathfrak{A})$ equals the closed unit disc. The same discussion as that given in the proof of Theorem 7.23 shows then that the spectrum of $\psi(f)$ in $\mathfrak{M}_r(\mathfrak{A})$ equals the spectrum of f in H^∞ for every $f \in H^\infty$. In particular, $\nu_r(\varphi(f)) = \|f\|$ for every $f \in H^\infty$. If $U \in H^\infty$ and if $\varphi(U)$ is pseudo-bounded, then $\sup_{f \in U} \nu_r(\varphi(f)) < \infty$, so it follows from the above equality that U is bounded in H^∞.

This achieves the proof of the theorem.

We now define an "extended spectrum" for arbitrary Banach algebras with dense principal ideals.

DEFINITION 7.25. Let $\mathfrak{A}$ be a commutative Banach algebra such that $[a\mathfrak{A}]^- = \mathfrak{A}$ for some $a \in \mathfrak{A}$. Put $\mathfrak{A}^* = \mathfrak{A}/L$ where $L = \{x \in \mathfrak{A} \mid x\mathfrak{A} = 0\}$. The carrier space of $\mathfrak{M}_r(\mathfrak{A}^*)$ will be denoted by $\hat{\hat{\mathfrak{A}}}$ and called the <u>extended spectrum</u> of $\mathfrak{A}$. The normed algebra obtained by equipping the quotient of $\mathfrak{M}_r(\mathfrak{A}^*)$ by its radical with the norm $\|.\|_r$ induced by ν_r is called the extended <u>Gelfand transform</u> of $\mathfrak{A}$, and it is denoted by $\hat{\hat{\mathcal{G}}}(\mathfrak{A})$.

Note that, if $[a\mathfrak{A}]^- = \mathfrak{A}$ for some $a \in \mathfrak{A}$, and if $x\mathfrak{A} \subset L$, then $xa^2 = 0$, so that $x\mathfrak{A} = 0$ because $a^2\mathfrak{A}$ is dense in $\mathfrak{A}$. This shows that $y\mathfrak{A}^* \neq 0$ for every nonzero $y \in \mathfrak{A}^*$, and hence $\mathfrak{M}_r(\mathfrak{A}^*)$ is well defined and the above definition is meaningful.

COROLLARY 7.26. Let $\mathfrak{A}$ be a commutative, radical Banach algebra with b.a.i. Then the extended Gelfand transform $\hat{\hat{\mathcal{G}}}(\mathfrak{A})$ contains an isometric, spectrum-preserving copy of H^∞, and $\hat{\hat{\mathfrak{A}}}$ can be mapped continuously onto the carrier space of H^∞.

<u>Proof</u>. The first assertion follows immediately from the first assertion of

Theorem 7.24.

Let $\varphi: H^\infty \to \mathfrak{M}_r(\mathfrak{a}^*)$ be the corresponding homomorphism from H^∞. For every $\chi \in \hat{\hat{\mathfrak{a}}}$, $\chi \circ \varphi$ belongs to the carrier space Ω of H^∞. Put $T = \varphi(\alpha)$, where α is the position function $z \mapsto z$, and let t be an element of the open unit disc. There exists $\chi \in \hat{\hat{\mathfrak{a}}}$ such that $\chi(T) = t$, so that $(\chi \circ \varphi)(\alpha) = t$. If f is any element of H^∞, $f - f(t)1$ vanishes at t, so $f - f(t)1 = (\alpha - t)g$ for some $g \in H^\infty$. We obtain

$$(\chi \circ \varphi)(f) = (\chi \circ \varphi)[f - f(t)1 + f(t)1] = [(\chi \circ \varphi)(\alpha) - t][(\chi \circ \varphi)(g)] + f(t) = f(t).$$

This shows that the image of the open unit disc in the carrier space of H^∞ is contained in the image of $\hat{\hat{\mathfrak{a}}}$ by the map $\chi \mapsto \chi \circ \varphi$. But the image of $\hat{\hat{\mathfrak{a}}}$ is compact, and the open disc is dense in the carrier space of H^∞. (This is the famous Corona theorem of Carleson [16]; see [36], Appendix for a shorter proof.)

This proves the corollary.

Note that there is no reason to suppose that $\hat{\hat{\mathfrak{g}}}(\mathfrak{a})$ is complete, and so the fact that it necessarily contains H^∞ is rather surprising.

Corollary 7.26 shows that $\hat{\hat{\mathfrak{a}}}$ is very complicated if $\mathfrak{a}$ is a commutative radical Banach algebra with a b.a.i., or more generally a commutative radical Banach algebra $\mathfrak{R}$ which possesses a continuous semigroup $(a^t)_{t>0}$ satisfying $[\bigcup_{t>0} a^t\mathfrak{R}]^- \mathfrak{R}$ (such algebras are similar to a Banach algebra with a b.a.i.). The question whether or not any commutative radical Banach algebra $\mathfrak{R}$ such that $[x\mathfrak{R}]^- = \mathfrak{R}$ for some $x \in \mathfrak{R}$ possesses such real continuous semigroups is the main open problem raised in the other paper by the author in these Proceedings [25]. It was shown in [24] that these radical Banach algebras actually possess semigroups $(a^t)_{t>0}$ such that $[a^t\mathfrak{R}]^- = \mathfrak{R}$ for every $t > 0$, but these semigroups have no reason to be continuous. Zouakia[51] has just obtained a nice version of the heavy construction of [24] which is analogous to his version of the theory of Cohen elements given in these Proceedings [50], but the question of the continuity of these semigroups remains unclear.

We conclude this section with a few remarks.

REMARK 7.27. (1) In defining the preorder on the set Λ we arranged the injections $\theta_{\beta,\alpha}$ to be norm-decreasing. It is not always possible to make such an arrangement for inductive limits of non-commutative Banach algebras, as is shown by Raouyane in his thesis [42].

(2) In the case where $\mathfrak{a}_\alpha$ possesses a b.a.i., $\mathfrak{a}_\alpha$ is isomorphic as a Banach algebra with its image in $\mathfrak{M}(\mathfrak{a}_\alpha)$. So the discussion of the isomorphism between $\mathfrak{M}_r(\mathfrak{a})$ and $\varinjlim_{\alpha} \mathfrak{M}(\mathfrak{a}_\alpha)$ becomes much simpler when $\mathfrak{a}$ is similar to a

Banach algebra with b.a.i.

(3) The argument used to establish the second assertion of Remark 7.14 shows that, if $\varphi : \mathfrak{A} \to \mathfrak{B}$ and $\psi : \mathfrak{B} \to \mathfrak{C}$ are s-homomorphisms, then $\psi \circ \varphi$ is an s-homomorphism too. In fact, the subalgebras $\mathfrak{B}$ of $\mathfrak{M}_r(\mathfrak{A})$ containing $\mathfrak{A}$ which belong to the class $(\mathfrak{A}_\alpha)$ are exactly the Banach algebras $\mathfrak{B}$ containing $\mathfrak{A}$ for which the injection from $\mathfrak{A}$ into $\mathfrak{B}$ is an s-homomorphism.

(4) Koua [37] observed that $\mathfrak{M}_r(\mathfrak{A}) = \mathfrak{m}(\mathfrak{A})$ as a pseudo-Banach algebra if $\mathfrak{A}$ is a uniform algebra. It would be interesting to know exactly when the pseudo-Banach algebra $\mathfrak{M}_r(\mathfrak{A})$ is a Banach algebra.

(5) If $\mathfrak{A}$ is radical, then $\mathfrak{A}_r = \bigcup_{\alpha\in\Lambda} \mathfrak{A}_\alpha$ is contained in the radical of $\mathfrak{M}_r(\mathfrak{A})$. I see no reason why equality should hold. Also, if $\mathfrak{A}$ is a Banach algebra, the injection from $\varinjlim \mathfrak{A}_\alpha$ into $\mathfrak{M}_r(\mathfrak{A}) = \varinjlim \mathfrak{m}(\mathfrak{A}_\alpha)$ is "continuous" with respect to the pseudo-Banach algebras structures, but I see no reason for it to be "bicontinuous."

(6) If $\varphi : \mathfrak{D} \to \mathfrak{A}$ is as in Proposition 7.3 it is not true in general that $x\mathfrak{D}$ is dense in $\mathfrak{D}$ provided $x\mathfrak{D}$ is dense in $\mathfrak{A}$. To see this, put $\mathfrak{D} = C_*[0,1]$, $\mathfrak{A} = L^1_*[0,1]$, and $x(t) = 1$ for $t \in [0,1]$.

8. Applications to the closed ideal problem

We now give some applications of the above results to the closed ideal problem. The following theorems follow from Corollary 7.12 and Theorem 7.22.

THEOREM 8.1. If there exists an infinite-dimensional, commutative, topologically simple Banach algebra $\mathfrak{R}$ which possesses a nonzero continuous semigroup $(a^t)_{t>0}$ over the positive reals, then there exists an infinite-dimensional commutative, topologically simple Banach algebra with a bounded approximate identity.

THEOREM 8.2. If there exists an infinite-dimensional, commutative, topologically simple Banach algebra $\mathfrak{R}$ which possesses a nonzero continuous semigroup $(a^t)_{t>0}$ over the positive reals, then there exists a one-to-one representation $h \mapsto h(T)$ of H^∞ over a Banach space E which possesses the following properties:

(1) $\operatorname{Sp} T = \{1\}$;

(2) T does not possess any non-trivial hyperinvariant subspace.

THEOREM 8.3. If there exists an infinite-dimensional, commutative, topologically simple Banach algebra $\mathfrak{R}$ which possesses a nonzero continuous semigroup $(a^t)_{t>0}$ over the positive reals, then there exists an isometric representation $h \mapsto h(T)$ of H^∞ over a Banach space E which possesses the following properties:

(1) the left essential spectrum of T equals the closed unit disc;

(2) T has no proper hyperinvariant subspace.

Proof. It follows from Theorem 7.8 that every commutative Banach algebra with a b.a.i. which is similar to a topologically simple Banach algebra is topologically simple. So Theorem 8.1 follows from Corollary 7.12.

If $\mathfrak{a}$ if any commutative Banach algebra, and if $V \in \mathfrak{m}(\mathfrak{a})$, let F be a closed subspace of $\mathfrak{a}$ which is hyperinvariant for V. For $a \in \mathfrak{a}$, denote by $\tilde{a}$ the map $x \mapsto ax$. Then $(V \circ \tilde{a})(x) = V(ax) = a(Vx) = \tilde{a}V(x)$ for every $x \in \mathfrak{a}$, so that $V \circ \tilde{a} = \tilde{a} \circ V$. Thus $aV = \tilde{a}(V) \subset V$ for every $a \in \mathfrak{a}$. This shows that V is a proper closed ideal of $\mathfrak{a}$. Hence, Theorems 8.2 and 8.3 follow from Theorem 7.22.

We conclude by a reduction of the closed ideal problem to an invariant subspace problem in a special case.

THEOREM 8.4. If there exists an infinite-dimensional, commutative, topologically simple Banach algebra $\mathfrak{R}$ spanned by a continuous semigroup $(a^t)_{t>0}$ over the positive reals, then there exists a one-to-one representation $h \mapsto h(T)$ of H^∞ over a Banach space E which possesses the following properties:

(1) $\operatorname{Sp} T = \{1\}$;

(2) the operator T does not possess any proper invariant subspace.

Proof. Applying Corollary 7.12 to $\mathfrak{R}$, we obtain an s-homomorphism θ from $\mathfrak{R}$ into a commutative Banach algebra $\mathfrak{R}_1$ similar to $\mathfrak{R}$ such that $\sup_{t>0} \|\exp(-ct)\theta(a^t)\| \leq 1$, where c is a suitable positive real. By Remark 7.10, we have $[\theta(\mathfrak{R})]^- = \mathfrak{R}_1$, so the semigroup $(\theta(a^t))_{t>0}$ spans $\mathfrak{R}_1$. By considering the semigroup $(\exp(-ct)a^t)_{t>0}$ instead of $(a^t)_{t>0}$, we may assume in fact that $\sup_{t>0} \|\theta(a^t)\| \leq 1$. Also, $\mathfrak{R}_1$ is topologically simple (Theorem 7.8).

Note that Lemma 3.8 gives a one-to-one homomorphism ψ from H^∞ into $\mathfrak{M}_r(L^1(\mathbb{R}^+)$ such that $\psi(1-\alpha) \in L^1(\mathbb{R}^+)$. In fact, $\psi(1-\alpha)$ is the function $v : x \mapsto 2e^{-2x}$ $(x \geq 0)$. The map $\omega : f \mapsto \omega(f)$, where $\omega(f)(x) = 2f(2x)$ for $x \geq 0$, is an algebra homomorphism from $L^1(\mathbb{R}^+)$ into itself, and ω is in fact an isomorphism. We have $v = \omega(u)$, where u is the function $x \mapsto e^{-x}$. Since the powers of u span $L^1(\mathbb{R}^+)$, the powers of v span $L^1(\mathbb{R}^+)$ too. Denote by $\varphi : L^1(\mathbb{R}^+) \to \mathfrak{R}_1$ the Sinclair map $f \mapsto \int_0^\infty f(t)b^t dt$, where $b^t = \theta(a^t)$ $(t > 0)$. Note that, if $t > 0$, then $b^t = \lim_{n\to\infty} \varphi(a_{n,t})$, where $a_{n,t} = n\chi_{[t,t+1/n]}$ so that $b^t \in [\varphi(L^1(\mathbb{R}^+))]^-$ for every $t > 0$. We thus see that the powers of $(\varphi \circ \psi)(1-\alpha)$ span $\mathfrak{R}_1$.

Let S be the unit ball of H^∞. Then $(\hat{\varphi} \circ \psi)(S)$ is pseudo-bounded and stable under products. Applying Theorem 7.11 we obtain an s-homomorphism ρ from $\mathfrak{R}_1$ into a topologically simple Banach algebra $\mathfrak{R}_2$ such that $\hat{\rho} \circ \hat{\varphi} \circ \psi$ is a bounded homomorphism from H^∞ into $\mathfrak{M}(\mathfrak{R}_2)$. Also, $\mathfrak{R}_2$ has a b.a.i., and $(\hat{\rho} \circ \hat{\varphi} \circ \psi)(M_1) \subset [(\rho \circ \varphi \circ \psi)(1-\alpha)\mathfrak{M}(\mathfrak{R}_2)]^- \subset \mathfrak{R}_2$. Since $\rho(\mathfrak{R}_1)$ is dense in $\mathfrak{R}_2$, the powers of $y = (\rho \circ \hat{\varphi} \circ \psi)(1-\alpha)$ span $\mathfrak{R}_2$. Put $T = (\hat{\rho} \circ \hat{\varphi} \circ \psi)(\alpha)$, and set $h(T) = (\hat{\rho} \circ \hat{\varphi} \circ \psi)(h)$ for every $h \in H^\infty$. The map $T \mapsto h(T)$ is a bounded representation of H^∞. Let F be a closed subspace of $\mathfrak{R}_2$ which is invariant with respect to T. Since $\mathfrak{M}(\mathfrak{R}_2)$ has no non-trivial idempotent, we have $(\hat{\rho} \circ \hat{\varphi} \circ \psi)(1) = e$, so that $yF \subset F$, $y^n F \subset F$ for every $n \geq 1$ and finally $bF \subset F$ for every $b \in \mathfrak{R}_2$. So either $F = \{0\}$ or $F = \mathfrak{R}_2$ and T has no proper invariant subspace.

We need to show that $\hat{\rho} \circ \hat{\varphi} \circ \psi$ is one-to-one. Note that ρ and ψ are one-to-one. Domar's theorem [22] suggests that φ is necessarily one-to-one, but his theorem does not exactly show it. So assume that φ is not one-to-one. It follows from Lemma 3.6 that the set $L = \{z \in \mathbb{C} \mid \operatorname{Re} z \geq 0, \mathfrak{L}(f)(z) = 0$ for every $f \in \operatorname{Ker} \varphi\}$ is unbounded. Put $K = \{z/(z+2) : z \in L\}$. Then K is contained in the open unit disc. The homomorphism ψ constructed in Lemma 3.8 is such that $\psi(h - h\alpha) \in L^1(\mathbb{R}^+)$ and $(\mathfrak{L} \circ \psi)(h - h\alpha)(z) = 2/(z+2)\, h(z/(z+2))$ $(\operatorname{Re} z \geq 0)$ for every $h \in H^\infty$. Let $h \in H^\infty$ be such that $\psi(h) \in \operatorname{Ker} \hat{\varphi}$. Then $\psi(h - h\alpha) \in \operatorname{Ker} \varphi$, $h(z/(z+2)) = 0$ for every $z \in L$ and h vanishes on K. If h is not the zero function, then $h = BSF$ where B is a Blaschke product and where the singular and outer parts S and F of h have no zero in the open unit disc. Since K is infinite, we write $B = B_1 B_2$ where B_1 and B_2 are Blaschke products such that neither B_1 nor B_2 vanish over the whole set K. So $\psi(B_1) \notin \operatorname{Ker} \hat{\varphi}$, $\psi(B_2 SF) \notin \operatorname{Ker} \hat{\varphi}$, and $(\hat{\rho} \circ \hat{\varphi} \circ \psi)(B_1) \neq 0$, $(\hat{\rho} \circ \hat{\varphi} \circ \psi)(B_2 SF) \neq 0$. But $(\hat{\rho} \circ \hat{\varphi} \circ \psi)(B_1 B_2 SF) = \hat{\rho}(\hat{\varphi}(h)) = 0$ and $\mathfrak{M}(\mathfrak{R}_2)$, and hence $\mathfrak{R}_2$, is not an integral domain. This contradicts the fact that $\mathfrak{R}_2$ is topologically simple. So the representation $T \mapsto h(T)$ is one-to-one and the theorem is proved.

Note that the Brown-Chevreau-Pearcy theorem [15] shows that, if T is a contraction acting on a Hilbert space E whose spectrum is "rich" (which means that $\sup_{z \in \Delta} |h(z)| = \|h\|$ for every $h \in H^\infty$, where $\Delta = \{z \in Sp(T) \mid |z| < 1\}$, then T has a proper invariant subspace. It is not known whether their result extends to contractions on a Hilbert space whose spectrum contains the unit circle (see [30] for partial results in that direction). Apostol [6] has extended the Brown-Chevreau-Pearcy theorem to representations of H^∞ on some Banach spaces E satisfying additional hypothesis. So the reductions of the closed ideal problem which we have obtained lead to hyperinvariant or invariant subspace problems which are still unsolved even in the case of a Hilbert space.

Note also that, if Ω is any countable subset of the unit disc, we may assume that the operator T in Theorem 8.3 satisfies $\liminf_{n \to \infty} \|(T - \lambda e)^n x\|^{1/n} = 0$ for every $\lambda \in \Omega$ whenever $x \neq 0$ (use Theorem 6.3) and that $\limsup_{n \to \infty} \|(T - \mu e)^n y\|^{1/n} = 1 + |\mu|$ for every $\mu \in \mathbb{C}$ and every nonzero $y \in E$ (use Theorem 5.18). I do not know whether or not this gap between lim sup and lim inf can be of any help to get invariant or hyperinvariant subspaces for T.

9. Appendix

Melting pot of applications to operator theory.

In this section we give a more or less disconnected set of applications of some of the ideas involved in the paper to operator theory. One of the main ideas is that after a change of norm, as in the case of similar algebras, some operators acting on a Banach space give operators acting on another normed or Banach space which enjoy nicer properties. The first result, which might be new, is that $\lim_{n\to\infty} \|(e-T)T^n\| = 0$ for every contraction in a Banach algebra such that $\operatorname{Sp} T = \{1\}$. To show this we use the auxiliary seminorm $x \mapsto \lim_{n\to\infty} \|T^n(x)\|$ with respect to which T acts as an isometry. The second result gives inequalities of the type $\inf_{\|x\| \geq \rho} \|g(x)\| \geq \delta$ where g is an analytic function on a neighbourhood of the origin having a zero of order 1 at the origin. The inequality holds in any Banach algebra $\mathfrak{A}$ such that $ab \neq a$ for every nonzero $a \in \mathfrak{A}$ and every $b \in \mathfrak{A}$. The constants ρ and δ do not depend on $\mathfrak{A}$. The idea is analogous to the idea used in Section 4, and we use a classical theorem of Caratheodory ([44], Chapter 8, Theorem 6.11) related to the Bloch-Landau theorem ([44] Chapter 2, Theorem 1.1) to obtain a relation between ρ and δ. Note that Caratheodory's proof makes use of elliptic functions. This gives for example universal lower bounds for the rate of decrease of $\|(I-T)T^n\|$ if T is an operator whose spectrum equals $\{1\}$.

We then give an interpretation of a recent paper of Beauzamy [9] concerning hyperinvariant subspaces for contractions T on a Banach space E such that (T^n) does not strongly converge to zero and such that there exists a nonzero sequence (x_n) of elements of E satisfying $x_n = T(x_{n+1})$, $\|x_n\| \leq K\omega_n$ $(n \in \mathbb{N})$, where (ω_n) is an increasing sequence of positive numbers such that $\sum (\log \omega_n)/(1+n^2) < \infty$, $\omega_{n+m} \leq \omega_n \omega_m$ $(n,m \in \mathbb{N})$. In fact, we show that the first condition implies that T acts as an isometry over a Banach space obtained by completion of E with respect to a weaker norm, and that the second one implies that T acts as a "Wermer operator" on a subspace of E which is complete with respect to a stronger norm. A "Wermer operator" is an invertible operator T such that $\sum_{n\in\mathbb{Z}} (\log^+\|T^n\|)(1+n^2) < \infty$ and such that $\operatorname{Sp}(T)$ is not a singleton. Wermer's theorem [49] shows that these operators have a proper hyperinvariant subspace which can be obtained using a functional calculus involving the weighted algebra $\ell^1(\mathbb{Z}, \|T^n\|)$, as observed by Atzmon in [7], Section 6. (This observation was implicit in Davie's beautiful paper [21] about Bishop operators). In fact, Beauzamy's hyperinvariant subspaces can be obtained by taking the closure in E of the subspaces obtained in a subspace of E by Wermer's methods. Roughly speaking, Beauzamy's conditions ensure that the kernel of some possibly unbounded closed operator "commuting" with T (and with any operator commuting with T) does not reduce to zero.

We then turn to a discussion of results proved or announced by Atzmon in [7]. These results concern operators satisfying the condition $\|T^n\| = \underline{O}(n^k)$ as $n \to \infty$ for some $k > 0$. Atzmon shows in [7] that if $\|T^{-n}\| = \underline{O}\,[\exp(cn^{1/2})]$ as $n \to \infty$ for some $c > 0$ and if $Sp(T) = \{1\}$, then T has a proper hyperinvariant subspace (in fact an uncountable chain of hyperinvariant subspaces if T is not nilpotent). Also, he announces without proof that if $Sp(T) = \{1\}$ and if $|\ell[T^{-n}(x)]| = \underline{O}\,(\exp(cn^{1/2}))$ as $n \to \infty$ for some $c > 0$ and some nonzero $x \in E$ and some nonzero $\ell \in E^*$, then T has a proper invariant subspace, provided that E is reflexive.

We get here somewhat stronger results in the same direction, by a different method based upon a classical theorem of Paley and Wiener, applied to a variant of the classical resolvent formula for continuous semigroups. If $T^n = \underline{O}(n^k)$ as $n \to \infty$, there exists a continuous map $t \mapsto u_{t,k+2}(T)$ from $[0,\infty)$ into $\mathcal{L}(E)$ such that $(T - I)^{2k+3}\, u_{t+t',k+2}(T) = u_{t,k+2}(T)\, u_{t',k+2}(T)$ $(t,t' \geq 0)$. The formula $\int_0^\infty e^{-\lambda t}\, u_{t,k+2}(T)dt = (T - I)^{2k+2}[\lambda(T - I) - T - I]^{-1}$ holds for $\operatorname{Re}\lambda > 0$, and it follows from the Paley-Wiener theorem that, if $|\ell((T - I)^n(x))|^{1/n} = \underline{O}(1/n)$ as $n \to \infty$ then $\ell[u_{t,k+2}(T)(x)] = 0$ for $t \geq t_0$. This gives invariant subspaces for T if $x \neq 0$, $\ell \neq 0$. If $\|({}^tT - I)^n(\ell)\| = \underline{O}(1/n)$ as $n \to \infty$ for some nonzero $\ell \in E^*$, we obtain hyperinvariant subspaces for T. If $Sp(T) = \{1\}$, then, using a theorem of Cartwright, we see that Atzmon's condition $|\ell(T^{-n}(x))\| = \underline{O}(\exp cn^{1/2})$ for some $c > 0$ implies that $|\ell[(T - I)^n(x)]| = \underline{O}(1/n)$ as $n \to \infty$, and the result announced by Atzmon follows, without any assumption of reflexivity about E.

Results of this nature seem to have some interest for the closed ideal problem because they apply to Banach algebras which are integral domains. In fact, if $\|(e - x)^n\| = \underline{O}(n^k)$, and if $\|\ell \circ x^n\|^{1/n} = \underline{O}(1/n)$ for some nonzero $\ell \in \mathcal{A}$ and some nonzero $x \in \mathcal{A}$ this method gives proper closed ideals in $\mathcal{A}$, and it is possible to show that all closed ideals of the integral domain $L^1(\mathbb{R}^+, e^{-t^2})$ can be obtained by this method. A similar application of the Paley-Wiener theorem was given by the author in his UCLA postgraduate course in the spring 1979 for elements x in a Banach algebra such that $\|x(\lambda e - x)^{-1}\| \leq 1$ for every $\lambda > 0$ (see Sinclair's notes [47], Chapter 6), but the condition $\|(e - x)^n\| = \underline{O}(n^k)$ as $n \to \infty$ seems to be much more tractable.

We conclude the appendix and the paper by a very short proof of an extension of a spectral mapping theorem of Foias-Mlak [28] which ensures that, if $\varphi : f \mapsto f(T)$ is the representation of H^∞ associated to a completely non-unitary contraction on a Hilbert space such that $\lambda \in Sp(T)$, then $h(\lambda) \in Sp[\varphi(h)]$ for every $h \in H^\infty$ such that $h(\varphi) = \lim\{h(z) : z \to \lambda, |z| < 1\}$ does exist. Our proof uses only the fact that $M_\lambda = \{f \in \mathcal{A}(D) \mid f(\lambda) = 0\}$ possesses a nice bounded approximate identity, and it works for representations of H^∞ on arbitrary Banach spaces.

THEOREM 9.1. Let $\mathfrak{A}$ be a unital Banach algebra and let a be an element of $\mathfrak{A}$ of norm 1. If $Sp(a) = \{1\}$, then $\|a^n - a^{n+1}\| \to 0$ as $n \to \infty$.

Proof. Since $a - e$ is quasinilpotent, we may define $\log a = \log(e + (a - e))$ by the usual series, and $\log a$ is quasinilpotent. This shows that the function $z \mapsto a^z = \exp(z \log a)$ is of zero exponential type.

Now put

$$p(x) = \lim_{n\to\infty} \|a^n x\| \ (x \in \mathfrak{A}).$$

We obtain a linear seminorm over $\mathfrak{A}$, and $p(x) \leq \|x\|$ for every $x \in \mathfrak{A}$. Also, if $t \in \mathbb{R}$ we have

$$p(a^t) = \lim_{n\to\infty} \|a^{n+t}\| \leq \|a^{t-[t]}\| \lim_{n\to\infty} \|a^{n+[t]}\| \leq \|a^{t-[t]}\| \leq \sup_{0 \leq s \leq 1} \|a^s\| < \infty .$$

So if ℓ is any linear form over $\mathfrak{A}$ continuous with respect to p the function $\ell : z \mapsto \ell(a^z)$ is an entire function of zero exponential type which is bounded over the real line. It follows from [12], Chapter 1, Theorem 1.4.3 that ℓ is constant, so $\ell(a) = \ell(a^1) = \ell(a^0) = \ell(e)$ for every such ℓ. So $\lim_{n\to\infty} \|a^{n+1} - a^n\| = p(a - e) = 0$, which proves the theorem.

COROLLARY 9.2. Let E be a Banach space, and let T be a linear contraction with finite spectrum acting on E. If $T^n(x) \not\to 0$ as $n \to \infty$ for some $x \in E$, then either T is a scalar multiple of the identity or T has a proper hyper-invariant subspace.

Proof. If $Sp(T)$ contains more than one point, then it is disconnected and the result follows from the well-known fact that the closed unital subalgebra of $\mathcal{L}(E)$ generated by T contains a nontrivial idempotent. So assume that the spectrum of T is a singleton, say $\{\lambda\}$. Since $\|T^n\| \not\to 0$ as $n \to \infty$, we must have $|\lambda| = 1$. It follows from the theorem that $\|(T - \lambda I)T^n\| \to 0$ as $n \to \infty$.

Put $F = \{x \in E \mid T^n(x) \to 0 \text{ as } n \to \infty\}$. Then F is clearly a closed hyper-invariant subspace for T, and $F \neq E$. Since $Im(T - \lambda I) \subset F$, we obtain $F \neq \{0\}$ if $T \neq \lambda I$, which proves the corollary.

REMARK 9.3. (1) Let $\mathfrak{A}$ be a unital Banach algebra, and let a be an element of $\mathfrak{A}$ of norm 1 such that the spectrum of a is a finite set $\{\lambda_1, \ldots, \lambda_k\}$. Then $\|(T - \lambda_1 e) \ldots (T - \lambda_k e)T^n\| \to 0$ as $n \to \infty$. To see this, note that the closed unital subalgebra $\mathfrak{B}$ of $\mathfrak{A}$ generated by a possesses a finite family $\{e_1, \ldots, e_k\}$ of idempotents such that $e = e_1 + \cdots + e_k$ and such that $Sp(e_i) \subset \{0, \lambda_i\}$ for every $i \leq k$. So $Te_i - \lambda_i e_i$ is quasinilpotent for every i, and it follows from the theorem that

$$\lim_{n\to\infty} \|(Te_i - \lambda_i e)T^n\| = \lim_{n\to\infty} \|(Te_i - \lambda_i e_i)(Te_i)^n\| = 0 .$$

Since

$$\begin{aligned}(T-\lambda_1 e)\dots(T-\lambda_k e)T^n &= (e_1 + \cdots + e_k)(T-\lambda_1 e)\dots(T-\lambda_k e)T^n \\ &= \sum_{i=1}^{k} \Big[\prod_{j\neq i} (T-\lambda_j e)\Big] [(Te_i - \lambda_i e_i)T^n] ,\end{aligned}$$

the result follows.

(2) Let E be a Banach space, and let T be a linear contraction acting on E such that $T^n(x_0) \not\to 0$ as $n \to \infty$ for some $x_0 \in E$. For $x \in E$, set $p(x) = \lim_{n\to\infty} \|T^n(x)\|$. We obtain a linear seminorm over E which is not identically zero, and $p(S(x)) \leq \|S\|p(x)$ for every $x \in E$ and every $S \in \mathfrak{U} = \{V \in \mathcal{L}(E) \mid VT = TV\}$. Denote by G the completion of the linear space $E/\text{Ker } p$ with respect to the quotient norm $\bar{p}$ induced by p. The surjection $\pi : E \to E/\text{Ker } p$ is a norm-decreasing homomorphism from E into G, and for every $S \in \mathfrak{U}$ there exists a unique linear map $\hat{S}$ from $E/\text{Ker } p$ into itself satisfying $\hat{S} \circ \pi = \pi \circ S$. Then $\hat{S}$ is continuous with respect to $\hat{p}$, and we may consider $\hat{S}$ as an element of $\mathcal{L}(G)$. The map $S \mapsto \hat{S}$ is norm-decreasing from $\mathfrak{U}$ into $\mathcal{L}(F)$, and $\hat{T}$ is an isometry. Using this observation, it is possible to deduce Theorem 9.1 and the well-known fact that an isometry whose spectrum equals $\{1\}$ is the identity map.

We now turn to a result related to Theorem 4.1.

THEOREM 9.4. Let g be an analytic function defined on a neighborhood of the origin such that $g(0) = 0$ and $g'(0) \neq 0$. Let R be the radius of convergence of the power series expansion of g at the origin, and let $\rho < R$ be a positive real number such that g is univalent over the disc $D_\rho = \{z \in \mathbb{C} \mid |z| < \rho\}$. If $\mathcal{G}$ is a Banach algebra such that $pq \neq 0$ for every nonzero elements p and q in $\mathcal{G}$, we have

$$\inf\{\|g(x)\| : \nu(x) < R,\ \|x\| > \rho\} \geq \frac{|f'(0)|\rho}{32} .$$

<u>Proof</u>. Put $h(z) = g(\rho z)/\rho g'(0)$. Then h is analytic over the open unit disc D, $h(0) = 0$ and $h'(0) = 1$. Also h is univalent over D, so $h(z) \neq 0$ if $z \neq 0$. It follows from a classical theorem of Carotheodory ([44], Chapter VIII, Theorem 6.11) that the image of D under h contains an open disc of radius $1/16$ centered at the origin (this result is related to the Bloch-Landau-Ahlfors theorems [44], Chapter II, Theorem 1.1 and Theorem 3.3). So the image of D_ρ under g contains the open disc $\Delta = \{z \in \mathbb{C} \mid |z| < \rho|g'(0)|/16\}$. There exists a univalent analytic function $f : \Delta \to D_\rho$ such that $g(f(z)) = z$ for every $z \in \Delta$.

Note that $f(0) = 0$.

Let $f(z) = \sum_{n\geq 1} a_n z^n$ be the power series expansion of f over Δ. Using Cauchy's inequalities we see that $|a_n| r^n \leq \rho$ for every $r < \rho|g'(0)|/16$, and hence for every $r \leq \rho|g'(0)|/16$.

Now let $\mathfrak{A}$ be any Banach algebra, and let x be an element of $\mathfrak{A}$ such that $\|x\| \leq |g'(0)|/32$. We define $f(x)$ to be the sum of the absolutely convergent series $\sum_{n\geq 1} a_n x^n$, and we obtain

$$\|f(x)\| \leq \sum_{n\geq 1} |a_n| \|x^n\| \leq \sum_{n\geq 1} |a_n| \|x\|^n < \sum_{n\geq 1} |a_n| \frac{\rho^n |g'(0)|^n}{32^n}$$
$$\leq \rho\Big(\sum_{n\geq 1} 2^{-n}\Big) = \rho .$$

Note also that, if u is an element in a Banach algebra satisfying $\|u\| < \rho|g'(0)|/16$, then $\nu(u) < \rho|g'(0)|/16$, and so we can put $f(u) = \sum_{n\geq 1} a_n u^n$ and certainly $\nu(f(u)) \leq \sup_{z\in\Delta} |f(z)| \leq R$. Thus, we can consider $g[f(u)] = \sum_{n\geq 1} b_n [f(u)]^n$, where $\sum_{n\geq 1} b_n z^n$ is the power series expansion of g at the origin. Also, $g[f(u)] = \sum_{n\geq 1} b_n (\sum_{m\geq 1} a_m u^m)^n = u$, since $g(f(z)) = \sum_{n\geq 1} b_n (\sum_{m\geq 1} a_m z^m)^m = z$ for every $z \in \Delta$.

Now let $\mathfrak{A}$ be a Banach algebra satisfying the condition of the theorem, and let x be an element of $\mathfrak{A}$ such that $\nu(x) < R$ and $\|g(x)\| < \rho|g'(0)|/32$. Put $y = f[g(x)]$. We have $\|y\| < \rho$, and $g(y) = g[f(g(x))] = g(x)$. So $\sum_{n\geq 1} b_n x^n = \sum_{n\geq 1} b_n y^n$ and $(x-y)(b_1 e + \sum_{n\geq 2} b_n [x^{n-1} + x^{n-2} y + \cdots + y^{n-1}]) = 0$. Since $b_1 = g'(0) \neq 0$, we obtain $(x-y)(e-v) = 0$, where $v \in \mathfrak{A}$. Our hypothesis implies that $x = y$. So $\|x\| < \rho$, which proves the theorem.

COROLLARY 9.5. Let $\mathfrak{A}$ be a unital Banach algebra, and let a be an element of $\mathfrak{A}$ such that $\mathrm{Sp}(a) = \{1\}$. If $a \neq e$ then $\liminf_{n\to\infty} n\|a^n - a^{n+1}\| \geq 1/96$.

Proof. Put $b = e - a$. Then b is nonzero and b is quasinilpotent, so the closed subalgebra $\mathfrak{B}$ of $\mathfrak{A}$ generated by b is radical and hence satisfies the conditions of the theorem.

Denote by g_n the polynomial $X(1-X)^n$, and denote by P_n the polynomial $\sum_{k=1}^n (k+1)\binom{k}{n} X^k$. Then P_n is the derivative of $Q_n = \sum_{k=1}^n \binom{k}{n} X^{k+1} = X(1+X)^n - X$, and so $P_n = (1+X)^n - 1 + nX(1+X)^{n-1}$. Note that all the coefficients of P_n are positive. We have $\lim_{n\to\infty} P_n(1/3n) = \exp(1/3) - 1 + 1/3 \exp(1/3) < 1$, so $P_n(1/3n) < 1$ eventually, and, if n is large enough, we have $|P_n(z)| < 1$ if $|z| < 1/3n$.

Now $g_n(0) = 0$, $g_n'(0) = 1$, and, if $z, z' \in \mathbb{C}$,

$$g_n(z) - g_n(z') = \sum_{k=0}^{n} (-1)^k \binom{k}{n} (z^{k+1} - z'^{k+1})$$

$$= (z - z')\Big[1 + \sum_{k=1}^{n} (-1)^k \binom{k}{n} (z^k + z^{k-1}z' + \cdots + z'^k) \Big]$$

$$= (z - z')(1 + u), \quad \text{where} \quad |u| \le P_n[\sup(|z|, |z'|)] .$$

This shows that, if n is large enough, then the function $z \mapsto g_n(z)$ is univalent over the open disc $D_{1/3n} = \{z \in \mathbb{C} \mid |z| < 1/3n\}$. It follows from the theorem that we have in $\mathfrak{B}$, for n large enough,

$$\inf\{ \|x(e - x)^n\| \mid \|x\| \ge 1/3n\} \ge \frac{1}{3.32n} = \frac{1}{96n} .$$

Since $b \ne 0$, $\|b\| \ge 1/3n$ eventually, and we obtain $\liminf_{n\to\infty} n\|a^n - a^{n+1}\| = \liminf_{n\to\infty} n\|b(e - b)^n\| \ge 1/96$. This proves the corollary.

REMARK 9.6. In the proof of Theorem 9.4 we used Caratheodory's theorem, which shows that, if g is analytic over the open unit disc and satisfies $g(0) = 0$, $g(z) \ne 0$ if $0 < |z| < 1$, then the image of the open unit disc under g contains the open disc $D_{1/16} = \{z \in \mathbb{C} \mid |z| < 1/16\}$. This theorem is sharp in the sense that there exists a function g satisfying the above condition such that $g(z) \ne 1/16$ for $0 \le |z| < 1$, but the value $1/16$ is of course rather crude if we work with a concrete function like the functions g_n introduced above. Also, in the proof of the theorem we used Cauchy's inequalities to get estimates for the function f associated to g in the proof of the theorem, but if all Taylor coefficients of f are positive it is possible to get much better estimates as in the example of Section 4. So the value $1/96$ in Corollary 9.5 is certainly very crude. A study of estimates for $g_n(x)$ when $x \in C_0[0,1]$, the algebra of continuous functions over $[0,1]$ vanishing at 0, suggests that the best possible estimate for $\liminf_{n\to\infty} n\|a^n - a^{n+1}\|$ in Corollary 9.5 could be $\exp(-1)$. (Note that, if a is any invertible element in a unital Banach algebra such that $a \ne e$, then the spectral radius formula gives $\liminf_{n\to\infty} \|xa^n - xa^{n+1}\|^{1/n} \ge \nu(a^{-1}) - 1$ if $x \in \mathfrak{G}$, $ax \ne x$.) We will not go here into further investigations concerning Theorem 9.4 and Corollary 9.5.

We now give a rather short proof of a "hyperinvariant subspace theorem" due to Beauzamy.

THEOREM 9.7. (Beauzamy [9]) Let E be a Banach space, and let T be a linear contraction acting on E. Assume that $\|T^n(x_0)\| \not\to 0$ as $n \to \infty$ for some

$x_0 \in E$, and assume that there exists an increasing sequence $(\omega_n)_{n\geq 1}$ of positive real numbers and a sequence $(x_n)_{n\geq 1}$ of elements of E, with $x_1 \neq 0$, satisfying $\omega_{n+m} \leq \omega_n\omega_m$ $(n,m \in \mathbb{N})$, $\sum_{n=1}^{\infty} \log \omega_n/(1+n^2) < \infty$ and $x_n = T(x_{n+1})$ $(n \in \mathbb{N})$. Then either T is a scalar multiple of the identity, or T has a proper hyperinvariant subspace.

<u>Proof</u>. Put $\omega_n = 1$ if $n \leq 0$. Since $\omega_1 \leq \omega_2 \leq \omega_1^2$, $\omega_1 \geq 1$ and the function $n \mapsto \omega_n$ is increasing over $\mathbb{Z}$. Also, $\sum_{n\in\mathbb{Z}} \log \omega_n/(1+n^2) < \infty$, and $\omega_{n+m} \leq \omega_n\omega_m$ if $n,m \in \mathbb{Z}$.

Since we are looking for hyperinvariant subspaces, we may assume without loss of generality that T is one-to-one. Also, $F = \{x \in E \mid \lim_{n\to\infty} \|T^n x\| = 0\}$ is a closed hyperinvariant subspace for T which is different from E, so we may assume that $F = \{0\}$.

Since T is one-to-one, there exists for every $x \in T^n(E)$ a unique $y \in E$ such that $T^n(x) = y$, and we may denote this element y by $T^{-n}(x)$. So T^n is well defined for every $n \in \mathbb{Z}$ as a linear operator from $\bigcap_{n\geq 1} T^n(E)$ into itself, and $T^{n+m}(x) = T^n(T^m(x))$ for every $x \in \bigcap_{n\geq 1} T^n(E)$ and every $n,m \in \mathbb{Z}$. Put

$$G = \{x \in \bigcap_{n\geq 1} T^n(E) \mid \sup_{n\geq 1} \frac{\|T^{-n}(x)\|}{\omega_n} < \infty\}.$$

Then G is clearly a linear subspace of E. Let

$$|||x||| = \sup_{n\geq 0} \frac{\|T^{-n}(x)\|}{\omega_n} \quad (x \in G).$$

Since $\omega_0 = 1$, $|||x||| \geq \|x\|$ for every $x \in G$. Also, $|||x||| = \sup_{n\in\mathbb{Z}} \|T^{-n}(x)\|/\omega_n$, as T is a contraction and as $\omega_n = 1$ if $n < 0$, and $(G, |||.|||)$ is a normed space. It follows from the hypothesis that $G \neq \{0\}$.

Denote by L the set of all sequences $(y_n)_{n\geq 0}$ of elements of E satisfying $\sup_{n\geq 0} \|y_n\|/\omega_n < \infty$. Then L is a Banach space with respect to the norm $(y_n)_{n\geq 0} \mapsto \sup_{n\geq 0} \|y_n\|/\omega_n$. Also, $M = \{(y_n)_{n\geq 0} \in L \mid y_n = T(y_{n-1})\ (n \geq 1)\}$ is a closed subspace of L, and $y_n = T^n(y_0)$ for every $n \geq 1$. The map $(y_n)_{n\in\mathbb{N}} \mapsto y_0$ is an isometry from L onto G, which shows that $(G, |||.|||)$ is a Banach space.

Denote by $\mathfrak{U}$ the commutant of T in $\mathfrak{L}(E)$. If $R \in \mathfrak{U}$, then $R(x) \in \bigcap_{n\in\mathbb{N}} T^n(E)$ for every $x \in \bigcap_{n\in\mathbb{N}} T^n(E)$ and $T^n[R(x)] = R[T^n(x)]$ for every $n \in \mathbb{Z}$. This shows that $R(G) \subset G$ for every $R \in U$, and $|||\hat{R}||| \leq \|R\|$, where $\hat{R} = R \mid G$ and where $|||S||| = \sup_{|||x||| \leq 1} |||S(x)|||$ for $S \in \mathfrak{L}(G)$. So $\hat{T}$ is a contraction over G.

Note that

$$\sup_{n\geq 0} \frac{\|T^{-n-p}(x)\|}{\omega_n} = \sup_{n\geq 0} \frac{\|T^{-n-p}(x)\|}{\omega_{n+p}} \times \frac{\omega_{n+p}}{\omega_n} \leq \omega_p \sup_{n\in\mathbb{Z}} \frac{\|T^{-n}(x)\|}{\omega_n}$$

$$\leq \omega_p \, |||x||| \quad (x \in h,\ b \in \mathbb{Z}).$$

So $T^{-1}(x) \in G$ for every $x \in G$, $\tilde{T}$ is invertible in $\mathcal{L}(G)$, and $|||\hat{T}^p||| \leq \omega_{-p}$ for every $p \in Z$.

If $Sp(\hat{T})$ were a singleton λ, we would have $|||(\hat{T} - \lambda I)\hat{T}^n||| \to 0$ as $n \to \infty$ by Theorem 9.1. (The fact is obvious for $|\lambda| < 1$, and is given by Theorem 9.1 applied to $\bar{\lambda}T$ if $|\lambda| = 1$.) So $|||T^n(T - \lambda I)x||| \to 0$, as $n \to \infty$ and hence $\|T^n(T - \lambda I)x\| \to 0$ as $n \to \infty$ for every $x \in G$. Thus, $(T - \lambda I)(G) \subset F = \{0\}$. So in that case λ is an eigenvalue for T, and T has a proper hyperinvariant subspace if $T \neq \lambda I$. Hence, we may assume that $Sp(\hat{T})$ is not a singleton.

A classical theorem of Wermer [49] now shows that $\hat{T}$ has a proper hyperinvariant subspace, because $\|\hat{T}^p\| \leq \omega_{-p}$ $(p \in \mathbb{Z})$ with $\sum_{n\in\mathbb{Z}} \log \omega_n/(1+n^2) < \infty$. More precisely, as indicated by Atzmon in [7], this hyperinvariant subspace can be obtained as follows. The algebra $\ell^1(\mathbb{Z}, \omega) = \{\sum_{n\in\mathbb{Z}} \lambda_n X^n \mid \sum_{n\in\mathbb{Z}} |\lambda_n|\omega_n < \infty\}$ is a completely regular Banach algebra in the sense of Šilov. This implies that, if χ_1, χ_2 are two distinct characters of $\ell^1(\mathbb{Z}, \omega)$, there exist two elements a and b of $\ell^1(\mathbb{Z}, \omega)$ such that $\chi_1(a) = 1$, $\chi_2(b) = 1$, $ab = 0$. (Here, products are "convolution" products, $(\sum_{n\in\mathbb{Z}} \lambda_n X^n)(\sum_{n\in\mathbb{Z}} \mu_n X^n) = \sum_{n\in\mathbb{Z}} (\sum_{p\in\mathbb{Z}} \lambda_p \mu_{n-p})X^n$.) The map $\varphi : \sum_{n\in\mathbb{Z}} \lambda_n X^n \mapsto \sum_{n\in\mathbb{Z}} \lambda_n \hat{T}^n$ is a continuous algebra homomorphism from $\ell^1(\mathbb{Z}, \omega)$ into $\mathcal{L}(G)$, and since $Sp(\hat{T})$ is not a singleton there exist at least two distinct characters χ_1 and χ_2 of $\ell^1(\mathbb{Z}, \omega)$ such that $\operatorname{Ker}\varphi \subset \operatorname{Ker}\chi_1$ and $\operatorname{Ker}\varphi \subset \operatorname{Ker}\chi_2$. Taking a and b as above, we obtain $\varphi(a)\varphi(b) = \varphi(ab) = 0$, $\varphi(a) \neq 0$, $\varphi(b) \neq 0$. In particular, $V = \operatorname{Ker}\varphi(a)$ is a closed subspace of G satisfying $V \neq \{0\}$, $V \neq G$. Since $\varphi(a) = \lim_{m\to\infty} \sum_{|n|\leq m} \lambda_n \hat{T}^n$, where $(\lambda_n)_{n\in\mathbb{Z}}$ is a suitable family of complex numbers, we have $\hat{R}\varphi(a) = \varphi(a)\hat{R}$, so that $R(V) = \hat{R}(V) \subset V$ for every $R \in \mathfrak{A}$.

Denote by Ω the closure of V in E. Then Ω is hyperinvariant with respect to T, and $\Omega \neq \{0\}$. We assumed that the linear seminorm $p : x \mapsto \lim_{n\to\infty} \|T^n(x)\|$ is actually a norm over E. Denote by H the completion of E with respect to p, and put $p(S) = \sup_{p(x)\leq 1} p(S(x))$ for $S \in \mathcal{L}(H)$. According to Remark 9.3 every element R of $\mathfrak{A}$ is continuous over E with respect to p, and can be extended to an element $\tilde{R}$ of $\mathcal{L}(H)$ satisfying $p(\tilde{R}) \leq \|R\|$. Also, $\tilde{T}$ is an isometry, and, since we assumed that $T(E)$ is dense in E, we see a fortiori that $\tilde{T}(H)$ is dense in H, and hence equals H. Thus, $\tilde{T}$ is an invertible isometry.

The map $\psi : \sum_{n\in\mathbb{Z}} \lambda_n X^n \mapsto \sum_{n\in\mathbb{Z}} \lambda_n \tilde{T}^n$ is an algebra homomorphism from $\ell^1(\mathbb{Z})$ into $\mathcal{L}(H)$. Of course, $\tilde{T}^{-1} | G = \hat{T}^{-1}$, and since $(\omega_n)_{n\geq 1}$ is increasing we have $\ell^1(\mathbb{Z}, \omega) \subset \ell^1(\mathbb{Z})$. If $c = \sum_{n\in\mathbb{Z}} \lambda_n X^n \in \ell^1(\mathbb{Z}, \omega)$, we have $|||\varphi(c)(x) - \sum_{|n|\leq m} \lambda_n \hat{T}^n(x)||| \to 0$, as $n \to 0$, and hence $p[\varphi(c)(x) - \sum_{|n|\leq m} \lambda_n \hat{T}^n(x)] \to 0$ as $n \to \infty$. Thus, $\varphi(c)(x) = \psi(c)(x)$ for every $x \in G$.

Let a be as above, and let $y \in G$ with $\varphi(a)(y) \neq 0$. Then $\psi(a)(y) \neq 0$. If $\Omega = E$, then $V = \operatorname{Ker} \varphi(a)$ would be dense in E with respect to the given norm, hence with respect to p, and we would have $\operatorname{Ker} \psi(a) \supset V$, so that $\operatorname{Ker} \psi(a) = H$, $\psi(a) = 0$, a contradiction. This achieves the proof of the theorem.

COROLLARY 9.8. (Colojoara-Foias [17], p. 136, Corollary 1.10.) Let T be a weakly compact contraction acting over a Banach space E such that $\|T^n(x)\| \not\to 0$ as $n \to \infty$ for some $x \in E$, and such that $\|\ell \circ T^n\| \not\to 0$ as $n \to \infty$ for some $\ell \in E^*$. Then either T is a scalar multiple of the identity, or T possesses a proper hyperinvariant subspace.

<u>Proof</u>. The sequence $(\|\ell \circ T^n\|)_{n\in\mathbb{N}}$ is decreasing since T is a contraction, so there exists $\alpha > 0$ such that $\|\ell \circ T^n\| \geq 2\alpha$ for every n, and there exists a sequence $(x_n)_{n\in\mathbb{N}}$ of elements of E of norm 1 satisfying $|\ell[T^n(x_n)]| \geq \alpha$ for every n.

For $p \geq 0$, denote by S_p the weak closure of the set $\{T^{n-p}(x_n)\}_{n\geq p+1}$. Since $\|T^k(x_n)\| \leq 1$ for every $k \geq 0$ and every $n \in \mathbb{N}$, S_p is weakly compact for every p. Also, $T(S_{p+1}) \subset S_p$ for every p. Denote by Δ the topological cartesian product $\prod_{p\geq 0} S_p$. Tychonoff's theorem shows that Δ is compact, and since T is weakly continuous the set $U_q = \{(y_p)_{p\geq 0} \in \Delta \mid y_p = T(y_{p+1}),$ $p = 0,1,\dots,q\}$ is closed in Δ for every $q \geq 0$. The sequence (U_q) is decreasing. Now fix $q \geq 0$ and put $y_p = T^{q+2-p}(x_{q+2})$ if $p \leq q + 1$, $y_p = T(x_{p+1})$ if $p > q + 1$. Then the sequence $(y_p)_{p\geq 0}$ belongs to U_q, and so $U_q \neq \emptyset$ for every q. Thus, $\bigcap_{q\in\mathbb{N}} U_q \neq \emptyset$, and we may find a sequence $(u_p)_{p\geq 0}$ of elements of E satisfying $u_p = T(u_{p+1})$ and $u_p \in S_p$ for every $p \geq 0$. Since $\|x_n\| \leq 1$ for every n, we have $\|u_p\| \leq 1$ for every $p \geq 0$. Also, $|\ell(u)| \geq \alpha$ for every $u \in S_0$, and certainly $u_0 \neq 0$. The result then follows immediately from the theorem (just take $\omega_n = 1$ for every $n \geq 1$).

Beauzamy actually gives in [9] two theorems, Theorem 1 and Theorem 2, which seem more general than Theorem 9.7. In fact, these two theorems reduce to Theorem 9.7. Rome notices in [43] that the operator U introduced in Theorem 1 of [9] must equal T if $\lim_{n\to\infty} \|T^n(x)\| > 0$ for every $x \neq 0$, (which is immediate), so this theorem reduces to Theorem 9.7. Theorem 2 of [9] involves another operator U equal to T and a rather complicated technical condition, but the argument used in the proof of Corollary 9.8 shows that, if this condition is satisfied for $\varepsilon = 1/2$, then T satisfies in fact the conditions of Theorem 9.7. Note that, in the case of Corollary 9.8, the proof of Theorem 9.7 associates to the weakly compact contraction T two Banach spaces G and H, where G is continuously embedded in E and E is continuously embedded in H, such that T admits a restriction to G and an extension to H which are isometries. In fact, it is possible to define $f(T)$ for $f \in \ell^1(\mathbb{Z})$ to be the smallest closed extension of $f(\hat{T})$, where $\hat{T} = T \mid G$, and to obtain a functional calculus involving unbounded closed operators. The fact that $f(\hat{T})$ has a closed extension to E follows from the relationships between $\hat{T}$ and the extension $\tilde{T}$ of T to H.) Similar remarks hold for $\ell^1(\mathbb{Z}, \omega)$ if T satisfies the conditions of Theorem 9.7, and the hyperinvariant subspace for T is given by the non-trivial kernel of the possibly unbounded closed operator $f(T)$ for some $f \in \ell^1(\mathbb{Z}, \omega)$.

We now wish to improve some results proved or announced by Atzmon in [7]. Denote by $\mathcal{A}_k$ the algebra of all analytic functions f over the open unit disc such that $f, f', \ldots, f^{(k)}$ admit a continuous extension to the closed unit disc $\bar{D}$. The algebra $\mathcal{A}_k$ (equipped with poinwise product) is, as is well known, a Banach algebra with respect to the norm

$$f \mapsto \|f\|_k = \sum_{p=0}^{k} \sup_{|z|<1} \frac{|f^{(p)}(z)|}{p!}.$$

We first need an easy lemma, suggested by the classical resolvent formula for continuous semigroups.

We set

$$u_{t,k}(z) = (z-1)^{2k+1} \exp\left(\frac{t(z+1)}{(z-1)}\right) \qquad (|z| < 1,\ t \geq 0).$$

LEMMA 9.9. The function $u_{t,k}$ belongs to $\mathcal{A}_k$ for every $t \geq 0$ and $k \in \mathbb{N}$. The map $t \mapsto u_{t,k}$ is continuous from $[0,\infty)$ into $\mathcal{A}_k$, and $\|u_{t,k}\|_k = \underline{O}(t^k)$ as $t \to \infty$ for every $k \geq 0$. Moreover, if $\operatorname{Re} \lambda > 0$, then $\lambda(\alpha-1) - \alpha - 1$ is invertible in $\mathcal{A}_k$ (where α is the position function $z \mapsto z$) and $(\alpha-1)^{2k+2}[\lambda(\alpha-1)-\alpha-1]^{-1}$ equals the Bochner integral $\int_0^\infty \exp(-\lambda t) u_{t,k} dt$ (the integral being computed in $\mathcal{A}_k$).

<u>Proof</u>. The function $u_{t,k}$ is in fact analytic on $\mathbb{C} - \{1\}$, and an easy induction shows that

$$(u_{t,k})^{(p)} = (z-1)^{2k+1-2p} Q_p(z,t) \exp\left(t \frac{z+1}{z-1}\right) \text{ if } z \neq 1,$$

where $Q_p(z,t)$ is a polynomial in two variables whose degree with respect to t is less than or equal to p. Since the function $z \mapsto \exp(t(z+1)/(z-1))$ is bounded by 1 over $\overline{D}\setminus\{1\}$, we see that $u_{t,k}^{(p)} \in \mathfrak{A}(D)$ for every $p \le k$, so that $u_{t,k} \in \mathfrak{A}_k$. Since the degree of $Q_p(z,t)$ with respect to t is less or equal to p, we have $\|u_{t,k}\|_k = \underline{O}(t^k)$ as $t \to \infty$.

If $\delta > 0$, put $\Omega_\delta = \{z \in \overline{D} \mid |z-1| \ge \delta\}$. Clearly, if $t_0 > 0$ is fixed, then

$$\sup_{z\in\Omega_\delta} |u_{t_0,k}^{(p)}(z) - u_{t,k}^{(p)}(z)| \to 0 \text{ as } t \to t_0 \text{ with } t > 0,$$

for every $\delta > 0$ and every $p \le k$, and, if $s > 0$ and $\varepsilon > 0$ are given, there exists $\delta > 0$ such that $|u_{t,k}^{(p)}(z)| < \varepsilon$ for every $z \in D\setminus\{1\}$ such that $|z-1| < \delta$, for every $p \le k$ and for every $t \in [0,s]$. It follows that the map $t \mapsto u_{t,k}$ is continuous from $[0,\infty)$ into $\mathfrak{A}_k$ for every $k \ge 0$.

The characters of $\mathfrak{A}_k$ are the maps $\chi_z : f \mapsto f(z)$ where $|z| \le 1$. Since $\mathrm{Re}((z+1)/(z-1)) \le 0$ if $|z| \le 1$, $z \ne 1$, the function $\lambda(\alpha-1)-\alpha-1$ is invertible in $\mathfrak{A}_k$ if $\mathrm{Re}\,\lambda > 0$. Since $\|u_{t,k}\| = \underline{O}(t^k)$ as $t \to \infty$, the Bochner integral $\int_0^\infty \exp(-\lambda t)u_{t,k}dt$ exists in $\mathfrak{A}_k$ for every $k \ge 0$ provided that $\mathrm{Re}\,\lambda > 0$. Since Bochner integrals commute with continuous linear forms, we obtain, for $z \in D$,

$$\begin{aligned}\left[\int_0^\infty \exp(-\lambda t)u_{t,k}dt\right](z) &= \int_0^\infty \exp(-\lambda t)\,\chi_z(u_{t,k})dt \\ &= \int_0^\infty (z-1)^{2k+1} \exp\left[-\lambda t + t\left(\frac{z+1}{z-1}\right)\right]dt \\ &= \frac{(z-1)^{2k+1}}{\lambda - \left(\frac{z+1}{z-1}\right)} = \frac{(z-1)^{2k+2}}{\lambda(z-1)-(z+1)} \\ &= \left[(\alpha-1)^{2k+2}[\lambda(\alpha-1)-\alpha-1]^{-1}\right](z).\end{aligned}$$

So

$$\int_0^\infty \exp(-\lambda t)u_{t,k}dt \quad\text{and}\quad (\alpha-1)^{2k+2}[\lambda(\alpha-1)-\alpha-1]^{-1}$$

agree over D, and hence are equal, which achieves the proof of the lemma.

LEMMA 9.10. Let $\mathfrak{A}_k$, α and $u_{t,k}$ be as in Lemma 9.9. Then $u_{t,k}$ belongs to the closed span of the set $\{(1-\alpha)^{n+p}\, u_{t,k}\}_{n\ge 0}$ in $\mathfrak{A}_k$ for every $p \ge 0$ and every $t \ge 0$.

Proof. We have $u_{t,k}^{(m)} = (\alpha-1)^{2k-2m}\, v_m$ for every $m \le k$, where $v_m \in \mathfrak{M}_1 = \{f \in \mathfrak{A}(D) \mid f(1) = 0\}$. Put $e_q = (\alpha-1)(\alpha-1-1/q)^{-1}$ $(q \ge 1)$. Clearly, $\|e_q\| \le 1$ and e_q converges uniformly to 1 over every compact subset of $\overline{D}\setminus\{1\}$, so that $\|fe_q - f\| \to 0$ as $q \to \infty$ for every $f \in \mathfrak{M}_1$. Also $e_q \in \mathfrak{A}_k$.

If $1 \leq r \leq m \leq k$, we have

$$u_{t,k}^{(m-r)} e_q^{(r)} = \frac{(-1)^r r!}{q} \left(\alpha - 1 - \frac{1}{q} \right)^{-r-1} v_m(\alpha - 1)^{2k-2m+2r} .$$

Since $2k - 2m + r - 1 \geq 0$, $\|u_{t,k}^{(m-r)} e_q^{(r)}\| \to 0$ as $q \to \infty$, and, using the Leibniz rule, we see that $\|(u_{t,k} e_q)^{(m)} - u_{t,k}^{(m)}\| \to 0$ as $q \to \infty$ for every $m \leq k$. So $\|u_{t,k} e_q - u_{t,k}\|_k \to 0$ as $q \to \infty$.

Also the power series expansion of $(\alpha - 1 - 1/q)^{-1}$ has a radius of convergence strictly larger than 1, and so does the power series expansion of all its derivatives. So $(\alpha - 1 - 1/q)^{-1} = \sum_{n \geq 0} a_n(q) \alpha^n$ where the series is convergent in $\mathcal{G}_k$, and $(\alpha - 1 - 1/q)^{-1} \in \overline{\text{lin}\{\alpha^n\}_{n \geq 0}}, e_q \in \overline{\text{lin}\{(\alpha - 1)^n\}_{n \geq 1}}$ for every q. So $u_{t,k} \in \overline{\text{lin}\{(\alpha - 1)^n u_{t,k}\}_{n \geq 1}}$, and $(\alpha - 1) u_{t,k} \in \overline{\text{lin}\{(\alpha - 1)^n u_{t,k}\}_{n \geq 2}}$. An easy induction shows then that $\overline{\text{lin}\{(\alpha - 1)^n u_{t,k}\}_{n \geq p}} = \overline{\text{lin}\{(\alpha - 1)^n u_{t,k}\}_{n \geq 1}}$, and the lemma is proved.

The following lemma is a simple reformulation of a classical theorem of Paley and Wiener about elements of $L^2(\mathbb{R})$ which are the restriction to the real line of entire functions of exponential type.

LEMMA 9.11. Let f be a measurable function over $[0, \infty)$ such that $\sup_{t \geq 0} |f(t)|/|1 + t|^k < \infty$ for some $k \geq 0$, and let F be the Laplace transform of f. If $\sup_{n \in \mathbb{N}} |F^{(n+p)}(1)|^{1/n} < \infty$ for some $p \in \mathbb{Z}$, then $\inf\{\delta \geq 0 \mid f \equiv 0$ a.e. over $[\delta, \infty)\} = \limsup_{n \to \infty} |F^{(n+q)}(1)|^{1/n}$ for every $q \in \mathbb{Z}$.

<u>Proof</u>. An easy verification that we omit shows that $\limsup_{n \to \infty} (u_n)^{1/n} = \limsup_{n \to \infty} (u_{n+p})^{1/n}$ for every $p \in \mathbb{Z}$ and every sequence (u_n) of positive reals. So $\limsup_{n \to \infty} |F^{(n+q)}(1)|^{1/n} = \limsup_{n \to \infty} |F^{(n)}(1)|^{1/n}$ $(q \in \mathbb{Z})$. Put $c = \limsup_{n \to \infty} |F^{(n)}(1)|^{1/n}$. Since c is finite, F is in fact an entire function, and an elementary argument given in [12], Theorem 2.2.10 and 2.2.11 shows that F is of exponential type c (which means that $c = \inf\{r > 0 \mid \sup_{z \in \mathbb{C}} |F(z)| e^{-r|z|} < \infty\}$).

Now put $G(z) = F(z + 1)$ $(z \in \mathbb{C})$. Then $G(iy) = \int_0^\infty e^{-t} f(t) e^{-ity} dt$ $(y \in \mathbb{R})$, so the function $g : y \mapsto G(iy)$ is the Fourier transform of an L^2-function, hence belongs to $L^2(\mathbb{R})$. A classical theorem of Paley and Wiener ([12], Theorem 6.8.1) shows that there exists an L^2-function φ over $[-c, +c]$ such that $G(iz) = \int_{-c}^{c} \varphi(t) e^{-itz} dt$ $(z \in \mathbb{C})$ since $z \mapsto G(iz)$ is of exponential type c. The uniqueness theorem for Fourier transforms shows that $\varphi(t) = 0$ a.e. if $t \leq 0$,

$\varphi(t) = e^{-t}f(t)$ a.e. if $0 \leq t \leq c$, and $f(t) = 0$ a.e. over $[c,\infty)$. Moreover, $f(t)$ cannot vanish a.e. over $[c',\infty)$ for every $c' < c$, because otherwise F would be of exponential type strictly less than c. This proves the lemma.

If $x \in E$ and $\ell \in E^*$ (where E is a Banach space and E^* its dual), we will denote by $\langle x,\ell\rangle$ the complex number $\ell(x)$. Also, if $T \in \mathcal{L}(E)$ we will denote as usual by ${}^tT$ the map $\ell \mapsto \ell \circ T$ $(\ell \in E^*)$. If $T \in \mathcal{L}(E)$, $x \in E$, we will write Tx instead of $T(x)$. We have the following theorem, strongly connected with Theorems 1 and 1* in Atzmon's paper [7].

THEOREM 9.12. Let E be a Banach space, and let $T \in \mathcal{L}(E)$ be such that $\|T^n\| = \underline{O}(n^k)$ as $n \to \infty$ for some integer $k \geq 0$.

(1) If there exists a nonzero $x \in E$ and a nonzero $\ell \in E^*$ such that $|\langle (T-I)^n x,\ell\rangle|^{1/n} = \underline{O}(1/n)$ as $n \to \infty$ then T has a proper invariant subspace, and, if $\langle (T-I)^{2k+3}x, \ell\rangle \neq 0$, then T has an uncountable chain of proper invariant subspaces.

(2) If there exists a nonzero $x \in E$ such that $\|(T-I)^n x\|^{1/n} = \underline{O}(1/n)$ as $n \to \infty$, and if $T \neq I$, then T has a proper hyperinvariant subspace. If, further $(T-I)^{2k+3}x \neq 0$, then T has an uncountable chain of hyperinvariant subspaces.

(3) If there exists a nonzero $\ell \in E^*$ such that $\|({}^tT-I)^n\ell\|^{1/n} = \underline{O}(1/n)$ as $n \to \infty$, and if $T \neq I$, then T has a proper hyperinvariant subspace. If, further, $({}^tT-I)^{2k+3}\ell \neq 0$, then T has an uncountable chain of hyperinvariant subspaces.

<u>Proof</u>. If $f : z \mapsto \sum_{n=0}^{\infty} a_n z^n$ $(|z| < 1)$ belongs to $\mathcal{G}_{k+2}$, Cauchy's inequalities imply that $\sup_{n\geq 0} n^{k+2}|a_n| < \infty$. More precisely, there exists a constant M independent of f such that $\sup_{n\geq 0} (n+1)^{k+2}|a_n| \leq M\|f\|_{k+2}$. So $\sum_{n=0}^{\infty} \|a_n T^n\| < \infty$, and the series $\sum_{n=0}^{\infty} a_n T^n$ converges in $\mathcal{L}(E)$ to an operator that we denote by $f(T)$. The map $\varphi : f \mapsto f(T)$ is a continuous algebra homomorphism from $\mathcal{G}_{k+2}$ into $\mathcal{L}(E)$, and, if $\alpha : z \mapsto z$ is the position function, we have of course $\alpha(T) = T$.

Since k is fixed, we will write u_t for $u_{t,k+2}$ to simplify notations. Also the operator $T-I$ will be denoted by S.

Since $u_t u_{t'} = (\alpha-1)^{2k+3}u_{t+t'}$, we have $u_t(T)u_{t'}(T) = S^{2k+3}u_{t+t'}$ $(t,t' \geq 0)$. If $x \in E$ and $t \geq 0$, put $F_{x,t} = \overline{\mathrm{lin}\{T^n u_t(T)x\}}_{n\geq 0}$. The closed linear space $F_{x,t}$ is invariant for T (but may equal E or reduce to $\{0\}$). We have of course, $F_{x,t} = \overline{\mathrm{lin}(S^n u_t(T)(x))}_{n\geq 0}$. Since $u_{t'}(T) \in \overline{\mathrm{lin}\{T^n\}}_{n\geq 0}$ in $\mathcal{L}(E)$, we have $F_{u_{t'}(T)x,t} \subset F_{x,t}$ $(t,t' \geq 0)$. Now if $t' < t$, put $s = t-t'$. Then $u_s(T)u_{t'}(T)\cdot x = S^{2k+3}u_t(T)\cdot x$. Since $u_s(T)\,u_{t'}(T)\cdot x \in F_{x,t'}$, we obtain $S^{2k+3}u_t(T)\cdot x \in F_{x,t'}$ and $S^p u_t(T)\cdot x \in F_{x,t'}$ for every $p \geq 2k+3$. It follows from Lemma 9.10 that $F_{x,t} = \overline{\text{lin}\{S^p \cdot u_t(T)\cdot x\}}_{p \geq 2k+1}$, so that $F_{x,t} \subset F_{x,t'}$ if $t' \leq t$.

Now let $t \geq 0$, $s > 0$ and assume that $F_{x,t} = F_{x,t+s}$. Then $u_t(T)\cdot x \in F_{x,t+s}$, so

$$S^{2k+3}u_{t+t'}(T)\cdot x \in u_{t'}(T)(F_{x,t+s}) = \overline{\text{lin}\{S^n u_{t'}(T)u_{t+s}(T)\cdot x\}}_{n \geq 0}$$

$$= \overline{\text{lin}\{S^{2k+3+n}u_{t'+t+s}(T)\cdot x\}}_{n \geq 0} \subset F_{t+t'+s,x}\,.$$

Since $F_{t+t',x} = \overline{\text{lin}\{S^n u_{t+t'}(T)\cdot x\}}_{n \geq 2k+3}$ (here we use again Lemma 9.10), we obtain $F_{t+t'+s,x} = F_{t+t',x}$ for every $t' \geq 0$. We thus see that if there exists a sequence (t_n) such that $t_n \to t$ as $n \to \infty$, $t_n \leq t$ and $u_{t_n}(T)\cdot x \notin F_{x,t}$ for every n, then $F_{s,x} \neq F_{r,s}$ if $s \leq t$, $r \leq t$, $r \neq s$.

Now put $G_t = \text{Ker}\, u_t(T)$ $(t \geq 0)$. It follows from Lemma 9.10 that $G_t = \text{Ker}\, S^p u_t(T)$ for every $p \geq 0$. Let $t,s \geq 0$ and let $x \in G_t$. Then $u_s(T)u_t(T)\cdot x = 0$, so $S^{2k+3}u_{t+s}(T)\cdot x = 0$ and $x \in \text{Ker}\, S^{2k+3}u_{t+s}(T) = G_{t+s}$. So $G_{t'} \subset G_t$ if $t,t' \geq 0$, $t' \leq t$.

Let $t \geq 0$, $s > 0$ and assume that $G_{t+s} = G_t$. Let $t' > 0$, and let $x \in G_{t+t'+s}$. Then $u_{t+s}(T)u_{t'}(T)\cdot x = S^{2k+3}u_{t+t'+s}(T)\cdot x = 0$, so that $u_{t'}(T)\cdot x \in G_{t+s} = G_t$, $u_t(T)\,u_{t'}(T)\cdot x = 0$, $x \in \text{Ker}\, u_t(T)\,u_{t'}(T) = \text{Ker}\, S^{2k+3}u_{t+t'}(T) = G_{t+t'}$, and $G_{t+t'+s} \subset G_{t+t'}$, hence $G_{t+t'} = G_{t+t'+s}$. Thus if $t \geq 0$ and if there exists a sequence (t_n) of positive reals such that $t_n \to t$, with $t_n \geq t$ and $G_{t_n} \neq G_t$ for every n, then $G_r \neq G_s$ if $r \leq t$, $s \leq t$, $r \neq s$.

Since Bochner integrals commute with continuous linear maps, we have, for $x \in E$, $\ell \in E^*$, and $\text{Re}\,\lambda > 0$:

$$\langle S^{2k+4}[\lambda S - S - 2I]^{-1}\cdot x, \ell\rangle = \int_0^\infty e^{-\lambda t}\,\langle u_t(T)\cdot x, \ell\rangle dt\,.$$

Put

$$g_{x,\ell}(\lambda) = \langle S^{2k+4}[\lambda S - S - 2I]^{-1}\cdot x, \ell\rangle \quad (\text{Re}\,\lambda > 0)\,.$$

Then

$$g_{x,\ell}(\lambda) = -\frac{1}{2}\left\langle S^{2k+4}\left[I - \frac{(\lambda-1)}{2}\,S\right]^{-1}\cdot x, \ell\right\rangle$$

$$= -\frac{1}{2}\sum_{n \geq 0} \langle S^{2k+4+n}\cdot x, \ell\rangle\;\frac{(\lambda-1)^n}{2^n} \qquad (|\lambda-1| < 1)\,.$$

So $g_{x,\ell}^{(n)}(1) = -(n!/2^{n+1})\langle S^{2k+4+n}\cdot x,\ell\rangle$ $(n \geq 0)$, and, if we put $x_p = S^p\cdot x$, we have more generally $g_{x_{p,\ell}}^{(n)}(1) = -(n!/2^{n+1})\langle S^{p+n+2k+4}\cdot x,\ell\rangle$ for every $n \geq 0$ and every $p \geq 0$, so $|g_{x_{p,\ell}}^{(n-2k-4-p)}(1)|^{1/n} \sim n|\langle S^n\cdot x,\ell\rangle|^{1/n}/2e$ as $n \to \infty$.

Now assume that x, ℓ and T satisfy the first condition of the Theorem. Put $\delta = \limsup_{n\to\infty} n|\langle S^n\cdot x,\ell\rangle|^{1/n}$. Since $g_{x_{p,\ell}}$ is the Laplace transform of the function $t \to \langle u_t(T)\cdot x_p,\ell\rangle$, it follows from Lemma 9.11 that $\langle u_t(T)\cdot x_{p,\ell}\rangle = 0$ for every $t \geq \delta/2e$, and that there exists a sequence (t_n) such that $0 \leq t_n < \delta/2e$, $\langle u_{t_n}(T)\cdot x,\ell\rangle \neq 0$ for every $n \in \mathbb{N}$ and $t_n \to \delta/2e$ as $n \to \infty$ in the case where $\delta > 0$.

If $\delta = 0$, then $\langle T^p(T-I)^{2k+3}\cdot x,\ell\rangle = \Sigma_{j=0}^{p} \binom{j}{p} \langle u_0(T)S^j\cdot x,\ell\rangle = 0$ for every $p \geq 0$. If $(T-I)^{2k+3}\cdot x \neq 0$, the set $\{T^p\cdot(T-I)^{2k+3}\cdot x\}_{p\geq 0}$ is a proper invariant subspace for T, as $\ell \neq 0$. If $(T-I)^{2k+3}\cdot x = 0$, then $\mathrm{Ker}(T-I)^{2k+3} \neq 0$, and T possesses a proper hyperinvariant subspace if $T \neq I$ (and of course, an invariant subspace if $T = I$).

Now assume that $\delta > 0$, (which is certainly the case if $\langle (T-I)^{2k+3}\cdot x,\ell\rangle \neq 0$), put $t_0 = \delta/2e$, and let (t_n) be the sequence introduced above. Then $F_{x,t_0} = \mathrm{lin}\{T^p u_{t_0}(T)\cdot x\}_{p\geq 0} = \mathrm{lin}\{u_{t_0}(T)\cdot S^p\cdot x\}_{p\geq 0} \subset \mathrm{Ker}\,\ell$, and $u_{t_n}(T)\cdot x \notin \mathrm{Ker}\,\ell$ for every $n \in \mathbb{N}$. So $F_{x,t_n} \neq F_{x,t_0}$ for every $n \in \mathbb{N}$, and the family $(F_{x,t})_{0\leq t<t_0}$ gives an uncountable chain of closed invariant subspaces for T. This proves the first assertion.

Now assume that there exists a nonzero $x \in E$ such that $\|(T-I)^n\cdot x\|^{1/n} = \underline{O}(1/n)$ as $n \to \infty$. If $u_0(T)\cdot x = (T-I)^{2k+3}\cdot x = 0$, then $\mathrm{Ker}(T-I)$ is a proper hyperinvariant subspace for T. If $u_0(T)\cdot x \neq 0$ then there exists $s > 0$ such that $u_s(T)\cdot x \neq 0$ since the map $t \mapsto u_t(T)$ is continuous. Now put $c = \limsup n\|(T-I)^n\cdot x\|^{1/n}/2e$. It follows from the above discussion that $\langle u_c(T)\cdot x,\ell\rangle = 0$ for every $\ell \in E^*$, so that $u_c(T)\cdot x = 0$. Put $t_0 = \inf\{t > 0 \mid x \in \mathrm{Ker}\,u_t(T)\}$. Since $\mathrm{Ker}\,u_t(T) \subset \mathrm{Ker}\,u_{t'}(T)$ if $t < t'$, we have $c \geq t_0 > 0$, and, as the map $t \mapsto u_t(T)$ is continuous, we have $x \in \mathrm{Ker}\,u_{t_0}(T)$. Then $u_{t_0+1/n}(T)\cdot x \neq 0$, $\mathrm{Ker}\,u_{t_0+1/n}(T) \neq \mathrm{Ker}\,u_{t_0}(T)$ for every $n \in \mathbb{N}$, and the family $\{\mathrm{Ker}\,u_t(T)\}_{0\leq t\leq t_0}$ gives an uncountable chain of closed hyperinvariant subspaces for T.

Assume next that there exists a nonzero $\ell \in E^*$ such that $\|({}^tT-I)^n\ell\|^{1/n} = \underline{O}(1/n)$ as $n \to \infty$. If $({}^tT-I)^{2k+3}\ell = 0$, then $\mathrm{Im}(T-I)^{2k+3} \subset \mathrm{Ker}\,\ell$

and $[\mathrm{Im}(T-I)]^-$ gives a proper hyperinvariant subspace for T if $T \neq I$. If not, there exists $t_0 > 0$ such that $\mathrm{Ker}\ {}^t(u_t(T)) \subsetneq \mathrm{Ker}\ {}^t(u_{t'}(T))$ if $0 \leq t \leq t' \leq t_0$. Since $\mathrm{Ker}\ {}^t(u_t(T)) = [\mathrm{Im}\ u_t(T)]^\perp$, we see that the family $[\mathrm{Im}\ u_t(T)]^-_{0 \leq t \leq t_0}$ gives an uncountable chain of hyperinvariant subspaces for T. (Recall that $u_t(T)$ is a uniform limit of polynomials in T.) This achieves the proof of the theorem.

Theorems 1 and 1* in Atzmon's paper [7] involve operators T such that $\mathrm{Sp}(T) = \{1\}$, $\|T^n\| = \underline{O}(n^k)$ as $n \to \infty$ for some $k \geq 0$, and $\|T^{-n}\| = \underline{O}(\exp(cn^{1/2}))$ (or $|\langle T^{-n}x, \ell\rangle| = \underline{O}(\exp(cn^{1/2})))$ as $n \to \infty$ for some constant $c > 0$. The bridge between Theorem 9.12 and these results is given by the following proposition.

PROPOSITION 9.13. Let T be a bounded operator acting on a Banach space E, let $x \in E$, and let $\ell \in E^*$. If $\mathrm{Sp}(T) = \{1\}$, then

$$\limsup_{n\to\infty} n|\langle (T-I)^n\cdot x, \ell\rangle|^{1/n} \leq e^2 \limsup_{|n|\to\infty} |n|^{1/2} \log|\langle T^n\cdot x, \ell\rangle| .$$

Similar inequalities hold for $\limsup_{n\to\infty} n\|(T-I)^n\cdot x\|^{1/n}$ and $\limsup_{n\to\infty} n\|(T-I)^n\|^{1/n}$.

<u>Proof</u>. Put $U = \sum_{n\geq 1} (-1)^{n+1} (T-I)^n/n$. Then $T^n = \exp(nU)$ $(n \in \mathbb{Z})$. Also, $U = (T-I)[\sum_{n\geq 0} (-1)^n (T-I)^n/(n+1)]$, so U is a quasinilpotent and $\sup_{z\in\mathbb{C}} \exp(-\varepsilon|z|)\|\exp zU\| < \infty$ for every $\varepsilon > 0$.

Note that $T - I = \exp U - I$, so $(T-I)^n = U^n + \sum_{p\geq n+1} \alpha_{p,n} U^p/p!$, where $\alpha_{p,n} = \sum_{k=0}^n (-1)^{n-k} \binom{k}{n} k^p$ is the pth derivative at the origin of the function $z \mapsto (\exp z - 1)^n$. Thus,

$$|\langle (T-I)^n\cdot x, \ell\rangle| \leq |\langle U^n\cdot x, \ell\rangle| + \sum_{p\geq n+1} \frac{2^n n^p |\langle U^p\cdot x, \ell\rangle|}{p!}$$

$$\leq (2e)^n \sup_{p\geq n} |\langle U^p\cdot x, \ell\rangle| .$$

Let $c > \limsup_{n\to\infty} n|\langle U^n\cdot x, \ell\rangle|^{1/n}$. Then $|\langle U^n\cdot x, \ell\rangle| \leq c^n/n^n$ eventually. Since the sequence (c^n/n^n) is eventually decreasing, $\sup_{p\geq n} |\langle U^p\cdot x, \ell\rangle| \leq c^n/n^n$, and $|\langle (T-I)^n\cdot x, \ell\rangle|^{1/n} \leq 2ec/n$ for n large enough. So $\limsup_{n\to\infty} n|\langle (T-I)^n\cdot x, \ell\rangle|^{1/n} \leq 2e \limsup_{n\to\infty} n|\langle U^n\cdot x, \ell\rangle|^{1/n}$. Now put $\delta = \limsup_{|n|\to\infty} |n|^{1/2} \log|\langle T^n\cdot x, \ell\rangle|^{1/n}$, and let $d > \delta$. Put $T^z = \exp(zU)$ $(z \in \mathbb{C})$. If $\mathrm{Re}\ z \geq 0$, denote by $z^{1/2}$ the square root of z which belongs to the right-hand half-plane. Then $|z|^{-1} \log|\exp[-dz^{1/2}]\cdot\langle T^z\cdot x, \ell\rangle| \to 0$ as $|z| \to \infty$ uniformly over the closed right-hand half-plane. Since $d > \delta$, the function $\exp[-dz^{1/2}]\langle T^z\cdot x, \ell\rangle$ is bounded over the positive integers. A classical theorem of Cartwright ([12], Chapter 10, Theorem 10.2.1) shows that $\sup_{t\geq 0} \exp[-dt^{1/2}]|\langle T^t\cdot x,\ell\rangle| < \infty$. Applying the same argument to T^{-t}, we see that $\sup_{t\in\mathbb{R}} \exp[-d|t|^{1/2}]|\langle T^t\cdot x,\ell\rangle| < \infty$. The function $g : z \mapsto \exp[-d\sqrt{2}\, z^{1/2}]|\langle T^{iz}\cdot x,\ell\rangle|$ is continuous for $\operatorname{Re} z \geq 0$, analytic for $\operatorname{Re} z > 0$, and of zero exponential type in the right-hand half-plane. Also, g is bounded over the vertical axis, and a refinement of the Phragmén-Lindelöf theorem given in [12], Chapter I, Theorem 1.4.3 shows that $\sup_{\operatorname{Re} z\geq 0} |g(z)| < \infty$. The same argument applied to T^{-iz} shows then that there exists $M > 0$ such that $|\langle T^z\cdot x,\ell\rangle| \leq M \exp[d\sqrt{2}\,|z|^{1/2}]$ for every $z \in \mathbb{C}$. This shows that the entire function $z \mapsto \langle T^z\cdot x,\ell\rangle$ is of order $1/2$ and of the type $\delta\sqrt{2}$ in the usual sense of [12], Chapter 1, Definition 2.1.3. Using [12], Theorem 2.2.10, we see that $\limsup_{n\to\infty} (n/e)|\langle U^n\cdot x,\ell\rangle|^{1/n} = \delta^2/2$, so that

$$\limsup_{n\to\infty} n|\langle (T-I)^n\cdot x,\ell\rangle|^{1/n} \leq e^2\delta^2 = e^2 \limsup_{|n|\to\infty} |n|^{1/2} \log|\langle T^n\cdot x,\ell\rangle| .$$

The inequalities

$$\limsup_{n\to\infty} n\|(T-I)^n\cdot x\|^{1/n} \leq e^2 \limsup_{|n|\to\infty} |n|^{-1/2} \log\|T^n x\| \quad (x \in E),$$

$$\limsup_{n\to\infty} n\|(T-I)^n\|^{1/n} \leq e^2 \limsup_{|n|\to\infty} |n|^{-1/2} \log\|T^n\|$$

follow easily. (The last inequality can be proved without using Cartwright's theorem.)

Using Proposition 9.13 we see that the first assertion of Theorem 9.12 improves Theorem 1* of [7], announced without proof. Theorem 9.12 gives an uncountable chain of hyperinvariant subspaces, and the space E is not assumed to be reflexive, as it is in [7]. Theorem 1 of [7] ensures that $(T-I)^{2k+2}\cdot u_t(T) = 0$ where $t = 2c^2$ if $\|T^n\| = \underline{O}(n^k)$ as $n \to \infty$ and if $\|T^{-n}\| = \underline{O}(\exp c|n|^{1/2})$ as $n \to -\infty$, where $c > 0$. Using assertion 2 of Theorem 9.12 and Proposition 9.13, we obtain $u_t(T) = 0$ if $t = e^2c^2$ under the same conditions, but the constant e^2 can certainly be improved because our estimate of $\langle (T-I)^n\cdot x,\ell\rangle$ was crude in the proof of Proposition 9.13. In fact, it is likely that there exists a constant λ such that

$$\limsup_{n\to\infty} n|\langle (T-I)^n\cdot x,\ell\rangle|^{1/n} = \lambda \limsup_{|n|\to\infty} |n|^{1/2} \log|\langle T^n\cdot x,\ell\rangle|^{1/n} \quad (x \in E, \ell \in E^*)$$

if $Sp(T) = \{1\}$, but we will not enter into this here.

Note that the last assertion of Theorem 9.12 suggests a way to attack the closed ideal problem. If $\mathfrak{A}$ is a radical Banach algebra with b.a.i., we can find a Banach algebra $\mathfrak{B}$ similar to $\mathfrak{A}$ such that the set $\{x \in \mathfrak{B} \mid \|e - x\| = 1\}$ is very large. If one can find in that set an element x such that $\|\ell \circ x^n\| = \underline{O}(1/n!)$ as $n \to \infty$ for some nonzero $\ell \in \mathfrak{A}^*$, then $\mathfrak{B}$, and hence $\mathfrak{A}$, possesses a proper closed ideal.

Note also that, if $(a^t)_{t>0}$ is a continuous semigroup in a Banach algebra $\mathfrak{A}$ satisfying $\sup_{t>0} \|a^t\| < \infty$, and if $u = \int_0^\infty \exp(-t)a^t dt$, it is possible to show, using standard estimates about Laguerre polynomials, that $\|u(e - u)^n\|$ is bounded. This implies that $\|(e - u)^n\| = \underline{O}(n)$ as $n \to \infty$. So the above theory can be used to find bounded continuous semigroups $(a^t)_{t>0}$ in Banach algebras such that $[a^t\mathfrak{A}]^- \neq [a^{t'}\mathfrak{A}]^-$ for some $t' \neq t$.

It is also possible to weaken the condition $\|T^n\| = \underline{O}(n^k)$ by the condition $\|VT^n\| = \underline{O}(n^k)$ as $n \to \infty$, where V commutes with T (or commutes with every operator which commutes with T). Using this idea it would be possible to control by some refinement of the last assertion of Theorem 9.12 all continuous (possibly unbounded at the origin) semigroups in Banach algebras such that $[a^t\mathfrak{A}]^- \neq [a^{t'}\mathfrak{A}]^-$ for some $t' \neq t$, but we will not enter into this here. A similar application of the Paley-Wiener theorem was given by the author in his UCLA postgraduate course to obtain hyperinvariant subspaces for operators S such that $\|S(\lambda e - S)^{-1}\| \leq 1$ for every $\lambda > 0$, (see Sinclair's notes [47], Chapter 6), but Atzmon's condition $\|(I - S)^n\| = \underline{O}(n^k)$ as $n \to \infty$ seems easier to handle. (We will not discuss here the relationship between these conditions.)

We conclude the appendix and the paper with a very short proof of an extension of a classical spectral mapping theorem of Foiaş and Mlak [28] concerning representations of H^∞ associated to a completely non-unitary contraction on a Hilbert space.

THEOREM 9.14. Let $\mathfrak{A}$ be a subalgebra of H^∞ containing $\mathfrak{A}(D)$, and let $\varphi : f \mapsto f(T)$ be a continuous representation of $\mathfrak{A}$ on a Banach space, where T is the image of the position function $\alpha : z \mapsto z$. If $\lambda \in Sp(T)$, $|\lambda| = 1$, and if $\lim\{h(z) : z \to \lambda, |z| < 1\}$ exists, then $h(\lambda) = \lim\{h(z) : z \to \lambda, |z| < 1\} \in Sph(T)$.

Proof. Put $M_\lambda = \{f \in \mathfrak{A}(D) \mid f(\lambda) = 0\}$. Then M_λ possesses a bounded approximate identity given, for example, by the sequence $e_n = (1 - \overline{\lambda}\alpha)^{1/n}$. (To define $(1 - \overline{\lambda}z)^{1/n}$ for $|z| \leq 1$, $z \neq \lambda$, take the determination of the argument which belongs to $[-\pi/2, \pi/2]$.) We have $\|e_n\| \leq 2^{1/4}$ $(n \in \mathbb{N})$ and $e_n(z) \to 0$ as $n \to \infty$ uniformly over compact subsets of $\overline{D} \setminus \{\lambda\}$. It follows immediately from these facts that $\|[h - h(\lambda) \cdot 1]e_n - h + h(\lambda) \cdot 1\| \to 0$ as $n \to \infty$.

Put $J = \varphi(1)$. Then $J^2 = J$ and $J \cdot g(T) = g(T) \cdot J = g(T)$ for every $g \in \mathfrak{A}$.

We have $\|h(T)e_n(T) - h(\lambda)e_n(T) - h(T) + h(\lambda)J\| \to 0$ as $n \to \infty$, so that $\|[h(T) - h(\lambda)I][e_n(T) - J]\| \to 0$ as $n \to \infty$. If $e_n(T) \to J$ as $n \to \infty$, then $e_n(T)$ is invertible, for n large enough, in the algebra $\mathfrak{D} = [\varphi(\mathfrak{a})]^-$ which has J as a unit element. So $J - \bar{\lambda}T = [e_n(T)]^n$ is invertible too in $\mathfrak{D}$. We obtain, denoting by B the inverse of $T - \lambda J$ in $\mathfrak{D}$ (note that $(T - \lambda I)B = TB - \lambda B = (T - \lambda J)B = J$)

$$\left[B + \frac{J - I}{\lambda}\right][T - \lambda I] = [T - \lambda I]\left[B + \frac{J - I}{\lambda}\right] = J + \frac{T(J - I)}{\lambda} + I - J = I.$$

This contradicts the fact that $\lambda \in Sp(T)$.

Thus $\limsup_{n\to\infty} \|e_n(T) - J\| > 0$. (In fact the same argument shows that $\liminf_{n\to\infty} \|e_n(T) - J\| > 0$.) Since $\|(h(T) - h(\lambda)I)(e_n(T) - J)\| \to 0$ as $n \to \infty$, $h(T) - h(\lambda)I$ is not invertible, and the theorem is proved.

References

[1] M. Akkar, Etude spectrale et structure des algèbres topologiques et bornologiques complètes, Thèse de Doctorat d'Etat de Mathématiques, Bordeaux, 1976.

[2] G. R. Allan, Elements of finite closed descent in a Banach algebra, J. London Math. Soc., 7 (1973), 462-466.

[3] __________, Ideals of rapidly growing functions, Proceedings International Symposium on functional analysis and its applications, Ibadan, Nigeria (1977), 85-109.

[4] G. R. Allan, H. G. Dales and J. P. McClure, Pseudo-Banach algebras, Studia Math., 40 (1971), 55-69.

[5] M. Altman, Contracteurs dans les algèbres de Banach, C. R. Acad. Sci. Paris Ser A-B, 274 (1972), 399-400.

[6] C. Apostol, Functional calculus and invariant subspaces, J. Operator Theory, 4 (1980), 159-190.

[7] A. Atzmon, Operators which are annihilated by analytic functions and invariant subspaces, Acta Math., 144 (1980), 27-63.

[8] W. G. Bade and H. G. Dales, Norms and ideals in radical convolution algebras, J. Functional Analysis, 41 (1981), 77-109.

[9] B. Beauzamy, Sous espaces invariants de type fonctionnel dans les espaces de Banach, Acta Math., 144 (1980), 65-82.

[10] __________, Sous espaces invariants dans les espaces de Banach, Seminaire N. Bourbaki, 549, (Février, 1980).

[11] M. Berkani, Inegalités dans les algèbres de Banach et proprietés spectrales, Thèse de 3e cycle, Bordeaux (in preparation).

[12] R. P. Boas, Entire Functions, Academic Press, New-York, 1954.

[13] F. F. Bonsall and J. Duncan, Complete Normed Algebras, Springer-Verlag, Berlin, 1973.

[14] __________, Numerical Ranges II, London Math. Soc. Lecture Note Series 10, Cambridge University Press, 1973.

[15] S. Brown, B. Chevreau and C. Pearcy, Contractions with rich spectrum have invariant subspaces, J. Operator Theory, 1 (1979), 123-126.

[16] L. Carleson, Interpolation by bounded analytic functions and the corona problem, Annals of Math., 76 (1962), 547-559.

[17] I. Colojoara and C. Foias, Theory of Generalized Spectral Operators, Gordon and Breach, New-York, 1968.

[18] P. J. Cohen, Factorization in group algebras, Duke Math. J., 26 (1959), 199-206.

[19] H. G. Dales, Automatic continuity: a survey, Bull. London Math. Soc., 10 (1978), 129-183.

[20] H. G. Dales, Convolution algebras on the real line, this Volume.

[21] A. M. Davie, Invariant subspaces for Bishop's operators, Bull. Lond. Math. Soc., 6 (1974), 343-348.

[22] Y. Domar, A solution of the translation-invariant subspace problem for weighted L^p on $\mathbb{R}$, $\mathbb{R}^+$ or $\mathbb{Z}$, this Volume.

[23] P. Enflo, On the Invariant Subspace Problem in Banach Spaces, Mittag-Leffler Institute Publications.

[24] J. Esterle, Universal properties of some commutative radical Banach algebras, J. für die Reine und ang. Math., 321 (1981), 1-24.

[25] ___________, Elements for a classification of commutative radical Banach algebras, this Volume.

[26] W. Feller, On the generation of unbounded semigroups of bounded linear operators, Annals of Math., (2) 58 (1953), 166-174.

[27] S. Fisher, The convex hull of finite Blaschke products, Bull. Amer. Math. Soc., 74 (1968), 1128-1129.

[28] C. Foias and W. Mlak, The extended spectrum of completely nonunitary contractions and the spectral mapping theorem, Studia Math., 26 (1966), 239-245.

[29] B. Sz-Nagy and C. Foias, Analyse Harmonique des Operateurs de L'espace de Hilbert, Akademiai Kiadó, Budapest, 1967.

[30] C. Foias, C. M. Pearcy and B. Sz-Nagy, Contractions with spectral radius one and invariant subspaces, Acta Sci. Math., 43 (1981), 273-280.

[31] E. Hille and R. S. Phillips, Functional Analysis and Semigroups, Amer. Math. Soc. Colloquium Publications 31, Providence, 1957.

[32] H. Hogbe-Nlend, Theorie des Bornologies et Applications, Springer-Verlag Lecture Notes, 273, 1971.

[33] ___________, Les fondements de la theorie spectrale des algèbres bornologiques, Bol. Soc. Brasil Math., 3 (1972).

[34] B. E. Johnson, Continuity of centralizers on Banach algebras, J. Lond Math. Soc., 41 (1966), 639-640.

[35] Ju Kwei Wang, Multipliers of commutative Banach algebras, Pacific J. Math., 11 (1961), 1131-1149.

[36] P. Koosis, Introduction to H_p Spaces, London Math Soc. Lecture Note Series 40, Cambridge University Press, 1980.

[37] K. Koua, Multiplicateurs et quasimultiplicateurs dans les algèbres de Banach commutatives non-unitaires à unité approchée bornée, Thèse de 3e cycle, Bordeaux, Juin 1982.

[38] ___________, Un exemple d'algèbre de Banach commutative et radicale à unité approchée bornee sans multiplicateur non-trivial, submitted.

[39] V. J. Lomonosov, Invariant subspaces for operators commuting with a compact operator, Funk. Anal. i Prilozen, 7 (1973), 55-56.

[40] B. Nyman, On the one dimensional translation group on certain function spaces, Thesis, Uppsala, 1960.

[41] H. Radjavi and P. Rosenthal, Invariant Subspaces, Springer-Verlag, New York, 1973.

[42] M. Raouyane, Image numerique et elements hermitiens dans les algèbres bornologiques multiplicativement convexes complètes, Thèse de 3e cycle, Bordeaux, Décembre 1981.

[43] M. Rome, Dilatations isometriques d'operateurs et le problème des sous espaces invariants, J. Operator Theory, 6 (1981), 39-50.

[44] 1. Segal, Nine Introductions in Complex Analysis, Notas de Matematica 80, North Holland Math. Studies, North Holland, Amsterdam, 1981.

[45] A. M. Sinclair, Bounded approximate identities, factorization, and a convolution algebra, J. Functional Analysis, 29 (1978), 308-318.

[46] __________, Cohen's factorization methods using an algebra of analytic functions, Proc. London Math. Soc., 39 (1979), 451-468.

[47] __________, Continuous Semigroups in Banach Algebras, London Math. Soc. Lecture Note Series 63, Cambridge University Press, 1982.

[48] N. Th. Varopoulos, Continuité des formes lineaires positives sur une algèbre de Banach avec involution, C. R. Acad. Sci. Paris Serie A-B, 258 (1964), 1121-1124.

[49] J. Wermer, The existence of invariant subspaces, Duke Math. J., 19 (1952), 615-622.

[50] F. Zouakia, The theory of Cohen elements, this Volume.

[51] __________, Semigroupes reels dans certaines algèbres de Banach commutatives radicales, Thèse de 3e cycle, Bordeaux, Juin, 1980.

U.E.R. de Mathématiques et Informatique
Université de Bordeaux I
351 Cours de la Libération
33405 Talence, France

THE THEORY OF COHEN ELEMENTS

F. Zouakia

Introduction

It is shown here that for each commutative, separable Banach algebra A with a bounded approximate identity, and for each totally ordered group H of cardinality $\aleph_1$, there is an injection φ of H^+ into A satisfying $\overline{\varphi(x)A} = A$ and $\varphi(x+x') = \varphi(x)\varphi(x')$ for each $x,x' \in H^+$.

This result is one of the key steps of the method of J. Esterle for the construction of a discontinuous homomorphism from $C(K)$, the algebra of continuous, complex-valued functions on an infinite compact space K ([5], [4], and [3]). It was obtained in [5] by a complicated method involving the calculation of some infinite products. The method used here, based on the property of Baire and on the theorem of Mittag-Leffler ([1, II, 3,5]), is much more direct than that given in [5] and avoids all explicit recourse to infinite products.

In Section 1 a proof of a variant of a theorem of Mittag-Leffler is given.

In Section 2, the notion of weak Cohen element is introduced. This is a weaker notion than that of Cohen element introduced in [5]. We show, using the property of Baire, that the set $\mathbb{C}$ of weak Cohen elements is a dense G_δ in $\overline{G} \cap A$, where $G = \exp A^\#$ [7]. This will allow us to construct a projective limit U which is a complete metric space and a unital abelian monoid. The image of the set $\mathbf{S}$ of non-invertible elements of U by the first projection coincides with the set of Cohen elements of A in the sense of [5].

In Section 3, we show that the positive rationals operate on U and we construct subsets $\alpha\mathbf{S}$, (e,α), and (α,β) of U which are complete metric spaces with respect to suitable metrics. The theorem of Mittag-Leffler allows us to prove that the intersections of each decreasing sequence $(\beta_n\mathbf{S})$ [respectively, $((e,\alpha_n))$, $((\alpha_n,\beta_n))$] is non-empty.

In Section 4, we adopt additive notation in U, and we make it symmetric. We thus obtain a group Δ which has the structure of a rational vector space and a preorder. The maximal totally ordered vector subspaces of Δ are "of η_1-type", and we conclude by using a theorem of Hausdorff [6].

I must thank J. Esterle for numerous fruitful discussions which I have had with him, discussions from which this article originates.

1. The Mittag-Leffler theorem

The basis of the present work is the following theorem.

THEOREM 1.1. Let (E_n) be a sequence of complete metric spaces such that, for each $n \in \mathbb{N}$, there is a continuous map $\theta_n : E_{n+1} \to E_n$ such that $\theta_n(\theta_{n+1}(E_{n+2}))$ is dense in $\theta_n(E_{n+1})$. Let $E = \varprojlim E_n$, and let π_1 be the canonical map $E \to E_1$. Then $\pi_1(E)$ is dense in $\theta_1(E_2)$.

This theorem is a variant of the following Mittag-Leffler theorem:

THEOREM 1.2. ([1, II, 3,5].) Let (E_n, f_{nm}) be a projective system of complete metric spaces, where the maps f_{nm} are uniformly continuous. Suppose that, for each $n \in \mathbb{N}$, there exists $m \geq n$ such that

$$\text{for each } p \geq m,\ f_{np}(E_p) \text{ is dense in } f_{nm}(E_m). \qquad (*)$$

Let $E = \varprojlim E_n$, and let π_n be the canonical map $E \to E_n$. Then, for each $n \in \mathbb{N}$ and each $m \geq n$ which satisfies (*), $\pi_n(E)$ is dense in $f_{nm}(E_m)$.

Let us remark that in Theorem 1.1 we only suppose that the maps θ_n are continuous, whereas Bourbaki supposes in Theorem 1.2 that the maps f_{nm} are uniformly continuous. We now give a proof of Theorem 1.1.

<u>Proof</u>. Let d_n denote a metric which defines a complete topology on E_n. Let $x \in \theta_1(E_2)$, and take $\varepsilon > 0$. Define by induction a family $(x_n)_{n \geq 2}$ such that $x_n \in E_n$ for all n, such that $x = \theta_1(x_2)$, and such that

$$d_k(\theta_k \circ \cdots \circ \theta_n(x_{n+1}), \theta_k \circ \cdots \circ \theta_{n-1}(x_n)) < \frac{\varepsilon}{2^n} \quad (k < n).$$

To construct such a family, first take $x_2 \in E_2$ such that $x = \theta_1(x_2)$. Now suppose that $x_2, x_3, \ldots, x_n$ have been chosen. Since $\theta_{n-1}(\theta_n(E_{n+1}))$ is dense in $\theta_{n-1}(E_n)$, there is a sequence (y_p) in E_{n+1} such that

$$\theta_{n-1}(\theta_n(y_p)) \to \theta_{n-1}(x_n) \text{ in } E_{n-1} \text{ as } p \to \infty.$$

Thus

$$\theta_k \circ \cdots \circ \theta_{n-1} \circ \theta_n(y_p) \to \theta_k \circ \cdots \circ \theta_{n-1}(x_n) \text{ in } E_k \text{ as } p \to \infty$$

for each $k < n$. By choosing $x_{n+1} = y_p$ for p sufficiently large, we obtain the desired inequalities.

For $k \geq 1$ and $n > k$, set

$$\gamma_{k,n} = \theta_k \circ \cdots \circ \theta_{n-1}(x_n),$$

so that $\gamma_{k,k+1} = \theta_k(x_{k+1})$. Then, for all $n > k$, $d_k(\gamma_{k,n+1}, \gamma_{k,n}) < \varepsilon/2^n$. The sequence $(\gamma_{k,n})_{n=k+1}^{\infty}$ is a Cauchy sequence in the complete space E_k, and so it converges. Set $y_k = \lim_{n\to\infty} \gamma_{k,n}$. Then

$$\begin{aligned} d_1(y_1,x) &= d_1(y_1,\theta_1(x_2)) \\ &= \lim_{p\to\infty} d_1(\gamma_{1,p},\theta_1(x_2)) \\ &\le \lim_{p\to\infty} [d_1(\gamma_{1,p},\gamma_{1,p-1}) + \cdots + d_1(\gamma_{1,3},\theta_1(x_2))] \\ &< \sum_{n=2}^{\infty} \frac{\varepsilon}{2^n} < \varepsilon . \end{aligned}$$

For each $n > k + 1$, we have

$$\gamma_{k,n} = \theta_k \circ \cdots \circ \theta_{n-1}(x_n) = \theta_k[\theta_{k+1} \circ \cdots \circ \theta_{n-1}(x_n)] = \theta_k(\gamma_{k+1,n}) .$$

Since θ_k is continuous, we have $\lim_{n\to\infty} \gamma_{k,n} = \theta_k(\lim_{n\to\infty} \gamma_{k+1,n})$, and so, for each k, $y_k = \theta_k(y_{k+1})$. Consequently, $(y_k) \in \varprojlim E_k$. This completes the proof.

We easily obtain the following corollary.

COROLLARY 1.3. Let (E_n) be a decreasing sequence of complete metric spaces such that, for each n, the injections $E_{n+1} \to E_n$ are continuous, and E_{n+2} is dense in E_{n+1} with respect to the topology of E_n. Then $\bigcap_n E_n \neq \emptyset$ (and $\bigcap_n E_n$ is dense in E_2 for the topology of E_1).

We note that Baire's theorem is a consequence of Corollary 1.3. Indeed, if (U_n) is a sequence of dense open subsets of a complete metric space E, and if $V_n = U_1 \cap \cdots \cap U_n$, then (V_n) is a decreasing sequence of dense open subsets of E, and each V_n is homeomorphic to a complete metric space.

2. Weak Cohen elements

In the sequel, A will denote a complex, commutative, separable Banach algebra, $A^\# = A \oplus \mathbb{C}e$ will be the algebra obtained by adjoining an identity to A, and $G = \exp A^\#$ (see [7]). By [7, §1.4], G is the component of the set $\mathrm{Inv}\, A^\#$ of invertible elements of $A^\#$ which contains the identity element, so that G is open and closed in $\mathrm{Inv}\, A^\#$, and hence open in $A^\#$.

An approximate identity for A is a sequence $(e_n) \subset A$ such that $xe_n \to x$ as $n \to \infty$ for each $x \in A$.

We shall later use the following remark.

REMARK 2.1. Let A be a commutative Banach algebra, and let $\alpha \in A$.

(a) If A has an approximate identity and if $\overline{\alpha A} = A$, then the map $x \mapsto \alpha x$ is injective on A.

(b) If $\overline{\alpha A} = A$, and if (f_n) is a bounded sequence in A such that $\alpha f_n \to \alpha$

as $n \to \infty$, then for each β in A it is also true that $\beta f_n \to \beta$ as $n \to \infty$.

Proof. Take $x \in A$ with $\alpha x = 0$. Then $x\alpha A = 0$, and so $xA = x\overline{(\alpha A)} = 0$. Let (e_n) be an approximate identity for A. Then $x = \lim xe_n = 0$. This proves (a), and the proof of (b) is easy.

DEFINITION 2.2. Let A be a commutative, separable Banach algebra with approximate identity bounded by 1, $A^{\#} = A \oplus \mathbb{C}e$, and $G = \exp A^{\#}$. An element x of A is a weak Cohen element if and only if there exists a sequence (x_n) of elements of G with the following properties:

(a) $x_n \to x$ as $n \to \infty$;

(b) $\|xx_n^{-1}\| < 1$ for all n;

(c) $\overline{xA} = A$.

The notion of weak Cohen element is weaker than that of Cohen element introduced by J. Esterle [5, 2.5], where the sequence (x_n) of G is taken such that the positive rationals operate on the x_n, and, in addition to conditions (a), (b), (c) of Definition 2.2, $x_n^{1/i}$ converges for each $i \in \mathbb{N}$. Let us remark that, if $x \in A$ is a weak Cohen element, then (a) and (b) imply that

$$x \cdot xx_n^{-1} - x = xx_n^{-1}(x - x_n) \to 0 \quad \text{as} \quad n \to \infty,$$

and from (c) and Remark 2.1 (b) it follows that $\beta xx_n^{-1} \to \beta$ as $n \to \infty$ for each $\beta \in A$.

Let C be the set of weak Cohen elements of A. We have the following proposition.

PROPOSITION 2.3. The set $\mathbb{C}$ is a dense G_δ in $\overline{G} \cap A$.

To prove this result, we shall use the following lemma.

LEMMA 2.4. Let A be a commutative, separable Banach algebra with approximate identity bounded by 1. Let $A^{\#} = A \oplus \mathbb{C}e$ and $G = \exp A^{\#}$. Then there exists a sequence (e_p) of elements of $\overline{G} \cap A$ such that $\|e_p\| < 1$ for each p and such that $\|\alpha e_p - \alpha\| \to 0$ as $p \to \infty$ for each element α of A.

Proof. Sinclair (see [8, page 465]) has proved that, for such an algebra A, there is a continuous semi-group $t \mapsto a^t$, $(0,\infty) \to A$, such that, for each $t > 0$, one has $\|a^t\| < 1$, $\|a^t x - x\| \to 0$ as $t \to 0+$ for each $x \in A$, and

$$sp(a^t) \subset \{z \in \mathbb{C} : 0 < |z| \leq 1 \text{ and } |\arg z| \leq t\}.$$

Here, $sp(a^t)$ denotes the spectrum of a^t.

Set $e_p = a^{\pi/2p}$ for $p \in \mathbb{N}$. We have that $-1/n \notin sp(e_p)$ for each $n,p \in \mathbb{N}$, and so $e_p + \frac{1}{n} e \in \text{Inv } A^{\#}$, the set of invertible elements of $A^{\#}$. Also,

$$\text{sp}\left(e_p + \frac{1}{n} e \right) \subset \frac{1}{n} + \left\{ z \in \mathbb{C} : 0 \leq |z| \leq 1 \text{ and } |\arg z| \leq \frac{\pi}{2p} \right\}.$$

Thus, 0 is an element of the connected, unbounded component of $\mathbb{C} \setminus \text{sp}(e_p + \frac{1}{n} e)$, and, from [2, I, Exercice 4.1], $e_p + \frac{1}{n} e \in G$. Since $e_p = \lim_{n\to\infty} (e_p + \frac{1}{n} e)$ for each $p \in \mathbb{N}$, it follows that $e_p \in \overline{G} \cap A$ for each $p \in \mathbb{N}$. This completes the proof.

Proof of Proposition 2.3. Let (f_m) be a dense sequence in A. For $n \in \mathbb{N}$ and $m \in \mathbb{Z}^+$, set

$$\Omega_{n,m} = \left\{ x \in \overline{G} \cap A : \text{there exists } y \in G \text{ such that } \|x - y\| < 1/n, \right.$$
$$\left. \|xy^{-1}\| < 1, \text{ and } \|f_m - f_m x^2 y^{-2}\| < 1/n \right\}.$$

The sets $\Omega_{n,m}$ are finite intersections of open sets of $\overline{G} \cap A$, and so are open in $\overline{G} \cap A$.

Let $x \in \bigcap_{n,m} \Omega_{n,m}$. For each $m \in \mathbb{Z}^+$ and $n \in \mathbb{N}$, there exists $y_n \in G$ such that $\|x - y_n\| < 1/n$, $\|xy_n^{-1}\| < 1$, and $\|f_m - f_m x^2 y_n^{-2}\| < 1/n$. Then $y_n \to x$ as $n \to \infty$, $\|xy_n^{-1}\| < 1$, and $f_m = \lim_{n\to\infty} f_m x^2 y_n^{-2}$ for each $m \in \mathbb{Z}^+$. But the set $\{f_m\}$ is dense in A, and so $y_n \to x$, $\|xy_n^{-1}\| < 1$, and $\overline{xA} = A$, whence $x \in C$.

Let $x \in C$, and take a sequence (x_n) of elements of G such that $x_n \to x$ as $n \to \infty$, $\|xx_n^{-1}\| < 1$, and $\overline{xA} = A$. Since $(x^2 x_n^{-2})$ is a bounded sequence in A,

$$x^2(x^2 x_n^{-2}) - x^2 = x^2 x_n^{-2}(x^2 - x_n^2) \to 0 \text{ as } n \to \infty,$$

and since $\overline{x^2 A} = A$, it follows from Remark 2.1 (b) that

$$f_m x^2 x_n^{-2} \to f_m \text{ as } n \to \infty$$

for each $m \in \mathbb{Z}^+$. For each $n \in \mathbb{N}$, $m \in \mathbb{Z}^+$, we see that, for p sufficiently large,

$$\|x_p - x\| < 1/n, \ \|xx_p^{-1}\| < 1, \text{ and } \|f_m - f_m x^2 x_p^{-2}\| < 1/n,$$

and so $x \in \bigcap_{n,m} \Omega_{n,m}$. We have shown that $C = \bigcap_{n,m} \Omega_{n,m}$, and so C is a G_δ in $\overline{G} \cap A$.

It remains to show that C is dense in $\overline{G} \cap A$. For this, it is sufficient to prove that each $\Omega_{n,m}$ is dense in $\overline{G} \cap A$, and to apply Baire's theorem (see [1]). Let $n \in \mathbb{N}$, $m \in \mathbb{Z}^+$ be fixed, and take $u \in \overline{G} \cap A$. For each $\varepsilon > 0$, there exists $v \in G$ such that $\|u - v\| < \varepsilon/2$. Let (e_p) be the sequence constructed in Lemma 2.4. For p sufficiently large, we have

$$\begin{aligned}
\|ve_p - v\| &< \min\{1/n, \varepsilon/2\}, \\
\|ve_p v^{-1}\| &< 1, \\
\|f_m - f_m e_p^2\| &\le \|e_p\| \|f_m e_p - f_m\| + \|f_m e_p - f_m\| \\
&\le 2\|f_m e_p - f_m\| < 1/n .
\end{aligned}$$

Set $z = ve_p$. Then $z \in \overline{G} \cap A$ because $e_p \in \overline{G} \cap A$, and, by the preceding calculation, $z \in \Omega_{n,m}$ and

$$\|z - u\| = \|ve_p - u\| \le \|ve_p - v\| + \|v - u\| < \varepsilon .$$

This completes the proof.

The set G is open in $A^{\#}$ and so $\mathcal{C} \cup G$ is a G_δ. It is thus homeomorphic to a complete metric space.

PROPOSITION 2.5. Let $\mathcal{C}$ be as above.

(a) If $x \in \mathcal{C}$ and $y \in \mathcal{C}$, then $xy \in \mathcal{C}$.

(b) If $x \in \mathcal{C}$ and $y \in G$, then $xy \in \mathcal{C}$.

Proof. If $x,y \in \mathcal{C}$, there are two sequences $(x_n),(y_n) \subset G$ such that $x_n \to x$ and $y_n \to y$ as $n \to \infty$, such that $\|xx_n^{-1}\| < 1$ and $\|yy_n^{-1}\| < 1$ for all n. Also, $\overline{xA} = \overline{yA} = A$. Then $(x_n y_n)$ is a sequence in G such that $x_n y_n \to xy$ and $\|xy(x_n y_n)^{-1}\| < 1$ for all n. Take $t \in A$ and $\varepsilon > 0$. Since $\overline{xA} = A$, there exists $\alpha \in A$ such that $\|t - x\alpha\| < \varepsilon/2$. But $\overline{yA} = A$, and so there exists $\beta \in A$ such that $\|\alpha - y\beta\| < \varepsilon/2\|x\|$, whence $\|t - xy\beta\| < \varepsilon$. Thus, we have $\overline{xyA} = A$, and this proves (a).

If $x \in \mathcal{C}$ and $y \in G$, there is a sequence $(x_n) \subset G$ such that $x_n \to x$, $\|xx_n^{-1}\| < 1$ for all n, and $\overline{xA} = A$. The sequence $(x_n y)$ is a sequence in G which satisfies $x_n y \to xy$ as $n \to \infty$ and $\|(xy)(x_n y)^{-1}\| < 1$ for all n. Take $t \in A$ and $\varepsilon > 0$. There exists $\alpha \in A$ such that $\|t - x\alpha\| < \varepsilon$. Set $z = y^{-1}\alpha$. Then $z \in A$ and $\|t - xyz\| < \varepsilon$. We have thus shown that $\overline{xyA} = A$, which proves (b).

Since G is stable for multiplication, it follows from 2.5 that $\mathcal{C} \cup G$ is stable for multiplication. For each $n \in \mathbb{N}$, consider the map $\theta_n : x \mapsto x^{n+1}$, $\mathcal{C} \cup G \to \mathcal{C} \cup G$. Let

$$U = \varprojlim(\mathcal{C} \cup G, \theta_n) = \{(x_n) \in (\mathcal{C} \cup G)^{\mathbb{N}} : x_{n+1}^{n+1} = x_n\} .$$

If (x_n) is an element of U, then we have:

<u>either</u> $x_1 \in G$, and hence $x_n \in G$ for all n because $x_1 = x_n^{n!}$, $C \subset A$, and $A \cap G = \emptyset$;

<u>or</u> $x_1 \in C$, and hence $x_n \in C$ for all n because $x_1 = x_n^{n!}$, G is stable for multiplication, and $A \cap G = \emptyset$.

It follows that $U = \mathcal{S} \cup \mathcal{G}$, where

$$\mathcal{S} = \{(x_n) \in U : x_1 \in C\},\ \mathcal{G} = \{(x_n) \in U : x_1 \in G\}.$$

THEOREM 2.6. The set $\mathcal{S}$ is nonempty.

<u>Proof</u>. Note that $(e,e,e,\ldots,e,e,\ldots) \in \mathcal{G}$, so that $\mathcal{G}$ is non-empty. We have $\mathcal{S} = \varprojlim (C,\theta_p)$, where C is homeomorphic to a complete metric space and the application θ_p is continuous for all p. To show that $\mathcal{S}$ is non-empty, it is sufficient to show that $\theta_p(C)$ is dense in C for all p and to then apply Theorem 1.1.

Let $x \in C$ and $p \in \mathbb{N}$. Then there exists a sequence (x_n) of elements of G such that $x_n \to x$ as $n \to \infty$, $\|xx_n^{-1}\| < 1$ for all n, and $\beta xx_n^{-1} \to \beta$ as $n \to \infty$ for all $\beta \in A$.

We have

$$\beta x^p(x_n^{-1})^p - \beta = \beta x^p(x_n^{-1})^p - \beta x^{p-1}(x_n^{-1})^{p-1} + \cdots + \beta x^2(x_n^{-1})^2 - \beta xx_n^{-1} + \beta xx_n^{-1} - \beta,$$

and $\|\beta x^p(x_n^{-1})^p - \beta\| < p\|\beta xx_n^{-1} - \beta\|$ for all $\beta \in A$. So $\beta x^p(x_n^{-1})^p \to \beta$ as $n \to \infty$ for all $\beta \in A$. In particular this holds for $\beta = x$. Put $z_n = x^{p+1}(x_n^{-1})^p$. We have $z_n \to x$ as $n \to \infty$ and by using Proposition 2.5 (b), we see that $x \cdot (x_n^{-1})^{1/(p+1)} \in C$, so that $z_n = \theta_p(x(x_n^{-1})^{1/(p+1)}) \in \theta_p(C)$. Thus $\theta_p(C)$ is dense in C, as required.

3. Properties of subsets of U

If $(x_n),(y_n) \in U$, we set $(x_n)(y_n) = (x_ny_n)$. It is easily checked that U is stable for this product, and that U is a unital abelian monoid with an identity element $(e,e,\ldots)$. If $(x_n) \in U$ is invertible, then x_1 is in G and hence $(x_n) \in \mathcal{G}$. Conversely, each element (x_n) of $\mathcal{G}$ is invertible; its inverse is (x_n^{-1}).

Let $X = (x_n) \in U$. We see by induction that, for all $n \in \mathbb{N}$,

$$x_n = x_{nj}^{(nj)!/n!} \quad (j \in \mathbb{N}).$$

Let $\mathbb{Q}^+$ be the set of strictly positive rational numbers. We can define $X^{i/j}$

for $i/j \in \mathbb{Q}^+$ in the following fashion:

$$X^{i/j} = \left(x_{nj}^{(i/j)(nj)!/n!} \right).$$

The element $X^{i/j}$ is well defined, for if $i/j = p/q$, with $j < q$, say, then

$$x_{nj} = x_{nq}^{(nj+1)\dots(nq)} \quad \text{for all } n,$$

and so

$$x_{nj}^{(i/j)(nj)!/n!} = x_{nq}^{(p/q)(nq)!/n!}$$

for each n, whence $X^{i/j} = X^{p/q}$.

We have the following proposition.

PROPOSITION 3.1.
(a) For $r \in \mathbb{Q}^+$ and $X \in U$, $X^r \in U$.
(b) For $r,s \in \mathbb{Q}^+$ and $X \in U$, $X^r X^s = X^{r+s}$.
(c) For $r \in \mathbb{Q}^+$ and $X,Y \in U$, $X^r Y^r = (XY)^r$.
(d) For $r,s \in \mathbb{Q}^+$ and $X \in U$, $(X^r)^s = X^{rs}$.

Proof. (a) Let $X = (x_n) \in U$ and $r = i/j \in \mathbb{Q}^+$. Then $X^r = (\alpha_n)$, where $\alpha_n = x_{nj}^{(i/j)(nj)!/n!}$. The α_n are elements of $\mathbb{C} \cup G$, and for each n,

$$\begin{aligned} \alpha_{n+1}^{n+1} &= x_{(n+1)j}^{(i/j)((n+1)j)!(n+1)/(n+1)!} \\ &= x_{(n+1)j}^{(i/j)(nj)!(nj+1)\dots(nj+j)/n!}. \end{aligned}$$

But $x_{nj+1}^{nj+1} = x_{nj}$, $x_{nj+2}^{(nj+1)(nj+2)} = x_{nj}, \dots,$ and so $\alpha_{n+1}^{n+1} = x_{nj}^{(i/j)(nj)!/n!} = \alpha_n$ for all n. Thus, $X^r \in U$.

(b) Let $X = (x_n) \in U$, $r = i/j \in \mathbb{Q}^+$, and $s = p/q \in \mathbb{Q}^+$. Then

$$X^r = \left(x_{nj}^{(i/j)(nj)!/n!} \right), \quad X^s = \left(x_{nq}^{(p/q)(nq)!/n!} \right).$$

For each n, we have

$$x_{nj} = x_{njq}^{(nj+1)\dots(njq)} \quad \text{and} \quad x_{nq} = x_{njq}^{(nq+1)\dots(nqj)},$$

and so

$$x_{nj}^{(i/j)(nj)!/n!}\, x_{nq}^{(p/q)(nq)!/n!} = x_{njq}^{N},$$

where

$$N = \frac{i}{j}\frac{(nj)!}{n!}(nj+1)\dots(njq) + \frac{p}{q}\frac{(nq)!}{n!}(nq+1)\dots(njq)$$
$$= \left(\frac{i}{j}+\frac{p}{q}\right)\frac{(njq)!}{n!}.$$

Hence, $X^rX^s = X^{r+s}$, as required.

(c) Let $X = (x_n) \in U$, $Y = (y_n) \in U$, and $r = i/j \in \mathbb{Q}^+$. Then

$$X^rY^r = \left(x_{nj}^{(i/j)(nj)!/n!}\, y_{nj}^{(i/j)(nj)!/n!}\right)$$
$$= \left((x_{nj}y_{nj})^{(i/j)(nj)!/n!}\right) = (XY)^r,$$

as required.

(d) Let $X = (x_n) \in U$, $r = i/j \in \mathbb{Q}^+$, and $s = p/q \in \mathbb{Q}^+$. Then $X^r = (\alpha_n)$, where $\alpha_n = x_{nj}^{(i/j)(nj)!/n!}$, and so

$$(X^r)^s = \left(\alpha_{nq}^{(p/q)(nq)!/n!}\right) = \left(x_{njq}^N\right),$$

where

$$N = \frac{i}{j}\frac{(nqj)!}{(nq)!}\frac{p}{q}\frac{(nq)!}{n!} = \frac{i}{j}\frac{p}{q}\frac{(nqj)!}{n!},$$

and so $(X^r)^s = X^{rs}$, as required.

The set U, being closed in $(\mathbb{C} \cup G)^{\mathbb{N}}$, is a complete metric space. The set $\mathbb{S}$ is closed in U because it is closed in $\mathbb{C}^{\mathbb{N}}$, and hence in $(\mathbb{C} \cup G)^{\mathbb{N}}$. We write d for the metric on U induced by the metric on $(\mathbb{C} \cup G)^{\mathbb{N}}$.

REMARK 3.2.

(a) If $\beta \in U$ and $\gamma \in \beta\mathbb{S}$, then γ has a unique representation in the form $\gamma = \beta\alpha$ with $\alpha \in \mathbb{S}$.

(b) If $\alpha \in \mathbb{S}$, there exists $\beta \in \mathbb{S}$ with $\alpha \in \beta\mathbb{S}$.

(c) If $\alpha, \beta \in \mathbb{S}$ are such that $\beta \in \alpha\mathbb{S}$, then there exists $\delta \in \mathbb{S}$ such that $\delta \in \alpha\mathbb{S}$ and $\beta \in \delta\mathbb{S}$.

<u>Proof</u>. Let $\beta = (\beta_n) \in U$, $\alpha = (\alpha_n) \in \mathbb{S}$, and $\alpha' = (\alpha'_n) \in \mathbb{S}$ with $\beta\alpha = \beta\alpha'$. There are two cases: either (i) $\beta \in G$, and so $\alpha = \alpha'$ because β is invertible; or (ii) $\beta \in \mathbb{S}$, and $\beta_n\alpha_n = \beta_n\alpha'_n$ for all n, the elements β_n being in $\mathbb{C}$. In the latter case, $\overline{\beta_n A} = A$ for all n, and so, by 2.1 (a), $\alpha_n = \alpha'_n$ for all n. Again, this shows that $\alpha = \alpha'$, and (a) is proved.

For (b), it is sufficient to take $\beta = \alpha^{1/2}$.

For (c), we have $\beta = \alpha\gamma$ with $\gamma \in \mathbf{S}$, and $\delta = \alpha\gamma^{1/2}$ satisfies the required conditions.

If $x \in U$ and $\alpha \in \mathbf{S}$ with $x \in \alpha\mathbf{S}$, we denote by x/α the unique element β of $\mathbf{S}$ such that $x = \alpha\beta$. Remark 3.2 allows us to define for $\alpha \in \mathbf{S}$ the set

$$(e,\alpha) = \{x \in \mathbf{S} : \alpha \in x\mathbf{S}\},$$

and for $\alpha,\beta \in \mathbf{S}$ such that $\beta \in \alpha\mathbf{S}$ the set

$$(\alpha,\beta) = \{x \in \mathbf{S} : x \in \alpha\mathbf{S} \text{ and } \beta \in x\mathbf{S}\}.$$

We give to $\beta\mathbf{S}$, (e,α), and (α,β) the following metrics:

$$d_1(\beta x,\beta y) = d(x,y) \quad \text{for } x,y \in \mathbf{S};$$

$$d_2(x,y) = d(\alpha/x,\alpha/y) + d(x,y) \quad \text{for } x,y \in (e,\alpha);$$

$$d_3(x,y) = d(x/\alpha,y/\alpha) + d(\beta/x,\beta/y) \quad \text{for } x,y \in (\alpha,\beta).$$

PROPOSITION 3.3. The metric spaces $\beta\mathbf{S}$, (e,α), and (α,β), given the respective metrics d_1, d_2, and d_3, are complete.

Proof. Let $(\gamma_n) = (\beta x_n)$ be a Cauchy sequence in $\beta\mathbf{S}$. Then (x_n) is a Cauchy sequence in $\mathbf{S}$, and so it converges to an element, say x, of $\mathbf{S}$. It follows that (γ_n) converges to βx in $\beta\mathbf{S}$.

Let (α_n) be a Cauchy sequence in (e,α). Then (α/α_n) and (α_n) are Cauchy sequences in $\mathbf{S}$. Hence, there exist $\gamma \in \mathbf{S}$ and $\delta \in \mathbf{S}$ such that $\alpha/\alpha_n \to \delta$ and $\alpha_n \to \gamma$ as $n \to \infty$. Thus, $\alpha = \delta\gamma$, and $d_2(\alpha_n,\gamma) = d(\alpha/\alpha_n,\delta) + d(\alpha_n,\gamma) \to 0$ as $n \to \infty$.

Finally, let $\alpha,\beta \in \mathbf{S}$ with $\beta \in \alpha\mathbf{S}$, and let (x_n) be a Cauchy sequence in (α,β). Then (x_n/α) and (β/x_n) are Cauchy sequences in $\mathbf{S}$, and so there exist two elements of $\mathbf{S}$, say γ and δ, such that $x_n/\alpha \to \gamma$ and $\beta/x_n \to \delta$ as $n \to \infty$. We have that $x_n \to \alpha\gamma$ and $\beta = \alpha(x_n/\alpha)(\beta/x_n) \to \alpha\gamma\delta$ as $n \to \infty$, and so $\alpha\gamma \in (\alpha,\beta)$ and

$$d_3(x_n,\alpha\gamma) = d(x_n/\alpha,\gamma) + d(\beta/x_n,\delta) \to 0 \quad \text{as } n \to \infty.$$

PROPOSITION 3.4. For each $\alpha \in \mathbf{S}$, there exists a sequence (α_n) of elements of $\mathcal{G}$ such that $\alpha_n \to \alpha$ and $\beta\alpha\alpha_n^{-1} \to \beta$ as $n \to \infty$ for each $\beta \in \mathbf{S}$.

Proof. Let $\alpha = (u_n) \in \mathbf{S}$. For each n, $u_{n+1}^{n+1} = u_n$, and there exists a sequence $(z_{p,n})_{p=1}^{\infty} \subset G$ such that $\|u_n z_{p,n}^{-1}\| < 1$ for each p, $z_{p,n} \to u_n$ as

$p \to \infty$, and $\delta u_n z_{p,n}^{-1} \to \delta$ as $p \to \infty$ for each $\delta \in A$.

Set $u_{n,n} = z_{p,n}$ for p sufficiently large so that

$$\|u_n - u_{n,n}\| < 1/n \quad \text{and} \quad \|u_m - u_{n,n}^{n!/m!}\| < 1/n \quad \text{for} \quad m < n .$$

Take $v \in A^{\#}$ such that $u_{n,n} = \exp v$. For each m, set

$$u_{n,m} = \exp((n!/m!)v) .$$

Then, for each m, $u_{n,m} \in G$ and $u_{n,m+1}^{m+1} = u_{n,m}$, whence $(u_{n,m})_{m=1}^{\infty} \in \mathcal{G}$.

Let us consider $(\alpha_n) = ((u_{n,m})_{m=1}^{\infty})_{n=1}^{\infty}$. This is a sequence of elements of $\mathcal{G}$, and $\alpha_n \to \alpha$ as $n \to \infty$ because, for each m, $\|u_m - u_{n,m}\| \to 0$ as $n \to \infty$ since $\|u_m - u_{n,m}\| < 1/n$ whenever $m < n$.

Let $\beta = (\beta_n) \in \mathcal{S}$. For each m, $\|u_m u_{n,m}^{-1}\| < 1$ whenever $n > m$, and so $u_m u_m u_{n,m}^{-1} - u_m = u_m u_{n,m}^{-1}(u_m - u_{n,m}) \to 0$ as $n \to \infty$. It follows from Remark 2.1 (b) that $\beta_n u_m u_{n,n}^{-1} \to \beta_m$ as $n \to \infty$ for each m, and hence that $\beta\alpha\alpha_n^{-1} \to \beta$ as $n \to \infty$.

This completes the proof.

PROPOSITION 3.5.

(a) For each $\alpha,\beta,\alpha',\beta' \in \mathcal{S}$ such that $\beta \in \alpha\mathcal{S}$, $\alpha \in \alpha'\mathcal{S}$, $\beta' \in \beta\mathcal{S}$, and for each $x \in (\alpha,\beta)$, $x\mathcal{G}$ is dense in (α,β) for the topology of (α',β').

(b) For each $\beta_1,\beta_2 \in \mathcal{S}$ such that $\beta_2 \in \beta_1\mathcal{S}$, and for each $x \in (e,\beta_1)$, $x\mathcal{G}$ is dense in (e,β_1) for the topology of (e,β_2).

(c) For each $\alpha,\alpha' \in \mathcal{S}$ such that $\alpha \in \alpha'\mathcal{S}$, and for each $x \in \alpha\mathcal{S}$, $x\mathcal{G}$ is dense in $\alpha\mathcal{S}$ for the topology of $\alpha'\mathcal{S}$.

Proof. Let us first prove (a). Take $\rho,\tau \in \mathcal{S}$ such that $\alpha = \alpha'\rho$ and $\beta' = \beta\tau$. Let $x,y \in (\alpha,\beta)$. Then there exist $\gamma,\delta,u,v \in \mathcal{S}$ such that $x = \alpha\gamma$, $\beta = x\delta$, $y = \alpha u$, and $\beta = yv$. We have $\alpha\gamma\delta = \alpha uv$, and so, by 3.2 (a), $\gamma\delta = uv$.

By Proposition 3.4, there exist two sequences (u_n) and (γ_n) in $\mathcal{G}$ such that $u_n \to u$, $\gamma_n \to \gamma$, and, for each $\beta \in \mathcal{S}$, $\beta u u_n^{-1} \to \beta$ and $\beta\gamma\gamma_n^{-1} \to \beta$ as $n \to \infty$. We can construct by induction two strictly increasing sequences $(q(n))$ and $(p(n))$ in $\mathbb{N}$ such that

$$d(\tau\delta\gamma u_{q(n)}^{-1}, \tau v) < 1/n$$

and

$$d(\tau\delta\gamma_{p(n)} u_{q(n)}^{-1}, \tau\delta\gamma u_{q(n)}^{-1}) < 1/2n .$$

Let us consider the sequence $(\omega_n) = (u_{q(n)}\gamma_{p(n)}^{-1})$ in $\mathcal{G}$. Now

$$x\omega_n/\alpha' = \rho\gamma u_{q(n)}\gamma^{-1}_{p(n)} \to \rho u = y/\alpha'$$

and

$$\beta'/x\omega_n = \delta\tau\omega_n^{-1} = \delta\tau\gamma_{p(n)}u^{-1}_{q(n)} \to \tau v = \beta'/y$$

as $n \to \infty$, and so $x\omega_n \to y$ as $n \to \infty$ for the topology of (α',β'), thus proving (a).

Now take $\tau \in \mathbf{S}$ such that $\beta_2 = \beta_1\tau$, and take $x,y \in (e,\beta_1)$. There exist $\gamma,u \in \mathbf{S}$ such that $\beta_1 = x\gamma$ and $\beta_1 = yu$. By Proposition 3.4, there exist two sequences (u_n) and (γ_n) in $\mathbf{G}$ such that $u_n \to u$, $\gamma_n \to \gamma$, and, for each $\beta \in \mathbf{S}$, $\beta u u_n^{-1} \to \beta$ and $\beta\gamma\gamma_n^{-1} \to \beta$ as $n \to \infty$. We can construct by induction two strictly increasing sequences $(q(n))$ and $(p(n))$ in $\mathbb{N}$ such that

$$d(x\gamma u^{-1}_{q(n)},y) < 1/2n$$

and

$$d(x\gamma_{p(n)}u^{-1}_{q(n)},x\gamma u^{-1}_{q(n)}) < 1/2n\,.$$

Let us consider the sequence $(\omega_n) = (\gamma_{p(n)}u^{-1}_{q(n)})$ in $\mathbf{G}$. Now

$$x\omega_n = x\gamma_{p(n)}u^{-1}_{q(n)} \to y$$

and

$$\beta_2/x\omega_n = (\beta_2/x)\gamma^{-1}_{p(n)}\, u_{q(n)} = \gamma\tau\gamma^{-1}_{p(n)}\, u_{q(n)} \to \tau u = \beta_2/y$$

as $n \to \infty$, and so $x\omega_n \to y$ in $(e_1\beta_2)$ as $n \to \infty$, thus proving (b).

Finally, take $\alpha,\alpha' \in \mathbf{S}$ such that $\alpha \in \alpha'\mathbf{S}$. Take $x,y \in \alpha\mathbf{S}$, and set $\beta = xy$, $\beta' = (xy)^2$. By (a), there is a sequence (ω_n) in $\mathbf{G}$ such that $x\omega_n \to y$ as $n \to \infty$ for the topology of (α',β'), and hence $x\omega_n \to y$ for the topology of $\alpha'\mathbf{S}$, thus proving (c).

COROLLARY 3.6.

(a) If $((\alpha_n,\beta_n))$ is a decreasing sequence such that $\alpha_{n+1} \in \alpha_n\mathbf{S}$ for each n, then $\bigcap_n (\alpha_n,\beta_n) \neq \emptyset$.

(b) If $((e,\beta_n))$ is a decreasing sequence, then $\bigcap_n (e,\beta_n) \neq \emptyset$.

(c) If (α_n) is a sequence in $\mathbf{S}$ such that $\alpha_{n+1} \in \alpha_n\mathbf{S}$, then $\bigcap_n \alpha_n\mathbf{S} \neq \emptyset$.

Proof. Let us prove (a). Note first that the injections $(\alpha_{n+1},\beta_{n+1}) \mapsto (\alpha_n,\beta_n)$ are continuous because the maps $t \mapsto t\rho$, $\mathbf{C} \to \mathbf{C}$ (where ρ is fixed in $\mathbf{C}$) are continuous and hence the maps $\alpha \to \alpha\gamma$, $\mathbf{S} \to \mathbf{S}$ (where γ is fixed in $\mathbf{S}$) are continuous. The injections are even uniformly continuous (see [1, II, page 11]). By Proposition 3.5, $(\alpha_{n+1},\beta_{n+1})$ is dense in (α_n,β_n) for the topology of $(\alpha_{n-1},\beta_{n-1})$. It follows from Corollary 1.3 that $\bigcap_n (\alpha_n,\beta_n) \neq \emptyset$.

An analogous proof gives (b) and (c).

4. Injective homomorphisms of totally ordered groups in $\mathbb{C}$

Let us now adopt additive notation for the group operation in U. We have seen that U is a unital abelian monoid on which the positive rationals operate.

We consider on $U \times U$ the relation:

$$(x,y) \sim (x',y') \text{ if and only if } x+y' = x'+y\,.$$

Introduce on $U \times U$ coordinatewise addition, and multiplication by a positive rational by the formula

$$\alpha(x,y) = (\alpha x,\alpha y) \quad (\alpha \in \mathbb{Q}^+, x\ y \in U)\,.$$

Then $\sim$ is an equivalence relation which is compatible with addition and with multiplication by a positive rational.

The set $\Delta = U \times U/\sim$ is a group. The identity element is $\overline{(0,0)}$ (where $\bar{a}$ denotes the equivalence class of an element a), and $\overline{(x,y)} + \overline{(y,x)} = \overline{(0,0)}$ for $x,y \in U$. The map $x \mapsto \overline{(x,0)}$, $U \to \Delta$, is an injection, and this allows us to identify U with $\{\overline{(x,0)} : x \in U\}$. For each $\overline{(x,y)} \in \Delta$, we have $\overline{(x,y)} = \overline{(x,0)} - \overline{(y,0)} \in U - U$. If $\alpha \in \mathbb{Q}$ with $\alpha < 0$, define $\alpha\overline{(x,y)} = (-\alpha)\overline{(y,x)} = ((-\alpha)y,(-\alpha)x)$ for $\overline{(x,y)} \in \Delta$. Then, with this addition and multiplication by a rational, Δ is a vector space over $\mathbb{Q}$.

Consider the relation $>$ on Δ defined by

$$x > y \text{ if and only if } x - y \in \mathcal{S}\,.$$

The relation $>$ is a preorder which is compatible with addition and with multiplication by a positive rational.

Let E be a maximal, totally ordered, rational vector subspace of Δ.

DEFINITION 4.1. A totally ordered vector space X is said to be of η_1-type if, for each sequence (x_n) of elements of X, there are two elements a and b of X such that $a > x_n > b$ for all n, and if, for each pair of sequences (x_n) and (y_n) of elements of X such that $x_m > y_n$ for all n, m, there exists c in X with $x_m > c > y_n$ for all n, m.

The notation of a set of η_1-type was introduced by Hausdorff in [6].

LEMMA 4.2. The set E is of η_1-type.

To prove Lemma 4.2, we use the following lemma.

LEMMA 4.3. If E is a totally ordered, rational vector subspace of a preordered space Δ, and if $E \cup \{\gamma\}$ is totally ordered for each $\gamma \in \Delta$, then the vector subspace generated by $E \cup \{\gamma\}$ is totally ordered.

Proof of Lemma 4.3. Let $x = \lambda\gamma + y$ with $\lambda \in \mathbb{Q}$, $y \in E$. If $\lambda = 0$, then $x > 0$ or $x < 0$ or $x = 0$. If $\lambda < 0$, then $-x = (-\lambda)\gamma + (-y)$, and so we need only consider the case in which λ is positive. Suppose in fact that $\lambda = 1$. Then $x = \gamma + y$. Since $E \cup \{\gamma\}$ is totally ordered, $x > 0$ or $x < 0$ or $x = 0$. The lemma is proved.

Proof of Lemma 4.2. Let (α_n) be a sequence in E which can be supposed to be increasing. If $\alpha_n \leq 0$ for all n, each positive element of E majorizes (α_n). If $\alpha_n > 0$ from some point, then $\alpha_n \in \mathcal{S}$ from that point, and we can suppose that this holds for all elements of the sequence. There are two cases: either (α_n) is eventually stationary, say from $n = m$, and then $2\alpha_m > \alpha_m \geq \alpha_n$ for all n; or, if not, one can suppose that (α_n) is strictly increasing, and then Corollary 3.6 (c) shows us that $\bigcap_n (\alpha_n + \mathcal{S}) \neq \emptyset$. Thus, there exists $\gamma \in \mathcal{S}$ which majorizes (α_n). If $\gamma \notin E$, there exists an element of E which majorizes (α_n), for otherwise $E \cup \{\gamma\}$ would be totally ordered and, by 4.3, it would follow that the subspace generated by $E \cup \{\gamma\}$ would be totally ordered, a contradiction of the maximality of E. We have shown that each sequence of E is majorized by an element of E. Each sequence (α_n) of E is minimized by an element of E because the sequence $(-\alpha_n)$ is majorized by an element of E.

Let (α_n) and (β_n) be two sequences of elements of E such that $\alpha_m < \beta_n$ for each n,m. We can suppose that (α_n) is increasing and that (β_n) is decreasing. If $\beta_n < 0$ from some point, it can be supposed that $\beta_n < 0$ for all n, and then we can bring this to the case when all elements are positive. If $\beta_n > 0$ and $\alpha_n < 0$ for all n, we have $\alpha_n < 0 < \beta_n$ for all n. If $\beta_n > 0$ and $\alpha_n = 0$ from some point, two cases arise: either (β_n) is stationary from some point, say from $n = m$, and so there exists $\gamma \in \mathcal{S}$ such that $0 < \gamma < \beta_m$, whence $\alpha_n < \gamma < \beta_n$ for all n; or, if not, one can suppose that (β_n) is strictly decreasing, and then Corollary 3.6 (b) shows us that $\bigcap_n (0,\beta_n) \neq \emptyset$. This implies that there exists $\gamma \in \mathcal{S}$ such that

$$\alpha_n \leq 0 < \gamma < \beta_n \quad \text{for all } n .$$

If $\gamma \notin E$, there is an element y of E such that $\alpha_n < y < \beta_n$ for all n, for otherwise $E \cup \{\gamma\}$ would be totally ordered and the subspace generated by $E \cup \{\gamma\}$ would be totally ordered, a contradiction.

If $\beta_n > 0$ for all n and if $\alpha_n > 0$ from some point, then it can be supposed that $\alpha_n > 0$, and hence that $\alpha_n \in \mathcal{S}$, for all n. Three cases arise (i) (α_n) and (β_n) are eventually stationary, and then there exists $\gamma \in \mathcal{S}$

such that $\alpha_n < \gamma < \beta_n$ for all n.

(ii) (α_n) is stationary from some point, but (β_n) is not. It can be supposed that $\alpha_n = 0$ from some point and that (β_n) is strictly decreasing. Then $\bigcap_n (\alpha_n, \beta_n) \neq \emptyset$, and so there exists $\gamma \in \mathbf{S}$ such that $\alpha_n < \gamma < \beta_n$ for all n.

(iii) (α_n) is not stationary from some point, and so can be supposed to be strictly increasing, (β_n) is stationary from some point or can be supposed to be strictly decreasing. Proposition 3.6 (a) shows us that $\bigcap_n (\alpha_n, \beta_n) \neq \emptyset$. Hence, there exists $\gamma \in \mathbf{S}$ such that $\alpha_n < \gamma < \beta_n$ for all n. If $\gamma \notin E$, there is an element y of E such that $\alpha_n < y < \beta_n$ for all n, for otherwise $E \cup \{\gamma\}$ would be totally ordered and the subspace generated by $E \cup \{\gamma\}$ would be totally ordered, a contradiction.

This completes the proof of Lemma 4.2.

Let ω_1 be the smallest uncountable ordinal [3]. Lemma 4.2 and Hausdorff's theorem (see [6], or [3, Remarks 3.4 and 3.6]) give us:

THEOREM 4.4. Let H be a totally ordered group such that $H = \bigcup\{H_\xi : \xi < \omega_1\}$, where (H_ξ) is an increasing family of totally ordered subgroups of H such that each subset of H_ξ has a cofinal sequence. Then there is a homomorphism $\psi : H \to E$ which preserves the order.

We have the following corollary.

COROLLARY 4.5. If H is a totally ordered group of cardinality $\aleph_1$, there exists an injective homomorphism $\varphi : H^+ \to C$.

It follows easily from Theorem 4.4 that there exist real semi-groups $(a^t)_{t>0}$ in C, but a much stronger result is found in the article of Sinclair [8]: there exist in A analytic semi-groups $(a^t)_{\mathrm{Re}\, t>0}$ such that $\overline{a^t A} = A$ for each t with $\mathrm{Re}\, t > 0$ and such that $\sup_{t \in \mathbb{R}} \|a^t\| = 1$.

References

[1] N. Bourbaki, Topologie générale, Act. Sci. Ind. (Hermann, Paris 1971).

[2] __________, Théories spectrales, Act. Sci. Ind. (Hermann, Paris 1967).

[3] J. Esterle, Solution d'un probléme d'Erdös, Gillman et Henriksen et application à l'etude des homomorphismes de C(K), Acta. Math. Acad. Sci. Hungar., 30 (1977), 113-127.

[4] __________, Sur l'existence d'un homomorphisme discontinu de C(K), Proc. London Math. Soc., (3) 36 (1978), 46-58.

[5] __________, Injection des semigroupes divisibles dans les algèbres de convolution et construction d'homomorphismes discontinus de C(K), Proc. London Math. Soc., (3) 36 (1978), 59-85.

[6] F. Hausdorff, Grundzüge der Mengenlehre, (Leipzig, 1914).

[7] C. E. Rickart, General Theory of Banach Algebras, (Van Nostrand, New York, 1960).

[8] A. M. Sinclair, Cohen's factorization method using an algebra of analytic functions, Proc. London Math. Soc., (3) 39 (1979), 451-468.

U. E. R. de Mathématiques et Informatique
Université de Bordeaux I
351 Cours de la Libération
33405 Talence, France

[Translated by H. G. Dales]

II. EXAMPLES OF RADICAL BANACH ALGEBRAS

CONVOLUTION ALGEBRAS ON THE REAL LINE

H. G. Dales[†]

1. Introduction

This expository paper gives an introduction to the theory of some Banach algebras consisting of locally integrable functions on the real line, $\mathbb{R}$. The functions of the algebras are multiplied by convolution. My aim is to provide an exposition of the basic results, from which more advanced work can grow. Most of what I have to say will be well known, but there is some reworking of the material, and in particular I shall give details of some results that are not easily available.

Some of this work is based on a manuscript which it is hoped will grow into a research monograph, to be written jointly with Jean Esterle.

Elementary treatments of some of the material here can be found in the books of Gelfand, Raikov, and Šilov ([18]), and of Hille and Phillips ([25]). An approach which centres on the theory of semi-groups in these algebras is given in a forthcoming book ([38]) by Allan Sinclair in the 'Lecture Notes in Mathematics' series of the London Mathematical Society.

Perhaps I can begin by explaining how I became interested in the algebras that are to be discussed here. At what may now be taken to be the first conference in a series, held in Los Angeles in 1974, I talked about my interest in constructing a discontinuous algebra homomorphism from the Banach algebra $C(X)$ of continuous functions on an infinite compact space X into a Banach algebra. Bade and Curtis had shown that one must look for a non-zero homomorphism $\theta : M \to R$, where M is a maximal ideal of $C(X)$ and R is a radical Banach algebra. It seemed to be a good idea to take R to be a radical Banach algebra $L^1(\omega)$, the only such algebras that I knew of that did not obviously fail. In fact it turned out to be a good choice, as the following theorem of Esterle ([14]) shows.

THEOREM 1.1. (CH) Let $\mathcal{I}$ be the class of integral domains, without identity, of cardinality c, and let ω be a radical weight on $\mathbb{R}^+$. Then $L^1(\omega)$ is universal for $\mathcal{I}$, in the sense that $L^1(\omega)$ belongs to $\mathcal{I}$, and that, if A belongs to $\mathcal{I}$, then there is an embedding $A \to L^1(\omega)$.

COROLLARY 1.2. (CH) Let X be an infinite compact space, let M be a maximal ideal of $C(X)$, and let ω be a radical weight. Then there is a non-zero (and hence discontinuous) homomorphism $M \to L^1(\omega)$.

These theorems are discussed in the survey article [10], and their dependence

† Supported by NATO Grant No. RG 073.81.

on the continuum hypothesis, CH, is explained. In fact, a great deal more is now known about the radical Banach algebras which can be the closures of the ranges of discontinuous homomorphisms from maximal ideals of $C(X)$: see [15] and [17]. A proof of the above theorem which is complete, considerably shorter, more comprehensive, and much more pleasing than the originals is now emerging, and it is hoped that it will appear in the monograph referred to above.

The organization of this paper is as follows.

In §2, I introduce the locally integrable functions on $\mathbb{R}$, weight functions on $\mathbb{R}^+$, and the algebras $L^1(\omega)$.

A theme of this paper is the use of complex function theory in this area. Basic notations and the elementary theory of the Laplace transform are given in §3, together with the deduction of Titchmarsh's convolution theorem from the Ahlfors-Heins theorem.

In §4, I discuss polynomial generators of the algebras $L^1(\omega)$. Let $u(t) = 1$ for $t \in \mathbb{R}^+$. Then it is shown that u is a polynomial generator, and this allows a rapid identification of the character space of $L^1(\omega)$. An element which is not a polynomial generator is given: this example uses Schwartz's theory of exponential sums.

In §5, Beurling algebras are discussed, and I show when spectral analysis holds for these algebras.

In §6, a proof of Nyman's theorem on the closed ideals of $L^1(\mathbb{R}^+)$ is given. The proof uses Kreĭn's theorem; that theorem is deduced from an easy special case of the Ahlfors-Heins theorem in §7.

2. Locally integrable functions

We denote by $L^1_{loc}(\mathbb{R})$ the set of locally integrable functions on $\mathbb{R}$. Thus, f belongs to $L^1_{loc}(\mathbb{R})$ if f is a complex-valued, Lebesgue measurable function on $\mathbb{R}$ such that $\int_K |f(t)|dt < \infty$ for each compact subset K of $\mathbb{R}$. Clearly, $L^1_{loc}(\mathbb{R})$ is a vector space with respect to the pointwise operations. Similarly, we can define $L^1_{loc}(\mathbb{R}^+)$, where $\mathbb{R}^+ = [0,\infty)$. Throughout, functions defined on subsets of $\mathbb{R}$ will be implicitly extended to $\mathbb{R}$ by setting them equal to 0 on the complement of their domain, and in this way we regard $L^1_{loc}(\mathbb{R}^+)$ as a subspace of $L^1_{loc}(\mathbb{R})$.

If $f,g \in L^1_{loc}(\mathbb{R})$, their <u>convolution product</u> is $f * g$, where

$$(f * g)(t) = \int_{-\infty}^{\infty} f(t-s)g(s)ds \quad (t \in S), \tag{2.1}$$

and S is the subset of $\mathbb{R}$ on which the integral is absolutely convergent. If $f,g \in L^1_{loc}(\mathbb{R}^+)$, then $f * g$ is defined almost everywhere on $\mathbb{R}^+$, and the formula becomes

$$(f*g)(t) = \int_0^t f(t-s)g(s)ds \quad (t \in \underset{\sim}{R}^+).$$ (2.2)

Now let ω be a real-valued, measurable function on $\underset{\sim}{R}^+$ such that

$$\omega(s) > 0, \quad \omega(s+t) \leq \omega(s)\omega(t) \quad (s,t \in \underset{\sim}{R}^+).$$ (2.3)

Then ω is a _weight function_ on $\underset{\sim}{R}^+$: it is a _radical_ weight function if, further,

$$\underset{t>0}{\inf} \ \omega(t)^{1/t} = 0.$$ (2.4)

The following elementary properties of weight functions can be found in Chapter 7 of the book ([25]) of Hille and Phillips, for example.

LEMMA 2.1. Let ω be a weight function. Then:
(i) $\sup \omega(K) < \infty$ for each compact $K \subset (0,\infty)$;
(ii) $\inf \omega(K) > 0$ for each compact $K \subset \underset{\sim}{R}^+$;
(iii) $\underset{t\to\infty}{\lim} \ \omega(t)^{1/t} = \underset{t>0}{\inf} \ \omega(t)^{1/t}.$

In general, weight functions are not necessarily continuous, and they may not be bounded in any neighbourhood of the origin. However, it is quite often assumed that ω is continuous on $\underset{\sim}{R}^+$.

DEFINITION 2.2. Let ω be a weight function on $\underset{\sim}{R}^+$. Then

$$L^1(\omega) = \{f \in L^1_{loc}(\underset{\sim}{R}^+) : \|f\| = \int_0^\infty |f(t)|\omega(t)dt < \infty\}.$$

The following theorem is easily checked: use Lemma 2.1 (ii) for the first part.

THEOREM 2.3. Let ω be a weight function. Then $L^1(\omega)$ is a subset of $L^1_{loc}(\underset{\sim}{R}^+)$, and $(L^1(\omega), \|\cdot\|)$ is a commutative Banach algebra with respect to the convolution product defined in equation (2.2).

We shall see later that, if ω is a radical weight function, then $L^1(\omega)$ is a radical Banach algebra.

The algebra $L^1(\omega)$ does not have an identity. If we assume that the weight function ω is bounded in a neighbourhood of 0, then $L^1(\omega)$ has a bounded approximate identity: this is an important distinction from the maximal ideals $M_1(\omega)$ of the algebras $\ell^1(\omega)$, discussed elsewhere in this volume ([3]), which do not have bounded approximate identities. For example, we can take

$$e_n(t) = \begin{cases} n & (0 \leq t \leq 1/n), \\ 0 & (t > 1/n). \end{cases}$$

Then $(\|e_n\|)$ is bounded in $L^1(\omega)$, and $f * e_n \to f$ as $n \to \infty$ for each f in $L^1(\omega)$. The algebras $L^1(\omega)$ also contain several interesting, classical analytic and continuous semi-groups: for a discussion of these semi-groups, see [38]. Of course, if $L^1(\omega)$ has a bounded approximate identity, then it factors.

It is probable that interesting counter-examples in the theory of radical Banach algebras can be constructed by considering the algebras $L^1(\omega)$ when ω is not bounded near 0.

EXAMPLES 2.4.

(i) Let $\omega(t) = 1$ $(t \in \mathbb{R}^+)$. Then $L^1(\omega)$ is the well-known Banach algebra that we denote by $L^1(\mathbb{R}^+)$. We shall discuss this algebra further in §6.

(ii) For $\sigma \geq 0$, let $\omega_\sigma(t) = e^{-\sigma t}$. Then $L^1(\omega_\sigma)$ is an algebra which is isometrically isomorphic to $L^1(\mathbb{R}^+)$: if we define $(\theta f)(t) = f(t)e^{-\sigma t}$ $(t \in \mathbb{R}^+$, $f \in L^1(\mathbb{R}^+))$, then $\theta : L^1(\mathbb{R}^+) \to L^1(\omega_\sigma)$ is an isometric isomorphism.

(iii) Let $\omega(t) = \exp(-t^\gamma)$, where $\gamma > 1$, or let $\omega(t) = \exp(-t(\log(1+t))^\gamma)$, where $\gamma \geq 1$. Then ω is a continuous, radical weight function.

(iv) Let $\omega(t) = \exp((1-t^3)/t)$. Then ω is a radical weight function, but ω is not bounded in any neighbourhood of 0.

DEFINITION 2.5. For $f \in L^1_{loc}(\mathbb{R})\setminus\{0\}$, let

$$\alpha(f) = \inf \operatorname{supp} f ,$$

and take $\alpha(0) = \infty$.

It may be that $\alpha(f) = -\infty$. If $\alpha(f) > -\infty$, then

$$\alpha(f) = \sup\{\delta : f = 0 \text{ almost everywhere on } (-\infty, \delta)\} .$$

If $\delta \geq 0$, we set

$$M_\delta(\omega) = \{f \in L^1(\omega) : \alpha(f) \geq \delta\} .$$

It is easy to check that each $M_\delta(\omega)$ is a closed ideal in the algebra $L^1(\omega)$. These ideals, together with the zero ideal, are the <u>standard ideals</u> of $L^1(\omega)$. It is an important question whether or not, in the case that ω is a radical weight function, all closed ideals of $L^1(\omega)$ are standard: it was this question which attracted a lot of attention to the algebras $L^1(\omega)$. A most remarkable positive solution to this question (for each radical weight function ω satisfying certain regularity and rate-of-growth conditions) will be presented later in this Volume by Professor Yngve Domar ([12]). Earlier partial results were obtained by G. R. Allan ([2]) and by Domar.

There are two other remarks that I should like to make at this stage about the radical algebras $L^1(\omega)$.

Firstly, the standard ideals of $L^1(\omega)$ can often be characterized by the rate of growth of the sequence $(\|f^{*n}\|^{1/n})$.

THEOREM 2.6. ([4], Theorem 3.8)

Let ω be a radical weight function such that $\sup\{\omega(s+t+\delta)/\omega(s+\delta)\omega(t+\delta) : s,t \in \underset{\sim}{R}^+\}$ is finite for each $\delta \geq 0$. Then the following are equivalent for $f \in L^1(\omega)$ and $\delta > 0$:

(a) $\alpha(f) \geq \delta$, i.e., $f \in M_\delta(\omega)$;

(b) $\lim(\|f^{*n}\|/\omega(\delta n))^{1/n} = 0$;

(c) $\lim\sup(\|f^{*n}\|/\omega(\delta n))^{1/n} < \infty$.

Secondly, we can identify the derivations on $L^1(\omega)$.

THEOREM 2.7. Let ω be a radical weight function, and let $D : L^1(\omega) \to L^1(\omega)$ be a derivation.

(i) ([26, 3(a)]) D is automatically continuous.

(ii) ([19, Theorem 2.5].) There is a locally finite measure μ such that

$$\|D\| = \sup_{t \in \underset{\sim}{R}^+} t \int_0^\infty \frac{\omega(s+t)}{\omega(t)} \, d|\mu|(s) < \infty$$

and

$$D(f)(t) = t(f * \mu)(t) \qquad (f \in L^1(\omega),\ t \in \underset{\sim}{R}^+).$$

(iii) ([19, Theorem 2.6].) The algebra $L^1(\omega)$ has a non-zero derivation if and only if there exists $b > 0$ such that $\sup\{t\omega(t+b)/\omega(t) : t \in \underset{\sim}{R}^+\} < \infty$.

3. The Laplace transform

We now introduce some tools from the theory of one complex variable that we shall use to study the algebras described above.

The idea of using complex-function theory to solve problems in what is essentially real-variable analysis is very old. Here is an example. Let (λ_n) be a sequence of positive integers. Then the set of functions $\{\exp(i\lambda_n t)\}$ is complete in the Banach space $L^2[0,1]$ if and only if there is no function $f \in L^2[0,1]$ with $f \neq 0$ and $(\mathcal{L}f)(\lambda_n) = 0$ (and hence if and only if $\Sigma\, 1/\lambda_n = \infty$). Most of this result was already proved by Szász in 1916 ([39, §4]; cf [34, 15.26]). A seminal work is, of course, Paley and Wiener ([34]), and another important early contribution is due to Carleman ([8]).

Throughout, we write $z = x + iy = re^{i\theta}$ for a complex number z.

Take $\sigma \in \underset{\sim}{R}$. We write

$$\Pi_\sigma = \{ z \in \mathbb{C} : x > \sigma \}, \qquad {}_\sigma\Pi = \{ z \in \mathbb{C} : x < \sigma \},$$

so that Π_σ and ${}_\sigma\Pi$ are, respectively, open right-hand and left-hand half-planes in $\mathbb{C}$. We shall often write Π for Π_0.

DEFINITION 3.1. Take $\sigma \in \mathbb{R}$. Then

$$A(\overline{\Pi}_\sigma) = \{F \text{ continuous and bounded on } \overline{\Pi}_\sigma, \text{ analytic on } \Pi_\sigma\},$$
$$A_0(\overline{\Pi}_\sigma) = \{F \in A(\overline{\Pi}_\sigma) : F(z) \to 0 \text{ as } z \to \infty \text{ in } \overline{\Pi}_\sigma\}.$$

We write $|\cdot|_\sigma$ for the uniform norm on $\overline{\Pi}_\sigma$. Then it is clear that $A(\overline{\Pi}_\sigma)$ and $A_0(\overline{\Pi}_\sigma)$ are uniform algebras on $\overline{\Pi}_\sigma$ with respect to the norm $|\cdot|_\sigma$, and that each character on $A_0(\overline{\Pi}_\sigma)$ is given by evaluation at a point of $\overline{\Pi}_\sigma$.

DEFINITION 3.2. If $f \in L^1_{loc}(\mathbb{R})$, then

$$F(z) = (\mathcal{L}f)(z) = \int_{-\infty}^{\infty} f(t)e^{-zt}dt \quad (z \in \mathcal{D}_F), \tag{3.1}$$

where $\mathcal{D}_F$, the domain of F, is the set of numbers z for which the integral is absolutely convergent. The function F is the <u>Laplace transform</u> of f.

The elementary properties of the Laplace transform are given in many books written with applications in mind, but these works do not usually give the results that we require in sufficient detail. A comprehensive account is given by Widder in [41].

The restriction of F to a vertical line in $\mathcal{D}_F$ is just the Fourier transform of an appropriate function in $L^1(\mathbb{R})$, and so many of the properties of the Laplace transform are immediate consequences of properties of the Fourier transform that can be found in [36], for example.

The domain $\mathcal{D}_F$ of F always has the form $I \times \mathbb{R}$, where I is an interval of $\mathbb{R}$. The function F is continuous on $\mathcal{D}_F$ and analytic on $\operatorname{int} \mathcal{D}_F$, the interior of $\mathcal{D}_F$. Further, by the Riemann-Lebesgue lemma, $F(\sigma + iy) \to 0$ as $|y| \to \infty$ for each σ with $(\sigma, 0) \in \mathcal{D}_F$.

If $f, g \in L^1_{loc}(\mathbb{R})$, and if $z \in \mathcal{D}_{\mathcal{L}f} \cap \mathcal{D}_{\mathcal{L}g}$, then $f * g$ is defined on $\mathbb{R}$, and

$$\mathcal{L}(f * g)(z) = (\mathcal{L}f)(z)(\mathcal{L}g)(z). \tag{3.2}$$

We shall need an elementary lemma.

LEMMA 3.3. Let $f \in L^1_{loc}(\mathbb{R})$, and suppose that $0 \in \operatorname{int} \mathcal{D}_F$. If $\int_{-\infty}^{\infty} t^n f(t)dt = 0$ $(n \in \mathbb{N})$, then $f = 0$.

<u>Proof</u>. If $z \in \operatorname{int} \mathcal{D}_F$, then we can differentiate under the integral sign in equation (3.1) to see that

$$F^{(n)}(z) = \int_{-\infty}^{\infty} (-t)^n f(t) e^{-zt} dt \qquad (n \in \underset{\sim}{Z}^+).$$

Apply this at $z = 0$ to see that F is constant. Since $F(iy) \to 0$ as $|y| \to \infty$, $F = 0$, and so $f = 0$.

A basic theorem on the Laplace transform is the following: it is easily checked.

THEOREM 3.4. For each $\sigma \geq 0$, the map $\mathcal{L} : L^1(\omega_\sigma) \to A_0(\overline{\Pi}_\sigma)$ is a continuous embedding, where ω_σ is as in Example 2.4 (ii).

We now relate the value $\alpha(f)$ for $f \in L^1_{loc}(\underset{\sim}{R})$ to properties of the Laplace transform of f. I shall sketch some of the proofs: this will show how far one can get by quite elementary means.

A sector of the complex plane is a set of the form $S = \{z : a < \arg z < b\}$, where $b - a \leq 2\pi$. We also regard the whole plane as a sector. In the former case, the frontier, ∂S, of S consists of the two rays $\{\arg z = a\}$ and $\{\arg z = b\}$, together with the origin.

DEFINITION 3.5. Let S be a sector. A function F is of exponential type on S if F is analytic on S, if F is continuous on $\overline{S}$, and if there exist constants $A, B > 0$ with

$$|F(z)| \leq A \exp(B|z|) \qquad (z \in S). \tag{3.3}$$

The infimum of the B's that will serve in this inequality is the type of F.

We shall use the following form of the Phragmén-Lindelöf principle. The proof is given in [5, 1.4.3], but a hypothesis is omitted in the statement of that result.

LEMMA 3.6. Let F be a function of exponential type on Π. Suppose that $|F(iy)| \leq M$ and that, for some $\varphi \in (-\pi/2, \pi/2)$, $|F(re^{i\varphi})| = \underline{O}(e^{cr \cos \varphi})$ as $r \to \infty$. Then

$$|F(z)| \leq Me^{cx} \quad (z \in \Pi).$$

THEOREM 3.7. Let $f \in L^1_{loc}(\underset{\sim}{R})$, let $F = \mathcal{L}f$, and suppose that $\mathcal{D}_F \supset \overline{\Pi}$. Then the following are equivalent for $a \in \underset{\sim}{R}$:

(a) $\alpha(f) \geq a$;

(b) $|F(z)| = \underline{O}(e^{-ax})$ as $z \to \infty$ with $z \in \overline{\Pi}$;

(c) $\alpha(f) > -\infty$, and, for some $\varphi \in (-\pi/2, \pi/2)$, $|F(re^{i\varphi})| = \underline{O}(e^{-ar \cos \varphi})$ as $r \to \infty$.

Proof. Suppose that (a) holds. If $z \in \overline{\Pi}$, then

$$|F(z)| \le \int_a^\infty |f(t)|e^{-xt}dt \le e^{-ax} \int_a^\infty |f(t)|dt \,.$$

The integral $\int_a^\infty |f(t)|dt$ is finite because $0 \in \mathcal{D}_F$. Thus, (b) holds.

Next, suppose that (b) holds. Set $f_1 = f|(-\infty,a]$, $f_2 = f|[a,\infty)$, $F_1 = \mathcal{L}f_1$, and $F_2 = \mathcal{L}f_2$, so that $F_1 + F_2 = F$. Then $|F_2(z)| = \underline{O}(e^{-ax})$ on $\overline{\Pi}$ by what we have proved, and so $|F_1(z)| = \underline{O}(e^{-ax})$ on $\overline{\Pi}$. On the other hand, if $z \in {}_0\overline{\Pi}$, then

$$|F_1(z)| \le \int_{-\infty}^a |f(t)|e^{-xt}dt \le e^{-ax} \int_{-\infty}^a |f(t)|dt \,,$$

so that $\mathcal{D}_{F_1} \supset {}_0\overline{\Pi}$ and $|F_1(z)| = \underline{O}(e^{-ax})$ on ${}_0\overline{\Pi}$. This shows that $F_1(z)e^{az}$ is a bounded entire function. By Liouville's theorem, it is a constant. But $F_1(iy) \to 0$ as $|y| \to \infty$, and so $F_1 = 0$, whence $f_1 = 0$. This shows that $\alpha(f) = \alpha(f_2) \ge a$, proving (a).

Clearly, (a) and (b) imply (c).

Finally, suppose that (c) holds. Because $\alpha(f) > -\infty$, F is of exponential type on Π, and it is also true that F is bounded on the imaginary axis. Thus, (b) follows from Lemma 3.6.

COROLLARY 3.8. The following are equivalent for $f \in L^1_{loc}(\underset{\sim}{R})$ and $a \in \underset{\sim}{R}^+$:

(a) $\operatorname{supp} f \subset [-a,a]$;

(b) $\mathcal{D}_F = \underset{\sim}{C}$, and F is an entire function of exponential type a.

Proof. If (a) holds, then $|F(z)| = \underline{O}(e^{ax})$ as $z \to \infty$ with $z \in \overline{\Pi}$. Similarly, $|F(z)| = \underline{O}(e^{-ax})$ as $z \to \infty$ with $z \in {}_0\overline{\Pi}$. Hence, F is of exponential type a.

Conversely, suppose that (b) holds. Take $\varepsilon > 0$. Then $|F(x)| = \underline{O}(e^{(a+\varepsilon)x})$ as $x \to \infty$, and, by 3.6, $|F(z)| = \underline{O}(e^{(a+\varepsilon)x})$ as $z \to \infty$ with $z \in \overline{\Pi}$. Hence $\alpha(f) \ge -a-\varepsilon$. Since this holds for each $\varepsilon > 0$, $\alpha(f) \ge -a$. The same argument applied to the function $\check{f} : t \mapsto f(-t)$ shows that $\alpha(\check{f}) \ge -a$, and so $\operatorname{supp} f \subset [-a,a]$.

COROLLARY 3.9. Let $f \in L^1_{loc}(\underset{\sim}{R}) \setminus \{0\}$ with $\alpha(f) > -\infty$, and suppose that $\mathcal{D}_F \supset \overline{\Pi}$. Then

$$\lim_{r\to\infty} \frac{1}{r} \log|F(re^{i\theta})| = -\alpha(f) \cos\theta$$

for almost all $\theta \in (-\pi/2,\pi/2)$.

Proof. By the Ahlfors-Heins theorem, Theorem 7.1(i), there is a constant $c \in \underset{\sim}{R}$ such that

$$\lim_{r\to\infty} \frac{1}{r} \log|F(re^{i\theta})| = c \cos\theta$$

for almost all $\theta \in (-\pi/2, \pi/2)$. Let $\alpha = \alpha(f)$. Then $|F(re^{i\theta})| = \underline{O}(e^{-\alpha r \cos\theta})$ for each θ, and so $c \leq -\alpha$. Now take φ such that $r^{-1}\log|F(re^{i\varphi})| \to c\cos\varphi$, and take $\varepsilon > 0$. Then $|F(re^{i\varphi})| = \underline{O}(e^{(c+\varepsilon)r\cos\varphi})$, and so $\alpha \geq -c-\varepsilon$. It follows that $c = -\alpha$, as required.

THEOREM 3.10. (Titchmarsh's convolution theorem.)

Let $f, g \in L^1_{loc}(\underset{\sim}{R}) \setminus \{0\}$ with $\alpha(f), \alpha(g) > -\infty$. Then

$$\alpha(f * g) = \alpha(f) + \alpha(g). \tag{3.4}$$

Proof. Since $(f*g)(t)$ only depends on the values of $f(s)$ and $g(s)$ for $s \leq t$, we can suppose that f and g have compact support, and hence that their transforms, F and G, are entire functions. By 3.9, there exists φ such that $r^{-1}\log|F(re^{i\varphi})| \to -\alpha(f)\cos\varphi$, $r^{-1}\log|G(re^{i\varphi})| \to -\alpha(g)\cos\varphi$, and $r^{-1}\log|(FG)(re^{i\varphi})| \to -\alpha(f*g)\cos\varphi$, where we are using (3.2). This gives the result.

COROLLARY 3.11. $L^1_{loc}(\underset{\sim}{R}^+)$ and its subalgebras are integral domains.

There are many other proofs of Titchmarsh's convolution theorem. See [27], [32], and [35], for example. It is an extension of Titchmarsh's theorem that is at the heart of Domar's new result, [12].

4. Polynomial generators

Let $\mathfrak{A}$ be a commutative Banach algebra without identity, and let a be an element of $\mathfrak{A}$.

DEFINITION 4.1. The subalgebra polynomially generated by a in $\mathfrak{A}$ is the smallest closed subalgebra of $\mathfrak{A}$ containing a: it is written $\overline{P(a)}$. The element a is a polynomial generator of $\mathfrak{A}$ if $\overline{P(a)} = \mathfrak{A}$.

Clearly, $\overline{P(a)}$ is the closure of the set of polynomials (with zero constant term) in a.

We wish to determine the polynomial generators of the radical Banach algebras $L^1(\omega)$. We shall need to know that the dual space of $L^1(\omega)$ is the Banach space $L^\infty(\omega^{-1})$, where

$$L^\infty(\omega^{-1}) = \{\varphi : \|\varphi\|_\infty = \text{ess sup}|\varphi(t)|/\omega(t) < \infty\}.$$

The duality between $L^1(\omega)$ and $L^\infty(\omega^{-1})$ is implemented by the formula

$$\langle f, \varphi\rangle = \int_0^\infty f(t)\varphi(t)dt \qquad (f \in L^1(\omega),\ \varphi \in L^\infty(\omega^{-1})). \tag{4.1}$$

Note that $L^\infty(\omega^{-1}) \subset L^1_{loc}(\underset{\sim}{R}^+)$ if and only if $\int_0^\delta \omega < \infty$ for some $\delta > 0$; if this is the case, we can take the Laplace transform of elements of $L^\infty(\omega^{-1})$.

DEFINITION 4.2. Let $u(t) = 1 \ (t \in \mathbb{R}^+)$.

The function u is of importance because convolution multiplication by u is the operation of indefinite integration:

$$(u * f)(t) = \int_0^t f(s)ds \qquad (f \in L^1_{loc}(\mathbb{R}^+)) .$$

Note that $(\mathcal{L}u)(z) = 1/z \ (z \in \Pi)$, and that

$$u^{*n}(t) = \frac{t^{n-1}}{(n-1)!} \qquad (n \in \mathbb{N},\ t \in \mathbb{R}^+) .$$

PROPOSITION 4.3. Suppose that ω is a weight function such that $\int_0^\infty \omega < \infty$. Then u is a polynomial generator of $L^1(\omega)$.

<u>Proof</u>. Take $\varphi \in L^\infty(\omega^{-1})$ with

$$\langle u^{*n}, \varphi \rangle = 0 \qquad (n \in \mathbb{N}) , \tag{4.2}$$

and let $\Phi = \mathcal{L}\varphi$. Since $\int \omega < \infty$, $0 \in \operatorname{int} \mathcal{D}_\Phi$, and, by (4.1) and (4.2), $\int t^n \varphi(t)dt = 0 \ (n \in \mathbb{N})$. By Lemma 3.3, $\varphi = 0$, and so, by the Hahn-Banach theorem, the linear span of the elements u^{*n} is dense in $L^1(\omega)$. This shows that u is a polynomial generator of $L^1(\omega)$.

Note the use of complex analysis and functional analysis in this simple result.

Now let ω be any weight function on $\mathbb{R}^+$, and set $\rho = \lim_{t\to\infty} \omega(t)^{1/t}$. If $\rho > 0$, set $\sigma = -\log \rho$. Recall [6, 16.5] that the character space of a commutative Banach algebra is identified with the space of maximal modular ideals.

THEOREM 4.4. Let ω be a weight function.

(i) If $\rho > 0$, then $L^1(\omega)$ is a semi-simple Banach algebra, and the character space of $L^1(\omega)$ is $\overline{\Pi}_\sigma$.

(ii) If $\rho = 0$, so that ω is a radical weight, then $L^1(\omega)$ is a radical Banach algebra.

<u>Proof</u>. (i) Here, $\omega(t) \geq \omega_\sigma(t) \ (t \in \mathbb{R}^+)$ by Lemma 2.1(iii), and so $L^1(\omega) \hookrightarrow L^1(\omega_\sigma)$. By Theorem 3.4, $\mathcal{L} : L^1(\omega_\sigma) \to A_0(\overline{\Pi}_\sigma)$ is a continuous embedding. For $\zeta \in \overline{\Pi}_\sigma$, let $\varphi_\zeta(f) = (\mathcal{L}f)(\zeta) \ (f \in L^1(\omega))$. Then $\iota : \zeta \mapsto \varphi_\zeta$ is a continuous map from $\overline{\Pi}_\sigma$ into the character space of $L^1(\omega)$. By considering the cases $f = \chi_{[a,b]}$, where $0 < a < b < \infty$, we see that ι is injective. We shall show that ι is also surjective.

Firstly, suppose that $\int \omega < \infty$. If $z \in \mathbb{C}$, then

$$\sum_{n=1}^{\infty} (zu)^{*n}(t) = \sum_{n=1}^{\infty} \frac{z^n t^{n-1}}{(n-1)!} = ze^{zt} \quad (t \in \underset{\sim}{R}^+),$$

and so the series $\sum_{n=1}^{\infty} (zu)^{*n}$ converges in $L^1(\omega)$ if and only if $z \notin \overline{\Pi}_\sigma$. Thus, if $z \notin \overline{\Pi}_\sigma$, then $-\sum_{n=1}^{\infty} (zu)^{*n}$ is the quasi-inverse of zu in $L^1(\omega)$. Let $sp(u)$ be the spectrum of u. Then

$$sp(u) \subset \{0\} \cup \{z : z^{-1} \in \overline{\Pi}_\sigma\} = \{0\} \cup (\mathcal{L}u)(\overline{\Pi}_\sigma).$$

Since u is a polynomial generator of $L^1(\omega)$, it follows from [6, 19.2] that ι is a surjection in this case.

The general case can be deduced from this case by straightforward technicalities.

(ii) This is immediate from (i).

There is a simple direct proof that, if ω is a radical weight, then $L^1(\omega)$ is a radical algebra. One shows by direct calculation that the elements $\chi_{[a,b]}$ are quasi-nilpotents whenever $0 < a < b$, and this is sufficient. A different way from the above of identifying the character space of $L^1(\omega)$ is given in [25, §4.4] and in [18].

The above calculation suggests the question of identifying the polynomial generators of, say, a radical Banach algebra $L^1(\omega)$. This has become a more feasible project now that Domar has solved the closed ideal problem, but it still seems to be intractable at the above level of generality. The natural guess is that, if ω is a sufficiently nice radical weight, and if $f \in L^1(\omega)$ with $\alpha(f) = 0$, then f is a polynomial generator of $L^1(\omega)$. The following example shows that this is not the case; it is a development of an example given by Ginsberg and Newman in [20].

EXAMPLE 4.5. A proper closed subalgebra of $L^1(\omega)$ which contains elements h with $\alpha(h) = 0$.

<u>Construction</u>. In this example, we take ω to be a continuous, radical weight function; we can suppose that ω is monotone decreasing ([4, 1.1]).

Define h by the formula

$$h(t) = \frac{1}{\pi^{\frac{1}{2}} t^{3/2}} \exp\left(-\frac{1}{t}\right) \quad (t \in \underset{\sim}{R}^+).$$

Then $h \in L^1(\underset{\sim}{R}^+) \subset L^1(\omega)$. We need to know the Laplace transform, H, of h.

In fact, if $x > 0$, then

$$H(x) = \frac{1}{\pi^{1/2}} \int_0^\infty \frac{1}{t^{3/2}} \exp\left(-\frac{1}{t} - tx\right) dt = \left(\frac{x}{\pi}\right)^{1/2} \int_0^\infty \frac{1}{s^{3/2}} \exp\left(-s - \frac{x}{s}\right) ds .$$

Let $F(p) = \int_0^\infty s^{-\frac{1}{2}} \exp(-s - p^2/s)ds$ for $p \geq 0$. Then F is continuous on $\mathbb{R}^+$, and we can differentiate under the integral sign for $p > 0$ to obtain

$$F'(p) = -2p \int_0^\infty \frac{1}{s^{3/2}} \exp\left(-s - \frac{p^2}{s}\right) ds \qquad (p > 0). \tag{4.3}$$

Setting $s = p^2/u$, we see that

$$F'(p) = -2 \int_0^\infty \frac{1}{u^{1/2}} \exp\left(-u - \frac{p^2}{u}\right) du \qquad (p > 0),$$

and so $F'(p) + 2F(p) = 0$ for $p > 0$, whence $F(p) = Ce^{-2p}$ $(p \geq 0)$, where C is a constant. Since $F(0) = \Gamma(\frac{1}{2}) = \pi^{\frac{1}{2}}$, $C = \pi^{\frac{1}{2}}$, and it follows from (4.3) that

$$-2\pi^{\frac{1}{2}}e^{-2p} = -2p \int_0^\infty \frac{1}{s^{3/2}} \exp\left(-s - \frac{p^2}{s}\right) ds \qquad (p > 0).$$

Hence, $H(x) = \exp(-2x^{\frac{1}{2}})$ $(x > 0)$, and so $H(z) = \exp(-2z^{\frac{1}{2}})$ $(z \in \overline{\Pi})$.

Now let $h_n(t) = n\pi^{-\frac{1}{2}}t^{-3/2}\exp(-n^2/t)$. Then

$$\begin{aligned}(\mathcal{L}h_n)(z) &= \int_0^\infty \frac{n}{\pi^{1/2}t^{3/2}} \exp\left(-\frac{n^2}{t} - zt\right) dt \\ &= \int_0^\infty \frac{1}{\pi^{1/2}t^{3/2}} \exp\left(-\frac{1}{t} - zn^2t\right) dt \\ &= (\mathcal{L}h)(n^2z) \\ &= \exp(-2nz^{1/2}) \\ &= (\mathcal{L}h)(z)^n,\end{aligned}$$

and so $h^{*n} = h_n$. This shows that $\overline{P(h)}$, the closed subalgebra of $L^1(\omega)$ which is polynomially generated by h, is equal to $\overline{\text{lin}}\{h_n : n \in \mathbb{N}\}$.

Let $\mathfrak{A} = \overline{P(h)}$, and take $f \in \mathfrak{A}$. For each $\varepsilon > 0$ and $N > 1$, there exist $a_1, \ldots, a_k \in \mathbb{C}$ with

$$\int_0^\infty \left| f(t) - \sum_1^k a_j t^{-3/2} \exp(-j^2/t) \right| \omega(t)dt < \varepsilon N^{-1/2}\omega(N).$$

Then

$$\int_0^N \left| f(t) - \sum_1^k a_j t^{-3/2} \exp(-j^2/t) \right| dt < \varepsilon N^{-1/2}.$$

Let $g(t) = t^{-3/2}f(1/t)$ $(t > 0)$. Then

$$\int_{1/N}^\infty s^{-1/2} \left| g(s) - \sum_1^k a_j \exp(-j^2 s) \right| ds < \varepsilon N^{-1/2},$$

and so

$$\int_{1/N}^{N} |g(s) - \sum_{1}^{k} a_j \exp(-j^2 s)|ds < \varepsilon .$$

We have now made contact with the theory of exponential sums expounded in the thesis of L. Schwartz ([37]), for example. Take a,b with $0 \le a < b \le \infty$, and let $\Lambda = (\lambda_j)$ be an increasing sequence of positive integers. Then $\bar{A}^1(\Lambda,a,b)$ is the closure in the Banach space $L^1[a,b]$ of the linear span of the elements $\exp(-2\pi\lambda_j s)$. We have shown that $g \in A^1((j^2/2\pi),1/N,N)$, and so, by [37, p. 58], $g \in A^1((j^2/2\pi),1/N,\infty)$. The point is that $\Sigma\, 1/j^2$ is convergent, and so this last space is a proper subspace of $L^1[1/N,\infty)$. Moreover, we can characterize g more precisely. By the 'Théorème Fondamental I' of [37], g has an analytic extension, also g, to $\Pi_{1/N}$, and g has a Dirichlet series: there are constants c_n such that

$$g(z) = \sum_{n=1}^{\infty} c_n \exp(-n^2 z) \qquad (x > 1/N), \tag{4.4}$$

the series being normally convergent in each Π_σ with $\sigma > 1/N$. Since N was arbitrary, g is analytic in Π, and the series in (4.4) is normally convergent in Π_σ for each $\sigma > 0$.

Since $f(t) = t^{-3/2}g(1/t)$ $(t > 0)$, f extends to an analytic function, also f, on Π, and the series in the expression

$$f(z) = \sum c_n z^{-3/2} \exp(-n^2/z) \qquad (z \in \Pi)$$

converges normally to f in each disc of the form $\{(x,y) : (x-q)^2 + y^2 < q^2\}$ for $q > 0$.

We see that with each $f \in \mathfrak{A}$ we can associate a unique series

$$\pi^{\frac{1}{2}} \sum_{n=1}^{\infty} \frac{c_n}{n} h_n = \pi^{\frac{1}{2}} \sum_{n=1}^{\infty} \frac{c_n}{n} h^{*n}, \tag{4.5}$$

and hence we can associate a formal power series $\Sigma_{n=1}^{\infty} \tilde{c}_n X^n$, where $\tilde{c}_n = \pi^{1/2} c_n/n$. The map $f \mapsto \Sigma\, \tilde{c}_n X^n$ is clearly an embedding of $\mathfrak{A}$ into $\mathbb{C}[[X]]$.

It follows from [37, p.34] that the coefficients c_n are continuous functions of f as f varies in $\mathfrak{A}$, and so we have identified $\mathfrak{A}$ as a Banach algebra of power series in the sense of [3]. (Since each formal series has zero constant term, and since the polynomials are dense in $\mathfrak{A}$, $\mathfrak{A}$ is also a Banach algebra of power series in the sense of Grabiner, [21].)

Note that, if $f,g \in \mathfrak{A}$, and if f,g, and $f*g$ have extensions F,G, and $F*G$ respectively, to Π, then

$$(F * G)(z) = \int_0^z F(z - \zeta)G(\zeta)d\zeta \qquad (z \in \Pi), \tag{4.6}$$

where the integral is evaluated along a contour in Π from 0 to z. This is true because the right-hand side of (4.6) is an analytic function of z which agrees with $(f * g)(z)$ for $z \in \mathbb{R}^+$.

It is clear that the algebra $\mathfrak{A}$ in the above example can bear further investigation. It seems to be an interesting example of a Banach algebra of power series: compare the algebras in Chapter 13 of [21]. I have not given an intrinsic characterization of $\mathfrak{A}$, and it is not obvious to me whether or not the series given in (4.5) converges to f in $\mathfrak{A}$.

From the characterization of $\mathfrak{A}$ that we do have, we see that, if $g \in \mathfrak{A}$ and $\alpha(g) > 0$, then $g = 0$. So there are no non-trivial standard ideals of the form $\{f \in \mathfrak{A} : \alpha(f) \geq \delta\}$. However, we do have the 'standard ideals' that arise from the representation of $\mathfrak{A}$ as a Banach algebra of power series: they are the ideals $\mathfrak{A}^n$ for $n \in \mathbb{N}$. Is each non-zero closed ideal of $\mathfrak{A}$ of this form?

The above represents all that I know about the polynomial generators of $L^1(\omega)$, and it leaves open an obvious problem.

Problem. Let ω be a radical weight. Characterize the polynomial generators of $L^1(\omega)$. In particular, if f is a polynomial generator of $L^1(\omega)$, and if $g \in L^1(\omega)$ is such that $\alpha(f - g) > 0$, is g also a polynomial generator of $L^1(\omega)$? What if $f = u$?

5. Beurling algebras

In this section, we consider algebras on $\mathbb{R}$ rather than on $\mathbb{R}^+$.

Let φ be a non-negative, measurable function on $\mathbb{R}$ such that

$$\varphi(s + t) \leq \varphi(s) + \varphi(t) \qquad (s,t \in \mathbb{R}). \tag{5.1}$$

Such a function is bounded above on compact subsets of $\mathbb{R}$.

DEFINITION 5.1. Let

$$L^1_\varphi = \{f \in L^1_{loc}(\mathbb{R}) : \|f\| = \int_{-\infty}^{\infty} |f(t)|e^{\varphi(t)}dt < \infty\}.$$

Then L^1_φ is a commutative Banach algebra without identity. These algebras were introduced by Beurling in 1938, and are now called Beurling algebras. Note that $L^1_\varphi \subset L^1(\mathbb{R})$, so that, as opposed to $L^1(\omega)$, L^1_φ contains only functions that are "small at infinity".

Condition (5.1) ensures the existence of the finite limits $\alpha = \lim_{t\to\infty} \varphi(t)/t$

and $\beta = \lim_{t\to-\infty} \varphi(t)/t$, and it follows easily from Theorem 4.4(i) that the character space of L^1_φ is the strip

$$\Phi = \{z : -\alpha \leq x \leq -\beta\}.$$

If $\alpha = \beta$, then Φ is a line. By replacing φ by $t \to \varphi(t) - \alpha t$, we can suppose that $\alpha = 0$. For $f \in L^1_\varphi$, $\mathfrak{L}f$ is continuous on Φ and analytic on $\operatorname{int}\Phi$. We denote by J_φ the important integral

$$J_\varphi = \int_{-\infty}^{\infty} \frac{\varphi(t)}{1+t^2}\,dt.$$

The following classification was introduced by Beurling:

(i) φ is <u>analytic</u> if $\beta < \alpha$;

(ii) φ is <u>quasi-analytic</u> if $\beta = \alpha = 0$ and if $J_\varphi = \infty$;

(iii) φ is <u>non-quasi-analytic</u> if $\beta = \alpha = 0$ and if $J_\varphi < \infty$.

The finiteness of the integral J_φ is important in a number of problems.

Let I be a proper closed ideal of L^1_φ, and let

$$Z(I) = \{z \in \Phi : (\mathfrak{L}f)(z) = 0 \quad (f \in I)\}.$$

Then $Z(I)$ is a closed subset of Φ, and I is contained in a maximal modular ideal of L^1_φ if and only if $Z(I) \neq \emptyset$. Let $\mathfrak{A} = L^1_\varphi/I$. Then $\mathfrak{A}$ is a radical Banach algebra if and only if $Z(I) = \emptyset$.

DEFINITION 5.2. Let L^1_φ be a Beurling algebra. Then <u>spectral analysis holds</u> for L^1_φ if each proper closed ideal of L^1_φ is contained in a maximal modular ideal.

I should like to discuss the following theorem.

THEOREM 5.3. Let L^1_φ be a Beurling algebra. Then the following conditions on φ are equivalent:

(a) spectral analysis holds for L^1_φ;

(b) $I = L^1_\varphi$ for each closed ideal with $Z(I) = \emptyset$;

(c) L^1_φ is a regular Banach algebra;

(d) φ is non-quasi-analytic.

It is trivial that (a) and (b) are equivalent. This shows that to prove that spectral analysis holds is to prove a Tauberian theorem, albeit, as Loomis ([31, p.85]) remarks, 'in a disguise that the reader may find perfect.' See Loomis for an elucidation: the result for the case in which $\varphi \equiv 0$ is Wiener's Tauberian theorem.

It was originally shown by Beurling that (d) implies (a).

Modern proofs of this proceed as follows. The heart of the matter is that, if $J_\varphi < \infty$, then

$$F(z) = \exp\left(- \frac{1}{\pi} \int_{-\infty}^{\infty} \left(\frac{tz+i}{t+iz}\right) \frac{\varphi(t)dt}{1+t^2} \right)$$

defines a function F which is analytic on Π, which is such that $|F(z)| \leq 1$, and which is such that $\lim_{x\to 0+} |F(x+iy)| = e^{-\varphi(y)}$ for almost all y. One easily then proves that L^1_φ is regular ([11]), and so (d) implies (c). That (c) implies (a) is particularly easy: if I is a closed ideal of L^1_φ with $Z(I) = \emptyset$, then I contains each $f \in L^1_\varphi$ such that $\mathcal{L}f$ has compact support, and, because L^1_φ is regular, these elements are dense in L^1_φ.

Until recently, the fact that (a) implies (d) was only known when φ satisfied some extra conditions, but the full result has now been proved by Domar, and it is discussed in this Volume ([13]).

I now describe a new proof [11], in the spirit of radical Banach algebra theory, that (d) implies (a). This proof does not involve the regularity of L^1_φ.

Let us start from the Herglotz representation. This says that, if u is a harmonic function on the open unit disc, Δ, with $u(z) \leq 0$ $(z \in \Delta)$, then there is a finite, positive measure μ on the circle such that

$$u(re^{i\theta}) = -\frac{1}{2\pi} \int_{-\pi}^{\pi} P_r(\theta - t)d\mu(t) \qquad (re^{i\theta} \in \Delta), \tag{5.2}$$

where $\{P_r\}$ is the Poisson kernel. (This result follows by applying [36, 11.19] to $-u$.) Regard μ as a function of bounded variation. In [24], Hayman and Korenblum consider the possibility of a representation (5.2) in which μ is a bounded function, but not necessarily of bounded variation. Let k be a positive, continuous, increasing function on $[0,1)$, and let u be a harmonic function such that

$$u(re^{i\theta}) \leq k(r) \quad (re^{i\theta} \in \Delta).$$

Then Hayman and Korenblum prove that a representation analogous to (5.2) holds provided that

$$\int_0^1 \left(\frac{k(r)}{1-r}\right)^{1/2} dr < \infty, \tag{5.3}$$

and they show that the rate of growth implied by (5.3) is exactly best possible, in that the representation need not hold if the integral in (5.3) diverges. The proof in [24] uses the Ahlfors distortion theorem, and it is not easy. I am grateful to Professor Korenblum for telling me that a more direct proof will be given in [29].

Let $\zeta = \rho e^{i\psi} \in \Pi$, and let $R = \rho/\cos\psi$. A conformal map from Δ onto Π shows that a representation analogous to (5.2) holds for harmonic functions U on Π such that $U(\rho e^{i\psi}) \le K(R)$ if and only if

$$\int_1^\infty \left(\frac{K(R)}{R^3}\right)^{1/2} dR < \infty . \tag{5.4}$$

A consequence of this representation is that various results which were known to hold for bounded, analytic functions on Π in fact hold for analytic functions F on Π such that $\log|F(\rho e^{i\psi})| \le K(R)$ $(\rho e^{i\psi} \in \Pi)$, provided that K satisfies (5.4). Note, however, that we cannot apply the representation directly to $\log|F(\zeta)|$, for this is in general a subharmonic, and not a harmonic, function. We require a further theorem of Hayman and Korenblum ([24, Theorem 5]): this shows that such an F has 'locally bounded Nevanlinna characteristic', and in particular it implies that $\limsup_{\rho\to\infty} \rho^{-1} \log|F(\rho)| > -\infty$ unless $F \equiv 0$.

We now turn to the proof that (d) implies (a). It follows a beautiful idea of Jean Esterle ([16]). Let

$$a^\zeta(t) = \frac{1}{(\pi\zeta)^{1/2}} \exp\left(-\frac{t^2}{\zeta}\right) \quad (t \in \mathbb{R},\ \zeta \in \Pi).$$

Then $a^\zeta \in L^1_\varphi$, and the map $\zeta \mapsto a^\zeta$, $\Pi \to L^1_\varphi$, is an analytic semi-group. (The semi-group (a^ζ) is the <u>Gaussian</u> semi-group. It is discussed in [25, 21.4] and in [38, 2.15].) We must calculate $\|a^\zeta\|$ in L^1_φ when $\zeta = \rho e^{i\psi}$. Let $R = \rho/\cos\psi$. Then elementary calculus given in [11] shows that $\log\|a^\zeta\| \le K(R)$ for a certain function K which depends on φ, and that K satisfies (5.4) if and only if $J_\varphi < \infty$, i.e., if φ is non-quasi-analytic. Thus, we have established a surprisingly intimate connection between Beurling's condition that $J_\varphi < \infty$, the condition (5.3) of Hayman and Korenblum, and the rate of growth of $\|a^\zeta\|$, where (a^ζ) is the Gaussian semi-group. The result also shows the sense in which the growth of $\|a^\zeta\|$ for $\zeta \in \Pi$ is minimal among analytic semi-groups over Π.

Let I be a closed ideal in L^1_φ with $Z(I) = \emptyset$, and let $\mathfrak{A} = L^1_\varphi/I$, a radical algebra. Let $[a^\zeta]$ be the coset of a^ζ in $\mathfrak{A}$ take $\lambda \in \mathfrak{A}'$, and set

$$F(\zeta) = \langle [a^\zeta], \lambda \rangle \quad (\zeta \in \Pi).$$

Then F is analytic over Π, and, by our calculation, $\log|F(\zeta)| \le K(R)$, where K is a function satisfying (5.4). Since $\mathfrak{A}$ is radical, $\rho^{-1}\log|F(\rho)| \to -\infty$ as $\rho \to \infty$, and so $F \equiv 0$. It follows that $[a^\zeta] = 0$, that $a^\zeta \in I$ $(\zeta \in \Pi)$, and hence that $I = L^1_\varphi$, as required.

6. The closed ideals of $L^1(\mathbb{R}^+)$

Throughout this section we write A for $L^1(\mathbb{R}^+)$, so that

$$A = \{f : \|f\| = \int_0^\infty |f(t)|dt < \infty\} .$$

Then A is a semi-simple Banach algebra, and its character space is $\overline{\Pi}$.

For $f \in A$, let $Z(f) = \{z \in \overline{\Pi} : (\mathcal{L}f)(z) = 0\}$, and, if I is a closed ideal of A, let

$$Z(I) = \bigcap \{Z(f) : f \in I\} .$$

Then the quotient Banach algebra A/I is radical if and only if $Z(I) = \emptyset$. We slightly extend the notation of Definition 2.5 by setting

$$\alpha(I) = \inf\{\alpha(f) : f \in I\} .$$

We are interested in conditions which imply that I is equal to A. It is clear that there are two necessary conditions: if $I = A$, then $Z(I) = \emptyset$ and $\alpha(I) = 0$. In this section it will be proved that these two conditions are also sufficient.

This result was first proved by Nyman in 1950 ([33]). However, this work is somewhat inaccessible. The result has also been proved by Gurariĭ in [22, Theorem 1], a recommended paper. In fact, Gurariĭ deals with more general (but not all — there remain open cases) semi-simple algebras $L^1(\omega)$, and he gives considerable details of the structure of the family of closed ideals.

It is the aim of this section to give greater prominence to this important theorem. Our proof is somewhat shorter and simpler than those of Nyman and Gurariĭ, although it has much in common with those works. The ideas in the first part of the proof are quite standard: they will recur in the proof of Domar in [12] and were used by Allan in [2, §5]. I have aimed to minimize the dependence on results from complex function theory: the one result used, Kreĭn's theorem, will be fully proved in §7.

THEOREM 6.1. Let I be a closed ideal of A. Then $I = A$ if and only if (1) $Z(I) = \emptyset$ and (2) $\alpha(I) = 0$.

Proof. We must prove that (1) and (2) imply that $I = A$.

It is convenient to identify the dual space, A', of A with $L^\infty(\mathbb{R}^-)$. In this case, the duality is implemented by the formula

$$\langle f, \varphi\rangle = \int_0^\infty f(s)\varphi(-s)ds = (\varphi * f)(0) \qquad (f \in A,\ \varphi \in A') .$$

Now take $\varphi \in A'$ with $\langle f, \varphi\rangle = 0$ $(f \in I)$. Our aim is to show that $\varphi = 0$, for, by the Hahn-Banach theorem, this will imply that $I = A$.

Let S_t denote the operation of translation to the right by t units, so that $(S_t f)(s) = f(s-t)$. If $e_n = n\chi_{[t,t+1/n]}$, then $f * e_n \to S_t f$ in A as $n \to \infty$, and so $S_t(I) \subset I$ $(t \geq 0)$.

Choose an element f of $I \setminus \{0\}$ and let

$$h = \varphi * f, \tag{6.1}$$

so that $h \in L^\infty(\underset{\sim}{R})$. For each $t \geq 0$, $\langle S_t f, \varphi \rangle = 0$ because $S_t f \in I$. But this says exactly that $h(t) = 0$ for $t \leq 0$. Thus, we see that

$$\varphi \in L^\infty(\underset{\sim}{R}^-),\ f \in L^1(\underset{\sim}{R}^+), \quad h \in L^\infty(\underset{\sim}{R}^+).$$

We now take the Laplace transforms of φ, f, and h: say that they are Φ, F, and H, respectively. We must be careful to check the properties of these functions. Certainly, F belongs to $A_0(\overline{\Pi})$. We cannot say that H belongs to $A_0(\overline{\Pi})$, but H is analytic on Π, and also

$$|H(x+iy)| \leq \int_0^\infty |h(t)|e^{-xt}dt \leq \|h\|_\infty/x,$$

so that $x|H(x+iy)|$ is bounded on Π. Similarly, Φ is analytic and $|x|\,|\Phi(x+iy)|$ is bounded on ${}_0\Pi$.

Define

$$\Psi(z) = \begin{cases} \Phi(z) & (z \in {}_0\Pi), \\ \dfrac{H(z)}{F(z)} & (z \in \Pi \setminus Z(f)). \end{cases} \tag{6.2}$$

We shall show that Ψ extends to be an entire function of exponential type. Let us suppose for a moment that we have done this. Then Φ has exponential type, and so, by Corollary 3.8, φ has compact support. Thus, we can apply Titchmarsh's convolution theorem to deduce from (6.1) that

$$\alpha(h) = \alpha(\varphi) + \alpha(f).$$

By hypothesis (2), $\inf\{\alpha(f) : f \in I\} = 0$, and we know that $\alpha(h) \geq 0$, and hence $\alpha(\varphi) \geq 0$. Thus, $\varphi = 0$, as required.

The above is the standard part of the proof.

Write V for the imaginary axis. To complete the proof, our first problem is to show that we can extend Ψ to be analytic on $V \setminus Z(f)$. Our technique for doing this is related to a technique due to Carleman and explained in [28, VI.8]. A somewhat different technique, due to Beurling, is used by Nyman. It is convenient first to give a lemma from [28].

LEMMA 6.2. Let J be an interval of V, and let D be the disc with J as diameter. Suppose that (Ψ_n) is a sequence of analytic functions on D such that

(i) $(|x\Psi_n(x+iy)|)$ is uniformly bounded on D,

(ii) $\Psi_n(z) \to \Psi(z)$ as $n \to \infty$ $(z \in D \setminus V)$.

Then Ψ extends to an analytic function on D.

Proof of lemma. Let Ω be an open disc concentric with D and of smaller diameter, and let L be a compact subset of Ω. Let $\partial\Omega$ meet J at z_1 and z_2. By (i), the sequence $((z-z_1)(z-z_2)\Psi_n(z))$ is uniformly bounded on $\partial\Omega$, and so, by the maximum modulus theorem, it is uniformly bounded on Ω. Thus, (Ψ_n) is uniformly bounded on L. This shows that (Ψ_n) is a normal family on Ω and hence there is a subsequence of (Ψ_n) which converges uniformly on compact subsets of Ω. The limit function is analytic on Ω, and, by (ii), it extends Ψ.

We shall also require some standard results from the theory of Fourier transforms. Write $C(\mathbb{R})$ for the continuous functions, and $D(\mathbb{R})$ for the infinitely differentiable functions of compact support on $\mathbb{R}$.

If $\lambda \in L^1(\mathbb{R})$, its Fourier transform is $\hat{\lambda}(y) = \int \lambda(t)e^{-ity}dt$. The inversion theorem shows that, if $\lambda \in L^1(\mathbb{R}) \cap C(\mathbb{R})$ and if $\hat{\lambda} \in L^1(\mathbb{R})$, then

$$\lambda(t) = \frac{1}{2\pi}\int \hat{\lambda}(y)e^{ity}dy \qquad (t \in \mathbb{R}) . \tag{6.3}$$

This implies that, if $\Lambda \in D(\mathbb{R})$, then there exists $\lambda \in L^1(\mathbb{R}) \cap C(\mathbb{R})$ with $\hat{\lambda} = \Lambda$, for we can take $\lambda(t) = \hat{\Lambda}(t)/2\pi$.

Let E be a closed set in $\mathbb{R}$, and let $I_E = \{\lambda \in L^1(\mathbb{R}) : \hat{\lambda} \mid E = 0\}$. Since $L^1(\mathbb{R})$ is regular, the character space of $L^1(\mathbb{R})/I_E$ is E. Thus, if $\lambda \in L^1(\mathbb{R})$ and $\hat{\lambda}(y) \neq 0$ $(y \in E)$, then $\hat{\lambda} + I_E$ is invertible in $L^1(\mathbb{R})/I_E$. If $\lambda_k \to \lambda_0$ in $L^1(\mathbb{R})$, and if $\hat{\lambda}_k(y) \neq 0$ $(y \in E,\ k \in \mathbb{Z}^+)$, then, by using the continuity of inversion in a Banach algebra, we see that there exist $\mu_k, \mu_0 \in L^1(\mathbb{R})$ with $(\lambda_k * \mu_k)\hat{}(y) = 1$ $(y \in E)$ and $\mu_k \to \mu_0$.

Finally, we require a triviality on convergence.

LEMMA 6.3. Let $\lambda_k \to \lambda$ in $L^1(\mathbb{R})$, let $\mu_k, \mu \in L^\infty(\mathbb{R})$ with $\|\mu_k\|_\infty \leq \|\mu\|_\infty$ and $\mu_k \to \mu$ pointwise. Then $(\lambda_k * \mu_k)(0) \to (\lambda * \mu)(0)$.

Proof of lemma. Clearly,

$$\begin{aligned}|(\lambda*\mu)(0) - (\lambda_k*\mu_k)(0)| &\leq \int |\lambda(s)\mu(-s) - \lambda_k(s)\mu_k(-s)|ds \\ &\leq \int |\lambda(s)||\mu(-s) - \mu_k(-s)|ds + \int |\mu_k(-s)||\lambda(s) - \lambda_k(s)|ds .\end{aligned}$$

In the first integral, the integrand is bounded by $2\|\mu\|_\infty|\lambda(s)|$, and it converges pointwise to 0, so the integral converges by the dominated convergence theorem.

The second integral is bounded by $\|\mu\|_\infty \|\lambda_k - \lambda\|$, and this converges to 0 as $k \to \infty$.

We now return to the proof of Theorem 6.1. For $\eta > 0$, let $\Delta_\eta = \{z : |z| < \eta\}$. Suppose first that $F(0) \neq 0$, and take $\delta > 0$ so that $F(z) \neq 0$ $(z \in \Delta_{2\delta} \cap \overline{\Pi})$. If $\Lambda \in D(\underset{\sim}{R})$ with $\operatorname{supp} \Lambda \subset (-\delta, \delta)$, we can set

$$(\Psi \cdot \Lambda)(\zeta) = \int \Psi(\zeta + iy)\Lambda(y)dy \quad (\zeta \in \Delta_\delta \setminus V). \tag{6.4}$$

Clearly, $\Psi \cdot \Lambda$ is analytic on $\Delta_\delta \setminus V$, and we claim that $\Psi \cdot \Lambda$ has an analytic extension to Δ_δ. It is sufficient to show that $\Psi \cdot \Lambda$ has a continuous extension to Δ_δ. In fact, we shall show that there exists $c \in \underset{\sim}{R}$ such that $(\Psi \cdot \Lambda)(\zeta_k) \to c$ for each $(\zeta_k) \subset \Pi$ with $\zeta_k \to 0$ and for each $(\zeta_k) \subset {}_0\Pi$ with $\zeta_k \to 0$, which is still sufficient.

First take $(\zeta_k) \subset \Pi$ with $\zeta_k \to 0$. Take $\lambda \in L^1(\underset{\sim}{R}) \cap C(\underset{\sim}{R})$ with $\hat{\lambda} = \Lambda$. For $\zeta \in \Pi$, let $h_\zeta(t) = e^{-\zeta t}h(t)$ and $f_\zeta(t) = e^{-\zeta t}f(t)$ $(t \in \underset{\sim}{R})$. Then $h_\zeta, f_\zeta \in L^1(\underset{\sim}{R})$. Note that $\|h_\zeta\|_\infty \leq \|h\|_\infty$, that $h_\zeta \to h$ and $f_\zeta \to f$ pointwise on $\underset{\sim}{R}$, and that $f_\zeta \to f$ in $L^1(\underset{\sim}{R})$ as $\zeta \to 0$. Take $f_0 = f$. Then for $\zeta \in \Pi \cup \{0\}$, we have $f_\zeta \in L^1(\underset{\sim}{R})$ and $\hat{f}_\zeta(y) = F(\zeta + iy)$, and hence, if $y \in [-\delta, \delta]$ and $\zeta \in \Pi \cup \{0\}$ with $|\zeta| < \delta$, $\hat{f}_\zeta(y) \neq 0$. Take $(g_{\zeta_k}) \subset L^1(\underset{\sim}{R})$ such that $(f_{\zeta_k} * g_{\zeta_k})^\wedge(y) = 1$ $(y \in [-\delta, \delta])$ and $g_{\zeta_k} \to g_0$ in $L^1(\underset{\sim}{R})$. Then

$$(g_{\zeta_k} * h_{\zeta_k} * \lambda)^\wedge(y) = H(\zeta_k + iy)\Lambda(y)/F(\zeta_k + iy) = \Psi(\zeta_k + iy)\Lambda(y)$$

for $y \in \underset{\sim}{R}$, noting that $\Lambda(y) = 0$ for $y \notin [-\delta, \delta]$. By the inversion theorem,

$$2\pi(g_{\zeta_k} * h_{\zeta_k} * \lambda)(0) = \int \Psi(\zeta_k + iy) \ \Lambda(y)dy = (\Psi \cdot \Lambda)(\zeta_k),$$

and so, by Lemma 6.3,

$$(\Psi \cdot \Lambda)(\zeta_k) \to 2\pi(g_0 * h * \lambda)(0) \quad \text{as} \quad \zeta_k \to 0. \tag{6.5}$$

Secondly, take $(\zeta_k) \subset {}_0\Pi$ with $\zeta_k \to 0$. For $\zeta \in {}_0\Pi$, let $\varphi_\zeta(t) = e^{-\zeta t}\varphi(t)$ $(t \in \underset{\sim}{R})$. Then $\varphi_\zeta \in L^1(\underset{\sim}{R})$, and

$$(g_0 * \varphi_{\zeta_k} * f * \lambda)^\wedge(y) = \Phi(\zeta_k + iy)\Lambda(y) = \Psi(\zeta_k + iy)\Lambda(y)$$

for $y \in \underset{\sim}{R}$. By the inversion theorem,

$$2\pi(g_0 * \varphi_{\zeta_k} * f * \lambda)(0) = (\Psi \cdot \Lambda)(\zeta_k),$$

and so

$$(\Psi \cdot \Lambda)(\zeta_k) \to 2\pi(g_0 * \varphi * f * \lambda)(0) \quad \text{as} \quad \zeta_k \to 0 . \tag{6.6}$$

The convergence that we are seeking now follows from (6.1), (6.5), and (6.6), and so $\Psi \cdot \Lambda$ has an analytic extension to Δ_δ, as claimed.

Let J be the open interval $(-i\delta, i\delta)$ of V. For $n \in \mathbb{N}$, let $\Lambda_n \in D(\mathbb{R})$ be a positive function with $\operatorname{supp} \Lambda_n \subset [-1/n, 1/n]$ and $\int \Lambda_n = 1$, and set $\Psi_n(z) = (\Psi \cdot \Lambda_n)(z)$ $(z \in \Delta_\delta \setminus V)$. For n sufficiently large, Ψ_n is analytic on Δ_δ. Clearly, $(|x\Psi_n(x+iy)|)$ is uniformly bounded on Δ_δ, and we easily check that $\Psi_n(z) \to \Psi(z)$ $(z \in \Delta_\delta \setminus V)$. By Lemma 6.2, Ψ extends to be an analytic function on Δ_δ.

We have proved that, for each $f \in I$, the function Ψ defined in (6.2) extends to be analytic on $\mathbb{C} \setminus Z(f)$. Since all such functions agree on ${}_0\Pi$, we see from hypothesis (1) that Ψ extends to be an entire function.

Let Ψ have the representation (6.2) for some $f \in I \setminus \{0\}$. It remains to prove that Ψ has exponential type, and to do this we call in aid Kreĭn's theorem, which we discuss in the next section as Theorem 7.4. Clearly, hypothesis (i) of Theorem 7.4 is satisfied (for each $\sigma > 0$). Since $|\Phi(z)| \le \|\varphi\|_\infty / |x|$ $(z \in {}_0\Pi)$,

$$\iint_{{}_0\Pi} \frac{\log^+ |\Phi(z)|}{1 + |z|^2} \, dxdy \le \|\varphi\|_\infty \int_{\mathbb{R}} \frac{dy}{1 + |y|^2} \int_0^1 \log\left(\frac{1}{x}\right) dx < \infty ,$$

and a similar estimate holds for the integral involving $\log^+ |H(z)|$ over Π. Thus hypothesis (ii) of 7.4 holds by Proposition 7.5.

This concludes the proof of the theorem.

COROLLARY 6.4. Let $f \in A$ with $Z(f) = \emptyset$ and $\alpha(f) = 0$. Then $\overline{f * A} = A$.

7. The Ahlfors-Heins theorem and Kreĭn's theorem

In this section, I first wish to discuss the Ahlfors-Heins theorem. This result has, perhaps rather surprisingly, been a key ingredient of several recent results in the theory of radical Banach algebras. I give a version that is fairly close to the best possible: usually, weaker forms suffice for the applications.

For $\delta \in (0, \pi/2)$, set $S_\delta = \{z \in \mathbb{C} : -\delta < \arg z < \delta\}$, and set $-S_\delta = \{z : -z \in S_\delta\}$.

THEOREM 7.1. (Ahlfors-Heins)

Let F be a function of exponential type in Π, and suppose that

$$\int_{-\infty}^{\infty} \frac{\log^+ |F(iy)|}{1 + y^2} \, dy < \infty , \tag{7.1}$$

and that $F \not\equiv 0$. Then

$$c = \lim_{r\to\infty} \frac{2}{\pi r} \int_{-\pi/2}^{\pi/2} \log|F(re^{i\theta})| \cos\theta \, d\theta \tag{7.2}$$

exists in $\underset{\sim}{R}$. Further:

(i) for almost all $\theta \in (-\pi/2,\pi/2)$, $\lim_{r\to\infty} r^{-1} \log|F(re^{i\theta})| = c \cos\theta$;

(ii) there is an open subset E of $\underset{\sim}{R}^+$ such that E has finite logarithmic length (i.e., $\int_E dt/t < \infty$) and such that

$$\lim_{\substack{r\to\infty \\ r\notin E}} \frac{1}{r} \log|F(re^{i\theta})| = c \cos\theta \quad \text{uniformly for} \quad |\theta| < \pi/2 ;$$

(iii) if $\delta \in (0,\pi/2)$, if F has no zeros on S_δ, and if $\eta < \delta$, then $\lim_{r\to\infty} r^{-1}\log|F(re^{i\theta})| = c \cos\theta$ uniformly for $|\theta| \leq \eta$.

This theorem has a long history. Quite a lot was already proved by Cartwright in 1935 ([9, Theorem III]). Essentially the above result was given by Ahlfors and Heins in 1949 ([1]): in their result, $\log|F|$ is replaced by an arbitrary subharmonic function. A proof of most of the result is given by Boas in [5, §7.2], but it requires some following back to sort out what everything means: the constant c of Theorem 7.2.6 is defined in (6.5.7). A version of the theorem is proved by Sinclair in [38, Appendix 1].

If F is bounded on the imaginary axis (as is usually the case in applications), then (7.1) is satisfied, and in this case $c \leq 0$. A discussion of variants of (7.1) is given by Boas.

The excluded values of θ in (i) are actually contained in an open set of outer capacity zero.

Only a weaker form of (ii) is proved in Boas. The full form follows from a theorem of Hayman ([23, Theorem 2]). That theorem is proved with $\log|F|$ replaced by an arbitrary non-positive subharmonic function, but we can deduce our form of (ii) from it. Let F satisfy (7.1), and set

$$v(z) = \frac{1}{\pi} \int_{-\infty}^{\infty} \frac{x \log^+|F(iy)|}{|iy - z|^2} \, dy \qquad (z \in \overline{\Pi}) .$$

Then v is a non-negative harmonic function on Π, continuous on $\overline{\Pi}$, and $\log|F(iy)| \leq v(iy)$. Thus, we can apply [23, Theorem 2] to $\log|F(z)| - v(z)$ and to $-v(z)$, and subtract to obtain our result. (I am grateful to Professor Hayman for this remark.) It is also shown in [23] that 'finite logarithmic length' can not be weakened in (ii).

A first step in the proof of Theorem 7.1 could be the proof of Proposition 7.2, below. This is very easy, and I give it because it is all that is needed to give a full proof of Krein's theorem. A proof of Theorem 7.1 must involve an analysis of the Blaschke product formed from the zeros of F.

PROPOSITION 7.2. Let F be a bounded analytic function on Π such that $F(z) \neq 0$ $(z \in \Pi)$. Then there is a constant c such that, for each $\delta \in (0,\pi/2)$, $\lim_{r\to\infty} r^{-1} \log|F(re^{i\theta})| = c \cos\theta$ uniformly for $|\theta| \leq \delta$.

<u>Proof</u>. Let f be a bounded analytic function on the unit disc, Δ, say $|f(\zeta)| \leq 1$ $(\zeta \in \Delta)$. Suppose that f has no zeros in Δ. Then $f(\zeta) = \exp(-h(\zeta))$, where h is analytic and $\operatorname{Re} h(\zeta) \geq 0$ on Δ. Let $h = u + iv$, where u and v are real-valued functions on Δ. Then u is a positive harmonic function, and so we have the Herglotz representation (e.g. [36, 11.19])

$$u(\zeta) = \int_{-\pi}^{\pi} \operatorname{Re}\left(\frac{e^{i\varphi}+\zeta}{e^{i\varphi}-\zeta}\right) d\mu(\varphi) \qquad (\zeta \in \Delta)$$

where μ is a finite, positive Borel measure on the circle. Thus,

$$f(\zeta) = \exp\left(-\int_{-\pi}^{\pi} \frac{e^{i\varphi}+\zeta}{e^{i\varphi}-\zeta}\, d\mu(\varphi)\right) \qquad (\zeta \in \Delta).$$

Set $\mu = -c\delta_1 + \mu'$, where δ_1 is the point mass at $\zeta = 1$ and μ' is a finite measure on the remainder of the circle. Then

$$f(\zeta) = \exp\left(c\left(\frac{1+\zeta}{1-\zeta}\right)\right) \exp\left(-\int_{-\pi}^{\pi} \frac{e^{i\varphi}+\zeta}{e^{i\varphi}-\zeta}\, d\mu'(\varphi)\right) \qquad (\zeta \in \Delta). \tag{7.3}$$

We transform this to an equation on Π. Let $z = (1+\zeta)/(1-\zeta)$ for $|\zeta| < 1$, so that $\zeta = (z-1)/(z+1)$. On the boundary, the map is given by $e^{i\varphi} = (it-1)/(it+1)$, and we note that

$$\frac{e^{i\varphi}+\zeta}{e^{i\varphi}-\zeta} = \frac{tz+i}{t+iz}.$$

Now let F be as specified in the Proposition: we can suppose that $|F(z)| \leq 1$ $(z \in \Pi)$. Then, from (7.3),

$$F(z) = e^{cz} \exp\left(-\int_{-\infty}^{\infty} \left(\frac{tz+i}{t+iz}\right) d\nu(t)\right) \qquad (z \in \Pi),$$

where ν is a finite, positive measure on the imaginary axis. Thus, if $z = re^{i\theta} \in \Pi$, then

$$\log|F(re^{i\theta})| = cr\cos\theta - \int_{-\infty}^{\infty} \operatorname{Re}\left(\frac{tz+i}{t+iz}\right) d\nu(t).$$

Take $\delta \in (0,\pi/2)$. To prove the result, it clearly suffices to show that

$$\frac{1}{|z|}\int_{-\infty}^{\infty}\left|\frac{tz+i}{t+iz}\right| d\nu(t) \to 0 \quad \text{as} \quad |z| \to \infty \quad \text{with} \quad z \in \overline{S}_\delta . \tag{7.4}$$

But, if $t \in \underset{\sim}{R}$ and $z \in \overline{S}_\delta$ with $|z| \geq 1$, then $|t + iz| \geq |t| \cos \delta$ and $|t + iz| \geq \sin \delta$, and so there is a constant K such that

$$\left|\frac{tz+i}{z(t+iz)}\right| \leq K \quad (t \in \underset{\sim}{R},\ z \in \overline{S}_\delta,\ |z| \geq 1) .$$

Also, for each $t \in \underset{\sim}{R}$. $|(tz + i)/z(t + iz)| \to \infty$. Thus, (7.4) follows from the dominated convergence theorem.

The second theorem to be discussed in this section is Kreĭn's theorem. It is taken from a paper written in Russian in 1947 by M. G. Kreĭn ([30]). As far as I am aware, a proof of the theorem has not been given in English, but the proof is set as Problem 37 in [7], where a somewhat different approach from ours is indicated. A proof of a weaker version of the theorem, a version that is not sufficient for the application in §6, is given in [38, Appendix 1].

We shall require the fact that a bounded analytic function F on Π can be factored as $F = BG$, where B is the Blaschke product formed from the zeros of F and G is a bounded analytic function on Π with no zeros and such that $\|G\|_\infty = \|F\|_\infty$: see [36, 17.9], for example.

We shall also require the following standard form of the Phragmén-Lindelöf principle.

LEMMA 7.3. Let S be a sector of angle π/α, and let F be analytic on S and continuous on $\overline{S}$. Suppose that F is bounded on ∂S and that $|F(z)| = \underline{O}(\exp(r^\beta))$ as $r \to \infty$ for some $\beta < \alpha$. Then F is bounded on $\overline{S}$.

THEOREM 7.4. Let F be an entire function. Suppose that:

(i) for some $\sigma \geq 0$, F is equal to a quotient of bounded analytic functions separately on each of ${}_{-\sigma}\Pi$ and Π_σ;

(ii) for some $m \geq 0$,

$$\iint_{\underset{\sim}{R}^+} \frac{\log^+|F(z)|}{1 + |z|^m}\, dx\, dy < \infty . \tag{7.5}$$

Then F is of exponential type.

<u>Proof</u>. We can suppose that $m \geq 2$. Let $u(z) = \log^+|F(z)|$ $(z \in \underset{\sim}{C})$. Since F is entire, u is subharmonic on $\underset{\sim}{C}$, and so

$$u(\zeta) \leq \frac{1}{2\pi}\int_{-\pi}^{\pi} u(\zeta + re^{i\theta})d\theta \quad (r > 0,\ \zeta \in \underset{\sim}{C}) .$$

(See [36, Chapter 17]). For $\zeta \neq 0$, let K_ζ be the disc, centre ζ, radius $|\zeta|$. Then

$$\int_0^{|\zeta|} u(\zeta)\, r dr \leq \frac{1}{2\pi} \iint_{K_\zeta} u(z) dx\, dy,$$

and so

$$u(\zeta) \leq \frac{1}{\pi|\zeta|^2} \iint_{K_\zeta} u(z) dx\, dy \qquad (\zeta \in \underset{\sim}{C} \setminus \{0\}).$$

On the disc K_ζ, we have $|z| \leq 2|\zeta|$, and so

$$u(\zeta) \leq \frac{1+2^m|\zeta|^m}{\pi|\zeta|^2} \iint_{K_\zeta} \frac{u(z) dx\, dy}{1+|z|^m} \qquad (\zeta \in \underset{\sim}{C} \setminus \{0\}),$$

noting that $u(z) \geq 0$ for $z \in \underset{\sim}{C}$. It follows from (7.5) that $|u(\zeta)| = \underline{O}(|\zeta|^{m-2})$ as $|\zeta| \to \infty$ with $\zeta \in \underset{\sim}{C}$, and so

$$|F(\zeta)| = \underline{O}(\exp(|\zeta|^m) \quad \text{as} \quad \zeta \to \infty. \tag{7.6}$$

Now take $\delta \in (0,\pi/2)$ with $\pi/(\pi - 2\delta) > m$, and take $\eta \in (\delta,\pi/2)$. Let $\tilde{F}(z) = F(z + \sigma)$. Then $\tilde{F}$ is a quotient of bounded analytic functions on Π, and so $\tilde{F} = BG_1/G_2$, where B is a Blaschke product and G_1, G_2 are bounded analytic functions on Π with no zeros. By 7.2, there are constants $c_1, c_2 \in \underset{\sim}{R}$ such that $\lim_{r\to\infty} r^{-1} \log|G_j(re^{i\theta})| = c_j \cos\theta$ uniformly for $|\theta| < \eta$ $(j = 1,2)$. Thus, there exist c and R_δ such that

$$|\tilde{F}(re^{i\theta})| \leq \exp(cr\cos\theta) \qquad (r \geq R_\delta,\ |\theta| < \eta).$$

This shows that there is a constant M such that

$$|F(z)| = \underline{O}(\exp(Mr)) \quad \text{as} \quad r \to \infty \quad \text{with} \quad z \in \overline{S}_\delta.$$

We can suppose that the same estimate holds if $r \to \infty$ with $z \in -\overline{S}_\delta$.

Let $S = \{z : \delta < \arg z < \pi - \delta\}$, and let $G(z) = F(z)\exp(iMz \operatorname{cosec}\delta)$ for $z \in \overline{S}$. Then G is bounded on ∂S, and, by (7.6), $|G(z)| = \underline{O}(\exp(r^m))$ as $r \to \infty$. Since $m < \pi/(\pi - 2\delta)$, G is bounded on $\overline{S}$. Hence

$$|F(re^{i\theta})| = \underline{O}(\exp(Mr \operatorname{cosec}\delta)) \quad \text{as} \quad r \to \infty$$

uniformly for $\delta \leq \theta \leq \pi - \delta$. Again, we can suppose that the same estimate holds uniformly for $-\pi + \delta \leq \theta \leq -\delta$.

It is now clear that F is of exponential type, as required.

In the applications of Theorem 7.4, the following fact is useful.

PROPOSITION 7.5. Let F be a bounded analytic function on Π with $F \not\equiv 0$. Then

$$\iint_{\Pi} \frac{\log^{+}|1/F(z)|}{1+|z|^{4}}\, dx\, dy < \infty . \tag{7.7}$$

Proof. Let f be a bounded analytic function on Δ. Suppose that f has a zero of order k at 0, and let $g(\zeta) = f(\zeta)/\zeta^{k}$. By Jensen's formula ([36, 15.19]),

$$\frac{1}{2\pi}\int_{-\pi}^{\pi} \log|f(re^{i\varphi})|\,d\varphi \geq \log|g(0)| + k\log r \qquad (0 < r < 1).$$

Hence, $\iint_{\Delta} \log^{+}|1/f(re^{i\varphi})|\,r\,dr\,d\varphi < \infty$. We write this area integral as

$$\left| \iint_{\Delta} \log^{+}|1/f(\zeta)|\,d\zeta \wedge d\bar{\zeta} \right| .$$

Transform the integral to be over Π by setting $\zeta = (z-1)/(z+1)$. Then $d\zeta = 2dz/(z+1)^{2}$ and $d\bar{\zeta} = 2d\bar{z}/(\bar{z}+1)^{2}$, so that

$$\left| \iint_{\pi} \frac{\log^{+}|1/F(z)|}{|1+z|^{4}}\, dz \wedge d\bar{z} \right| < \infty .$$

Since $|1+z|^{4}/(1+|z|^{4})$ is bounded on Π, inequality (7.7) follows.

We may combine this with Theorem 7.4 to obtain the original version of Kreǐn's theorem.

THEOREM 7.6. (Kreǐn)

Let F be an entire function. Suppose that F is a quotient of bounded analytic functions on each of ${}_{0}\Pi$ and Π. Then F is of exponential type.

Theorem 7.6 may also be deduced directly from 7.1(ii) in a way that gives a little more information.

Suppose that $F = F_1/F_2$ on Π and that $F = F_3/F_4$ on ${}_{0}\Pi$, where $F_1,\ldots,F_4$ are bounded analytic functions. Let $c_1,\ldots,c_4$ be the constants associated with $F_1,\ldots,F_4$, respectively, as in (7.2). Let $a = c_1 - c_2$, $b = c_3 - c_4$, and take $\varepsilon > 0$. It follows from Theorem 7.1(ii) that, for each sufficiently large ρ, there exists $r \in [\rho,(1+\varepsilon)\rho]$ such that $|r^{-1}\log|F(re^{i\theta})| - a\cos\theta| < \varepsilon$ $(|\theta| < \pi/2)$, and similarly on ${}_{0}\Pi$. Let $K = \max\{a,b\}$, and let $M(r) = \sup\{|F(re^{i\theta})|\}$. Then, for sufficiently large ρ,

$$\log M(\rho) \leq \log M(r) \leq (K+\varepsilon)r \leq (K+\varepsilon)(1+\varepsilon)\rho .$$

Thus, $\limsup[\log M(\rho)]/\rho \leq K$, and so F is of exponential type at most K.

The above proof was pointed out to me by Professor Hayman.

References

[1] L. Ahlfors and M. Heins, Questions of regularity connected with the Phragmén-Lindelöf principle, Ann. of Math., (2) 50 (1949), 341-346.

[2] G. R. Allan, Ideals of rapidly growing functions, Proceedings International Symposium on Functional Analysis and its Applications, Ibadan, Nigeria, 1977.

[3] W. G. Bade, Multipliers of weighted ℓ^1 algebras, this Volume.

[4] W. G. Bade and H. G. Dales, Norms and ideals in radical convolution algebras, J. Functional Analysis, 41 (1981), 77-109.

[5] R. P. Boas, Jr., Entire Functions, Academic Press, New York, 1954.

[6] F. F. Bonsall and J. Duncan, Complete Normed Algebras, Springer-Verlag, New York, 1973.

[7] L. de Branges, Hilbert Spaces of Entire Functions, Prentice Hall, New Jersey, 1968.

[8] T. Carleman, L'Integrale de Fourier et Questions Qui s'y Rattachent, Almquist and Wiksells, Uppsala, 1944.

[9] M. L. Cartwright, On functions which are regular and of finite order in an angle, Proc. London Math. Soc., (2) 38 (1935), 158-179.

[10] H. G. Dales, Automatic continuity: a survey, Bull. London Math. Soc., 10 (1978), 129-183.

[11] H. G. Dales and W. K. Hayman, Esterle's proof of the Tauberian theorem for Beurling algebras, Ann. Inst. Fourier, Grenoble, 31 (1981), 141-150.

[12] Y. Domar, A solution of the translation-invariant subspace problem for weighted L^p on $\mathbb{R}$, $\mathbb{R}^+$ or $\mathbb{Z}$, this Volume.

[13] __________, Bilaterally translation-invariant subspaces of weighted $L^p(\mathbb{R})$, this Volume.

[14] J. Esterle, Homomorphismes discontinus des algèbres de Banach commutatives séparables, Studia Math., 66 (1979), 119-141.

[15] __________, Universal properties of some commutative radical Banach algebras, J. Reine Angew. Math., 321 (1981), 1-24.

[16] __________, A complex-variable proof of the Wiener Tauberian theorem, Ann. Inst. Fourier, Grenoble, 30 (1980), 91-96.

[17] __________, Elements for a classification of commutative radical Banach algebras, this Volume.

[18] I. M. Gelfand, D. A. Raikov, and G. E. Šilov, Commutative Normed Rings, Chelsea, New York, 1964.

[19] F. Ghahramani, Homomorphisms and derivations on weighted convolution algebras, J. London Math. Soc., (2) 21 (1980), 149-161.

[20] J. I. Ginsberg and D. J. Newman, Generators of certain radical algebras, J. Approximation Theory, 3 (1970), 229-235.

[21] S. Grabiner, Derivations and automorphisms of Banach algebras of power series, Mem. Amer. Math. Soc., 146 (1974).

[22] V. P. Gurariĭ, Harmonic analysis in spaces with a weight, Trans. Moscow Math. Soc., 35 (1979), 21-75.

[23] W. K. Hayman, Questions of regularity connected with the Phragmén-Lindelöf principle, J. Math. Pures et Appliquées, 35 (1956), 115-126.

[24] W. K. Hayman and B. Korenblum, An extension of the Riesz-Herglotz formula, Annales Academiae Scientiarum Fennicae, Series A1, Mathematica, 2 (1976), 175-201.

[25] E. Hille and R. S. Phillips, Functional Analysis and Semi-groups, Colloquium Publications Series, Vol. 31, Amer. Math. Soc., Providence, Rhode Island, 1957.

[26] N. P. Jewell and A. M. Sinclair, Epimorphisms and derivations on $L^1[0,1]$ are continuous, Bull. London Math. Soc., 8 (1976), 135-139.

[27] G. K. Kalisch, A functional analysis proof of Titchmarsh's theorem on convolution, J. Math. Anal. and Appl., 5 (1962), 176-183.

[28] Y. Katznelson, An Introduction to Harmonic Analysis, Dover, New York, 1976.

[29] B. Korenblum and K. Samotij, in preparation.

[30] M. G. Kreĭn, A contribution to the theory of entire functions of exponential type, Izvestiya Akad. Nauk SSSR, 11 (1947), 309-326 (Russian).

[31] L. H. Loomis, An Introduction to Abstract Harmonic Analysis, van Nostrand, New Jersey, 1953.

[32] J. Mikusiński, Operational Calculus, Pergamon Press, Oxford, 1959.

[33] B. Nyman, On the one-dimensional translation group and semi-group in certain function spaces, Uppsala Thesis, 1950.

[34] R. E. A. C. Paley and N. Wiener, Fourier Transforms in the Complex Domain, Colloquium Publications Series, Vol. 19, Amer. Math. Soc., New York, 1934.

[35] J. R. Ringrose, Compact Non-Self-Adjoint Operators, van Nostrand-Reinhold, New York, 1971.

[36] W. Rudin, Real and Complex Analysis, McGraw-Hill, New York, 1966.

[37] L. Schwartz, Étude des Sommes d'exponentielles Réelles, Hermann, Paris, 1943.

[38] A. M. Sinclair, Continuous Semigroups in Banach Algebras, London Mathematical Lecture Note Series, 63, Cambridge University Press, 1982.

[39] P. Szász, Uber die Approximation stetiger Funktionen durch lineare Aggregate von Potenzen, Math. Ann., 77 (1916), 482-496.

[40] E. C. Titchmarsh, The Theory of Functions, Oxford University Press, Oxford, 1939 (2nd edition).

[41] D. V. Widder, The Laplace Transform, Princeton University Press, Princeton, 1941.

School of Mathematics
University of Leeds
Leeds, LS2 9JT
England

BILATERALLY TRANSLATION-INVARIANT SUBSPACES OF WEIGHTED $L^p(\mathbb{R})$

Y. Domar

Let w be a positive Lebesgue measurable function on $\mathbb{R}$ such that

$$\sup_{y\in\mathbb{R}} \frac{w(x+y)}{w(y)} \tag{1}$$

is bounded on every compact interval. For $1 \leq p \leq \infty$, $L^p(w)$ denotes the Banach space of complex-valued functions f on $\mathbb{R}$ with $fw \in L^p$, provided with the induced norm. Then the translation operator T_y, defined by

$$T_y f(x) = f(x-y), \quad x \in \mathbb{R},$$

is a bounded operator on $L^p(w)$, for every $y \in \mathbb{R}$.

Our aim is to construct a certain closed non-trivial subspace K of $L^p(w)$, invariant under the family of all translations.

In the case when, for some $b \in \mathbb{R}$,

$$e^{bx}/w(x) \in L^q, \tag{2}$$

where q is the conjugate of p, we have that

$$\hat{f}(t) = \int_{-\infty}^{\infty} f(x)\, e^{-itx} dx \tag{3}$$

converges absolutely, for $f \in L^p(w)$, $\mathrm{Im}(t) = b$, by Hölder's inequality. Given t_0 with $\mathrm{Im}(t_0) = b$, the subspace $A(t_0)$ of all $f \in L^p(w)$ for which (3) vanishes at $t = t_0$, is closed and translation-invariant. It is then of interest to compare our space K with $A(t_0)$. It turns out that it is possible to show that $A(t_0)$ does not include K, if

$$\int_{-\infty}^{\infty} \frac{\log w(x) - bx}{1+x^2}\, dx = \infty. \tag{4}$$

In particular this proves the implication (a) $\Rightarrow$ (d) of Theorem 5.3 in the survey [1] by H. G. Dales.

A subspace with these properties was first constructed, for special weights w, by Nyman [9]. A different type of construction was later developed and investigated by Korenblum [7,8], Geĭsberg and Konjuhovski [5,6], and Vretblad [11]. All these papers have additional assumptions on w, apart from (2) and (4), and are restricted to the case $p = 1$. In particular, the above mentioned statement by Dales is not covered by these results.

In [4] it was sketched how a non-trivial translation-invariant subspace can be constructed quite generally, even without assuming (2) or (4). This was done by modifying some of Vretblad's arguments. Here we shall give a more detailed exposition of the construction in [4], and after that we will explain why (2) and (4) imply that the obtained subspace is not included in $A(t_0)$ if $\mathrm{Im}(t_0) = b$.

Let w_0 be the continuous function which coincides with w on $\mathbb{Z} \subset \mathbb{R}$ and has $\log w_0$ linear in the complementary intervals. By the boundedness condition (1), w/w_0 and w_0/w are bounded on $\mathbb{R}$, and the derivative of $\log w_0$ is bounded on $\mathbb{R}\setminus\mathbb{Z}$. A simple modification leads to a twice-differentiable function w_1, with $w_1(0) = 1$, with w/w_1 and w_1/w bounded, and with the derivative of $\log w_1$ bounded. $L^p(w)$ and $L^p(w_1)$ then have translation-invariant subspace structures which are isomorphic in the obvious way. We can therefore assume, without loss of generality, that w has from the outset all the properties just stated for w_1.

For every s, $1 \leq s \leq \infty$, we use the notation H^s for the (closed) subspace of all $f \in L^s(\mathbb{R})$, such that $\hat{f}$, defined in the distributional sense in accordance to the formal relation (3), has $\mathrm{supp}(\hat{f}) \subseteq \mathbb{R}^+ = [0,\infty)$.

By Poisson's formula for the half-plane, we can find a bounded continuous function u on the closed upper half-plane $\Pi \subseteq \mathbb{C}$, with values equal to $\frac{d}{dx}(\log w(x))$ on the real axis, and harmonic in the interior Π^0. The differentiability assumptions on w show that u has continuous partial derivatives in Π.

By an elementary method, we can find a conjugate harmonic function v of u, extendable to a continuous function in Π. The function

$$f(z) = u(z) + iv(z)$$

is then continuous in Π and analytic in Π^0. Let us define

$$F(z) = \int_\gamma f(w)dw,$$

$z \in \Pi$, where γ is any integration path in Π from 0 to z. F is well defined, continuous in Π and analytic in Π^0.

For every $a \in \mathbb{R}$ we have

$$F(z+a) - F(z) = \int_0^a u(z+t)dt + i\int_0^a v(z+t)dt, \tag{5}$$

for $z \in \Pi$. By (5) and the boundedness of u, the function

$$z \mapsto \mathrm{Re}(F(z+a) - F(z))$$

is bounded in Π. Defining $G(z) = \exp\{-F(z)\}$, we thus find that

$$z \mapsto G(z+a)/G(z)$$

is bounded in Π, for every fixed $a \in \mathbb{R}$.

Let g be the restriction of G to the real axis. Using the earlier introduced translation symbol T_a, $a \in \mathbb{R}$, we thus know that the functions $T_a g/g$ are restrictions to the real axis of functions, continuous and bounded in Π, analytic in Π^0. Elementary distribution theory shows that this implies that $T_a g/g \in H^\infty$, for every $a \in \mathbb{R}$.

The relation (5), for $z = 0$, and the assumption $w(0) = 1$, show that

$$\operatorname{Re} F(a) = \operatorname{Re}(F(a) - F(0)) = \int_0^a u(t)dt = \log w(a) - \log w(0) = \log w(a)$$

for $a \in \mathbb{R}$, hence $|g| = 1/w$ on $\mathbb{R}$.

Let us now define K as the set of all fg, where $f \in H^p$. Thus, K is the image of the isometric mapping from H^p into $L^p(w)$ given by $f \mapsto fg$. Hence K is a closed subspace of $L^p(w)$, and it is obviously non-trivial. Let us consider the relation

$$T_a(fg) = T_a f(T_a g/g)g\,. \tag{6}$$

If $f \in H^p$, then $T_a f \in H^p$. As proved above, $(T_a g/g) \in H^\infty$. By elementary distribution theory, the product of functions in H^p and H^∞ belongs to H^p. Hence the right-hand member of (6) belongs to K, and the translation invariance of K is proved.

Let us finally assume that (2) and (4) hold, and that $K \subseteq A(t_0)$, for a certain t_0, with $\operatorname{Im}(t_0) = b$. We shall prove that this implies a contradiction.

By the definition of K, the function $h(x) = (x+i)^{-2} g(x)$ belongs to K, and so do the functions $h(x)\exp(iax)$, for every $a \in \mathbb{R}^+$. (1) and (2) imply that $g(x)e^{bx}$ is bounded, and hence $h(x)e^{bx} \in L^1 \cap L^\infty$. The assumption that $K \subseteq A(t_0)$ gives

$$\int_{-\infty}^{\infty} h(x)\, e^{iax} \cdot e^{-it_0 x}\, dx = 0\,,$$

$a \in \mathbb{R}^+$. But this shows that the function

$$\ell(x) = h(x)e^{-it_0 x}$$

has a Fourier transform which vanishes on $(-\infty, 0]$. Furthermore, $\ell \in L^2(\mathbb{R})$ and, by (4)

$$\int_{-\infty}^{\infty} \frac{\log|\ell(x)|}{1+x^2}\,dx \leq \int_{-\infty}^{\infty} \frac{\log|g(x)| + bx}{1+x^2}\,dx = -\int_{-\infty}^{\infty} \frac{\log w(x) - bx}{1+x^2}\,dx = -\infty\,.$$

By the famous Theorem XII in Paley-Wiener [10], ℓ vanishes identically, a contradiction.

If $w = \{w_n\}_{-\infty}^{\infty}$ is a positive sequence with $\{w_{n+1}/w_n\}$ and $\{w_n/w_{n+1}\}$ bounded, we can define, in an analogous way, Banach spaces $\ell^p(w)$, $0 \le p \le \infty$, for which right and left translations are bounded operators.

Problem: Does $\ell^p(w)$, $1 \le p < \infty$, contain a non-trivial closed bilaterally translation-invariant subspace?

For some positive partial results, see [3]. It should also be mentioned that there exist sequences $w = \{w_n\}$, with $w_n \ge 1$, with w growing to infinity arbitrarily slowly as $|n| \to \infty$, and for which $\ell^1(w)$ contains a non-trivial closed invariant subspace, not contained in any of the subspaces characterized by the vanishing of the Fourier series at a fixed point (see [2]).

References

[1] H. G. Dales, Convolution algebras on the real line, this Volume.

[2] Y. Domar, Spectral analysis in spaces of sequences summable with weights, J. Functional Analysis, 5 (1970), 1-13.

[3] ___________, On the existence of non-trivial or non-standard translation invariant subspaces in weighted ℓ^p and L^p, 18th Scandinavian Congress of Mathematicians, Proceedings, 1980, Progress in Mathematics, Vol. 11 (1981), 226-235.

[4] ___________, Translation invariant subspaces of weighted ℓ^p and L^p spaces, Math. Scand., 49 (1981), 133-144.

[5] S. P. Geĭsberg, Completeness of function shifts in certain spaces, Dokl. Akad. Nauk SSSR, 173 (1967), 741-744.

[6] S. P. Geĭsberg and V. S. Konjuhovskiĭ, Description of certain systems of primary ideals in rings of functions integrable with weight, Leningrad. Gos. Ped. Inst. Učen. zap., 404 (1971), 349-361.

[7] B. Korenblum, A generalization of Wiener's tauberian theorem and spectrum of fast-growing functions, Trudy Moskov. Mat. Obšč., 7 (1958), 121-148.

[8] ___________, Phragmén-Lindelöf type theorems for quasi-analytic classes of functions, Issled. sovremen. Probl. Teor. Funkciĭ Kompleks. Peremen., IV. vsesojuz. Konf. Moskov. Univ., 1958 (1961), 510-514.

[9] B. Nyman, On the one-dimensional translation group and semi-group in certain function spaces, Thesis, Uppsala 1950.

[10] R. E. A. C. Paley and N. Wiener, Fourier Transforms in the Complex Domain, American Math. Soc., New York, 1934.

[11] A. Vretblad, Spectral analysis in weighted L^1 spaces on $\mathbb{R}$, Ark. Mat., 11 (1973), 109-138.

University of Uppsala
Thunbergsvägen 3
S-752 38 Uppsala, Sweden

A SOLUTION OF THE TRANSLATION-INVARIANT SUBSPACE PROBLEM FOR WEIGHTED L^p ON $\mathbb{R}$, $\mathbb{R}^+$ OR $\mathbb{Z}$

Yngve Domar

1. Introduction

From a Banach algebra point of view, the most significant result in this paper is a theorem on the closed ideal structure in radical convolution algebras of type $L^p(w,\mathbb{R}^+)$ (Theorem 7). We will, however, discuss a more general theory, dealing with a class of convolution equations on $\mathbb{R}$, with the radical algebra result as a corollary. This greater generality does not complicate the deduction of the special result; rather the converse, since it provides the set-up with a certain symmetry. The general theory has moreover an independent interest. Thus it has applications in other directions as well.

The paper can be considered as a survey and a complement of the more technical and detailed papers [7] and [8]. We present three basic theorems (Theorem 1, 2, and 3), give the main ideas of their proofs, make applications, and mention the most important open problems. Section 8 contains a result, not to be found in the earlier papers (Theorem 10).

We start with a weight w on $\mathbb{R}$, a positive continuous function such that $\log w$ is ultimately convex, as $x \to -\infty$, and ultimately concave, as $x \to \infty$, and fulfills the condition

$$u(x) = -x^{-1} \log w(x) \to \infty,$$

as $|x| \to \infty$. μ and ν are assumed to be complex Borel measures on $\mathbb{R}$, satisfying

$$\int_{-\infty}^{\infty} w(x)\,|d\mu(x)| < \infty, \quad \int_{-\infty}^{\infty} (w(-x))^{-1}\,|d\nu(x)| < \infty .$$

We may observe that w_1, defined by $w_1(x) = (w(-x))^{-1}$, $x \in \mathbb{R}$, satisfies the same regularity as w, and also the relation $w(x) = (w_1(-x))^{-1}$. Thus μ and ν are in a sense interchangeable, and this is the symmetry which was referred to above.

Let $\mathbb{R}^-$ denote the interval $(-\infty,0)$. Defining measure convolution in the usual way, and using the fact that w decreases outside some finite interval, we find by Fubini's theorem that $\mu * \nu$ is a well-defined (bounded) Borel measure on $\mathbb{R}^-$.

Let us consider the two following properties for the pair (μ,ν):

A. $\mu * \nu = 0$ on $\mathbb{R}^-$.

B. $\mathrm{supp}(\mu) \subseteq [a,\infty)$, $\mathrm{supp}(\nu) \subseteq [-a,\infty)$, for some $a \in \mathbb{R} \cup \{-\infty\} \cup \{\infty\}$. Here the intervals are interpreted as $\emptyset$ or $\mathbb{R}$, if a is infinite.

Obviously B implies A. Does the converse hold? The answer is yes, if both supports are bounded below. This is just a version of the Titchmarsh convolution theorem [4,5]. Our results are more general, and they can thus be regarded as extensions of the Titchmarsh theorem. Here are the basic theorems.

THEOREM 1. $A \Rightarrow B$, if

$$\varliminf_{x\to-\infty} \frac{u(x)}{|x|^{\alpha}} > 0, \quad \varliminf_{x\to\infty} \frac{u(x)}{x^{\beta}} > 0,$$

where $\alpha > 0$, $\beta > 0$, $\alpha\beta = 1$, and where at least one of the lower limits is infinite.

THEOREM 2. If $\mathrm{supp}(\mu)$ is bounded below, then $A \Rightarrow B$, if

$$\lim_{x\to\infty} \frac{\log u(x)}{\sqrt{\log x}} = \infty .$$

THEOREM 3. If $\mathrm{supp}(\mu) \subseteq \mathbb{Z}$, then $A \Rightarrow B$, if

$$\varliminf_{x\to\infty} \,(u(x) - \alpha \log|x|) > -\infty, \quad \varliminf_{x\to\infty} \,(u(x) - \beta \log x) > -\infty,$$

where $\alpha > 0$, $\beta > 0$, $\alpha + \beta = 2$, and where at least one of the lower limits is infinite.

2. Preliminaries

The conditions of Theorem 1 and Theorem 3 imply that

$$\varlimsup_{|x|\to\infty} \frac{\log x}{u(x)} < \infty .$$

The corresponding relation holds, as $x \to \infty$, in the case of Theorem 2. Since the values of w on $\mathbb{R}^-$ are irrelevant in that case, we can assume that the relation holds as well for $x \to -\infty$. Then we can, in all three cases, define N as the smallest positive integers, for which

$$N > \varlimsup_{|x|\to\infty} \frac{\log x}{u(x)} . \tag{1}$$

In order to prove any of the three theorems, it suffices, by the Titchmarsh theorem, to find a contradiction if (μ,ν) satisfies A, while $\mathrm{supp}(\mu)$ and $\mathrm{supp}(\nu)$ are non-empty, and at least one of them is unbounded from below. It is then possible to introduce certain additional assumptions, without loss of generality. In the case of Theorem 1 or Theorem 2, we will assume the following:

1°. $\log w$ is convex on $(-\infty,0]$, concave on $[0,\infty)$, $w(0) = 1$, and $w'(0)$ exists and has the value 0.

2°. Assuming 1° fulfilled, we can define, for every $\sigma > 0$, real numbers $x_1 = x_1(\sigma) > 0$, and $x_2 = x_2(\sigma) > 0$, such that $\sigma x + \log w(x)$ increases in $[-x_1,x_2]$, and decreases in the complementary intervals. If σ is large enough we assume

$$x_1(\sigma) \le e^{N\sigma},\ x_2(\sigma) \le e^{N\sigma},\ \int_0^\infty e^{\sigma x}\ \frac{w(x_2+x)}{w(x_2)}\ dx \le 2e^{N\sigma},$$

$$\int_0^\infty e^{\sigma x}\ \frac{w(-x_1)}{w(-x_1-x)}\ dx \le 2e^{N\sigma}.$$

3°. μ and ν are absolutely continuous with respect to the Lebesgue measure, with derivatives f and g, respectively, satisfying

$$\int_{-\infty}^{\infty} |f(x)|w(x)dx \le 1, \qquad \int_{-\infty}^{\infty} |g(x)|(w(-x))^{-1}dx \le 1,$$

$$|f(x)| \le 1/w(x), \qquad |g(x)| \le w(-x),\ x \in \mathbb{R}.$$

4°. $\inf(\operatorname{supp}(f)) < -N$, $\inf(\operatorname{supp}(g)) < -N$.

These additional assumptions can be justified by diverse modifications of w, μ, and ν, involving regularizations and normalizations. Here we shall only make a few explanatory remarks. One is that the families of weights w fulfilling the conditions of Theorem 1 or Theorem 2, are invariant under right or left translations on $\mathbb{R}$, and so is, too, the associated constant N, defined by (1). This makes it possible to place the measures in accordance with 4°. As for 2°, a circumstance that prevents the integral inequalities to be valid quite generally, is the fact that $\log w$ may be linear, or close to linear, on some long intervals, giving rise to integrands with values close to 1 on such intervals. This can, however, be overcome by first using (1) to prove the inequality

$$|x|w(x+N) \le w(x),$$

valid for large $|x|$. By a translation argument we can then move into a situation where the weight equals $|x|^{-1}w(x)$, as $x \to -\infty$, and $xw(x)$, as $x \to \infty$, and then the corresponding inequalities can be proved. 3° can be legitimated by convoluting μ and ν with continuous functions with small support, and the remaining properties cause no difficulties to justify.

3. Transfer to a problem in complex function theory

In the proofs of Theorem 1 and Theorem 2 we shall thus assume property A of Section 1, that is the convolution equation

$$f * g(x) = 0, \quad x \in \mathbb{R}^-,$$

where w, f, g fulfill the additional assumptions 1° - 4° of Section 2.

We introduce f_0 and g_0, truncations at 0 of f and g, respectively, by the relations

$$f_0(x) = \begin{cases} f(x), & x \leq 0, \\ 0, & x > 0, \end{cases}$$

$$g_0(x) = \begin{cases} g(x), & x \leq 0, \\ 0, & x > 0. \end{cases}$$

For f_0 and g_0 we can introduce their Fourier-Laplace transforms F_0 and G_0, defined by

$$F_0(s) = \int_{-\infty}^{\infty} f_0(x)e^{-sx}dx = \int_{-\infty}^{0} f(x)e^{-sx}dx,$$

$$G_0(s) = \int_{-\infty}^{\infty} g_0(x)e^{-sx}dx = \int_{-\infty}^{0} g(x)e^{-sx}dx.$$

By 3° and the growth conditions on w, F_0 and G_0 are entire functions. We will study them only in the closed right half-plane

$$\Pi = \{s \in \mathbb{C} : \mathrm{Re}(s) \geq 0\},$$

and will show that there is a constant $C > 0$ such that

$$\inf(|F_0(s)|, |G_0(s)|) \leq Ce^{N\sigma}, \tag{2}$$

for $s = \sigma + it \in \Pi$. To this end it turns out to be convenient — and this is in fact the crucial point of the proof — to study instead

$$F_\sigma(s) = \int_{-\infty}^{\infty} f_\sigma(x)e^{-sx}dx = \int_{-\infty}^{x_2} f(x)e^{-sx}dx,$$

$$G_\sigma(s) = \int_{-\infty}^{\infty} g_\sigma(x)e^{-sx}dx = \int_{-\infty}^{x_1} g(x)e^{-sx}dx,$$

where

$$f_\sigma(x) = \begin{cases} f(x), & x \leq x_2(\sigma), \\ 0, & x > x_2(\sigma) \end{cases}$$

$$g_\sigma(x) = \begin{cases} g(x), & x \leq x_1(\sigma), \\ 0, & x > x_1(\sigma), \end{cases}$$

with x_1 and x_2 taken from assumption 2° of Section 2.

By elementary Fourier analysis

$$|F_\sigma(s)G_\sigma(s)| \le \int_{-\infty}^{\infty} |f_\sigma * g_\sigma(x)| e^{-\sigma x} dx , \tag{3}$$

and thus we are left to estimate $|f_\sigma * g_\sigma|$ on $\mathbb{R}$.

The support of $f_\sigma * g_\sigma$ is included in $(-\infty, x_1 + x_2]$. For $x \in [0, x_1 + x_2]$ we have

$$|f_\sigma * g_\sigma(x)| \le e^{\sigma x} ,$$

using the additional assumptions. On $\mathbb{R}^-$ we can use the representation

$$f_\sigma * g_\sigma = (f_\sigma - f) * (g_\sigma - g) + (f_\sigma - f) * g + f * (g_\sigma - g) + f * g .$$

The first and the last term of the sum vanish on $\mathbb{R}^-$, and estimating the rest, using the additional assumptions, we obtain

$$|f_\sigma * g_\sigma(x)| \le \frac{w(x_2 - x)}{w(x_2)} + \frac{w(-x_1)}{w(x - x_1)} , \quad x \in \mathbb{R}^- .$$

These estimates give, using (3), and 2° of the additional assumptions,

$$|F_\sigma(s)G_\sigma(s)| \le 6e^{N\sigma} ,$$

if σ is large enough. After a generous estimate we obtain, for the same values of σ,

$$\inf(|F_\sigma(s)|, |G_\sigma(s)|) \le 6e^{N\sigma} .$$

The additional assumptions give

$$|F_\sigma(s) - F_0(s)| \le 1, \quad |G_\sigma(s) - G_0(s)| \le 1 ,$$

if $\sigma > 0$, and thus

$$\inf(|F_0(s)|, |G_0(s)|) \le 7e^{N\sigma} ,$$

if σ is large enough. By the assumptions on w, $f(x)$ and $g(x)$ decrease to 0 faster than exponentially as $x \to -\infty$. It follows from this fact that F_0 and G_0 are uniformly bounded on any vertical strip in $\mathbb{C}$ of finite width. Hence there is a $C > 0$ such that (2) holds in Π.

In order to set the stage for the final part of the proof of Theorem 1 and Theorem 2, let us form

$$\Phi_1(s) = e^{-Ns} F_0(s), \quad \Phi_2(s) = e^{-Ns} G_0(s) ,$$

and consider these functions in Π. By (2) we have

$$\inf\{|\Phi_1(s)|, |\Phi_2(s)|\} \leq C,$$

for $s \in \Pi$. Furthermore, Φ_1 and Φ_2 are bounded on the imaginary axis. The additional assumption 4° and simple Paley-Wiener theory show that both Φ_1 and Φ_2 are unbounded in Π (see for instance [5]). In the next section we will show that these properties of the analytic functions Φ_1 and Φ_2, together with growth restrictions on Φ_1 and Φ_2 at infinity caused by the growth assumptions on w, give us the desired contradiction.

4. Final part of the proof of Theorem 1 and Theorem 2

We shall use the following very informative lemma, essentially given by formula (107) in Beurling [3]. [8] contains a second proof, based on Ahlfors' distortion theorem [1].

LEMMA. Let L_m, $1 \leq m \leq n$, be simple, non-intersecting Jordan curves in $\mathbb{C}$, extending from 0 to ∞ and splitting $\mathbb{C}$ into n simply-connected regions D_m. For every m, u_m is subharmonic and unbounded above in D_m, but uniformly bounded above at the finite part of the boundary of D_m. Put

$$M_m(r) = \sup_{z \in D_m, |z|=r} u_m(z).$$

Then there is a constant $c > 0$ such that

$$\sum_{m=1}^{n} \frac{1}{\log M_m(r)} < \frac{2}{\log r - c},$$

if r is sufficiently large.

In the situation, described at the end of the preceding section, it is possible to prove that there is a Jordan curve L, in the interior Π^0 of Π, from 0 to ∞, with Φ_1 and Φ_2 bounded on L, and such that L splits Φ into two parts, one where Φ_1 is unbounded, and one where Φ_2 is unbounded. After an elementary conformal mapping of Π^0 onto $\mathbb{C}$, cut along the negative real axis, we can apply the lemma to the subharmonic functions $\log|\Phi_1|$ and $\log|\Phi_2|$. Putting

$$M_i(r) = \sup_{s \in \Pi, |s|=r} \log|\Phi_i(s)|, \qquad i = 1,2,$$

we obtain

$$\frac{1}{\log M_1(r)} + \frac{1}{\log M_2(r)} < \frac{1}{\log r - c}, \tag{4}$$

for some $c > 0$ if r is sufficiently large.

Simple estimates show that the growth conditions of Theorem 1 imply that

$$\overline{\lim_{r\to\infty}} (\log M_1(r) - (\beta+1)\log r) < \infty,$$

$$\overline{\lim_{r\to\infty}} (\log M_2(r) - (\alpha+1)\log r) < \infty,$$

with at least one of these upper limits equal to $-\infty$. This contradicts (4). In the case of Theorem 2, the lower boundedness of supp(f) and the growth condition on w given

$$\overline{\lim_{r\to\infty}} (\log M_1(r) - \log r) < \infty,$$

$$\log M_2(r) = \underline{o}(\log r)^2, \quad \text{as} \quad r \to \infty,$$

and this, too, contradicts (4). Hence Theorem 1 and Theorem 2 are proved.

5. Proof of Theorem 3

With $\text{supp}(\mu) \subseteq \mathbb{Z}$, it is easy to show that we can assume $\text{supp}(\nu) \subseteq \mathbb{Z}$ as well, without losing generality in the proof of Theorem 3. Then it is possible to reformulate the problem in terms of sequences in the following way. Let $\{a_n\}_{-\infty}^{\infty}$ and $\{b_n\}_{-\infty}^{\infty}$ be complex sequences with

$$\sum_{-\infty}^{\infty} |a_n| w(n) < \infty, \quad \sum_{-\infty}^{\infty} |b_n| (w(-n))^{-1} < \infty,$$

and consider the convolution equation

$$\sum_{-\infty}^{\infty} a_{m-n} b_n = 0, \quad m < 0 .$$

We can regularize and normalize in a way analogous to the discussion in the continuous case. In particular we can assume, without loss of generality, that neither a_n nor b_n vanishes for all $n < -N$, attempting to find a contradiction.

The earlier proof can be adopted very easily to our new situation. This time we have

$$\Phi_1(s) = e^{-Ns} \sum_{-\infty}^{0} a_n e^{-sn},$$

$$\Phi_2(s) = e^{-Ns} \sum_{-\infty}^{0} b_n e^{-sn},$$

with Φ_1 and Φ_2 having exactly the same properties as before, in particular fulfilling (2), for some $C > 0$.

But Φ_1 and Φ_2 are not only analytic in s, but also analytic in $z = e^s$. To be more precise, putting

$$\Psi_i(z) = \Psi_i(e^s) = \Phi_i(s) ,$$

$i = 1,2$, we find that Ψ_i are analytic in $\{z \in \mathbb{C} : |z| \geq 1\}$, where they satisfy

$$\inf(|\Psi_1(z)|, |\Psi_2(z)|) \leq C .$$

Ψ_i are unbounded in the region considered, the exclusion of the unit circle is irrelevant, and we can, in a similar way as before, apply the lemma, this time in its original version, for $\mathbb{C}$ itself, with $n = 2$. We obtain

$$\frac{1}{\log M_1(r)} + \frac{1}{\log M_2(r)} < \frac{2}{\log r - c} , \tag{5}$$

for some $c > 0$ if r is sufficiently large, where

$$M_i(r) = \sup_{|z|=r} \log|\Psi_i(z)| , \qquad i = 1,2 .$$

The growth assumptions of Theorem 3 give

$$\overline{\lim_{r\to\infty}} \left(\log M_1(r) - \frac{1}{\alpha} \log r\right) < \infty ,$$

$$\overline{\lim_{r\to\infty}} \left(\log M_2(r) - \frac{1}{\beta} \log r\right) < \infty ,$$

where at least one of the upper limits equals $-\infty$. This contradicts (5), and Theorem 3 is proved.

6. On the necessity of the growth conditions on w

Let us keep the basic assumptions on w, namely that $\log w$ is what may be called convex-concave, and that $u(x) \to \infty$, as $x \to \infty$. In this section we will see whether the three basic theorems fail if the growth conditions are sufficiently weakened. We have two results, pertaining to Theorem 1 and Theorem 3, respectively.

THEOREM 4. The conclusion of Theorem 1 is false with $0 < \alpha < 1$, $0 < \beta < 1$.

Proof. For every positive decreasing function w on $\mathbb{R}$ such that

$$\int_{-\infty}^{\infty} \frac{d(\log w(x))}{1 + x^2} > -\infty , \tag{6}$$

it is possible to find a not identically vanishing function G, analytic in the upper half-plane, with $|G|$ having boundary values w on the real axis, and such that $x \to |G(x+iy)|$ decreases, for every fixed $y > 0$.

If α and β are as stated, we can find an admissible, decreasing weight, satisfying (6). With G as above, let $h \not\equiv 0$ be continuous, belong to $L^1(\mathbb{R})$, and be extendable to a function, bounded and analytic in the upper half-plane. Let us define

$$f(x) = (G(x))^{-1}h(x),\ g(x) = G(-x)h(-x)\,,$$

$x \in \mathbb{R}$. Then $t \mapsto f(x-t)g(t)$ is, for every fixed $x \in \mathbb{R}^-$, the restriction to $\mathbb{R}$ of a function, analytic and bounded in the upper half-plane, and such that the boundary function on the real axis is in $L^1(\mathbb{R})$. Hence (see for instance [5]) its integral over $\mathbb{R}$ vanishes. Thus $f * g = 0$ on $\mathbb{R}^-$, giving the desired counter-example.

REMARKS. For $w(x) = \exp(-|x|^\gamma \operatorname{sign} x)$, $1 < \gamma < 2$, we can define G explicitly, namely by

$$G(z) = \exp\left\{\frac{-i(-iz)^\gamma}{\sin\frac{\pi\gamma}{2}}\right\},$$

where the power takes its principal value.

It should be mentioned that the condition (6) is in fact necessary for the existence of a function G with the mentioned properties, and thus this type of counter-example can unfortunately not be used for other values of α and β.

THEOREM 5. For $\alpha = \beta = 1$, it is not possible to remove in Theorem 3 the condition that at least one of the lower limits is infinite.

Proof. Counter-examples are easily found by expanding $\exp(c/z)$ and $\exp(-c/z)$ in powers of $1/z$, and considering the identity

$$\exp(c/z)\exp(-c/z) = 1\,.$$

7. Applications

For $a \geq 0$, right translation T_a of a function f on $\mathbb{R}$, $\mathbb{R}^+$, or $\mathbb{Z}$ is defined by the relation

$$T_a f(x) = f(x-a)\,,$$

where, in the case of $\mathbb{R}^+$, we think of f as extended with the values 0 on $\mathbb{R}^-$. For linear spaces of functions on any of these semigroups, the standard (invariant) subspaces are the spaces of all functions vanishing for all $x < a$, with a in the semigroup, and the two trivial subspaces.

Let us first consider $\mathbb{R}$. For $1 \leq p \leq \gamma$, and with w as any weight, we define $L^p(w,\mathbb{R})$ as the Banach space of functions f, with $fw \in L^p(\mathbb{R})$, under the induced norm. Right translation is a bounded operator, since $w(x)$ decreases for large $|x|$, and the standard subspaces are closed. Let w_1 be defined as in Section 1. Elementary functional analysis shows, in case $1 \leq p < \infty$, that no other closed right translation invariant subspaces exist if and only if the only solutions of

$$f * g(x) = 0, \quad x \in \mathbb{R}^-,$$

with $0 \not\equiv f \in L^p(w,\mathbb{R})$, $0 \not\equiv g \in L^q(w_1,\mathbb{R})$, $1/p + 1/q = 1$, are pairs (f,g), for which both f and g belong to some proper standard subspace of their respective space. By a simple translation argument we can restrict to the case when $f \in L^1(w,\mathbb{R})$, $g \in L^1(w_1,\mathbb{R})$, and then obtain the following from Theorem 1.

THEOREM 6. Under the conditions of Theorem 1, all closed right translation invariant subspaces of $L^p(w,\mathbb{R})$, $1 \leq p < \infty$, are standard.

Theorem 2 can be applied in a similar way to the Banach spaces $L^p(w,\mathbb{R}^+)$ of functions f on $\mathbb{R}^+$ with $fw \in L^p(\mathbb{R}^+)$, $1 \leq p < \infty$. By a result of Grabiner [9], the growth condition in Theorem 2 implies that $L^p(w,\mathbb{R}^+)$ is a Banach algebra under convolution. The notion of closed ideal coincides with the notion of closed right translation-invariant subspace, and thus we obtain the following result.

THEOREM 7. Under the conditions of Theorem 2, the only closed ideals of $L^p(w,\mathbb{R}^+)$, $1 \leq p < \infty$, are the standard subspaces.

Let us assume that the positive sequence $\{w_n\}_{-\infty}^{\infty}$ has the property that $\{\log w_n\}$ is ultimately convex, as $n \to -\infty$, ultimately concave, as $n \to \infty$. Interpolating linearly between the integer points, we obtain a function w on $\mathbb{R}$. Theorem 3 can then be applied, as above, to give the following theorem, where $\ell^p(w,\mathbb{Z})$, $1 \leq p < \infty$, is the Banach space of functions f on $\mathbb{Z}$ with $\{f(n)w(n)\} \in \ell^p(\mathbb{Z})$.

THEOREM 8. All right translation invariant subspaces in $\ell^p(w,\mathbb{Z})$, $1 \leq p < \infty$, are standard if

$$\varliminf_{n\to-\infty} \left(\frac{\log w_n}{|n|} - \alpha \log|n| \right) > -\infty,$$

$$\varliminf_{n\to-\infty} \left(- \frac{\log w_n}{n} - \beta \log n \right) > -\infty,$$

where $\alpha > 0$, $\beta > 0$, $\alpha + \beta = 2$, and where at least one of the limits is infinite.

It is easy to see that Theorem 4 holds with "Theorem 1" exchanged to "Theorem 6", and that Theorem 5 holds with "Theorem 3" exchanged to "Theorem 8". Let us compare the second of these two assertions with Theorem 8 itself, in the case

$\alpha = \beta = 1$. In terms of the notion of bilateral weighted shift (see [10]) we then obtain the following theorem, which gives an answer to Question 17 of [10].

THEOREM 9. Let the positive sequence $\{\lambda_n\}_{-\infty}^{\infty}$ be increasing, for $n \leq 0$, and decreasing, for $n \geq 0$. Let T be the corresponding weighted shift on $\ell^2(\mathbb{Z})$. If $n\lambda_n \to 0$ as $|n| \to \infty$, all closed invariant subspaces are standard (equivalently, the weighted shift is unicellular). This is never the case, if $\varliminf_{|n| \to \infty} |n|\lambda_n > 0$.

8. A partial result

The theorem below gives a partial information, related to Theorem 7, in the case of a weight w for which we have just the very weak assumption (7) below. The result should be compared with Theorem 4 and its corollary in Allan [2], and with Theorem 3 in [6].

THEOREM 10. Let w be a weight with

$$\varliminf_{x \to \infty} \frac{u(x)}{\log x} > 0\,, \tag{7}$$

and let $f_0 \in L^1(w, \mathbb{R}^+)$, $g_0 \in L^\infty(w_1, \mathbb{R}^-)$, with $\inf(\mathrm{supp}(f_0)) = 0$, and

$$f_0 * g_0(x) = 0, \quad x \in \mathbb{R}^-.$$

(Here w_1 is as in Section 1, and convolution is defined after extending f_0 by defining it to be 0 on $\mathbb{R}^-$.)

For a fixed $d > 0$, we form the set E of all $s = \sigma + it$ in the right half-plane Π, for which

$$\left| \int_0^{d+1} f_0(x) e^{-sx} dx \right| < e^{-d\sigma}.$$

Then g_0 vanishes (almost everywhere) on $\mathbb{R}^-$, if the points s in every unbounded component of E satisfy $\sigma/\sqrt{|s|} \to 0$, as $s \to \infty$.

The proof follows along the same lines as the proof of Theorem 1 and Theorem 2, and we shall just sketch it briefly. We can make the same kind of regularizations and normalizations as those we did in the mentioned proof, and it is then possible to transfer the discussion to the problem of finding a contradiction in the case when we have two entire functions Φ_1 and Φ_2 with the following properties:

Φ_1 is bounded on the imaginary axis, but unbounded (of exponential growth) in Π. For every $C > 0$, the set $E(C)$ of all $s \in \Pi$ with $|\Phi_1(s)| < C$ has the property that $\sigma/\sqrt{|s|} \to 0$, as $s \to \infty$, in every unbounded component of $E(C)$.

Φ_2 is bounded on the imaginary axis, but unbounded in Π, where it satisfies

$$\varlimsup_{s \to \infty} \sigma^{-1} \log^+ \log^+ |\Phi_2(s)| < \infty.$$

(The last inequality follows from (7).)

Finally $\inf(|\Phi_1|,|\Phi_2|)$ is bounded in Π.

As in the proof of Theorem 1 and Theorem 2 we can split Π into two parts E_1 and E_2, such that Φ_i is unbounded in E_i, $i = 1,2$, but bounded on ∂E_i. This time the splitting can be chosen in such a way that $\sigma/\sqrt{|s|} \to 0$, as $s \to \infty$ in E_2. Hence

$$|s|^{-1/2} \log^+\log^+|\Phi_2(s)| \to 0,$$

as $s \to \infty$ in E_2. An application of the Ahlfors distortion theorem [1] for $\log|\Phi_2|$ in E_2 (see the proof of the lemma of [8] for a similar argument), shows that the last relation gives a contradiction.

9. Open problems

Problem 1. Is there a weight w on $\mathbb{R}^+$ with $\log w$ concave and

$$\lim_{x\to\infty} x^{-1} \log w(x) \to -\infty,$$

such that $L^1(w,\mathbb{R}^+)$ has non-standard ideals?

The corresponding question for $\ell^1(w,\mathbb{Z}^+)$ has the answer no (see for instance [6]). But this fact does not speak too strongly in favour of the same answer to Problem 1. For Theorem 3 and Theorem 5 indicate a very clear distinction between the needed growth conditions in the continuous and the discrete case in the questions under discussion.

When attempting to find an affirmative answer to Problem 1, Theorem 10 and its proof may perhaps give some indication on what type of functions are required to provide counterexamples.

Problem 2. Is there a weight w on $\mathbb{R}$, with $\log w$ convex-concave,

$$\lim_{|x|\to\infty} x^{-1} \log w(x) = -\infty,$$

and

$$\int_{-\infty}^{\infty} \frac{d(\log w(x))}{1+x^2} = -\infty,$$

for which $L^1(w,\mathbb{R})$ has non-standard right invariant subspaces?

Of particular interest are positive examples where $w(x)$ tends to infinity very rapidly, as $x \to -\infty$. By a limit procedure this might give a positive answer of Problem 1 as well. It is believable that examples for Problem 1 and Problem 2, if they exist, can be constructed by means of explicit functions taken from complex theory.

Problem 3. How relevant is the concavity assumption on $\log w$ in Theorem 7?

Problem 4. Let w_n be defined as $1/n!$, for $n \geq 0$, as $|n|!$, for $n < 0$. Is it then possible to obtain a complete description of the closed right translation-invariant subspaces of $\ell^1(w)$ or $\ell^2(w)$?

References

[1] L. Ahlfors, Untersuchungen zur Theorie der Konformen Abbildung und der ganzen Funktionen, Acta Soc. Sci. Fenn., 1, 9 (1930).

[2] G. R. Allan, Ideals of rapidly growing functions, Proceedings International Symposium on Functional Analysis and its Applications, Ibadan, Nigeria, 1977.

[3] A. Beurling, Études sur un problème de majoration, Uppsala 1933.

[4] R. P. Boas, Entire Functions, Academic Press, New York, 1954.

[5] H. G. Dales, Convolution algebras on the real line, this Volume.

[6] Y. Domar, Cyclic elements under translation in weighted L^1 spaces on $\mathbb{R}^+$, Ark. Mat., 19 (1981), 137-144.

[7] __________, Translation invariant subspaces of weighted ℓ^p and L^p spaces, Math. Scand., 49 (1981), 133-144.

[8] __________, Extensions of the Titchmarsh convolution theorem with applications in the theory of invariant subspaces, Proc. London Math. Soc., to appear.

[9] S. Grabiner, Analogies between Banach algebras of power series and weighted convolution algebras, this Volume.

[10] A. L. Shields, Weighted shift operators and analytic function theory, Amer. Math. Soc. Mathematical Surveys, 13 (1974), 49-128.

Department of Mathematics
University of Uppsala
Thunbergsvägen 3
S-752 38 Uppsala, Sweden

MULTIPLIERS OF WEIGHTED ℓ^1-ALGEBRAS

William G. Bade

Introduction

This paper is a report on joint work with H. G. Dales and K. B. Laursen concerning a new class of Banach algebras of power series. These algebras are the multiplier algebras of certain radical weighted convolution algebras.

A Banach algebra of power series is a subalgebra of the algebra $\mathbb{C}[[X]]$ of all formal power series in one variable X which has a complete algebra norm with respect to which the coordinate functionals are continuous. Let ω be a radical weight function, a positive submultiplicative function on $\mathbb{Z}^+$ with $\inf \omega(n)^{1/n} = 0$. Denote by $\ell^1(\omega)$ the subalgebra of $\mathbb{C}[[X]]$ consisting of those formal power series $x = \Sigma x(n)X^n$ for which $\|x\| = \sum_{n=0}^{\infty} |x(n)|\omega(n) < \infty$. Multiplication of formal power series corresponds to convolution of sequences of coefficients. If ω is a radical weight function, then $\ell^1(\omega)$ is a local algebra, and its unique maximal ideal, $M = \{x : x(0) = 0\}$, is a radical algebra. We study the multiplier algebra $\mathfrak{m}(M)$ of M, and prove that $\mathfrak{m}(M)$ may be naturally identified with another Banach algebra of power series containing $\ell^1(\omega)$ as a subalgebra. For general radical weights ω we explore the connection between properties of ω and those of the multiplier algebra $\mathfrak{m}(M)$. Since the algebras $\ell^1(\omega)$ are local algebras, the closure $\mathfrak{L}(M)$ of $\ell^1(\omega)$ in the operator norm of $\mathfrak{m}(M)$ is always a local algebra. We show, however, that $\mathfrak{m}(M)$ need not be local, and indeed we describe an example in which $\mathfrak{m}(M)$ contains a multiplier whose spectrum has non-empty interior.

For the broad class of weights ω for which the maps $y \mapsto x + y; \ell^1(\omega) \mapsto \ell^1(\omega)$ are compact for each $x \in \ell^1(\omega)$, we prove that there is an interesting equivalence between various algebraic properties associated with the left shift $L\omega$ of ω and Banach space properties of the Banach algebra $\mathfrak{m}(M)$. For example, $\mathfrak{m}(M)$ is separable if and only if some multiple of $L\omega$ is submultiplicative. When $\mathfrak{m}(M)$ is non separable, it contains subspaces isomorphic to ℓ^∞, and $\mathfrak{m}(M)/\mathfrak{L}(M)$ contains subspaces isometric to ℓ^∞/c_0. When $\mathfrak{L}(M) \neq \mathfrak{m}(M)$, $\mathfrak{L}(M)$ is uncomplemented in $\mathfrak{m}(M)$.

Our initial interest in studying the multiplier algebras $\mathfrak{m}(M)$ arose from an important problem concerning the algebras $\ell^1(\omega)$. For $k \in \mathbb{N}$, let $M_k = \{(x(j)) : x(j) = 0, 0 \leq j < k\}$. These are closed ideals in $\ell^1(\omega)$. Together with $\{0\}$ and $\ell^1(\omega)$, they are the standard ideals of $\ell^1(\omega)$. No example is known of a weight ω for which $\ell^1(\omega)$ has non-standard closed ideals. In [10] Nikolskii claims to give an example of a radical weight ω such that $\ell^1(\omega)$ contains non-standard ideals, but his proof is in error. (However his arguments do yield nonstandard closed translation invariant subspaces in certain Banach spaces $\ell^1(\omega)$ where ω is not submultiplicative.)

It has been a difficult problem to construct specific weights for which $\ell^1(\omega)$ might contain nonstandard closed ideals. We show in this work how to construct weights ω for which $\mathfrak{M}(M)$ has unusual properties. It seems likely that for such ω the algebras $\ell^1(\omega)$ contain nonstandard closed ideals, but we have not been able to prove this.

A full account of the details of this work will appear in [2]. However, the present paper contains some new results not included there. In Section 2 we give the full details of the construction of a class of weights ω for which $\mathfrak{L}(M)$ is a proper subalgebra of $\mathfrak{M}(M)$. These examples are much simpler than the difficult example in [2] of a weight ω for which there are multipliers in $\mathfrak{M}(M)$ whose spectrum has nonempty interior. It is not known even for the examples constructed in this paper whether $\ell^1(\omega)$ contains only standard closed ideals.

In the discussion of Banach space properties of $\mathfrak{M}(M)$ it is proved that $\mathfrak{M}(M)$ is always a dual space. Following a suggestion of G. Bachelis, we constructed in [2] a predual for $\mathfrak{M}(M)$ as a quotient of a certain projective tensor product. In this paper we give a simple and natural construction of another predual, by very different methods. It has been pointed out to us by G. Bachelis that these two preduals are in fact isometrically isomorphic, thus raising the question of when $\mathfrak{M}(M)$ has a unique predual.

Two lectures in this volume concern topics especially close to those discussed here. In [8], K. B. Laursen discusses an important special class of weights, the Domar weights. For a Domar weight, all closed ideals in $\ell^1(\omega)$ are standard, and $\mathfrak{M}(M)$ is local, although it can properly contain $\mathfrak{L}(M)$. For a Domar weight $\mathfrak{M}(M)$ is always nilpotent. In his paper [12], M. Thomas establishes the weakest conditions presently known that insure for a weight ω that all closed ideals in $\ell^1(\omega)$ are standard.

1. Weights and multipliers

A real-valued function ω on $\mathbb{Z}^+$ is a weight function if $\omega(n) > 0$ $(n \in \mathbb{Z}^+)$ and if

$$\omega(m+n) \leq \omega(m)\omega(n) \quad (m,n \in \mathbb{Z}^+). \tag{1.1}$$

It is often convenient to set

$$\omega(n) = \exp(-\eta(n)) \quad (n \in \mathbb{Z}^+), \tag{1.2}$$

and write $\omega = e^{-\eta}$. Thus if ω is a weight function, then

$$\eta(m) + \eta(n) \leq \eta(m+n) \quad (m,n \in \mathbb{Z}^+). \tag{1.3}$$

Conversely if η is a nonnegative function satisfying (1.3), then (1.2) defines a weight function ω. We say ω is a radical weight if

$$\lim_{n\to\infty} \omega(n)^{1/n} = 0.$$

Equivalently ω is radical if

$$\lim_{n\to\infty} \frac{\eta(n)}{n} = \infty.$$

If $\omega = (\omega(n))$ is any sequence of positive numbers, we denote by $\ell^1(\omega)$ the set of complex-valued functions x on $\mathbb{Z}^+$ for which $\|x\| = \sum_{n=0}^{\infty} |x(n)|\omega(n) < \infty$. The dual space of $\ell^1(\omega)$ is $\ell^\infty(\omega^{-1})$, the Banach space of complex-valued functions y on $\mathbb{Z}^+$ for which $|y|/\omega$ is bounded. The norm is

$$\|y\|_\infty = \sup\{|y(n)|/\omega(n) : n \in \mathbb{Z}^+\}$$

and the duality is implemented by $\langle x,y\rangle = \sum_{n=0}^{\infty} x(n)y(n)$.

Let ω be a weight function, and let $x,y \in \ell^1(\omega)$. We define the convolution product $x * y$ by setting

$$(x*y)(n) = \sum_{i=0}^{n} x(i)y(n-i) \quad (n \in \mathbb{Z}^+).$$

Then $\|x*y\| \leq \|x\|\|y\|$ $(x,y \in \ell^1(\omega))$.

If ω is a radical weight, then $\ell^1(\omega)$ is a local algebra. Its unique character is the functional $(x(n)) \to x(0)$ $(x \in \ell^1(\omega))$, and its unique maximal ideal is $M_1(\omega)$, where, for $n \geq 1$, we write

$$M_n(\omega) = \{x \in \ell^1(\omega) : x(i) = 0,\ i < n\} \quad (n \in \mathbb{N}) \tag{1.4}$$

for the standard ideals.

We shall suppose hereafter that ω is a radical weight and write M for $M_1(\omega)$. Then M is the radical of $\ell^1(\omega)$.

For $j \geq 0$, let $e_j = \{\delta_{jn} : n \geq 0\}$. Then e_0 is the identity of the algebra $C[[X]]$. Note that $e_j * e_k = e_{j+k}$ $(j,k \geq 0)$. Moreover, for each $x \in \ell^1(\omega)$

$$(e_j * x)(n) = \begin{cases} 0 & (n < j), \\ x(n-j) & (n \geq j), \end{cases}$$

so that multiplication by e_1 effects a right shift on $\ell^1(\omega)$. Also, if $x \in \ell^1(\omega)$, then $x = \sum x(n)\, e_1^{*n} = \sum x(n)X^n$, with the series converging to x

in $\ell^1(\omega)$.

Our purpose is to study the multiplier algebra [14] of the ideal M in $\ell^1(\omega)$. Recall that for a commutative Banach algebra A, a linear map $T : A \to A$ is a <u>multiplier</u> if

$$T(ab) = aTb \qquad (a,b \in A).$$

The set $\mathfrak{M}(A)$ of all continuous multipliers of A is a strongly closed subalgebra of the Banach algebra $\mathfrak{B}(A)$ of all bounded linear operators, and $\mathfrak{M}(A)$ contains the identity operator. Via the regular representation $x \to T_x$, where $T_x(y) = xy$ $(y \in A)$, we can regard A as an ideal in $\mathfrak{M}(A)$. We shall use $|||\cdot|||$ to denote the operator norm in $\mathfrak{B}(A)$, and note $|||a||| \leq \|a\|$ $(a \in A)$.

Taking $A = M \subseteq \ell^1(\omega)$, it follows easily from the closed graph theorem that every multiplier on M continuous. We shall see next that every multiplier of M is given by convolution by some sequence, and that $\mathfrak{M}(M)$, with the operator norm, is a Banach algebra of power series. For convenience we define the left and right shifts on sequences. If $x = (x(0),x(1),x(2),\ldots)$ is a sequence, then

$$Lx = (x(1),x(2),\ldots)$$

and

$$Rx = (0,x(0),x(1),\ldots).$$

1.1. THEOREM. Let ω be a radical weight function.

(i) If $T \in \mathfrak{M}(M)$, define $\alpha = LTe_1$. Then $Tx = \alpha * x$ for all $x \in M$.

(ii) A sequence $\alpha = (\alpha(j))$ determines a multiplier T on M by convolution if and only if

$$\sup_{n\geq 1} \sum_{j=0}^{\infty} |\alpha(j)| \frac{\omega(n+j)}{\omega(n)} < \infty, \tag{1.5}$$

and then this supremum is the operator norm $|||T|||$ of the multiplier.

(iii) The map $T \to \alpha$ defines an isometric isomorphism of $\mathfrak{M}(M)$ onto a Banach algebra of power series.

<u>Proof</u>. To prove (i), let $T \in \mathfrak{M}(M)$ and define $\alpha = LTe_1$. Then, since RL is the identity on M, $Te_1 = R\alpha = e_1 * \alpha$ and $T(e_n) = T(e_{n-1} * e_1) = e_{n-1} * Te_1 = e_n * \alpha$ $(n \geq 2)$. Let $x = \sum_{n=1}^{\infty} x(n)e_n$ be an element of M. Then $Tx = \sum_{n=1}^{\infty} x(n)Te_n = \sum_{n=1}^{\infty} x(n)(\alpha * e_n)$, and so

$$(Tx)(j) = \sum_{n=1}^{j} x(n)(\alpha * e_n)(j) = \sum_{n=1}^{j} x(n)\alpha(j-n) = (\alpha * x)(j) \qquad (j \in \mathbb{N}).$$

This proves that $Tx = \alpha * x$ for all $x \in M$.

For (ii), let $\alpha = (\alpha(n))$ be a sequence and, for each $x \in M$, let Tx be the sequence $\alpha * x$. Then $\|Te_n\| = \sum_{j=0}^{\infty} |\alpha(j)|\omega(n+j)$ $(n \in \mathbb{N})$. If T is a multiplier, then $\sup_{n \geq 1} \|Te_n\| / \|e_n\| < \infty$, that is, $\sup_{n \geq 1} \sum_{j=0}^{\infty} |\alpha(j)|\omega(n+j)/\omega(n) < \infty$. For the converse, let $C = \sup_{n \geq 1} \sum_{j=0}^{\infty} |\alpha(j)|\omega(n+j)/\omega(n)$. If $x = \sum_{n=1}^{\infty} x(n)e_n \in M$, then

$$\sum_{n=1}^{\infty} |(\alpha * x)(n)|\omega(n) \leq \sum_{n=1}^{\infty} |x(n)| \left(\sum_{j=0}^{\infty} |\alpha(j)|\omega(n+j) \right)$$
$$\leq \sum_{n=1}^{\infty} |x(n)|\omega(n) < \infty ,$$

and so $\alpha * x \in M$, showing that T is a multiplier.

The rest of (ii) is clear, as is (iii).

It follows from (i) that $\ell^1(\omega) \subseteq \mathfrak{m}(M) \subseteq L(M)$. We make some observations on when these inclusions may be equalities. Note that it is impossible that $L(M) = \ell^1(\omega)$. For if $L(M) = \ell^1(\omega)$, then R maps $\ell^1(\omega)$ onto M. Since R is continuous $(|||R||| \leq \omega(1))$, L must be continuous by the open mapping theorem. Then we have $\omega(0) = \|e_0\| = \|L^k e_k\| \leq |||L|||^k \omega(k)$, so $\omega(k)^{1/k} \geq \omega(0)/|||L|||$ $(k \in \mathbb{N})$, contradicting the assumption that ω is a radical weight.

The next result deals with the possibility that $L(M) = \mathfrak{m}(M)$.

1.2. THEOREM. Let ω be a radical weight function. The following are equivalent.

(a) $L(M) = \mathfrak{m}(M)$,

(b) $L(M)$ is an algebra under convolution,

(c) There exists a constant C such that

$$\omega(m+n+1) \leq C\omega(m+1)\omega(n+1) \qquad (m,n \in \mathbb{Z}^+), \tag{1.6}$$

and $\mathfrak{m}(M) = \ell^1(\omega')$ (where $\omega'(n) = C\omega(n+1)$ for $n \in \mathbb{N}$), the two norms being equivalent.

<u>Proof</u>. Certainly (a) implies (b). The equivalence of (b) and (c) is due to Grabiner [5]. We give the simple proof that holds in our setting. Suppose that (b) holds. Then $L(M)$ is a Banach space for the norm $\|x\| = \sum |x(n)|\omega(n+1)$, and it is an algebra in which multiplication is separately continuous (this follows from the closed graph theorem). Hence there exists a constant C such that $\|x * y\| \leq C\|x\|\|y\|$ $(x,y \in L(M))$. Taking $x = e_m$ and $y = e_n$, we obtain (1.6). Thus ω' is a weight and $\mathfrak{m}(M) = \ell^1(\omega')$, the two norms being equivalent.

Finally, let (c) holds, let $x \in M$, and let $\alpha = Lx$. Then $\omega(n+j)/\omega(n) \leq C\omega(j+1)$ by assumption, so that

$$\sum_{j=0}^{\infty} |\alpha(j)| \frac{\omega(n+j)}{\omega(n)} \leq C \sum_{j=0}^{\infty} |\alpha(j)|\omega(j+1)$$

$$= C \sum_{k=1}^{\infty} |x(k)|\omega(k) < \infty .$$

It follows from Theorem 1.1 (ii) that $\alpha \in \mathfrak{M}(M)$ and that (a) holds. This completes the proof of the equivalences.

Grabiner proves that a sufficient condition for (c) to hold is that $\{\omega(n+1)/\omega(n)\}$ be eventually decreasing to zero [5, p. 646].

We use $\mathcal{L}(M)$ to denote the operator-norm closure of $\ell^1(\omega)$ in $\mathfrak{M}(M)$. Clearly, if $L(M) = \mathfrak{M}(M)$, then $\mathcal{L}(M) = \mathfrak{M}(M)$, since by Theorem 1.2 (c), the polynomials are dense in $\mathfrak{M}(M)$. Thus the "normal" situation is that

$$\ell^1(\omega) \subsetneq \mathcal{L}(M) = \mathfrak{M}(M) = L(M) .$$

This is the case with the weight function given by $\omega(n) = \exp(-n^2)$ $(n \in \mathbb{Z}^+)$. However, the next example shows that it is possible to have $\ell^1(\omega) = \mathfrak{M}(M)$.

1.3. EXAMPLE. Define $\omega(0) = 1$, $\omega(2^p) = (2^{p+1}!)^{-1}$ $(p \in \mathbb{Z}^+)$, and $\omega(2^p + k) = \omega(2^p)\omega(k)$ $(0 \leq k < 2^p)$. It is clear that $\omega(n)^{1/n} \to 0$ as $n \to \infty$. Now for each $j \in \mathbb{N}$ let P_j be the statement that $\omega(m+n) \leq \omega(m)\omega(n)$ whenever $m,n \geq 0$ and $m + n = j$. Clearly P_1 is true. Suppose P_i is true for all $i < j = 2^p + k$, where $0 \leq k < 2^p$. Let $j = m + n$, with $m \leq n$. Then we have $2^{p-1} \leq n < 2^p + k$. There are two cases to consider. The first is when n has the form $\underline{n = 2^p + r}$, where $0 \leq m = k - r$. Then by the induction hypothesis.

$$\omega(m)\omega(n) = \omega(k-r)\omega(2^p+r) = \omega(k-r)\omega(2^p)\omega(r)$$
$$\geq \omega(k)\omega(2^p) = \omega(m+n) .$$

In the other case $\underline{n = 2^{p-1}+s}$, where $0 \leq s < 2^{p-1}$, and $m = 2^{p-1} + (k-s)$, where $(k-s) < s$. Then

$$\omega(m)\omega(n) = \omega(2^{p-1} + (k-s))\omega(2^{p-1}+s) = \omega(2^{p-1})^2\omega(k-s)\omega(s)$$
$$\geq \omega(2^p)\omega(k) = \omega(m+n) .$$

It follows that P_j is true. Thus ω is a radical weight. Moreover,

$$\omega(k) = \frac{\omega(2^p+k)}{\omega(2^p)} \qquad (0 \leq k < 2^p) . \tag{1.7}$$

If $\alpha \in \mathfrak{M}(M)$ then

$$\sum_{k=0}^{2^p-1} |\alpha(k)|\omega(k) = \sum_{k=0}^{2^p-1} |\alpha(k)| \frac{\omega(2^p+k)}{\omega(2^p)} \leq |||\alpha||| .$$

Since p is arbitrary, $\|\alpha\| \leq |||\alpha|||$. But the reverse inequality holds, so $\|\alpha\| = |||\alpha|||$ $(\alpha \in \mathfrak{M}(M))$. Thus $\mathfrak{M}(M) = \ell^1(\omega)$, so we have

$$\ell^1(\omega) = \mathfrak{L}(M) = \mathfrak{M}(M) \subsetneqq L(M) .$$

The strict inclusion results from our earlier remark that one never has $\ell^1(\omega) = L(M)$ for a radical weight.

This elegant example was shown to us at the Conference by H. Kamowitz. A much more complicated example with similar properties is given in [1].

The elements of $\mathfrak{L}(M)$ are just those multipliers whose series converge in the operator norm. If $\alpha \in \mathfrak{M}(M)$, define

$$S_m(\alpha) = (0,0,\ldots,0,\alpha(m),\alpha(m+1),\ldots) \qquad (m \in \mathbb{Z}^+) .$$

Then $S_m(\alpha) \in \mathfrak{M}(M)$ and

$$|||S_m(\alpha)||| = \sup_{n \geq 1} \sum_{j=m}^{\infty} |\alpha(j)| \frac{\omega(n+j)}{\omega(n)} .$$

The proof of the next lemma is elementary.

1.4. LEMMA. A multiplier α belongs to $\mathfrak{L}(M)$ if and only if $\lim_{m\to\infty} |||S_m(\alpha)||| = 0$. In fact, $\lim_{m\to\infty} |||S_m(\alpha)||| = \text{dist}(\alpha,\mathfrak{L}(M))$ for every $\alpha \in \mathfrak{M}(M)$.

We next describe those multipliers which are compact, as operators on M. The compact multipliers form a closed ideal $\mathfrak{K}$ in $\mathfrak{M}(M)$. An element $x \in \ell^1(\omega)$ is <u>compact</u> if the operator of multiplication by x is compact. Thus an element of $\ell^1(\omega)$ is compact if and only if it is compact as a multiplier on M. It is easily seen that the compact elements form a closed ideal $\mathcal{K}$ in $\ell^1(\omega)$. A compact multiplier must have countable spectrum containing zero. Since $\mathfrak{M}(M)$ is an integral domain, it follows from Silov's idempotent theorem that each element of $\mathfrak{M}(M)$ has connected spectrum. Consequently the ideal $\mathfrak{K}$ of compact multipliers is a radical Banach algebra.

The existence of compact elements and multipliers is connected with a property of weights.

1.5. DEFINITION. Let ω be a weight function. We say that ω is <u>regulated at</u> k if

$$\lim_{n\to\infty} \frac{\omega(n+k)}{\omega(n)} = 0 .$$

1.6. THEOREM. Let ω be a weight function

(i) A multiplier $\alpha \in \mathfrak{M}(M)$ is compact if and only if

$$\lim_{n\to\infty} \sum_{j=0}^{\infty} |\alpha(j)| \frac{\omega(n+j)}{\omega(n)} = 0 .$$

(ii) Let $\bar{k}$ be the smallest $k \geq 1$ such that ω is regulated at k. Then the ideal $\mathcal{K}$ of compact elements equals the standard closed ideal $M_{\bar{k}}$.

(iii) Every compact multiplier lies in $\mathfrak{L}(M)$. In fact $\mathcal{H}$ is the operator closure of $\mathcal{K}$.

Clearly $\bar{k}$ is the smallest integer k such that e_k is compact element of $\ell^1(\omega)$. We remark that for ω to be regulated at 1 (or, equivalently, that every element of M be compact) it is sufficient that $\omega(n)^{1/n}$ decrease monotonically to zero. On the other hand there exist weights ω for which zero is the only compact multiplier.

For the weight ω of Example 1.3, there are no nonzero compact multipliers, since by (1.7) ω is not regulated at any k.

We have noted that for any radical weight, the equivalent conditions of Theorem 1.2 imply that $\mathfrak{M}(M) = \mathfrak{L}(M)$. It is an interesting fact that the converse holds, when ω is regulated at 1. By Theorem 1.2 (c) we must prove that if $\mathfrak{M}(M) = \mathfrak{L}(M)$, then there is a constant C such that

$$\omega(m+j+1) \leq C\omega(m+1)\omega(j+1) \qquad (j,m \in \mathbb{N}) . \tag{1.8}$$

In this connection it is convenient to introduce the notation

$$\mu_n(j) = \frac{\omega(n+j)}{\omega(n)} \qquad (n \in \mathbb{N},\ j \in \mathbb{Z}^+), \tag{1.9}$$

and for a sequence α,

$$\|\alpha\|_n = \sum_{j=0}^{\infty} |\alpha(j)|\mu_n(j) .$$

Thus, $\alpha \in \mathfrak{M}(M)$ if and only if

$$|||\alpha||| = \sup_{n\geq 1} \|\alpha\|_n < \infty ,$$

while $\alpha \in \mathfrak{L}(M)$ if and only if $\alpha \in \mathfrak{M}(M)$ and

$$\lim_{m\to\infty} \sup_{n\geq 1} \|S_m(\alpha)\|_n = 0 .$$

Taking $m + 1 = n$ and $C\omega(1) = K$, we see that (1.8) can be expressed in the equivalent form

$$K\mu_1(j) \geq \mu_n(j) \qquad (j \in \mathbb{N}, n \geq 2),$$

which states that a multiple of μ_1 majorizes all the other sequences μ_n for $n \geq 2$.

1.7. THEOREM. Let ω be a weight which is regulated at 1. Then the following are equivalent

(a) $\mathfrak{m}(M) = L(M)$

(b) $\mathfrak{m}(M) = \mathfrak{L}(M)$.

The proof that (b) $\to$ (a) proceeds by showing that if no multiple of μ_1 majorizes the set $\{\mu_n : n \geq 2\}$, then there exists a sequence $\beta \in \mathfrak{m}(M) \sim \mathfrak{L}(M)$. The details may be found in [2]. The assumption that ω is regulated at 1 is used in the form that $\lim_{n\to\infty} \mu_n(j) = 0$ $(j \in \mathbb{N})$. This is equivalent to the statement that

$$\lim_{n\to\infty} \|\alpha\|_n = 0 \tag{1.10}$$

whenever $\alpha \in \mathfrak{L}(M)$ and $\alpha(0) = 0$.

The implication (b) $\to$ (a) may be false without the assumption that ω is regulated at 1. Let ω be the weight of Example 3.1. It follows from [1, Lemma 2.1 ii] that there exists an increasing sequence (j_i) such that $\omega(j_i+1)/\omega(j_i) \to 0$ as $i \to \infty$. Let $n_i = 2^{j_i}$ $(i \in \mathbb{N})$. By (1.7) we have $\omega(n_i + j_i) = \omega(n_i)\omega(j_i)$, so that

$$\lim_{i\to\infty} \frac{\mu_{n_i}(j_i)}{\mu_1(j_i)} = \lim_{i\to\infty} \frac{\omega(1)\omega(j_i)}{\omega(j_i+1)} = \infty .$$

This shows that no multiple of μ_1 can majorize the set $\{\mu_n : n \geq 2\}$. However, for this ω one has $\mathfrak{L}(M) = \mathfrak{m}(M)$, since $\mathfrak{m}(M) = \ell^1(\omega)$.

2. Examples

In this section we construct a class of examples of radical weights for which $\mathfrak{L}(M)$ is properly contained in $\mathfrak{m}(M)$. It is convenient to construct these weights by constructing the function η in the formula $\omega = \exp(-\eta)$. It is required of η that it satisfy $\eta(m) + \eta(n) \leq \eta(m+n)$ $(m,n \in \mathbb{N})$, and that $\lim_{n\to\infty} \eta(n)/n = \infty$. For $p \geq 0$ let $j_p = 2^p$ and

$$k_p = j_p + j_{p-1}.$$

2.1. DEFINITION. We define the function $\eta : \mathbb{Z}^+ \to \mathbb{Z}^+$ inductively as follows.

(i) Let $\eta(0) = 0$, $\eta(1) = 1$.

(ii) If $p \geq 1$ and η has been defined on $[0, j_p)$, define $\eta(j_p)$ to be a number satisfying

$$\eta(j_p) \geq 2\eta(j_p - 1) + j_{p+2} \log(p+1) \tag{2.1}$$

(iii) If $n = k_p$, define

$$\eta(k_p) = \eta(j_p) + \eta(j_{p-1}) . \tag{2.2}$$

(iv) If $n \in (j_p, k_p) \cup (k_p, j_{p+1})$ define

$$\eta(n) = 2 \log p + \max\{\eta(m_1) + \eta(n_1) : m_1, n_1 \in \mathbb{N},\ m_1 + n_1 = n\} . \tag{2.3}$$

2.2. LEMMA. (i) The function η satisfies the conditions

$$\eta(m+n) \geq \eta(m) + \eta(n) \qquad (m,n \in \mathbb{N}) , \tag{2.4}$$

and if $m + n \in [j_p, k_p) \cup (k_p, j_{p+1})$, then

$$\eta(m+n) \geq \eta(m) + \eta(n) + 2 \log p . \tag{2.5}$$

Proof. We suppose that $p \geq 1$, and assume for purposes of an inductive proof that (2.4) holds when $m + n < j_p$. If $m + n = j_p$, then $m,n < j_p$, so (2.4) follows from (2.1). Also (2.5) is trivially satisfied if $m + n \in (j_p, k_p) \cup (k_p, j_{p+1})$. It remains to prove (2.4) in the case that $m + n = k_p$. Again if $m,n < j_p$, then

$$\eta(k_p) > \eta(j_p) > \eta(m) + \eta(n) .$$

If $n = j_p$, then $m = j_{p-1}$, and we have additivity. Thus we can suppose

$$n = j_p + s,\ m = j_{p-1} - s ,$$

where $1 \leq s < j_{p-1}$. Then from (2.1)

$$\eta(m) \leq \frac{1}{2} [\eta(j_{p-1}) - j_{p+1} \log p] = \frac{1}{2} \eta(j_{p-1}) - j_p \log p ,$$

and there exist m_1, n_1 with $m_1 \leq n_1$, $n = m_1 + n_1$, and

$$\eta(n) = \eta(m_1) + \eta(n_1) + 2 \log p .$$

As before we can suppose $n_1 \geq j_p$. If $n_1 = j_p$ then $m_1 = s$ and

$$\eta(m) + \eta(n) = \eta(m) + \eta(j_p) + \eta(s) + 2 \log p \leq \frac{1}{2} \eta(j_{p-1}) - j_p \log p$$
$$+ \eta(j_p) + \frac{1}{2} \eta(j_{p-1}) < \eta(k_p) .$$

If $n_1 > j_p$, we apply the definition again to get

$$\eta(n) = \eta(m_2) + \eta(n_2) + \eta(m_1) + 4 \log p .$$

After $k \leq s$ repetitions we obtain

$$\eta(n) = \eta(j_p) + \eta(m_1) + \cdots + \eta(m_k) + 2k \log p .$$

and, since $m_1 + m_2 + \cdots + m_k = s$, the induction hypothesis yields

$$\eta(n) \leq \eta(j_p) + \eta(s) + 2s \log p$$
$$\leq \eta(j_p) + \frac{1}{2} \eta(j_{p-1}) ,$$

and we have again that $\eta(m) + \eta(n) < \eta(k_p)$. This completes the proof.

2.3. THEOREM. Let $\omega = \exp(-\eta)$, where η is one of the functions of Definition 2.1. Then ω satisfies the following conditions

$$\text{(i)} \quad \omega(m+n) \leq \omega(m)\omega(n) \qquad (m,n \in \mathbb{N}) , \tag{2.7}$$

while if $m + n \in [j_p,k_p) \cup (k_p,j_p)$ then

$$\omega(m+n) \leq \frac{\omega(m)\omega(n)}{p^2} . \tag{2.8}$$

(ii) If $n = k_p = j_p + j_{p-1}$, then

$$\omega(k_p) = \omega(j_p)\omega(j_{p-1}) . \tag{2.9}$$

(iii) The function ω is a radical weight which is regulated at 1.

Proof. Statements (i) and (ii) are immediate from Lemma 2.2. To show that ω is a radical weight, we must show that $\lim_{n\to\infty} \eta(n)/n = \infty$. However, if $j_p \leq n < j_{p+1}$, then

$$\frac{\eta(n)}{n} \geq \frac{\eta(j_p)}{j_{p+1}} > 2 \log(p+1) \to \infty .$$

The weight ω is regulated at 1 if $\lim_{n\to\infty} \omega(n+1)/\omega(n) = 0$. Thus it is necessary to show that $\lim_{n\to\infty} (\eta(n) - \eta(n-1)) = \infty$. It follows from the definitions that $\eta(n) - \eta(n-1) \geq 2 \log(p+1)$ if $n \in [j_p,j_{p+1}) \sim \{k_p\}$. The argument of Lemma 2.2 to prove formula (2.6) shows that we have the inequality

$$\eta(k_p - 1) \le \eta(j_p) + \eta(s) + 2s \log p ,$$

where s is an integer satisfying $1 \le s < j_{p-1}$. From (2.1) for $p-1$ we get

$$\begin{aligned}\eta(k_p - 1) &\le \eta(j_p) + (\tfrac{1}{2})\eta(j_{p-1}) - (j_p - 2s) \log p \\ &\le \eta(j_p) + (\tfrac{1}{2})\eta(j_{p-1}) .\end{aligned}$$

Hence $\eta(k_p) - \eta(k_p - 1) > j_p \log p$.

We next investigate the multiplier algebras $\mathfrak{M}(M)$ for the weights ω we have constructed.

2.4. DEFINITION. Let $\alpha(0) = 0$, $\alpha(1) = 1$, and

$$\alpha(j) = \begin{cases} \dfrac{1}{\omega(j_p)} & (j = j_p,\ p \ge 1) \\ 0 & (\text{otherwise}) . \end{cases}$$

2.5. THEOREM. The sequence α belongs to $\mathfrak{M}(M)$, but not to $\mathfrak{L}(M)$.

Proof. By Theorem 1.1 we must find a bound which is independent of n for the sum

$$\sum_{j=0}^{\infty} \alpha(j) \frac{\omega(n+j)}{\omega(n)} = \frac{\omega(n+1)}{\omega(n)} + \sum_{p=1}^{\infty} \frac{\omega(n+j_p)}{\omega(j_p)\omega(n)} .$$

Let n be fixed with $j_q \le n < j_{q+1}$, where $q \ge 2$. For large values of p such that $j_{p-1} > n$ we have by (2.8), that

$$\frac{\omega(n+j_p)}{\omega(n)\omega(j_p)} \le \frac{1}{p^2} .$$

Also for small values of p, when $p \le q-1$, $n + j_p$ has the form k_r only when $r = q$, and this happens for at most one value of p. If $n + j_p = k_r$ for some r, we use the fact that the ratio is bounded by one. These remarks allow us to estimate the sum when $p \le q-1$ or $p \ge q+2$. Thus we get

$$\begin{aligned}\frac{\omega(n+1)}{\omega(n)} + \sum_{p=1}^{\infty} \frac{\omega(n+j_p)}{\omega(n)\omega(j_p)} &\le 1 + \sum_{p \le q-1} \frac{\omega(n+j_p)}{\omega(n)\omega(j_p)} \\ &+ \frac{\omega(n+j_q)}{\omega(n)\omega(j_q)} + \frac{\omega(n+j_{q+1})}{\omega(n)\omega(j_{q+1})} + \sum_{p \ge q+2} \frac{\omega(n+j_p)}{\omega(n)\omega(j_p)} \\ &\le 4 + \sum_{p=1}^{\infty} \frac{1}{p^2} .\end{aligned}$$

This shows $\alpha \in \mathfrak{M}(M)$.

Recall that $\alpha \in \mathfrak{L}(M)$ if and only if $\lim_{m\to\infty} |||S_m\alpha||| = 0$, where $S_m\alpha = (0,0,\ldots,\alpha_{m+1},\ldots)$. Let q be an integer such that $j_{q-1} > m$. Then

$$|||S_m\alpha||| = \sup_n \sum_{j=m+1}^{\infty} \alpha_j \frac{\omega(n+j)}{\omega(n)\omega(j)}$$

$$\geq \frac{\omega(j_q + j_{q-1})}{\omega(j_q)\omega(j_{q-1})} = 1 .$$

Since this is true for every m, it follows that $\alpha \notin \mathfrak{L}(M)$.

It seems likely that for each of the weights ω constructed above the algebra $\mathfrak{M}(M)$ is local, although we have not proved this. Using the same general method, we have constructed in the paper [2] a much more complicated example of a weight ω and a multiplier $\alpha \in \mathfrak{M}(\omega)$ whose spectrum contains a neighborhood of zero. Since $\mathfrak{L}(\omega)$ is local, it follows in particular that $\mathfrak{L}(M) \subsetneq \mathfrak{M}(M)$. We describe below the construction of this weight, referring to [2] for the proof of its properties. As before, it is convenient to define $\eta = -\log \omega$.

Let $j_p = 2^p (p \in \mathbb{Z}^+)$. If $p \geq 1$ divide the interval $J_p = [j_p, j_{p+1})$ into subintervals of length $j_{[p/2]}$. The ends of the smaller intervals, together with the points j_p, are called <u>endpoints</u>. Each endpoint e in J_p has the form $e = j_p$, or

$$e = j_p + j_{p_1} + \cdots + j_{p_r} ,$$

where $p > p_1 > \cdots > p_r \geq [p/2]$. Let E_p denote the set of endpoints in J_p.

2.6. DEFINITION. We define the function $\eta : \mathbb{Z}^+ \to \mathbb{R}^+$ inductively as follows.

(i) Let $\eta(0) = 0$, $\eta(1) = 1$.

(ii) If $p \geq 1$ and η has been defined on $[0, j_p)$, define $\eta(j_p)$ to be a number satisfying

$$\eta(j_p) \geq 7\eta(j_p - 1) + j_{2p+1} \log(2p+1) .$$

(iii) If $n = j_p + j_{p_1} + \cdots + j_{p_r}$ (where $p_r \geq [p/2]$) is an endpoint in (j_p, j_{p+1}), define

$$\eta(n) = \eta(j_p) + \eta(j_{p_1}) + \cdots + \eta(j_{p_r}) .$$

(iv) If n lies in J_p, but is not an endpoint, and if $\eta(m)$ has been defined for all $m < n$, define

$$\eta(n) = \log p + \max\{\eta(m_1) + \eta(n_1) : m_1, n_1 \in \mathbb{N},\ m_1 + n_1 = n\} .$$

Define $\omega = \exp(-\eta)$. Then it can be proved that

$$\omega(m+n) \leq \omega(m)\omega(n) \qquad (m,n \in \mathbb{N}),$$

while if $m + n \in J_p$, but $m + n \notin E_p$, then

$$\omega(m+n) \leq \frac{\omega(m)\omega(n)}{p}.$$

If $n = j_p + j_{p_1} + \cdots + j_{p_r}$ is an endpoint of J_p, then

$$\omega(n) = \omega(j_p)\omega(j_{p_1}) \ldots \omega(j_{p_r}).$$

Further the function ω is a radical weight which is regulated at 1.

We now define a multiplier α by $\alpha(0) = 0$, $\alpha(1) = 1$, and

$$\alpha(j) = \begin{cases} \dfrac{1}{p\omega(j_p)} & (j = j_p,\ p \geq 1) \\ 0 & (\text{otherwise}). \end{cases}$$

It is proved in [2] that $|||\alpha||| < \infty$, and that $\sigma(\alpha)$ contains the disk about the origin of radius $1/2$.

2.7. THEOREM. There exists a separable Banach algebra of power series which is neither local nor semi-simple.

Proof. Let α be the multiplier defined above, and let $\mathfrak{A}$ be the smallest closed subalgebra of $\mathfrak{m}(M)$ containing $\mathfrak{L}(M)$ and α. Then $\alpha \notin \operatorname{rad}\mathfrak{A}$, but $\operatorname{rad}\mathfrak{A}$ contains all those multipliers in $\mathfrak{L}(M)$ whose first coordinate vanishes. Clearly $\mathfrak{A}$ is separable, so $\mathfrak{A}$ is the required example.

This example answers a question raised by R. J. Loy. It would be interesting to identify the radical of $\mathfrak{A}$.

3. Banach space structure of $\mathfrak{m}(M)$

In this section we examine the Banach space structure of $\mathfrak{m}(M)$ and of its subspace $\mathfrak{L}(M)$.

The Banach algebra $\mathfrak{L}(M)$ is, of course, separable since the polynomials are dense in it. We shall see that, if $\mathfrak{m}(M)$ properly contains $\mathfrak{L}(M)$, then $\mathfrak{m}(M)$ is always non-separable, and $\mathfrak{L}(M)$ is uncomplemented in $\mathfrak{m}(M)$. Moreover, in this case, $\mathfrak{m}(M)$ contains subspaces isomorphic to ℓ^∞, and $\mathfrak{m}(M)/\mathfrak{L}(M)$ contains subspaces isometric to ℓ^∞/c_0.

We note that $\mathfrak{m}(M)$ is a module over the Banach algebra ℓ^∞ with respect to the action

$$(y \cdot \alpha)(j) = y(j)\alpha(j) \qquad (j \in \mathbb{N},\ \alpha \in \mathfrak{M}(M),\ y \in \ell^\infty).$$

Clearly, $\mathfrak{L}(M)$ is a submodule and

$$|||y \cdot \alpha||| \leq \|y\|_\infty \, |||\alpha||| \qquad (\alpha \in \mathfrak{M}(M),\ y \in \ell^\infty).$$

If $\alpha \in \mathfrak{M}(M)$, let $[\alpha]$ denote the coset in $\mathfrak{M}(M)/\mathfrak{L}(M)$ containing α. Recall from Lemma 1.4 that the quotient norm is given by

$$|||[\alpha]||| = \lim_{m\to\infty} |||S_m(\alpha)|||, \tag{3.1}$$

where

$$|||S_m(\alpha)||| = \sup_{n \geq 1} \sum_{j=m+1}^{\infty} |\alpha(j)| \frac{\omega(n+j)}{\omega(n)}; \tag{3.2}$$

the latter is non-increasing as a function of m. Then $\mathfrak{M}(M)/\mathfrak{L}(M)$ is a module over ℓ^∞/c_0 if we define

$$[y] \cdot [\alpha] = [y \cdot \alpha] \qquad (y \in \ell^\infty,\ \alpha \in \mathfrak{M}(M)).$$

Here, $[y]$ denotes the coset in ℓ^∞/c_0 containing y. This action is well defined since $y \cdot \alpha \in \mathfrak{L}(M)$ whenever $y \in c_0$ and $\alpha \in \mathfrak{M}(M)$. Moreover,

$$|||[y] \cdot [\alpha]||| \leq \|[y]\| \, |||[\alpha]||| \qquad (y \in \ell^\infty,\ \alpha \in \mathfrak{M}(M)).$$

Let $\alpha \in \mathfrak{M}(M) \sim \mathfrak{L}(M)$ and $\{\varepsilon_n\}$ be a positive sequence with $\varepsilon_n \to 0$. By (3.1) we can find an increasing sequence $\{m_n\}$ such that $|||S_{m_n}(\alpha)||| > (1-\varepsilon_n)|||[\alpha]|||$. Moreover the norm $|||S_m(\alpha)|||$ is a supremum of infinite sums. By a careful choice of the sequence $\{m_n\}$, we can approach each supremum by a convenient partial sum. An inductive argument yields the following result.

3.1. LEMMA. Let $\alpha \in \mathfrak{M}(M) \sim \mathfrak{L}(M)$, and let $\{\varepsilon_n\}$ be a decreasing sequence of positive numbers with $\varepsilon_1 < 1$ and $\lim_{n\to\infty} \varepsilon_n = 0$. Then there exist strictly increasing sequences $\{m_n\}$ and $\{k_n\}$ of integers, with $k_1 \geq 1$ and $m_1 = 0$, such that

$$\sum_{j=m_n+1}^{m_{n+1}} |\alpha(j)| \frac{\omega(k_n+j)}{\omega(k_n)} > |||[\alpha]|||(1-\varepsilon_n) \qquad (n \in \mathbb{N}). \tag{3.3}$$

We now suppose that we have a fixed $\alpha \in \mathfrak{M}(M) \sim \mathfrak{L}(M)$, a sequence $\{\varepsilon_n\}$ and the sequences $\{k_n\}$ and $\{m_n\}$ provided by the lemma. Consider the decomposition of $\mathbb{N}$ into the union of the disjoint finite sets F_n, where

$$F_n = \{k : m_n < k \leq m_{n+1}\}.$$

Let A be the subalgebra of ℓ^∞ consisting of those $y \in \ell^\infty$ which are constant on each of the sets F_n. We can consider $\mathfrak{m}(M)$ as a module over A. Since A is isomorphic to ℓ^∞, we can define $T : \ell^\infty \to \mathfrak{m}(M)$ by

$$T(y)(j) = y(n)\alpha(j) \qquad (j \in F_n,\ n \in \mathbb{N},\ y \in \ell^\infty). \tag{3.4}$$

Thus $T(y) = z \cdot \alpha$, where $z \in A$, and $z(j) = y(n)$ $(j \in F_n)$. Note that T carries the constant function 1 in ℓ^∞ onto α.

3.2. THEOREM. Under the above assumptions, the map T defines an isomorphism of ℓ^∞ onto a subspace of $\mathfrak{m}(M)$ containing α. Moreover

$$(1-\varepsilon_1)\,|||[\alpha]|||\ \|y\|_\infty \leq |||T(y)||| \leq |||\alpha|||\ \|y\|_\infty \qquad (y \in \ell^\infty).$$

Proof. Using Lemma 3.1, we see that

$$\begin{aligned}(1-\varepsilon_1)\|y\|_\infty\ |||[\alpha]||| &\leq \sup_{n\geq 1} |y(n)|(1-\varepsilon_n)\,|||[\alpha]||| \\ &\leq \sup_{n\geq 1} |y(n)| \sum_{j\in F_n} |\alpha(j)| \frac{\omega(k_n+j)}{\omega(k_n)} \\ &\leq \sup_{k\geq 1} \sum_{n=1}^{\infty} \sum_{j\in F_n} |y(n)\alpha(j)| \frac{\omega(k+j)}{\omega(k)} \\ &= |||T(y)||| \\ &\leq \|y\|_\infty\ |||\alpha||| .\end{aligned}$$

A similar computation yields the following inequalities.

$$\begin{aligned}(1-\varepsilon_p)\,|||[\alpha]|||\left(\sup_{n\geq p} |y(n)|\right) &\leq |||S_{m_p}(T(y))||| \\ &\leq |||S_{m_p}(\alpha)|||\left(\sup_{n\geq p} |y(n)|\right) \qquad (p \in \mathbb{N}).\end{aligned} \tag{3.5}$$

3.3. THEOREM. Let $\alpha \in \mathfrak{m}(M) \sim \mathfrak{L}(M)$ with $|||[\alpha]||| = 1$. Then there exists an isometric linear map of ℓ^∞/c_0 into $\mathfrak{m}(M)/\mathfrak{L}(M)$ which carries the coset $[1]$ containing 1 in ℓ^∞ onto the coset $[\alpha]$.

Proof. Suppose that $|||[\alpha]||| = 1$, and let $\{\varepsilon_n\}$, $\{m_n\}$ and $\{k_n\}$ be as in Lemma 3.1. We define the map $\hat{T} : \ell^\infty/c_0 \to \mathfrak{m}(M_1)/\mathfrak{L}(M_1)$ by

$$\hat{T}([y]) = [T(y)] \qquad (y \in \ell^\infty).$$

Letting $p \to \infty$ in (3.5), we get

$$|||[\alpha]|||\ \|[y]\| \le |||[T(y)]||| \le |||[\alpha]|||\ \|[y]\|$$

using (3.1). The rest is clear.

3.4. THEOREM. The Banach space $\mathfrak{m}(M)$ is separable if and only if $\mathfrak{m}(M) = \mathfrak{L}(M)$.

Proof. We have seen that if $\mathfrak{m}(M)$ contains elements not in $\mathfrak{L}(M)$, then the quotient algebra contains subspaces isometric to ℓ^∞/c_0, and so in this case $\mathfrak{m}(M)$ cannot be separable. The converse is clear.

When $\mathfrak{m}(M)$ is strictly larger than $\mathfrak{L}(M)$, the subspace $\mathfrak{L}(M)$ bears an analogy to the subspace c of ℓ^∞. The following theorem is proved in [2].

3.5. THEOREM. If $\mathfrak{m}(M)$ properly contains $\mathfrak{L}(M)$, then $\mathfrak{L}(M)$ is uncomplemented in $\mathfrak{m}(M)$.

The proof uses the imbedding of ℓ^∞ into $\mathfrak{m}(M)$ of Theorem 3.2, and follows standard proofs of Phillips' theorem [11, p. 539] that c is uncomplemented in ℓ^∞.

We now give a theorem which includes the results obtained in Theorems 1.2, 1.7, and 3.4.

3.6. THEOREM. For each radical weight ω the following are equivalent.

(a) $\mathfrak{m}(M) = L(M)$,

(b) $L(M)$ is an algebra under convolution,

(c) there exists a constant C such that

$$\omega(m+n+1) \le C\omega(m+1)\omega(n+1) \qquad (m,n \in \mathbb{N}),$$

and $\mathfrak{m}(M) = \ell^1(\omega')$ (where $\omega'(n) = C\omega(n+1)$ for $n \in \mathbb{N}$), the two norms being equivalent.

Also the following conditions are equivalent.

(d) $\mathfrak{m}(M)$ is separable,

(e) $\mathfrak{m}(M) = \mathfrak{L}(M)$.

Moreover, (c) implies (d) and, if ω is regulated at 1, then (a),...,(e) are all equivalent, and equivalent to the condition

(f) $\mathfrak{L}(M)$ is weakly sequentially complete.

Proof. It remains to prove that (f) implies (e) under the assumption that ω is regulated at 1. For this we will need certain auxillary spaces. Let $X_n = \ell^1(\mu_n)$, $\mu_n(j) = \omega(n+j)/\omega(n)$ $(n \in \mathbb{N}, j \in \mathbb{Z}^+)$. Denote by $\mathfrak{D}$ the space of all sequences $\vec{x} = \{x_n\}$, where $x_n \in X_n$, $\|\vec{x}\| = \sup_{n\ge 1} \|x_n\|_n < \infty$, and $\lim_{n\to\infty} \|x_n\|_n = 0$. Then $\mathfrak{D}$ is a Banach space whose dual is the space $\mathfrak{R}$ of all sequences $\vec{y} = \{y_k\}$, where $y_k \in X_k^* = \ell^\infty(\mu_n^{-1})$,

$$\|y_k\|_k^* = \sup_{n \geq 0} \{|y_k(j)|/\mu_{n(j)} : j \in \mathbb{Z}^+\}$$

and $\|\vec{y}\| = \sum_{k=1}^\infty \|y_k\|_k^* < \infty$. The pairing is given by

$$\langle \vec{y}, \vec{x} \rangle = \sum_{k=1}^{\infty} \langle y_k, x_k \rangle_k ,$$

where $\langle y_k, x_k \rangle_k = \sum_{j=0}^\infty y_k(j)x_k(j)$ (see [4], p. 34). It follows from (1.10) that the map $\beta \to \{\beta,\beta,\beta,\ldots\}$ imbeds $\mathfrak{L}_1(M) = \{\beta \mid \beta \in \mathfrak{L}(M), \beta(0) = 0\}$ isometrically into $\mathfrak{D}$.

Now let $\alpha \in \mathfrak{M}(M)$. We suppose $\alpha(0) = 0$. Define $\alpha_n(j) = \alpha(j)$ $(0 \leq j \leq n)$, $\alpha_n(j) = 0$ $(j > n)$. Then $\alpha_n \in \mathfrak{L}_1(M)$ $(n \in \mathbb{N})$. Let F be any element of $\mathfrak{L}_1(M)^*$. By the Hahn-Banach Theorem there exists $\vec{y} = \{y_k\}$ in $\mathfrak{R}$ such that

$$F(\beta) = \sum_{k=1}^{\infty} \langle y_k, \beta \rangle_k \qquad (\beta \in \mathfrak{L}_1(M)) .$$

Hence

$$\lim_{n\to\infty} F(\alpha_n) = \lim_{n\to\infty} \sum_{k=1}^{\infty} \sum_{j=0}^{n} y_k(j)\alpha(j)$$

exists, since

$$\sum_{k=1}^{\infty} \sum_{j=0}^{\infty} |y_k(j)||\alpha(j)| \leq \sum_{k=1}^{\infty} \|y_k\|_k^* \sum_{j=0}^{\infty} |\alpha(j)|\mu_k(j)$$
$$\leq |||\alpha||| \; \|\vec{y}\| < \infty .$$

Since $\mathfrak{L}_1(M)$ is weakly sequentially complete, there is an element γ in $\mathfrak{L}_1(M)$ such that

$$\lim_{n\to\infty} F(\alpha_n) = F(\gamma) \qquad (F \in \mathfrak{L}_1(M)^*) .$$

However, since $\alpha(j) = \lim_{n\to\infty} \alpha_n(j) = \gamma(j)$ for each $j \in \mathbb{Z}^+$, it follows that $\alpha = \gamma$, so $\alpha \in \mathfrak{L}_1(M)$.

We close by proving that $\mathfrak{M}(M)$ is always a dual space for any radical weight ω. Let $A_n = c_0(\mu_n^{-1})$ $(n \in \mathbb{N})$, where $\mu_n(j) = \omega(n+j)/\omega(n)$ $(n \in \mathbb{N}, j \in \mathbb{Z}^+)$. Denote by $\mathfrak{A}$ the class of all sequences $\vec{x} = \{x_n\}$, where $x_n \in A_n$, $\|\vec{x}\| = \sum_{n=1}^\infty \|x_n\|_n < \infty$, and

$$\|x_n\|_n = \sup_{j\in\mathbb{Z}^+} \frac{|x_n(j)|}{\mu_n(j)} \qquad (n \in \mathbb{N}) .$$

Let $\mathfrak{B}$ be the class of all sequences $\vec{y} = \{y_n\}$ such that $y_n \in A_n^* = \ell^1(\mu_n)$ $(n \in \mathbb{N})$ and

$$\|\vec{y}\| = \sup_n \|y_n\|_n < \infty .$$

In a way similar to the last proof one sees that $\mathfrak{A}$ and $\mathfrak{B}$ are Banach spaces and that $\mathfrak{B} = \mathfrak{A}^*$, for the pairing

$$\langle \vec{y},\vec{x} \rangle = \sum_{k=1}^{\infty} \langle y_k, x_k \rangle_k \qquad (\vec{y} \in \mathfrak{B},\ \vec{x} \in \mathfrak{A}) .$$

(See [4, page 31]). If $y \in \mathfrak{m}(M)$, define $\vec{y} = \{y,y,y,\ldots\}$. Then $\vec{y} \in \mathfrak{B}$ and $\|\vec{y}\| = |||y|||$. The correspondence $y \to \vec{y}$ imbeds $\mathfrak{m}(M)$ onto a closed subspace, say $\mathfrak{C}$, of $\mathfrak{B}$, and $\mathfrak{C}$ consists of all elements $\vec{y} = \{y_1,y_2,\ldots\}$ in $\mathfrak{B}$ such that $y_1(j) = y_2(j) = \cdots$ for each $j \in \mathbb{Z}^+$. We denote by $\mathfrak{C}^\top$ the set $\{\vec{x} \in \mathfrak{A} : \langle \vec{y},\vec{x} \rangle = 0$ for all $\vec{y} \in \mathfrak{C}\}$.

3.7. LEMMA. $\mathfrak{C}$ is a weak-star closed subspace of $\mathfrak{B} = \mathfrak{A}^*$. Consequently $\mathfrak{C}$ is the dual of $\mathfrak{A}/\mathfrak{C}^\top$.

<u>Proof</u>. Let $\{\vec{y}^{(\alpha)}\}$ be a net in $\mathfrak{C}$ which converges weak-star to $\vec{y}$ in $\mathfrak{B}$. Then for each $\vec{x} \in \mathfrak{A}$

$$\langle \vec{y}^{(\alpha)}, \vec{x} \rangle = \sum_{j=0}^{\infty} y^{(\alpha)}(j)\left(\sum_{k=1}^{\infty} x_k(j) \right) \to \sum_{j=0}^{\infty} \sum_{k=1}^{\infty} y_k(j)x_k(j) .$$

Taking $\vec{x} = \{x_k\}$, where $x_k = 0,\ k \neq n$, and $x_k = e_p$ yields

$$y^{(\alpha)}(p)x_n(p) = y^{(\alpha)}(p) \to y_n(p)x_n(p) = y_n(p) \qquad (n,p \in \mathbb{N}) ,$$

so $y_1(p) = y_2(p) = \cdots\ (p \in \mathbb{Z}^+)$, showing $\vec{y} \in \mathfrak{C}$. Hence $\mathfrak{C}$ is weak-star closed. The duality follows from the bipolar theorem.

It follows from the isomorphism of $\mathfrak{m}(M)$ with $\mathfrak{C}$ that $\mathfrak{m}(M)$ is a dual space. However, a more concrete description of the predual is possible. It follows from the definition of $\mathfrak{C}$ that

$$\mathfrak{C}^\top = \{\vec{x} \in \mathfrak{A} : \sum_{n=1}^{\infty} x_n(j) = 0 \qquad (j \in \mathbb{Z}^+)\} .$$

Thus two elements $\vec{x}$ and $\vec{y}$ of $\mathfrak{A}$ lie in the same coset modulo $\mathfrak{C}^\top$ if and only if

$$\sum_{n=1}^{\infty} x_n(j) = \sum_{n=1}^{\infty} y_n(j) \qquad (j \in \mathbb{Z}^+) .$$

3.8. THEOREM. Let $\mathfrak{m}(M)_*$ be the Banach space of all sequences $\alpha = (\alpha(j))$, where for some $\vec{x} \in \mathfrak{U}$, $\alpha(j) = \sum_{n=1}^{\infty} x_n(j)$ $(j \in \mathbb{Z}^+)$, and

$$\|\alpha\| = \inf\{\|\vec{x}\| : \sum_{n=1}^{\infty} x_n(j) = \alpha(j) \quad (j \in \mathbb{Z}^+)\}.$$

Then $\mathfrak{m}(M)_*$ is a predual for $\mathfrak{m}(M)$ for the duality

$$\langle \beta,\alpha \rangle = \sum_{j=0}^{\infty} \beta(j)\alpha(j) \qquad (\beta \in \mathfrak{m}(M),\ \alpha \in \mathfrak{m}(M)_*).$$

In [2] another predual for $\mathfrak{m}(M)$ was constructed involving tensor products by an argument suggested by G. Bachelis. He has recently informed us these two preduals of $\mathfrak{m}(M)$ are actually isometric.

References

[1] W. G. Bade and H. G. Dales, Norms and ideals in radical convolution algebras, J. Functional Analysis, 41 (1981), 77-109.

[2] W. G. Bade, H. G. Dales and K. B. Laursen, Multipliers of radical Banach algebras of power series, to appear in Memoirs Amer. Math. Soc.

[3] F. F. Bonsall and J. Duncan, Complete Normed Algebras, Springer-Verlag New York, 1973.

[4] M. M. Day, Normed Linear Spaces, Third edition, Springer-Verlag, New York, 1973.

[5] S. Grabiner, A formal power series operational calculus for quasi-nilpotent operators, Duke Math. J., 38 (1971), 641-658.

[6] ________, Derivations and automorphisms of Banach algebras of power series, Memoirs Amer. Math. Soc., 146 (1974), 1-124.

[7] ________, Weighted shifts and Banach algebras of power series, Amer. J. Math., 97 (1974), 16-42.

[8] K. B. Laursen, Ideal structure in radical sequence algebras, this Volume.

[9] R. J. Loy, Banach algebras of power series, J. Austral. Math. Soc., 17 (1974), 263-273.

[10] N. K. Nikolskii, Selected problems of weighted approximation and spectral analysis, Trudy Mat. Inst. Steklov, 120 (1974), 1-270 = Proc. Steklov Inst. Math., 120 (1974), 1-278 (A.M.S. Translation).

[11] R. S. Phillips, On linear transformations, Trans. Amer. Math. Soc., 48 (1940), 516-541.

[12] M. Thomas, Closed ideals and biorthogonal systems in radical Banach algebras of power series, to appear in Proc. Edinburgh Math. Soc.

[13] ________, Approximation in the radical algebra $\ell^1(\omega_n)$ when $\{\omega_n\}$ is star-shaped, this Volume.

[14] M. Thomas, Closed ideals of $\ell^1(\omega)$ when $\{\omega_n\}$ is star-shaped, to appear in Pacific J. Math.

Department of Mathematics
University of California
Berkeley, CA 94720

IDEAL STRUCTURE IN RADICAL SEQUENCE ALGEBRAS

K. B. Laursen[†]

In this lecture we report on joint work with W. G. Bade and H. G. Dales, [3].

All terms used here are defined in W. G. Bade's paper [1], to which we refer. Some of the motivation behind the work that I shall describe here may also be found in Bade's paper: he presents a class of examples of weights ω for which the multiplier algebra $\mathfrak{m}(M)$ of the maximal ideal M in $\ell^1(\omega)$ is much larger than the local algebra $\mathfrak{L}(M)$ (the operator norm closure of $\ell^1(\omega)$ in $\mathfrak{m}(M)$): there are weights ω for which $\mathfrak{m}(M)$ contains elements with spectra with non-empty interior.

Here we shall present a class of weights for which $\mathfrak{m}(M)$ is local, yet properly contains $\mathfrak{L}(M)$. As this class also yields some interesting examples related to the standard ideal question, we shall begin with this aspect.

Recall that an linear operator $T : X \to X$ on a Banach space X is <u>unicellular</u> if the lattice $\mathrm{Lat}(T)$ of closed T-invariant subspaces of X is totally ordered (with respect to the inclusion order). In $\ell^1(\omega)$ the right shift R is multiplication by $e_1 = (0,1,0,\ldots)$ and it is easily seen that $\mathrm{Lat}(R)$ equals the collection of closed ideals of $\ell^1(\omega)$. Moreover, a moment's reflection will reveal that R is unicellular if and only if $\mathrm{Lat}(R) = \{M_k\}$, where

$$M_k = \{x \in \ell^1(\omega) \mid x(j) = 0,\ j < k\} \quad (k \in \mathbb{N})$$

(and $M_0 = \ell^1(\omega)$, $M_\infty = \{0\}$). It is entirely natural, then, to say that a weight ω is <u>unicellular</u> if and only if every closed ideal of $\ell^1(\omega)$ is standard.

It is known [3, Theorem 1.17] that a sufficient condition for unicellularity of ω is that ω be a <u>basis</u> weight (see Definition 1 below). Here we show that this condition is not necessary. This answers a question raised by Nikolskii [7].

Recall that a weight ω is <u>radical</u> if $\omega(n)^{1/n} \to 0$ as $n \to \infty$ and that ω is <u>submultiplicative</u> if there is a constant C for which $\omega(m+n) \le C\omega(m)\omega(n)$ $(m,n \in \mathbb{N})$.

Suppose ω is a radical and submultiplicative weight. We may even assume that ω is monotonically decreasing [2]. Define, for $k \in \mathbb{N}$:

$$\omega_k(n) = \omega(k+n) \quad (n \in \mathbb{Z}^+)\,.$$

[†] Supported by the Danish Science Research Council.

Then it is easy to see that each ω_k is radical. However (see below), no ω_k need be submultiplicative.

DEFINITION 1. If ω is a radical and submultiplicative weight and if each ω_k is submultiplicative, then ω is called a basis weight.

Sufficient conditions for unicellularity have also been given by Grabiner [6]. To state these we need the following notion:

With ω and ω_k as above consider the Banach spaces $\ell^1(\omega_k)$ and the maximal linear subspaces

$$M(\omega_k) = \{x \in \ell^1(\omega_k) \mid x(0) = 0\} .$$

Define

$$M(\omega_\infty) = \bigcup_{k=1}^{\infty} M(\omega_k) .$$

Letting e_k denote the kth canonical basis element it is easy to see that $x \in M(\omega_k)$ if and only if $x * e_k \in \ell^1(\omega)$. From this it follows readily that $M(\omega_\infty)$ is a convolution algebra.

We shall say that an element $x \in A$, where A is an algebra, is quasi-invertible if there is an element $y \in A$ for which $x + y = xy = yx$. Then A is radical if every $x \in A$ is quasi-invertible. We may then quote from [6], Theorem 3.15:

THEOREM 2. Let ω be submultiplicative and radical. The following are equivalent:

(a) $M(\omega_\infty)$ is radical.

(b) Each non-zero ideal in $\ell^1(\omega)$ contains e_k for some $k \in \mathbb{Z}^+$.

(c) For each $x \in M(\omega_\infty)$ there is an integer k such that $x^{*k} \in M$.

COROLLARY 3. If $M(\omega_\infty)$ is radical, then ω is unicellular.

Proof. A closed non-zero ideal I in $\ell^1(\omega)$ is a standard ideal if and only if $e_k \in I$ for some $k \in \mathbb{Z}^+$.

COROLLARY 4. If $M(\omega_\infty)$ is radical then $\mathfrak{m}(M)$ is local.

Proof. If $\alpha \in \mathfrak{m}(M)$ then $\alpha * e_1 \in M$, so $\alpha \in M(\omega_1)$, if $\alpha(0) = 0$. Hence $\alpha \in M(\omega_\infty)$ and so, by (c) above $\alpha^{*k} \in M$ for some k. But then α is quasi-nilpotent.

For any positive sequence $(\xi(0), \xi(1), \ldots)$, where $\xi(0) = 1$, it is possible to find a largest submultiplicative (in the strict sense, i.e. with the constant $C = 1$) sequence $(\omega(0),\omega(1),\ldots)$ for which $\omega(n) \le \xi(n)$ $(n \in \mathbb{Z}^+)$. In fact it is routine to check that the following formula holds:

PROPOSITION 5. Let ω be any positive sequence of numbers with $\omega(0) = 1$. Then the largest submultiplicative minorant $\overline{\omega}$ of ω is given by

$$\overline{\omega}(n) = \inf\{\omega(n_1)\ldots\omega(n_k) : n_1,\ldots,n_k \ge 0,\ n_1 + \cdots + n_k = n\}\ (n \in \mathbb{Z}^+).$$

We may then form a set analogous to $M(\omega_\infty)$, using the submultiplicative weights $\overline{\omega}_k$ (corresponding to $\omega_k/\omega_k(0)$): for each $k \in \mathbb{N}$ consider the local algebra $\ell^1(\overline{\omega}_k)$ with unique maximal ideal $M(\overline{\omega}_k)$ and let

$$M(\overline{\omega}_\infty) = \bigcup_{k=1}^{\infty} M(\overline{\omega}_k).$$

The following is then immediate from the definitions.

PROPOSITION 6. The set $M(\overline{\omega}_\infty)$ is a radical algebra containing $M(\omega_\infty)$.

Obviously, equality of these two algebras enables us to invoke Theorem 2 above. It turns out that a reasonably simple necessary and sufficient condition on ω for this equality to hold may be given.

DEFINITION 7. Let ω be radical and submultiplicative and let $k \in \mathbb{N}$ be given. Then ω is a D_k-weight if there are numbers $N \ge k$ and $C > 0$ such that $\omega_N \le C\overline{\omega}_k$ (that is, $\omega(N+n) \le C\overline{\omega}_k(n)$ $(n \in \mathbb{Z}^+)$). If ω is a D_k-weight for each $k \in \mathbb{N}$ then ω is a Domar weight.

The choice of name here is due to the fact that Y. Domar suggested the possible relevance of such a condition in this context.

We then have the following

PROPOSITION 8. A weight ω is a Domar weight if and only if $M(\omega_\infty) = M(\overline{\omega}_\infty)$.

Proof. Suppose first that ω is a D_k-weight for some k. Then ω_N is dominated by $\overline{\omega}_k$, for some N. Hence $M(\overline{\omega}_k) \subseteq M(\omega_N)$. Since k is arbitrary, $M(\overline{\omega}_\infty) \subseteq M(\omega_\infty)$.

To prove the converse, let $A = \bigcup_{n=1}^{\infty} M(\omega_n) = \bigcup_{n=1}^{\infty} M(\overline{\omega}_n)$, and let A be given the inductive limit topology induced by the spaces $M(\overline{\omega}_n)$, $n = 1,2,\ldots$. Let m be fixed. The canonical map of $M(\overline{\omega}_m)$ into A is continuous. Similarly, since each $\overline{\omega}_n$ is dominated by ω_n, the canonical maps of $M(\omega_n)$ into A are all continuous. These remarks make [4], Theorem 6.5.1 directly applicable: we conclude that there is an integer k such that $M(\overline{\omega}_m) \subset M(\omega_k)$ and that the unit ball of $M(\omega_k)$ contains a neighborhood of zero in $M(\overline{\omega}_m)$. This is exactly the statement that ω is a D_m-weight. Since m is arbitrary, this completes the proof.

REMARK. We owe the suggestion (and an outline of the proof) that the condition $M(\omega_\infty) = M(\overline{\omega}_\infty)$ is sufficient to make ω a Domar weight to M. Neumann.

It is not hard to give examples of Domar weights; clearly, any basis weight is a Domar weight, and so, in particular, the weight $\omega(n) = \exp(-n^2)$ is a Domar weight.

To obtain more illuminating examples we turn to a class of weights studied by Grabiner [5].

Let ω be any sequence of positive real numbers for which the sequence ω_1 is dominated by ω. Let

$$L_\omega = \lim\inf[\omega_1(n)/\omega(n)]^{1/n},$$

and let

$$U_\omega = \lim\sup[\omega_1(n)/\omega(n)]^{1/n},$$

and so $0 \leq L_\omega \leq U_\omega \leq 1$.

PROPOSITION 9. Let ω be a strictly decreasing weight for which $0 < L_\omega \leq U_\omega < 1$. Then ω is a Domar weight.

<u>Proof</u>. Let $m \geq 2$ be a given integer. Since

$$\overline{\omega}_m(n) = \inf\{\omega_m(n_1)\dots\omega_m(n_p)/\omega_m(0)^p : n_1,\dots,n_p \geq 0,\ n_1 + \dots + n_p = n\}\ (n \in \mathbb{Z})$$

(Proposition 5) we have to show that the negative logarithm η of ω satisfies the following: for each $m \in \mathbb{N}$, there are integers K_m and N_m such that

$$\eta(n_1 + m) + \dots + \eta(n_k + m) - \eta(n + N_m) - k\eta(m) \leq K_m,$$

whenever $k \in \mathbb{N}$, $n, n_1,\dots,n_k \in \mathbb{Z}^+$, and $n_1 + \dots + n_k = n$.

Choose $\delta_0 > 0$ so that $0 < L_\omega - \delta_0$ and $U_\omega + \delta_0 < 1$. Then, for each sufficiently large n,

$$L_\omega - \delta_0 < (\omega(n+1)/\omega(n))^{1/n} < U_\omega + \delta_0.$$

Let $c_1 = -\log(L_\omega - \delta_0)$, $c_2 = -\log(U_\omega + \delta_0)$. Then $0 < c_2 < c_1$ and, for sufficiently large n,

$$c_2 < \frac{1}{n}\,(\eta(n+1) - \eta(n)) < c_1 .$$

Since η is strictly increasing, we can suppose that

$$c_2 n < \eta(n+1) - \eta(n) < c_1 n \quad (n \in \mathbb{N}) . \tag{*}$$

Let $n = n_1 + \cdots + n_k$ and let $N \in \mathbb{N}$ be given. Then by repeated use of (*) we obtain

$$\begin{aligned} &\eta(n_1+m) + \eta(n_2+m) + \cdots + \eta(n_k+m) - (n+N) - k\eta(m) \leq \\ &\quad \eta(n_1) + c_1(n_1+m-1+n_1+m-2+\cdots+n_1) + \cdots + \eta(n_k) \\ &\quad + c_1(n_k+m-1+n_k+m-2+\cdots+n_k) \\ &\quad - \eta(n) - c_2(n+N-1+n+N-2+\cdots+n) - k\eta(n) \\ &\leq c_1(2n_1+m-1)m/2 + c_1(2n_2+m-1)m/2 + \cdots + c_1(2n_k+m-1)m/2 \\ &\quad - c_2(2n+N-1)N/2 - k\eta(m) \\ &= c_1 nm - c_2 nN + k(c_1 m(m-1)/2 - \eta(m)) - c_2 N(N-1)/2 . \end{aligned} \tag{**}$$

In the last inequality we used the superadditivity of η.

Now, consider the coefficient of k and let $A = \max[0, c_1 m(m-1)/2 - \eta(m)]$. Since $k \leq n$ we obtain

$$k(c_1 m(m-1)/2 - \eta(m)) \leq n\,A ,$$

and consequently

$$\begin{aligned} &\eta(n_1+m) + \cdots + \eta(n_k+m) - \eta(n+N) - k\eta(m) \\ &\quad \leq n(c_1 m + A - c_2 N) - c_2 N(N-1)/2 . \end{aligned}$$

Clearly we may choose N so large that

$$c_1 m + A - c_2 N \leq 0 .$$

This establishes the boundedness of (**) and completes the proof.

We shall use Proposition 9 to construct a Domar weight ω for which ω_1 is not submultiplicative. Thus this weight is a unicellular, but not a basis, weight. At the same time, the algebra $\ell^1(\omega)$ provides an example for which $\mathfrak{M}(M)$ is local (Corollary 4), yet $\mathfrak{M}(M) \neq \mathfrak{L}(M)$ ([3], Theorem 1.4). Below (Theorem 14), we shall give a more general result by showing that, if ω is a D_m-weight for some $m \geq 2$, then each element in the quotient algebra $\mathfrak{M}_1(M)/\mathfrak{L}_1(M)$ is nilpotent.

Our aim now is to find a weight ω to which we can apply Proposition 9. This means that, for certain constants $c_1, c_2 \in (0,\infty)$, we want to choose $\eta = -\log \omega$ such that

$$c_2 n < \eta(n+1) - \eta(n) < c_1 n \quad (n \in \mathbb{N}). \tag{10}$$

At the same time we want to ensure that ω_1 is not submultiplicative. In terms of η this means that

$$\eta(m+1) + \eta(n+1) - \eta(n+m+1)$$

must be unbounded. One way to achieve this is to make η additive at certain points. To be precise, we seek sequences of distinct integers (a_k), (b_k) for which

$$\eta(a_k + b_k) = \eta(a_k) + \eta(b_k). \tag{11}$$

For values of m and n for which $m+1 = a_k$ and $n+1 = b_k$ we then obtain

$$\begin{aligned}\eta(m+1) + \eta(n+1) - \eta(m+n+1) &= \eta(a_k) + \eta(b_k) - \eta(a_k + b_k - 1)\\ &= \eta(a_k + b_k) - \eta(a_k + b_k - 1) \geq c_2(a_k + b_k - 1) \to \infty \quad \text{as} \quad k \to \infty.\end{aligned}$$

Our objective then is to construct a weight ω for which (10) and (11) are satisfied.

EXAMPLE. For $p \in \mathbb{Z}^+$ let $j_p = 2^p$ and let $k_p = j_p + j_{p-1}$. Note that k_p is the midpoint of the interval $[j_p, j_{p+1}]$.

We shall define the function η inductively.

DEFINITION 12. Let $\eta(0) = \eta(1) = 1$ and for $p \in \mathbb{N}$, let

$$\eta(j_p) = j_{2p-1}$$

and

$$\eta(k_p) = \eta(j_p) + n(j_{p-1}).$$

If $n \in (j_p, k_p)$ and if $\eta(0),\ldots,\eta(n-1)$ have been defined, let

$$\eta(n) = \eta(n-1) + \frac{1}{5} n;$$

if $n \in (k_p, j_{p+1})$ and if $\eta(0),\ldots,\eta(n-1)$ have been defined, let

$$\eta(n) = \eta(n-1) + \frac{11}{7} n.$$

It is a tedious task to check that the weight ω thus defined is indeed submultiplicative [3]. We omit the details here and content ourselves with a statement of the relevant properties of the above weight.

THEOREM 13. There exists a Domar weight which is not a basis weight.

Proof. Let η be the function that we just constructed. It follows that

$$\frac{n}{6} \le \eta(n) - \eta(n-1) \le \frac{11}{7} n \quad (n \in \mathbb{N}).$$

Hence $\omega = \exp(-\eta)$ is a submultiplicative weight (because η is superadditive) for which

$$0 < e^{-11/7} \le L_\omega \quad \text{and} \quad U_\omega \le e^{-1/6} < 1 .$$

By Proposition 9, ω is a Domar weight. In particular, ω is unicellular. On the other hand, since $\eta(k_p) = \eta(j_p) + \eta(j_{p-1})$ for every $p \in \mathbb{N}$, it follows from the preceding discussion, that ω_1 is not submultiplicative, and hence that ω is not a basis weight.

REMARK. Marc Thomas, in [8], has also given an example of a unicellular weight φ which is not a basis weight. Since he shows that for φ, the algebra $M(\varphi_\infty)$ is not radical (he does this by showing that Theorem 2(b) is not true), it follows from Proposition 8, that φ is not a Domar weight.

We conclude this paper by giving a stronger form of Corollary 4. Our aim is to show that, if ω is a D_m-weight for some $m \ge 2$, then the multiplier algebra $\mathfrak{M}(M)$ is local. Recall that $\mathfrak{M}_1(M) = \{\alpha \in \mathfrak{M}(M) : \alpha(0) = 0\}$ and that $\mathfrak{L}_1(M) = \{\alpha \in \mathfrak{L}(M) : \alpha(0) = 0\}$.

THEOREM 14. Let ω be a D_m-weight for some $m \ge 2$, and let $N \in \mathbb{N}$ be chosen such that ω_N is dominated by $\overline{\omega}_m$. Then $\mathfrak{M}_1(M)/\mathfrak{L}_1(M)$ is nilpotent of order at most $N+1$.

For the proof of this theorem we shall need the following inequality.

LEMMA 15. Let ω be a radical weight and let $N \ge m \ge 2$ be integers such that ω_N is dominated by $\overline{\omega}_m$. Let s be any non-negative integer greater than $N - 2m$. Then there exists a positive constant C such that

$$\frac{\omega(j+n)}{\omega(n)} \le \frac{C}{\omega(m)^{k+2}} \, \omega(k-s)\omega(j_1+1)\dots\omega(j_k+1),$$

when $n \ge m$ and $j = j_1 + \dots + j_k$, where $j_1,\dots,j_k \ge m-1$ and $k > s+m$.

Proof. By replacing ω by an equivalent weight (see the argument of [2], Lemma 1.1) we may assume, without loss of generality, that ω is decreasing.

Choose a non-negative $s > N - 2m$, and let $k > 2+m$. Let $n \geq m$. By Proposition 5, there exists a constant C so that

$$\omega(t+N) \leq \frac{C}{\omega(m)^{k+2}} \omega(n_1+m) \ldots \omega(n_{k+2}+m)$$

whenever $t = n_1 + \cdots + n_{k+2}$. Now make the choices $n_1 + m = n$, $n_2 + m = k - s$, $n_3 + m = j_1 + 1, \ldots, n_{k+2} + m = j_k + 1$. We have

$$t = n + j + 2k - s - (k+2)m ,$$

where $j = j_1 + \cdots + j_k$, and therefore $t + N < n + j$ because

$$2k - s - (k+2)m + N < 0 .$$

Since ω is decreasing, we obtain

$$\omega(n+j) \leq \omega(t+N) \leq \frac{C}{\omega(m)^{k+2}} \omega(n)\omega(k-s)\omega(j_1+1) \ldots \omega(j_k+1)$$

which is the claim we are making.

<u>Proof of Theorem 14</u>. Let $\alpha = (0,\alpha(1),\alpha(2),\ldots)$ be an element of $\mathfrak{m}_1(M)$, and let $\beta = \alpha(1)e_1 + \cdots + \alpha(m-1)e_{m-1}$. Let $\gamma = \alpha - \beta$. Then, for any $k \in \mathbb{N}$, we have

$$\alpha^{*k} = (\beta+\gamma)^{*k} = \sum_{j=0}^{k} \binom{k}{j} \beta^{*j} * \gamma^{*(k-j)} = \gamma^{*k} + \delta ,$$

say, where $\delta \in M \subset \mathcal{L}_1(M)$. Hence $\alpha^{*k} \in \mathcal{L}_1(M)$ if and only if $\gamma^{*k} \in \mathcal{L}_1(M)$. Thus there is no loss of generality if we assume that $\alpha(1) = \cdots = \alpha(m-1) = 0$. By [3], Lemma 1.6, we have that $\alpha^{*k} \in \mathcal{L}_1(M)$ if and only if $\lim_{t\to\infty} S_t(\alpha^{*k}) = 0$, where $S_t x = (0,0,\ldots,x_t,x_{t+1},\ldots)$. Since

$$|||S_t(\alpha^{*k})||| = \sup_{n\geq 1} \|S_t(\alpha^{*k}) * \bar{e}_n\| ,$$

where $\bar{e}_n = e_n/\omega(n)$, we have that

$$|||S_t(\alpha^{*k})||| = \max_{1\leq j\leq m-1} (\|S_t(\alpha^{*k}) * \bar{e}_j\|, \sup_{p\geq m} \|S_t(\alpha^{*k}) * \bar{e}_p\|) .$$

It is clear from the definition of the norm in $\ell^1(\omega)$ that $\lim_{t\to\infty} \|S_t(\alpha^{*k}) * \bar{e}_j\| = 0$ for $j = 1,\ldots,m-1$, and hence we need only concern ourselves with the term

$$\lim_{t\to\infty} \sup_{n\geq m} \|S_t(\alpha^{*k}) * \bar{e}_n\| ,$$

where $k > N$. We have

$$\sup_{n \geq m} \|S_t(\alpha^{*k}) * \bar{e}_n\| = \sup_{n \geq m} \sum_{j \geq t} \left| \sum_{j_1+\cdots+j_k=j} \alpha(j_1)\alpha(j_2)\ldots\alpha(j_k) \right| \frac{\omega(j+n)}{\omega(n)} .$$

Since $\alpha(1) = \cdots = \alpha(m-1) = 0$, we may assume that $j_1,\ldots,j_k \geq m$. Take s of Lemma 15 to be $N-m$, and fix $k > N$. We obtain a constant $C(m,k)$ such that

$$\sup_{n \geq m} \|S_t(\alpha^{*k}) * \bar{e}_n\|$$

$$\leq C(m,k) \sum_{j \geq t} \sum_{j_1+\cdots+j_k=j} |\alpha(j_1)\ldots\alpha(j_k)|\omega(j_1+1)\ldots\omega(j_k+1) .$$

Since $j_1 + \cdots + j_k \geq t$, one of the numbers $j_1,\ldots,j_k$ is at least t/k, and consequently we may continue our estimate to obtain

$$\sup_{n \geq m} \|S_t(\alpha^{*k}) * \bar{e}_n\| \leq C(m,k)\left(\sum_{p \geq t/k} |\alpha(p)|\omega(p+1)\right)\left(\sum_{p \geq m-1} |\alpha(p)|\omega(p+1)\right)^{k-1}$$

$$= C(m,k)\|S_{t/k}(\alpha * e_1)\| \, \|\alpha * e_1\|^{k-1} ,$$

provided that $k \mid t$. Changing notation slightly (and assuming, as we may, that $\|\alpha * e_1\| \leq 1$) we finally obtain

$$\sup_{n \geq m} \|S_{kt}(\alpha^{*k}) * \bar{e}_n\| < C(m,k)\|S_t(\alpha * e_1)\| \quad (t \in \mathbb{N}) .$$

Since the right-hand side converges to zero as $t \to \infty$ the Theorem follows.

REMARK. We have not been able to decide on the localness of $\mathfrak{M}(M)$ when ω is assumed only to be a D_1-weight. More generally, it would be interesting to know more about the radical of $\mathfrak{M}(M)$ for arbitrary radical ω.

References

[1] W. G. Bade, Multipliers of weighted ℓ^1-algebras, this Volume.

[2] W. G. Bade and H. G. Dales, Norms and ideals in radical convolution algebras, J. Functional Analysis, 41 (1981), 77-109.

[3] W. G. Bade, H. G. Dales and K. B. Laursen, Multipliers of radical Banach algebras of power series, Mem. Amer. Math. Soc., to appear.

[4] R. E. Edwards, Functional Analysis, Holt, Rinehart & Winston, New York 1965.

[5] S. Grabiner, A formal power series operational calculus for quasi-nilpotent operators, Duke Math. J., 38 (1971), 641-658.

[6] ___________, A formal power series operational calculus for quasi-nilpotent operators II, J. Math. Analysis and Appl., 43 (1973), 170-192.

[7] N. K. Nikolskii, Selected problems of weighted approximation and spectral analysis, Proc. Steklov Inst. Math., 120 (1974), 1-270, Amer. Math. Soc. Transl., (1976), 1-278.

[8] M. Thomas, Closed ideals of $\ell^1(\omega_n)$ when $\{\omega_n\}$ is star-shaped, Pacific J. Math., to appear.

Mathematics Institute
Copenhagen University
Copenhagen, Denmark

APPROXIMATION IN THE RADICAL ALGEBRA $\ell^1(\omega_n)$ WHEN $\{\omega_n\}$ IS STAR-SHAPED

Marc P. Thomas

Let $A = \ell^1(\omega_n)$ be a radical Banach algebra of power series where the weight $\{\omega_n\}$ is star-shaped. Let T be the operator of right translation on A. We show that all closed ideals of A are <u>standard</u> provided $n(\omega_n)^{1/n} \to 0$. Equivalently, if $x = \sum \zeta_j z^j$ is a non-zero element of A, we show that $\overline{\text{span}}\,\{T^n x : n = 0,1,2,\ldots\}$ contains a power of z. This is the approximation problem for x. The technical device used is a κ-function. This enables the use of inductive and recursive techniques and shows specifically how such an approximation can be made.

1. Introduction

Let $\mathbb{C}[[z]]$ be the algebra of formal power series over the complex field $\mathbb{C}$. In this paper we study the subspace $\ell^1(\omega_n)$ of $\mathbb{C}[[z]]$ defined as follows:

$$\ell^1(\omega_n) = \left\{ \sum_{n=0}^{\infty} \alpha_n z^n : \sum_{n=0}^{\infty} |\alpha_n|\omega_n \equiv \left\| \sum_{n=0}^{\infty} \alpha_n z^n \right\| < \infty \right\}.$$

If the following conditions on the weight $\{\omega_n\}$ are satisfied, $\ell^1(\omega_n)$ is a Banach algebra under convolution of formal power series:

$$\omega_0 = 1 \tag{1.1}$$

$$0 < \omega_n \le 1, \quad \text{all } n \tag{1.2}$$

$$\omega_{m+n} \le \omega_m \omega_n, \quad \text{all } m \text{ and } n \tag{1.3}$$

$$\lim_{n\to\infty} (\omega_n)^{1/n} = 0 . \tag{1.4}$$

Such a weight is called a <u>radical</u> weight since $\ell^1(\omega_n)$ is then a commutative radical Banach algebra with unit adjoined. We shall generally shorten this to <u>radical Banach algebra</u>. Much of the structure we shall build up does not require algebra multiplication. If we replace (1.3) above by the weaker condition:

$$\omega_{n+1} \le \omega_n , \quad \text{all } n , \tag{1.3$'$}$$

then $\ell^1(\omega_n)$ is still a Banach space and the operator T of right translation is continuous on $\ell^1(\omega_n)$. We shall observe the terminology in the literature that a closed T-invariant <u>subspace</u> K of $\ell^1(\omega_n)$ is <u>standard</u> provided K is the zero subspace or

$$K = \left\{ f \in \ell^1(\omega_n) : f = \sum_{n=n(1)}^{\infty} \alpha_n z^n \right\} \tag{1.5}$$

for some $n(1)$. All other closed T-invariant subspaces are referred to as <u>non-standard</u>. In the case that $\ell^1(\omega_n)$ is an algebra (using (1.3)), the closed T-invariant subspaces are precisely the closed <u>ideals</u> of $\ell^1(\omega_n)$. At present, it remains an open question whether there exists some weight $\{\omega_n\}$ satisfying (1.1), (1.2), (1.3) and (1.4) such that $\ell^1(\omega_n)$ contains a non-standard ideal. We list some general results concerning standard T-invariant subspaces and standard ideals. We refer the reader to [1], [2, p. 188 §3.2] and [4, Introduction] for a general discussion of these and similar results.

(1.6) If $A = \ell^1(\omega_n)$ is an algebra then the closed ideal $(Ax)^-$ generated by an element x in A is equal to $\overline{\text{span}}\,\{T^n x : n = 0,1,2,\ldots\}$.

(1.7) Let x be a non-zero element of $X = \ell^1(\omega_n)$, assuming the weaker (1.3)'. Then $\overline{\text{span}}\,\{T^n x : n = 0,1,2,\ldots\}$ is standard if and only if it contains a power of z [1, Lemma 4.5].

(1.8) If for each r there is C_r such that $\omega_{m+n+r} \le C_r \omega_{m+r} \omega_{n+r}$, all m, n, then $\ell^1(\omega_{n+r})$ is an algebra, each r. Also if x is non-zero in $A = \ell^1(\omega_n)$, the <u>algebraic</u> ideal, Ax, contains a power of z. Hence $(Ax)^-$ is standard [2, Theorems 1 and 2, pp. 191-193]. Such a weight is called a <u>basis</u> weight.

(1.9) Let x be a non-zero element of $X = \ell^1(\omega_n)$, assuming the weaker (1.3)'. Then $K = \overline{\text{span}}\,\{T^n x : n = 0,1,2,\ldots\}$ is standard if and only if $K^\perp$ is finite dimensional in $X^* = \ell^\infty(1/\omega_n)$. Since $K^\perp$ is weak-star closed and since the only weak-star closed subspaces of $c_0(1/\omega_n) \subseteq \ell^\infty(1/\omega_n)$ are finite dimensional, it follows that K is standard if and only if $K^\perp \subseteq c_0(1/\omega_n)$

In Section two we assume only (1.1), (1.2), (1.3)', and (1.4) and work in the Banach space $X = \ell^1(\omega_n)$. We fix a non-zero x in X and recall the definition of the associated sequence $\{c_n\}$ from [3]. Basically this sequence is defined inductively so that

$$\left(\sum_{n=0}^{\infty} c_n z^n \right) * x = z^{n(1)} \quad \underline{\text{in } \mathbb{C}[[z]],} \tag{1.10}$$

where $n(1)$ is the first non-zero index of x. We also assume without loss of generality that the first term of x is one. We then define a κ-function (Definition 2.2) and the resulting derived set $S = S(x,\kappa)$ (Definition 2.4). Intuitively, one expects an index n to be in the derived set S if there is an "appreciable drop" in the weight $\omega_{n(1)+n}$ relative to the coefficients in the associated sequence $\{c_n\}$. The main results in this section are Lemma 2.5, which relates S to the norms of the biorthogonal polynomials $\{\chi_n^*\}$, and Theorem 2.7, which asserts that the cardinality of S is infinite. From the latter follows the technical Corollary 2.9, which is the basis for the approximation in Section three.

In Section three we specialize to the case of a star-shaped weight (Definition 3.1). This is a stronger requirement than (1.1), (1.2), (1.3) and (1.4), so $A = \ell^1(\omega_n)$ is an algebra. Geometrically, the star-shaped condition means that the region below the graph of $y = \ln(\omega_n)$ is "illuminated" by the origin. Earlier in [4] we had shown that certain star-shaped weights $\{\omega_n\}$ had the following property: All closed ideals of $\ell^1(\omega_n)$ are standard but there are elements x in $\ell^1(\omega_n)$ such that the algebraic ideal, Ax, contains no power of z [4, Proposition 4.2]. This should be contrasted to the case where $\ln \omega_n$ is <u>concave</u>, in which case (1.8) holds. In Theorem 3.5 we assume that $\{\omega_n\}$ is star-shaped and for some b, $n(\omega_n)^{b/n} \to 0$. Then if $x = \sum_{j=n(1)}^{\infty} \zeta_j z^j$, $\zeta_{n(1)} = 1$, is in $A = \ell^1(\omega_n)$ we show specifically how $z^{2n(1)+b}$ can be approximated by linear combinations of translates of x. In particular, if $\{\omega_n\}$ is star-shaped and $n(\omega_n)^{1/n} \to 0$, all closed ideals of $\ell^1(\omega_n)$ are standard. This is a powerful extension of our results in [4] where we had to assume the weight was "induced" from certain subsequences with nice properties to conclude all closed ideals were standard.

Finally, the question arises whether these techniques can be applied to other spaces. The continuous analogue of $\ell^1(\omega_n)$ is

$$L^1(\omega, \mathbb{R}^+) = \left\{ f : \int_0^\infty |f(t)|\omega(t)dt < \infty \right\},$$

where the weight function $\omega(\cdot)$ satisfies conditions similar to (1.1), (1.2), (1.3) and (1.4). The major difficulty is the inability to normalize and make the leading term one as one can do in the discrete case. However, if f is smooth and $f(0) \neq 0$ similar results can be obtained.

2. κ-Functions

We shall not need any algebra structure in this section. We shall only require that $\{\omega_n\}$ is a weight satisfying the conditions (1.1), (1.2), (1.3)' and (1.4) of the introduction. This makes $\ell^1(\omega_n)$ into a Banach space where the operator T of right translation is continuous. Hence convolution by polynomials is also continuous although in general $\ell^1(\omega_n)$ need not be an algebra (i.e. $f, g \in \ell^1(\omega_n)$ doesn't imply $f * g \in \ell^1(\omega_n)$). Let $x = \sum_{j=n(1)}^{\infty} \zeta_j z^j$ be a <u>fixed</u> element of $X = \ell^1(\omega_n)$ with $\zeta_{n(1)} = 1$. As in [3] we define the associated sequence $\{c_n\}$ as follows:

DEFINITION 2.1. Let x be fixed as above. Let

$$c_0 = \frac{1}{\zeta_{n(1)}} = 1,$$

and if $c_0, c_1, c_2, \ldots, c_{n-1}$ have been chosen, let

$$c_n = -\sum_{k=0}^{n-1} c_k \zeta_{n(1)+n-k}.$$

We shall refer to $\{c_n\}$ as the associated sequence for x.

We require the following elementary results [3, Lemma 4.2]. First, since the dual of $X = \ell^1(\omega_n)$ is $X^* = \ell^\infty(1/\omega_n)$, let the canonical dual weak-star basis be denoted by $\{e_n^*\}$ (i.e. $e_m^*(z^n) = \delta_{m,n}$). It easily follows that

(2.1) $\sum_{n=0}^m c_n T^n x$ agrees with $z^{n(1)}$ on $[0, n(1)+m]$ where $Tf = z * f$ as before, $f \in X$.

(2.2) $(\sum_{n=0}^\infty c_n z^n) * x = z^{n(1)}$ as formal power series in $\mathbb{C}[[z]]$.

(2.3) If $\chi_n^* = \sum_{k=0}^n c_k e_{n(1)+n-k}^*$, then $\chi_n^*(T^m x) = \delta_{m,n}$ (biorthogonality), and

$$\|\chi_n^*\| = \sup_{0 \le k \le n} \frac{|c_k|}{\omega_{n(1)+n-k}},$$

in $X^* = \ell^\infty(1/\omega_n)$.

We next define the concept of a κ-function for x. The intuitive idea is that $\kappa(n)$ should be an index where χ_n^* carries "appreciable mass."

DEFINITION 2.2. Let x be fixed as above and let $\{c_n\}$ be the associated sequence for x. A κ-function for x is a mapping κ from the non-negative integers to itself satisfying the following conditions:

(i) For all n, $0 \le \kappa(n) \le n$.

(ii) There exists an $n_0 \ge 1$ such that $n \ge n_0$ implies $\kappa(n) < n$.

(iii) If $\kappa(n) = 0$ then $\|\chi_n^*\| = 1/\omega_{n(1)+n}$.

(iv) There exists a constant c, $0 < c \le 1$ such that

$$c\|\chi_n^*\| \le \frac{|c_{\kappa(n)}|}{\omega_{n(1)+n-\kappa(n)}}, \quad \text{all } n.$$

It is clear that the main difficulty in constructing a κ-function is to simultaneously satisfy (ii) and (iv).

LEMMA 2.3. Let x be fixed as above. Then there exists a κ-function for x.

Proof. Let $\mu(n)$ be the smallest index such that

$$\|x_n^*\| = \frac{|c_{\mu(n)}|}{\omega_{n(1)+n-\mu(n)}} .$$

If $\mu(n) = n$ for only finitely many n, let $c = 1$, $\kappa(n) = \mu(n)$ and the result follows. Hence we may suppose that $\mu(n) = n$ for infinitely many n. Suppose for each $c > 0$ there is $n(c)$, arbitrarily large, with $\mu(n(c)) = n(c)$ satisfying

$$c\|x_{n(c)}^*\| > \sup_{0<k<n(c)} \left(\frac{|c_k|}{\omega_{n(1)+n(c)-k}} \right) . \tag{2.4}$$

Then we may pick a sequence $\{c(i)\}$ tending to zero and obtain an increasing sequence $\{n_i = n(c(i))\}$ with $\mu(n_i) = n_i$ such that

$$\lim_{i\to\infty} \left[\frac{\sup_{0<k<n_i} \left(\frac{|c_k|}{\omega_{n(1)+n_i-k}} \right)}{\|x_{n_i}^*\|} \right] = 0 . \tag{2.5}$$

But equation (2.5) and the fact that $\|x_{n_i}^*\| = |c_{n_i}|(\omega_{n(1)})^{-1}$ imply that

$$\frac{\bar{c}_{n_i} x_{n_i}^*}{|c_{n_i}|\|x_{n_i}^*\|} \xrightarrow{\text{weak-star}} \omega_{n(1)} e_{n(1)}^* ,$$

a contradiction, since any such weak-star limit would annihilate x and all its translates. Clearly, $\omega_{n(1)} e_{n(1)}^*$ does not annihilate x. It must follow, then, that there is a c, $0 < c \leq 1$, and an $n_0 \geq 1$ such that if $n \geq n_0$ and $\mu(n) = n$, then

$$c\|x_n^*\| \leq \sup_{0<k<n} \left(\frac{|c_k|}{\omega_{n(1)+n-k}} \right) . \tag{2.6}$$

For such n, let $\kappa(n)$ be any index k, $0 < k < n$ in equation (2.6) such that

$$c\|x_n^*\| \leq \frac{|c_{\kappa(n)}|}{\omega_{n(1)+n-\kappa(n)}} .$$

If $n < n_0$, or if $n \geq n_0$ but $\mu(n) < n$, let $\kappa(n) = \mu(n)$. It is now easily verified that κ satisfies (i)-(iv) of Definition 2.2.

We remark that if the latter case ($\mu(n) = n$ infinitely many times) actually occurs above, this forces $\overline{\text{span}}\, \{T^n x : n = 0,1,2,\ldots\}$ to be a non-standard closed T-invariant subspace. This follows because the set $\{x_n^*/\|x_n^*\| : \mu(n) = n\}$ then has a weak-star cluster point in X^* which annihilates x and all its translates and doesn't vanish at the $n(1)$st place. We need one more definition concerning

κ-functions.

DEFINITION 2.4. Let x be fixed as above. Let $\{c_n\}$ be the associated sequence for x and let κ be any κ-function for x. The derived set $S = S(x,\kappa)$ is the set of all positive integers n satisfying the following:

(i) $\kappa(n) = 0$

(ii) If $0 < k < n$, then

$$(\omega_{n(1)+n-k})^{1/n(1)+n-k} \geq (\omega_{n(1)+n})^{1/n(1)+n}.$$

(iii) If $0 < k < n$, then

$$|c_k|(\omega_{n(1)+n})^{(n(1)+k)/(n(1)+n)} \leq 1.$$

At this point it is difficult to motivate the choice of the above three conditions. The reasons will become clear later. Roughly speaking, one expects n to be in the derived set S if $\|x_n^*\|$ is assumed at the $(n(1)+n)$th place and there is an "appreciable drop" in the weight ω_m when $m = n(1)+n$. We first relate the derived set to the norms of the $\{x_n^*\}$.

LEMMA 2.5. Let x be a fixed element of $X = \ell^1(\omega_n)$ as above. Let κ be any κ-function for x and let $S = S(x,\kappa)$ be the derived set. Then there exist constants $m_0 \geq 2$ and $C \geq 1$ such that the following holds: Let n be any positive integer. There exist rationals $r_i > 0$ and integers s_i satisfying

(1) $r_1 s_1 + r_2 s_2 + \cdots + r_j s_j \leq n$

(2) $1 \leq s_i \leq n$, $s_i \in (S \cup \{1,2,3,\ldots,m_0\})$

(3) $\ln\|x_n^*\| \leq \sum_{i=1}^{j} r_i \ln\|x_{s_i}^*\| + ((\sum_{i=1}^{j} r_i) - 1) \ln C.$

Proof. Let $m_0 = \max\{n_0, 2\}$, n_0 as in Definition 2.2. Let $C = \sup_{0 \leq k \leq m_0} c^{-1}\|x_k^*\|$, c as in Definition 2.2. It is clear that the result holds for $1 \leq n \leq m_0$. We proceed by induction. Suppose the result holds for $1,2,\ldots,m_0$, $m_0+1,\ldots,n-1$. We consider $\|x_n^*\|$ and have several cases.

Case I. $\kappa(n) \neq 0$. Since $n > m_0 \geq n_0$ of Definition 2.2, it follows that $0 < \kappa(n) < n$ and $0 < (n - \kappa(n)) < n$. Hence

$$\|x_n^*\| \leq \frac{|c_{\kappa(n)}|}{c\omega_{n(1)+n-\kappa(n)}} \leq \frac{1}{c} \frac{|c_{\kappa(n)}|}{\omega_{n(1)}} \frac{|c_0|}{\omega_{n(1)+n-\kappa(n)}}$$

$$\leq \frac{1}{c} \|x_{\kappa(n)}^*\| \, \|x_{n-\kappa(n)}^*\| .$$

Thus

$$\ln\|\chi_n^*\| \leq \ln\|\chi_{\kappa(n)}^*\| + \ln\|\chi_{n-\kappa(n)}^*\| + \ln C .$$

Applying the inductive hypothesis and collecting terms yields the result.

Case II. $\kappa(n) = 0$, but condition (ii) of Definition 2.4 fails. Hence there is k, $0 < k < n$, satisfying

$$(\omega_{n(1)+n-k})^{1/(n(1)+n-k)} < (\omega_{n(1)+n})^{1/(n(1)+n)} .$$

Let $m = (n-k)$ and the above is easily seen to imply

$$\frac{1}{\omega_{n(1)+n}} < \left(\frac{1}{\omega_{n(1)+m}} \right)^{(n(1)+n)/(n(1)+m)} .$$

Hence by condition (iii) of Definition 2.2

$$\begin{aligned} \|\chi_n^*\| = \frac{1}{\omega_{n(1)+n}} &\leq \left(\frac{1}{\omega_{n(1)+m}} \right)^{(n(1)+n)/(n(1)+m)} \leq \|\chi_m^*\|^{(n(1)+n)/(n(1)+m)} \\ &\leq \|\chi_m^*\|^{n/m} . \end{aligned} \tag{2.7}$$

Since $1 \leq m \leq n-1$, we can apply the inductive hypothesis to obtain

$$\ln \|\chi_m^*\| \leq \sum_{i=1}^{j} r_i \ln\|\chi_{s_i}^*\| + \left(\left(\sum_{i=1}^{j} r_i\right) - 1\right) \ln C$$

with $r_1 s_1 + \cdots + r_j s_j \leq m$, etc. From (2.7) it follows that

$$\begin{aligned} \ln\|\chi_n^*\| \leq \frac{n}{m} \ln\|\chi_m^*\| &\leq \sum_{i=1}^{j} \frac{nr_i}{m} \ln\|\chi_{s_i}^*\| + \left(\left(\sum_{i=1}^{j} \frac{nr_i}{m}\right) - \frac{n}{m}\right) \ln C \\ &\leq \sum_{i=1}^{j} t_i \ln\|\chi_{s_i}^*\| + \left(\left(\sum_{i=1}^{j} t_i\right) - 1\right) \ln C \end{aligned}$$

where $t_i = nr_i/m$. This is easily seen to complete Case II.

Case III. $\kappa(n) = 0$, but condition (iii) of Definition 2.4 fails. Hence there is k, $0 < k < n$, satisfying

$$|c_k| (\omega_{n(1)+n})^{(n(1)+k)/(n(1)+n)} > 1 .$$

This implies that

$$\frac{1}{\omega_{n(1)+n}} < |c_k|^{(n(1)+n)/(n(1)+k)} .$$

Hence

$$\|\chi_n^*\| = \frac{1}{\omega_{n(1)+n}} < |c_k|^{(n(1)+n)/(n(1)+k)}$$
$$\le \|\chi_k^*\|^{(n(1)+n)/(n(1)+k)} \le \|\chi_k^*\|^{n/k}. \tag{2.8}$$

Since $1 \le k \le n-1$, we can apply the inductive hypothesis to obtain

$$\ln\|\chi_k^*\| \le \sum_{i=1}^{j} r_i \ln\|\chi_{s_i}^*\| + \left(\left(\sum_{i=1}^{j} r_i\right) - 1\right) \ln C$$

with $r_1 s_1 + \cdots + r_j s_j \le k$, etc. From (2.8) it follows that

$$\ln\|\chi_n^*\| \le \frac{n}{k} \ln\|\chi_k^*\| \le \sum_{i=1}^{j} \left(\frac{nr_i}{k}\right) \ln\|\chi_{s_i}^*\| + \left(\left(\sum_{i=1}^{j} \frac{nr_i}{k}\right) - \frac{n}{k}\right) \ln C$$
$$\le \sum_{i=1}^{j} t_i \ln\|\chi_{s_i}^*\| + \left(\left(\sum_{i=1}^{j} t_i\right) - 1\right) \ln C$$

where $t_i = nr_i/k$. This is easily seen to complete Case III.

We now have the final case.

Case IV. $\kappa(n) = 0$, and conditions (ii) and (iii) of Definition 2.4 hold. Hence $n \in S$ and we can simply write

$$\ln\|\chi_n^*\| \le \ln\|\chi_n^*\| + (1-1) \ln C .$$

This completes the proof of the lemma.

LEMMA 2.6. Let x be a fixed element of $X = \ell^1(\omega_n)$ as above. Let κ be any κ-function for x and let $S = S(x,\kappa)$ be the derived set.

(1) If s_1, s_2 are elements of S and $s_1 < s_2$, then

$$\ln\|\chi_{s_1}^*\| \le \frac{s_1}{s_2} \ln\|\chi_{s_2}^*\| .$$

(2) There exists a constant $B \ge 1$ such that if n is a positive integer

$$\ln\|\chi_n^*\| \le \frac{n}{s} \ln\|\chi_s^*\| + n \ln B$$

for some $s \le n$, $s \in (S \cup \{1\})$.

Proof. We first prove (1). Using condition (ii) of Definition 2.4, it follows that

$$(\omega_{n(1)+s_1})^{1/(n(1)+s_1)} \ge (\omega_{n(1)+s_2})^{1/(n(1)+s_2)}$$

since $1 \le s_1 \le (s_2 - 1)$. Hence

$$\|\chi^*_{s_1}\| = \frac{1}{\omega_{n(1)+s_1}} \leq \left(\frac{1}{\omega_{n(1)+s_2}}\right)^{(n(1)+s_1)/(n(1)+s_2)}$$
$$\leq \|\chi^*_{s_2}\|^{s_1/s_2} .$$

Upon taking logarithms of both sides, (1) follows. To prove (2), let C and m_0 be as in Lemma 2.5. Choose $D \geq C$ sufficiently large so that

$$\sup_{0 \leq m \leq m_0} (\ln\|\chi^*_m\|) \leq \ln D .$$

If n is a positive integer Lemma 2.5 implies that

$$\ln\|\chi^*_n\| \leq r_1 \ln\|\chi^*_{s_1}\| + \cdots + r_j \ln\|\chi^*_{s_j}\| + (n-1) \ln C ,$$

where $s_i \in (S \cup \{1,2,\ldots,m_0\})$. If no s_i is in S it is easily seen that

$$\ln\|\chi^*_n\| \leq n \ln D + (n-1) \ln C ,$$

from (2.9). Otherwise let

$$s = (\max_{s_i \in S} s_i) \leq n .$$

We may also suppose, upon re-indexing, that $s_1, s_2, \ldots, s_t$ are in S and $s_{t+1}, s_{t+2}, \ldots, s_j$ are not. Then using (1) and (2.9) we obtain

$$\ln\|\chi^*_n\| \leq \frac{(r_1 s_1 + \cdots + r_t s_t)}{s} \ln\|\chi^*_s\| + (r_{t+1} + \cdots + r_j) \ln D + (n-1) \ln C$$
$$\leq \frac{n}{s} \ln\|\chi^*_s\| + n \ln D + (n-1) \ln C .$$

In any case we may let $B = DC$ to obtain

$$\ln\|\chi^*_n\| \leq \frac{n}{s} \ln\|\chi^*_s\| + n \ln B ,$$

where $s \in S$ or, if no s_i is in S, then $s = 1$. This completes the proof of the lemma.

We remark in passing that assertion (2) of Lemma 2.6 is still sharp. This can be seen upon considering examples in [4, Section 4]. We now obtain our main result concerning the derived set.

THEOREM 2.7. Let x be a fixed element of $X = \ell^1(\omega_n)$ as above. Let κ be any κ-function for x and let $S = S(x,\kappa)$ be the derived set. Then the cardinality of S is infinite.

Proof. If the result fails, $S \cup \{1\}$ is a finite set and assertion (2) of Lemma 2.6 implies that there is a constant $H \geq B$ such that

$$\ln\|\chi_n^*\| \leq n \ln H, \quad n \geq 1,$$

or

$$\|\chi_n^*\| \leq H^n \leq H^{n(1)+n}.$$

But $(\omega_{n(1)+n})^{-1} \leq \|\chi_n^*\|$, all n. Thus

$$(\omega_{n(1)+n})^{1/n(1)+n} \geq \frac{1}{H}, \quad \text{all } n,$$

a contradiction, since $(\omega_m)^{1/m} \to 0$. The result follows.

Although we have worked so far with a fixed x in $X = \ell^1(\omega_n)$ we could fix $n(1)$ and look at several such x's. We could pick a κ-function for each and ask whether the derived sets $S(x,\kappa)$ had any common elements. One result in this direction is the following proposition.

PROPOSITION 2.8. Let $x_1, x_2, \ldots, x_m$ be a finite number of elements of $X = \ell^1(\omega_n)$. Suppose each has initial non-zero term one at the $n(1)$st place. Let $\kappa_1, \kappa_2, \ldots, \kappa_m$ be κ-functions for the respective elements above. Then the cardinality of $\bigcap_{i=1}^m S(x_i,\kappa_i)$ is infinite.

Proof. Suppose the result fails. Let $\|\chi(i)_n^*\|$ be the norm of the nth biorthogonal polynomial of the ith element x_i. Let

$$F_n \equiv \max_{1 \leq i \leq m} \|\chi(i)_n^*\|.$$

Apply Lemma 2.5 to each element and without loss of generality assume that C and m_0 are the same for each of the elements. There must be some index $m_1 \geq m_0 \geq 2$ such that $n > m_1$ implies $n \notin \bigcap_{i=1}^m S(x_i,\kappa_i)$. Pick E sufficiently large so that

$$\ln F_n \leq n \ln E - \ln C \tag{2.10}$$

for $n = 1,2,3,\ldots,m_1$. We prove by induction that (2.10) holds for all n. Suppose (2.10) holds for $1,2,\ldots,m_1,\ldots,n-1$. We consider F_n and have two cases.

Case I. $F_n = \|\chi(i)_n^*\|$ and $n \notin S(x_i,\kappa_i)$. By Lemma 2.5 there exist rationals r_t and integers s_t $1 \leq s_t \leq n$ such that $r_1 s_1 + \cdots + r_j s_j \leq n$ and

$$\ln\|\chi(i)_n^*\| \le \sum_{t=1}^{j} r_t \ln\|\chi(i)_{s_t}^*\| + \left(\left(\sum_{t=1}^{j} r_t\right) - 1\right) \ln C$$
$$\le \sum_{t=1}^{j} r_t \ln F_{s_t} + \left(\left(\sum_{t=1}^{j} r_t\right) - 1\right) \ln C .$$

Applying the inductive hypothesis, the above is

$$\le \sum_{t=1}^{j} r_t s_t \ln E - \sum_{t=1}^{j} r_t \ln C + \left(\left(\sum_{t=1}^{j} r_t\right) - 1\right) \ln C$$

$$\le n \ln E - \ln C .$$

Thus, (2.10) holds in this case.

Case II. $F_n = \|\chi(i)_n^*\|$ and $n \in S(x_i,\kappa_i)$. Since $n > m_1$ there is $j \neq i$ such that $n \notin S(x_j,\kappa_j)$. But then

$$F_n = \|\chi(i)_n^*\| = \frac{1}{\omega_{n(1)+n}} ,$$

since x_i has first non-zero term at $n(1)$st place. The above is then

$$\le \|\chi(j)_n^*\| \le F_n .$$

Hence $F_n = \|\chi(j)_n^*\|$ also and this case reduces to the first case above.

We then obtain that (2.10) holds for all $n \ge 1$. But

$$\omega_{n(1)+n} \ge \|\chi(i)_n^*\|^{-1}, \quad \text{all } i,$$
$$\ge (F_n)^{-1} \ge E^{-n}C \ge E^{-(n(1)+n)}C .$$

Then

$$(\omega_{n(1)+n})^{1/(n(1)+n)} \ge \frac{C^{1/(n(1)+n)}}{E}, \quad \text{all } n ,$$

a contradiction since $(\omega_n)^{1/n} \to 0$. We then conclude that the cardinality of $\bigcap_{i=1}^{m} S(x_i,\kappa_i)$ is infinite.

The above proposition shows that there is considerable overlap among derived sets. An open question is whether infinite intersections contains a common subsequence. We conclude this section with a technical corollary which will be used to solve the main approximation problem in the next section.

COROLLARY 2.9. Let x be a fixed element of $X = \ell^1(\omega_n)$ as above. Let κ be any κ-function for x and let $S = S(x,\kappa)$ be the derived set. Suppose, for some $b > 0$ that $m(\omega_m)^{b/m} \to 0$. Then the following hold:

$$\text{(i)} \qquad \lim_{n\in S} \sum_{k=0}^{n-1} |c_k| (\omega_{n(1)+n})^{(n(1)+k+b)/(n(1)+n)} = 0 .$$

$$\text{(ii)} \qquad \liminf_{n\to\infty} \sum_{k=0}^{n-1} |c_k| (\omega_{n(1)+n})^{(n(1)+k+b)/(n(1)+n)} = 0 .$$

Proof. Theorem 2.7 implies that the cardinality of S is infinite so (i) makes sense. Clearly (ii) follows from (i). To prove (i) let $n \in S$. Using condition (iii) of Definition 2.4 it follows that

$$\sum_{k=0}^{n-1} |c_k| (\omega_{n(1)+n})^{(n(1)+k)/(n(1)+n)} \le |c_0| (\omega_{n(1)+n})^{(n(1))/(n(1)+n)} + (n-1) \tag{2.11}$$

and if n is sufficiently large, (2.11) is less than or equal to n. Since $n(\omega_n)^{b/n} \to 0$, it follows that given $\varepsilon > 0$

$$\sum_{k=0}^{n-1} |c_k| (\omega_{n(1)+n})^{(n(1)+k+b)/(n(1)+n)} < \varepsilon$$

if n is in S and n is sufficiently large. The result then follows.

3. Approximation in $\ell^1(\omega_n)$

As before let $x = \sum_{j=n(1)}^{\infty} \zeta_j z^j$ be a fixed element of $\ell^1(\omega_n)$ with $\zeta_{n(1)} = 1$. The requirement that the leading term be one is only for convenience. We will henceforth specialize to the case when $\{\omega_n\}$ is star-shaped.

DEFINITION 3.1. We say that the weight $\{\omega_n\}$ is star-shaped provided the following hold:

(i) $\omega_0 = 1$

(ii) $(\omega_n)^{1/n} \to 0$ as $n \to \infty$

(iii) $m > n$ implies $(\omega_m)^n \le (\omega_n)^m$.

It is elementary then that conditions (1.1), (1.2), (1.3) and (1.4) of the introduction are satisfied. Hence $A = \ell^1(\omega_n)$ is a radical Banach algebra. Intuitively, condition (iii) implies that the region below the graph of $y = \ln \omega_n$ is "illuminated" by the origin.

DEFINITION 3.2. If $f = \sum_{j=0}^{\infty} \alpha_j z^j \in A = \ell^1(\omega_n)$ we define the projection Q_m as follows:

$$Q_m f = \sum_{j=m}^{\infty} \alpha_j z^j .$$

We have the following basic lemma for star-shaped domains.

LEMMA 3.3. Let $y = \sum_{i=0}^{\infty} \alpha_i z^i$ be <u>any</u> element of $A = \ell^1(\omega_n)$ where $\{\omega_n\}$ is <u>star-shaped</u>. Let m be fixed. Then

$$\|Q_m T^h y\| \leq (\omega_m)^{h/m}\|y\|,$$

for all non-negative integers h.

<u>Proof</u>. If $h = 0$ the result is obvious and we may suppose $h \geq 1$. Let $Q = Q_m$. Then

$$\|QT^h y\| = \|Q \sum_{i=0}^{\infty} \alpha_i z^{i+h}\| = \| \sum_{i=s}^{\infty} \alpha_i z^{i+h}\|,$$

where $s = \max\{0, m-h\}$. The above is then

$$= \sum_{i=s}^{\infty} |\alpha_i|\omega_{i+h}$$

$$= \sum_{i=s}^{m-1} |\alpha_i|\omega_{i+h} + \sum_{i=m}^{\infty} |\alpha_i|\omega_{i+h}.$$

In the first term $(i+h) \geq m$ and in the second term $(i+h) \geq i$. Since $\{\omega_n\}$ is star-shaped the above is

$$\leq \sum_{i=s}^{m-1} |\alpha_i|(\omega_m)^{(i+h)/m} + \sum_{i=m}^{\infty} |\alpha_i|(\omega_i)^{(i+h)/i}$$

$$\leq (\omega_m)^{h/m} \sum_{i=s}^{m-1} |\alpha_i|(\omega_m)^{i/m} + \sum_{i=m}^{\infty} |\alpha_i|\omega_i(\omega_i)^{h/i}.$$

In the first term $m > i$ whereas in the second term $i \geq m$. Since $\{\omega_n\}$ is star-shaped the above is

$$\leq (\omega_m)^{h/m} \sum_{i=s}^{m-1} |\alpha_i|\omega_i + (\omega_m)^{h/m} \sum_{i=m}^{\infty} |\alpha_i|\omega_i$$

$$\leq (\omega_m)^{h/m} \sum_{i=s}^{\infty} |\alpha_i|\omega_i$$

$$\leq (\omega_m)^{h/m}\|y\|,$$

and the result follows.

We require one more lemma.

LEMMA 3.4. Let $A = \ell^1(\omega_n)$ where $\{\omega_n\}$ is <u>star-shaped</u>, and let $x \in A$ where $x = \sum_{j=n(1)}^{\infty} \zeta_j z^j$, $\zeta_{n(1)} = 1$. Let $\{c_n\}$ be the associated sequence for x and let $r \geq 1$ be fixed. If n is a positive integer, then

$$\|z^{n(1)+r} - \sum_{k=0}^{n-1} c_k T^{r+k} x\| \leq \sum_{k=0}^{n-1} |c_k|(\omega_{n(1)+n})^{(r+k)/(n(1)+n)} \|x\|.$$

Proof. Using (2.1) it is clear that $z^{n(1)}$ and $\sum_{k=0}^{n-1} c_k T^k x$ agree on $[n(1), n(1)+n-1]$. Hence $z^{n(1)+r}$ and $\sum_{k=0}^{n-1} c_k T^{r+k} x$ agree on $[n(1), n(1)+n+r-1]$. Thus

$$\begin{aligned}\|z^{n(1)+r} - \sum_{k=0}^{n-1} c_k T^{r+k} x\| &= \|Q_{n(1)+n+r}(z^{n(1)+r} - \sum_{k=0}^{n-1} c_k T^{r+k} x)\| \\ &= \|\sum_{k=0}^{n-1} c_k Q_{n(1)+n+r} T^{r+k} x\| \\ &\leq \sum_{k=0}^{n-1} |c_k| \|Q_{n(1)+n+r} T^{r+k} x\| \\ &= \sum_{k=0}^{n-1} |c_k| \|Q_{n(1)+n} T^{r+k} x\| \\ &\leq \sum_{k=0}^{n-1} |c_k| (\omega_{n(1)+n})^{(r+k)/(n(1)+n)} \|x\|,\end{aligned}$$

by Lemma 3.3, and the result follows.

We now have our main result, which shows that $z^{2n(1)+b}$ can be approximated by the specific linear combination $(\sum_{k=0}^{n-1} c_k T^{n(1)+k+b} x)$, $n \in S$, provided $m(\omega_m)^{b/m} \to 0$ for some fixed b.

THEOREM 3.5. Let $A = \ell^1(\omega_n)$ where $\{\omega_n\}$ is <u>star-shaped</u> and let $x \in A$ where $x = \sum_{j=n(1)}^{\infty} \zeta_j z^j$, $\zeta_{n(1)} = 1$. Let $\{c_n\}$ be the associated sequence for x. Let κ be any κ-function for x and $S = S(x,\kappa)$ be the derived set. Let b be some positive integer satisfying $m(\omega_m)^{b/m} \to 0$, then

$$\lim_{n \in S} \|z^{2n(1)+b} - \sum_{k=0}^{n-1} c_k T^{n(1)+k+b} x\| = 0 .$$

Proof. Apply Lemma 3.4 with $n \in S$ and $r = n(1) + b$. Corollary 2.9, assertion (i), implies the result.

We finally obtain

COROLLARY 3.6. Let $A = \ell^1(\omega_n)$ where $\{\omega_n\}$ is <u>star-shaped</u>. If for some b, $n(\omega_n)^{b/n} \to 0$, then all closed ideals of A are <u>standard</u>.

Proof. It suffices to show $(Ax)^-$ is standard for all $x \in A$. Since $(Ax)^-$ is standard if and only if $(A(\lambda x))^-$ is standard, $\lambda \neq 0$, we may assume the leading coefficient is one, i.e., $x = \sum_{j=n(1)}^{\infty} \zeta_j z^j$, $\zeta_{n(1)} = 1$. Theorem 3.5 implies $(Ax)^- = \overline{\text{span}}\{T^n x : n = 0,1,2,\ldots\}$ contains a power of z. Hence by (1.7), $(Ax)^-$ is standard and the result follows.

References

[1] S. Grabiner, Weighted shifts and Banach algebras of power series, Amer. J. Math., 97 (1975), 16-42.

[2] N. K. Nikolskii, Selected problems of weighted approximation and spectral analysis, Proc. Steklov Inst. Math., 120 (1974) (AMS translation, 1976).

[3] M. P. Thomas, Closed ideals and biorthogonal systems in radical Banach algebras of power series, Proc. Edinburgh Math. Soc., 25 (1982), to appear.

[4] __________, Closed ideals of $\ell^1(\omega_n)$ when $\{\omega_n\}$ is star-shaped, Pacific J. Math., to appear.

Mathematics Department
California State College
Bakersfield, CA 93309

A CLASS OF UNICELLULAR SHIFTS WHICH CONTAINS NON-STRICTLY CYCLIC SHIFTS

Sandy Grabiner† and Marc P. Thomas

Let $\{\omega_n\}_0^\infty$ be a sequence of positive scalars. For each p, $1 \leq p < \infty$, $\ell^p(\omega_n)$ is the Banach space of formal power series $f = \Sigma_0^\infty \lambda_n z^n$ for which the norm

$$\|f\| \equiv \left[\sum_0^\infty |\lambda_n \omega_n|^p\right]^{1/p}$$

is finite. When $\{\omega_{n+1}/\omega_n\}_0^\infty$ is bounded, $\ell^p(\omega_n)$ is closed under multiplication by z, and multiplication by z is similar to the weighted shift on $\ell^p = \ell^p(1)$ with weights ω_{n+1}/ω_n. As many papers in this volume indicate, there has been renewed interest in the question whether multiplication by z is <u>unicellular</u>, that is, the only non-zero closed invariant subspaces are the spaces

$$\ell_k^p(\omega_n) \equiv \left\{\sum_0^\infty \lambda_n z^n \in \ell^p(\omega_n) : \lambda_0 = \lambda_1 = \cdots = \lambda_{k-1} = 0\right\}.$$

Particular interest has focused upon the question how the unicellularity of multiplication by z is related to the question whether $\ell^1(\omega_n)$ or $\ell^p(\omega_n)$ is an algebra. If $\ell^p(\omega_n)$ is an algebra, the weighted shift on ℓ^p which is similar to multiplication by z on $\ell^p(\omega_n)$ is said to be <u>strictly cyclic</u> [5, Proposition 32, p. 96]. We will study a class of spaces $\ell^p(\omega_n)$ for which multiplication by z is always unicellular (Theorem 1), but need not be strictly cyclic (Theorem 2). Theorem 2 provides an answer in all ℓ^p simultaneously to Shield's [5, Question 14, p. 100] and to the latter half of [5, Question 20, p. 106].

We say that the sequence $\{\omega_n\}_0^\infty$ is <u>quasi star-shaped</u> if there is a non-negative integer k and a positive constant c such that the sequence $\{(c\omega_{n-k})^{1/n}\}$ is eventually non-increasing. This implies that we can define a sequence $\{\overline{\omega}_n\}_0^\infty$ with $\overline{\omega}_0 = 1$, $\{(\overline{\omega}_n)^{1/n}\}_0^\infty$ non-increasing and $\overline{\omega}_n = c\omega_{n-k}$ for sufficiently large n. Hence $\overline{\omega}_{m+n} \leq \overline{\omega}_m \overline{\omega}_n$ [3, Theorem 7.2.4(i), p. 239], $\ell^1(\overline{\omega}_n)$ is an algebra, and all $\ell^p(\omega_n)$ are closed under multiplication by z.

THEOREM 1. If $\{\omega_n\}_0^\infty$ is quasi star-shaped and there is an $\varepsilon > 0$ for which $(\omega_n)^{1/n} = \underline{O}(1/n^\varepsilon)$, then multiplication by z is unicellular on each $\ell^p(\omega_n)$.

<u>Proof</u>. Let c, k, and $\{\overline{\omega}_n\}_0^\infty$ be as above. Then $\ell^1(\overline{\omega}_n)$ is a Banach algebra for which multiplication by z is unicellular [7, Theorem 3.5]. It is also clear that $z^k(\ell^1(\omega_n)) \subseteq \ell^1(\omega_n)$. Since $\{(\overline{\omega}_n)^{1/n}\}_0^\infty$ is non-increasing, it is easy to show

†This work was supported in part by NSF Grant MCS 8002923.

that $(\bar{\omega}_{n+1}/\bar{\omega}_n) \leq (\bar{\omega}_n)^{1/n}$ for all n, so that $(\omega_{n+1}/\omega_n) = \underline{O}(1/n^{\varepsilon})$. Let j be a positive integer for which $\varepsilon j \geq 1$. Then $(\omega_{n+j}/\omega_n) = \underline{O}(1/n)$ and hence $\{\omega_{n+j}/\omega_n\}_0^\infty$ belongs to ℓ^q where $\frac{1}{p} + \frac{1}{q} = 1$, $1 \leq p < \infty$. An easy calculation now shows that $z^j(\ell^p(\omega_n)) \subseteq \ell^1(\omega_n)$, hence

$$z^{j+k}(\ell^p(\omega_n)) \subseteq \ell^1(\bar{\omega}_n) \subseteq \ell^p(\omega_n).$$

Let L be a closed non-zero subspace of $\ell^p(\omega_n)$ with L invariant under multiplication by z. Then $L \cap \ell^1(\bar{\omega}_n)$ is a closed non-zero subspace of $\ell^1(\bar{\omega}_n)$ and is invariant under multiplication by z. Hence $L \cap \ell^1(\bar{\omega}_n) = \ell^1_m(\bar{\omega}_n)$ for some m. Therefore L contains z^m and hence $L \supseteq \ell^p_m(\omega_n)$. We now have the situation that whenever a series f belongs to L, a polynomial g with the same leading coefficients also belongs to L. Hence there is an integer i for which $L \subseteq \ell^p_i(\omega_n)$ and a polynomial $g = z^i(1+h)$ in L for which h has zero constant term. Then h belongs to the radical Banach algebra $\ell^1_1(\bar{\omega}_n)$, and $z^i = g(1+h)^{-1} \in L \cap \ell^1(\bar{\omega}_n) \subseteq \ell^p(\omega_n)$. Hence $L = \ell^p_i(\omega_n)$, completing the proof of the theorem.

Recall that multiplication by z on $\ell^p(\omega_n)$ is said to be a basis operator if $\ell^p(\omega_{n+k})$ is an algebra for all $k > 0$. This is equivalent to saying that the weighted shift on ℓ^p which is similar to multiplication by z on $\ell^p(\omega_n)$ is strongly strictly cyclic [5, p. 98]. To simplify the statement of the following theorem, we will consider a weighted shift T with a fixed sequence of weights as acting on all ℓ^p.

THEOREM 2. There exist weighted shifts T and U which are unicellular on all ℓ^p, $1 \leq p < \infty$, and for which:

(a) T is not strictly cyclic on any ℓ^p;

(b) U is strictly cyclic on all ℓ^p but is not strongly strictly cyclic on any ℓ^p.

Proof. We will prove the theorem by considering multiplication by z on appropriate weighted ℓ^p spaces. Recall that if $\ell^1(\omega_n)$ is not an algebra, then no $\ell^p(\omega_n)$ can be an algebra [2, Theorem 6.9, p. 38], [5, Proposition 32, p. 96]. By [6, Proposition 2.3 and Lemma 2.4] we can find a sequence $\{\omega_n\}_0^\infty$ of positive scalars for which $\omega_0 = 1$, $\{(\omega_n)^{1/n}\}_0^\infty$ is non-increasing, $(\omega_n)^{1/n} = \underline{O}(1/n)$, and $\ell^1(\omega_{n+1})$ is not an algebra. Let $\tilde{\omega}_n = \omega_{n+1}$, all n. Let T be the shift on ℓ^p with weights $\tilde{\omega}_{n+1}/\tilde{\omega}_n$. Then T is unicellular on all ℓ^p by Theorem 1. But T is not even strictly cyclic on ℓ^1 since $\ell^1(\tilde{\omega}_n)$ is not an algebra.

Now define $\{\bar{\omega}_n\}_0^\infty$ so that $\bar{\omega}_n = \omega_{n-2}$ for $n \geq 2$. Let U be the shift on ℓ^p with weights $\bar{\omega}_{n+1}/\bar{\omega}_n$. Since $\{(\bar{\omega}_n)^{1/n}\}_0^\infty$ is eventually non-increasing it follows from Theorem 1 that multiplication by z is unicellular on each $\ell^p(\overline{\omega}_n)$. It is clear from the definitions that f belongs to the space $\ell^1(\omega_n)$ if and only if fz^2 belongs to $\ell^1(\overline{\omega}_n)$. It follows, as in the proof of Theorem 1, that $z\ell^p(\overline{\omega}_n) \subseteq \ell^1(\overline{\omega}_n)$. Hence by [2, Theorem 3.4, p. 24] or remarks preceding [1, Lemma 3.15, p. 650] it follows that $\ell^p(\overline{\omega}_n)$ is an algebra for all p, but $\ell^1(\overline{\omega}_{n+3}) = \ell^1(\omega_{n+1})$ is not an algebra. Thus U is a unicellular strictly cyclic shift on all ℓ^p but is not strongly strictly cyclic on any ℓ^p. This completes the proof of the theorem.

It is easy to extend Theorem 1 to shifts with respect to a <u>normalized M-basis</u> (Markushevich Basis) $\{x_n, x_n^*\}$ of a Banach space X [2, Definition 2.1, p. 18] provided the basis is <u>bounded</u>, that is, $\{\|x_n^*\|\}$ is bounded. Note that every separable Banach space has a bounded normalized M-basis [4, Theorem 1.f.4, p. 44].

COROLLARY 3. Suppose that $\{x_n, x_n^*\}$ is a bounded normalized M-basis of the Banach space X, and that $\{\omega_n\}$ is quasi star-shaped. If $\Sigma_0^\infty (\omega_n)^{1/n}$ converges, then the weighted shift for the M-basis with weights ω_{n+1}/ω_n is unicellular.

<u>Proof</u>. A summable decreasing sequence is always $\underline{O}(1/n)$, so $\{\omega_n\}$ satisfies the hypotheses of Theorem 1. We can consider X as a Banach space of formal power series by letting $z^n = \omega_n x_n$ (cf. [2, Definition 2.5, pp. 19-20]). Since $\{\omega_{n+1}/\omega_n\}$ is summable $zX \subseteq \ell^1(\omega_n)$. The rest of the proof is the same as in Theorem 1.

Theorem 2 above suggests the following problem which is an extension of [5, Question 20, p. 106].

<u>Question</u>. Is there a sequence $\{\omega_n\}$ for which multiplication by z is unicellular on all $\ell^p(\omega_n)$, or at least on $\ell^1(\omega_n)$, but no $\ell^1(\omega_{n-k})$ is an algebra?

References

[1] S. Grabiner, A formal power series operational calculus for quasi-nilpotent operators, <u>Duke Math. J.</u>, 38 (1971), 641-658.

[2] __________, Weighted shifts and Banach algebras of power series, <u>Amer. J. Math.</u>, 97 (1975), 16-42.

[3] E. Hille and R. Phillips, <u>Functional Analysis and Semigroups</u>, AMS Colloquium Publication No. 31, 1957, 2nd rev. ed.

[4] J. Lindenstrauss and L. Tzafriri, <u>Classical Banach Spaces</u>, Vol. I, Springer-Verlag, Berlin, 1973.

[5] A. Shields, Weighted shifts operators and analytic function theory, <u>Topics in Operator Theory</u>, (ed. C. Pearcy) Amer. Math. Soc. Surveys, No. 13, Providence, RI, 1974.

[6] M. Thomas, Closed ideals of $\ell^1(\omega_n)$ when $\{\omega_n\}$ is star-shaped, to appear in Pacific J. Math., to appear.

[7] ___________, Approximation in the radical algebra $\ell^1(\omega_n)$ when $\{\omega_n\}$ is star-shaped, this Volume.

Department of Mathematics
Pomona College
Claremont, CA 91711

and

Department of Mathematics
California State College
Bakersfield, CA 93309

AN INEQUALITY INVOLVING PRODUCT MEASURES

G. R. Allan†

The inequality, proved in this note, arose in answer to two questions:

1. If K is a compact subset of $\mathbb{R}^3$ with volume V and if the orthogonal projections of K onto the three coordinate planes have areas A_1, A_2, A_3, then is it true that $V^2 \le A_1A_2A_3$?

2. If f is a non-negative function in $L^1[0,1]$ and we set $\omega(n) = \|f^{*n}\|_1$ ($n = 1,2,\ldots; f^{*n}$ denotes the nth convolution power of f, restricted to $[0,1]$), then is ω a 'star-shaped' weight, in the sense of Marc Thomas (see [1]), i.e. is it true that $\omega(n+1)^{1/(n+1)} \le \omega(n)^{1/n}$ $(n \ge 1)$?

In fact both these questions have affirmative answers, being special cases of the inequality to be proved. The first question was posed to me by J. B. Boyling (Leeds, England) in 1978, in connection with work in statistical physics; the second question arose in discussion at this conference.

Suppose now that $\mu_1,\ldots,\mu_n$ $(n \ge 2)$ are positive Borel measures, each defined on a copy of the real line $\mathbb{R}$; in the ith copy we take co-ordinate x_i. Let $\mu = \mu_1 \times \mu_2 \times \cdots \times \mu_n$ be the corresponding product Borel measure on $\mathbb{R}^n$; for $i = 1,\ldots,n$ let $\hat{\mu}_i = \mu_1 \times \cdots \times \hat{\mu}_i \times \cdots \times \mu_n$ be the $(n-1)$-dimensional measure on the co-ordinate hyperplane $x_i = 0$; similarly $\hat{\mu}_{ij}$ is the corresponding $(n-2)$-dimensional measure on $x_i = x_j = 0$ $(1 \le i, j \le n;\ i \ne j)$.

Note that, if B is a Borel subset of $\mathbb{R}^n$ and if $\hat{\pi}_i : \mathbb{R}^n \to \mathbb{R}^{n-1}$ is projection onto the hyperplane $x_i = 0$, then $\hat{\pi}_i(B)$ is analytic, hence universally measurable; in particular $\hat{\pi}_i(B)$ is $\hat{\mu}_i$-measurable, so that $\hat{\mu}_i(\hat{\pi}_i(B))$ is well defined, where $\hat{\mu}_i$ now denotes the completion of the Borel measure $\hat{\mu}_i$ as defined above.

THEOREM. Let B be a Borel subset of $\mathbb{R}^n$ $(n \ge 2)$ and let $V = \mu(B)$, $A_i = \hat{\mu}_i(\hat{\pi}_i(B))$, in the above notation. Then

$$V^{n-1} \le A_1A_2\ldots A_n\,.$$

[Note: on the right-hand side, we take $0\cdot\infty = 0$, where necessary.]

Proof. The proof is by induction on n. We may assume each $A_i > 0$, since otherwise B is contained in a product set of measure zero and so $V = 0$. We may also assume that $0 < A_i < \infty$, the case $A_i = \infty$ being trivial.

† Partially supported by NATO Grant No. RG 073.81.

For $n = 2$ the result is trivial, because then

$$B \subset \hat{\pi}_2(B) \times \hat{\pi}_1(B),$$

and so

$$\begin{aligned}\mu(B) &\le \mu(\hat{\pi}_2(B) \times \hat{\pi}_1(B)) \\ &= \mu_1(\hat{\pi}_2(B)) \cdot \mu_2(\hat{\pi}_1(B)) \\ &= \hat{\mu}_2(\hat{\pi}_2(B)) \cdot \hat{\mu}_1(\hat{\pi}_1(B)),\end{aligned}$$

i.e.,

$$V \le A_1 A_2 .$$

Let $n \ge 3$ and suppose that the result holds for smaller values of n.

For any $x \in \mathbb{R}$, let

$A(x) = \hat{\mu}_n$-measure of the section B_x of B through $x_n = x$ (projected onto the plane $x_n = 0$), i.e.

$$A(x) = \hat{\mu}_n\{(x_1,\dots,x_{n-1},0) : (x_1,\dots,x_{n-1},x) \in B\} .$$

Trivially,

$$A(x) \le A_n \quad (\text{all } x \in \mathbb{R}) . \tag{1}$$

For $i = 1,\dots,n-1$, let $B_i(x)$ be the $\hat{\mu}_{i,n}$-measure of the projection $\hat{\pi}_{i,n}(B_x)$ (into the 'plane' $x_i = x_n = 0$). From Fubini's theorem:

$$V = \int_{\mathbb{R}} A(x_n)\, \mu_n(dx_n) , \tag{2}$$

and

$$A_i = \int_{\mathbb{R}} B_i(x_n)\, \mu_n(dx_n) \quad (i = 1,\dots,n-1), \tag{3}$$

while the induction assumption applied to the section B_x (a Borel set) gives

$$A(x)^{n-2} \le B_1(x) \dots B_{n-1}(x) \quad (x \in \mathbb{R}) . \tag{4}$$

Thus, from (2):

$$\begin{aligned}V &= \int_{\mathbb{R}} A(x_n)^{1/(n-1)}\, A(x_n)^{(n-2)/(n-1)}\, \mu_n(dx_n) \\ &\le A_n^{1/(n-1)} \int_{\mathbb{R}} (B_1(x_n) \dots B_{n-1}(x_n))^{1/(n-1)}\, \mu_n(dx_n) \\ &\qquad\qquad\qquad\qquad \text{(by (1), (4))} \\ &\le A_n^{1/(n-1)} \left(\int_{\mathbb{R}} B_1(x_n)\mu_n(dx_n)\right)^{1/(n-1)} \dots \\ &\qquad\qquad \dots \left(\int_{\mathbb{R}} B_{n-1}(x_n)\mu_n(dx_n)\right)^{1/(n-1)}\end{aligned}$$

(by generalized Hölder's inequality)

$$= (A_1 \ldots A_n)^{1/(n-1)}, \quad \text{by (3)}.$$

Thus

$$V^{n-1} \leq A_1 \ldots A_n ,$$

which completes the proof.

For the questions at the beginning of the note, it is clear that Question 1 is answered by taking $n = 3$ and each of μ_1, μ_2, μ_3 to be 1-dimensional Lebesgue measure.

For Question 2, we take $\mu_1 = \mu_2 = \cdots = \mu_n = f(x)dx$ and

$$B = \{(x_1,\ldots,x_n) \in \mathbb{R}^n : x_1 + \cdots + x_n \leq 1,\ x_i \geq 0 \quad (i = 1,\ldots,n)\},$$

a compact set. Then

$$V = \mu(B) = \int_B \cdots \int f(x_1) \ldots f(x_n) dx_1 \ldots dx_n = \|f^{*n}\|_1 ,$$

while, for $i = 1,\ldots,n$,

$$A_i = \int_{\substack{x_1 + \cdots + \hat{x}_i + \cdots + x_n \leq 1,\\ x_j \geq 0}} \cdots \int f(x_1) \ldots \widehat{f(x_i)} \ldots f(x_n) dx_1 \ldots \widehat{dx_i} \ldots dx_n$$

$$= \|f^{*(n-1)}\|_1 .$$

Thus, the inequality $V^{n-1} \leq A_1 \ldots A_n$ is precisely $\|f^{*n}\|^{n-1} \leq \|f^{*(n-1)}\|^n$, i.e. $\omega(n)^{1/n} \leq \omega(n-1)^{1/(n-1)}$ $(n \geq 2)$.

Problem. Given a star-shaped weight ω on $\mathbb{Z}^+$, does there exist a positive function f in $L^1[0,1]$ such that $\omega(n) = \|f^{*n}\|_1$ $(n \geq 1)$?

Reference

[1] M. Thomas, Approximation in the radical algebra $\ell^1(\omega_n)$ when (ω_n) is star-shaped, this Volume.

Department of Pure Mathematics and Mathematical Statistics
16 Mill Lane
Cambridge CB2 1SB, U.K.

Note: Please see also the following article, [W.1].

THE NORMS OF POWERS OF FUNCTIONS IN THE VOLTERRA ALGEBRA

G. A. Willis

This note provides an example of a weight function (ω_n) with $\omega_{m+n} \leq \omega_m \omega_n$ and $\lim_n \omega_n^{1/n} = 0$, such that there is no function $f \in L^1(0,1)$ with $(\|f^{*n}\|)$ equivalent to (ω_n). It thus provides an answer to the problem in [1].

To begin with, recall from [2] that a weight (ω_n) is said to be <u>regulated at p</u> if $\lim_n \sup (\omega(p+n)/\omega(n)) = 0$. If (ω_n) is regulated at p, then it is regulated at q for every $q \geq p$. It is shown in [2] how to construct, for any $p > 1$, a weight which is regulated at p but not at $p-1$. For example, if we let ω_n be defined by putting $\omega(0) = \omega(1) = 1$ and $\omega(2n) = 2^{-n^2}$, $\omega(2n+1) = 2^{-(n^2+1)}$ for $n = 1,2,3,\ldots$, then (ω_n) is a radical weight function which is regulated at 2 but is not regulated at 1.

Now let R be a commutative Banach algebra. For each non-nilpotent $f \in R$ define a seminorm, $|\cdot|_f$, on R by

$$|a|_f = \lim_n \sup \frac{\|af^n\|}{\|f^n\|} \qquad (a \in R).$$

Then it is immediate that for every $a,b \in R$:

(i) $|a+b|_f \leq |a|_f + |b|_f$;

(ii) $|a|_f \leq \|a\|$; and

(iii) $|ab|_f \leq \|a\| \, |b|_f$.

Hence, if we define $J_f = \{a \in R \mid |a|_f = 0\}$, then J_f is a closed ideal in R.

LEMMA. Let R be a Banach algebra and f be an element of R with $f \in [f^2R]^-$. Then $(\|f^n\|)$ is either regulated at one or is not regulated at any p.

<u>Proof</u>. Suppose that $f^p \notin J_f$ for every $p = 1,2,3,\ldots$. Then

$$0 \neq |f^p|_f = \lim_n \sup \frac{\|f^{p+n}\|}{\|f^n\|} \quad (p = 1,2,3,\ldots).$$

Hence $(\|f^n\|)$ is not regulated at any p.

Suppose on the other hand that $f^p \in J_f$ for some p. Then, since J_f is a closed ideal in R, $[f^pR]^-$ is contained in J_f. Now an induction argument using the hypothesis that $f \in [f^2R]^-$ shows that $f \in [f^pR]^-$ for every p. Hence, $f \in J_f$ and so

$$0 = |f|_f = \limsup_n \frac{\|f^{n+1}\|}{\|f^n\|} .$$

That is, $(\|f^n\|)$ is regulated at one.

COROLLARY. Let (ω_n) be a weight function which for some $p > 1$ is regulated at p but not at $p-1$. Then there is no function $f \in L^1(0,1)$ such that $(\|f^{*n}\|)$ is equivalent to (ω_n).

Proof. The property of being regulated at p is preserved under equivalence of weights and so it will suffice to show that there is no $f \in L^1(0,1)$ such that $(\|f^{*n}\|)$ is equal to (ω_n).

Let f be in $L^1(0,1)$. If $\alpha(f) = \inf(\mathrm{supp}(f)) > 0$, then f is nilpotent and $(\|f^{*n}\|)$ is not equal to (ω_n). Hence we may suppose that $\alpha(f) = 0$. It follows, by Titchmarsh's convolution theorem, that $\alpha(f^{*2}) = 0$. Hence, by theorem 7.9 (i) of [3], $[f^{*2} * L^1(0,1)]^- = L^1(0,1)$ and so $f \in [f^{*2} * L^1(0,1)]^-$. Therefore, by the lemma, $(\|f^{*n}\|)$ is either regulated at one or is not regulated at any p and is thus not equal to (ω_n).

It is shown by G. R. Allan in [1] that if f is a positive function on $(0,1)$, then $(\|f^{*n}\|)$ is star-shaped, i.e., $(\|f^{*n}\|^{1/n})$ is monotonically decreasing. Hence, by [2] Cor. 2.8, $(\|f^{*n}\|)$ is regulated at one if f is positive. In view of this fact and the above corollary, it seems likely that $(\|f^{*n}\|)$ is regulated at one for every f in $L^1(0,1)$.

References

[1] G. R. Allan, An inequality involving product measures, this Volume.

[2] W. G. Bade and H. G. Dales, Norms and ideals in radical convolution algebras, J. Functional Analysis, 41 (1981), 77-109.

[3] H. G. Dales, Automatic continuity: a survey, Bull. London Math. Soc., 10 (1978), 129-183.

Mathematics Department
The University of New South Wales
P.O. Box 1
Kensington, New South Wales 2033
Australia

WEIGHTED CONVOLUTION ALGEBRAS AS ANALOGUES OF BANACH ALGEBRAS OF POWER SERIES

Sandy Grabiner†

By a <u>weight</u> we mean a positive Borel function $\omega(x)$ on $\mathbb{R}^+$ which is essentially bounded and essentially bounded away from 0 on compact subsets of $(0,\infty)$ and is also essentially bounded away from 0 at the origin. For $1 \le p < \infty$, $L^p(\omega)$ is the Banach space of (equivalence classes of) functions f in $L^1_{loc}(\mathbb{R}^+)$ for which the norm

$$\|f\| = \|f\|_p = \|f\|_{p,\omega} = \left(\int_0^\infty |f\omega|^p \right)^{1/p}$$

is finite. In this paper, we will describe various results about $L^p(\omega)$ under the convolution product $f * g(x) = \int_0^x f(x-t)g(t)dt$.

Almost everyone who has studied $L^p(\omega)$ under convolution has noticed analogies with the space $\ell^p(\omega)$ given by a <u>discrete weight</u>, that is, a sequence $\{\omega_n\}_0^\infty$ of positive numbers. It is usually convenient to represent $\ell^p(\omega)$ as the Banach space of formal power series $f = \sum_0^\infty \lambda_n z^n$ with norm $\|f\|_{p,\omega} = [\sum_0^\infty |\lambda_n \omega_n|^p]^{1/p}$ finite. Then convolution multiplication just becomes formal power series multiplication.

For sufficiently well-behaved discrete weights, the structure of $\ell^p(\omega)$ and similar spaces of power series has been well understood for many years [6], [7], [8], [14, Section 3.2, pp. 188-204], and known results about $\ell^p(\omega)$ have served as a source of fruitful conjectures about $L^p(\omega)$. In this paper we study questions of the form: when is $L^p(\omega) * L^q(\omega) \subseteq L^r(\omega)$? And, in particular, when is $L^1(\omega)$ or $L^p(\omega)$ an algebra? In answering those questions, not only results about $\ell^p(\omega)$ but even some proof techniques have been useful.

1. $L^1(\omega)$ as an algebra

In this section we discuss necessary and sufficient conditions, and also convenient sufficient conditions, for $L^1(\omega)$ to be an algebra. We say that the weight $\omega(x)$ is an <u>algebra weight</u> if there is a positive number M for which $\omega(x+y) \le M\,\omega(x)\omega(y)$ for almost every (x,y) in $\mathbb{R}^+ \times \mathbb{R}^+$. Analogously, $\{\omega_n\}$ is a <u>discrete algebra weight</u> if $\omega_{n+m}/\omega_n\omega_m$ is bounded. It is easy to see that $L^1(\omega)$ is an algebra when $\omega(x)$ is an algebra weight. An easy calculation shows that

$$|f*g|(x)\omega(x) \le M(|f\omega| * |g\omega|)(x) \qquad (1)$$

---

†Research partly supported by NSF Grant No. MCS-8002923.

for almost every x. Formula (1) implies not only that $L^1(\omega)$ is an algebra, but also that all the usual convolution formulas for unweighted L^p-spaces [11, p. 298], [19, pp. 37-38] hold for $L^p(\omega)$. The discrete analogue of formula (1) is even easier to prove and gives analogous results for $\ell^p(\omega)$. We thus have implications in one direction for each of the following theorems.

THEOREM 1. Suppose that $\omega(x)$ is a weight and that $1 \leq p < \infty$. Then the following are equivalent:

(A) $\omega(x)$ is an algebra weight;

(B) $L^1(\omega)$ is an algebra;

(C) $L^1(\omega) * L^p(\omega) \subseteq L^p(\omega)$.

THEOREM 1'. Suppose that $\{\omega_n\}$ is a discrete weight and that $1 \leq p < \infty$. Then the following are equivalent:

(A) $\{\omega_n\}$ is a discrete algebra weight;

(B) $\ell^1(\omega)$ is an algebra;

(C) $\ell^1(\omega) * \ell^p(\omega) \subseteq \ell^p(\omega)$.

In each case we need only prove that (C) implies (A). In the discrete case this is quite easy. When (C) holds, it follows from the closed graph theorem that power series multiplication is separately continuous from $\ell^1(\omega) \times \ell^p(\omega)$ to $\ell^p(\omega)$, and hence is a bounded bilinear operator [16, Th. 2.17, p. 51]. If M is the bound of this bilinear operator, then

$$\omega_{n+m} = \|z^{n+m}\|_p \leq M\|z^n\|_1 \; \|z^m\|_p = M \, \omega_n \omega_m ,$$

for all m and n. In the non-discrete case, it is still true that convolution, when defined, is a bounded bilinear operator. But the analogue of z^n is the point mass δ_a for a in $\mathbb{R}^+$; but the point masses are not functions and cannot belong to $L^p(\omega)$. To prove that (C) implies (A) in Theorem 1, we need to "represent" point masses by translates of approximate identities and these approximate identities need to be chosen carefully enough to give almost everywhere convergence rather than just the usual norm convergence. Since Theorem 1 is given in [10, Th. (2.1)], we prove the following more general result, which reduces to Theorem 1 when p and/or q equals 1.

THEOREM 2. Suppose that $\omega(x)$ is a weight and that p, q and r in $[1,\infty)$ satisfy $\frac{1}{p} + \frac{1}{q} = 1 + \frac{1}{r}$. Then $\omega(x)$ is an algebra weight if and only if $L^p(\omega) * L^q(\omega) \subseteq L^r(\omega)$.

In the proof, we will need the following definition (cf. [10, Def. (2.3)] or [19, p. 65]).

DEFINITION. The number $a \geq 0$ is a <u>right Lebesgue point</u> of the locally integrable function f if

$$\lim_{h\to 0^+} \frac{1}{h}\int_0^h |f(a+x) - f(a)|\,dx = 0 .$$

Almost every point is a right Lebesgue point of f [19, p. 65], as is every point of right continuity.

<u>Proof of Theorem 2</u>. That $L^p(\omega) * L^q(\omega) \subseteq L^r(\omega)$ when $\omega(x)$ is an algebra weight follows from formula (1) and [11, Th. (20.18), p. 296].

Suppose therefore that $L^p(\omega) * L^q(\omega) \subseteq L^r(\omega)$. As indicated above, there is an $M > 0$ for which $\|f * g\|_r \leq M\|f\|_p\|g\|_q$ for all f in $L^p(\omega)$ and g in $L^q(\omega)$. Since almost all points are right Lebesgue points of a fixed function, it follows from the Fubini theorem that we need only prove that there is a constant C for which $\omega(a+b) \leq CM\,\omega(a)\omega(b)$ whenever a is a right Lebesgue point of $\omega(x)^p$, b is a right Lebesgue point of $\omega(x)^q$, and $a+b$ is a right Lebesgue point of $\omega(x)^r$.

For each $h > 0$ let φ_h be the characteristic function of $[0,h)$ and let $K_h(x) = \varphi_h * \varphi_h(x) = (h - |x-h|)^+$. Then for each h we have

$$\frac{\|\delta_{a+b} * K_h\|_r}{h^{1+1/r}} \leq M \frac{\|\delta_a * \varphi_h\|_p}{h^{1/p}} \; \frac{\|\delta_b * \varphi_h\|_q}{h^{1/q}} .$$

It follows from the definition of right Lebesgue point that $\lim_{h\to 0^+} \|\delta_a * \varphi_h\|_p / h^{1/p} = \omega(a)$, and similarly for $\omega(b)$. Also notice that $\int_0^\infty (K_h(x))^r dx = \int_0^{2h} (K_h(x))^r dx = 2h^{r+1}/(r+1)$, and that $0 \leq K_h(x)^r \leq h^r$ for all x. Therefore as in the proof of [10, Lemma (2.4)], it follows from the fact that $a+b$ is a right Lebesgue point of $\omega(x)^r$ that

$$\lim_{h\to 0^+} \frac{\|\delta_{a+b} * K_h\|_r^r}{h^{r+1}} = \frac{2}{r+1}\,\omega(a+b)^r .$$

This completes the proof of Theorem 2, and also of Theorem 1.

In a technical sense Theorems 1 and 1' completely solve the problem of when $L^1(\omega)$ or $\ell^1(\omega)$ is an algebra. In practice, however, it is often difficult to determine directly when a weight is an algebra weight; moreover, additional conditions on $\omega(x)$ and $\{\omega_n\}$ are often needed to describe the structure of $L^1(\omega)$ or $\ell^1(\omega)$ in detail. Two conditions which are sufficient for $\omega(x)$ to be an algebra weight have been known for a long time [12, p. 239]. $\omega(x)$ is an algebra weight if there is a $C > 0$ for which $(C\omega(x))^{1/x}$ is eventually non-increasing or if $\omega(x)$ is logarithmically concave (or equivalently, [15, Th. C, p. 215], if $\omega(x+a)/\omega(x)$ is non-increasing for each $a > 0$). The weights considered by Allan

[1] and Domar [3], [4] are logarithmically concave; and discrete weights $\{\omega_n\}$ with $\omega_n^{1/n}$ non-increasing have been extensively studied by Thomas [17], [18]. The following two theorems give conditions which we have found useful and which are sufficient for $L^1(\omega)$ and $\ell^1(\omega)$, respectively, to be algebras.

THEOREM 3. Suppose that $\omega(x)$ is a weight satisfying:

(i) There is an $a > 0$ for which $\omega(x+a)/\omega(x)$ is eventually non-increasing;

(ii) $\omega(x+y)/\omega(x)\omega(y)$ is essentially bounded for x and y in some neighborhood of the origin;

Then, $\omega(x)$ is an algebra weight.

THEOREM 3'. The discrete weight $\{\omega_n\}$ is a discrete algebra weight if there is an integer k for which ω_{n+k}/ω_n is eventually non-increasing.

We omit the proofs, which can be found in [10, Th. (2.2)] and [6, Th. (2.10), p. 645], respectively.

In studying $L^p(\omega)$, one is of course primarily interested in the spaces and not the weights. Thus it is often convenient to replace $\omega(x)$ with an <u>equivalent weight</u> $\overline{\omega}(x)$, that is, a weight $\overline{\omega}(x)$ with $\omega(x)/\overline{\omega}(x)$ and $\overline{\omega}(x)/\omega(x)$ both essentially bounded; so that $L^p(\omega) = L^p(\overline{\omega})$ for all p. The following special case of [10, Th. (3.1)] (compare [2, Lemma 1.1]) is adequate for many purposes. We omit the proof.

THEOREM 4. Suppose that $\omega(x)$ is an algebra weight, and that there is a positive number C for which $\text{ess. sup}(\omega(x+a)/\omega(x)) \leq C$ for all sufficiently small a. Then there is a weight $\omega(x)$ equivalent to $\overline{\omega}(x)$ and satisfying:

(A) $\overline{\omega}(x)$ is right continuous;

(B) $\overline{\omega}(x+y) \leq \overline{\omega}(x)\overline{\omega}(y)$ for all x and y;

(C) If $\omega(x)$ is bounded, then $\overline{\omega}(x)$ is decreasing.

When $\omega(x)$ is an algebra weight, $\omega(x+a)/\omega(x) \leq M\,\omega(a)$, so the added assumption on $\omega(x)$ above is satisfied, in particular, when $\omega(x)$ is bounded near the origin. This condition is precisely equivalent to the right translation semigroup, which is given by convolution with $\{\delta_t\}_{t\geq 0}$, being strongly continuous at $t = 0$ [2, Lemma 1.6] [10, Lemma (2.8)c)].

2. $L^p(\omega)$ as an algebra

In this section we prove the following theorem and examine some of its consequences.

THEOREM 5. Suppose that $\omega(x)$ is a weight satisfying hypotheses (i) and (ii) of Theorem 3. If there is a $b > 0$ for which $\omega(x+b)/\omega(x)$ belongs to $L^q(\mathbb{R}^+)$ for some $1 \leq q < \infty$, then all $L^p(\omega)$ are algebras.

In Section 5 of [10], there are a number of convolution formulas for various $L^p(\omega)$, of which Theorem 5 is essentially [10, Th. (5.1)]. By concentrating on proving only that $L^p(\omega)$ is an algebra, the proof is somewhat simplified, and the hypothesis is weakened.

In many applications [3], [4], growth conditions are more naturally given for $\omega(x)^{1/x}$ than for $\omega(x+b)/\omega(x)$. The following simple proposition shows that Theorem 5 applies in this case.

PROPOSITION 6. Suppose that $\omega(x)$ is a weight which is bounded away from 0 on compact subsets of $[0,\infty)$. If $\omega(x+a)/\omega(x)$ is eventually non-increasing, then $\omega(x+a)/\omega(x) = O(\omega(x)^{a/x})$.

Proof of Proposition 6. Suppose $\omega(x+a)/\omega(x)$ is non-increasing for $x+a \geq d$. Since $\omega(x)$ is bounded below on $[0,d]$ there is a $C > 0$ for which $\omega(x_0)^a \geq C[\omega(x)/\omega(x-a)]^{x_0}$ whenever $x \geq a \geq x_0$. Suppose $x \geq a$ and let $x = x_0 + na$ with $x_0 < d \leq x_0 + a$. Then

$$\omega(x)^a = \left[\frac{\omega(x)}{\omega(x-a)} \frac{\omega(x-a)}{\omega(x-2a)} \cdots \frac{\omega(x_0+a)}{\omega(x_0)}\right]^a \omega(x_0)^a$$

$$\geq C\left[\frac{\omega(x)}{\omega(x-a)}\right]^{na} \left[\frac{\omega(x)}{\omega(x-a)}\right]^{x_0} = C\left[\frac{\omega(x)}{\omega(x-a)}\right]^x .$$

This completes the proof.

In both the discrete and nondiscrete cases we prove that spaces of power series or spaces of locally integrable functions are algebras by considering spaces of left translates [6, Def. (2.6), p. 644], [9, Formula (6.6), p. 37], [10, Def. (4.1) and Formula (4.2)]. For a discrete weight $\{\omega_n\}$ and a non-negative integer k, we define

$$S_{-k}(\ell^p(\omega_n)) = \{f \in \mathbf{C}[[z]] : fz^k \in \ell^p(\omega_n)\} = \ell^p(\omega_{n+k}).$$

For a continuous weight $\omega(x)$ and an $a \geq 0$, we define

$$\mathbf{s}^{-a}(L^p(\omega)) = \{f \in L^1_{loc}(\mathbb{R}^+) : \delta_a * f \in L^p(\omega(x))\} = L^p(\omega(x+a)).$$

Now we are ready to prove Theorem 5.

Proof of Theorem 5. Since $\omega(x+c)/\omega(x)$ is always essentially bounded, we can replace b by a multiple of b, if necessary, and assume that $\omega(x+b)/\omega(x)$ belongs to $L^{p'}(\mathbb{R}^+)$, where p' is the Holder conjugate of p. Then

$L^p(\omega) \subseteq \mathcal{S}^{-b}(L^1(\omega))$ [10, Lemma (5.3)]. Let $\omega_1(x) = \omega(x+b)$. Then $\omega_1(x)$ satisfies hypotheses (i) and (ii) of Theorem 3, so that $\omega_1(x)$ is an algebra weight.

Now suppose that f and g belong to $L^p(\omega)$ and let $g = g_1 + \delta_b * g_2$ where $g_1 \in L^p(\omega)$ has bounded support. Since g_1 has bounded support, it belongs to $L^1(\omega)$, so that $f * g_1 \in L^p(\omega)$, by Theorem 1 (C). Also

$$g_2 \in \mathcal{S}^{-b}(L^p(\omega)) = L^p(\omega_1) \quad \text{and} \quad f \in L^p(\omega) \subseteq \mathcal{S}^{-b}(L^1(\omega)) = L^1(\omega_1).$$

Since ω_1 is an algebra weight, Theorem 1 (C) implies that $f * g_2 \in \mathcal{S}^{-b}(L^p(\omega))$ and hence that $f * (\delta_b * g_2) \in L^p(\omega)$. So $f * g = f * g_1 + f * \delta_b * g_2$ belongs to $L^p(\omega)$. This completes the proof.

The above proof doesn't really use the monotonicity of $\omega(x+a)/\omega(x)$ but only the fact that $\mathcal{S}^{-b}(L^1(\omega))$ is an algebra, so we have proved (cf [10, Th. (5.2)]):

THEOREM 7. Suppose that $\omega(x)$ is an algebra weight. If $\mathcal{S}^{-b}(L^1(\omega))$ is an algebra and if $L^p(\omega) \subseteq \mathcal{S}^{-b}(L^1(\omega))$, then $L^p(\omega)$ is an algebra.

The following discrete analogue is proved in essentially the same way (cf. [6, Lemma (3.6), p. 648]). But by [6, Lemma (3.15), p. 650] we don't need to assume that $\{\omega_n\}$ is an algebra weight.

THEOREM 7'. Suppose that $\{\omega_n\}$ is a discrete weight. If $S_{-k}(\ell^1(\omega_n))$ is an algebra and $\ell^p(\omega_n) \subseteq S_{-k}(\ell^1(\omega_n))$, then $\ell^p(\omega_n)$ is an algebra.

3. Questions suggested by power series analogies

In this paper and [10] we showed that $L^1(\omega)$ and $L^p(\omega)$ are algebras under analogous hypotheses to those which imply that $\ell^1(\omega)$ and $\ell^p(\omega)$ are algebras. In [3], [4] Domar showed that under sufficiently strong hypotheses on $\omega(x)$, the closed ideals on $L^1(\omega)$ and $L^p(\omega)$ are exactly what one would expect from analogies with $\ell^1(\omega)$ and $\ell^p(\omega)$. In [5] Ghahramani decribed all the derivations of $L^1(\omega)$ under analogous hypotheses to those used in describing the derivation of Banach algebras of power series [8, Section 4]; in fact even earlier the derivations were described for the convolution algebra $L^1[0,1]$ by Kamowitz and Scheinberg [13]. In this section, we will draw attention to two areas in which our knowledge of Banach algebras of power series like $\ell^1(\omega)$ is essentially complete, when $\{\omega_n\}$ is sufficiently well-behaved, but our knowledge about convolution algebras like $L^1(\omega)$, or even $L^1[0,1]$ is minimal.

Under appropriate hypotheses on $\{\omega_n\}$, one can give explicit descriptions of all the automorphisms of Banach algebras of power series like $\ell^1(\omega_n)$ [8, pp. 30-32, 47]. In particular one can show that if $f = \Sigma_1^\infty \lambda_n z_n$ and $g = \Sigma_1^\infty \mu_n z^n$ belong to

$\ell^p(\omega_n)$ and if $\lambda_1 \neq 0$, then there is an automorphism T of $\ell^p(\omega_n)$ with $Tf = g$ if and only if $|\lambda_1| = |\mu_1|$. This suggests the following:

Question 1. If $\omega(x)$ is a sufficiently well-behaved radical algebra weight, characterize those f and g in $L^1(\omega)$ with $\alpha(f) = \alpha(g) = 0$ for which there is an automorphism T of $L^1(\omega)$ with $Tf = g$. In particular, what is the orbit of $u(x) \equiv 1$ under the automorphism group of $L^1(\omega)$? Can one answer these questions at least for $L^1[0,1]$?

To have enough automorphisms one will presumably need to assume that $x\omega(x+a)/\omega(x)$ is bounded for some or all a. Also one will presumably need to assume that all $S^{-a}(L^1(\omega))$ are algebras or even that $\omega(x)$ is logarithmically concave.

Even a partial answer to Question 1 will have interesting applications similar to those given by the characterization of automorphisms for Banach algebras of power series. For instance, if $Tu = g$ then g generates $L^1(\omega)$ as an algebra (cf. [8, pp. 41-42]), and convolution by g is similar to the Volterra integral operator on $L^1(\omega)$ (cf. [8, Th. (12.11), p. 92]).

In [9] we showed that every separable Banach space can be turned into a Banach algebra of power series all of whose closed ideals are standard. This suggests:

Question 2. Which separable Banach spaces can be continuously imbedded as a subalgebra of $L^1_{loc}(\mathbb{R}^+)$ generated as an algebra by $u(x) \equiv 1$? Can this be done so that the algebra has only standard closed ideals (so that convolution by $u(x)$ is unicellular)?

References

[1] G. R. Allan, Ideals of rapidly growing functions, Proceedings International Symposium on Functional Analysis and its Applications, Ibadan, Nigeria (1977).

[2] W. G. Bade and H. G. Dales, Norms and ideals in radical convolution algebras, J. Functional Analysis, 41 (1981), 77-109.

[3] Y. Domar, Extensions of the Titchmarsh convolution theorem with applications in the theory of invariant subspaces, Proc. London Math. Soc., to appear.

[4] __________, A solution of the translation-invariant subspace problem for weighted L^1 on $\mathbb{R}$ and $\mathbb{R}^+$, this Volume.

[5] F. Ghahramani, Homomorphisms and derivations on weighted convolution algebras, J. London Math. Soc., 21 (1980), 149-161.

[6] S. Grabiner, A formal power series operational calculus for quasinilpotent operators, Duke Math. J., 38 (1971), 641-658.

[7] __________, A formal power series operational calculus for quasinilpotent operators, II, J. Math. Anal. Appl., 43 (1973), 170-192.

[8] ___________, Derivations and automorphisms of Banach algebras of power series, Mem. Amer. Math. Soc., 146 (1974).

[9] ___________, Weighted shifts and Banach algebras of power series, Amer. J. Math., 97 (1975), 16-42.

[10] ___________, Weighted convolution algebras on the half line, J. Math. Anal. Appl., 83 (1981), 531-553.

[11] E. Hewitt and K. A. Ross, Abstract Harmonic Analysis, I, Springer-Verlag, Berlin, 1963.

[12] E. Hille and R. S. Phillips, Functional Analysis and Semi-groups, American Mathematical Society, Providence, R.I., 1957.

[13] H. Kamowitz and S. Scheinberg, Derivations and automorphisms of $L^1(0,1)$, Trans. Amer. Math. Soc., 135(1969), 415-427.

[14] N. K. Nikolskii, Selected problems of weighted approximation and spectral analysis, Proc. Steklov Inst. Math., 120 (1974).

[15] A. W. Roberts and D. E. Varberg, Convex Functions, Academic Press, New York, 1973.

[16] W. Rudin, Functional Analysis, McGraw-Hill, New York, 1973.

[17] M. P. Thomas, Closed ideals of $\ell^1(\omega_n)$ when $\{\omega_n\}$ is star-shaped, Pacific J. Math., to appear.

[18] ___________, Approximation in the radical algebra $\ell^1(\omega_n)$ when $\{\omega_n\}$ is star-shaped, this Volume.

[19] A. Zygmund, Trigonometric Series, Volume 1, Cambridge University Press, Cambridge, England, 1959.

Department of Mathematics
Pomona College
Claremont, CA 91711

COMMUTATIVE BANACH ALGEBRAS WITH POWER-SERIES GENERATORS

G. R. Allan[†]

The subject of this paper arose from the question, much discussed at the conference, of the existence, or non-existence, of a non-standard closed ideal I in some algebra $\ell^1(\omega)$, with ω a radical weight. The suggested approach is to study properties of the quotient algebra $\ell^1(\omega)/I$. It is, of course, conceivable that no such non-standard closed ideal exists; in that case there might be a proof by reductio ad absurdum from consideration of the quotient. However, it seems most probable that non-standard ideals exist in abundance. In any case, if we do not restrict attention to radical weights, then non-standard closed ideals certainly exist.

Recall that, if ω is a sub-multiplicative weight function on $\mathbb{Z}^+$, then the algebra $\ell^1(\omega)$ is semi-simple if $\lim_{n\to\infty} \omega(n)^{1/n} > 0$ and is radical (with an identity adjoined) if $\lim_{n\to\infty} \omega(n)^{1/n} = 0$. We shall write $\ell_0^1(\omega)$ for the maximal ideal of $\ell^1(\omega)$ consisting of all sequences $x = (x_n)_{n\geq 0}$ with $x_0 = 0$. In the 'radical case', $\ell_0^1(\omega)$ is the radical of $\ell^1(\omega)$. We shall normally identify the sequence (x_n) with the formal power series $\sum_{n\geq 0} x_n X^n$.

Let A be a (commutative) complex Banach algebra, not necessarily with identity. We say that the element x of A is a <u>power-series generator</u> (p.s.g.) for A, if and only if every element y of A can be written in the form

$$y = \sum_{n\geq 0} \alpha_n x^n,$$

where $(\alpha_n)_{n\geq 0}$ is a complex sequence with $\sum |\alpha_n| \|x^n\| < \infty$; of course $\alpha_0 \neq 0$ is allowed only if A has an identity.

A p.s.g. x for A is, of course, also a generator in the ordinary sense of Banach algebra theory; in particular the maximal ideal space Φ_A is homeomorphic to $Sp_A(x)$ (or to $Sp_A(x)\setminus\{0\}$ in case A has no identity) and A is necessarily separable and commutative.

Clearly, if A has a p.s.g. and if I is a closed ideal of A, then A/I has a p.s.g. Also, for any weight ω, the algebras $\ell^1(\omega)$, $\ell_0^1(\omega)$ have the obvious p.s.g. $X = (0,1,0,0,\ldots)$; thus any quotient algebra of $\ell^1(\omega), \ell_0^1(\omega)$ also has a p.s.g.

But also, if A has a p.s.g. x and we define $\omega(n) = \|x^n\|$ $(n \geq 0)$ then ω is a sub-multiplicative weight on $\mathbb{Z}^+$ and the mapping $\theta : \ell^1(\omega) \to A$ (or

---

[†]Partially supported by NATO Grant No. RG 073.81.

$\ell^1_0(\omega) \to A$ if A has no identity), defined by

$$\theta\left(\sum_{n\geq 0} \lambda_n x^n\right) = \sum_{n\geq 0} \lambda_n x^n,$$

is a continuous epimorphism. Thus, the class of Banach algebras with a p.s.g. is precisely the class of all quotient algebras (trivial or otherwise) of algebras $\ell^1(\omega)$, $\ell^1_0(\omega)$.

THEOREM 1. Let A be a commutative Banach algebra having a p.s.g. x. Then:

either $A \cong \ell^1(\omega)$, $\ell^1_0(\omega)$ for some weight ω,

or Φ_A is totally disconnected.

[If ω is not a radical weight, then these possibilities are mutually exclusive, since the maximal ideal space of $\ell^1(\omega)$ is then a disc.]

Proof. Suppose that $\sum_{n\geq 0} \lambda_n x^n = 0$ in A, where (λ_n) is a complex sequence such that $\sum|\lambda_n|\,\|x^n\| < \infty$. Then, if $r(x)$ is the spectral radius of x in A, $\sum|\lambda_n|r(x)^n < \infty$, so that the power series $\sum_{n\geq 0} \lambda_n z^n$ converges absolutely on the closed disc $\Delta(0,r(x))$, while the sum function vanishes on $Sp_A(x)$. Hence if Φ_A, and hence also $Sp_A(x) \cong \Phi_A$ (or $\Phi_A \cup (0)$), is not totally disconnected, then we deduce $\lambda_n = 0$ all n.

Thus, setting $\omega(n) = \|x^n\|$, the continuous epimorphism $\theta : \ell^1(\omega) \to A$ (or $\ell^1_0(\omega) \to A$), defined by $\theta(\sum\mu_n X^n) = \sum\mu_n x^n$, is an isomorphism.

COROLLARY 1. Let A be a commutative Banach algebra with Φ_A not totally disconnected. If A has a p.s.g., then A is semi-simple and Φ_A is homeomorphic to a disc (or a punctured disc if A has no identity).

COROLLARY 2. Let A be a non-radical Banach algebra without a 1 having a p.s.g. x. The following are equivalent:

(i) x has finite closed descent;
(ii) A contains a non-zero element of finite closed descent;
(iii) Φ_A is totally disconnected.

Proof. (i) $\Rightarrow$ (ii): is trivial;

(ii) $\Rightarrow$ (iii): follows from Theorem 1, since no non-zero element of $\ell^1_0(\omega)$ has finite closed descent, so that $A \not\cong \ell^1_0(\omega)$ (for all weights ω);

(iii) $\Rightarrow$ (i): for then, since A is non-radical, we cannot have $A \cong \ell^1_0(\omega)$, since the maximal ideal space of $\ell^1_0(\omega)$ is not totally disconnected. Thus A is a proper quotient of $\ell^1_0(\omega)$, $\omega(n) = \|x^n\|$ $(n \geq 1)$. But then, as in the proof of

Theorem 1, x satisfies a non-trivial equation $\sum_{n\geq 1} \lambda_n x^n = 0$, where $\sum |\lambda_n| \|x^n\| < \infty$.

If k is the first non-zero coefficient we easily deduce $\overline{Ax^k} = \overline{Ax^{k+1}}$, so that x has finite closed descent.

COROLLARY 3. Let A be a uniform algebra on a compact space X that is not totally disconnected. Then A does not have a p.s.g.

Proof. Since X is homeomorphically embedded in Φ_A, then Φ_A is not totally disconnected. The result now follows from Theorem 1, since no algebra $\ell^1(\omega)$ is isomorphic to a uniform algebra.

We are indebted to Garth Dales for pointing out an improvement incorporated in the following result.

PROPOSITION. Let X be a compact Hausdorff space. Then $C(X)$ has a power series generator if and only if X is a totally disconnected metric space.

Proof. If $C(X)$ has a p.s.g., then X is totally disconnected by Corollary 3 and X is metrizable, since $C(X)$ is separable.

Conversely, if X is a compact, totally disconnected metric space, then X is homeomorphic to a closed subset of the Cantor set K ([3], Corollary 2.99, page 100). But K may be realized homemorphically as a Helson set in the circle $\mathbb{T}$, which is then also an interpolation set for the algebra $A^+(\mathbb{T})$ of absolutely convergent Taylor series ([4], Chapitre IV, §7). Hence also X is homeomorphic to an interpolation set for $A^+(\mathbb{T})$ and $C(X)$ is isomorphic to a quotient of $A^+(\mathbb{T})$. The latter algebra certainly has a p.s.g. and so, therefore, does $C(X)$.

We note however, in contrast, that if M is a maximal ideal in the Shilov boundary of $A^+(\mathbb{T})$, then M does not have a p.s.g., for, it is easily seen that M has an element of finite closed descent, yet Φ_M is certainly not totally disconnected.

We now give a result for the radical case. It is just a sample, to show that the assumption of the existence of a p.s.g. can have interesting consequences. We hope to produce a more systematic study in the near future. Recall [1] that a weight ω on $\mathbb{Z}^+$ is called regulated at k (for some $k = 1,2,\ldots$) if $\omega(n+k)/\omega(n) \to 0$ as $n \to \infty$. It is shown in [1] that ω is regulated at k if and only if multiplication by X^k is a compact operator on $\ell^1(\omega)$.

THEOREM 2. Let R be a non-zero commutative radical Banach algebra, having a p.s.g. x, with x not a zero-divisor. Suppose that the sequence $\omega(n) = \|x^n\|$ $(n \geq 1)$ is regulated at some $k(\geq 1)$. Then:

(i) the map $y \mapsto x^k y \ (y \in R)$ is compact;

(ii) R is a dual space and multiplication in R is separately weak $*$-continuous;

(iii) R does not have a bounded approximate identity;

(iv) $x^n/\|x^n\| \to 0$ (weak $*$); in particular $\{x^n/\|x^n\| : n = 1,2,\ldots\}$ is a discrete set in the norm topology.

<u>Proof</u>. Set $\omega(n) = \|x^n\|$, $A = \ell_0^1(\omega)$ and define $\theta : A \to R$ by $\theta(\sum_{n\geq 1}\lambda_n X^n) = \sum_{n\geq 1} \lambda_n x^n$, so that θ is a continuous epimorphism.

(i) By the result of [1] mentioned above, $T_k : f(X) \mapsto X^k f(X)$ is a compact mapping $A \to A$.

Since $R \cong A/\ker\theta$ and T_k induces the mapping $y \mapsto x^k y$ on R, it follows that this latter mapping is also compact.

(ii) For $\varphi \in R^*$, define $\varphi_k \in R^*$ by $\varphi_k(y) = \varphi(x^k y) \ (y \in R)$; let

$$L = \{\varphi_k : \varphi \in R^*\}.$$

Since x is not a zero-divisor, L separates the points of R, so that the weak topology $\sigma(R,L)$ is Hausdorff. Now, for any $\varphi_k \in L$, $\theta^*\varphi_k \in A^*$ is given by the $\ell_0^\infty(\omega^{-1})$-sequence $(\varphi(x^{n+k}))_{n\geq 1}$; but $|\varphi(x^{n+k})| \leq \|\varphi\|\|x^{n+k}\| = o(\|x^n\|)$, since ω is regulated at k. Thus $\theta^*\varphi_k$ is given by a $c_0(\omega^{-1})$-sequence.

It follows that θ is $\sigma(A, c_0(\omega^{-1})) \to \sigma(R,L)$ continuous; since $\sigma(R,L)$ is Hausdorff, $\ker\theta$ is weak $*$-closed in A, so that $R \cong A/\ker\theta$ is a dual space and θ is weak $*$-continuous.

We next note that multiplication in $\ell^1(\omega)$ is separately weak $*$-continuous ([2]). Now let $f \in R_*$ (pre-dual of R) and let $z \in R$; define $f_z \in R^*$ by $f_z(y) = f(yz) \ (y \in R)$. We shall show that $f_z \in R_*$, which will imply the separate weak $*$-continuity of multiplication in R. For that, we show $\ker(f_z)$ weak $*$-closed; by a well-known result, it suffices to show $B \cap \ker(f_z)$ to be weak $*$-closed, where B is the closed unit ball of R. Thus, let (y_α) be a net in $B \cap \ker(f_z)$, $y_\alpha \to y$ (weak $*$). Since θ is norm-continuous, we can find (h_α) in A with $\theta(h_\alpha) = y_\alpha$, $\|h_\alpha\| \leq 2$ (all α). Then (h_α) has a weak $*$-convergent subnet (h_β) with limit h, say. But θ is weak $*$-continuous, so that $\theta(h) = y$. Choose any k A with $\theta(k) = z$; then $f_z(y) = f(yz) = (f\circ\theta)(hk) = \lim_\beta (f\circ\theta)(h_\beta k) = \lim_\beta f(y_\beta z) = 0$. Thus $y \in \ker f_z$, so that $\ker f_z$ is weak $*$-closed, and the proof of (ii) is complete.

(iii) It follows from (ii) that R does not have a bounded approximate identity. For suppose that (e_α) were such a b.a.i. Then, from (ii), there is a weak $*$-convergent subnet (e_β) say, $e_\beta \to e$ (weak $*$). But then, again by (ii). $e_\beta x \to ex$ (weak $*$) for all x in A. However $e_\beta x \to x$ in norm, so that $ex = x$ $(x \in A)$, contradicting R being a radical algebra.

(iv) By (ii), the bounded sequence $(x^n/\|x^n\|)$ has a weak *-cluster point, say y. But then, by weak *-continuity of multiplication, every weak *-neighbourhood U of $x^k y$ contains $x^{N+k}/\|x^N\|$ for infinitely many N. But $\|x^{N+k}\|/\|x^N\| \to 0$ $(N \to \infty)$, so that $x^k y = 0$. Since x is not a zero-divisor, $y = 0$; i.e. 0 is the unique weak *-cluster point of $(x^n/\|x^n\|)_{n \geq 1}$ and so $x^n/\|x^n\| \to 0$ (weak *).

If z were a cluster-point in the norm topology, for the same sequence, then $\|z\| = 1$; but z is also then a weak *-cluster point, so that $z = 0$. This contradiction proves that $\{x^n/\|x^n\| : n \geq 1\}$ is a discrete set in the norm topology.

References

[1] W. G. Bade, Multipliers of weighted ℓ^1-algebras, this Volume.

[2] F. Gharamani, Homomorphisms and derivations on weighted convolution algebras, J. London Math. Soc., (2), 21 (1980), 149-161.

[3] J. G. Hocking and G. S. Young, Topology, (Addison-Wesley 1961).

[4] J.-P. Kahane, Séries de Fourier Absolument Convergentes, (Springer-Verlag, 1970).

Department of Pure Mathematics and Mathematical Statistics
16 Mill Lane
Cambridge CB2 1SB, U.K.

WEIGHTED DISCRETE CONVOLUTION ALGEBRAS

Niels Grønbæk

In this note I shall discuss a class of commutative Banach algebras, the weighted discrete convolution algebras. These algebras are analogous to the more familiar weighted convolution algebras on the half-line of the type $L^1(\omega)$, and most of the questions concerning the structure of the latter have their counterpart in the discrete case, e.g., existence of elements of finite closed descent and the standard ideal problem.

A weighted discrete convolution algebra is defined as follows: Let S be a commutative semigroup and let $\omega : S \to \mathbb{R}^+$ be a function satisfying $\omega(s+t) \leq \omega(s)\omega(t)$ for all $s,t \in S$. The weighted discrete convolution algebra $\ell^1(S,\omega)$ has the usual weighted ℓ^1-space as its underlying Banach space and the algebra product is given by convolution, i.e., if $f,g \in \ell^1(S,\omega)$, then fg is the function given by

$$(fg)(s) = \begin{cases} \sum_{t+u=s} f(t)g(u), & \text{if } t+u=s \text{ has solutions} \\ 0 & \text{otherwise}. \end{cases}$$

It is easily checked that this indeed defines a commutative Banach algebra. If $\omega(ns)^{1/n} \to 0$ for all $s \in S$, then the algebra $\ell^1(S,\omega)$ is radical. It is often convenient to write a function in $\ell^1(S,\omega)$ as a formal power series with exponents in S, so that with $f \in \ell^1(S,\omega)$ is associated the power series $\sum_{s\in S} \lambda_s X^s$, $\lambda_s = f(s)$. The product then is formal power series multiplication.

In the present note, I shall only discuss convolution algebras associated with subsemigroups of $\mathbb{R}^+$. For a slightly more general treatment, see [5]. First I give a description of elements of closed descent zero in $\ell^1(\mathbb{Q}^+,\omega)$, where ω is a radical weight, and next I give an example of a subsemigroup S of $\mathbb{R}^+$ and a non-standard ideal in $\ell^1(S,\omega)$. I close with some remarks on factorization in Banach algebras.

1. Elements of closed descent zero in $\ell^1(S,\omega)$

If S is a subsemigroup of $\mathbb{R}^+$, then a necessary condition for an element $f \in \ell^1(S,\omega)$ to have closed descent zero is that $\alpha(f) = 0$, where $\alpha(f) = \inf \operatorname{supp} f$. When ω is a radical weight the natural conjecture is that this is also a sufficient condition. The theorem to follow states that, if $\alpha(f) = 0$, then there is a restriction of f, whose closed descent is zero. This restriction can be chosen close to f in two senses, the norm sense, and a sense that, loosely speaking, says that the support of the restriction includes points where f assumes large values. The theorem is slightly more general than Theorem 4.1 of [5].

THEOREM 1. Let S be a subsemigroup of $\mathbb{R}^+$ such that $(S-S)\cap\mathbb{R}^+\subseteq S$ and $\inf S = 0$. Let $\omega: S\to\mathbb{R}^+$ be a radical weight and put $\mathfrak{R} = \ell^1(S,\omega)$. Then $U = \{f\in\mathfrak{R} : (f\mathfrak{R})^- = \mathfrak{R}\}$ is a dense G_δ-set. Moreover, let $f\in\mathfrak{R}$ with $\alpha(f) = \inf\operatorname{supp} f = 0$, and let $(H_n)_{n\in\mathbb{N}}$ be a family of functions $H_n : S\to\mathbb{R}^+$ such that $\sum_{s\in S}|f(s)|H_n(s) = \infty$ for all $n\in\mathbb{N}$. Then to each $\varepsilon > 0$ there exists a restriction f_1 of f such that $\|f-f_1\| < \varepsilon$, $\sum_{s\in S}|f_1(s)|H_n(s) = \infty$ for all $n\in\mathbb{N}$, and $(f_1\mathfrak{R})^- = \mathfrak{R}$.

<u>Proof</u>. We only prove the last statement in the case $S = \mathbb{Q}^+$. The proof easily generalizes to give the theorem in full. Thus, let $f\in\ell^1(\mathbb{Q}^+,\omega)$ such that $\alpha(f) = 0$ and $\sum_{q\in\mathbb{Q}^+}|f(q)|H_n(q) = \infty$ for all $n\in\mathbb{N}$, and let $\mathfrak{F} = \{g\in\mathfrak{R} \mid g$ is a restriction of $f\}$.

Then clearly $\mathfrak{F}$ is closed in $\mathfrak{R}$. Put

$$U_n = \{g\in\mathfrak{F} : \inf_{h\in\mathfrak{R}}\|gh - X^{1/n}\| < \frac{1}{n},\quad \sum_{q\in\mathbb{Q}^+}|g(q)|H_m(q) > n \text{ for } m\le n\}.$$

Obviously U_n is open in $\mathfrak{F}$. We want to prove that U_n is dense in $\mathfrak{F}$ for all $n\in\mathbb{N}$. Let $\delta > 0$ and let $g\in\mathfrak{F}$. We choose $k\in\mathfrak{F}$ with finite support such that

(i) $$\|g-k\| < \delta$$

(ii) $$\alpha(k) < \frac{1}{n}$$

(iii) $$\sum_{q\in\mathbb{Q}^+}|k(q)|H_m(q) > n \text{ for } m\le n.$$

Such k exists: First we take $k_1\in\mathfrak{F}$ with finite support such that $\|g-k_1\| < \delta/3$. Next, since $f(q)\omega(q)\to 0$ as $q\to 0$, we can find $q_0 < 1/n$ such that $0 < |f(q_0)\omega(q_0)| < \delta/3$. Finally since $\sum_{q\in\mathbb{Q}^+}|f(q)|H_n(q) = \infty$ for all $n\in\mathbb{N}$, there is a finite restriction k_2 of f such that $\|k_2\| < \delta/3$ and $\sum_{q\in\mathbb{Q}^+}k_2(q)H_m(q) > n$ for all $m\le n$. The function k is then restriction of f with $\operatorname{supp} k = \operatorname{supp} k_1\cup\operatorname{supp} k_2\cup\{q_0\}$. One verifies easily the properties (i), (ii), and (iii).

Now $k = X^q b$ for some $q < 1/n$, where $b\in\operatorname{Inv}(\mathfrak{R}\oplus\mathbb{C})$. Put $h = X^{(1/n)-q}b^{-1}$. Then $kh = X^{1/n}$, so $k\in U_n$.

By the Baire category theorem $\bigcap_{n\in\mathbb{N}} U_n$ is a dense G_δ-set in $\mathfrak{F}$, which yields the assertion of the theorem.

REMARK. In [1] G. R. Allan proves that, if the function $t \mapsto f(t)e^{-Kt}$ is absolutely integrable for some $K > 0$ and $\alpha(f) = 0$, then f has closed descent zero in $L^1(\omega)$ for certain nice weights. With minor changes the proof of the above theorem can be adapted to give elements of closed descent zero in $L^1(\omega)$ with rapid growth at infinity, e.g. we can find an element f in $L^1(\omega)$ with closed descent zero, but such that $\int_{\mathbb{R}^+} |f(t)|e^{-nt}dt = \infty$ for all $n \in \mathbb{N}$.

2. A subsemigroup S of $(\mathbb{R}^+, +)$ and a non-standard ideal of $\ell^1(S,\omega)$. Let S be a subsemigroup of $\mathbb{R}^+$ and let $\omega : S \to \mathbb{R}^+$ be a weight. Let $t \in \mathbb{R}^+$. Then there are two closed ideals associated with t in $\ell^1(S,\omega)$:

$$M_{(t)} = \{f \in \ell^1(S,\omega) \mid \operatorname{supp} f \subseteq (t,\infty)\}$$
$$M_{[t]} = \{f \in \ell^1(S,\omega) \mid \operatorname{supp} f \subseteq [t,\infty)\} .$$

Clearly $M_{(t)} \neq M_{[t]}$ if and only if $t \in S$. If ω is a radical weight these ideals correspond to the standard ideals in $L^1(\omega)$ and will therefore also be termed standard ideals. For no radical weight $\omega : \mathbb{Q}^+ \to \mathbb{R}^+$ is it known whether the standard ideals in $\ell^1(\mathbb{Q}^+,\omega)$ are the only closed ideals.

The following example shows that the algebraic structure of the underlying semigroup may play an important role for the structure of the closed ideals in the ℓ^1-case, and therefore that the standard ideal problem in $\ell^1(S,\omega)$ is of a different nature than the usual standard ideal problem.

EXAMPLE 2. Let $(t_i)_{i\in I}$ be a family of $\mathbb{Q}$-independent positive reals less than 1, and let $S = \{\sum_{i\in I} q_i t_i : q_i \in \mathbb{Q}^+$, all but finitely many q_i's equal zero$\}$ be the subsemigroup of $\mathbb{R}^+$ generated by $(t_i)_{i\in I}$, using positive rational coefficients. Let $\omega : \mathbb{R}^+ \to \mathbb{R}^+$ be any bounded weight. We define a semigroup homomorphism $\varphi : S \to \mathbb{Q}^+$ by $\varphi(\sum_{i\in I} q_i t_i) = \sum_{i\in I} q_i$. This homomorphism determines an epimorphism $\theta : \ell^1(S,\omega) \to \ell^1(\mathbb{Q}^+,\omega)$ by

$$\theta\left(\sum_{s\in S} \lambda_s X^s\right) = \sum_{s\in S} \lambda_s X^{\varphi(s)} .$$

The inequalities $t_i \leq 1$ for all $i \in I$ imply that $\varphi(s) \geq s$ for all $s \in S$. Hence $\omega(\varphi(s)) \leq \omega(\varphi(s) - s)\omega(s) \leq M\omega(s)$ for all $s \in S$, where M is the bound of ω and we for convenience have put $\omega(0) = 1$. Using this, one checks easily that θ indeed is an epimorphism from $\ell^1(S,\omega)$ onto $\ell^1(\mathbb{Q}^+,\omega)$. It is obvious that $\ker\theta$ is a non-standard closed ideal of $\ell^1(S,\omega)$. Note that there are elements $f \in \ker\theta$ with $\alpha(f) = 0$.

This example, however, cannot be modified to give non-standard ideals if $S = \mathbb{Q}^+$ or $S = \mathbb{R}^+$, and for these semigroups the question, whether all closed ideals are standard, remains open. The example, though, indicates that it may be possible to find a non-standard ideal by means of some algebraic relation in the semigroup.

3. Factorization in separable Banach algebras.

One of the questions that is naturally posed concerning the weighted discrete convolution algebras, is: Can every element be written as a product of two elements? The only condition known to me that ensures factorization in separable Banach algebras is the existence of a bounded approximate identity. (H. G. Dales has brought to my attention examples of non-separable commutative Banach algebras with factorization, but without a bounded approximate identity. One example is described on p. 166 of [4]. Another example goes as follows: Let $H^\infty(\Delta)$ be the uniform algebra of bounded analytic functions on the disc. Take a point x which is a one-point part off the Silov boundary in the character space. Since x is not a strong boundary point, the maximal ideal $M_x = \{f \mid f(x) = 0\}$ does not have a bounded approximate identity. But M_x factors (see [6].) Of course, the Banach algebras which are the subject of this paper do not have bounded approximate identities, and the question of factorization in these algebras is open, e.g., is there factorization in $\ell^1(\mathbb{Q}^+, e^{-t^2})$, $\ell^1(\mathbb{Q}^+, \chi_{[0,1]})$, etc.?

Recall that, if a Banach algebra has a bounded approximate identity, then not only do we have factorization, but we also get some estimates of the norms of the factors. The following theorem, by J. P. R. Christensen and myself ([3]), goes in the converse direction and may serve as a means to sort out separable Banach algebras with factorization.

THEOREM 3. Let A be a separable Banach algebra (not necessarily commutative), and assume that there is factorization in A, i.e., every element of A is a product of two elements of A. Then there exists $C > 0$ such that each $a \in A$ can be written

$$a = x_1y_1 + x_2y_2, \quad x_i, y_i \in A, \quad i = 1,2,$$

where $\|x_i\| \leq 1$ and $\|y_i\| \leq C\|a\|$, $i = 1,2$.

Proof. The proof of the theorem utilizes Pettis's lemma in a manner much similar to [2]. Let $P = \{xy \mid \|x\| < 1, \|y\| < 1\}$. Since A is separable, P is analytic as a subset of A. From the existence of factorization it follows that P is absorbing, and therefore P must be of the second category in A. Hence the Pettis lemma applies to give that $P - P$ is a neighborhood of zero. It is easy to see that this implies the conclusion of the theorem.

COROLLARY 4. There is not factorization in separable topologically nilpotent Banach algebras.

Proof. Recall that a topologically nilpotent Banach algebra A (by definition) has the property that the nth root of e_n, defined by $e_n = \sup\{\|x_1 \cdots x_n\| \mid \|x_i\| < 1, i = 1,\dots,n\}$ tends to zero. However, with C as in the theorem, we easily obtain the inequality $e_n \leq 2C\, e_{n+1}$, from which we conclude that the above mentioned limit is at least $1/2C$. Hence, if A has factorization, it is not topologically nilpotent.

Problem List

1. Let $\omega : \mathbb{Q}^+ \to \mathbb{R}^+ \cup \{0\}$ be a weight. Is there factorization in $\ell^1(\mathbb{Q}^+, \omega)$? Of special interest are the weights

$$\omega_1(t) = e^{-t^2}$$

$$\omega_2(t) = \begin{cases} 1 & \text{if } t \leq 1 \\ 0 & \text{if } t > 1 \end{cases}$$

$$\omega_3(t) = 1$$

Conjecture. No, in all three cases.

2. What are the closed ideals of $\ell^1(\mathbb{Q}^+, \omega)$, ω a radical weight? In particular: are all closed ideals standard?

Conjecture. There exists non-standard ideals for all weights.

3. It is clear that, if $\ell^1(S)$ is an integral domain, then S is a cancellation semigroup. On the other hand $\ell^1(\mathbb{Z})$ is not an integral domain, whereas $\ell^1(\mathbb{N})$ is. Give conditions on semigroups and/or weights such that $\ell^1(S, \omega)$ is an integral domain.

4. The following are equivalent for a Banach algebra A:

(i) There exists a constant $C > 0$ such that every $f \in A$ is a product

$$f = ag,$$

where $a, g \in A$, $\|a\| \leq C$, $\|g\| \leq \|f\|$.

(ii) There is factorization in $c_0(A)$, the Banach algebra of sequences of elements from A, tending to zero.

(iii) The set $S = \{ab \mid \|a\| \leq 1, \|b\| \leq 1\}$ is a neighborhood of zero in A.

Suppose there is factorization in A. Does (i), and hence (ii) and (iii) hold? One may assume that A is commutative and separable.

References

[1] G. R. Allan, Ideals of rapidly growing functions, Proceedings International Symposium on Functional Analysis and its Applications, Ibadan, Nigeria, 1977.

[2] J. P. R. Christensen, Codimensions of some subspaces in a Fréchet algebra, Proc. Amer. Math. Soc., 57 (1976), 276-278.

[3] J. P. R. Christensen and N. Grønbæk, Factorization in Banach Algebras, (København Universitet, Matematisk Institut, Preprint Series, 3, 1981).

[4] R. S. Doran and J. Wichmann, Approximate Identities and Factorization in Banach Modules, Lecture Notes in Mathematics, 768, Springer-Verlag, New York, 1979.

[5] N. Grønbæk, Radical weighted discrete convolution algebras, Trans. Amer. Math. Soc., to appear.

[6] K. Hoffman, Banach Spaces of Analytic Functions, Prentice-Hall, New Jersey, 1962.

Refsnaesgade 43, st. tv.
DK-2200 Copenhagen N
Denmark

SOME RADICAL QUOTIENTS IN HARMONIC ANALYSIS

Gregory F. Bachelis

Introduction

In this paper we are concerned with several phenomena in harmonic analysis which give rise to commutative radical quotient Banach algebras. We survey some known results about these phenomena and examine them from the standpoint of radical Banach algebras; we also raise several questions which arise naturally as a consequence of this examination.

Before proceeding further, we need some notation: G will always denote an infinite locally compact abelian group, and Γ its dual group. $A(G)$ denotes the Fourier algebra of G, that is, the algebra of Fourier transforms of integrable functions on Γ. $A(G)$ is a Banach algebra under pointwise multiplication, with the norm given by

$$\|\hat{f}\|_A = \|f\|_1, \quad f \in L^1(\Gamma).$$

For $G = \mathbb{T}$, the circle group, $A(G)$ is the Wiener algebra of absolutely converging Fourier series. $M(G)$ denotes the finite regular Borel measures on G; it is a Banach algebra under convolution and the total variation norm. $L^1(G)$ is identified in the usual way with the ideal of measures absolutely continuous with respect to Haar measure. $\text{Rad}\, L^1(G)$ denotes the kernel of the hull of $L^1(G)$ in $M(G)$, that is, the intersection of the maximal ideals of $M(G)$ which contain $L^1(G)$.

The phenomena we consider in which radical quotients arise are as follows:

1. For G non-discrete, the existence of closed sets $E \subset G$ not of synthesis for $A(G)$, that is, the existence of closed ideals not the kernel of their hull.

2. For G discrete, the existence of closed subalgebras of $A(G)$ not the closed span of their idempotents.

3. For G non-discrete, the fact that $\text{Rad}\, L^1(G)$ properly contains $L^1(G)$.

We also consider phenomena analogous to (3) when the measure algebra is replaced by certain multiplier algebras.

For general notions of Banach algebras and abstract harmonic analysis, the reader is referred to one of the standard texts, e.g. [7] or [29], and [16], [20], [28] or [30]. As for specifics, we give some references as we go along, and also in the concluding remarks. However, our survey is of necessity selective and does not pretend to give a full historical account.

1. Sets of non-synthesis for $A(G)$

Let E be a closed subset of G and let I_E denote the kernel of E:

$$I_E = \{f \in A(G) : f = 0 \text{ on } E\}.$$

Let J_E denote the closure of the set of functions in $A(G)$ which are each zero on some neighborhood of E. Then I_E is the largest closed ideal with hull equal to E and J_E is the smallest. We say that E is of spectral synthesis if $I_E = J_E$. There are equivalent definitions involving pseudomeasures which we shall not go into. When E is not of synthesis, then I_E/J_E is a commutative radical Banach algebra.

The first example of a set of non-synthesis was given by L. Schwartz [34], who showed that the unit sphere S^2 in the Euclidean space $\mathbb{R}^3$ is not of synthesis. In 1959, P. Malliavin [25] showed that every non-discrete G contains a set of non-synthesis. One version of Malliavin's criterion for the existence of such a set is as follows [20, page 231]:

THEOREM. Let A be a commutative, semi-simple, regular Banach algebra with identity, considered as functions on its maximal ideal space. Suppose there exist $f \in A$ and $0 \neq F \in A^*$, the dual space of A, and a positive integer n such that

$$\int_{-\infty}^{\infty} \|e^{iuf}F\|_{A^*} |u|^n du < \infty .$$

Then there exists $a \in \mathbb{R}$ such that the functions $f - a, \ldots, (f - a)^{n+1}$ generate distinct closed ideals, and thus $f^{-1}(a)$ is not of synthesis.

Here A^* is an A-module in the canonical way.

Whenever E is not of synthesis, there always exist closed ideals strictly between J_E and I_E ([15], [31]), and in fact one can find continuum many such ideals ([35], [6, page 194], [28, page 35]).

It is possible for the algebras I_E/J_E to be nilpotent. For example, if $E = S^{n-1} \subset \mathbb{R}^n$, then $I_E^{[(n+1)/2]} = J_E$, $n \geq 3$ [38]. On the other hand, for any non-discrete G there always exists $f \in A(G)$ such that all powers of f generate distinct closed ideals [6, page 193].

More recently, examples have been given in any non-discrete G of compact sets which are at once not of synthesis and a Helson set [22], [21], [32]. A compact set E is a Helson set if the monomorphism $f + I_E \mapsto f\,|\,E$, from $A(G)/I_E$ into $C(E)$, is surjective, that is, if every continuous function on E is the restriction to E of a function in $A(G)$. If E is a Helson set which is not of synthesis and if $B = A(G)/J_E$, then the radical N of B is I_E/J_E, and B/N is isometrically isomorphic to $A(G)/I_E$, which is isomorphic to $C(E)$. One can ask as in [5] whether B is (strongly) decomposable, that is, does there exist a (closed) subalgebra M of B such that $B = M \oplus N$? By [5, Thm. 4.1], if B is decomposable, then it is strongly decomposable and the strong decomposition is unique. So we have:

Question 1. If E is a Helson set which is not of synthesis and if $B = A(G)/J_E$ with radical $N = I_E/J_E$, so that $B/N \simeq C(E)$, is there a subalgebra M of B such that $B = M \oplus N$?

If E is totally disconnected and N is nilpotent, then B is strongly decomposable [5, Thm. 4.2]. We conjecture that N is not nilpotent and that B is never decomposable.

2. Closed subalgebras of A(G) not generated by their idempotents

If G is discrete then every closed ideal in A(G) is the kernel of its hull. In this case, however, one can obtain radical quotients by examining the structure of closed subalgebras. Given a closed subalgebra S, one considers the constancy sets,

$$E(x) = \{y \in G : f(y) = f(x),\ f \in S\},\ x \in G .$$

(In this context constancy sets are often called Rudin classes.) The trivial constancy set E_0 is the one on which each $f \in S$ vanishes. If $E(x) \neq E_0$ then $E(x)$ is finite, and its characteristic function is an idempotent which is in S [30, page 232]. Thus there exists a smallest and a largest closed subalgebra of A(G) with the same nontrivial constancy sets as S. The smallest S_m, is the closed span of the idempotents of S. The largest, S_M, is the set of functions in A(G) which are constant on each $E(x)$ and zero on E_0. By [12, Thm. 2.1], S_m and S_M have the same maximal ideal space, so that whenever $S_m \neq S_M$ then S_M/S_m is a radical algebra.

In [18], Kahane showed that $A(\mathbb{Z})$ contains closed subalgebras S such that $S_m \neq S_M$, where $\mathbb{Z}$ denotes the integers. In Theorem 3 of [18] an example is given of such an S whose non-trivial constancy sets have bounded cardinality. If one takes the function f constructed in the proof of that theorem, then it is not difficult to see that none of the powers of f are in S_m. Thus in this case S_M/S_m is not nilpotent. We do not know if all the powers of f generate distinct closed ideals in some closed subalgebra of S_M containing f and S_m, although we conjecture that this is the case. If this is not the case, then S_M/S_m contains a copy of the formal power series with zero constant term [1]. The examples in [18] carry over to any discrete G, as do our comments.

We mention parenthetically that it is possible to have a commutative regular semi-simple Banach algebra with a discrete maximal ideal space having a non-empty subset E not of synthesis. Mirkil constructed such an example in [26]. Here, as above, a set E is of synthesis if I_E, the kernel of E, is the only closed ideal with hull equal to E. The smallest closed ideal with hull equal to E, J_E, is the closed span of the idempotents whose Gelfand transform has support contained in the complement of E. It is easy to see, for Mirkil's set E not of synthesis, that $I_E^2 \subset J_E$. Recently, Atzmon showed that this E can be written as the union of two sets of synthesis [2]. The problem for $A(G)$, G non-discrete, of whether the union of two sets of synthesis is necessarily of synthesis remains open.

3. Singular measures with absolutely continuous convolution powers

Wiener and Wintner gave the first example of a singular measure on $\mathbb{T}$ whose convolution square is absolutely continuous [39], and Hewitt and Zuckerman exhibited the same phenomenon for any non-discrete G [17]. In fact, if G is non-discrete, then for each positive integer n there exist measures whose nth convolution power is singular but whose $(n+1)$st power is absolutely continuous [13, Cor. 7.2.4]. Since $\mathrm{Rad}\, L^1(G)$ is, by definition, the kernel of the hull of $L^1(G)$ as an ideal in $M(G)$, we see that $\mathrm{Rad}\, L^1(G)/L^1(G)$ is a non-nilpotent radical algebra. We do not know if the nilpotent elements are dense in this algebra, so we can ask:

Question 2. Does there exists a measure in $\mathrm{Rad}\, L^1(G)$ which is not in the closure of the measures which are nilpotent modulo $L^1(G)$?

Each nilpotent element of $\mathrm{Rad}\, L^1(G)/L^1(G)$ of order $\leq n$ generates an ideal whose elements obviously have the same property. However, Saeki has shown that the set of all nilpotent elements of order $\leq n$ is not a subspace [33, Cor. 2.5].

It is easy to see that $\mathrm{Rad}\, L^1(G) \subset M_0(G)$, the set of measures μ whose Fourier-Stieltjes transform $\hat{\mu}$ vanishes at infinity. It is known that these two algebras are different, and in fact that only entire functions operate in $M_0(G)$ [37]. By this latter we mean that if $F : [-1,1] \to \mathbb{C}$ is such that $F \circ \hat{\mu} \in M_0(G)^\wedge$ for all $\mu \in M_0(G)$ with $\hat{\mu}(\Gamma) \subset [-1,1]$, then F coincides with an entire function in a neighborhood of zero. The fact that only entire functions operate in $M_0(G)$ implies that its maximal ideal space is larger than Γ, the maximal ideal space of $\mathrm{Rad}\, L^1(G)$ and $L^1(G)$.

4. Multipliers of L^p vanishing at infinity which are not compact

One can also look for radical quotients when $M(G)$ is replaced by the multiplier algebras $M^p(G)$, $1 < p < \infty$, $p \neq 2$, and $L^1(G)$ by $m^p(G)$, the closure of $L^1(G)$ in $M^p(G)$. For $1 \leq p < \infty$, $M^p(G)$ can be defined as the space of pseudo-measures on G which convolve $L^p(G)$ into itself. When such a convolution operator is defined, it is necessarily continuous, and the norm in $M^p(G)$ is the operator norm. The multiplication is convolution, which corresponds to composition of the corresponding operators. Among the basic facts about these algebras are that $M^p(G) = M^q(G)$ when $\frac{1}{p} + \frac{1}{q} = 1$ and $p \neq 1$, that $M^1(G) = M(G)$, and that $M^2(G) = PM(G)$, the set of all pseudomeasures on G. Thus $m^1(G) = L^1(G)$ and $m^2(G) \simeq C_0(\Gamma)$. The maximal ideal space of $m^p(G)$ is Γ.

For simplicity, we will confine the remainder of our discussion to the case when G is compact. Then $m^p(G)$ is an ideal in $M^p(G)$ and consists of those L^p-multipliers which are compact as operators on $L^p(G)$. We denote by $M_0^p(G)$ the set of L^p-multipliers whose (generalized) Fourier transform vanishes at infinity. Then $m^p(G) \subset \text{Rad}\, m^p(G) \subset M_0^p(G)$. If $p = 2$ then $m^p(G) = M_0^p(G)$. For $p \neq 2$, $1 < p < \infty$, Figà-Talamanca and Gaudry gave the first example of a multiplier in $M_0^p(G)$ but not in $m^p(G)$ [11]. Their example has the property that its square is in $m^p(G)$, which for $p < 2$ follows from the fact that it convolves $L^p(G)$ into $L^2(G)$. Thus $\text{Rad}\, m^p(G) \neq m^p(G)$, $p \neq 2$.

More recently, Zafran has shown that, when G is the n-torus $\mathbb{T}^n$, the maximal ideal space of $M_0^p(G)$ is larger than Γ [40], and in fact that only entire functions operate in $M_0^p(G)$, $1 < p < \infty$, $p \neq 2$ [41]. The arguments in [40] can be used to obtain, for any positive integer j, a multiplier in $M_0^p(\mathbb{T}^n)$ whose jth power is not in $m^p(\mathbb{T}^n)$ but whose $(j+1)$st power is. The author is indebted to Professor Zafran for pointing this out. To see this, consider, for each j, the multiplier operator T_j produced in [40, Comment 4.3] from a sequence of measures $\{\nu_{n,j}\}_{n=1}^{\infty}$, which has the property that $T_j \in M_0^p(\mathbb{T})$ and $T_j^j \notin m^p(\mathbb{T})$. Using [40, Lemma 3.5] and the proof of [40, Lemma 3.6], one can show that $\sum_{n=1}^{\infty} 2^{(j+1)n/p'} \|\nu_{n,j}^{j+1}\|_{M^p} < \infty$, from which it follows that $T_j^{j+1} \in m^p(\mathbb{T})$. These operators can be "lifted" to $\mathbb{T}^n$, and evidently to any compact abelian group whose dual contains $\mathbb{Z}$. Thus in these cases $\text{Rad}\, m^p(G)/m^p(G)$ is not nilpotent. One can ask as in §3 whether the nilpotent elements are dense.

5. Concluding remarks

(i) The major open question concerning commutative radical Banach algebras is whether there exists a topologically simple one, that is, one with no non-trivial closed ideals. Now the examples of radical quotients we have considered all have many closed ideals; however, one could look for a closed subalgebra which is topologically simple.

Given a commutative Banach algebra B, a hyperinvariant subspace for multiplication by an element of B is in particular an ideal. Thus if multiplication by some element of B is compact and non-zero as an operator on B, then B cannot be topologically simple by Lomonosov's Theorem ([24], [27]).

Now the compact multipliers of $M(G)$ or $L^1(G)$ consist of: convolution by an L^1-function when G is compact, and only the zero operator when G is non-compact. Thus the quotients considered in §§1 and 3 might contain closed topologically simple subalgebras. In §3 such containment is possible when G is compact since, in considering Rad $L^1(G)/L^1(G)$ we are dividing out by the compact multipliers. The quotients considered in §2 cannot have closed topologically simple subalgebras, since when multiplication by every element of a Banach algebra is compact as an operator on that algebra, the same is true for quotients of closed subalgebras by closed ideals.

When G is compact, the compact multipliers of $M^p(G)$ or $m^p(G)$ are given by multiplication by an element of $m^p(G)$. This can be proved using the fact that $m^p(G)$ has a bounded approximate identity consisting of trigonometric polynomials. Thus the situation here is analogous to that of M(G).

(ii) There are of course many instances in harmonic analysis besides those we have considered in which one has a closed ideal not the kernel of its hull and a radical quotient obtains. For example, there are semi-simple weighted L^1-algebras which have non-maximal closed ideals contained in a unique maximal modular ideal, that is, for which there are singleton sets not of synthesis. See [8], [9], [10], and [14].

(iii) For more information about spectral synthesis, the reader is referred to [6], [13], and [19], in addition to the standard harmonic analysis texts; about closed subalgebras of A(G) (and other algebras) not generated by their idempotents, to [4] and the references given there; about measure algebras, to [36] and [13]; and about multipliers, to [23] and [13]. Compact multipliers are treated in [3].

The author would like to thank a number of his colleagues, including David Salinger and Jan Stegeman, for their helpful comments during the preparation of this paper.

References

[1] G. R. Allan, Embedding the algebra of formal power series in a Banach algebra, Proc. London Math. Soc., (3), 25 (1972), 329-340.

[2] A. Atzmon, On the union of sets of synthesis and Ditkin's condition in regular Banach algebras, Bull. Amer. Math. Soc., (N.S.), 2 (1979), 317-320.

[3] G. F. Bachelis and J. E. Gilbert, Banach spaces of compact multipliers and their dual spaces, Math. Z., 125 (1972), 285-297.

[4] ________, Banach algebras with Rider subalgebras, Bull. Inst. Math. Academia Sinica (Taiwan), 7 (1979), 339-347.

[5] W. G. Bade and P. C. Curtis, Jr., The Wedderburn decomposition of commutative Banach algebras, Amer. J. Math., 82 (1960), 851-866.

[6] J. Benedetto, Spectral Synthesis, Academic Press, New York, 1975.

[7] F. F. Bonsall and J. Duncan, Complete Normed Algebras, Springer-Verlag, New York, 1973.

[8] Y. Domar, Closed primary ideals in a class of Banach algebra, Math. Scand., 7 (1959), 109-125.

[9] __________, On the ideal structure of certain Banach algebras, Math. Scand., 14 (1964), 197-212.

[10] __________, Primary ideals in Beurling algebras, Math. Z., 126 (1972), 361-367.

[11] A. Figà-Talamanca and G. I. Gaudry, Multipliers of L^p which vanish at infinity, J. Functional Analysis, 7 (1971), 475-486.

[12] S. Friedberg, Closed subalgebras of group algebras, Trans. Amer. Math. Soc., 147 (1970), 117-125.

[13] C. C. Graham and O. C. McGehee, Essays in Commutative Harmonic Analysis, Springer-Verlag, New York, 1979.

[14] V. P. Gurariĭ, Harmonic analysis in spaces with a weight, Trans. Moscow Math. Soc., (English translation), (1979, issue 1), 21-75.

[15] H. Helson, On the ideal structure of group algebras, Ark. Mat., 2 (1952), 83-86.

[16] E. Hewitt and K. A. Ross, Abstract Harmonic Analysis, Volumes I and II, Springer-Verlag, New York, 1963 and 1970.

[17] E. Hewitt and H. S. Zuckerman, Singular measures with absolutely continuous convolution squares, Proc. Cambridge Philos. Soc., 62 (1966), 399-420. Corrigendum, ibid., 63 (1967), 367-368.

[18] J.-P. Kahane, Idempotents and closed subalgebras of A(Z), Function Algebras, Scott Foresman, Chicago 1966, 198-207.

[19] __________, Séries de Fourier Absolument Convergentes, Springer-Verlag, Berlin, 1970.

[20] Y. Katznelson, An Introduction to Harmonic Analysis, John Wiley & Sons, New York, 1968.

[21] R. Kaufman, M-sets and distributions, Astérisque 5 (1973), 225-230.

[22] T. W. Körner, A pseudofunction on a Helson set, I and II, Astérisque 5 (1973), 3-224 and 231-239.

[23] R. Larsen, An Introduction to the Theory of Multipliers, Springer-Verlag, Berlin, 1971.

[24] V. J. Lomonosov, Invariant subspaces for operators commuting with compact operators, Functional Anal. and Appl., 7 (1973), 55-56.

[25] P. Malliavin, Impossibilité de la synthèse spectrale sur les groupes abéliens non compacts, Publ. Math. Inst. Hautes Études Sci. Paris, (1959), 61-68.

[26] H. Mirkil, A counterexample to discrete spectral synthesis, Compositio Math., 14 (1959/60), 269-273.

[27] C. Pearcy and A. L. Shields, A survey of the Lomonosov technique in the theory of invariant subspaces, Topics in Operator Theory, Amer. Math. Soc. Surveys No. 13, (1974), 219-229.

[28] H. Reiter, Classical Harmonic Analysis and Locally Compact Groups, Clarendon, Oxford, 1968.

[29] C. Rickart, Banach Algebras, Van Nostrand, New York, 1960.

[30] W. Rudin, Fourier Analysis on Groups, Interscience, New York, 1962.

[31] S. Saeki, An elementary proof of a theorem of Henry Helson, Tôhoku Math. J., 20 (1968), 244-247.

[32] __________, Helson sets which disobey spectral synthesis, Proc. Amer. Math. Soc., 47 (1975), 371-377.

[33] __________, Singular measures having absolutely continuous convolution powers, Illinois J. Math., 21 (1977), 395-412.

[34] L. Schwartz, Sur une propriété de synthèse spectrale dans les groupes non compacts, C. R. Acad. Sci. Ser. A-B, 227 (1948), A424-A426.

[35] J. D. Stegeman, Extension of a theorem of H. Helson, Proc. Int. Cong. Math. Abstracts, Section 5 (1966), 28.

[36] J. L. Taylor, Measure Algebras, Regional Conference Series in Math., No. 16, Amer. Math. Soc., Providence, 1972.

[37] N. Th. Varopoulos, The functions that operate on $B_0(\Gamma)$ of a discrete group, Bull. Soc. Math. France, 93 (1965), 301-321.

[38] __________, Spectral synthesis on spheres, Proc. Camb. Philos. Soc., 62 (1966), 379-387.

[39] N. Wiener and A. Wintner, Fourier-Stieltjes transforms and singular infinite convolutions, Amer. J. Math., 60 (1938), 513-522.

[40] M. Zafran, The spectra of multiplier transformations on the L^p spaces, Ann. Math., 103 (1976), 355-374.

[41] __________, The functions operating on multiplier algebras, J. Functional Analysis, 26 (1977), 289-314.

Department of Mathematics
Wayne State University
Detroit, MI 48202

A BANACH ALGEBRA RELATED TO THE DISK ALGEBRA

Y. Domar

Richard J. Loy asked in [4] whether it is possible to find a multiplicative norm in the space of complex polynomials in one variable such that the completion is a Banach algebra with non-nilpotent radical, and such that the corresponding spectrum of the polynomial z is the unit disk. Here such a construction is given. However, the example is not a Banach algebra of power series, and so it leaves open the construction of a non-local, non-semi-simple, singly generated Banach algebra of power series: an example with two generators is discussed in [1, Theorem 2.7].

Let $w = \{w_n\}_0^\infty$ be a positive sequence such that

$$w_{m+n} \leq w_m w_n, \; m,n \in \mathbb{Z}^+, \tag{1}$$

such that

$$n^{-1} \log w_n + \log n \to -\infty, \quad \text{as} \quad n \to \infty. \tag{2}$$

Then $\ell^1(w)$ is the Banach space of complex sequences $c = \{c_n\}_0^\infty$ with $cw = \{c_n w_n\}_0^\infty \in \ell^1$, and $\ell^1(w)$ has the induced norm. By (1), $\ell^1(w)$ is a commutative Banach algebra under sequence convolution, and (2) implies, with a wide margin, that its only non-trivial complex homomorphism is the mapping $c \mapsto c_0$.

Let D be the closed unit disk in $\mathbb{C}$. Then $A(D)$ denotes the disk algebra, that is the Banach algebra of all complex-valued functions on D which are continuous in D and analytic in its interior, with pointwise multiplication as operation, and equipped with the uniform norm. Its only non-trivial complex homomorphisms are the point evaluations on D.

We form the Banach space B of pairs (f,c), where $f \in A(D)$, $c \in \ell^1(w)$, $f(1) = c_0$, and with the norm defined by

$$\|(f,c)\| = \|f\|_{A(D)} + \|c\|_{\ell^1(w)} . \tag{3}$$

We easily see that B is a Banach algebra with respect to componentwise operations. A complex homomorphism of B induces a complex homomorphism on each of its two subalgebras $A(D)$ and $\ell^1(w)$, and from this we find easily that the non-trivial complex homomorphisms of B are given by

$$(f,c) \mapsto f(z_0), \; z_0 \in D .$$

Hence D is the spectrum of the element $b = (a,d)$, where a denotes the function $z \mapsto z$ in $A(D)$, and

$$d = \left\{ \frac{1}{n!} \right\}_0^\infty .$$

The radical of B consists of the elements $(0,c)$, $c \in \ell^1(w)$, $c_0 = 0$, and hence the radical contains non-nilpotent elements. Indeed, it has no non-zero nilpotents. Thus polynomials in b, normed in accordance to (3), give the desired example, if it can be proved that these polynomials form a dense subspace of B.

To show the assertion last mentioned, let us first observe that an arbitrary bounded linear functional on B is given by a mapping

$$(f,c) \mapsto F(f) + \sum_1^\infty c_n g_n , \tag{4}$$

where F is a bounded linear functional on $A(D)$, and $\{g_n\}_1^\infty$ is a complex sequence such that

$$\sup |g_n| w_n^{-1} < \infty . \tag{5}$$

Let us then assume that the functional (4) is annihilated by the elements b^m, $m \geq 0$.

Since convolution in $\ell^1(w)$ is given by formal multiplication of the corresponding power series, we obtain

$$d^m = \left\{ \frac{m^n}{n!} \right\}_0^\infty , \quad m \in \mathbb{Z}^+ .$$

Hence (4) gives

$$F(a^m) + \sum_1^\infty \frac{m^n}{n!} g_n = 0, \quad m \in \mathbb{Z}^+ . \tag{6}$$

But $\|a^m\|_{A(D)} \leq 1$, for every m, and therefore the entire function

$$G(z) = \sum_1^\infty \frac{z^n}{n!} g_n \tag{7}$$

is bounded at the points $z = m$, $m \in \mathbb{Z}^+$. The relations (2) and (5) show that to every $\varepsilon > 0$, there corresponds a constant C_ε such that

$$\frac{|g_n|}{n!} \leq C_\varepsilon \frac{\varepsilon^{2n}}{(2n)!} , \quad n \geq 0 .$$

Hence, for every $\varepsilon > 0$,

$$|G(z)| \leq C_\varepsilon \, e^{\varepsilon\sqrt{|z|}}, \quad z \in \mathbb{C}.$$

It follows that $H(z) = G(z^2)$ defines an entire function, satisfying, for every $\varepsilon > 0$,

$$|H(z)| \leq C_\varepsilon \, e^{\varepsilon |z|}, \quad z \in \mathbb{C}. \tag{8}$$

Since $\{H(m)\}_{-\infty}^{\infty} = \{G(m^2)\}_{-\infty}^{\infty}$ is a bounded sequence, a theorem of M. L. Cartwright [3]; see [2, 10.2.3] shows that H is bounded on the real axis. Equation (8) and a standard Phragmén-Lindelöf result then show that H is bounded. Thus H is a constant, by Liouville's theorem. Hence G is constant, and (7) shows that $g_n = 0$, $n \geq 1$. Returning to the relation (6) we find that $F(a^m) = 0$, $m \geq 0$. It is however well known that polynomials in a are dense in $A(D)$, and it follows from this that F vanishes identically.

Thus we have that a bounded linear functional on B vanishes identically if it vanishes for all b^n, $n \geq 0$. By the Hahn-Banach theorem, this implies that polynomials in b are dense in B, as required.

References

[1] W. G. Bade, Multipliers of weighted ℓ^1-algebras, this volume.

[2] R. P. Boas, Jr., Entire Functions, Academic Press, New York, 1954.

[3] M. L. Cartwright, On certain integral functions of order one, Quart. J. Math., 7 (1936), 46-55.

[4] R. L. Loy, Commutative Banach algebras with non-unique complete norm topology, Bull. Austr. Math. Soc., 10 (1974), 409-420.

Uppsala Universitet
Matematiska Institutionen
Thunbergsvägen 3
752 38 Uppsala
Sweden

III. AUTOMATIC CONTINUITY FOR HOMOMORPHISMS AND DERIVATIONS

AUTOMATIC CONTINUITY CONDITIONS FOR A LINEAR MAPPING FROM A BANACH ALGEBRA ONTO A SEMI-SIMPLE BANACH ALGEBRA

Bernard Aupetit†

For a commutative semi-simple Banach algebra it is easy to prove that every Banach algebra norm $\| \|_1$ on A is equivalent to the original norm $\| \|$. For twenty years the same problem for semi-simple non-commutative Banach algebras was unsolved. In 1967, B. E. Johnson gave a solution by intensively using irreducible representations and by reducing the question to one for primitive algebras.

First we give a generalization of a result of [2] and of B. E. Johnson's theorem. The proof uses a subharmonic technique and is purely internal.

Finally we give a partial solution to the problem of continuity of morphisms from a Banach algebra onto a dense subalgebra of a semi-simple Banach algebra.

We also mention that the results obtained in this paper can be extended to Banach Jordan algebras. In particular the analog of B. E. Johnson's theorem is true for Banach Jordan algebras; this implies that all involutions are continuous on Banach Jordan algebras which are semi-simple in the sense of McCrimmon (see [3]). This conjecture was unsolved until today.

1. Spectrally contractive mappings

We denote by $\rho(x)$ the spectral radius of an element x of a Banach algebra.

THEOREM 1. Let A be a complex Banach algebra and let B be a semi-simple Banach algebra. Suppose that T is a linear mapping from A onto B such that $\rho(Tx) \leq \rho(x)$ for every x in A. Then T is continuous.

Proof. We apply the closed graph theorem. Suppose that $x_n \to 0$ in A and that $Tx_n \to Ta$ in B when n goes to infinity. The problem is to prove that $Ta = 0$. Let x be an arbitrary element of A, and let λ be arbitrary in $\mathbb{C}$. Then $\lambda x_n + x \to x$ when $n \to +\infty$ and $\rho(T(\lambda x_n + x)) = \rho(\lambda Tx_n + Tx) \leq \rho(\lambda x_n + x)$ by hypothesis. So

$$\overline{\lim_{n\to\infty}}\, \rho(\lambda Tx_n + Tx) \leq \overline{\lim_{n\to\infty}}\, \rho(\lambda x_n + x) \leq \rho(x)$$

because the spectral radius is upper semi-continuous on A. We put $\varphi_n(\lambda) = \rho(\lambda Tx_n + Tx)$ and we set $\varphi(\lambda) = \overline{\lim}_{n\to\infty}\, \varphi_n(\lambda)$. By E. Vesentini's theorem (see

---

† Supported by Natural Sciences and Engineering Research Council of Canada Grant A7668.

[1], p. 9), φ_n is subharmonic on $\mathbb{C}$. Consequently φ satisfies the mean inequality on $\mathbb{C}$, but in general it is not upper semi-continuous. We put

$$\psi(\lambda) = \overline{\lim_{\mu\to\lambda}}\ \varphi(\mu).$$

Of course $\varphi(\lambda) \leq \psi(\lambda) \leq \rho(x)$ for $\lambda \in \mathbb{C}$. Also it is easy to see that ψ is upper semi-continuous and satisfies the mean inequality, so it is subharmonic and bounded on $\mathbb{C}$. By Liouville's theorem for bounded subharmonic functions (see [1], p. 166) we conclude that ψ is a constant. So

$$\rho(Tx) = \varphi(0) \leq \psi(0) = \psi(\lambda) \quad \text{for every} \quad \lambda \in \mathbb{C}.$$

By the upper semi-continuity of the spectral radius on B, we have:

$$\varphi(\lambda) \leq \rho(\lambda Ta + Tx).$$

So $\psi(\lambda) \leq \overline{\lim}_{\mu\to\lambda}\, \rho(\mu Ta + Tx) \leq \rho(\lambda Ta + Tx)$. Consequently we have $\rho(Tx) \leq \rho(Ta + Tx)$ for every $x \in A$. Taking $x = y - a$ in this formula we obtain

$$\rho(Ty - Ta) \leq \rho(Ty)$$

for every y in A. The linear mapping T being onto we conclude that $\rho(Ta + u) = 0$ for every u quasi-nilpotent in B. But by Zemánek's characterization of the radical (see [1], p. 23) we have Ta in the radical of B, and so $Ta = 0$.

REMARK 1. If in Theorem 1 we replace the condition "T onto" by "T(A) is dense in B" the same conclusion is not necessarily true. To see this we take for A a Banach algebra containing a countable linear basis $(f_n)_{n\geq 1}$ such that $\|f_n\| = 1/n$, and for B the example given by P. G. Dixon (see [1], pp. 39-41). In this example B has a dense subalgebra B_0, which is linearly generated by a countable family $(m_n)_{n\geq 1}$ of elements verifying $\|m_n\| = 1$ and is such that $\rho(v) = 0$ for v in B_0. We define T as follows. Let H be a Hamel basis for A containing $\{f_n\}$, set $Tf_n = m_n$, set $Tx = 0$ for each $x \in H\setminus\{f_n\}$, and extended T linearly to A. Of course, $T(A) = B_0$, and so T(A) is dense in the semi-simple Banach algebra B. Also, $\rho(Tx) \leq \rho(x)$ for every x in A because $\rho(Tx) = 0$. But T is not continuous because

$$\frac{\|Tf_n\|}{\|f_n\|} = n.$$

REMARK 2. If we suppose only that T is an epimorphism from A onto B, of course, the condition that $\rho(Tx) \leq \rho(x)$ is verified, and in this case the use of Zemánek's result is not necessary. By formula (*), with $Ty = 0$, we obtain

$\rho(Ta) = 0$. In this situation it is easy to show that $I = \{Ta \in B : \text{there exists } X_n \to 0,\ Tx_n \to Ta\}$ is a two-sided ideal of B. But this two-sided ideal is contained in the set of quasi-nilpotent elements of B, so is in the Jacobson radical, and then $Ta = 0$.

COROLLARY 1. (B. E. Johnson's theorem). A semi-simple Banach algebra has a unique complete norm topology, and every epimorphism from a Banach algebra onto a semi-simple Banach algebra is continuous.

COROLLARY 2. On a semi-simple Banach algebra every involution is continuous.

COROLLARY 3. Let A be a complex Banach algebra with identity and let B be a semi-simple Banach algebra with identity. Suppose that T is a linear mapping from A onto B which sends each invertible element to an invertible element. Then T is continuous.

Proof. Of course we have $\mathrm{Sp}\, Tx \subset \mathrm{Sp}\, x$ and so we can apply Theorem 1.

If these conditions are satisfied is it true that T is a Jordan morphism (see [2])?

2. Morphisms with dense range

Is a morphism from a Banach algebra onto a dense subalgebra of a semi-simple Banach algebra continuous? This problem is mentioned in [4], p. 140, in [5], p. 43, and elsewhere in this Volume. The only partial result obtained in this direction is Theorem 2 which is in some sense a converse to a result of T. Kato which implies the following: if T is a continuous morphism from A into B with the range of T of finite or countable codimension then T is onto.

THEOREM 2. Let A be a complex Banach algebra and B be a semi-simple Banach algebra. Suppose that T is a morphism from A into B and suppose that its range $T(A)$ is dense in B with finite or countable codimension. Then T is continuous and onto.

Proof. Let $I = \{b \in B : \text{there exists } x_n \to 0 \text{ with } Tx_n \to b\}$. Because $T(A)$ is dense, I is a closed two-sided ideal of B. The argument used in the proof of Theorem 1 shows that every element of $I \cap T(A)$ is quasi-nilpotent. Let y be in I. Of course $e^{\alpha y} - 1$ is in I for every complex number α. The cosets corresponding to these elements in the quotient linear space $B/T(A)$ are linearly dependent because $T(A)$ is of finite or countable codimension. Hence there exist $\alpha_1,\dots,\alpha_n$, $\beta_1,\dots,\beta_n$ different from 0 such that

$$u = \beta_1(e^{\alpha_1 y} - 1) + \cdots + \beta_n(e^{\alpha_n y} - 1) \in T(A).$$

Of course $u \in I \cap T(A)$ so $\rho(u) = 0$. By the holomorphic functional calculus the spectrum of y is included in the set of zeros of the entire function

$$f(z) = \beta_1(e^{\alpha_1 z} - 1) + \cdots + \beta_n(e^{\alpha_n z} - 1),$$

which is not identically zero. So the spectrum of y is finite. By Newburgh's continuity theorem (see [1], p. 8), the spectrum function is continuous at y, and so $\rho(y) = 0$. Hence $I \subset \mathrm{Rad}\, B = \{0\}$. So T is continuous.

References

[1] B. Aupetit, Propriétés spectrales des algèbres de Banach, Lecture Notes in Mathematics 735 (Springer-Verlag, Berlin, 1979).

[2] __________, Une généralisation du théorème de Gleason-Kahane-Żelazko pour les algèbres de Banach, Pacific J. Math., 85 (1979), 11-17.

[3] __________, The uniqueness of the complete norm topology in Banach algebras and Banach Jordan algebras, J. Functional Analysis, to appear.

[4] H. G. Dales, Automatic continuity: a survey, Bull. London Math. Soc., 10 (1978), 129-183.

[5] A. M. Sinclair, Automatic Continuity of Linear Operators, Cambridge University Press, Cambridge, 1976.

Département de Mathématiques
Université Laval
Québec, G1K 7P4
Canada

THE UNIQUENESS OF NORM PROBLEM IN BANACH ALGEBRAS WITH FINITE DIMENSIONAL RADICAL

Richard J. Loy

Let A be a Banach algebra with radical R. If R is trivial it is well known (see, for example, [2] Theorem 25.9) that A has unique topology as a Banach algebra, while if R is nontrivial this property may fail even in the commutative case with $\dim R = 1$, $AR = 0$ (see [1]). This latter situation was considered in [8] and more generally for finite dimensional annihilator radicals in [14]. In the present paper we extend and modify the results of [14] and investigate how the methods of [13] can be used in the area.

For a Banach algebra A with norm $\|\cdot\|$ denote by π the map from $A \otimes A$ to A induced by multiplication: $\pi(\Sigma x_i \otimes y_i) = \Sigma x_i y_i$. Taking the projective norm $\|\cdot\|_\gamma$ on $A \otimes A$ let $\|\cdot\|_\pi$ be the quotient norm on $A^2 = \pi(A \otimes A)$:

$$\begin{aligned} \|z\|_\pi &= \inf\{\|w\|_\gamma : \pi(w) = z\} \\ &= \inf\{\Sigma \|x_i\| \cdot \|y_i\| : \Sigma x_i y_i = z\}. \end{aligned}$$

Clearly $\|\cdot\|_\pi \geq \|\cdot\|$ on A^2 and these two norms are equivalent on A^2 if and only if π is open. Now let T denote the convex hull of the set $\{xy : \|x\| \leq 1, \|y\| \leq 1\}$. In [8] and [14] A is said to have property (S) if the gauge function of T is equivalent to $\|\cdot\|$ on A^2. But this gauge function is precisely $\|\cdot\|_\pi$, so that A has (S) if and only if π is open. Property (S) is used in [8] to give necessary and sufficient conditions for strong decomposability in commutative Banach algebras with one dimensional radical, and in [9] to the same end for certain Banach algebras with general nil radical. A similar condition (in view of the results of [13]) that the maximal ideals M satisfy M^n closed and cofinite, is used in [4] for the same type of result. There are also relevant remarks in [6].

The same construction can be done for the n-fold tensor product and the resulting quotient norm on A^n will again be denoted by $\|\cdot\|_\pi$. Lack of explicit usage of n will simplify symbolism and should cause no difficulties. For similar reasons the annihilator condition $A^k R = 0$ for some $k \geq 1$ will be used to indicate the triviality of <u>all</u> $(k+1)$-fold products with one factor in R, not merely those with <u>last</u> factor in R.

1. We begin with the following result which is the exact parallel of [14] Lemma 1.

LEMMA 1. Let A be an algebra with (Jacobson) radical R satisfying $A^{n-1}R = 0$ for some $n \geq 2$. Then for any two Banach algebra norms $\|\cdot\|$, $\|\cdot\|'$ on

A the corresponding norms $\|\cdot\|_\pi$, $\|\cdot\|'_\pi$ on A^n are equivalent.

Proof. Since A/R is semisimple the two quotient norms $|\cdot|$, $|\cdot|'$ thereon are equivalent, so there is a constant $K > 0$ such that $|\cdot|' \leq K|\cdot|$. Now let $x \in A^n$ and $\varepsilon > 0$. Then there are $x_{ij} \in A^n$, $1 \leq i \leq k$, $1 \leq j \leq n$ such that

$$x = \sum_{i=1}^{k} x_{i1}x_{i2}\cdots x_{in}, \quad \|x\|_\pi \geq \sum_{i=1}^{k} \|x_{i1}\| \cdot \|x_{i2}\| \cdots \|x_{in}\| - \varepsilon .$$

Thus

$$\begin{aligned} \|x\|_\pi &\geq \sum_{i=1}^{k} |x_{i1}| \cdot |x_{i2}| \cdots |x_{in}| - \varepsilon \\ &\geq K^{-n} \sum_{i=1}^{k} |x_{i1}|' \cdot |x_{i2}|' \cdots |x_{in}|' - \varepsilon \\ &\geq K^{-n} \sum_{i=1}^{k} \|x_{i1} + r_{i1}\|' \cdot \|x_{i2} + r_{i2}\|' \cdots \|x_{in} + r_{in}\|' - 2\varepsilon \end{aligned}$$

for suitable $r_{ij} \in R$. But then

$$\begin{aligned} \|x\|_\pi &\geq K^{-n} \left\| \sum_{i=1}^{k} (x_{i1} + r_{i1})(x_{i2} + r_{i2}) \cdots (x_{in} + r_{in}) \right\|'_\pi - 2\varepsilon \\ &= K^{-n}\|x\|'_\pi - 2\varepsilon \end{aligned}$$

by the hypothesis $A^{n-1}R = 0$. It follows that $\|\cdot\| \geq K^{-n}\|\cdot\|'_\pi$ and the same argument with $\|\cdot\|$, $\|\cdot\|'$ interchanged completes the proof.

Another step in the argument of [14] is the $n = 2$, $\dim R = 1$ case of the following.

LEMMA 2. Let A be a Banach algebra with norm $\|\cdot\|$ and radical R satisfying $\dim(R \cap A^n) < \infty$ for some $n \geq 2$. Then $\|\cdot\|$ and $\|\cdot\|_\pi$ are equivalent on A^n if $|\cdot|$ and $|\cdot|_\pi$ are equivalent on $(A/R)^n$.

Proof. Let $K > 0$ be a constant such that $|\cdot| \leq |\cdot|_\pi \leq K|\cdot|$ on $(A/R)^n$. Take a sequence $\{x_k\} \subset A^n$ with $\|x_k\| \to 0$. Then certainly $|x_k + R| \to 0$ and hence $|x_k + R|_\pi \to 0$. Now for each fixed k choose x_{ij}, $(1 \leq j \leq m,\ 1 \leq j \leq n)$ such that

$$x_k - \sum_{i=1}^{m} x_{i1}\cdots x_{in} \in R, \quad |x_k + R|_\pi > \sum_{i=1}^{m} |x_{i1} + R| \cdot |x_{i2} + R| \cdots |x_{in} + R| - \frac{1}{k}.$$

and then choose y_{ij} $(1 \leq i \leq m,\ 1 \leq j \leq n)$ such that $x_{ij} - y_{ij} \in R$ and

$$\begin{aligned} |x_k + R|_\pi &> \sum_{i=1}^{m} \|y_{i1}\| \cdot \|y_{i2}\| \cdots \|y_{in}\| - \frac{1}{k} \\ &\geq \left\| \sum_{i=1}^{m} y_{i1}y_{i2}\cdots y_{in} \right\|_\pi - \frac{1}{k}. \end{aligned}$$

Thus $r_k = x_k - \sum_{i=1}^{m} y_{i1}y_{i2}\cdots y_{in} \in R \cap A^n$ satisfies

$$|x_k + R|_\pi \geq \|x_k - r_k\|_\pi - \frac{1}{k}.$$

But $|x_k + R|_\pi \to 0$ whence $\|x_k - r_k\|_\pi \to 0$ and so $\|x_k - r_k\| \to 0$. Since $\|x_k\| \to 0$ we have $\|r_k\| \to 0$, and hence $\|r_k\|_\pi \to 0$ since $\dim(R \cap A^n) < \infty$. It then follows that $\|x_k\|_\pi \to 0$ as required.

The lemma clearly holds for any closed two-sided ideal in place of the radical; the significance of using the radical is that then the equivalence or otherwise of $|\cdot|$ and $|\cdot|_\pi$ is independent of the choice of norm since A/R is semisimple. The converse is discussed in §2 below; it does not hold in general.

THEOREM 1. Let A be a Banach algebra satisfying $A^{n-1}R = 0$, $\dim(R \cap A^n) < \infty$ for some $n \geq 2$. If $|\cdot|$ and $|\cdot|_\pi$ are equivalent on $(A/R)^n$ then all Banach algebra norms on A are equivalent on A^n.

<u>Proof</u>. Let $\|\cdot\|$ be the given norm on A, and $\|\cdot\|'$ any other Banach algebra norm on A. By semisimplicity of A/R the norms $|\cdot|$ and $|\cdot|'$ are equivalent and so Lemma 2 is applicable to both $\|\cdot\|$ and $\|\cdot\|'$. But by Lemma 1 $\|\cdot\|_\pi$ and $\|\cdot\|'_\pi$ are equivalent on A^n and we are done.

COROLLARY 1. Let A be a separable Banach algebra such that $\dim R < \infty$, and $\operatorname{codim} A^{n+1} < \infty$, $A^nR = 0$ for some $n \geq 1$. Then A has unique topology as a Banach algebra.

<u>Proof</u>. Since $A^nR = 0$ it is easily seen that $(A/R)^{n+1} \cong A^{n+1}$ so that $\operatorname{codim}(A/R)^{n+1} \leq \operatorname{codim} A^{n+1} < \infty$. Applying Theorems 1.1 and 1.2 of [13] to A/R we deduce that $|\cdot|$ and $|\cdot|_\pi$ are equivalent on $(A/R)^n$. Thus by Theorem 1 all Banach algebra norms on A are equivalent on A^{n+1}. By [13], A^{n+1} is closed in the given (and hence any) Banach algebra norm on A, and the result is now immediate.

Alternatively, by the uniqueness of the Banach algebra topology on A/R, and $\dim R < \infty$, A is separable under any Banach algebra norm, so [13] directly applies to A and an appeal to Lemma 1 would complete the argument. The first proof given shows the hypotheses imply (and are easily seen to be equivalent to) $\dim R < \infty$ and $\operatorname{codim}(A/R)^{n+1} < \infty$, $A^nR = 0$ for some $n \geq 1$, paralleling Theorem 1 of [14]. Theorem 2 of [14] is a converse to Corollary 1 in the case $n = 1$, but the proof given is unfortunately invalid. (The isomorphism F of Lemma 5 of [14] is continuous since $\|\cdot\|_\pi \geq \|\cdot\|$ and so is bicontinuous by the open mapping theorem. This contradicts the subsequent step in the argument. As well the converse to Lemma 2 is invoked.) We give a correct version in the next theorem. The proof attempted in [14] uses an extension of $\|\cdot\|_\pi$ to the whole algebra and such an approach could be used here. Our proof, based on ideas of [8] and [9], is a little more direct.

THEOREM 2. Let A be a Banach algebra with nontrivial annihilator radical R.

1. In order that all Banach algebra norms on A be equivalent on A^2 it is necessary that A have property (S). If, further, A is decomposable, it is necessary that A/R have property (S).

2. In order that A have a unique topology as a Banach algebra it is necessary that A have property (S) and $\text{codim } A^2$ be finite. If, further, A is decomposable, it is necessary that A/R have property (S), and that both $\text{codim}(A/R)^2$ and $\dim R$ be finite.

Proof. Let $\|\cdot\|$ be the given norm on A. Supposing (S) fails in A let φ be a linear functional on A which is $\|\cdot\|$-discontinuous on A^2 yet $|\varphi(x)| \leq \|x\|_\pi$ for $x \in A^2$; such φ exist since the $\|\cdot\|$-unit ball of A^2 is $\|\cdot\|_\pi$-unbounded. If $\text{codim } A^2$ is infinite let φ be a $\|\cdot\|$-discontinuous linear functional vanishing on A^2. In either case, by taking trivial multiplication in the second summand $A \oplus \underset{\sim}{C}$ is a Banach algebra under the norm

$$\|(x,\alpha)\|_1 = 2(\|x\| + |\varphi(x) - \alpha|)$$

which is an algebra norm because of the choice of φ:

$$\|(x,\alpha) \cdot (y,\beta)\|_1 = 2(\|xy\| + |\varphi(xy)|) \leq 4\|x\| \cdot \|y\|$$
$$\leq \|(x,\alpha)\|_1 \cdot \|(y,\beta)\|_1 .$$

Now take $r \in R \setminus \{0\}$ and define $\Phi : A \oplus \underset{\sim}{C} \to A$ by $\Phi(x,\alpha) = x - \alpha r$. Then Φ is an epimorphism of $A \oplus \underset{\sim}{C}$ onto A with kernel $\underset{\sim}{C}(r,1)$ which is closed in $A \oplus \underset{\sim}{C}$. Denote by $\|\cdot\|'$ the quotient norm induced on A. Since $(0,1) \notin \underset{\sim}{C}(r,1)$ we have

$$\|r\|' = \inf\{\|(0,1) - \lambda(r,1)\|_1 : \lambda \in \underset{\sim}{C}\} > 0 .$$

Now take $(x_n) \subset A$ such that $\|x_n\| \to 0$ and $\|(x_n,1)\|_1 \to 0$. In the case when (S) fails in A our choice of φ ensures such (x_n) can be chosen in A^2. Then

$$\|x_n\|' = \inf\{\|x_n + \lambda(r,1)\|_1 : \lambda \in \underset{\sim}{C}\}$$
$$\geq \inf\{\|\lambda(r,1) - (0,1)\|_1 - \|(x_n,1)\|_1 : \lambda \in \underset{\sim}{C}\}$$
$$\geq \tfrac{1}{2} \|r\|'$$

if n is sufficiently large. Thus $\|\cdot\|$ and $\|\cdot\|'$ are inequivalent (on A^2 in the case when (S) fails).

If we have A decomposable with (S) failing in A/R or $\text{codim}(A/R)^2$ infinite then proceed as above with A/R in place of A as far as the construction of $\|\cdot\|_1$ on $A/R \oplus \underset{\sim}{C}$. Then on $A/R \oplus \underset{\sim}{C}$ we also have the Banach algebra norm

$$\|(x,\alpha)\|_2 = |x| + |\alpha|$$

inequivalent to $\|\cdot\|_1$ (on $(A/R)^2$ if (S) fails). Now take $r_0 \in R\setminus\{0\}$ and let R_0 be a $\|\cdot\|$-closed subspace of R with $R = \underset{\sim}{C}r_0 \oplus R_0$. Since $A \cong A/R \oplus \underset{\sim}{C}r_0 \oplus R_0$ we can define two inequivalent Banach algebra norms on A via this isomorphism:

$$\|(x,\alpha,r)\|_1 = \|(x,\alpha)\|_1 + \|r\|$$
$$\|(x,\alpha,r)\|_2 = \|(x,\alpha)\|_2 + \|r\|.$$

Once again inequivalence holds on A^2 in the case when (S) fails.

It remains to show that dim R finite is necessary in the decomposable situation. But if dim R is infinite take two inequivalent Banach space norms $\|\cdot\|_1$, $\|\cdot\|_2$ on R. Then via the isomorphism $A \cong A/R \oplus R$ we have two inequivalent Banach algebra norms on A:

$$\|(x,r)\|_1 = |x| + \|r\|_1$$
$$\|(x,r)\|_2 = |x| + \|r\|_2.$$

We remark that the use of a $\|\cdot\|$-discontinuous, $\|\cdot\|_\pi$-continuous functional is exactly the idea of the Feldman example [7], [1]. A similar technique was used in [12].

COROLLARY 2. Let A be a decomposable Banach algebra with $1 \leq \dim R < \infty$ and $AR = 0$. Then all Banach algebra norms on A are equivalent on A^2 if and only if A/R has property (S) if and only if A is strongly decomposable.

<u>Proof</u>. Theorems 1 and 2 give the first equivalence, Corollary 5.3 of [9] the second.

COROLLARY 3. Let A be a separable Banach algebra with $1 \leq \dim R < \infty$, $AR = 0$. Then A has unique topology as a Banach algebra if and only if $\text{codim}\, A^2 < \infty$. If $\text{codim}\, A^2 < \infty$ and A is decomposable then A is strongly decomposable.

<u>Proof</u>. Corollaries 1, 2 and Theorem 2.

We would like similar results for more general radicals if possible. In fact perusal of the proof of Theorem 2 shows that for the result on property (S) in A we can relax the hypothesis on R to merely requiring R contain a (non zero) annihilating element. It is by no means clear, however, that the same weakening of hypothesis works for the A/R result (why should $\|\cdot\|_1$, $\|\cdot\|_2$ be algebra norms?); a converse to Lemma 2 would of course help here. On the other hand property (S) is trivially satisfied in any algebra with identity, and there are Banach algebras A with unique topology as Banach algebras in which (S) fails and A^2 has infinite codimension. In such algebras, which range from semisimple to radical (see §2) the proof of Theorem 2 gives a discontinuous embedding of A into a larger Banach algebra with an annihilating element.

There is one approach which yields results at least for finite-dimensional radicals. For a Banach algebra A with radical R set

$$N = \{x \in A : xR = Rx = 0\},$$

the annihilator of R in A. Then N is a Banach algebra with radical $N \cap R$ which is annihilating, so we are in the situation considered above.

COROLLARY 4. Let A be a separable Banach algebra with nontrivial finite-dimensional radical. Then A has unique topology as a Banach algebra if and only if $\operatorname{codim} N^2 < \infty$.

<u>Proof</u>. If $k > 0$ is the least integer with $R^k = 0$, then $0 \neq R^{k-1} \subset N \cap R$. Thus N satisfies the hypotheses of Corollary 3 and so has unique topology as a Banach algebra if, and only if, N^2 has finite codimension in N, equivalently in A since N has finite codimension in A. But N is a closed subalgebra of A for any Banach algebra norm on A and so the result follows.

In the case A commutative with identity, and with singly generated radical, $(N+R)/R$ is a maximal ideal in A/R (see [14]). Since $N/N \cap R$ is topologically isomorphic to $(N+R)/R$ we have the following version of Theorem 3 of [14].

COROLLARY 5. Let A be a commutative unital Banach algebra with singly generated finite dimensional radical. If A is separable and the maximal ideals M of A/R satisfy $\operatorname{codim} M^2 < \infty$ then A has unique topology as a Banach algebra.

<u>Proof</u>. $(N/N \cap R)^2$ has finite codimension in $N/N \cap R$ whence N^2 has finite codimension in N. Thus Corollary 4 applies.

The hypothesis here on <u>all</u> maximal ideals of A is stronger than necessary, but ensures A is of class $\mathfrak{B}$. Related results for such algebras are given in §5 of [4].

2. We now give some examples and counterexamples relevant to the results of §1 and raise some open problems.

To begin with, consider the situation of Lemma 2. Let A be a Banach algebra with radical R satisfying $\dim(R \cap A^2) < \infty$. Then Lemma 2 shows that property (S) holds in A if it holds in A/R. The following elegant example of Dr. George Willis shows the converse fails.

For each positive integer m set $A_m = \underset{\sim}{C} \oplus \underset{\sim}{C}$ with product and norm given by

$$(\lambda,\mu)(\lambda',\mu') = (\lambda\lambda',0), \ \|(\lambda,\mu)\| = m|\lambda-\mu| + m^{1/2}|\mu| .$$

The norm is submultiplicative since

$$\begin{aligned} \|(\lambda,\mu)(\lambda',\mu')\| &= m|\lambda\lambda'| \\ &\leq m^{1/2}(|\lambda-\mu| + |\mu|)\cdot m^{1/2}(|\lambda'-\mu'| + |\mu'|) \\ &\leq (m|\lambda-\mu| + m^{1/2}|\mu|)\cdot(m|\lambda'-\mu'| + m^{1/2}|\mu'|) \\ &= \|(\lambda,\mu)\|\cdot\|(\lambda',\mu')\| . \end{aligned}$$

Then $\|(1,0)\| \leq \|(1,0)\|_\pi \leq \|(1,1)\|\cdot\|(1,1)\| = \|(1,0)\|$ so that $\|\cdot\| = \|\cdot\|_\pi$ on $A_m^2 = \underset{\sim}{C} \oplus 0$. But on $A_m/R_m = \underset{\sim}{C}$

$$\begin{aligned} |(1,0)+R_m| &= \inf\{m|1-\mu| + m^{1/2}|\mu|\} = m^{1/2} \\ |(1,0)+R_m|_\pi &= \inf\{\Sigma|(\lambda_i,0)+R_m|\cdot|(\lambda_i',0)+R_m| : \Sigma\lambda_i\lambda_i' = 1\} = m . \end{aligned}$$

Now let A be the c_0-sum of the A_m, $m \geq 1$:

$$\begin{aligned} A &= \{(\lambda_m,\mu_m) : m|\lambda_m-\mu_m| + m^{1/2}|\mu_m| \to 0\} \\ \|(\lambda_m,\mu_m)\| &= \sup_m\{m|\lambda_m-\mu_m| + m^{1/2}|\mu_m|\} . \end{aligned}$$

Then $A^2 = \{(\lambda_m,0) : m\lambda_m \to 0\}$, $R = \{(0,\mu_m) : m\mu_m \to 0\}$, so $A^2 \cap R = 0$. Because of the norm on A and the fact that $\|\cdot\| = \|\cdot\|_\pi$ on each A_m^2 it follows that $\|\cdot\| = \|\cdot\|_\pi$ on A^2. On the other hand A/R contains a copy of A_m/R_m for each $m \geq 1$ and so $|\cdot|$ and $|\cdot|_\pi$ cannot be equivalent on $(A/R)^2$. /

This example makes essential use of an infinite-dimensional radical; what about finite-dimensional radicals? Suppose A is a Banach algebra with property (S) with $\|\cdot\| \leq \|\cdot\|_\pi \leq K\|\cdot\|$ on A^2. Then if $x \in A^2$

$$\begin{aligned} |x+R| \leq |x+R|_\pi &= \inf\{\Sigma|x_{i1}+R|\cdot|x_{i2}+R| : \Sigma x_{i1}x_{i2} - x \in R\} \\ &= \inf\{\Sigma\|x_{i1}\|\cdot\|x_{i2}\| : \Sigma x_{i1}x_{i2} - x \in R\} \\ &= \inf\{\|y\|_\pi : y - x \in R \cap A^2\} \\ &\leq K\inf\{\|y\| : y - x \in R \cap A^2\} . \end{aligned}$$

If $R \cap A^2 = R$ the right side here is just $K|x+R|$ and we see A/R has property (S). At the other extreme, if $R \cap \overline{A^2} = 0$ and $\dim R < \infty$, there is K' such that if $x \in A^2$, $|x+R| \leq \|x\| \leq K'|x+R|$ whence $|x+R| \leq |x+R|_\pi \leq KK'|x+R|$ so that again A/R has property (S). Indeed, if $R \cap A^2 = 0$ and A/R has property (S) with $|\cdot| \leq |\cdot|_\pi \leq L|\cdot|$ on $(A/R)^2$ then for $x \in A^2$, $|x+R| \leq \|x\| \leq \|x\|_\pi \leq L|x+R|$. In particular if $\{x_k\} \subset A^2$ with $x_k \to r \in R$ then $|x_k + R| \to 0$ so $\|x_k\| \to 0$ and hence $r = 0$. Thus $R \cap \overline{A^2} = 0$. For the case $\dim R = 1$ one of the cases $R \cap A^2 = R$, $R \cap \overline{A^2} = 0$ must occur and so the problem reduces to whether $R \cap A^2 = 0$ necessitates $R \cap \overline{A^2} = 0$. If $R \cap A^2 = 0$ then $RA = 0$ and here the converse to Lemma 2 is stated as obvious in [14].

<u>Problem 1</u>. If $\dim R < \infty$ and A has property (S) does A/R necessarily have property (S)? Indeed, if $\dim R = 1$ and $R \cap A^2 = 0$ is $R \cap \overline{A^2} = 0$? We conjecture the answer to both is in the negative. [If we consider a one-dimensional ideal J then $A^2 \cap J = 0$, $\overline{A^2} \cap J = J$ is possible in the absence of property (S). The construction of [15] yields such an example.]

The convolution algebras $\ell^1(\omega)$ where ω is a submultiplicative weight on the non-negative integers receive much attention elsewhere in this volume. By [11] these algebras always have unique topology as Banach algebras. It is thus of some interest to confirm the following (expected) result; a more general theorem is given in [3].

PROPOSITION. If ω is a radical weight, M the maximal ideal in $\ell^1(\omega)$, then property (S) fails in M and codim $(M)^2$ is uncountable.

<u>Proof</u>. Consider the standard basis element e_p for some $p \geq 2$ and suppose $e_p = \sum_{n=1}^k x_n y_n$ in M. Then

$$\sum_{n=1}^{k} \|x_n\| \cdot \|y_n\| \geq \sum_{n=1}^{k} \sum_{r=1}^{p-1} |x_n(r)\omega_r| \cdot |y_n(p-r)\omega_{p-r}|$$

$$\geq \sum_{n=1}^{k} \sum_{r=1}^{p-1} |x_n(r) y_n(p-r)| \Omega_p$$

where $\Omega_p = \min\{\omega_r \omega_{p-r} : 1 \leq r \leq p-1\}$. Thus

$$\sum_{n=1}^{k} \|x_n\| \cdot \|y_n\| \geq \left| \sum_{n=1}^{k} \sum_{r=1}^{p-1} x_n(r) y_n(p-r) \right| \Omega_p = \Omega_p$$

since $\sum_{n=1}^k \sum_{r=1}^{p-1} x_n(r) y_n(p-r) = e_p(p) = 1$. It follows that $\|e_p\|_\pi = \Omega_p$.

If property (S) held in M with $\|x\| \leq \|x\|_\pi \leq K\|x\|$ for $x \in (M)^2$, then in particular $K\omega_p \geq \Omega_p$ for each $p \geq 2$. But then a simple inductive argument shows $\omega_p \geq K^{1-p}\omega_1^p$ whence $\underline{\lim}\ \omega_p^{1/p} \geq K^{-1}\omega_1 > 0$, so that ω is not a radical weight. Since M is separable the last part of the proposition follows from [13].

Corollary 5 of §1 is a weak version of Theorem 3 of [14]. We now give counter-examples to the result stated in [14]. Let $A = \mathbb{C} \oplus \ell^1 \oplus \mathbb{C}$ with direct sum norm and product

$$(\alpha,x,\beta)\cdot(\alpha',x',\beta') = (\alpha\alpha',\alpha x'+\alpha' x+xx',\alpha\beta'+\alpha'\beta+x(0)\beta'+x'(0)\beta)$$

where ℓ^1 has pointwise product. Thus $(1,0,0)$ is an identity, $R = 0 \oplus 0 \oplus \mathbb{C}$ and $A/R = \mathbb{C} \oplus \ell^1$. The maximal ideals of A/R are $M_0 = 0 \oplus \ell^1$ and $M_{n+1} = \{(\lambda,x) : \lambda = -x(n)\}$, $n \geq 0$. Consider M_1; its elements can be written in the form $(\lambda,-\lambda,x)$ for $\lambda \in \mathbb{C}$, $x \in \ell^1$. But

$$(\lambda,-\lambda,x) = (\lambda^{1/2},-\lambda^{1/2},0)^2 + (1,-1,0)\cdot(0,0,x).$$

Thus $M_1^2 = M_1$ and $\|\cdot\|_\pi \leq 2\|\cdot\|$ thereon. It follows that $M_n^2 = M_n$ for $n \geq 1$ and they all have property (S). Further M_0 has property (S) and satisfies M_0^2 properly dense in M_0. Thus property (S) holds in all the maximal ideals of the unital algebra A/R. But in A

$$\begin{aligned} N &= \{(\alpha,x,\beta) : (\alpha,x,\beta)R = 0\} \\ &= 0 \oplus \{x \in \ell^1 : x(0) = 0\} \oplus \mathbb{C} \end{aligned}$$

so that N^2 has infinite codimension. Corollary 4 now shows that A does <u>not</u> have unique topology as a Banach algebra.

Again, let B be a commutative separable semisimple Banach algebra with identity which contains maximal ideals I, J such that $\operatorname{codim} I^2 < \infty$ and J does not satisfy property (S). Define $A = B \oplus \mathbb{C}$ as above:

$$(x,\alpha)\cdot(y,\beta) = (xy,\alpha\hat{y}(I) + \beta\hat{x}(I)).$$

Then $N = I \oplus \mathbb{C}$ and so by Corollary 4 A has unique topology as a Banach algebra. But by hypothesis J is a maximal ideal in $B = A/R$ with property (S). An example of such a B is given by

$$B = \{f \in C[-1,1] : f \mid [0,1] \in C^1[0,1]\}$$

$$\|f\| = \max_{-1\leq t\leq 1} |f(t)| + \max_{0\leq t\leq 1} |f'(t)|$$

with $I = \{f \in B : f(-1) = 0\}$ (which has a bounded approximate identity so $I^2 = I$) and $J = \{f \in B : f(1) = 0\}$ (which fails to have property (S)).

These two examples refute both parts of Theorem 3 of [14].

To circumvent the unital hypothesis in Corollary 5 one could adjoin an identity and work in the resulting algebra A_+ (as is done in [8]). This raises the following question.

Problem 2. Is there any relation between property (S) in a maximal ideal M of A_+ and in the corresponding maximal modular ideal $M \cap A$ of A? As shown above by ℓ^1_+, $M^2 = M$ yet codim $(M \cap A)^2 = \infty$ is possible; but note that $(M \cap A)^2 = M \cap A$ implies $M^2 = M$ [10].

Finally, we have used separability to obtain property (S) and A^2 closed from the hypothesis codim $A^2 < \infty$.

Problem 3. In the situations considered in this paper, is separability necessary? (Compare with the pathologies demonstrated in [5] and [13].)

ACKNOWLEDGEMENT. We would like to thank Dr. R. S. Anderssen for his help with translation of portions of [14].

References

[1] W. G. Bade and P. C. Curtis, Jr., Homomorphisms of commutative Banach algebras, Amer. J. Math., 82 (1960), 589-608.

[2] F. F. Bonsall and J. Duncan, Complete Normed Algebras, Springer-Verlag, New York, 1973.

[3] N. Grønbaek, Weighted discrete convolution algebras, this Volume.

[4] H. G. Dales and J. P. McClure, Continuity of homomorphisms into certain commutative Banach algebras, Proc. London Math. Soc., 26 (1973), 69-81.

[5] P. G. Dixon, Nonseparable Banach algebras whose squares are pathological, J. Functional Analysis, 26 (1977), 190-200.

[6] J. Esterle, An algebra norm on ℓ^2 whose restriction to ℓ^1 is the usual ℓ^1-norm, Proc. Amer. Math. Soc., 69 (1978), 303-307.

[7] C. J. Feldman, The Wedderburn principal theorem in Banach algebras, Proc. Amer. Math. Soc., 2 (1951), 771-777.

[8] A. J. Helemskii, Description of commutative annihilator Banach algebras, Soviet Math. Dokl., 5 (1964), 902-904.

[9] B. E. Johnson, The Wedderburn decomposition of Banach algebras with finite dimensional radical, Amer. J. Math., 90 (1968), 866-876.

[10] R. J. Loy, Commutative Banach algebras with idempotent maximal ideals, J. Austral. Math. Soc., 9 (1969), 275-286.

[11] __________, Uniqueness of the complete norm topology and continuity of derivations on Banach algebras, Tôhoku Math. J., 22 (1970), 371-378.

[12] R. J. Loy, Commutative Banach algebras with non-unique complete norm topology, Bull. Austral. Math. Soc., 10 (1974), 409-420.

[13] __________, Multilinear mappings and Banach algebras, J. London Math. Soc., (2), 14 (1976), 423-429.

[14] V. M. Zinde, The 'unique norm' property for commutative Banach algebras with finite dimensional radical, Vestnik Moskov. Univ. Ser. I Mat. Meh., 25 (1970), 3-8. (English translation, Moscow Univ. Math. Bull., 25 (1970).)

[15] P. G. Dixon, On the intersection of the principal ideal generated by powers in a Banach algebra, this Volume.

Department of Mathematics
Faculty of Science
The Australian National University
Canberra ACT 2600 Australia

DERIVATIONS IN COMMUTATIVE BANACH ALGEBRAS

Philip C. Curtis, Jr.

Let $\mathfrak{A}$ be a commutative Banach algebra with identity and D a derivation in $\mathfrak{A}$, that is, a linear map in $\mathfrak{A}$ satisfying the derivative identity, $D(ab) = aDb + (Da)b$, $a,b \in \mathfrak{A}$. In 1955 Singer and Wermer [7] showed that if D is a bounded derivation in $\mathfrak{A}$, then $D(\mathfrak{A}) \subset R$, the Jacobson radical of $\mathfrak{A}$, and they conjectured that $D(\mathfrak{A}) \subset R$ for an arbitrary derivation, bounded or not. Since then this conjecture has occupied the attention of various authors, [2,3,4,6]. In 1969 Johnson [4] showed that if $R = \{0\}$, then a derivation in $\mathfrak{A}$ would necessarily be bounded, hence trivial, which verified the Singer-Wermer conjecture in this case. He showed further that if D is a derivation in a not necessarily semisimple algebra $\mathfrak{A}$, then there exist finitely many orthogonal idempotents, $e_1,\ldots,e_n$ in $\mathfrak{A}$ such that if $e_0 = 1 - (e_1 + \cdots + e_n)$ then $D(e_0\mathfrak{A}) \subset R_0$, the radical of $e_0\mathfrak{A}$, and the algebras $e_i\mathfrak{A}$, $i = 1,\ldots,n$ each have a unique proper maximal ideal. Equivalently, the maximal ideal M_i or the unique complex homomorphism φ_i of $e_i\mathfrak{A}$ is isolated in the structure space $\Phi_{\mathfrak{A}}$ of $\mathfrak{A}$.

Necessarily $D(e_i\mathfrak{A}) \subset e_i\mathfrak{A}$, and if for some i, $D(e_i\mathfrak{A}) \not\subset R_i$, the radical of $e_i\mathfrak{A}$, then it is a theorem of J. Cusack that there exists a closed non-maximal prime ideal P of $e_i\mathfrak{A}$ and a closed ideal Q of $e_i\mathfrak{A}$ such that $Q \supset P$ and Q/P is radical and topologically simple. Thus the non-maximal prime ideal P is properly contained in the intersection of those closed ideals K of $e_i\mathfrak{A}$ properly containing P. A closed non-maximal prime ideal P of a commutative Banach algebra $\mathfrak{A}$ with this property we shall call <u>inaccessible</u>. Thus a closed prime ideal P is <u>accessible</u> if it is the intersection of those closed ideals properly containing it.

It is not known if inaccessible closed prime ideals exist in commutative Banach algebras. It is the purpose of this note to show that if a commutative Banach algebra $\mathfrak{A}$ has a unique maximal ideal M, then $D(\mathfrak{A}) \subset M$, provided that whenever closed non-maximal prime ideals exist in $\mathfrak{A}$, at least one of them is accessible. For general algebras $\mathfrak{A}$ this yields the following.

THEOREM. Let D be a derivation in a commutative Banach algebra $\mathfrak{A}$ with identity and with radical R. Then $D(\mathfrak{A}) \subset R$ if whenever there are isolated maximal ideals M of $\mathfrak{A}$ containing non-maximal closed prime ideals of $\mathfrak{A}$, then at least one of these prime ideals must be accessible.

1. Separating ideals and accessible primes

To study an unbounded derivation D in a commutative B-algebra $\mathfrak{A}$ with radical R one must consider the separating space

$$\mathfrak{S} = \mathfrak{S}(D) = \{a \in \mathfrak{A} : \text{there exists } \{x_n\} \subset \mathfrak{A},\ x_n \to 0,\ D x_n \to a\}.$$

The space $\mathfrak{S}$ is a closed ideal of $\mathfrak{A}$ which satisfies the following stability condition.

(1) For each sequence $\{a_n\} \subset \mathfrak{A}$ there exists an integer n_0 such that if $n > n_0$, then

$$a_1 \cdots a_n \mathfrak{S}^- = a_1 \cdots a_{n_0} \mathfrak{S}^- .$$

A closed ideal I of $\mathfrak{A}$ which satisfies (1) is called a separating ideal. See [3] and [5] for a discussion of these ideas and for proofs of the above assertions. It is an easy consequence of (1) that if I is a separating ideal of $\mathfrak{A}$, then $\dim(I/I \cap R) < \infty$. The proof is contained in [3].

1.1 LEMMA. If P is an accessible prime ideal of $\mathfrak{A}$ and I a separating ideal of $\mathfrak{A}$, then $I \subset P$, and consequently $\mathfrak{A}/P$ has no separating ideals.

Proof. If $P \not\supset I$, then since P is prime, $zI \neq 0$ for each $z \notin P$. Since I is separating, there exist $z_1, \dots, z_{n_0}$ not in P such that for each $z \notin P$

$$z_1 \cdots z_{n_0} z I^- = z_1 \cdots z_{n_0} I^- \neq 0 .$$

Set $z_0 = z_1 \cdots z_{n_0}$. Then $z_0 \notin P$. If K is a closed ideal properly containing P, pick $z \in K \setminus P$. Then $z_0 z I^- = z_0 I^- \subset K$. Therefore $z_0 I^- \subset \bigcap_{K \supsetneq P} K$, which equals P since P is accessible. This, however, is impossible since neither $z_0 \in P$ nor $I \subset P$. Lastly if I is a separating ideal of $\mathfrak{A}/P$, then $\rho^{-1}(I)$ is a separating ideal of $\mathfrak{A}$ containing P, where ρ is the natural map of $\mathfrak{A}$ onto $\mathfrak{A}/P$. Therefore $\mathfrak{A}/P$ can have no separating ideals.

Let $\mathfrak{M}$ be a Banach space which is a left Banach $\mathfrak{A}$-module, and assume $\|am\| \leq \|a\|\,\|m\|$, $a \in \mathfrak{A}$, $m \in \mathfrak{M}$. If S is a linear operator from $\mathfrak{A}$ to $\mathfrak{M}$, then the separating subspace $\mathfrak{S}(S)$ for S is defined exactly as for a derivation. It is shown in [1] p. 91-93, that $\mathfrak{S}(S)$ is a submodule of $\mathfrak{M}$ and satisfies the stability property (1) provided S is of class $\mathfrak{J}$; that is, for $a,b \in \mathfrak{A}$

$$S(ab) = aS(b) + L(a,b)$$

where L is a bilinear operator from $\mathfrak{A}$ to $\mathfrak{M}$ which is bounded in b for fixed a. A closed submodule of $\mathfrak{M}$ satisfying (1) we will call a separating submodule.

We need one further fact from automatic continuity theory (cf. [5] Lemma 1.3).

If T is a bounded operator from $\mathfrak{A}$ to a B-space Y, then TD is bounded if and only if $T(\mathfrak{S}(D)) = 0$. Hence if Q is the natural homomorphism of $\mathfrak{A}$ onto $\mathfrak{A}/\mathfrak{S}(D)$, QD is bounded, and when TD is bounded,

$$\|TD\| \leq \|QD\| \, \|T\| .$$

1.2 THEOREM. Let D be a derivation in $\mathfrak{A}$, K a closed ideal of $\mathfrak{A}$ contained in R. If $\mathfrak{A}/K$ contains no non-trivial separating ideals, then $D(\mathfrak{A}) \subset R$.

Proof. Let $\mathfrak{M} = \mathfrak{A}/K$. If $\mathfrak{M}$ is regarded as an $\mathfrak{A}$-module, then by hypothesis $\mathfrak{M}$ contains no separating sub-modules. If ρ is the natural map of $\mathfrak{A}$ onto $\mathfrak{A}/K$, set $\overline{D}_k(x) = \rho D^k(x)$. Since D is a derivation, $\overline{D}_1$ is an operator of class $\mathfrak{J}$. Since there are no non-trivial separating submodules in $\mathfrak{M}$, $\overline{D}_1$ is bounded and we set $M = \|\overline{D}_1\|$. The operator D^2 is not known to be an operator of class $\mathfrak{J}$ in $\mathfrak{A}$, therefore we may not assert that $\mathfrak{S}(D^2)$ is a separating ideal of $\mathfrak{A}$. However,

$$\begin{aligned}\overline{D}_2(xy) &= \rho D^2(xy) = \rho[xD^2(y) + 2Dx\, Dy + (D^2x)y] \\ &= \rho(x)\overline{D}_2(y) + 2\overline{D}_1 x\, \overline{D}_1 y + (\overline{D}_2 x)\rho(y) \\ &= \rho(x)\overline{D}_2(y) + L_1(x,y)\end{aligned}$$

where we set $L_1(x,y) = 2\overline{D}_1 x \overline{D}_1 y + \overline{D}_2 x\, \rho(y)$. Since $\overline{D}_1$ is bounded, the bilinear operator L is bounded in y for x fixed. Therefore $\overline{D}_2$ is of class $\mathfrak{J}$ and hence bounded since $\mathfrak{M}$ has no separating submodules. Therefore

$$\|\overline{D}_2\| = \|\overline{D}_1 D\| \leq \|\overline{D}_1\| \, \|QD\| = M\|QD\| ,$$

where $Q(\mathfrak{A}) = \mathfrak{A}/\mathfrak{S}(D)$. From the Leibnitz identity,

$$D^n(xy) = xD^n y + \sum_{k=1}^{n-1} \binom{n}{k} D^k x\, D^{n-k} y + (D^n x)y ,$$

it follows that

$$\begin{aligned}\overline{D}_n(xy) &= \rho(x)\overline{D}_n(y) + \sum_{k=1}^{n-1} \binom{n}{k} \overline{D}_K x \overline{D}_{n-k} y + (\overline{D}_n x)\rho(y) \\ &\equiv \rho(x)\overline{D}_n(y) + L_{n-1}(x,y) .\end{aligned}$$

By induction we may assume that the operators $\overline{D}_k$, $k = 1,\dots,n-1$, are bounded, and $\|\overline{D}_K\| \leq M\|QD\|^{k-1}$. As before, the bilinear operator L_{n-1} is now bounded in y for fixed x. Since $\mathfrak{M}$ has no separating submodules, $\overline{D}_n$ is bounded, and

$$\|\overline{D}_n\| = \|\overline{D}_{n-1} D\| \leq \|\overline{D}_{n-1}\| \, \|QD\| \leq M\|QD\|^{n-1} .$$

Lastly, to show $D(\mathfrak{A}) \subset R$ it is enough to prove that $\overline{D}(\mathfrak{A}) \subset R/K$. For this we adopt the argument of Allan Sinclair in [6]. Since $K \subset R$, the dual map ρ^*

takes $\Phi_{\mathfrak{M}}$ onto $\Phi_{\mathfrak{A}}$. For $\varphi \in \Phi_{\mathfrak{M}}$ we identify φ and $\rho^*\varphi$. We claim $\varphi(\overline{D}x) = 0$ for each $\varphi \in \Phi_{\mathfrak{M}}$, $x \in \mathfrak{A}$. By replacing x by $x - \varphi(x)1$ we may assume that $\varphi(x) = 0$. Now for $x \in \mathfrak{A}$,

$$D^n x^n = n!(Dx)^n + xa$$

for some $a \in \mathfrak{A}$. Hence

$$\overline{D}_n x^n = n!(\overline{D}x)^n + \rho(x)\rho(a).$$

Since $\varphi(x) = \varphi\rho(x) = 0$

$$|\varphi(\overline{D}x)|^n = \frac{1}{n!} |\varphi \overline{D}_n x^n| \le \frac{1}{n!} \|\overline{D}_n\| \|x^n\| \le \frac{1}{n!} M\|QD\|^{n-1} \|x\|^n .$$

Therefore taking nth roots and letting $n \to \infty$ we see that $\varphi(\overline{D}x) = 0$ which completes the proof.

The following corollary of the above result has been observed earlier by Johnson [4].

1.3 COROLLARY. If $\Phi_{\mathfrak{A}}$ has no isolated points, then $D(\mathfrak{A}) \subset R$.

Proof. If $\mathfrak{S}$ is a nontrivial separating ideal of $\mathfrak{A}/R$, then it follows easily from the stability property (*) that $\varphi_1(\mathfrak{S}) \neq 0$ for at most finitely many points $\varphi_i \in \Phi_{\mathfrak{A}}$, and hence these points must be isolated in $\Phi_{\mathfrak{A}}$.

A more general version of the next result due to Cusack is Lemma 2.3 of [3]. However, the proof is simpler and more transparent in the commutative case, so we include it here.

1.4 LEMMA. Let I be a non-nilpotent separating ideal of a commutative Banach algebra. Then there exists a minimal prime ideal P which is closed and does not contain P.

Proof. If I is not nilpotent, then there exists a prime ideal P of $\mathfrak{A}$ not containing I. We may assume by Zorn's lemma that P is minimal. It remains to show that P is closed.

If $z \in \mathfrak{A} \setminus P$, then $zI \neq 0$ since P is prime. Since I is separating, the same argument as in 1.1 yields that there exists $z_0 \in \mathfrak{A} \setminus P$ such that for all $b \in \mathfrak{A} \setminus P$

$$z_0 b\, I^- = z_0\, I^- .$$

Again since I is separating, either $z_0 y\, I = 0$ for all $y \in P$, or there exists $y_0 \in P$ such that $z_0 y_0\, I \neq 0$ and for each $y \in P$ either

$$\left.\begin{array}{l} z_0y_0yI^- = z_0y_0I^- \quad \text{or} \\ z_0y_0yI = 0 \end{array}\right\} \qquad (2)$$

and therefore (2) holds for each $y \in \mathfrak{A}$. Let $Q = (z_0y_0I)^\perp = \{y \in \mathfrak{A} : z_0y_0yI = 0\}$. The ideal Q is closed and by (2), Q is prime. We claim $Q \subset P$. For if $x \in Q$ and $x \notin P$, then

$$z_0xI^- = z_0I^- .$$

Consequently

$$z_0y_0xI^- = z_0y_0I^- \neq 0 .$$

But $x \in Q$ implies $z_0y_0xI = 0$, which is a contradiction. Therefore $Q = P$.

We remark that by 1.1 the prime ideal constructed in this lemma is either maximal or inaccessible.

Let M be a maximal ideal of $\mathfrak{A}$ with associated homomorphism φ and assume φ is isolated in $\Phi_{\mathfrak{A}}$. Then there exists an idempotent $e \in \mathfrak{A}$. such that $e(\psi) = 1$ if and only if $\psi = \varphi$. It is easily verified that the algebra $e\mathfrak{A}$ has a unique maximal ideal eM. Further more if Q is a nonmaximal closed prime in $e\mathfrak{A}$, then $Q' = \{x \in \mathfrak{A} : ex \in Q\}$ is a nonmaximal closed prime of $\mathfrak{A}$ contained in M. On the other hand if P is a nonmaximal closed prime contained in M, then eP is a nonmaximal closed prime in $e\mathfrak{A}$. Furthermore if P is accessible in $\mathfrak{A}$, then eP is accessible in $e\mathfrak{A}$. To see this let $x = ex \in eM$ and $x \notin eP$. Then $x \notin P$. Therefore there exists a closed ideal N of $\mathfrak{A}$, and $y \notin P$ such that $N \supset P$, $N \ni y$ and $N \not\ni x$. Then eN is a closed ideal of $e\mathfrak{A}$, $eN \supset eP$, $eN \ni ey$, and $eN \not\ni x$. Moreover, since P is prime, $ey \notin P$. Therefore $ey \notin eP$, and eP is accessible.

1.5 THEOREM. Let D be a derivation in a commutative Banach algebra $\mathfrak{A}$ with identity and with radical R. Then $D(\mathfrak{A}) \subset R$ if whenever there are isolated maximal ideals M of $\mathfrak{A}$ containing nonmaximal closed prime ideals of $\mathfrak{A}$ at least one of these prime ideals must be accessible.

Proof. By Johnson's theorem [4], Theorem 4, it suffices to show that if M is an isolated maximal ideal with associated idempotent e, and $D(e\mathfrak{A}) \subset e\mathfrak{A}$ then $D(e\mathfrak{A}) \subset eM$. If there are no closed nonmaximal primes in M, then by the discussion above there are none in $e\mathfrak{A}$. Consequently $\mathfrak{S}(D) \cap eM$ is nilpotent by Lemma 1.4 and $D(e\mathfrak{A}) \subset eM$ by [3] Lemma 4.2 and Corollary 4.7. On the other hand if M contains an accessible prime P, then eP is accessible in $e\mathfrak{A}$. Therefore $e\mathfrak{A}/eP$ contains no separating ideals by 1.1, and consequently $D(e\mathfrak{A}) \subset eM$ by 1.2, and we are done.

We remark in closing, that if there exists a commutative Banach algebra $\mathfrak{A}$ with a unique maximal ideal M and a derivation D in $\mathfrak{A}$ such that $D(\mathfrak{A}) \not\subset M$, then nonmaximal closed primes must exist and they all are inaccessible. These ideals in addition must satisfy a descending chain condition. This is clear since the intersection of an infinite decreasing collection of closed prime ideals of $\mathfrak{A}$ is necessarily prime and accessible.

References

[1] W. G. Bade and P. C. Curtis, Jr., Prime ideals and automatic continuity problems for Banach algebras, J. Functional Analysis, 29 (1978), 88-103.

[2] P. C. Curtis, Jr., Derivations of commutative Banach algebras, Bull. Amer. Math. Soc., 67 (1961), 271-273.

[3] J. Cusack, Automatic continuity and topologically simple radical Banach algebras, J. London Math. Soc., (2), 16 (1977), 493-500.

[4] B. E. Johnson, Continuity of derivations on commutative algebras, Amer. J. Math., 91 (1969), 1-10.

[5] A. M. Sinclair, Automatic continuity of linear operators, London Math. Soc. Lecture Note Series, 21 (C.U.P. Cambridge 1976).

[6] ________, Continuous derivations on Banach algebras, Proc. Amer. Math. Soc., 20 (1969), 166-170.

[7] I. M. Singer and J. Wermer, Derivations on commutative normed algebras, Math. Ann., 129 (1955), 260-264.

Department of Mathematics
University of California
Los Angeles, CA 90024

AUTOMATIC CONTINUITY OF HOMOMORPHISMS INTO BANACH ALGEBRAS

John C. Tripp

Given a homomorphism $\varphi : \mathcal{A} \to \mathcal{B}$ from a Banach algebra $\mathcal{A}$ into a Banach algebra $\mathcal{B}$, we show in Lemma 2, that φ is automatically continuous if there is an element b in $\varphi(\mathcal{A})$ such that $\bigcap_{n=1}^{\infty} \overline{b^n\mathcal{B}} = \{0\} = \{x \in \mathcal{B} : bx = 0\}$. Lemma 2 may be thought of as a generalization of the well-known fact that every homomorphism from a Banach algebra into a Banach algebra of power series is continuous [2, Theorem 5.1, p. 145]. In proving Lemmas 1 and 2 we use techniques similar to those used in proving Sinclair's Stability Lemma [6, Lemma 1.6, p. 11]. In fact, Lemma 2 can easily be proved as a corollary to Sinclair's Stability Lemma. In Theorem 3 we apply Lemma 2 to various examples of homomorphisms. In Theorem 5, we use Lemma 4 and results of Badé, Curtis, and Laursen to give necessary and sufficient conditions on a commutative radical Banach algebra $\mathcal{B}$ so that every homomorphism from a Banach algebra into $\mathcal{B}$ is automatically continuous. Throughout this paper we assume that all Banach algebras are complex. However, Lemma 1, Lemma 2, and Theorem 3 remain valid for homomorphisms between real Banach algebras.

LEMMA 1. Let $\varphi : \mathcal{A} \to \mathcal{B}$ be a discontinuous homomorphism from a Banach algebra $\mathcal{A}$ into a Banach algebra $\mathcal{B}$. Suppose that there exists an element b in $\varphi(\mathcal{A})$ such that $\bigcap_{n=1}^{\infty} \overline{b^n\mathcal{B}} = \{0\} = \{x \in \mathcal{B} \mid bx = 0\}$. Let a in $\mathcal{A}$ be such that $\varphi(a) = b$. Then given any positive integers k and N and any positive number ε, there exists an element x in $\mathcal{A}$ and a positive integer P such that $\|x\| < \varepsilon$ and $\|Q_P\varphi(a^kx)\| > N$, where $Q_P : \mathcal{B} \to \mathcal{B}/\overline{b^P\mathcal{B}}$ is the usual quotient map and $\|Q_P\varphi(a^kx)\|$ denotes the norm of $\varphi(a^kx) + \overline{b^P\mathcal{B}}$ in the quotient Banach space.

Proof. Since φ is discontinuous, by the closed graph theorem, there exists a sequence (x_n) in $\mathcal{A}$ such that $x_n \to 0$ and $\varphi(x_n) \to y \neq 0$. For a fixed positive integer k, we have as $n \to \infty$ that $a^kx_n \to 0$ and $\varphi(a^kx_n) = b^k\varphi(x_n) \to b^ky \neq 0$. Since $\bigcap_{n=1}^{\infty} \overline{b^n\mathcal{B}} = \{0\}$ and $b^ky \neq 0$, there is some integer P such that $b^ky \notin \overline{b^P\mathcal{B}}$. If $Q_P : \mathcal{B} \to \mathcal{B}/\overline{b^P\mathcal{B}}$ denotes the usual quotient map, then it follows that as $n \to \infty$, we have $x_n \to 0$ and $Q_P\varphi(a^kx_n) \to Q_Pb^ky \neq 0$. We choose x_m such that $\|x_m\| < \|Q_Pb^ky\|\varepsilon/(2N)$ and $\|Q_P\varphi(a^kx_m)\| > \|Q_Pb^ky\|/2$. Let $x = (2N/\|Q_Pb^ky\|)x_m$. Then $\|x\| < \varepsilon$ and $\|Q_P\varphi(a^kx)\| = \|Q_Pb^k\varphi(x)\| = (2N/\|Q_Pb^ky\|)\|Q_P\varphi(a^kx_m)\| >$ $(2N/\|Q_Pb^ky\|)\ (\|Q_Pb^ky\|/2) = N$.

REMARK. Suppose that the hypotheses of Lemma 1 are satisfied and that we have

$\|Q_P\varphi(a^k x)\| > N$. Let R be a positive integer such that $R \geq P$, and let $Q_R : \mathfrak{B} \to \mathfrak{B}/\overline{b^R\mathfrak{B}}$ be the usual quotient map. Since $\overline{b^R\mathfrak{B}} \subseteq \overline{b^P\mathfrak{B}}$, it follows that $\|Q_R\varphi(a^k x)\| \geq \|Q_P\varphi(a^k x)\| > N$.

LEMMA 2. Let $\varphi : \mathfrak{A} \to \mathfrak{B}$ be a homomorphism from a Banach algebra $\mathfrak{A}$ into a Banach algebra $\mathfrak{B}$. Suppose that there exists an element b in $\varphi(\mathfrak{A})$ such that $\bigcap_{n=1}^{\infty} b^n\mathfrak{B} = \{0\} = \{x \in \mathfrak{B} : bx = 0\}$. Then φ is continuous.

Proof. By taking an appropriate scalar multiple of b, if necessary, we may assume that we can choose a in $\mathfrak{A}$, such that $\varphi(a) = b$ and $\|a\| \leq 1$. Suppose that φ is discontinuous. By Lemma 1 and the remark following it, we can inductively choose a sequence (x_n) in $\mathfrak{A}$ and a function $k(n)$ from the set of nonnegative integers into the set of positive integers such that for each $n > 1$, we have $\|x_n\| < 2^{-n}$, $k(n) > k(n-1)$, and

$$\|Q_{k(n)}\varphi(a^{k(n-1)}x_n)\| > \left\|\varphi\left(\sum_{i=1}^{n-1} a^{k(i-1)}x_i\right)\right\| + n.$$

Let $x = \sum_{n=1}^{\infty} a^{k(n-1)}x_n$. We have for each n,

$$\begin{aligned}
\|\varphi(x)\| &\geq \|Q_{k(n)}\varphi(x)\| \\
&= \left\|Q_{k(n)}\varphi\left(\sum_{i=1}^{n-1} a^{k(i-1)}x_i\right) + Q_{k(n)}\varphi(a^{k(n-1)}x_n) + Q_{k(n)}\varphi\left(\sum_{i=n+1}^{\infty} a^{k(i-1)}x_i\right)\right\| \\
&= \left\|Q_{k(n)}\varphi\left(\sum_{i=1}^{n-1} a^{k(i-1)}x_i\right) + Q_{k(n)}\varphi(a^{k(n-1)}x_n) + Q_{k(n)}\varphi\left(a^{k(n)}\sum_{i=n+1}^{\infty} a^{k(i-1)-k(n)}x_i\right)\right\| \\
&= \left\|Q_{k(n)}\varphi\left(\sum_{i=1}^{n-1} a^{k(i-1)}x_i\right) + Q_{k(n)}\varphi(a^{k(n-1)}x_n) + Q_{k(n)}b^{k(n)}\varphi\left(\sum_{i=n+1}^{\infty} a^{k(i-1)-k(n)}x_i\right)\right\| \\
&= \left\|Q_{k(n)}\varphi\left(\sum_{i=1}^{n-1} a^{k(i-1)}x_i\right) + Q_{k(n)}\varphi(a^{k(n-1)}x_n)\right\| \\
&\geq \|Q_{k(n)}\varphi(a^{k(n-1)}x_n)\| - \left\|Q_{k(n)}\varphi\left(\sum_{i=1}^{n-1} a^{k(i-1)}x_i\right)\right\| \\
&\geq \|Q_{k(n)}\varphi(a^{k(n-1)}x_n)\| - \left\|\varphi\left(\sum_{i=1}^{n-1} a^{k(i-1)}x_i\right)\right\| > n.
\end{aligned}$$

Since $\|\varphi(x)\|$ must be a finite real number, this leads to a contradiction, and φ must be continuous.

REMARK. We see from Lemma 2 that the conditions of Lemma 1 can never be satisfied. Lemma 1 was used to simplify the proof of Lemma 2.

Before we state Theorem 3, we will need to develop some terminology. By a Banach algebra of power series $\mathfrak{B}$, we mean a subalgebra of the algebra of all formal power series in some number of commuting or noncommuting indeterminates with a norm defined such that $\mathfrak{B}$ is a Banach algebra and such that each coordinate linear functional is continuous. If x is a formal power series in $\mathfrak{B}$, then by the <u>order of x</u>, denoted $\mathrm{ord}(x)$, we mean the smallest degree of the degrees of the nonzero terms of x. We define $\mathrm{ord}(0) = \infty$. We have $\mathrm{ord}(xy) = \mathrm{ord}(x) + \mathrm{ord}(y)$ for any x and y in $\mathfrak{B}$.

By a <u>Banach algebra of Dirichlet series $\mathfrak{B}$</u>, we mean a sub-algebra of the algebra of all sequences of the form (a_n), $n = 1,2,\ldots$, with multiplication defined by $[(a_n)(b_n)]_m = \sum_{\substack{i,j \\ ij=m}} a_i b_j$, for each positive integer m, with a norm defined so that $\mathfrak{B}$ is a Banach algebra and such that for each k, the linear functional $(a_n) \to a_k$ is continuous. For (a_n) in $\mathfrak{B}$, we define $\mathrm{ord}(a_n) = \min\{n \mid a_n \neq 0\}$, with $\mathrm{ord}(0) = \infty$. We have $\mathrm{ord}[(a_n)(b_n)] = \mathrm{ord}(a_n)\ \mathrm{ord}(b_n)$ for $(a_n),(b_n) \in \mathfrak{B}$.

By a <u>Banach algebra of matrices $\mathfrak{B}$</u> we mean an algebra of infinite matrices (a_{ij}) with a norm defined so that $\mathfrak{B}$ is a Banach algebra and such that for each pair of positive integers k, ℓ, the coordinate linear functional $(a_{ij}) \to a_{k\ell}$ is continuous. If $B = (b_{ij})$ is an infinite matrix such that for some $k > 0$ we have $b_{j+k,j} \neq 0$ for each j and $b_{ij} = 0$ whenever $i - j < k$, then we call B a <u>shift matrix</u>. A shift matrix is a generalization of the notion of the matrix representation of a weighted shift operator.

Let ω be a positive continuous function on $\mathbb{R}^+ = [0,\infty)$ such that $\omega(s+t) \leq \omega(s)\omega(t)$ for every pair s, t in $\mathbb{R}^+$. Then the function ω is called a <u>weight function</u> and $L^1(\mathbb{R}^+, \omega)$ denotes the algebra of Lebesgue measurable functions f on $[0,\infty)$ with $\|f\| = \int_{\mathbb{R}^+} |f(t)|\omega(t) < \infty$ and with the usual convolution product. See [3] for more details. For f in $L^1(\mathbb{R}^+, \omega)$ we define $\alpha(f)$ to be the supremum of the set $\{x \in \mathbb{R}^+ \mid f = 0$ a.e. on $[0,x]\}$. We let $\alpha(0) = \infty$. By Titchmarsh's Convolution Theorem [3] we know that for f and g in $L^1(\mathbb{R}^+, \omega)$ we have $\alpha(fg) = \alpha(f) + \alpha(g)$.

THEOREM 3. Let $\varphi : \mathfrak{A} \to \mathfrak{B}$ be a homomorphism from a Banach algebra $\mathfrak{A}$ into a Banach algebra $\mathfrak{B}$. Each of the following conditions is sufficient for φ to be continuous.

(i) $\mathfrak{B}$ is a Banach algebra of power series in any number of commuting or noncommuting indeterminates.

(ii) $\mathfrak{B}$ is a Banach algebra of Dirichlet series.

(iii) $\mathfrak{B}$ is a Banach algebra of matrices and there is a matrix in $\varphi(\mathfrak{A})$ which is a shift matrix.

(iv) $\mathfrak{B} = L^1(\mathbb{R}^+, \omega)$ for some appropriate weight function ω and there is a function g in $\varphi(\mathfrak{A})$ with $\alpha(g) > 0$.

<u>Proof</u>. Our proof consists of showing that in each case, the hypotheses of Lemma 2 are satisfied.

(i) If $\varphi(\mathfrak{A}) = \{0\}$ or $\varphi(\mathfrak{A})$ is the linear span of an identity in $\mathfrak{B}$, then the continuity of φ follows from the automatic continuity of multiplicative linear functionals. If $\varphi(\mathfrak{A})$ is not contained in the linear span of an identity, then there is an element b in $\varphi(\mathfrak{A})$ with $\mathrm{ord}(b) \geq 1$. If $y = b^n c$ is an element of $b^n\mathfrak{B}$ we have $\mathrm{ord}(b^n c) = n\, \mathrm{ord}(b) + \mathrm{ord}(c) \geq n$, and the continuity of the coordinate linear functionals assures us that if $y \in \overline{b^n\mathfrak{B}}$, then we have $\mathrm{ord}(y) \geq n$. It follows that if $y \in \bigcap_{n=1}^{\infty} \overline{b^n\mathfrak{B}}$, then $\mathrm{ord}(y) = \infty$ and $y = 0$. Since $\mathfrak{B}$ is an integral domain, we have $\{x \in \mathfrak{B} : bx = 0\} = \{0\}$ and the hypotheses of Lemma 2 are satisfied.

(ii) The proof is analogous to the proof of (i). If $\varphi(\mathfrak{A})$ is not contained in the linear span of an identity for $\mathfrak{B}$, then we have an element b of $\varphi(\mathfrak{A})$ with order at least 2 and for y in $\overline{b^n\mathfrak{B}}$, we have $\mathrm{ord}(y) \geq 2^n > n$. It follows that $\bigcap_{n=1}^{\infty} \overline{b^n\mathfrak{B}} = \{0\}$. Since $\mathfrak{B}$ is an integral domain, it follows that $\{x \in \mathfrak{B} : bx = 0\} = \{0\}$.

(iii) Let $B = (b_{ij})$ be a shift matrix in $\varphi(\mathfrak{A})$ and k a positive integer such that $b_{j+k,j} \neq 0$ for $j = 1,2,\ldots$ and $b_{ij} = 0$ whenever $i - j < k$. Let $A = (a_{ij})$ be a matrix in $\mathfrak{B}$ and let p be the smallest positive integer such that $a_{pj} \neq 0$ for some j. Let $C = (c_{ij}) = BA$. For $m < p + k$ and any j we have $c_{mj} = \sum_{i=1}^{\infty} b_{mi}a_{ij} = 0$ since if $i < p$ then $a_{ij} = 0$ and if $i \geq p$ then $m - i \leq m - p < p + k - p = k$ and $b_{mi} = 0$. We also have $c_{p+k,j} = \sum_{i=1}^{\infty} b_{p+k,i}a_{ij} = \sum_{i=1}^{p-1} b_{p+k,i}a_{ij} + b_{p+k,p}a_{pj} + \sum_{i=p+1}^{\infty} b_{p+k,i}a_{ij} = b_{p+k,p}a_{pj}$, since if $i < p$, then $a_{ij} = 0$ and if $i > p$ then $(p+k) - i < (p+k) - p = k$ and $b_{p+k,i} = 0$. Simply stated, we have shown that left multiplication of a matrix by B increases the number of initial zero rows by k. We have also shown that left multiplication of a nonzero matrix by B does not annihilate the matrix. It follows by a routine induction that if D is an element of $B^n\mathfrak{B}$, then D has at least nk initial rows which are zero. Since the coordinate linear functionals are continuous, it also follows that if $D \in \overline{B^n\mathfrak{B}}$, then D has at least nk initial zero rows. We see that if $D \in \bigcap_{n=1}^{\infty} \overline{B^n\mathfrak{B}}$, then $D = 0$. Since $BA = 0$ implies $A = 0$, we have $\{A \in \mathfrak{B} : BA = 0\} = \{0\}$ and the conditions of Lemma 2 are satisfied.

(iv) Let $\alpha(b) = k > 0$. By Titchmarsh's Theorem, we have for any $y = b^n a$ in $b^n\mathfrak{B}$, that $\alpha(y) = \alpha(b^n a) = n\alpha(b) + \alpha(a) = nk + \alpha(a) \geq nk$. We also have for any y in $\overline{b^n\mathfrak{B}}$, that $\alpha(y) \geq nk$. Therefore if $y \in \bigcap_{n=1}^{\infty} \overline{b^n\mathfrak{B}}$ then $\alpha(y) = \infty$, and $y = 0$. Since $\mathfrak{B}$ is an integral domain, it follows that $\{x \in \mathfrak{B} : bx = 0\} = \{0\}$.

Theorem 3 (iii) cannot be extended to cover automatic continuity of all homomorphisms into Banach algebras of matrices. For example, if $\mathfrak{B}$ has nilpotent matrices and $\mathfrak{A}^2$ = linear span of $\{xy : x,y \in \mathfrak{A}\}$ has infinite codimension in $\mathfrak{A}$, then it is easy to construct a discontinuous homomorphism from $\mathfrak{A}$ into $\mathfrak{B}$ [4, Lemma 2.1, p. 123]. Theorem 3 (iv) cannot be extended to cover automatic continuity of all homomorphisms into $L^1(\mathbb{R}^+, \omega)$. Esterle [4, Corollaire 6.4, p. 139] has shown that if $\mathfrak{A}$ is a complex commutative radical Banach algebra which is not nilpotent, then (assuming the Continuum Hypothesis) there exists a discontinuous homomorphism from $\mathfrak{A}$ into $L^1(\mathbb{R}^+, \omega)$. Automatic continuity of φ does extend to the case where φ is a homomorphism from a Banach algebra into a convolution algebra $L^p(\mathbb{R}^+, \omega)$, $p > 1$, as defined by Grabiner in [5] if there is an element f in $\varphi(\mathfrak{A})$ with $\alpha(f) > 0$. The proof is the same as the proof of Theorem 3 (iv).

Finally, we consider the case of homomorphisms into commutative radical Banach algebras. Let b be an element of a Banach algebra $\mathfrak{B}$, and let X be a linear subspace of $\mathfrak{B}$. As in [1] we say that X is <u>b-divisible</u> if $(b - \lambda)X = X$ for every complex number λ. In the case where $\mathfrak{B}$ is a commutative radical Banach algebra, it is easy to show that X is the maximal b-divisible subspace of $\mathfrak{B}$ if and only if X is the maximal subspace of $\mathfrak{B}$ such that $bX = X$ [1, Remark 0.5, p. 162].

LEMMA 4. Let $\mathfrak{B}$ be a commutative radical Banach algebra with no nonzero nilpotent elements. Then $\mathfrak{B}$ has a nonzero b-divisible subspace if and only if $\bigcap_{n=1}^{\infty} b^n\mathfrak{B} \neq \{0\}$.

<u>Proof</u>. Let $Y = \bigcap_{n=1}^{\infty} b^n\mathfrak{B}$. If X is a b-divisible subspace of $\mathfrak{B}$, then $X = bX = \bigcap_{n=1}^{\infty} b^n X \subseteq \bigcap_{n=1}^{\infty} b^n\mathfrak{B} = Y$. Therefore, it is sufficient to prove that Y is a b-divisible subspace of $\mathfrak{B}$. We first show that the linear operator on the linear space $b\mathfrak{B}$ defined by $bx \mapsto b(bx) = b^2x$ is one-to-one. If $b(bx) = 0$, then $(bx)^2 = b^2x^2 = b(bx)x = 0x = 0$ and $bx = 0$, since we have no nonzero nilpotents in $\mathfrak{B}$. Since $Y \subseteq b\mathfrak{B}$, it follows that left multiplication by b is a one-to-one linear operator on Y. Let z be an element of $Y = \bigcap_{n=1}^{\infty} b^n\mathfrak{B}$. Then there is a sequence (y_n) in $\mathfrak{B}$ such that $z = by_1 = b^2y_2 = b^3y_3 = \cdots$. Since left multiplication by b is one-to-one on Y, it follows from $b(by_2) = b(b^2y_3)$ $b(b^3y_4) = \cdots$, that $by_2 = b^2y_3 = b^3y_4 = \cdots$ and that $by_2 \in \bigcap_{n=1}^{\infty} b^n\mathfrak{B} = Y$. We have $z = b(by_2) \in bY$. Hence $Y \subseteq bY$. Since $bY \subseteq Y$, it follows that $bY = Y$ and that Y is b-divisible.

An example, credited to Marc Thomas, of a commutative radical Banach algebra $\mathfrak{B}$ with $\bigcap_{n=1}^{\infty} b^n\mathfrak{B} \neq \{0\}$, but with no nonzero b-divisible subspaces is given in [1, Remark 2.14 (ii), p. 172]. Since Thomas's Banach algebra contains nonzero nilpotent elements, it does not contradict Lemma 4. In [1, Theorem 2.4, p. 167] Bade, Curtis, and Laursen prove that given a commutative Banach algebra $\mathfrak{B}$ with scattered spectrum, every homomorphism from a Banach algebra into $\mathfrak{B}$ is automatically continuous if and only if $\mathfrak{B}$ has no nonzero nilpotent elements and $\mathfrak{B}$ has no nonzero b-divisible subspaces for any element b in $\mathfrak{B}$. Since a commutative radical Banach algebra has scattered spectrum, the following theorem follows from the theorem of Bade, Curtis, and Laursen and Lemma 4.

THEOREM 5. Let $\mathfrak{B}$ be a commutative radical Banach algebra. Every homomorphism from a Banach algebra into $\mathfrak{B}$ is continuous if and only if $\mathfrak{B}$ has no nonzero nilpotent elements and $\bigcap_{n=1}^{\infty} b^n\mathfrak{B} = \{0\}$ for every element b of $\mathfrak{B}$.

Our results and comments lead naturally to the following questions.

Question 1. Suppose that b is an element of a Banach algebra $\mathfrak{B}$. Does $\bigcap_{n=1}^{\infty} b^n\mathfrak{B} = \{0\}$ imply $\bigcap_{n=1}^{\infty} \overline{b^n\mathfrak{B}} = \{0\}$?

Question 2. Let $\mathfrak{B}$ be a noncommutative Banach algebra. Suppose that every homomorphism from a Banach algebra into a commutative subalgebra of $\mathfrak{B}$ is continuous. Does it necessarily follow that every homomorphism from a Banach algebra into $\mathfrak{B}$ is continuous?

References

[1] W. G. Bade, P. C. Curtis, Jr., and K. B. Laursen, Divisible subspaces and problems of automatic continuity, Studia Math., 68 (1980), 159-186.

[2] H. G. Dales, Automatic continuity: a survey, Bull. London Math. Soc., 10 (1978), 129-183.

[3] ___________, Convolution algebras on the real line, this Volume.

[4] J. Esterle, Homomorphismes discontinus des algèbres de Banach commutatives séparables, Studia Math., 66 (1979), 119-141.

[5] S. Grabiner, Weighted convolution algebras on the half line, J. Math. Anal. Appl., 83 (1981), 531-553.

[6] A. M. Sinclair, Automatic Continuity of Linear Operators, London Math. Soc. Lecture Note Series 21, Cambridge Univ. Press, Cambridge, 1976.

Department of Mathematics
Southeast Missouri State University
Cape Girardeau, MO 63701

Note: Please see also the following article, [Di] .

ON THE INTERSECTION OF THE PRINCIPAL IDEALS GENERATED BY POWERS IN A BANACH ALGEBRA

P. G. Dixon

The following question is raised in the preceding paper of J. C. Tripp [1]: If b is an element of a Banach algebra $\mathfrak{B}$, does $\bigcap_{n=1}^{\infty} b^n\mathfrak{B} = \{0\}$ imply $\bigcap_{n=1}^{\infty} (b^n\mathfrak{B})^- = \{0\}$? We give an example to show that the answer is negative, even if $\mathfrak{B}$ is commutative and separable.

To construct the algebra $\mathfrak{B}$ for our example, we first define $\mathfrak{B}_0$ to be the complex commutative algebra generated by the set of formal symbols $\{a, b, c_{ij}, d_{ij} : 1 \leq i, j < \infty\}$ subject to the relations:

$$ab = ac_{ij} = ad_{ij} = bd_{ij} = c_{ij}c_{k\ell} = c_{ij}d_{k\ell} = 0 \qquad (\text{all } i,j,k,\ell),$$

$$b^i c_{ij} = a + d_{ij} \qquad (\text{all } i,j),$$

$$b^k c_{ij} = 0 \qquad (1 \leq i < k < \infty, \text{ all } j).$$

Thus a typical element of $\mathfrak{B}_0$ may be written, uniquely, as

$$x = \lambda a + \sum_{n=1}^{\infty} \mu_n b^n + \sum_{i,j=1}^{\infty} \sum_{k=0}^{i-1} \nu_{kij} b^k c_{ij} + \sum_{i,j=1}^{\infty} \pi_{ij} d_{ij}, \qquad (1)$$

with $\lambda, \mu_n, \nu_{kij}, \pi_{ij} \in \mathbb{C}$, each of the sums having only finitely many non-zero terms. We define a norm on $\mathfrak{B}_0$ by

$$\|x\| = |\lambda| + 2 \sum_{n=1}^{\infty} |\mu_n| + \sum_{i,j=1}^{\infty} \sum_{k=0}^{i-1} |\nu_{kij}| + \sum_{i,j=1}^{\infty} 2^{-j} |\pi_{ij}|,$$

where x is as in (1). It is easy to verify that this norm is submultiplicative.

The algebra $\mathfrak{B}$ is then defined as the completion of $\mathfrak{B}_0$ in this norm. Clearly, $\mathfrak{B}$ is a commutative, separable Banach algebra and a typical element x of $\mathfrak{B}$ may be written, uniquely, in the form (1), where the sums are no longer restricted to being finite.

Now $b^i c_{ij} \to a$ as $j \to \infty$, for each i, so $a \in \bigcap_{n=1}^{\infty} (b^n\mathfrak{B})^-$. We show that $\bigcap_{n=1}^{\infty} b^n\mathfrak{B} = \{0\}$. Suppose $x = b^N x'$, for some $N \geq 1$, where x is as in (1) but with unrestricted sums, and x' is expressed likewise in terms of $\lambda', \mu'_n, \nu'_{kij}, \pi'_{ij}$. Then

$$\lambda a + \sum \mu_n b^n + \sum_{k<i} \nu_{kij} b^k c_{ij} + \sum \pi_{ij} d_{ij}$$

$$= \sum \mu_n' b^{n+N} + \sum_{k+N<i} \nu_{kij}' b^{k+N} c_{ij} + \sum_{k+N=i} \nu_{kij}' (a + d_{ij}) . \tag{2}$$

Equating coefficients of $b^n, b^k c_{ij}, d_{ij}$ and a, we obtain, respectively,

$$\mu_n = 0 \qquad (1 \le n \le N), \tag{3}$$

$$\nu_{kij} = 0 \qquad (0 \le k < N;\ 1 \le i,\ j < \infty), \tag{4}$$

$$\pi_{ij} = 0 \qquad (1 \le i < N,\ 1 \le j < \infty), \tag{5}$$

$$\pi_{ij} = \nu'_{(i-N)ij} \qquad (N \le i < \infty,\ 1 \le j < \infty), \tag{6}$$

$$\lambda = \sum_{j=1}^{\infty} \sum_{i=N}^{\infty} \nu'_{(i-N)ij} . \tag{7}$$

From (6) and (7) we have

$$\lambda = \sum_{j=1}^{\infty} \sum_{i=N}^{\infty} \pi_{ij} . \tag{8}$$

Now if $x \in \bigcap_{N=1}^{\infty} b^N \mathfrak{B}$, then (3), (4), (5) and (8) hold for all N. Hence $\mu_n = 0$, $\nu_{kij} = 0$, $\pi_{ij} = 0$ and $\lambda = 0$ for all n, k, i, j; that is, $x = 0$. Thus $\bigcap_{N=1}^{\infty} b^N \mathfrak{B} = \{0\}$.

Reference

[1] J. C. Tripp, Automatic continuity of homomorphisms into Banach algebras, this Volume.

Mathematics Department
University of Sheffield
S3 7RH England, U.K.

AUTOMATIC CONTINUITY FOR OPERATORS OF LOCAL TYPE

E. Albrecht and M. Neumann

1. Introduction

In this note we survey some of the automatic continuity methods from [4] and [5] and give some new applications of this theory.

Let Ω be a compact Hausdorff space and let μ be a Borel measure on Ω without any point atoms. A linear operator $T : L^p(\Omega,\mu) \to L^p(\Omega,\mu)$ is called an operator of local type (cf. [18, p.401]) if for all closed $F, G \subset \Omega$ with $F \cap G = \emptyset$ the operator $P_F\, T\, P_G$ is compact, where $P_F, P_G : L^p(\Omega,\mu) \to L^p(\Omega,\mu)$ are the operators of multiplication by the characteristic function of F and G, respectively. Of course, the operators of multiplication by bounded measurable functions and also singular integral operators of the Calderon-Zygmund type are in this class (cf. [18]). As we shall see in this note, all operators of local type are automatically continuous. The methods of proof are those of [5]. They are very general and can be used to obtain continuity results for operators which are local in the sense of Peetre [20] and also for certain classes of intertwining operators (see [5] for details). Some applications to normal topological algebras of functions can be found in [7].

In the following section, we include some general remarks on the separating space of a linear map between topological vector spaces X and Y. These remarks show that this subspace of Y measures in a certain sense the distance of the map to the space $L(X,Y)$ of all continuous linear maps from X to Y. In the third section, we prove the main sliding hump lemma from [4] which will be our main tool in the last section and which will also be needed in [3].

2. The separating space

Thoughout this note, a topological vector space (TVS) is always assumed to be Hausdorff but not necessarily locally convex. For a linear map T from a TVS X to a TVS Y the separating space $\mathfrak{S}(T)$ is defined as

$$\mathfrak{S}(T) := \{y \in Y : \text{there is a net } (x_\alpha)_\alpha \text{ in } X \text{ with } x_\alpha \to 0 \text{ in } X \text{ and } Tx_\alpha \to y \text{ in } Y\}.$$

Let us recall some of the basic properties of $\mathfrak{S}(T)$ (cf. [16, 5]).

(a) $\mathfrak{S}(T)$ is a closed linear subspace of Y, and $\mathfrak{S}(T)$ is the null space if and only if T has closed graph.

(b) For every continuous linear map R from Y to a TVS Z we have

Both authors were supported by NATO Grant No. RG 073.81.

$R(\mathfrak{S}(T)) \subset \mathfrak{S}(RT)$.

(c) If $q : Y \to Y/\mathfrak{S}(T)$ denotes the canonical quotient map, then qT is a closed linear operator from X to $Y/\mathfrak{S}(T)$.

(b) and (c) can be strengthened if we impose some additional conditions on X and Y. Recall that a TVS Y is said to be countably boundedly generated if Y is the union of a countable family of bounded subsets. By [5, Prop. 2.2 and 2.3] we have:

2.1. PROPOSITION. Let T be a linear map from a complete metrizable TVS X to a TVS Y which is locally convex, countably boundedly generated, and sequentially complete. If $R : Y \to Z$ is a continuous linear operator from Y to a TVS Z then $RT : X \to Z$ is continuous if and only if $R(\mathfrak{S}(T)) = \{0\}$.

Given a linear map T from a TVS X to a TVS Y, one may ask if there exists a continuous linear map $S : X \to Y$ which does not differ "too much" from T. By (c) and Proposition 2.1, the operator $qT : X \to Y/\mathfrak{S}(T)$ is continuous in many cases. Hence, if qT admits a continuous lifting $S : X \to Y$ (i.e., a continuous linear map $S : X \to Y$ such that $qT = qS$) then the range of $T - S$ is contained in $\mathfrak{S}(T)$. Conversely, if $S : X \to Y$ is a continuous linear operator, then, by the definition of $\mathfrak{S}(T)$, the closure $\overline{(T-S)(X)}$ of the range of $T - S$ contains $\mathfrak{S}(T)$. Hence, $\overline{(T-S)(X)} = \mathfrak{S}(T)$ is the best we can hope for. As we shall see, the separating space is in many cases a rather small subspace of Y.

There are some obvious cases where a continuous linear lifting of qT is possible:

- If Y is a Hilbert space and if Q denotes the orthogonal projection onto $Y \ominus \mathfrak{S}(T)$, then QT is a continuous linear lifting of qT.
- If $X = \ell^1(I)$ for some index set I and if Y is a Banach space, then it is well known that a continuous linear lifting is always possible.
- If qT is known to be nuclear and if Y is a Banach space (or more general a TVS with a strict web in the sense of De Wilde [9]) then Hilfssatz 2.6 in [15] gives a continuous lifting.

Using the Liftingsatz of Floret [11] and the proof of the above mentioned lifting result of Gramsch we can now prove:

2.2. LEMMA. Let X and Y be as in Proposition 2.1 and let $\mathfrak{S}$ be a closed linear subspace of Y. Suppose that $T_o : X \to Y/\mathfrak{S}$ is a continuous linear map which can be factored in the following way

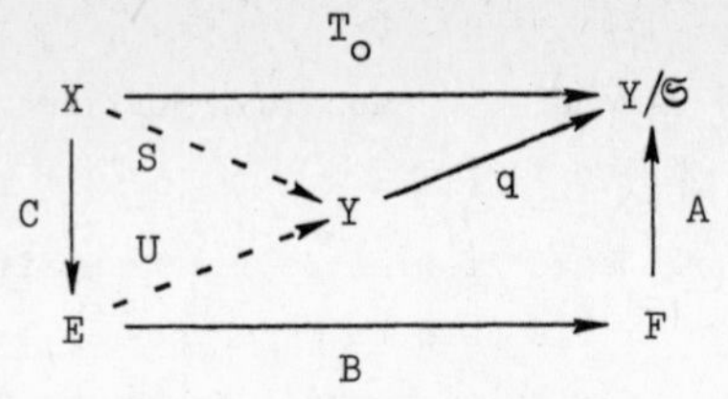

where E is an $\mathcal{L}_1$-space (in the sense of [17]), F is a complete, metrizable, and locally convex TVS, B is a compact linear map, A and C are continuous linear maps, and $T_o = ABC$. Then there exists a continuous linear operator $S : X \to Y$ such that $T_o = qS$.

Proof. (1) Of course, it is sufficient to construct a continuous linear lifting $U : E \to Y$ of AB, for $S := UC$ is then a continuous linear map with $T_o = qS$. Denote by E_1 the unit ball in E. As F is a Fréchet space and B is a compact map, the set $K := \overline{B(E_1)}$ is fast compact in F (cf. [9, Prop. III. 1.10]). Since A is continuous, A therefore maps fast convergent sequences of E into bounded sequences of $Y/\mathfrak{S}$, and we may apply [9, Prop. III.1.11] and obtain that $A(K)$ is fast compact in $Y/\mathfrak{S}$.

(2) Let us now prove that Y has a strict web in the sense of [9]. As Y is countably boundedly generated and locally convex, Y is the union of a countable family $(B_k)_{k=1}^{\infty}$ of closed, absolutely convex bounded subsets of Y. Because of [9, Prop. III.1.9] the sets B_k $(k \in \mathbb{N})$ are Banach discs, i.e., the linear hull Y_k of B_k, endowed with the Minkowski functional of B_k is a Banach space. Of course, Y_k is strictly webbed with respect to this Banach space topology (cf. [9, Prop. IV.3.1]) and therefore also with respect to the topology induced by Y (cf. [9, Prop. IV.4.2]). By [9, Prop. IV.4.5], Y is also strictly webbed.

(3) As Y is strictly webbed, we may apply [9, Prop. VI.3.7.(b)] and obtain a fast compact set $K_0 \subset Y$ with $A(K) \subset q(K_0)$. We may also choose K_0 to be absolutely convex (by [9, Prop. III.1.5]). Now we may apply the Liftingsatz of Floret [11] and obtain a continuous linear map $U : E \to Y''_{\nu}$ with $U(E) \subset Y$ (here, Y''_{ν} is the bidual of Y, endowed with the topology generated by the bipolars of neighborhoods of 0 in Y). As the canonical injection $Y \to Y''_{\nu}$ is continuous, the graph of $U : E \to Y$ is closed. Hence $U : E \to Y$ is continuous by [9, Prop. IV. 5.1].

Before giving some examples how to apply this lifting lemma, we prove:

2.3. LEMMA. Let $T : X \to Y$ be a continuous linear operator from a nuclear Fréchet space X to a countably boundedly generated locally convex TVS Y. Then T is nuclear.

Proof. Let $(B_k)_{k=1}^{\infty}$ be a sequence of closed absolutely convex bounded subsets of Y with $B_k \uparrow Y$. By the Baire theorem there is a $k \in \mathbb{N}$ and a continuous seminorm p on X such that $T(X)$ is contained in the linear hull Y_k of B_k and such that $\|Tx\|_k \le p(x)$ for all x in X, where $\|\cdot\|_k$ denotes the Minkowski functional for B_k on Y_k. As a continuous linear map from a nuclear space to a normed space, $T : X \to (Y_k, \|\cdot\|_k)$ is nuclear. Because of the continuity of the inclusion map $(Y_k, \|\cdot\|_k) \to Y$ we obtain the nuclearity of $T : X \to Y$.

2.4. PROPOSITION. Let $T : X \to Y$ be a linear map from a nuclear Fréchet space to a sequentially complete locally convex TVS Y which is countably boundedly generated. Then there exists a continuous linear map $S : X \to Y$ such that $(T - S)(X) \subset \mathfrak{S}(T)$.

Proof. (Cf. the proof of [15, Hilfssatz 2.6].) By Lemma 2.3 the continuous linear map $qT : X \to Y/\mathfrak{S}(T)$ is nuclear and hence has a representation

$$qT(x) = \sum_{k=1}^{\infty} \lambda_k < x, x_k' > y_k \qquad (x \in X),$$

where $\{y_k : k \in \mathbb{N}\}$ is bounded in $Y/\mathfrak{S}(T)$, $\{x_k' : k \in \mathbb{N}\}$ is an equicontinuous subset of X' and $\lambda_k > 0$, $\sum_{k=1}^{\infty} \lambda_k < \infty$. Then there is a sequence $(\mu_k)_{k=1}^{\infty}$ of positive real numbers such that $\sum_{k=1}^{\infty} \mu_k < \infty$ and $\lambda_k/\mu_k \to 0$ for $k \to \infty$. Consider now the weighted ℓ^1-spaces $E = \ell^1(\mu)$ and $F = \ell^1(\lambda)$, where for a sequence $\alpha = (\alpha_k)_{k=1}^{\infty}$ of positive real numbers $\ell^1(\alpha)$ is defined as

$$\ell^1(\alpha) = \left\{ u = (u_k)_{k=1}^{\infty} : u_k \in \mathbb{C},\ \|u\|_\alpha := \sum_{k=1}^{\infty} \alpha_k |u_k| < \infty \right\}.$$

We define $A : F \to Y/\mathfrak{S}(T)$, $B : E \to F$, and $C : X \to E$ by

$$Au := \sum_{k=1}^{\infty} \lambda_k u_k y_k \quad \text{for } u = (u_k)_{k=1}^{\infty} \in F,$$

$$Bu := u \quad \text{for } u \in E, \text{ and}$$

$$Cx := (\langle x, x_k' \rangle)_{k=1}^{\infty} \quad \text{for } x \in X.$$

Because $\lambda_k/\mu_k \to \infty$ as $k \to \infty$, the map B is compact (see e.g. [12, §19, Satz 4.2]). Hence we may apply Lemma 2.2 and obtain a continuous linear map $S : X \to Y$ such that $qS = qT$, i.e., such that the range of $T - S$ is contained in $\mathfrak{S}(T)$.

Let us give a special case of 2.4.

2.5. COROLLARY. Let $\mathbb{T} := \{z \in \mathbb{C} : |z| = 1\}$, $N \in \mathbb{N}$, and let Y be a TVS as in Proposition 2.4. Then, for every linear map $T : C^\infty(\mathbb{T}^N) \to Y$ there is a continuous linear operator $S : C^\infty(\mathbb{T}^N) \to Y$ with $(T - S)(C^\infty(\mathbb{T}^N)) \subset \mathfrak{S}(T)$.

Notice, that 2.4 cannot be applied immediately to linear maps $T: C^k(\mathbb{T}^n) \to Y$ for $k < \infty$. However, as an easy consequence of Corollary 2.5, one obtains the existence of some $m \geq k$, such that there is a continuous linear map $S: C^m(\mathbb{T}^N) \to Y$ with $(T-S)(C^m(\mathbb{T}^N)) \subset \mathfrak{S}(T)$. We now want to give an estimate for m.

First, we need some notations. For $N \in \mathbb{N}$ and $\alpha \geq 0$ we denote by W_N^α the space of all those continuous functions $f: \mathbb{T}^N \to \mathbb{C}$ which have a representation by a Fourier series

$$f(z) = \sum_{\nu \in \mathbb{Z}^N} f_\nu z^\nu$$

(where $z = (z_1,\ldots,z_N)$, $\nu = (\nu_1,\ldots,\nu_N)$, and $z^\nu = (z_1^{\nu_1},\ldots,z_N^{\nu_N})$) such that

$$\|f\|_{\alpha,N} := \sum_{\nu \in \mathbb{Z}^N} (1+|\nu|)^\alpha |f_\nu| < \infty$$

(with $|\nu| := (\nu_1^2 + \cdots + \nu_N^2)^{1/2}$). Endowed with the norm $\|\cdot\|_{\alpha,N}$, W_N^α is a Banach space which is topologically isomorphic to $\ell^1(\mathbb{Z}^N)$. Moreover, the canonical inclusion map $W_N^\alpha \to W_N^\beta$ for $\beta < \alpha$ is compact. This follows as in [12, p.90/91] by the corresponding version of the Kolmogoroff criterium for compact subsets of W_N^β.

2.6. PROPOSITION. Let Y be as in 2.4 and let $T: C^k(\mathbb{T}^N) \to Y$ be a linear map. Put $r := [N/2] + 1$. Then there is a continuous linear map $S: C^{k+r}(\mathbb{T}^N) \to Y$ such that $(T-S)(C^{k+r}(\mathbb{T}^N)) \subset \mathfrak{S}(T)$.

Proof. Put $E := W_N^\alpha$ with $\alpha := k + 1/4$ and $F := W_N^k$. Then the canonical injection $B: E \to F$ is compact and E is an $\mathcal{L}^1$-space. Denote by $M_j: E \to E$ the multiplication operators defined by $(M_j f)(z) := z_j f(z)$ for $f \in E$, $z \in \mathbb{T}^N$, $j = 1,\ldots,N$. An elementary calculation shows that

$$\|M_1^{\nu_1} \cdots M_N^{\nu_N}\| = \underline{O}(|\nu|^\alpha) \quad \text{for} \quad \nu \in \mathbb{Z}^N \quad \text{as} \quad |\nu| \to \infty.$$

Now we may apply [1, Satz 3.14] and conclude that $C^{\alpha+s}(\mathbb{T}^N) \subset W_N^\alpha$ with continuous inclusion map, whenever $s > N/2$. In particular, for $s := [N/2] + 3/4$ we obtain $C^{k+r}(\mathbb{T}^N) \subset W_N^\alpha = E$ with continuous inclusion map which we define to be C. Finally define $A: F \to Y/\mathfrak{S}(T)$ as $A = qTi$, where $q: Y \to Y/\mathfrak{S}(T)$ is the canonical epimorphism and $i: F \to C^k(\mathbb{T}^N)$ the (continuous) inclusion map. By Lemma 2.2 there is a linear map $S: C^{k+r}(\mathbb{T}^N) \to Y$ which is continuous with respect to the C^{k+r}-topology such that $qS = ABC = qT \mid C^{k+r}(\mathbb{T}^N)$. Hence $(T-S)(C^{k+r}(\mathbb{T}^N)) \subset \mathfrak{S}(T)$.

The remarks of this section show that in many cases the separating space measures the "distance" of a linear map to the set of all continuous linear maps. We shall see in the last section that, very often, the separating space is "rather small".

3. The sliding hump

In this section we prove one of the main sliding hump results of [4] (see also [19]) which will be needed in the last part.

3.1. THEOREM. Let $(X_n)_{n=0}^{\infty}$ and $(Y_n)_{n=0}^{\infty}$ be sequences of TVSs such that the spaces X_n, $n \in \mathbb{N}_0$, are complete and metrizable and such that Y_0 is countably boundedly generated. Let $(A_n)_{n=1}^{\infty}$ and $(\pi_n)_{n=1}^{\infty}$ be two sequences of continuous linear maps where $A_n : X_n \to X_{n-1}$ and $\pi_n : Y_0 \to Y_n$ $(n \in \mathbb{N})$. For $n \in \mathbb{N}$ put $S_n := A_1 \cdots A_n$. If $T : X_0 \to Y_0$ is a linear map such that $\pi_n T S_n : X_n \to Y_n$ is continuous for all $n \in \mathbb{N}$ then there is some $n \in \mathbb{N}$ such that $\pi_k T S_n : X_n \to Y_k$ is continuous for all $k \in \mathbb{N}$.

Proof. (a) By our assumptions, the maps $\pi_k T S_n$ are continuous for $k \leq n$. Assume now that the theorem is false. Then there is a strictly increasing sequence $(n(k))_{k=1}^{\infty}$ in $\mathbb{N}$ such that $\pi_{n(k+1)} T S_{n(k)}$ is discontinuous for all $k \in \mathbb{N}$. Fix a sequence $(B_m)_{m=1}^{\infty}$ of bounded subsets of Y_0 with $B_m \uparrow Y_0$ as $m \to \infty$, let $|\cdot|_n$ be an F-norm on X_n generating the topology of X_n, and put $n(0) := 0$, $\varepsilon_0 := 1$, and $S_{n(0)} = S_0 :=$ the identity on X_0. By induction we shall construct for every $k \in \mathbb{N}$ an integer $m(k) \geq k$, a neighborhood V_k of 0 in $Y_{n(k)}$, an element $x_k \in X_{n(k-1)}$, and an $\varepsilon_k \geq 0$ such that for all $k \geq 2$ the following conditions are fulfilled:

(1) $T\left(\sum_{r=1}^{k-1} S_{n(r-1)} x_r\right) \in B_{m(k)}$;

(2) $|x_k|_{n(k-1)} \leq 2^{-k}\varepsilon_{k-1}$ and $|A_{n(j)+1} \cdots A_{n(k-1)} x_k|_{n(j)} \leq 2^{-k}\varepsilon_j$ for $j = 0,1,\ldots,k-2$;

(3) $\pi_{n(k)} T S_{n(k-1)} x_k \notin \pi_{n(k)}(B_{m(k)} - B_{m(k)}) - V_k$;

(4) $\pi_{n(k)} T S_{n(k)} w \in V_k$ for all $w \in X_{n(k)}$ with $|w|_{n(k)} \leq \varepsilon_k$.

(b) For $k = 1$ we may take $m(1) := \varepsilon_1 := 1$ and choose an arbitrary neighborhood V_1 of 0 in $Y_{n(1)}$ and an arbitrary element $x_1 \in X_0$. Suppose now that $m(i)$, ε_i, V_i, and x_i have been constructed for $1 \leq i \leq k-1$, where $k \geq 2$ is fixed. First choose $m(k) \geq k$ such that (1) is fulfilled. By continuity, there exists $0 < \delta \leq 2^{-k}\varepsilon_{k-1}$ such that $|A_{n(j)+1} \cdots A_{n(k-1)} x|_{n(j)} \leq 2^{-k}\varepsilon_j$ for all $x \in X_{n(k-1)}$ with $|x|_{n(k-1)} \leq \delta$ and for $j = 0,\ldots,k-2$. By the discontinuity of $\pi_{n(k)} T S_{n(k-1)}$ there is a neighborhood U of 0 in $Y_{n(k)}$ such that for every $\eta > 0$ there is an element $z \in X_{n(k-1)}$ with $|z|_{n(k-1)} \leq \eta$ and $\pi_{n(k)} T S_{n(k-1)}(z) \notin U$. Let V be a neighborhood of 0 in $Y_{n(k)}$ with $V - V \subset U$. By the continuity of $\pi_{n(k)}$ there is a $p \in \mathbb{N}$ such that $\pi_{n(k)}(B_{m(k)} - B_{m(k)}) \subset pV$. Hence, for $\eta := \delta/p$ there is some $z \in X_{n(k-1)}$ such that $|z|_{n(k-1)} \leq \delta/p$ and $\pi_{n(k)} T S_{n(k-1)}(z) \notin U$.

Now, with $x_k := pz$ and $V_k := pV$ the conditions (2) and (3) are satisfied. Finally, by the continuity of $\pi_{n(k)}TS_{n(k)}$ there exists an $\varepsilon_k > 0$ such that (4) is fulfilled. This completes the inductive construction.

(c) Because of (2), the series $x = \sum_{i=1}^{\infty} S_{n(i-1)}x_i$ converges in X_0. Since $B_{m(k)} \uparrow Y_0$, there is some $k \geq 2$ with $Tx \in B_{m(k)}$ and therefore with

$$\pi_{n(k)}Tx \in \pi_{n(k)}B_{m(k)}. \tag{*}$$

Again by (2), the series

$$w = x_{k+1} + \sum_{i=k+2}^{\infty} A_{n(k)+1} \cdots A_{n(i-1)}x_i$$

converges in $X_{n(k)}$ and we have $|w|_{n(k)} \leq \varepsilon_k$. Hence, by (4) and by the continuity of $S_{n(k)}$, we have

$$\pi_{n(k)}T\left(\sum_{i=k+1}^{\infty} S_{n(i-1)}x_i\right) = \pi_{n(k)}TS_{n(k)}w \in V_k. \tag{**}$$

Now, (*), (**), and (1) imply

$$\pi_{n(k)}TS_{n(k-1)}x_k$$
$$= \pi_{n(k)}Tx - \pi_{n(k)}T\left(\sum_{i=1}^{k-1} S_{n(i-1)}x_i\right) - \pi_{n(k)}TS_{n(k)}w \in \pi_{n(k)}(B_{m(k)} - B_{m(k)}) - V_k$$

which is a contradiction of (3). Hence, the theorem must be true.

Sliding hump theorems of this type are useful instruments in the theory of automatic continuity. For some applications of Theorem 3.1 we refer to [3,4,5,19].

4. Local conditions which imply continuity properties

In the following let Ω be a regular Hausdorff space and denote by $\mathfrak{F}(\Omega)$ the family of all closed subsets of Ω. For a TVS X, denote by $\mathbf{S}(X)$ the collection of all closed linear subspaces of X.

4.1. DEFINITIONS. A map $\mathcal{E} : \mathfrak{F}(\Omega) \to \mathbf{S}(X)$ is called a

- precapacity if $\mathcal{E}(\emptyset) = \{0\}$ and if $\mathcal{E}(F) \subset \mathcal{E}(G)$ for all $F, G \in \mathfrak{F}(\Omega)$ with $F \subset G$.
- $\cap$-stable (respectively, countably $\cap$-stable) if for every family (resp. every countable family) $(F_i)_{i \in I}$ in $\mathfrak{F}(\Omega)$ we have

$$\mathcal{E}\left(\bigcap_{i \in I} F_i\right) = \bigcap_{i \in I} \mathcal{E}(F_i).$$

- spectral capacity (resp. a strongly spectral capacity) if $\mathcal{E}$ is countably

$\cap$-stable, if $\mathcal{E}(\emptyset) = \{0\}$ and $\mathcal{E}(\Omega) = X$, and if for every finite open covering $\{U_1,\dots,U_n\}$ of Ω, we have $X = \mathcal{E}(\overline{U_1}) + \cdots + \mathcal{E}(\overline{U_n})$ (resp. $\mathcal{E}(F) = \mathcal{E}(F \cap \overline{U_1}) + \cdots + \mathcal{E}(F \cap \overline{U_n})$ for every $F \in \mathfrak{F}(\Omega)$).

4.2. EXAMPLES. (a) The notion of a spectral capacity is taken from the theory of decomposable operators (cf. [8, 13, 14, 21]). A bounded linear operator T on a Banach space X is decomposable if and only if there is a spectral capacity $\mathcal{E}$ from $\mathfrak{F}(\mathbb{C})$ to the set of all those $Y \in \mathfrak{S}(X)$ which are invariant for T, such that $\sigma(T \mid \mathcal{E}(F)) \subset F$ for all $F \in \mathfrak{F}(\mathbb{C})$ (see [13]). This spectral capacity is uniquely determined. There exist decomposable operators such that their spectral capacity is not strongly spectral (cf. [2]).

(b) Let X be one of the spaces $L^p(\Omega)$, $L^p_{loc}(\Omega)$ $(0 < p \leq \infty)$, $C^k(\Omega)$ or its strong dual $(k = 0,1,\dots,\infty)$ where $\Omega \subset \mathbb{R}^N$ is open, or let X be $C(\Omega)$ for some normal topological space Ω, or let X be $L^p(\Omega,\mu)$ for some compact Hausdorff space Ω and some Borel measure μ on Ω. Endow X with its natural topology. Then a $\cap$-stable and strongly spectral capacity $\mathcal{E} : \mathfrak{F}(\Omega) \to \mathfrak{S}(X)$ is given by $\mathcal{E}(F) := \{x \in X : \operatorname{supp} x \subset F\}$ for $F \in \mathfrak{F}(\Omega)$.

We need one more notation. Let X and Y be TVSs with precapacities $\mathcal{E}_X : \mathfrak{F}(\Omega) \to \mathfrak{S}(X)$ and $\mathcal{E}_Y : \mathfrak{F}(\Omega) \to \mathfrak{S}(Y)$, and let $T : X \to Y$ be a linear operator. A point $t \in \Omega$ is called a <u>point of continuity for</u> T (with respect to $\mathcal{E}_X$ and $\mathcal{E}_Y$) if there are a closed neighborhood U and an open neighborhood V of t such that the map

$$p_{\Omega \setminus V} T \mid \mathcal{E}_X(U) : \mathcal{E}_X(U) \to Y/\mathcal{E}_Y(\Omega \setminus V)$$

is continuous (where $p_{\Omega \setminus V} : Y \to Y/\mathcal{E}_Y(\Omega \setminus V)$ is the canonical surjection). We write $\Lambda(T)$ for the set of all those points in Ω which are not points of continuity for T. The main result of this section is now:

4.3. THEOREM. Let X be a Fréchet space and suppose that Y is a countably boundedly generated TVS which is sequentially complete and locally convex. Let $\mathcal{E}_X : \mathfrak{F}(\Omega) \to \mathfrak{S}(X)$ and $\mathcal{E}_Y : \mathfrak{F}(\Omega) \to \mathfrak{S}(Y)$ be two precapacities and suppose that $T : X \to Y$ is a linear map such that

$$\text{for all closed } F,G \subset \Omega \text{ with } F \subset \operatorname{int} G, \text{ the map } p_G T \mid \mathcal{E}_X(F) : \mathcal{E}_X(F) \to Y/\mathcal{E}_Y(G) \text{ is continuous}, \tag{LC}$$

where $p_G : Y \to Y/\mathcal{E}_Y(G)$ denotes the canonical surjection.

(a) Then $\Lambda(T)$ is a finite subset of Ω.

(b) If $\mathcal{E}_Y(F) \cap \mathcal{E}_Y(G) = \{0\}$ for all closed $F,G \subset \Omega$ with $F \cap G = \emptyset$, then for every $t \in \Omega \setminus \Lambda(T)$ there exists a closed neighborhood U of t such that $T \mid \mathcal{E}_X(U) : \mathcal{E}_X(U) \to Y$ is continuous.

(c) If $\mathcal{E}_X$ is a strongly spectral capacity and if $\mathcal{E}_Y$ satisfies the condition from (b), then for every compact $K \subset \Omega \setminus \Lambda(T)$ there is a closed neighborhood W of K such that $T \mid \mathcal{E}_X(W)$ is continuous.

(d) If $\mathcal{E}_X(\overline{U}) + \mathcal{E}_X(\overline{V}) = X$ for all open $U, V \subset \Omega$ with $U \cup V = \Omega$, then for every $t \in \Omega \setminus \Lambda(T)$ there exists an open neighborhood V of t such that $p_{\Omega \setminus V} T : X \to Y/\mathcal{E}_Y(\Omega \setminus V)$ is continuous.

(e) If $\mathcal{E}_X$ is a spectral capacity and if $\mathcal{E}_Y$ is $\cap$-stable, then $\mathfrak{S}(T) \subset \mathcal{E}_Y(\Lambda(T))$.

(f) If $\mathcal{E}_X$ is a spectral capacity, if $\mathcal{E}_Y$ is countably $\cap$-stable, and if Ω is second countable, then $\mathfrak{S}(T) \subset \mathcal{E}_Y(\Lambda(T))$.

(g) If in the situation of (e) or (f) the space X is nuclear, then there exists a continuous linear map $S : X \to Y$ such that $(T - S)(X) \subset \mathcal{E}_Y(\Lambda(T))$.

(h) If in the situation of (e) or (f) we have moreover $\mathcal{E}_Y(\Lambda) = \{0\}$ for all finite subsets Λ of Ω, then T is continuous.

Proof. (a) Assume that $\Lambda(T)$ is not finite. Since Ω is a regular topological space, there is a sequence $(V_n)_{n=1}^{\infty}$ of open sets in Ω with mutually disjoint closures such that $V_n \cap \Lambda(T) \neq \emptyset$ for all $n \in \mathbb{N}$ (cf. [10, p.142]). For each $n \in \mathbb{N}$ we fix some $t_n \in V_n \cap \Lambda(T)$ and a closed neighborhood U_n of t_n satisfying $U_n \subset V_n$. Then we put

$$F_n := \overline{\bigcup_{k=n+1}^{\infty} U_k}, \quad W_n := \bigcup_{k=1}^{n} \operatorname{int} U_k, \quad X_n := \mathcal{E}_X(F_n), \quad Y_n := Y/\mathcal{E}_Y(\Omega \setminus W_n)$$

for all $n \in \mathbb{N}$ and $X_0 := X$, $Y_0 := Y$. We write A_n for the inclusion map from X_n into X_{n-1} and π_n for the canonical surjection from Y_0 onto Y_n for all $n \in \mathbb{N}$. Notice that $F_n \cap (U_1 \cup \cdots \cup U_n) = \emptyset$ and therefore $F_n \subset \operatorname{int}(\Omega \setminus W_n)$ for all $n \in \mathbb{N}$. Hence, because of (LC), the map $\pi_n T A_1 \cdots A_n = p_{\Omega \setminus W_n} T \mid \mathcal{E}_X(F_n)$ is continuous for every $n \in \mathbb{N}$. It now follows from Theorem 3.1 that for some suitable $n \in \mathbb{N}$ all the maps $\pi_k T A_1 \cdots A_n$ for $k \in \mathbb{N}$ are continuous. In particular, $\pi_{n+1} T A_1 \cdots A_n = p_{\Omega \setminus W_{n+1}} \mid \mathcal{E}_X(F_n)$ turns out to be continuous. Since F_n is a closed and W_{n+1} is an open neighborhood of t_{n+1}, we conclude that $t_{n+1} \notin \Lambda(T)$, which is a contradiction.

(b) Let $t \in \Omega$ be an arbitrary point of continuity for T and choose a closed neighborhood U and an open neighborhood V of t such that $p_{\Omega \setminus V} T \mid \mathcal{E}_X(U)$ is continuous. Because of the regularity of Ω, we then find closed neighborhoods W and W' of t such that $W \subset \operatorname{int} W' \subset \operatorname{int}(U \cap V)$. Thus $p_{\Omega \setminus V} T \mid \mathcal{E}_X(W)$ is continuous because of $\mathcal{E}_X(W) \subset \mathcal{E}_X(U)$, and $p_{W'} T \mid \mathcal{E}_X(W)$ is continuous because of (LC). Since $\Omega \setminus V$ and W' are disjoint, we conclude from Proposition 2.1 that

$\mathfrak{S}(T \mid \mathcal{E}_X(W)) \subset \mathcal{E}_Y(\Omega\setminus V) \cap \mathcal{E}_Y(W') = \{0\}$, and hence T is continuous on $\mathcal{E}_X(W)$.

(c) The proof is the same as that of [5, Theorem 3.6].

(d) Given an arbitrary $t \in \Omega\setminus\Lambda(T)$, we fix a closed neighborhood U and an open neighborhood V of t such that $p_{\Omega\setminus V}T \mid \mathcal{E}_X(U)$ is continuous. Next we choose open neighborhoods W and W' of t satisfying $\overline{W} \subset W' \subset \overline{W'} \subset \operatorname{int}(U \cap V)$. Then we claim that $p_{\Omega\setminus W}T$ is continuous on X. Using (LC) one easily sees that this mapping is continuous on each of the closed subspaces $\mathcal{E}_X(U)$ and $\mathcal{E}_X(\Omega\setminus W')$. But because $\operatorname{int} U \cup \operatorname{int}(\Omega\setminus W') = \Omega$, the assumption on $\mathcal{E}_X$ implies that the sum of these subspaces is the whole of X. Hence a standard application of the open mapping theorem (cf. [5, Lemma 3.1]) ensures the continuity of $p_{\Omega\setminus W}T$ on X.

(e) Because of (d), we obtain for every $t \in \Gamma := \Omega\setminus\Lambda(T)$ some open neighborhood $V(t)$ such that $p_{\Omega\setminus V(t)}T$ is continuous on X. By Proposition 2.1 we conclude that $\mathfrak{S}(T) \subset \mathcal{E}_Y(\Omega\setminus V(t))$ and hence

$$\mathfrak{S}(T) \subset \bigcap_{t\in\Gamma} \mathcal{E}_Y(\Omega\setminus V(t)) = \mathcal{E}_Y\Big(\bigcap_{t\in\Gamma} (\Omega\setminus V(t))\Big) = \mathcal{E}_Y(\Lambda(T)) .$$

(f) It is an easy consequence of Lindelöf's covering theorem that, in the situation of (f), the precapacity $\mathcal{E}_Y$ preserves even arbitrary intersections (cf. [5, Remark 3.8]). Consequently, the assertion is contained in (e).

(g) follows from (e) resp. (f) and Proposition 2.4.

(h) is an immediate consequence of (e) resp. (f) and (a).

4.4. REMARKS. (a) It should be noted that all parts of the theorem apply to the precapacity considered in Example 4.2.(b).

(b) Recall from [5] that the linear operator $T: X \to Y$ is called $(\mathcal{E}_X,\mathcal{E}_Y)$-<u>local</u> if $T(\mathcal{E}_X(F)) \subset \mathcal{E}_Y(F)$ holds for all closed $F \subset \Omega$. The class of these operators subsumes local operators in the classical sense, algebraic homomorphisms between certain topological algebras, and certain intertwining operators (cf. [5]). Now, every $(\mathcal{E}_X,\mathcal{E}_Y)$-local operator obviously satisfies condition (LC). Hence Theorem 4.3 generalizes some of the results from [5, Section 3].

In the following, we give some typical applications of Theorem 4.3 to operators between some of the usual spaces of analysis.

4.5. THEOREM. Let Ω be an open subset of $\mathbb{R}^N$ and let X and Y be two of the following Fréchet spaces:

$$L^p(\Omega),\ L^p_{loc}(\Omega)\ (1 \le p \le \infty),\ C^k(\Omega)\quad (k = 0,1,2,\dots,\infty).$$

For a linear map $T: X \to Y$ the following are equivalent.

(a) T is continuous.

(b) For all $\varphi,\psi \in C^\infty(\Omega)$ which are constant on $\Omega\setminus K$ for some compact subset $K = K(\varphi,\psi)$ and satisfy $\operatorname{supp}\varphi \cap \operatorname{supp}\psi = \emptyset$, the map $M_\varphi T M_\psi : X \to Y$ is continuous. Here M_φ, M_ψ are the operators of multiplication by φ (respectively, by ψ).

Proof. We have only to show that (b) implies (a). As, in all cases, $Y \subset L^1_{loc}(\Omega)$ with continuous inclusion map and as the closed graph theorem for the pair (X,Y) is valid, it is sufficient to give the proof for $Y = L^1_{loc}(\Omega)$. Let $(K_n)_{n=1}^\infty$ be a sequence of compact subsets of Ω with $K_n \subset \operatorname{int}(K_{n+1})$ for all $n \in \mathbb{N}$ and $\Omega = \bigcup_{n=1}^\infty K_n$ and denote by χ_n the characteristic function of K_n. $Y_n := M_{\chi_n} Y$ can be identified with $L^1(K_n)$ and is a Banach space in the topology induced by $Y = L^1_{loc}(\Omega)$. Let Ω^∞ be the one point compactification of Ω and define $\mathcal{E}_X : \mathfrak{F}(\Omega^\infty) \to \mathfrak{S}(X)$ and $\mathcal{E}_n : \mathfrak{F}(\Omega^\infty) \to \mathfrak{S}(Y_n)$ by $\mathcal{E}_X(F) := \{f \in X : \operatorname{supp} f \subset F \cap \Omega\}$ and $\mathcal{E}_n(F) := \{f \in Y_n : \operatorname{supp} f \subset F \cap \Omega\}$ for $F \in \mathfrak{F}(\Omega^\infty)$. Then $\mathcal{E}_X$ and $\mathcal{E}_n$ are spectral capacities and $\mathcal{E}_n(\Lambda) = \{0\}$ for all finite $\Lambda \subset \Omega^\infty$. Because of (b) the condition (LC) in Theorem 4.3 is fulfilled for $T_n := M_{\chi_n} T : X \to Y_n$. By 4.3(h), the maps $T_n : X \to Y_n \subset Y$ $(n \in \mathbb{N})$ are therefore continuous. As $T_n f \to f$ for $n \to \infty$ for all $f \in X$, we obtain by means of the Banach-Steinhaus theorem (cf. [12, p.51]) the continuity of T.

By means of Theorem 4.3 we are now able to prove the continuity result for operators of local type mentioned in the Introduction.

4.6. THEOREM. Suppose that Ω is a compact Hausdorff space and let μ be a Borel measure on Ω having no point atoms. Then every operator from $L^p(\Omega,\mu)$ to itself which is of local type is automatically continuous.

Proof. The map $\mathcal{E} : \mathfrak{F}(\Omega) \to \mathfrak{S}(L^p(\Omega,\mu))$ with $\mathcal{E}(F) := \{f \in L^p(\Omega,\mu) : \operatorname{supp} f \subset F\}$ for $F \in \mathfrak{F}(\Omega)$ is a $\cap$-stable strongly spectral capacity, and condition (LC) in 4.3 is satisfied with $X := Y := L^p(\Omega,\mu)$ and $\mathcal{E}_X := \mathcal{E}_Y := \mathcal{E}$. Moreover, $\mathcal{E}(\Lambda) = \{0\}$ for all finite subsets Λ of Ω. Hence, by Theorem 4.3(h), the operator T is continuous.

4.7. REMARK. Consider $\Omega := [-1,1]$, $\mu := \lambda + \delta_0$, where λ is the Lebesgue measure and δ_0 is the Dirac measure at 0. Then there exists a discontinuous local type operator $T : L^p(\Omega,\mu) \to L^p(\Omega,\mu)$. This shows that the condition for μ in 4.6 is necessary.

Proof. $X_0 := \{f \in L^p(\Omega,\mu) : f$ is constant in some neighborhood U_f of $0\}$ is a dense linear subspace of $X := L^p(\Omega,\mu)$. Let $L : X \to \mathbb{C}$ be a linear functional such that $L \mid X_0 \equiv 0$ and $L(g) = 1$, where $g(t) := t$ for all $t \in [-1,1]$. As X_0 is dense in X, L must be discontinuous. Hence, the linear operator

$T : X \to X$ given by $Tf := L(f)\chi_{\{0\}}$ for $f \in X$, is discontinuous. Moreover, T is of local type (it is even a local operator).

4.8. THEOREM. Let Ω be an open subset of $\mathbb{R}^N$ and suppose that $T : \mathcal{D}(\Omega) \to \mathcal{D}'(\Omega)$ is a linear operator satisfying the condition

$$M'_\varphi T M_\psi : \mathcal{D}(\Omega) \to \mathcal{D}'(\Omega) \text{ is continuous for all } \varphi, \psi \in \mathcal{D}(\Omega) \text{ with } \operatorname{supp}\varphi \cap \operatorname{supp}\psi = \emptyset. \tag{LC'}$$

Here, for $\varphi \in \mathcal{D}(\Omega)$, $M_\varphi : \mathcal{D}(\Omega) \to \mathcal{D}(\Omega)$ is the operator of multiplication with φ and $M'_\varphi : \mathcal{D}'(\Omega) \to \mathcal{D}'(\Omega)$ is its transpose.

(a) If the range of T is contained in $L^1_{loc}(\Omega)$, then T is continuous.

(b) There is a discrete subset Λ of Ω such that

$$T \mid \mathcal{D}(\Omega \setminus \Lambda) : \mathcal{D}(\Omega \setminus \Lambda) \to \mathcal{D}'(\Omega)$$

is continuous with respect to the topology of $\mathcal{D}(\Omega \setminus \Lambda)$.

Proof. (1) Let $(K_n)_{n=1}^\infty$ be as in the proof of 4.5. Then $\mathcal{D}(\Omega)$ is the inductive limit of the Fréchet spaces

$$\mathcal{D}(K_n) := \{\varphi \in C^\infty(\Omega) : \operatorname{supp}\varphi \subset K_n\}.$$

For each $n \in \mathbb{N}$ we fix a function $\varphi_n \in \mathcal{D}(K_{n+1})$ with $\operatorname{supp}(1 - \varphi_n) \cap K_n = \emptyset$. Let now Y be the strong dual of the Fréchet space $C^\infty(\Omega)$ in the general case and put $Y := L^1(\Omega)$ in the case that $T(\mathcal{D}(\Omega)) \subset L^1_{loc}(\Omega)$. Then the range of $T_n := M'_{\varphi_n} T$ is contained in Y in both cases. Let Ω^∞ be the one point compactification of Ω and define $\mathcal{E}_m(F) := \{\psi \in \mathcal{D}(K_m) : \operatorname{supp}\psi \subset F\}$, $\mathcal{E}_Y(F) := \{f \in Y : \operatorname{supp} f \subset F \cap \Omega^\infty\}$ for $F \in \mathfrak{F}(\Omega^\infty)$ and $m \in \mathbb{N}$. $\mathcal{E}_Y : \mathfrak{F}(\Omega^\infty) \to \mathfrak{S}(Y)$ and $\mathcal{E}_m : \mathfrak{F}(\Omega^\infty) \to \mathfrak{S}(\mathcal{D}(K_m))$ are strongly spectral capacities. Because of (LC') we see that for all $n, m \in \mathbb{N}$ the operator $T_n \mid \mathcal{D}(K_m) : \mathcal{D}(K_m) \to Y$ satisfies condition (LC) of Theorem 4.3.

(2) In the case that the range of T is contained in $L^1_{loc}(\Omega)$ we apply 4.3 (h) to $T_n \mid \mathcal{D}(K_m) : \mathcal{D}(K_m) \to Y$ and obtain the continuity of this operator. As the inclusion map $Y \to L^1_{loc}(\Omega)$ is continuous we see that $T_n \mid \mathcal{D}(K_m) : \mathcal{D}(K_m) \to L^1_{loc}(\Omega)$ is continuous. Now, $T_n\psi \to T\psi$ for all ψ in $\mathcal{D}(K_m)$. Therefore, by the Banach-Steinhaus theorem, $T \mid \mathcal{D}(K_m)$ is continuous for all $m \in \mathbb{N}$. This implies the continuity of T.

(3) We apply 4.3 to the maps $S_n := T_{n+1} \mid \mathcal{D}(K_n) : \mathcal{D}(K_n) \to Y$ and obtain that $\Lambda_n := \Lambda(S_n)$ is a finite set which is of course contained in $K_n (n \in \mathbb{N})$. Obviously, $\Lambda_n \subset \Lambda_{n+1}$ and $\Lambda_n \cap \operatorname{int} K_n = \Lambda_{n+1} \cap \operatorname{int} K_n$ for all $n \in \mathbb{N}$. Hence, $\Lambda := \bigcup_{n=1}^\infty \Lambda_n$ is a discrete subset of Ω. Let now K be an arbitrary compact subset of $\Omega \setminus \Lambda$. Then there is some $n_0 \in \mathbb{N}$ such that $K \subset K_n$ for all $n \geq n_0$ and $K \cap \Lambda_n = \emptyset$

for all n By 4.3 (c), $T_n | \mathcal{D}(K) : \mathcal{D}(K) \to Y$ and hence $T_n | \mathcal{D}(K) : \mathcal{D}(K) \to \mathcal{D}'(\Omega)$ is continuous for all $n \geq n_0$ (notice, that $\mathcal{D}(K) = \mathcal{E}_n(K)$ for all $n \geq n_0$). Again by the Banach-Steinhaus theorem we obtain the continuity of $T | \mathcal{D}(K)$, as $T_n\psi \to T\psi$ for all $\psi \in \mathcal{D}(K)$. This shows that $T | \mathcal{D}(\Omega \setminus \Lambda) : \mathcal{D}(\Omega \setminus \Lambda) \to \mathcal{D}'(\Omega)$ is continuous with respect to the topology of $\mathcal{D}(\Omega \setminus \Lambda)$.

Finally let us remark that of course, many of the results in this section can also be obtained for other spaces of functions, distributions or ultradistributions (see [6] in the case of local operators).

References

[1] E. Albrecht, Funktionalkalküle in mehreren Veränderlichen für stetige lineare Operatoren auf Banachräumen, Manuscripta Math., 14 (1974), 1-40.

[2] __________, On two questions of I. Colojoară and C. Foiaş, Manuscripta Math., 25 (1978), 1-15.

[3] E. Albrecht and H. G. Dales, Homomorphisms from C*-algebras and other Banach algebras, this Volume.

[4] E. Albrecht und M. Neumann, Automatische Stetigkeitseigenschaften einiger Klassen linearer Operatoren, Math. Ann., 240 (1979), 251-280.

[5] __________, Automatic continuity of generalized local linear operators, Manuscripta Math., 32 (1980), 263-294.

[6] __________, Local operators between spaces of ultradifferentiable functions and ultradistributions, Manuscripta Math., 38 (1982), 131-161.

[7] __________, Continuity properties of C^k-homomorphisms, this Volume.

[8] C. Apostol, Spectral decompositions and functional calculus, Rev. Roum. Math. Pures et Appl., 13 (1968), 1481-1528.

[9] M. De Wilde, Closed Graph Theorems and Webbed Spaces, Pitman, London-San Francisco-Melbourne, 1978.

[10] J. Dugundji, Topology, Allyn and Bacon, Boston, 1966.

[11] K Floret, $\mathcal{L}_1$-Räume und Liftings von Operatoren nach Quotienten lokalkonvexer Räume, Math. Z., 134 (1973), 107-117.

[12] K. Floret und J. Wloka, Einführung in die Theorie der lokalkonvexen Räume, Lecture Notes in Mathematics, No. 56, Springer-Verlag, Berlin-Heidelberg-New York, 1968.

[13] C. Foiaş, Spectral capacities and decomposable operators, Rev. Roum. Math. Pures et Appl., 13 (1968), 1539-1545.

[14] St. Frunză, The Taylor spectrum and spectral decompositions, J. Functional Analysis, 19 (1975), 390-421.

[15] B. Gramsch, Ein Schwach-Stark-Prinzip der Dualitätstheorie lokalkonvexer Räume als Fortsetzungsmethode, Math. Z., 156 (1977), 217-230.

[16] G. Köthe, Topological Vector Spaces II, Springer-Verlag, New York-Heidelberg-Berlin, 1979.

[17] J. Lindenstrauss and L. Tzafriri, Classical Banach Spaces, Lecture Notes in Mathematics, No. 338, Springer-Verlag, Berlin-Heidelberg-New York, 1973.

[18] S. G. Michlin und S. Prößdorf, Singuläre Integraloperatoren, Akademie-Verlag, Berlin, 1980.

[19] M. Neumann, Automatic Continuity of Linear Operators. In: Functional Analysis: Surveys and recent results II, K.-D. Bierstedt, B. Fuchssteiner (eds.), North-Holland Publ. Comp., Amsterdam-New York-Oxford, 1980, 269-296.

[20] J. Peetre, Rectification à l'article "Une caractérisation abstraite des opérateurs différentiels", Math. Scand., 8 (1960), 116-120.

[21] F.-H. Vasilescu, Analytic functional calculus and spectral decompositions, to appear.

Fachbereich Mathematik der
Universität des Saarlandes
D-6600 Saarbrücken
W. Germany

Fachbereich Mathematik der
Universität Essen-Gesamthochschule
D-4300 Essen
W. Germany

)

CONTINUITY PROPERTIES OF C^k-HOMOMORPHISMS

E. Albrecht and M. Neumann

1. Introduction

Let Ω be an open subset of $\mathbb{R}^n$ or $\mathbb{C}^n$, and for $k = 0,1,\dots,\infty$ let $C^k(\Omega)$ denote the usual Fréchet algebra of all k-times continuously differentiable functions $f:\Omega \to \mathbb{C}$. A central theme of this article will be the automatic continuity problem for homomorphisms from $C^k(\Omega)$ into some Banach algebra B. Discontinuous homomorphisms of this type can be easily constructed by adapting a standard technique from [7]; see also [8]. Moreover, at least if one assumes the continuum hypothesis, then these homomorphisms may be highly discontinuous in the sense that the restriction to $C^\infty(\Omega)$ is discontinuous with respect to the C^∞-topology [13, Th. 1]. On the other hand for a homomorphism $\varphi: C^k(\Omega) \to B$ which, for the sake of simplicity, we assume to have dense range, the last section of the present paper will establish the following positive results:

(a) There exists a finite subset Λ of Ω such that φ is continuous with respect to the C^{2k}-topology on the subalgebra of all $f \in C^{2k}(\Omega)$ vanishing on some neighborhood of Λ.

(b) φ is continuous on $C^{2k+1}(\Omega)$ for the C^{2k+1}-topology if and only if none of the operators of multiplication by $\varphi(Z_j)$ on B for $j = 1,\dots,n$ has a non-trivial divisible subspace, where $Z_1,\dots,Z_n : \Omega \to \mathbb{C}$ are the coordinate functions.

(c) φ is continuous on $C^\infty(\Omega)$ for the C^∞-topology if and only if φ is continuous on $C^{2k+1}(\Omega)$ for the C^{2k+1}-topology.

(d) If the radical of B is nil, then φ is necessarily continuous on $C^{2k+1}(\Omega)$ for the C^{2k+1}-topology.

(e) If the radical of B is finite dimensional, then φ is even continuous on $C^{2k}(\Omega)$ for the C^{2k}-topology.

Assertions (a) and (d) should be compared with classical results of Bade and Curtis [7] concerning homomorphisms on the Banach algebra C(K) of all continuous functions on a compact Hausdorff space K. It turns out that the situation for $C^\infty(\Omega)$ is quite similar to that for C(K), whereas some loss of orders of differentiation has to be taken into account for $C^k(\Omega)$ if $0 < k < \infty$. In fact, it follows from a result of Dales [14, Th. 3.1] that in (b) and in (c) the order $2k+1$ is best possible in general. We shall also prove corresponding results for homomorphisms on $C^k(\overline{\Omega})$, where Ω is an open and bounded subset of $\mathbb{R}^n$ or $\mathbb{C}^n$ satisfying some mild regularity condition at the boundary. These results are closely related to and actually an extension of some work of Bade, Curtis, Laursen, and

Both authors were supported by NATO Grant No. RG 073.81.

Ochoa [9; 10; 17]. But perhaps it should be noted that the present approach avoids tensor products and analytic spaces and that factorization arguments enter the scene only at the very end in connection with assertion (e). This paper is based on the automatic continuity theory for generalized local operators, and we shall occasionally use some results and special notions from [2] or [4].

Another essential ingredient will be the theory of non-analytic functional calculi from which our interest in this type of continuity problem arose. Recall from [12] that an operator $T \in L(X)$ is said to be generalized scalar if there exists a continuous homomorphism $\varphi : C^\infty(\mathbb{C}) \to L(X)$ such that $\varphi(1) = I$ and $\varphi(Z) = T$, where $L(X)$ is the Banach algebra of all continuous linear operators on a given Banach space X and I, Z denote the identity functions on X, $\mathbb{C}$ respectively. It is natural to ask to what extent the continuity assumption on φ is superfluous in this definition. We shall prove that a certain algebraic property of T is decisive for this question: An operator with some non-trivial divisible subspace cannot be generalized scalar, whereas for operators with no divisible subspaces different from $\{0\}$ any functional calculus on $C^\infty(\mathbb{C})$ is automatically continuous. We shall also obtain certain uniqueness results for functional calculi on $C^k(\Omega)$ and $C^k(\overline{\Omega})$.

The organization of this note is as follows. In section 2, we start to investigate the continuity properties of A-module homomorphisms $\varphi : X \to Y$, where A is an algebra admitting appropriate partitions of unity, X is complete metrizable, and Y has a fundamental sequence of bounded sets. The main idea is to regard such mappings as generalized local operators and then to apply the abstract theory from [2; 4]. Section 3 contains some straightforward, but useful consequences concerning algebra homomorphisms. These results are rather general and apply to various situations including, for example, algebras of Lipschitz functions or ultradifferentiable functions, but in this paper we shall concentrate on the algebras $C^k(\Omega)$ and $C^k(\overline{\Omega})$. In section 4, we obtain some basic continuity results for homomorphisms from these algebras into some topological algebra with a fundamental sequence of bounded sets. Here, no restriction will be imposed on the range of the homomorphisms. In section 5, we introduce some further techniques in the context of non-analytic functional calculi on a wide class of topological algebras. Results of this type have already proved to be useful in [3] and they give some insight into the role of divisible subspaces. In section 6, we finally derive the main theorems on C^k-homomorphisms including the above-mentioned results.

This paper is an extended and somewhat different version of our manuscript [5].

2. Module homomorphisms

We first have to fix some terminology. A topological vector space (TVS for short) is always supposed to be separated, but we do not assume local convexity in general. An algebra A endowed with some vector space topology is called a topological algebra if the multiplication in A is separately continuous. By an (F)-algebra we mean a complete metrizable topological algebra, neither assumed to be locally convex nor locally multiplicatively-convex. Here, a module over an algebra A is a TVS X together with an algebra homomorphism, say ρ, from A into the algebra $L(X)$ of all continuous linear operators on X; as usual, we simply write $fx := \rho(f)(x)$ for all $f \in A$, $x \in X$. Given a normal topological space Ω, an algebra A of functions $f : \Omega \to \mathbb{C}$ is said to be normal if for all disjoint closed subsets F and G of Ω there exists an $f \in A$ satisfying $f = 1$ on F as well as $f = 0$ on G. Using standard techniques on partitions of unity, one easily verifies that an algebra A of complex-valued functions on Ω is normal if and only if for every finite open covering $\{U_1, \dots, U_k\}$ of Ω there exist $f_j \in A$ with $\operatorname{supp} f_j \subseteq U_j$ for $j = 1, \dots, k$ such that $f_1 + \dots + f_k = 1$ on Ω. Finally recall that a linear operator between a pair of TVSs is said to be bounded if it maps some 0-neighborhood into a bounded set. Of course, boundedness in this sense implies continuity.

2.1. THEOREM. Consider an (F)-space X and a countably boundedly generated TVS Y, and assume that X and Y are unital modules over some normal algebra A of complex-valued functions on a normal topological space Ω. Then, for every A-module homomorphism $\varphi : X \to Y$, there exists a finite subset Λ of Ω such that every compact $K \subseteq \Omega \setminus \Lambda$ has an open neighborhood $W \subseteq \Omega$ such that φ is bounded on the submodule $\{fx : x \in X,\ f \in A \text{ with } \operatorname{supp} f \subseteq W\}$ of X.

Proof. We use the results and notations from [2] or [4]. For each closed subset F of Ω, let $\mathcal{E}_X(F) := \{x \in X : gx = 0 \text{ for all } g \in A \text{ with } F \cap \operatorname{supp} g = \emptyset\}$ and define $\mathcal{E}_Y(F) \subseteq Y$ in a similar way. Because A is normal, we thus obtain two strongly spectral and disjointly additive precapacities $\mathcal{E}_X : \mathfrak{F}(\Omega) \to \mathfrak{S}(X)$ and $\mathcal{E}_Y : \mathfrak{F}(\Omega) \to \mathfrak{S}(Y)$. Because of $\varphi(\mathcal{E}_X(F)) \subseteq \mathcal{E}_Y(F)$ for all closed $F \subseteq \Omega$, the general automatic continuity theory, see [2, Th. 3.6] or [4], supplies us with some finite $\Lambda \subseteq \Omega$ such that for every compact $K \subseteq \Omega \setminus \Lambda$ the restriction of φ to $\mathcal{E}_X(\overline{W})$ is continuous, where W is a suitable open neighborhood of K in Ω. Since $\mathcal{E}_X(\overline{W})$ is complete metrizable and since Y is the union of an increasing sequence of closed bounded sets, the Baire category theorem forces φ to be bounded on some 0-neighborhood of $\mathcal{E}_X(\overline{W})$. Now, the normality assumption implies that $\{fx : x \in X,\ f \in A \text{ with } \operatorname{supp} f \subseteq W\}$ is indeed a submodule of X, and it is clear that this submodule is contained in $\mathcal{E}_X(\overline{W})$. The assertion follows.

A disadvantage of the preceding result is that the continuity of φ is not under control if one considers a collection of compact sets $K \subseteq \Omega \setminus \Lambda$ approaching the singularity set Λ. We therefore supplement 2.1 by a considerably stronger continuity assertion in 2.4, where we have to strengthen the assumptions only slightly. Theorem 2.4 comes closer to previous work on automatic continuity and requires the following main boundedness type argument. The ancestor of this result is in [7], while the present proof is modelled after [16].

2.2. THEOREM. Consider a pair of (F)-spaces X_1 and X_2, a pair of TVSs Y_1 and Y_3 with fundamental sequences of bounded sets, and a pair of arbitrary TVSs X_3 and Y_2. Further, let $q : X_1 \times X_2 \to X_3$ and $r : Y_1 \times Y_2 \to Y_3$ denote separately continuous bilinear mappings, and for $j = 1,2,3$ let $T_j : X_j \to Y_j$ be linear operators (not assumed to be continuous) such that $r(T_1 \times T_2) = T_3 q$. Finally, assume that T_3 is continuous on some closed linear subspace M of X_3, and consider two bounded sequences $(a_n)_{n\in\mathbb{N}}$ and $(b_n)_{n\in\mathbb{N}}$ in X_1 and X_2 respectively such that $q(a_m,b_n) \in M$ whenever $m \neq n$. Then $(T_3 q(a_n,b_n))_{n\in\mathbb{N}}$ is bounded in Y_3.

Proof. (1) Let $(B_k)_{k\in\mathbb{N}}$ and $(C_k)_{k\in\mathbb{N}}$ denote fundamental sequences of bounded sets for Y_3 and Y_1 respectively, and let $|\cdot|_1$ and $|\cdot|_2$ be (F)-norms generating the topology of X_1 and X_2 respectively. $\mathfrak{P}(\mathbb{N})$ stands for the family of all subsets of $\mathbb{N}$. We shall need the following auxiliary result. We claim that a sequence $(y_j)_{j\in\mathbb{N}}$ in Y_3 is necessarily bounded if for some bounded set B in Y_3 and some pair of functions $\varphi : \mathfrak{P}(\mathbb{N}) \to Y_1$ and $\psi : \mathfrak{P}(\mathbb{N}) \to Y_2$ the condition

$$y_j - r(\varphi(L),\psi(K)) \in B \quad \text{for all } j \in \mathbb{N} \text{ and } K,L \subseteq \mathbb{N} \text{ with } K \cap L = \{j\}$$

is fulfilled. Suppose that this is false, choose a strictly increasing sequence of integers $k(m)$ such that $y_{k(m)} \notin B_m$ for all $m \in \mathbb{N}$, and group these integers into a sequence of sets as indicated by the following construction:

$$K_1 := \{k(1),k(2),k(5),\ldots\}$$
$$K_2 := \{k(4),k(3),k(6),\ldots\}$$
$$K_3 := \{k(9),k(8),k(7),\ldots\}$$
$$\ldots$$

We thus obtain a sequence of pairwise disjoint $K_m \in \mathfrak{P}(\mathbb{N})$ such that none of the sets $\{y_j : j \in K_m\}$ is bounded. On the other hand, the continuity assumption on r implies that each of the sets $r(C_m,\psi(K_m))$ is bounded in Y_3. Hence, for each $m \in \mathbb{N}$ there exists some $j(m) \in K_m$ such that $y_{j(m)} \notin r(C_m,\psi(K_m)) + B$. Let $L := \{j(m) : m \in \mathbb{N}\}$ and fix an $m \in \mathbb{N}$ such that $\varphi(L) \in C_m$. Then we have $K_m \cap L = \{j(m)\}$ and hence by assumption $y_{j(m)} \in r(\varphi(L),\psi(K_m)) + B \subseteq r(C_m,\psi(K_m)) + B$, a contradiction.

(2) Now assume that the theorem is false. Since $(a_n)_{n\in\mathbb{N}}$ and $(b_n)_{n\in\mathbb{N}}$ are bounded sequences, for each $k \in \mathbb{N}$ we obtain a real $t_k > 0$ such that $|t_k a_n|_1 \leq 2^{-k}$ and $|t_k b_n|_2 \leq 2^{-k}$ for all $n \in \mathbb{N}$. Then we choose by induction a strictly increasing sequence of integers $n(k)$ such that $T_3 q(a_{n(k)},b_{n(k)}) \notin t_k^{-2} B_k$

for all $k \in \mathbb{N}$. It is clear that, for each $K \subseteq \mathbb{N}$, the series

$$u_K := \sum_{k\in K} t_k a_{n(k)} \quad \text{and} \quad v_K := \sum_{k\in K} t_k b_{n(k)}$$

converge in the (F)-spaces X_1 and X_2 respectively, and it is easily seen that the sets $\{u_K : K \subseteq \mathbb{N}\}$ and $\{v_K : K \subseteq \mathbb{N}\}$ are bounded. Since q is separately continuous, the principle of uniform boundedness forces $\{q(u_K, v_L) : K, L \subseteq \mathbb{N}\}$ to be bounded in X_3. Furthermore, as $q(a_{n(k)}, b_{n(\ell)}) \in M$ whenever $k \neq \ell$, the assumptions on q and M imply that $q(u_K, v_L) \in M$ whenever $K, L \subseteq \mathbb{N}$ are disjoint. By the continuity of T_3 on M, we thus obtain some bounded $D \subseteq Y_3$ such that $T_3 q(u_K, v_L) \in D$ for all $K, L \subseteq \mathbb{N}$ with $K \cap L = \emptyset$. Now, given $j \in \mathbb{N}$ and $K, L \subseteq \mathbb{N}$ such that $K \cap L = \{j\}$ we conclude from

$$q(u_{\{j\}}, v_{\{j\}}) = q(u_K, v_L) - q(u_K, v_{L\setminus\{j\}}) - q(u_{K\setminus\{j\}}, v_{\{j\}})$$

that $T_3 q(u_{\{j\}}, v_{\{j\}}) \in r(T_1(u_K), T_2(v_L)) - D - D$. This enables us to apply the auxiliary result from (1): Let $\varphi : \mathfrak{P}(\mathbb{N}) \to Y_1$ and $\psi : \mathfrak{P}(\mathbb{N}) \to Y_2$ be given by $\varphi(K) := T_1(u_K)$ and $\psi(K) := T_2(v_K)$ for all $K \subseteq \mathbb{N}$, and define $y_k :=$ $T_3 q(t_k a_{n(k)}, t_k b_{n(k)})$ for all $k \in \mathbb{N}$. Then we know that $y_j - r(\varphi(K), \psi(L)) \in -D - D$ holds whenever $j \in \mathbb{N}$ and $K, L \subseteq \mathbb{N}$ satisfy $K \cap L = \{j\}$. Hence $(y_j)_{j\in\mathbb{N}}$ has to be bounded so that, for some suitable $k \in \mathbb{N}$, we arrive at $y_j \in B_k$ for all $j \in \mathbb{N}$. But $y_k \notin B_k$ according to the choice of $n(k)$. This contradiction completes the proof of the theorem.

2.3. COROLLARY. Consider an (F)-algebra A, an (F)-space X, and a TVS Y with a fundamental sequence of bounded sets. Let X and Y be modules over A, and assume that the module operation on X is separately continuous. Then, for every A-module homomorphism $\varphi : X \to Y$ and for all bounded sequences $(f_n)_{n\in\mathbb{N}}$ and $(x_n)_{n\in\mathbb{N}}$ in A resp. X such that $f_m x_n = 0$ whenever $m \neq n$, the sequence $(\varphi(f_n x_n))_{n\in\mathbb{N}}$ is bounded in Y.

Proof. Let $X_1 := X_3 := X, X_2 := A, M := \{0\}$ and consider the mapping $q : X \times A \to X$ given by $q(x,f) := fx$. Further, let $Y_1 := Y_3 := Y$, endow $Y_2 := A$ with the strongest vector space topology, and let $r : Y \times A \to Y$ be given by $r(y,f) := fy$. Then the assertion is immediate after 2.2.

2.4. THEOREM. Consider an (F)-space X and a TVS Y with a fundamental sequence of bounded sets, and assume that X and Y are unital modules over some normal (F)-algebra A of complex-valued functions on a normal topological space Ω. Furthermore, suppose that the module operation on X is separately continuous. Then, for every A-module homomorphism $\varphi : X \to Y$, there exist a finite subset Λ of Ω, a bounded set C in Y, and 0-neighborhoods U of X and V of Λ such that $\varphi(fgx) \in C$ holds for all $f, g \in \mathfrak{J}(\Lambda)$ and $x \in X$ satisfying

$f \in V, gx \in U$, where $\mathfrak{J}(\Lambda)$ denotes the ideal of all $f \in A$ having a compact support off Λ. In particular, the trilinear mapping given by $(f,g,x) \mapsto \varphi(fgx)$ is continuous on $\mathfrak{J}(\Lambda) \times \mathfrak{J}(\Lambda) \times X$.

Proof. (1) First note that the last assertion is an easy consequence of the preceding one, since the module operation on X is even jointly continuous by the uniform boundedness theorem. In the following, let $|\cdot|_X$ and $|\cdot|_A$ be (F)-norms generating the topologies of X and A respectively, let $(B_k)_{k\in\mathbb{N}}$ be a fundamental sequence of bounded sets for Y, and consider the finite singularity set $\Lambda \subseteq \Omega$ for φ given by 2.1.

(2) We now claim that there exist an $\varepsilon > 0$, a $k \in \mathbb{N}$, and a compact $K \subseteq \Omega\setminus\Lambda$ such that $\varphi(fgx) \in B_k$ for all $f,g \in A$ with compact support contained in $\Omega\setminus(\Lambda \cup K)$ and all $x \in X$ satisfying $|f|_A \le \varepsilon$ and $|gx|_X \le \varepsilon$. If this is false, for $n = 1,2,\ldots$ we may choose by induction $f_n, g_n \in A$ and $x_n \in X$ such that $\varphi(f_n g_n x_n) \notin B_n$, that $|f_n|_A \le 1/n$ and $|g_n x_n|_X \le 1/n$, and that $K_n := \operatorname{supp} f_n \cup \operatorname{supp} g_n$ is compact and disjoint to $\Lambda \cup K_1 \cup \cdots \cup K_{n-1}$, where of course $K_0 := \emptyset$. The sequences $(f_n)_{n\in\mathbb{N}}$ and $(g_n x_n)_{n\in\mathbb{N}}$ are in obvious contradiction to 2.3.

(3) Applying 2.1 to the compact set K given by (2), we obtain an open neighborhood W of K in Ω, a $\delta > 0$, and an $m \in \mathbb{N}$ such that $\varphi(hx) \in B_m$ for all $h \in A$, $x \in X$ satisfying $\operatorname{supp} h \subseteq W$ and $|hx|_X \le \delta$. Next, the normality of A supplies us with some $e \in A$ such that $\operatorname{supp} e \subseteq W$ and $e = 1$ on a neighborhood of K. Finally, from the continuity of the multiplication in A and the module operation on X, we obtain some $\gamma > 0$ such that the estimates $|u - eu|_A \le \varepsilon$, $|(v - ev)x|_X \le \varepsilon$, and $|e(2-e)uvx|_X \le \delta$ hold for all $u,v \in A$ and $x \in X$ satisfying $|u|_A \le \gamma$ and $|vx|_X \le \gamma$. Now the proof of 2.4 can be easily completed. Given arbitrary $u,v \in \mathfrak{J}(\Lambda)$ and $x \in X$ with $|u|_A \le \gamma$ and $|vx|_A \le \gamma$, we observe that the functions $f := u - eu$ and $g := v - ev$ have compact supports contained in $\Omega\setminus(\Lambda \cup K)$. And we have $uv = fg + h$, where the support of $h := e(2-e)uv$ is contained in W. Consequently $\varphi(uvx) \in B_k + B_m$, which proves the assertion.

3. Homomorphisms from normal (F)-algebras

We now specialize the results of the preceding section to the important case of algebra homomorphisms between certain topological algebras. Again, let Ω be a normal topological space and consider some normal (F)-algebra A of complex-valued functions on Ω. Then, for every homomorphism φ from A into some topological algebra B, we have:

3.1. THEOREM. Assume that B is countably boundedly generated. Then there exists a finite subset Λ of Ω such that each compact $K \subseteq \Omega\setminus\Lambda$ has a neighborhood $W \subseteq \Omega$ such that φ is bounded on the ideal $\{f \in A : \operatorname{supp} f \subseteq W\}$.

3.2. THEOREM. Assume that B has a fundamental sequence of bounded sets. Then there exists a finite subset Λ of Ω such that the bilinear mapping given by $(f,g) \mapsto \varphi(fg)$ is bounded on $\mathfrak{J}(\Lambda) \times \mathfrak{J}(\Lambda)$.

Proofs. Since B may be replaced by $\varphi(A)$, we may assume that $\varphi(1)$ is the unit element for B. Now regard B as a unital module over A by defining $fb := \varphi(f)b$ for all $f \in A$, $b \in B$. Then the assertions follow immediately from 2.1 and 2.4.

Given a pointwise bounded family of homomorphisms from A into some topological algebra with a fundamental sequence of bounded sets, similar automatic continuity properties can be derived outside some common finite singularity set $\Lambda \subseteq \Omega$; see [5] for details. We also note another typical consequence of 2.1, which has to be compared with the continuity properties of local operators in distribution theory [2, Ex. 3.11].

3.3. EXAMPLE. Let $\Omega \subseteq \mathbb{R}^n$ be open, and suppose that B is a countably boundedly generated topological algebra. Then, for every homomorphism $\varphi : \mathcal{D}(\Omega) \to B$, there exists a discrete subset Λ of Ω such that φ is continuous on $\mathcal{D}(\Omega \setminus \Lambda)$.

Proof. Consider a sequence of compact sets $K_j \subseteq \Omega$ such that $K_j \subseteq \overset{\circ}{K}_{j+1}$ and $K_j \uparrow \Omega$ as $j \to \infty$. For each $j \in \mathbb{N}$, fix an $e_j \in \mathcal{D}(\Omega)$ such that $e_j = 1$ on a neighborhood of K_j, let $X_j := \{f \in \mathcal{D}(\Omega) : \operatorname{supp} f \subseteq K_j\}$ and $Y_j := \varphi(X_j)$, and define $hb := \varphi(he_j)b$ for all $h \in C^\infty(\Omega)$, $b \in Y_j$. In a canonical way, X_j and Y_j are unital $C^\infty(\Omega)$-modules such that φ is a module homomorphism on X_j. Thus 2.1 supplies us with some finite $\Lambda_j \subseteq \Omega$ such that φ is continuous on $\{f \in \mathcal{D}(\Omega) : \operatorname{supp} f \subseteq K\}$ for all compact $K \subseteq K_j$ satisfying $K \cap \Lambda_j = \emptyset$. We may assume that $\Lambda_j \subseteq K_j$ and that $\Lambda_{j+1} \cap \overset{\circ}{K}_j \subseteq \Lambda_j$ for all $j \in \mathbb{N}$. Then the union Λ of the sets Λ_j for $j \in \mathbb{N}$ has the desired properties.

4. C^k-homomorphisms I

In this section, we investigate the continuity properties of homomorphisms from the usual algebras of differentiable functions into an arbitrary topological algebra B with a fundamental sequence of bounded sets. Given an open bounded subset Ω of $\mathbb{R}^n$, for each $k \in \mathbb{N}_0^* := \{0,1,\dots,\infty\}$ let $C^k(\overline{\Omega})$ denote the algebra of all functions $f : \Omega \to \mathbb{C}$ such that, for every multi-index $\alpha \in \mathbb{N}_0^n$ of order $|\alpha| \le k$, the partial derivative $D^\alpha f : \Omega \to \mathbb{C}$ exists as a continuous function on Ω and has a continuous extension to $\overline{\Omega}$, again denoted by $D^\alpha f$. For each $k < \infty$, $C^k(\overline{\Omega})$ is a Banach algebra with respect to the norm $\|\cdot\|_k$ given by

$$\|f\|_k := \sum_{|\alpha| \le k} \frac{1}{\alpha!} \sup\{|D^\alpha f(t)| : t \in \Omega\} \quad \text{for all } f \in C^k(\overline{\Omega}),$$

whereas $C^{\infty}(\overline{\Omega})$ is an (F)-algebra with respect to the sequence of these norms. Moreover, each of these algebras is a normal algebra of functions on the compact set $\overline{\Omega}$, so that the general theory from section 3 applies to this situation. In addition, we shall need suitable Taylor expansions which require some mild restriction on the boundary of Ω. Let us define Ω to be locally regular if for every $\lambda \in \overline{\Omega}$ there exist an open neighborhood U of λ in $\mathbb{R}^n$ and a constant $c > 0$ such that any two points $s,t \in \Omega \cap U$ can be joined by a rectifiable arc in Ω of length not greater than $c|s-t|$, where $|\cdot|$ stands for the Euclidean norm on $\mathbb{R}^n$. Conditions of this type are well-known from the extension theory for differentiable functions; see for instance [20;21]. Throughout this section, we assume $\Omega \subseteq \mathbb{R}^n$ to be open, bounded, and locally regular.

4.1. LEMMA. Given $\lambda \in \overline{\Omega}$ and $k \in \mathbb{N}_0$, let $M^k(\lambda)$ consist of all $f \in C^k(\overline{\Omega})$ such that $D^{\alpha}f(\lambda) = 0$ whenever $\alpha \in \mathbb{N}_0^n$ with $|\alpha| \leq k$. Further, for $m \in \mathbb{N}_0$ and $f \in M^{k+m}(\lambda)$, let $Sf : \Omega \to \mathbb{C}$ be given by $Sf(t) := f(t)|t-\lambda|^{-m}$ for $t \in \Omega\setminus\{\lambda\}$ and $Sf(t) := 0$ for $t \in \Omega \cap \{\lambda\}$. Then S is a continuous linear mapping from $M^{k+m}(\lambda)$ into $M^k(\lambda)$.

Proof. Choose U and c according to the local regularity of Ω at λ. Then one verifies as in [20, Lemma 2] that each $t \in U \cap \Omega$ can be joined to λ by an arc of length $\leq 2c|t-\lambda|$ which lies in Ω except possibly for the end point λ. Hence the Taylor remainder estimates from [20, Lemma 3] imply that

$$|D^{\alpha}f(t)| \leq M\varepsilon(f,t)|t-\lambda|^{k+m-|\alpha|}$$

for all $f \in M^{k+m}(\lambda)$, $t \in \Omega \cap U$, and $\alpha \in \mathbb{N}_0^n$ with $|\alpha| \leq k+m$, where $M > 0$ is a suitable constant and $\varepsilon(f,t) := \sup\{|D^{\beta}f(s)| : |\beta| = k+m,\ s \in \Omega$ with $|s-\lambda| \leq 2c|t-\lambda|\}$. On the other hand, an elementary calculation shows that

$$|D^{\alpha}g(t)| \leq N_{\alpha}|t-\lambda|^{-m-|\alpha|} \quad \text{for all } \alpha \in \mathbb{N}_0^n \text{ and } t \in \Omega,\ t \neq \lambda,$$

where $g(t) := |t-\lambda|^{-m}$ and $N_{\alpha} > 0$ is some constant. By the Leibniz rule, for every $f \in M^{k+m}(\lambda)$, $t \in \Omega \cap U$ with $t \neq \lambda$, and $\alpha \in \mathbb{N}_0^n$ with $|\alpha| \leq k$ we obtain

$$|D^{\alpha}(fg)(t)| \leq \sum_{\beta \leq \alpha} \binom{\alpha}{\beta} MN_{\beta}\varepsilon(f,t)|t-\lambda|^{k-|\alpha|}.$$

Because of $\varepsilon(f,t) \to 0$ as $t \to \lambda$ and $\varepsilon(f,t) \leq (k+m)!\|f\|_{k+m}$, the assertion follows.

For every finite subset $\Lambda = \{\lambda^1,\dots,\lambda^r\}$ of $\overline{\Omega}$ and $k \in \mathbb{N}_0^*$, let us introduce the ideal $\mathfrak{J}^k(\Lambda)$ of all $f \in C^k(\overline{\Omega})$ vanishing in some neighborhood of Λ. Further, if $k < \infty$, let $\varphi_{k,\Lambda} : \Omega \to \mathbb{R}$ be given by $\varphi_{k,\Lambda}(t) := |t-\lambda^1|^k \dots |t-\lambda^r|^k$ for all $t \in \Omega$, and consider on $\mathfrak{J}^k(\Lambda)$ a second norm $\|\cdot\|_{k,\Lambda}$ given by

$$\|f\|_{k,\Lambda} := \|f/\varphi_{k,\Lambda}\|_k \quad \text{for all} \quad f \in \mathfrak{J}^k(\Lambda).$$

By means of an appropriate partition of unity and by 4.1 we obtain immediately the following estimates:

4.2. LEMMA. For every finite $\Lambda \subseteq \overline{\Omega}$ and $k \in \mathbb{N}_0$, there exists some $c_{k,\Lambda} > 0$ such that $\|f\|_{k,\Lambda} \leq c_{k,\Lambda}\|f\|_{2k}$ for all $f \in \mathfrak{J}^{2k}(\Lambda)$.

4.3. THEOREM. Let $k \in \mathbb{N}_0$. For every homomorphism $\Phi : C^k(\overline{\Omega}) \to B$, there exists some finite $\Lambda \subseteq \overline{\Omega}$ such that Φ is bounded on the ideal $\mathfrak{J}^k(\Lambda)$ with respect to $\|\cdot\|_{k,\Lambda}$.

<u>Proof</u>. From 3.2 we obtain some finite $\Lambda \subseteq \overline{\Omega}$ and some bounded $C \subseteq B$ such that $\Phi(fg) \in C$ holds for all $f,g \in \mathfrak{J}^k(\Lambda)$ satisfying $\|f\|_k, \|g\|_k \leq 1$. By a standard construction [15, Lemma I.4.2], for every $\delta > 0$ there is a $g_\delta \in \mathfrak{J}^\infty(\Lambda)$ such that $g_\delta(t) = 1$ whenever $\operatorname{dist}(t,\Lambda) \geq \delta$ as well as $|D^\alpha g_\delta(t)| \leq c_\alpha \delta^{-|\alpha|}$ for all $t \in \Omega$ and $\alpha \in \mathbb{N}_0^n$, where $c_\alpha > 0$ does not depend on δ. From the Leibniz rule it follows that $\|g_\delta \varphi_{k,\Lambda}\|_k \leq M$ holds for some $M > 0$ and all $\delta > 0$. Now, given $f \in \mathfrak{J}^k(\Lambda)$ such that $\|f\|_{k,\Lambda} \leq 1/M$, for $\delta := \operatorname{dist}(\operatorname{supp} f,\Lambda) > 0$ we have $fg_\delta = f$ and therefore $\Phi(f) = \Phi(fg_\delta) = \Phi((f/\varphi_{k,\Lambda})(g_\delta\varphi_{k,\Lambda})) \in C$. This proves the theorem. Note that the local regularity of Ω is not needed for this proof.

4.4. COROLLARY. Let $k \in \mathbb{N}_0^*$. For every homomorphism $\Phi : C^k(\overline{\Omega}) \to B$, there exists some finite $\Lambda \subseteq \overline{\Omega}$ such that Φ is bounded on $\mathfrak{J}^{2k}(\Lambda)$ for the C^{2k}-topology.

<u>Proof</u>. For $k < \infty$, the assertion follows at once from 4.2 and 4.3. For $k = \infty$, the arguments are sufficiently similar and therefore omitted.

This corollary includes corresponding results from [9;17], but the technicalities are somewhat different. We also note that homomorphisms on $C^k(\overline{\Omega})$ may be discontinuous for the C^k-topology on every dense subalgebra [8]. Later, it will be useful to extract from a given (possibly discontinuous) homomorphism on $C^k(\overline{\Omega})$ some continuous homomorphism on a reasonable large subalgebra of $C^k(\overline{\Omega})$. To be more precise, for each $\Lambda \subseteq \overline{\Omega}$ and $k \in \mathbb{N}_0^*$, let $M^k(\Lambda)$, resp. $M_*^k(\Lambda)$ denote the closed subalgebra of all $f \in C^k(\overline{\Omega})$ satisfying $D^\alpha f = 0$ on Λ for all $\alpha \in \mathbb{N}_0^n$ with $0 \leq |\alpha| \leq k$, resp. $1 \leq |\alpha| \leq k$, and let $\mathfrak{J}_*^k(\Lambda)$ consist of all $f \in C^k(\overline{\Omega})$ which are constant on some neighborhood of Λ. Then we have:

4.5. COROLLARY. Assume now, that B is sequentially complete, let $\Phi : C^k(\overline{\Omega}) \to B$ be a homomorphism, and let $\Lambda \subseteq \overline{\Omega}$ denote its finite singularity set according to 4.4. If $k < \infty$, then there exists a continuous homomorphism $\psi : M_*^{2k}(\Lambda) \to B$ such that $\psi = \Phi$ on $\mathfrak{J}_*^{2k}(\Lambda)$. If $k = \infty$, then for some suitable $m \in \mathbb{N}_0$ there exists a continuous homomorphism $\psi : M_*^m(\Lambda) \to B$ such that $\psi = \Phi$

on $\mathfrak{I}_*^\infty(\Lambda)$.

Proof. Let $\Lambda = \{\lambda^1,\ldots,\lambda^r\}$ and choose a partition of unity $\{\varphi_1,\ldots,\varphi_r\}$ in $\mathfrak{I}_*^\infty(\Lambda)$ such that $\varphi_j = 0$ on a neighborhood of λ^k whenever $j \neq k$. Note that

$$\Phi(f) = \sum_{j=1}^{r} \Phi(\varphi_j(f - f(\lambda^j))) + \sum_{j=1}^{r} \Phi(f(\lambda^j)\varphi_j) \quad \text{for all} \quad f \in C^k(\overline{\Omega}).$$

Thus the continuity of the mappings $f \mapsto \Phi(\varphi_j(f - f(\lambda^j)))$ and $f \mapsto \Phi(f(\lambda^j)\varphi_j)$ on $\mathfrak{I}_*^{2k}(\Lambda)$ with respect to the C^{2k}-topology imply the same type of continuity for Φ. Now the construction of ψ is obvious, as for every $m \in \mathbb{N}_0$ the algebra $\mathfrak{I}_*^\infty(\Lambda)$ is dense in $M_*^m(\Lambda)$ with respect to $\|\cdot\|_m$.

The preceding results have natural counterparts concerning the usual (F)-algebra $C^k(\Omega_0)$ of all k-times continuously differentiable functions on some arbitrary open subset Ω_0 of $\mathbb{R}^n$. We restrict ourselves to the following typical result.

4.6. THEOREM. Let $\Omega_0 \subseteq \mathbb{R}^n$ be open and $k \in \mathbb{N}_0^*$. Then, for every homomorphism $\Phi : C^k(\Omega_0) \to B$, there exists some finite $\Lambda \subseteq \Omega_0$ such that Φ is bounded on $\mathfrak{I}^{2k}(\Lambda)$ with respect to the $C^{2k}(\Omega_0)$-topology, where the subalgebra $\mathfrak{I}^{2k}(\Lambda)$ consists of all $f \in C^{2k}(\Omega_0)$ having a compact support off Λ.

Proof. First let $k < \infty$. Then 3.2 supplies us with some finite $\Lambda \subseteq \Omega_0$, some compact $K \subseteq \Omega_0$, and some bounded $C \subseteq B$ such that $\Phi(fg) \in C$ holds for all $f,g \in \mathfrak{I}^k(\Lambda)$ satisfying $\|f \mid K\|_k, \|g \mid K\|_k \leq 1$. Let $\Omega \subseteq \mathbb{R}^n$ be open, bounded, and locally regular such that $K \cup \Lambda \subseteq \Omega \subseteq \overline{\Omega} \subseteq \Omega_0$. Then, by 4.2 and the proof of 4.3, we obtain some constant $M > 0$ such that $\Phi(f) \in C$ for all $f \in \mathfrak{I}^{2k}(\Lambda)$ satisfying $\|f \mid \Omega\|_{2k} \leq M$. For $k = \infty$ the proof is of course similar.

5. Non-analytic functional calculi

We now turn to the automatic continuity problem for a certain class of homomorphisms which naturally arises in the spectral theory of several commuting operators. In the following, the underlying scalar field is always assumed to be $\mathbb{C}$. Recall from [1;12] that an algebra A of complex-valued functions on some subset Ω of $\mathbb{C}^n$ is said to be admissible if the following conditions are fulfilled:

(a) $1, Z_1,\ldots,Z_n \in A$, where $Z_j : \Omega \to \mathbb{C}$ is given by $Z_j(z) := z_j$ for all $z = (z_1,\ldots,z_n) \in \Omega$ and $j = 1,\ldots,n$.

(b) For every finite open covering $\{U_1,\ldots,U_m\}$ of $\overline{\Omega}$ there are $f_1,\ldots, f_m \in A$ with $0 \leq f_j \leq 1$ and $\operatorname{supp} f_j \subseteq U_j$ for $j = 1,\ldots,m$ such that $f_1 + \cdots + f_m = 1$.

(c) For all $f \in A$ and $w \in \mathbb{C}^n \setminus \operatorname{supp} f$, there exist $f_1,\ldots,f_n \in A$ such that $(w_1 - z_1)f_1(z) + \cdots + (w_n - z_n)f_n(z) = f(z)$ for all $z \in \Omega$.

Here, supp f stands for the closure in $\mathbb{C}^n$ of $\{z \in \Omega : f(z) \neq 0\}$. Standard examples are the algebras $C^k(\overline{\Omega})$ and $C^k(\Omega)$. We now assume the following situation: Let A denote an admissible (F)-algebra on some $\Omega \subseteq \mathbb{C}^n$, let $T_1,\dots,T_n \in L(X)$ for some Banach space X, and let $\Phi : A \to L(X)$ be an algebraic homomorphism such that $\Phi(1) = I$ and $\Phi(Z_j) = T_j$ for $j = 1,\dots,n$. Thus, in the terminology of [1], $T = (T_1,\dots,T_n)$ is an A-scalar system, and Φ is an A-functional calculus for T.

5.1. THEOREM. There exists a finite subset Λ of the joint Taylor spectrum $\sigma(T,X) \subseteq \mathbb{C}^n$ of T such that the separating space $\mathfrak{S}(\Phi)$ is contained in $\{S \in L(X) : \Phi(f)S = 0$ for all $f \in A$ with $\Lambda \cap \operatorname{supp} f = \emptyset\}$ and such that Φ is continuous on $\{f \in A : \operatorname{supp} f \subseteq F\}$ for all closed $F \subseteq \mathbb{C}^n$ satisfying $F \cap \Lambda = \emptyset$.

Proof. We are going to apply the general theory from [2] or [4]; see also [3]. For each closed $F \subseteq \mathbb{C}^n$ let $\mathcal{E}_A(F) := \{f \in A : \operatorname{supp} f \subseteq F\}$. Then condition (b) on A implies that $\mathcal{E}_A(F)$ is closed and that $\mathcal{E}_A : \mathfrak{F}(\mathbb{C}^n) \to \mathfrak{S}(A)$ is a 2-spectral precapacity. Further, it is shown in [1, Th. 4] that the definition $\mathcal{E}_{L(X)}(F) := \{S \in L(X) : \Phi(f)S = 0$ for all $f \in A$ with $F \cap \operatorname{supp} f = \emptyset\}$ yields a countably $\cap$-stable precapacity $\mathcal{E}_{L(X)} : \mathfrak{F}(\mathbb{C}^n) \to \mathfrak{S}(L(X))$. In view of $\Phi(\mathcal{E}_A(F)) \subseteq \mathcal{E}_{L(X)}(F)$ for all closed $F \subseteq \mathbb{C}^n$, it follows from [2, Th. 3.7] or from [4] that the assertion holds true for some finite $\Lambda \subseteq \mathbb{C}^n$. But Φ vanishes on $\mathbb{C}^n \setminus \sigma(T,X)$ by the theorem of support [1, Th. 6]. Hence, by means of an $f \in A$ satisfying $(\Lambda \setminus \sigma(T,X)) \cap \operatorname{supp} f = \emptyset$ and $\sigma(T,X) \cap \operatorname{supp}(1-f) = \emptyset$, one easily deduces that Λ can be replaced by $\Lambda \cap \sigma(T,X)$ if necessary.

5.2. REMARK. If $A = C^k(\Omega)$ for some open $\Omega \subseteq \mathbb{R}^n$ or $\mathbb{C}^n$ and $k \in \mathbb{N}_0^*$, then $\sigma(T,X) \subseteq \Omega$ and hence $\Lambda \subseteq \Omega$ in 5.1. Similarly, if $A = C^k(\overline{\Omega})$ for some open bounded $\Omega \subseteq \mathbb{R}^n$ or $\mathbb{C}^n$ and $k \in \mathbb{N}_0^*$, then $\sigma(T,X) \subseteq \overline{\Omega}$ and hence $\Lambda \subseteq \overline{\Omega}$ in 5.1.

Proof. We concentrate on the first case only. Fix some $\lambda \in \mathbb{C}^n \setminus \Omega$ and let $f_j \in C^\infty(\Omega)$ be given by $f_j(z) := (\overline{\lambda}_j - \overline{z}_j)|\lambda - z|^{-2}$ for all $z \in \Omega$ and $j = 1,\dots,n$. Then we have $(\lambda_1 - Z_1)f_1 + \dots + (\lambda_n - Z_n)f_n = 1$ and consequently $(\lambda_1 I - T_1)Q + \dots + (\lambda_n I - T_n)Q = Q$, where Q denotes the commutant algebra of $T_1,\dots,T_n$ in $L(X)$. Thus λ is not in the joint commutant spectrum $\tilde{\sigma}(T,X)$ of T. This proves the first assertion, since under the present assumptions $\sigma(T,X) = \tilde{\sigma}(T,X)$ according to [1, Th. 6].

Next, we relate certain stronger continuity properties of Φ to some algebraic condition on $T_1,\dots,T_n$. Recall that a linear subspace Y of X is said to be divisible with respect to a linear operator S on X if $(S - \mu I)Y = Y$ for every $\mu \in \mathbb{C}$.

5.3. THEOREM. Assume that each of the operators $T_1,\dots,T_n$ has no divisible subspace different from $\{0\}$. Then, for $j = 1,\dots,n$, there exists a polynomial $0 \neq p_j \in \mathbb{C}[X]$ such that $p_j(T_j)\Phi : A \to L(X)$ is continuous. Moreover, p_j can be chosen so that all its roots are contained in $Z_j(\Lambda)$ where $\Lambda \subseteq \mathbb{C}^n$ is the finite set from 5.1.

Proof. Let Y consist of all $x \in X$ such that $\Phi(f)x = 0$ whenever $f \in A$ satisfies $\Lambda \cap \operatorname{supp} f = \emptyset$. Then Y is a closed linear subspace of X, invariant under $T_1,\dots,T_n$, and [1, Th. 4] ensures that $\sigma(T \mid Y,Y) \subseteq \Lambda$, where $T \mid Y :=$ $(T_1 \mid Y,\dots,T_n \mid Y)$. Now, fix some $j = 1,\dots,n$ and note that $\sigma(T_j \mid Y) \subseteq Z_j(\Lambda)$, by the projection property of the Taylor spectrum. Thus, if Y_∞ denotes the largest linear subspace of Y satisfying $(T_j - \mu I)Y_\infty = Y_\infty$ for all $\mu \in Z_j(\Lambda)$, an elementary observation [10, 0.5] confirms that $(T_j - \mu I)Y_\infty = Y_\infty$ holds true for all $\mu \in \mathbb{C}$. Hence, by assumption, $Y_\infty = \{0\}$. Next, observe that $T_j\mathfrak{S}(\Phi) \subseteq \mathfrak{S}(\Phi)$ and consider the larges linear subspace $\mathfrak{S}_\infty$ of $\mathfrak{S}(\Phi)$ satisfying $(T_j - \mu I)\mathfrak{S}_\infty = \mathfrak{S}_\infty$ for all $\mu \in Z_j(\Lambda)$. Clearly $Sx \in Y$ whenever $S \in \mathfrak{S}(\Phi)$ and $x \in X$, according to the choice of Λ. Thus, if W denotes the linear hull of $\{Sx : S \in \mathfrak{S}_\infty, x \in X\}$, then W is contained in Y and satisfies $(T_j - \mu I)W = W$ for all $\mu \in Z_j(\Lambda)$. We conclude that $W \subseteq Y_\infty$ so that $W = \{0\}$ and hence $\mathfrak{S}_\infty = \{0\}$. Now, since $Z_j(\Lambda)$ is finite, from [18, Lemma 1.2] we obtain some $0 \neq p_j \in \mathbb{C}[X]$ having all its roots in $Z_j(\Lambda)$ such that $\overline{p_j(T_j)\mathfrak{S}(\Phi)} = \overline{\mathfrak{S}_\infty}$ and hence $p_j(T_j)\mathfrak{S}(\Phi) = \{0\}$. In view of [2, Prop. 2.3], this condition is equivalent to the continuity of $p_j(T_j)\Phi$.

If $T_1,\dots,T_n$ are only supposed to have no common divisible subspace $\neq \{0\}$, then a similar reasoning leads to non-trivial $p_1,\dots,p_n \in \mathbb{C}[X]$ such that $p_1(T_1)\cdots p_n(T_n)\Phi$ is continuous on A. This condition, however does not suffice for the applications we have in mind.

6. C^k-homomorphisms II

Throughout this section, let X be a Banach space, let $k \in \mathbb{N}_0^*$, and consider an open, bounded, and locally regular subset Ω of $\mathbb{R}^n$ or $\mathbb{C}^n$. Recall that a system $T = (T_1,\dots,T_n) \in L(X)^n$ is said to be generalized scalar, if there exists a continuous homomorphism $\Theta : C^\infty(\mathbb{C}^n) \to L(X)$ such that $\Theta(1) = I$ and $\Theta(Z_j) = T_j$ for $j = 1,\dots,n$. Later, we shall see precisely to what extent the continuity hypothesis on the spectral distribution Θ is redundant. We start with the following basic result.

6.1. THEOREM. Consider a homomorphism $\Phi : C^k(\overline{\Omega}) \to L(X)$ such that $\Phi(1) = I$ and let $T_j := \Phi(Z_j)$ for $j = 1,\dots,n$. Then the following assertions are equivalent:

(a) Φ is continuous on $C^{2k+1}(\overline{\Omega})$ for the C^{2k+1}-topology.

(b) $T = (T_1,\dots,T_n)$ is a generalized scalar system.

(c) Each of the operators $T_1,\dots,T_n$ is a generalized scalar operator.

(d) There is no non-trivial linear subspace Y of X satisfying

$(\lambda_1 I - T_1)Y + \cdots + (\lambda_n I - T_n)Y = Y$ for all $\lambda = (\lambda_1, \ldots, \lambda_n) \in \mathbb{C}^n$.

(e) None of the operators $T_1, \ldots, T_n$ has a divisible subspace $\neq \{0\}$.

(f) Whenever $\psi : C^k(\overline{\Omega}) \to L(X)$ is a $\mathbb{C}[X_1, \ldots, X_n]$-module homomorphism with respect to $(T_1, \ldots, T_n)$ in the sense that ψ is linear and satisfies $T_j \psi(f) = \psi(Z_j f)$ for all $f \in C^k(\overline{\Omega})$ and $j = 1, \ldots, n$, there exists some $m \geq k$ such that ψ is continuous on $C^m(\overline{\Omega})$ for the C^m-topology.

Proof. (1) The implications (a) $\Rightarrow$ (b), (b) $\Rightarrow$ (c), (f) $\Rightarrow$ (b) are obvious, and (d) $\Rightarrow$ (e) is immediate, since the largest divisible subspace of any T_j is clearly invariant under $T_1, \ldots, T_n$. To prove that (b) implies (d), let Y be a linear subspace of X such that $(\lambda_1 I - T_1)Y + \cdots + (\lambda_n I - T_n)Y = Y$ for all $\lambda \in \mathbb{C}^n$, and fix an arbitrary $q \in \mathbb{N}$. By iterating the property of Y q^2 times, we obtain $(\lambda_1 I - T_1)^q Y + \cdots + (\lambda_n I - T_n)^q Y = Y$ for all $\lambda \in \mathbb{C}^n$. Thus [6, Th. 3.3] forces Y to be trivial. The proof of (c) $\Rightarrow$ (e) is a simple special case of the preceding argument. It remains to prove the implications (e) $\Rightarrow$ (a) and (b) $\Rightarrow$ (f).

(2) Assume that (e) holds and choose some finite subset $\Lambda = \{\lambda^1, \ldots, \lambda^r\}$ of $\overline{\Omega}$ as well as $p_1, \ldots, p_n \in \mathbb{C}[X]$ according to the general results from section 5. Next, fix $h_1, \ldots, h_r \in C^k(\overline{\Omega})$ such that $h_i = \delta_{im}$ on some neighborhood of λ^m and, moreover, such that the distance between the projections $Z_j(\operatorname{supp} h_l)$ and $Z_j(\operatorname{supp} h_m)$ is strictly positive whenever $\lambda_j^i \neq \lambda_j^m$, for $j = 1, \ldots, n$ and $i, m = 1, \ldots, r$. Finally, let $h_0 := 1 - h_1 - \cdots - h_r$ and for $i = 0, 1, \ldots, r$ consider the linear mapping $\Phi_i : C^k(\overline{\Omega}) \to L(X)$ given by $\Phi_i(f) := \Phi(fh_i)$ for all $f \in C^k(\overline{\Omega})$. Since h_0 vanishes on a neighborhood of Λ, Φ_0 is certainly continuous on $C^k(\overline{\Omega})$. Now fix some $i = 1, \ldots, r$ and some $j = 1, \ldots, n$. Then the continuity of $p_j(T_j)\Phi$ and the choice of $h_1, \ldots, h_r$ imply that $\Phi_i((Z_j - \lambda_j^i)^q \cdot)$ is continuous on $C^k(\overline{\Omega})$ for some sufficiently large integer q. Hence $\Phi_i(|Z_j - \lambda_j^i|^{2q} \cdot) = \Phi_i((Z_j - \lambda_j^i)^q (\overline{Z}_j - \overline{\lambda}_j^i)^q \cdot)$ is continuous on $C^k(\overline{\Omega})$. We claim that $2q$ can be replaced by $k+1$ provided $k < \infty$. To this end, let S_α be the closure of $\Phi(|Z_j - \lambda_j^i|^{\alpha+k+1/2})\mathfrak{S}(\Phi_i)$ for all real $\alpha > 0$. Since the proof of [9, Th. 2.1] can be easily adapted to the present situation, we obtain $S_\alpha = S_\beta$ for all $\alpha, \beta > 0$. We have of course $S_{2q} = \{0\}$ and consequently $S_{1/2} = \{0\}$. By a well-known property of the separating space, see [2, Prop. 2.3], we thus arrive at the desired continuity of $\Phi_i(|Z_j - \lambda_j^i|^{k+1} \cdot)$ on $C^k(\overline{\Omega})$. Given $f \in C^{2k+1}(\overline{\Omega})$, let $A_i f$ denote the Taylor polynomial for f of degree $2k+1$ at the point λ^i and define

$$B_i f := (f - A_i f)/(Z_1 - \lambda_1^i|^{k+1} + \cdots + |Z_n - \lambda_n^i|^{k+1}).$$

A slight variant of the Taylor type Lemma 4.1 implies that $B_i f \in C^k(\overline{\Omega})$ and $\|B_i f\|_k \leq c\|f\|_{2k+1}$ for some suitable constant $c > 0$. Applying Φ_i to the equation

$$f = A_i f + |Z_1 - \lambda_1^i| B_i f + \cdots + |Z_n - \lambda_n^i|^{k+1} B_i f$$

for all $f \in C^{2k+1}(\overline{\Omega})$, we conclude that each of the mappings Φ_i and hence $\Phi = \Phi_0 + \Phi_1 + \cdots + \Phi_r$ is continuous on $C^{2k+1}(\overline{\Omega})$ for the norm $\|\cdot\|_{2k+1}$. In the case $k = \infty$, the argument is of course similar and in fact easier, since a result like [9, Th. 2.1] is not needed. Thus (e) $\Rightarrow$ (a) is proved.

(3) To show that (b) implies (f), let Θ be a continuous $C^\infty(\mathbb{C}^n)$-functional calculus for T and consider a $\mathbb{C}[X_1,\ldots,X_n]$-module homomorphism $\psi : C^k(\overline{\Omega}) \to L(X)$. In [2, Th. 4.5] it is derived from the structure of the spectral maximal spaces of a generalized scalar system and from some general automatic continuity theory that, for a suitable finite subset Λ of $\mathbb{C}^n$, the separating space $\mathfrak{S}(\psi)$ is contained in $\{S \in L(X) : \Theta(f)S = 0 \text{ for all } f \in C^\infty(\mathbb{C}^n) \text{ with } \Lambda \cap \operatorname{supp} f = \emptyset\}$ and ψ is continuous on $\{f \in C^k(\overline{\Omega}) : \operatorname{supp} f \subseteq F\}$ for all closed $F \subseteq \mathbb{C}^n$ satisfying $F \cap \Lambda = \emptyset$. But then a close inspection of the proof of 5.3 shows that $p_j(T_j)\psi$ is continuous on $C^k(\overline{\Omega})$ for $j = 1,\ldots,n$, where $p_1,\ldots,p_n$ are appropriate polynomials. As before, we conclude in combination with 4.1 that ψ is continuous on $C^m(\overline{\Omega})$ for some $m \geq k$. This completes the proof of 6.1.

Next, we relate certain continuity properties of a C^k-homomorphism to algebraic properties of its separating space. Let B denote an arbitrary commutative Banach algebra.

6.2. THEOREM. A homomorphism $\Phi : C^\infty(\overline{\Omega}) \to B$ is continuous if and only if $\mathfrak{S}(\Phi)$ contains only nilpotent elements. In particular, Φ is necessarily continuous if the radical of B is nil.

6.3. THEOREM. Given $k \in \mathbb{N}_0$, for every homomorphism $\Phi : C^k(\overline{\Omega}) \to B$ the following assertions are equivalent:

(a) For all $s_1, s_2, s_3 \in \mathfrak{S}(\Phi)$ we have $s_1 s_2 s_3 = 0$.

(b) Each $s \in \mathfrak{S}(\Phi)$ is nilpotent.

(c) Φ is continuous on $C^\infty(\overline{\Omega})$ for the C^∞-topology.

(d) Φ is continuous on $C^{2k+1}(\overline{\Omega})$ for the C^{2k+1}-topology.

(e) There exists some finite $\Lambda \subseteq \overline{\Omega}$ such that the bilinear mapping given by $(f,g) \mapsto \Phi(fg)$ is continuous on $M^{k+1}(\Lambda) \times M^k(\Lambda)$.

(f) There exists some finite $\Lambda \subseteq \overline{\Omega}$ such that the trilinear mapping given by $(f,g,h) \mapsto \Phi(fgh)$ is continuous on $M^k(\Lambda) \times M^k(\Lambda) \times M^k(\Lambda)$.

In particular, Φ is necessarily continuous on $C^{2k+1}(\overline{\Omega})$ for the norm $\|\cdot\|_{2k+1}$, if the radical of B is nil.

<u>Proof of 6.2</u>. First note that each non-zero multiplicative linear functional on $C^\infty(\overline{\Omega})$ is given by evaluation at some point of $\overline{\Omega}$ and hence is continuous. $\mathfrak{S}(\Phi)$ is therefore contained in the radical of B so that the last assertion is indeed immediate from the first one. We also observe that actually $\mathfrak{S}(\Phi) = \{0\}$ if Φ is continuous. Now assume that $\mathfrak{S}(\Phi)$ consists only of nilpotents. It

suffices to establish the continuity of the homomorphism $\tilde{\Phi} : C^\infty(\Omega) \to L(X)$ given by $\tilde{\Phi} := L \circ \Phi$, where $L : X \to L(X)$ is the left regular representation on the Banach algebra $X := \overline{\Phi(C^\infty(\overline{\Omega}))}$. According to 6.1, this amounts to proving that for each fixed $j = 1,\dots,n$, the operator $T := \tilde{\Phi}(Z_j) \in L(X)$ has no divisible subspace different from $\{0\}$. For this purpose, we apply 4.5 to obtain some finite subset $\Lambda = \{\lambda^1,\dots,\lambda^r\}$ of $\overline{\Omega}$, some $m \in \mathbb{N}_0$, and a continuous homomorphism $\psi : M_*^m(\Lambda) \to X$ such that $\psi = \Phi$ on $\mathfrak{J}_*^\infty(\Lambda)$. Let $p := (X - \lambda_j^1)^{m+1} \dots (X - \lambda_j^r)^{m+1}$ and $q := (Z_j - \lambda_j^1)^{m+1} \dots (Z_j - \lambda_j^r)^{m+1}$ so that $p \in \mathbb{C}[X]$ and $q \in M_*^m(\Lambda)$. It is easily verified that $f \mapsto f \circ q$ defines a continuous homomorphism from $C^m(\mathbb{C})$ into $M_*^m(\Lambda)$. Consequently, the mapping $\Theta : C^m(\mathbb{C}) \to L(X)$ given by $\Theta(f) := L(\psi(f \circ q))$ for all $f \in C^m(\mathbb{C})$ is a continuous homomorphism such that $\Theta(1) = L(\psi(1)) = L(\Phi(1)) = I$. Hence $S := \Theta(Z) = L(\psi(q))$ is a generalized scalar operator on X, and [19, Th. 1.2] supplies us with some $\kappa \in \mathbb{N}$ such that the intersection of the spaces $(S - \mu I)^\kappa X$ over all $\mu \in \mathbb{C}$ is trivial. We next claim that the operator $N := S - p(T)$ is nilpotent. To this end, choose $f_i \in \mathfrak{J}_*^\infty(\Lambda)$ such that $f_i \to q$ in the C^∞-topology as $i \to \infty$. Then $\Phi(f_i - q) = \psi(f_i) - \Phi(q) \to \psi(q) - \Phi(q)$ as $i \to \infty$. Thus $\psi(q) - \Phi(q) \in \mathfrak{S}(\Phi)$ which forces $\psi(q) - \Phi(q)$ to be nilpotent. This implies $N^\nu = 0$ for some suitable $\nu \in \mathbb{N}$. Combining all these facts, we arrive at

$$(p(T) - \mu I)^{\kappa+\nu} = (S - \mu I - N)^{\kappa+\nu} = \sum_{j=0}^{\kappa+\nu} \binom{\kappa+\nu}{j} (S - \mu I)^{\kappa+\nu-j}(-N)^j$$
$$= (S - \mu I)^\kappa \sum_{j=0}^{\nu} \binom{\kappa+\nu}{j} (S - \mu I)^{\nu-j}(-N)^j$$

for all $\mu \in \mathbb{C}$ and hence

$$\bigcap_{\mu \in \mathbb{C}} (p(T) - \mu I)^{\kappa+\nu} X \subseteq \bigcap_{\mu \in \mathbb{C}} (S - \mu I)^\kappa X = \{0\}.$$

It follows that $p(T) \in L(X)$ has no non-trivial divisible subspace. Since each divisible subspace for T is certainly divisible for $p(T)$, we conclude that $\{0\}$ is the only divisible subspace for T. The assertion follows.

Proof of 6.3. (1) As before, the last assertion is clear because of $\mathfrak{S}(\Phi) \subseteq \mathrm{rad}(B)$. (a) $\Rightarrow$ (b) is trivial, and (b) $\Rightarrow$ (c) follows from 6.2. Now consider the Banach space $X := \Phi(C^k(\overline{\Omega}))$ and the homomorphism $\tilde{\Phi} : C^k(\overline{\Omega}) \to L(X)$ given by $\tilde{\Phi}(f)(x) := \Phi(f)x$ for all $f \in C^k(\overline{\Omega})$ and $x \in X$. If (c) holds true, then 6.1 implies that none of the operators $T_j := \tilde{\Phi}(Z_j)$ for $j = 1,\dots,n$ has a divisible subspace $\neq \{0\}$, and another application of 6.1 confirms the continuity of $\tilde{\Phi}$ on $C^{2k+1}(\overline{\Omega})$ with respect to $\|\cdot\|_{2k+1}$. Thus (c) and (d) are equivalent. It remains to show that (d) implies both (e) and (f) and that (e) $\Rightarrow$ (c) as well as (f) $\Rightarrow$ (a).

(2) Suppose that (d) is satisfied. Then 6.1 tells us that $T_1,\dots,T_n$ have no divisible subspaces different from $\{0\}$. Hence, using the notations and results

from part (2) in the proof of 6.1, we know that $\tilde{\Phi}_i(|z_j - \lambda_j^i|^{k+1})$ is continuous on $C^k(\overline{\Omega})$ for all $i = 1,\dots,r$ and $j = 1,\dots,r$. Exactly as in the proof of 4.1, one easily deduces from the Leibniz rule and from suitable Taylor estimates [20] that the bilinear mapping given by $(f,g) \mapsto fg/(|z_1 - \lambda_1^i|^{k+1} + \dots + |z_n - \lambda_n^i|^{k+1})$ maps $M^{k+1}(\Lambda) \times M^k(\Lambda)$ continuously into $M^k(\Lambda)$. It follows that for $i = 0,1,\dots,r$ the bilinear mapping $(f,g) \mapsto \tilde{\Phi}_i(fg)$ is continuous on $M^{k+1}(\Lambda) \times M^k(\Lambda)$. Because of $\tilde{\Phi} = \tilde{\Phi}_0 + \tilde{\Phi}_1 + \dots + \tilde{\Phi}_r$, we conclude that (e) holds true. The proof of (d) $\Rightarrow$ (f) is of course similar; here one has to use the fact that the trilinear mapping given by $(f,g,h) \mapsto fgh/(|z_1 - \lambda_1^i|^{k+1} + \dots + |z_n - \lambda_n^i|^{k+1})$ maps $M^k(\Lambda) \times M^k(\Lambda) \times M^k(\Lambda)$ continuously into $M^k(\Lambda)$.

(3) To prove that (e) implies (c), consider the function $g := \varphi_{k+1,\Lambda}$ from section 4. Then $g \in M^k(\Lambda)$, and 4.1 ensures that multiplication by $1/g$ defines a continuous linear mapping from $M^{2k+2}(\Lambda)$ into $M^{k+1}(\Lambda)$. Now, if $f_i \in M^{2k+2}(\Lambda)$ is such that $\|f_i\|_{2k+2} \to 0$ as $i \to \infty$, we conclude from (e) that $\Phi(f_i) = \Phi((f_i/g)g) \to 0$ as $i \to \infty$. Since $M^{2k+2}(\Lambda)$ is closed and of finite codimension in $C^{2k+2}(\overline{\Omega})$, Φ turns out to be continuous on $C^{2k+2}(\overline{\Omega})$ and in particular on $C^\infty(\overline{\Omega})$.

(4) We finally prove that (f) $\Rightarrow$ (a). Let $s_1, s_2, s_3 \in \mathfrak{S}(\Phi)$ be arbitrarily given. Since $M^k(\Lambda)$ is closed and of finite codimension on $C^k(\overline{\Omega})$, there exist f_i, g_i, h_i in $M^k(\Lambda)$ such that $\|f_i\|_k, \|g_i\|_k, \|h_i\|_k \to 0$ as well as $\Phi(f_i) \to s_1$, $\Phi(g_i) \to s_2$, $\Phi(h_i) \to s_3$ as $i \to \infty$. From $0 \leftarrow \Phi(f_i g_i h_i) = \Phi(f_i)\Phi(g_i)\Phi(h_i) \to s_1 s_2 s_3$ as $i \to \infty$, we conclude that $s_1 s_2 s_3 = 0$. This completes the proof of 6.3.

The preceding results include some work of Bade, Curtis, Laursen [10, Th. 3.3] and of Ochoa [17, Th. 5.29] and contain some further information, even if Ω is the unit interval as in [10] or the unit disc as in [17]. Moreover, the present approach may perhaps admit some additional insight into this kind of automatic continuity problem. We also note that, assuming the continuum hypothesis, Dales proved that homomorphisms on $C^k([0,1])$ for $k \in \mathbb{N}$ may be continuous on $C^{2k+1}([0,1])$, but discontinuous on $C^{2k}([0,1])$ for the respective topologies [14, Th. 3.1]. However, it is known from [10, Th. 3.11] and [17, Th. 5.31] that in certain situations C^{2k}-continuity can be obtained. These results carry over to our more general case. We restrict ourselves to the following typical result.

6.4. THEOREM. Consider the following assertions for a homomorphism $\Phi : C^k(\overline{\Omega}) \to B$, where $k \in \mathbb{N}$:

(a) $\mathfrak{S}(\Phi)$ is finite dimensional.

(b) For all $s_1, s_2 \in \mathfrak{S}(\Phi)$ we have $s_1 s_2 = 0$.

(c) There exists some finite $\Lambda \subseteq \overline{\Omega}$ such that the bilinear mapping given by $(f,g) \to \Phi(fg)$ is continuous on $M^k(\Lambda) \times M^k(\Lambda)$.

(d) Φ is continuous on $C^{2k}(\overline{\Omega})$ for the C^{2k}-topology.

Then (a) $\Rightarrow$ (b) $\Leftrightarrow$ (c) $\Rightarrow$ (d). In particular, Φ is necessarily continuous on

$C^{2k}(\overline{\Omega})$ with respect to $\|\cdot\|_{2k}$, if the radical of B is finite dimensional.

Proof. If (a) or (b) holds, then it follows exactly as in the proof of [10, Th. 3.11] that $\Phi(f\cdot)$ is continuous on $C^k(\overline{\Omega})$ for all $f \in M^k(\Lambda)$, where Λ is some finite subset of $\overline{\Omega}$. Hence (c) holds true. Conversely, the proof of (c) $\Rightarrow$ (b) is of course similar to part (4) in the proof of 6.3. Let us finally show that (c) implies (d). Following an argument due to Curtis, we introduce the algebra $A^k(\Lambda) := \{f/\varphi_{k,\Lambda} : f \in M^k(\Lambda)\}$ which is complete for the norm given by $|||f/\varphi_{k,\Lambda}||| := \|f\|_k$ for all $f \in M^k(\Lambda)$. One verifies that multiplication in $A^k(\Lambda)$ is continuous so that $A^k(\Lambda)$ becomes a Banach algebra for some equivalent norm. Moreover, one easily deduces from the Leibniz rule that the functions g_δ from the proof of 4.3 constitute a bounded approximate identity for $A^k(\Lambda)$. Now consider $f_i \in M^{2k}(\Lambda)$ such that $\|f_i\|_{2k} \to 0$ as $i \to \infty$. Then it follows from 4.1 that $g_i := f_i\varphi_{k,\Lambda}^{-2} \in A^k(\Lambda)$ and $|||g_i||| \to 0$ as $i \to \infty$. Hence, by a well-known consequence of the Cohen factorization theorem [11, §12, Cor. 12], there exist $h, h_i \in M^k(\Lambda)$ such that $g_i = hh_i\varphi_{k,\Lambda}^{-2}$ for all $i \in \mathbb{N}$ and $\|h_i\|_k \to 0$ as $i \to \infty$. Therefore (c) implies that $\Phi(f_i) = \Phi(hh_i) \to 0$ as $i \to \infty$. This proves (d), since $M^{2k}(\Lambda)$ is closed and of finite codimension in $C^{2k}(\overline{\Omega})$.

We finally turn to the automatic continuity problem for homomorphisms from $C^k(\Omega_0)$ into some Banach algebra, where Ω_0 is now an arbitrary open subset of $\mathbb{R}^N$ or $\mathbb{C}^n$. In this context, the basic observation is the following: Using a suitable left regular representation, such a homomorphism may be regarded as a $C^k(\Omega_0)$-functional calculus for some system $T = (T_1,\ldots,T_n)$ of operators on a Banach space X. Hence, by the theorem of support from [1] and by the inclusion $\sigma(T,X) \subseteq \Omega_0$ from 5.2, the homomorphism has to vanish on the complement of some compact subset of Ω_0. We conclude that 4.6 can be slightly sharpened in the present situation: Every homomorphism from $C^k(\Omega_0)$ into some Banach algebra is necessarily continuous for the $C^{2k}(\Omega_0)$-topology on the subalgebra of all $f \in C^{2k}(\Omega_0)$ vanishing on some neighborhood of a certain finite subset Λ of Ω_0. Moreover, we have:

6.5. THEOREM. All the results of this section remain valid for homomorphisms on the (F)-algebra $C^k(\Omega_0)$, where $k \in \mathbb{N}_0^*$ and Ω_0 is an arbitrary open subset of $\mathbb{R}^n$ or $\mathbb{C}^n$.

Sketch of proof. Given a homomorphism Φ from $C^k(\Omega_0)$ into some Banach algebra B, choose a compact $K \subseteq \Omega_0$ such that Φ vanishes on the complement of K and an open, bounded, and locally regular $\Omega \subseteq \mathbb{R}^n$, resp. $\mathbb{C}^n$ such that $K \subseteq \Omega \subseteq \overline{\Omega} \subseteq \Omega_0$. Next, let $g \in C^\infty(\Omega_0)$ be such that supp g is a compact subset of Ω and $g = 1$ on a neighborhood of K. Then it is clear that the definition $\tilde{\Phi}(f) := \Phi(fg)$ for all $f \in C^k(\overline{\Omega})$ yields a homomorphism $\tilde{\Phi} : C^k(\overline{\Omega}) \to B$ satisfying $\mathfrak{S}(\tilde{\Phi}) = \mathfrak{S}(\Phi)$. By means of this homomorphism and by an adaptation of the former arguments, the preceding results can be easily transferred to the $C^k(\Omega_0)$-situation.

We close with an application in operator theory which is similar to [10, Th. 3.16 and Cor. 3.17]. A corresponding result can be shown for generalized scalar operators whose spectrum lies on the unit circle.

6.6. COROLLARY. Let $T \in L(X)$ be a generalized scalar operator with real spectrum. Then T admits exactly one functional calculus on $C^\infty(\mathbb{R})$. This functional calculus is continuous of some order $k \in \mathbb{N}_0$. The natural functional calculus for T on $C^k(\mathbb{R})$ is unique up to the addition of nilpotents of order 3.

Proof. According to [12, Th. 5.4.5], there exists a continuous functional calculus Θ on $C^\infty(\mathbb{R})$ for T. Now the $C^\infty(\mathbb{R})$-version of 6.1 confirms that an arbitrary $C^\infty(\mathbb{R})$-functional calculus Φ for T is necessarily continuous. Since Θ and Φ coincide on all polynomials, we conclude that $\Theta = \Phi$ on $C^\infty(\mathbb{R})$. Let $\tilde{\Theta}$ be the natural extension of Θ to $C^k(\mathbb{R})$, and consider an arbitrary functional calculus $\tilde{\Phi}$ on $C^k(\mathbb{R})$ for T. Then $\tilde{\Phi} = \Theta$ on $C^\infty(\mathbb{R})$, which forces $\tilde{\Phi}$ to be continuous on $C^\infty(\mathbb{R})$. Hence the $C^k(\mathbb{R})$-version of 6.3 implies that $\mathfrak{S}(\tilde{\Phi})^3 = \{0\}$. But using an approximation by polynomials again, it is easily seen that $\tilde{\Phi}(f) - \tilde{\Theta}(f) \in \mathfrak{S}(\tilde{\Phi})$ holds for all $f \in C^k(\mathbb{R})$. The assertion follows.

References

[1] E. Albrecht and St. Frunză, Non-analytic functional calculi in several variables, Manuscripta Math., 18 (1976), 327-336.

[2] E. Albrecht and M. Neumann, Automatic continuity of generalized local linear operators, Manuscripta Math., 32 (1980), 263-294.

[3] ___________, On the continuity of non-analytic functional calculi, J. Operator Theory, 5 (1981), 109-117.

[4] ___________, Automatic continuity for operators of local type, this Volume.

[5] ___________, Stetigkeitsaussagen für Homomorphismen zwischen topologischen Algebren, preprint.

[6] E. Albrecht and F.-H. Vasilescu, Non-analytic local spectral properties in several variables, Czech. Math. J., 24 (1974), 430-443.

[7] W. G. Bade and P. C. Curtis, Homomorphisms of commutative Banach algebras, Amer. J. Math., 82 (1960), 589-608.

[8] ___________, The structure of module derivations of Banach algebras of differentiable functions, J. Functional Analysis, 28 (1978), 226-247.

[9] W. G. Bade, P. C. Curtis, and K. B. Laursen, Automatic continuity in algebras of differentiable functions, Math. Scand., 40 (1977), 249-270.

[10] ___________, Divisible subspaces and problems of automatic continuity, Studia Math., 68 (1980), 159-186.

[11] F. F. Bonsall and J. Duncan, Complete Normed Algebras, Springer, Berlin-Heidelberg-New York, 1973.

[12] I. Colojoară and C. Foiaş, Theory of Generalized Spectral Operators, Gordon and Breach, New York-London-Paris, 1968.

[13] H. G. Dales, Discontinuous homomorphisms from topological algebras, Amer. J. Math., 101 (1979), 635-646.

[14] __________, Eventual continuity in Banach algebras of differentiable functions, Studia Math., 70 (1980), 113-121.

[15] B. Malgrange, Ideals of Differentiable Functions, Oxford University Press, London, 1966.

[16] M. Neumann, Beschränktheitsaussagen für sublineare und bisublineare Funktionale, Arch. Math., 27 (1976), 539-548.

[17] J. C. Ochoa, Automatic continuity in algebras of differentiable functions of several variables, thesis, University of California, Berkeley, 1978.

[18] M. P. Thomas, Automatic continuity for linear functions intertwining continuous linear operators on Fréchet spaces, Canad. J. Math., 30 (1978), 518-530.

[19] P. Vrbová, Structure of maximal spectral spaces of generalized scalar operators, Czech. Math. J., 23 (1973), 493-496.

[20] H. Whitney, Functions differentiable on the boundaries of regions, Ann. of Math., 35 (1934), 482-485.

[21] __________, On the extension of differentiable functions, Bull. Amer. Math. Soc., 50 (1944), 76-81.

Fachbereich Mathematik
Universität des Saarlandes
D-6600 Saarbrücken
W.-Germany

Fachbereich Mathematik
Universität Essen GHS
D-4300 Essen
W.-Germany

CONTINUITY OF HOMOMORPHISMS FROM C*-ALGEBRAS AND OTHER BANACH ALGEBRAS

E. Albrecht and H. G. Dales

1. Introduction

In this article, we shall discuss three questions about the continuity of homomorphisms between Banach algebras, concentrating on the theory of homomorphisms from C*-algebras. We cannot solve any of the questions completely: essentially, we shall collect some known partial results, and we shall reorganize and extend them somewhat.

Let $\mathfrak{A}$ and $\mathfrak{B}$ be Banach algebras, and let $\theta : \mathfrak{A} \to \mathfrak{B}$ be an algebra homomorphism. Write $\mathfrak{S}(\theta)$ for the separating space of θ, so that

$$\mathfrak{S}(\theta) = \{b \in \mathfrak{B} : \text{there exists a sequence } (a_n) \subset \mathfrak{A} \text{ with } a_n \to 0 \text{ and } \theta(a_n) \to b\}.$$

Then $\mathfrak{S}(\theta)$ is a closed linear subspace of $\mathfrak{B}$, and it is a bi-ideal in $\mathfrak{B}$ if $\theta(\mathfrak{A})$ is dense in $\mathfrak{B}$. Of course, θ is continuous if and only if $\mathfrak{S}(\theta) = \{0\}$. Many authors have discussed automatic continuity questions in terms of $\mathfrak{S}(\theta)$, and we follow that tradition here. See the book of Allan Sinclair [22] for the basic properties of $\mathfrak{S}(\theta)$.

Denote by $\mathfrak{Q}(\mathfrak{B})$ the set of quasi-nilpotent elements in a Banach algebra $\mathfrak{B}$, and write $\operatorname{rad}\mathfrak{A}$ for the (Jacobson) radical of an algebra $\mathfrak{A}$.

Question I. Let $\theta : \mathfrak{A} \to \mathfrak{B}$ be a homomorphism between Banach algebras. Is $\mathfrak{S}(\theta)$ necessarily contained in $\mathfrak{Q}(\mathfrak{B})$?

The answer to Question I is certainly positive if $\mathfrak{A}$ is a commutative algebra. We shall be particularly interested in the case in which $\mathfrak{A}$ is a non-commutative C*-algebra. Note that, if $\mathfrak{B}$ is also a C*-algebra, and if θ is a *-homomorphism, then θ is continuous ([21, 1.5.7]), and so the answer is also positive in this case.

LEMMA 1.1. Let $\mathfrak{A}$ be a Banach algebra.

(i) If $\theta : \mathfrak{A} \to \mathfrak{B}$ is a homomorphism from $\mathfrak{A}$ into some Banach algebra $\mathfrak{B}$, then $\mathfrak{S}(\theta) \subset \mathfrak{Q}(\mathfrak{B})$ if and only if $\mathfrak{S}(\theta) \subset \operatorname{rad} \overline{\theta(\mathfrak{A})}$. Hence, if $\mathfrak{S}(\theta) \subset \mathfrak{Q}(\mathfrak{B})$ and if $\overline{\theta(\mathfrak{A})}$ is semi-simple, then θ is continuous.

(ii) If each homomorphism from $\mathfrak{A}$ onto a dense subalgebra of a semi-simple Banach algebra is continuous, then $\mathfrak{S}(\theta) \subset \mathfrak{Q}(\mathfrak{B})$ for each homomorphism θ from $\mathfrak{A}$ into a Banach algebra $\mathfrak{B}$.

Both authors were supported by NATO Grant No. RG 073.81.

Proof. (i) Certainly, $\operatorname{rad} \overline{\theta(\mathfrak{A})} \subset \mathfrak{Q}(\overline{\theta(\mathfrak{A})}) \subset \mathfrak{Q}(\mathfrak{B})$. Conversely, if $\mathfrak{S}(\theta) \subset \mathfrak{Q}(\mathfrak{B})$, then $\mathfrak{S}(\theta)$ is a closed bi-ideal in $\overline{\theta(\mathfrak{A})}$ consisting of quasi-nilpotent elements, and so $\mathfrak{S}(\theta) \subset \operatorname{rad}(\overline{\theta(\mathfrak{A})})$.

(ii) Let $\mathfrak{A}$ have the specified property, and let $\theta : \mathfrak{A} \to \mathfrak{B}$ be a homomorphism into a Banach algebra $\mathfrak{B}$. Denote by π the canonical homomorphism from $\overline{\theta(\mathfrak{A})}$ onto $\overline{\theta(\mathfrak{A})}/\operatorname{rad} \overline{\theta(\mathfrak{A})}$. Then $\pi \circ \theta$ is a homomorphism from $\mathfrak{A}$ such that $\overline{(\pi \circ \theta)(\mathfrak{A})}$ is semi-simple, and so $\pi \circ \theta$ is continuous. This implies that $\pi(\mathfrak{S}(\theta)) = \{0\}$, that is, that $\mathfrak{S}(\theta) \subset \operatorname{rad} \overline{\theta(\mathfrak{A})} \subset \mathfrak{Q}(\mathfrak{B})$.

Question I is an important question in automatic continuity theory. Lemma 1.1 shows that it is equivalent to the following question.

Question I'. If $\theta : \mathfrak{A} \to \mathfrak{B}$ is a homomorphism for which $\overline{\theta(\mathfrak{A})}$ is semi-simple, is θ automatically continuous?

We now come to our second question.

Question II. What is the class of Banach algebras for which each homomorphism from a member of this class into a Banach algebra is automatically continuous?

There are many C*-algebras which are known to belong to the above class. We shall recite and extend the known list below. On the other hand, if X is an infinite compact Hausdorff space, and if $C(X)$ is the Banach algebra of all continuous, complex-valued functions on X, then there are discontinuous homomorphisms from $C(X)$. See [9, §9]. Thus the class specified in Question II must exclude all infinite-dimensional commutative C*-algebras. (This last result requires the continuum hypothesis, a point discussed in [9]; we shall assume throughout the remainder of this article that the continuum hypothesis does hold.)

Question III. What is the class of Banach algebras for which each epimorphism from a member of this class onto a Banach algebra is automatically continuous?

Of course, a member of the above class has a unique complete norm topology, and there are elementary examples of Banach algebras which do not have this property ([9, p. 141]). We do not know whether or not each C*-algebra belongs to the above class, but the class does contain each C*-algebra in the class of Question II, together with all commutative C*-algebras. This last remark follows from an important theorem of Jean Esterle [15]: using that theorem, we shall prove in Theorem 4.1 that each closed bi-ideal of an AW*-algebra belongs to the class of Question III.

The organization of the paper is as follows. In §2, we describe the C*-algebras from which we know there is a discontinuous homomorphism into a Banach algebra, and we make the guess that this class is the complement in the class of C*-algebras of the class of Question II. In §3 we again reformulate Question I, and we obtain some positive results. §4 contains a complete solution of Questions I, II, and III for AW*M-algebras, a class of C*-algebras containing the class of closed (not necessarily proper) bi-ideals of AW*-algebras.

We thank Marc Rieffel, Chris Lance, Kjeld Laursen, and the participants of the "Study Period" in Long Beach, for some valuable comments.

2. Discontinuous homomorphisms from C*-algebras

We begin with a trivial lemma.

LEMMA 2.1. Let $\mathfrak{A}$ be a Banach algebra. Suppose that $\mathfrak{J}$ is a closed bi-ideal in $\mathfrak{A}$ such that there exists a discontinuous homomorphism from $\mathfrak{A}/\mathfrak{J}$ into a Banach algebra. Then there is a discontinuous homomorphism from $\mathfrak{A}$.

Proof. Let $\theta : \mathfrak{A}/\mathfrak{J} \to \mathfrak{B}$ be a discontinuous homomorphism, and let $\pi : \mathfrak{A} \to \mathfrak{A}/\mathfrak{J}$ be the canonical epimorphism. Then π is an open map, and so $\theta \circ \pi$ is a discontinuous homomorphism.

Already, this remark shows that several well-known C*-algebras occurring in analysis or operator theory admit discontinuous homomorphisms. For example, let $L^2(\mathbb{T})$ and $H^2(\mathbb{T})$ denote, respectively, the usual Lebesgue and Hardy spaces relative to the circle $\mathbb{T}$, and let P be the orthogonal projection from $L^2(\mathbb{T})$ onto $H^2(\mathbb{T})$. For $\varphi \in C(\mathbb{T})$, let $T_\varphi(f) = P(\varphi f)$ $(f \in H^2(\mathbb{T}))$, and let $\mathfrak{T}$ be the C*-subalgebra of $\mathfrak{L}(H^2(\mathbb{T}))$ generated by $\{T_\varphi : \varphi \in C(\mathbb{T})\}$. Then the symbol map $\Phi : \mathfrak{T} \to C(\mathbb{T})$ is an epimorphism ([17]). As we mentioned before, $C(\mathbb{T})$ admits a discontinuous homomorphism. Hence, by 2.1, there is a discontinuous homomorphism from $\mathfrak{T}$ into a Banach algebra.

Similar examples of this type, involving, for example, Toeplitz operators on the boundary of strongly pseudoconvex domains, or certain pseudo-differential operators of order 0, can be constructed in the same way: see [13, Chapter 1] for details and further references.

Now let H be a Hilbert space, and let $A \in \mathfrak{L}(H)$ be an essentially normal operator (so that $A^*A - AA^*$ is a compact operator). Let $\mathfrak{A}_A$ be the C*-subalgebra of $\mathfrak{L}(H)$ generated by A and by the set $\mathfrak{K}(H)$ of compact operators on H. By the spectral theorem, $\mathfrak{A}_A/\mathfrak{K}(H)$ is isomorphic to $C(\sigma_e(A))$. Here, $\sigma_e(A)$ denotes the essential spectrum of A, the spectrum of the coset of A in the quotient algebra $\mathfrak{L}(H)/\mathfrak{K}(H)$. Then we have the following result.

PROPOSITION 2.2. There is a discontinuous homomorphism from $\mathfrak{A}_A$ into a Banach algebra if and only if $\sigma_e(A)$ is infinite.

Proof. If $\sigma_e(A)$ is infinite, then $C(\sigma_e(A))$ admits a discontinuous homomorphism, and so, by 2.1, there is a discontinuous homomorphism from $\mathfrak{A}_A$. If

$\sigma_e(A)$ is finite, then $C(\sigma_e(A))$ is finite-dimensional, and so $\mathcal{K}(H)$ has finite codimension in $\mathfrak{A}_A$. Let θ be a homomorphism from $\mathfrak{A}_A$. By a well-known theorem of B. E. Johnson ([17]; see also §4), $\theta | \mathcal{K}(H)$ is continuous, and so θ itself is continuous.

In the remainder of this article, $M_n(\mathfrak{A})$ denotes the algebra of $n \times n$-matrices over an algebra $\mathfrak{A}$. Thus, $M_n(\mathfrak{A}) = M_n \otimes \mathfrak{A}$, where we write M_n for $M_n(\mathbb{C})$. If $\mathfrak{A}$ is a Banach algebra, then all cross-norms ([25, p. 188]) on $M_n(\mathfrak{A})$ are equivalent. If $\mathfrak{A}$ is a C*-algebra, then $M_n(\mathfrak{A})$ is a C*-algebra in the standard way, and the C*-norm on $M_n(\mathfrak{A})$ is uniquely determined.

LEMMA 2.3. Let $\mathfrak{A}$ be a Banach algebra which admits a discontinuous homomorphism θ into a Banach algebra $\mathfrak{B}$. Then $\theta_n : M_n(\mathfrak{A}) \to M_n(\mathfrak{B})$, given by $\theta((a_{ij})) = (\theta(a_{ij}))$ is a discontinuous homomorphism.

If we combine this lemma with Lemma 2.1 and the generalization of a theorem of Coburn given in [12, Theorem 1], we obtain a discontinuous homomorphism from the C*-algebra $\mathfrak{T}(C_{M_n}(\mathbb{T}))$ generated by the set of all Toeplitz operators with continuous matrix symbol.

We now try to characterize the C*-algebras which belong to the class of Question II in terms of their representations. We first require a lemma.

LEMMA 2.4. Let X and Y be infinite compact Hausdorff spaces, and let $\eta : X \to Y$ be a continuous surjection. Then there exists a unital Banach algebra $\mathfrak{B}$ and a unital monomorphism $\theta : C(X) \to \mathfrak{B}$ such that $\theta \circ \eta^* : C(Y) \to \mathfrak{B}$ is discontinuous.

Proof. Here, η^* is the dual map of η given by $\eta^*(f) = f \circ \eta$ $(f \in C(Y))$. Clearly, η^* is an embedding of $C(Y)$ in $C(X)$.

Let $\beta\mathbb{N}$ be the Stone-Čech compactification of $\mathbb{N}$, let $\mathfrak{p} \in \beta\mathbb{N} \setminus \mathbb{N}$, let $M_\mathfrak{p}$ be the maximal ideal of $C(\beta\mathbb{N})$ associated with $\mathfrak{p}$, and let $J_\mathfrak{p}$ be the ideal of all functions in $C(\beta\mathbb{N})$ which vanish in a neighbourhood of $\mathfrak{p}$ in $\beta\mathbb{N}$. Then there is a radical Banach algebra $\mathfrak{R}$ and a discontinuous homomorphism $\mu : M_\mathfrak{p} \to \mathfrak{R}$ with $\ker \mu = J_\mathfrak{p}$ ([8, Theorem 7.7] and [14]).

Let Y_0 be a countable subset of Y which is discrete in the relative topology. For $y \in Y_0$, choose $x_y \in X$ with $\eta(x_y) = y$, and let $X_0 = \{x_y : y \in Y_0\}$ so that X_0 is a countable discrete set in X. Let $\tau : \mathbb{N} \to X_0$ be a homeomorphism, and extend τ to be a continuous map $\tau : \beta\mathbb{N} \to \overline{X_0}$. Let $\mathfrak{B} = C(X) \oplus \mathfrak{R}^\#$, where $\mathfrak{R}^\#$ is the algebra $\mathfrak{R}$ with identity adjoined, and let $\theta(f) = (f, \mu(f \circ \tau))$ $(f \in C(X))$. Then $\theta : C(X) \to \mathfrak{B}$ is a unital monomorphism, and clearly $\theta \circ \eta^*$ is discontinuous.

THEOREM 2.5. Let $\mathfrak{A}$ be a C*-algebra. Suppose that, for some $n \in \mathbb{N}$, $\mathfrak{A}$ has infinitely many non-equivalent irreducible *-representations of dimension n. Then there exists a discontinuous homomorphism from $\mathfrak{A}$ into a Banach algebra.

Before we prove this theorem, we recall some notions from the theory of representations. A representation of an algebra $\mathfrak{A}$ is a pair (π, E) consisting of a vector space E and a homomorphism π of $\mathfrak{A}$ into the algebra $L(E)$ of all linear endomorphisms of E. If $\mathfrak{A}$ is a C*-algebra, then a *-representation is a representation (π, H), where H is a Hilbert space and π is a *-homomorphism from $\mathfrak{A}$ into $\mathfrak{L}(H)$. A representation (π, E) is irreducible if $\{0\}$ and E are the only linear subspaces of E which are invariant for $\pi(\mathfrak{A})$. The dimension of a representation (π, E) is the dimension of E.

Now let $\mathfrak{A}$ be a C*-algebra. Two *-representations (π_1, H_1) and (π_2, H_2) are (spatially) equivalent if there is an isometry u of H_1 onto H_2 such that $u\pi_1(a)u^* = \pi_2(a)$ $(a \in \mathfrak{A})$. The spectrum $\hat{\mathfrak{A}}$ of $\mathfrak{A}$ is the set of all equivalence classes of irreducible *-representations of $\mathfrak{A}$. The primitive spectrum $\operatorname{Prim} \mathfrak{A}$ of $\mathfrak{A}$ is the set of kernels of irreducible *-representations of $\mathfrak{A}$. There is a natural surjection of $\hat{\mathfrak{A}}$ onto $\operatorname{Prim} \mathfrak{A}$. The spaces $\hat{\mathfrak{A}}$ and $\operatorname{Prim} \mathfrak{A}$ can be given the Jacobson topologies as in [21, 4.1.4 and 4.1.12]; see also [10, 3.1]. If $\mathfrak{A}$ has an identity, then $\mathfrak{A}$ and $\operatorname{Prim} \mathfrak{A}$ are compact, but they are not necessarily Hausdorff ([10, 3.1.8]). Since two finite-dimensional irreducible *-representations with the same kernel are equivalent, the map from $\hat{\mathfrak{A}}$ to $\operatorname{Prim} \mathfrak{A}$ is injective over the finite-dimensional representations.

Proof of Theorem 2.5. We can suppose, without loss of generality, that $\mathfrak{A}$ has an identity. Let $\mathfrak{J}$ be the intersection of the kernels of the irreducible *-representations of dimension at most n. Then $\mathfrak{J}$ is a closed bi-ideal in $\mathfrak{A}$. Let $\mathfrak{B} = \mathfrak{A}/\mathfrak{J}$, let $X = \hat{\mathfrak{B}}$, let ${}_mX$ be the subset of X consisting of the equivalence classes of all irreducible *-representations of dimension at most m (for $m \in \mathbb{N}$), and let $U = {}_nX \setminus {}_{n-1}X$. Then X is homeomorphic to $\operatorname{Prim} \mathfrak{B}$, ${}_nX = X$, and, by [10, 3.6.3 and 3.6.4] or [21, 4.4.10], U is an open subset of X which is Hausdorff in the relative topology. (Note that X itself need not be Hausdorff.) By our hypothesis, U is infinite.

Let X_d denote X with its discrete topology, let βX_d denote its Stone-Čech compactification, and let $\mathfrak{C} = C(\beta X_d) \otimes M_{n!} = M_{n!}(C(\beta X_d))$. We construct a map $\rho : \mathfrak{B} \to \mathfrak{C}$ as follows. Let $x \in X$, and suppose that x corresponds to a representation of dimension d_x. For $a \in \mathfrak{B}$, let $\hat{a}(x)$ be the corresponding $d_x \times d_x$-matrix over $\mathbb{C}$, and place $n!/d_x$ copies of $\hat{a}(x)$ in blocks along the diagonal of $M_{n!}$ to obtain a matrix $a_\rho(x)$ in $M_{n!}$. If $\hat{a}_{ij}(x)$ is a component of the matrix $\hat{a}(x)$, then $|\hat{a}_{ij}(x)| \leq \|\hat{a}(x)\| \leq \|a\|$, and so $\hat{a}_{ij}$ is bounded on X_d, and hence defines an element of $C(\beta X_d)$ $(i,j = 1,2,\ldots,d_x)$. This shows that a_ρ is an element of $\mathfrak{C}$. Clearly, $\rho : \mathfrak{B} \to \mathfrak{C}$ is a *-monomorphism, and so, by [21, 1.5.7], ρ is an isometry.

Let $U^\star = U \cup \{\infty\}$ be the one-point compactification of U. (If U is closed in X, and hence compact, we still take $U^\star = U \cup \{\infty\}$.) Let f be a continuous function on $U^\star$. Then we can regard f as a continuous function on X by setting $f(x) = f(\infty)$ $(x \in X \setminus U)$. By the Dauns-Hofmann theorem [21, 4.4.8], there is an element a of (the centre of) $\mathfrak{B}$ such that $\hat{a}(x) = f(x)I_x$ $(x \in X)$. Here, I_x is the identity $d_x \times d_x$-matrix over $\mathbb{C}$. Then we see that $\rho(a) = f \otimes I \in C(U^\star) \otimes I$, where I is the identity of $M_{n!}$.

Let $\eta : \beta X_d \to U^\star$ be the natural surjection. Then η is continuous, and we have shown that $\eta^*(C(U^\star)) \otimes I \subset \rho(\mathfrak{B}) \subset \mathfrak{C}$. By Lemma 2.4, there is a unital Banach algebra $\mathfrak{K}$ and a unital monomorphism $\theta : C(X_d) \to \mathfrak{K}$ such that $\theta \circ \eta^* : C(U^\star) \to \mathfrak{K}$ is discontinuous. Then $\theta_{n!} \circ \rho : \mathfrak{B} \to \mathfrak{K} \otimes M_{n!}$ is a unital homomorphism. (See 2.3 for the definition of $\theta_{n!}$.) Take $(f_n) \subset C(U^\star)$ such that $f_n \to 0$ and $(\theta \circ \eta^*)(f_n) \not\to 0$, and take $(a_n) \subset \mathfrak{B}$ with $\rho(a_n) = \eta^*(f_n) \otimes I$. Then $a_n \to 0$ in $\mathfrak{B}$ (because ρ is an isometry), but $(\theta_{n!} \circ \rho)(a_n) = (\theta \circ \eta^*)(f_n) \otimes I \not\to 0$ in $\mathfrak{K} \otimes M_{n!}$. This shows that $\theta_{n!} \circ \rho$ is a discontinuous homomorphism from $\mathfrak{B}$ into a Banach algebra. The result follows from 2.1.

We are prepared to make the <u>conjecture</u> that, for each C*-algebra $\mathfrak{A}$ not satisfying the conditions of Theorem 2.5, every homomorphism from $\mathfrak{A}$ into a Banach algebra is necessarily continuous. In other words, we guess that the C*-algebras which fall into the class of Question II are exactly the C*-algebras which have only finitely many non-equivalent irreducible *-representations of dimension n for each $n \in \mathbb{N}$. We cannot prove this conjecture.

Let us recall a result that supports our conjecture. We start from a theorem of Johnson which is given as [23, 12.2]. Let $\theta : \mathfrak{A} \to \mathfrak{B}$ be a homomorphism from a C*-algebra $\mathfrak{A}$ onto a dense subalgebra of a Banach algebra $\mathfrak{B}$, and let

$$\mathfrak{J}(\theta) = \{a \in \mathfrak{A} : \theta(a)\mathfrak{S}(\theta) = \mathfrak{S}(\theta)\theta(a) = \{0\}\}$$

be the <u>continuity ideal</u> of θ. Then $\overline{\mathfrak{J}(\theta)}$ is a closed bi-ideal of finite codimension in $\mathfrak{A}$, and θ is continuous if and only if $\mathfrak{J}(\theta)$ is closed. Since the set of kernels of non-zero irreducible finite-dimensional *-representations of $\mathfrak{A}$ coincides with the set of maximal bi-ideals of finite codimension, we have immediately:

THEOREM 2.6. Let $\mathfrak{A}$ be a unital C*-algebra which has no non-zero irreducible finite-dimensional *-representations. Then each homomorphism from $\mathfrak{A}$ into a Banach algebra is automatically continuous.

We do not know how to extend this theorem to cover the case of a C*-algebra with identity which has finitely many (or even just one) non-equivalent irreducible finite-dimensional *-representations.

We shall prove in Theorem 4.5(b), below, that the above conjecture does hold in a rather large subclass of C*-algebras — this subclass includes all closed bi-ideals in AW*-algebras.

It might be thought that, if $\mathfrak{A}$ were a C*-algebra with infinitely many non-equivalent irreducible finite-dimensional *-representations, then there would necessarily be a discontinuous homomorphism from $\mathfrak{A}$. We shall show that this is not the case. We first require a trivial extension of Theorem 2.6.

THEOREM 2.7. Let $\mathfrak{A}$ be a C*-algebra such that each closed bi-ideal $\mathfrak{J}$ in $\mathfrak{A}$ of finite codimension has an identity in $\mathfrak{J}$ for $\mathfrak{J}$. Then each homomorphism from $\mathfrak{A}$ into a Banach algebra is continuous.

<u>Proof</u>. Let $\mathcal{J}(\theta)$ be the continuity ideal of a homomorphism θ from $\mathfrak{A}$. Then $\overline{\mathcal{J}(\theta)}$ is a Banach algebra with identity, and so $\overline{\mathcal{J}(\theta)} = \mathcal{J}(\theta)$. Hence, θ is continuous.

EXAMPLE 2.8. Let $\mathfrak{C}$ be the C*-product $\mathfrak{C} = \prod_{n=1}^{\infty} M_n$ ([21, 1.2.4]). Then every homomorphism from $\mathfrak{C}$ into a Banach algebra is necessarily continuous.

To see this, we shall note that $\mathfrak{C}$ satisfies the condition of 2.7. Regard each M_n as an ideal in $\mathfrak{C}$ in the obvious way, and let $\mathfrak{J}$ be a bi-ideal in $\mathfrak{C}$. Then $M_n \cap \mathfrak{J} = \{0\}$ or $M_n \cap \mathfrak{J} = M_n$ for each $n \in \mathbb{N}$. Thus a bi-ideal of finite codimension in $\mathfrak{C}$ has the form $\mathfrak{J} = \prod_{n \notin F} M_n$ for some finite subset of $\mathbb{N}$. Clearly, $\mathfrak{J}$ has an identity in $\mathfrak{J}$ for $\mathfrak{J}$, and so each homomorphism from $\mathfrak{C}$ is continuous.

An other example is the C*-direct sum $\mathfrak{A} = \oplus_{n=1}^{\infty} M_n$ ([21, 1.2.4]). Then $\mathfrak{A}$ is a closed bi-ideal of the von Neumann algebra $\mathfrak{C}$. It will follow from Theorem 4.5 in §4 that each homomorphism from $\mathfrak{A}$ into a Banach algebra is continuous. Notice that 2.7 is not directly applicable to $\mathfrak{A}$.

Obviously all irreducible *-representations of $\mathfrak{A}$ are finite-dimensional, whereas $\mathfrak{C}$ has irreducible *-representations of infinite dimension associated with the points of $\beta\mathbb{N} \setminus \mathbb{N}$ (cf. [25, p. 358]).

3. The separating space and quasi-nilpotents

We now consider Question I.

The radical of an algebra $\mathfrak{A}$ is the intersection of the maximal modular left ideals ([6, 24.14]). If $\mathfrak{A}$ is a Banach algebra, then these ideals are closed, and so rad $\mathfrak{A}$ is the intersection of the kernels of continuous irreducible representations $\pi : \mathfrak{A} \to \mathcal{L}(E)$, where E is a Banach space. Thus, we see from Lemma 1.1 that Question I is equivalent to the following question.

<u>Question I"</u>. Let $\theta : \mathfrak{A} \to \mathfrak{B}$ be a homomorphism from a Banach algebra $\mathfrak{A}$ onto a dense subalgebra of a Banach algebra $\mathfrak{B}$. Is $\mathfrak{S}(\theta)$ necessarily contained in

ker π for each continuous irreducible representation (π, E) of $\mathfrak{B}$?

In this section we shall consider continuous irreducible representations $\pi : \mathfrak{B} \to \mathfrak{L}(E)$ such that $\overline{\pi(\mathfrak{B})}$ contains a non-zero compact operator.

LEMMA 3.1. Let (π, E) be a continuous irreducible representation of a Banach algebra $\mathfrak{B}$. If T is a non-zero operator in $\mathfrak{L}(E)$ and if $0 \neq x \in E$, then there are $A, B \in \pi(\mathfrak{B})$ such that $ATBx = x$.

Proof. As $T \neq 0$, there are non-zero elements u, v in E with $Tu = v$. By Jacobson's density theorem ([6, 24.10]) there exist $A, B \in \pi(\mathfrak{B})$ with $Bx = u$ and $Av = x$. Hence, $ATBx = x$.

THEOREM 3.2. Let $\theta : \mathfrak{A} \to \mathfrak{B}$ be a homomorphism from a Banach algebra $\mathfrak{A}$ onto a dense subalgebra of a Banach algebra $\mathfrak{B}$. Let (π, E) be a continuous irreducible representation of $\mathfrak{B}$ such that $\overline{\pi(\mathfrak{B})}$ contains a non-zero compact operator. Then $\mathfrak{S}(\theta) \subset \ker \pi$.

Proof. Let $\mathfrak{S} = \mathfrak{S}(\pi \circ \theta)$, and denote by $\mathfrak{K} = \mathfrak{K}(E)$ the bi-ideal of all compact operators in $\mathfrak{L}(E)$. Clearly, $\pi(\theta(\mathfrak{A}))$ is dense in $\overline{\pi(\mathfrak{B})}$ and so $\mathfrak{S}$ is a closed bi-ideal in $\overline{\pi(\mathfrak{B})}$.

Assume first that $\mathfrak{S} \cap \mathfrak{K} \neq \{0\}$, i.e., that there is a non-zero compact operator K in $\mathfrak{S}$. Because of Lemma 3.1, there are $A, B \in \overline{\pi(\mathfrak{B})}$ such that 1 is an eigenvalue of AKB. Now, $\mathfrak{S} \cap \mathfrak{K}$ is a bi-ideal in $\overline{\pi(\mathfrak{B})}$, so that $AKB \in \mathfrak{S} \cap \mathfrak{K}$. As $1 \in \sigma(AKB)$, the spectrum of AKB fails to be a connected set containing 0. This contradicts a theorem of Barnes ([5], [23, 6.15]). Thus, our assumption was false, and we have proved that $\mathfrak{S} \cap \mathfrak{K} = \{0\}$.

In order to prove that $\mathfrak{S} = \{0\}$, take $T \in \mathfrak{S}$ and fix an arbitrary x in E. Now, $\overline{\pi(\mathfrak{B})}$ contains a non-zero compact operator K. By Lemma 3.1 there are $A, B \in \pi(\mathfrak{B})$ such that $AKBx = x$. Then $TAKB \in \mathfrak{S} \cap \mathfrak{K} = \{0\}$, so that $Tx = TAKBx = 0$, i.e., $T = 0$. This shows that $\mathfrak{S} = \{0\}$.

Since $\mathfrak{S} = \overline{\pi(\mathfrak{S}(\theta))}$ ([23, 1.3(iii)]), we now conclude that $\mathfrak{S}(\theta) \subset \ker \pi$, as required.

Theorem 3.2 will be the main tool of this section. As a first application, we obtain the following corollary.

COROLLARY 3.3. Let $\theta : \mathfrak{A} \to \mathfrak{B}$ be a homomorphism from a Banach algebra $\mathfrak{A}$ onto a dense subalgebra of a C*-algebra $\mathfrak{B}$ which is of type I. Then θ is continuous.

Proof. By [21, 6.1.5], $\pi(\mathfrak{B})$ contains $\mathfrak{K}(H)$ for each irreducible *-representation (π, H) of $\mathfrak{B}$. By Theorem 3.2, $\mathfrak{S}(\theta)$ is contained in the intersection of the kernels of such representations. By [21, 3.13.8], this intersection is

$\{0\}$. Hence, θ is continuous.

Our main application of Theorem 3.2 will require the notion of a polynomial identity. Let n be a natural number. Then an algebra $\mathfrak{A}$ satisfies the <u>standard polynomial identity</u> $S_n = 0$ if

$$S_n(a_1,\ldots,a_n) := \sum_{\tau} \operatorname{sgn}(\tau)\, a_{\tau(1)} \cdots a_{\tau(n)} = 0$$

for all $a_1,\ldots,a_n \in \mathfrak{A}$. Here, the sum runs over the symmetric group on n symbols. See [3] or [22].

LEMMA 3.4. Let $\mathfrak{A}$ be a Banach algebra satisfying $S_{2n} = 0$.

(i) Let $\theta : \mathfrak{A} \to \mathfrak{B}$ be a homomorphism from $\mathfrak{A}$ into a Banach algebra $\mathfrak{B}$. Then $\overline{\theta(\mathfrak{A})}$ satisfies $S_{2n} = 0$.

(ii) Each irreducible representation of $\mathfrak{A}$ has dimension at most n.

<u>Proof</u>. (i) is obvious.

(ii) Let (π, E) be an irreducible representation of $\mathfrak{A}$. Assume, if possible, that there is a subspace E_0 of E with $\dim E_0 = n+1$. Take $A_1,\ldots,A_{2n} \in \mathfrak{L}(E_0) \cong M_{n+1}$. By the Jacobson density theorem ([6, 24.10]), there exist $a_1,\ldots,a_{2n} \in \mathfrak{A}$ with $\pi(a_j)x = A_j x$ $(x \in E_0,\ j = 1,\ldots,2n)$. Hence,

$$S_{2n}(A_1,\ldots,A_{2n})(x) = \pi(S_{2n}(a_1,\ldots,a_{2n}))(x) = 0 \quad (x \in E_0).$$

But this says that M_{n+1} satisfies the standard polynomial identity $S_{2n} = 0$, a contradiction of the fact ([20, p. 338], [22, 1.4.5]) that each polynomial identity satisfied by M_{n+1} must have degree at least $2n+2$. It follows that $\dim E \leq n$.

THEOREM 3.5. Let $\mathfrak{A}$ be a Banach algebra satisfying the standard polynomial identity $S_{2n} = 0$ for some $n \in \mathbb{N}$, and let $\theta : \mathfrak{A} \to \mathfrak{B}$ be a homomorphism from $\mathfrak{A}$ into a Banach algebra $\mathfrak{B}$. Then $\mathfrak{S}(\theta) \subset \mathfrak{D}(\mathfrak{B})$.

<u>Proof</u>. We can suppose that $\mathfrak{B} = \overline{\theta(\mathfrak{A})}$. By Lemma 3.4, each continuous irreducible representation (π, E) of $\mathfrak{B}$ has dimension at most n, and so it satisfies the condition that $\mathfrak{K}(E) = \pi(\mathfrak{B})$. By Theorem 3.2, $\mathfrak{S}(\theta) \subset \ker \pi$. The result follows.

COROLLARY 3.6. Let $\mathfrak{A}$ be either (i) a semi-simple Banach algebra such that all continuous irreducible representations have dimension at most n, or (ii) a C*-algebra such that all irreducible *-representations have dimension at most n. Let $\theta : \mathfrak{A} \to \mathfrak{B}$ be a homomorphism from $\mathfrak{A}$ into a Banach algebra $\mathfrak{B}$. Then $\mathfrak{S}(\theta) \subset \mathfrak{D}(\mathfrak{B})$.

Proof. By the theorem it suffices to show that $\mathfrak{A}$ satisfies the identity $S_{2n} = 0$. Let $a_1,\dots,a_{2n} \in \mathfrak{A}$, and let (π,E) be a representation of dimension at most n. Using [3, Theorem 1] or [22, 1.4.1], we see that

$$\pi(S_{2n}(a_1,\dots,a_{2n})) = S_{2n}(\pi(a_1),\dots,\pi(a_{2n})) = 0\,.$$

In case (i), this shows that $S_{2n}(a_1,\dots,a_{2n})$ is in the intersection of the kernels of the continuous irreducible representations. Since $\mathfrak{A}$ is semi-simple, it follows that $S_{2n}(a_1,\dots,a_{2n}) = 0$. In case (ii), $S_{2n}(a_1,\dots,a_{2n})$ is in the intersection of the kernels of the irreducible *-representations, and so, by [21, 3.13.8], it again follows that $S_{2n}(a_1,\dots,a_{2n}) = 0$. In either case, we have shown that $\mathfrak{A}$ satisfies the identity $S_{2n} = 0$.

The next corollary will be used in the proof of Theorem 4.5.

COROLLARY 3.7. Let $\mathfrak{A}$ be a Banach subalgebra of $M_n(\mathfrak{C})$ for some commutative Banach algebra $\mathfrak{C}$. Let $\theta : \mathfrak{A} \to \mathfrak{B}$ be a homomorphism from $\mathfrak{A}$ into a Banach algebra $\mathfrak{B}$. Then $\mathfrak{S}(\theta) \subset \mathfrak{D}(\mathfrak{B})$.

Proof. It suffices to show that $M_n(\mathfrak{C})$, and hence also $\mathfrak{A}$, satisfies the identity $S_{2n} = 0$. But this is just the theorem of Amitsur and Levitzki, as given in [22, 1.4.1].

4. C*-algebras related to AW*-algebras

In this section, we consider 'sufficiently non-commutative' algebras having 'sufficiently good partitions of unity'. We obtain extensions of results due to B. E. Johnson and J. D. Stein. First, we develop a rather general method. Let us consider the following situation.

Situation 4.0.

(1) $\mathfrak{C}$ is a complex unital algebra.

(2) $\mathfrak{A}$ is a complex Fréchet algebra which is also a unital $\mathfrak{C}$-bimodule satisfying:

(i) $a((\alpha\beta)b) = (a\alpha)(\beta b) = (a(\alpha\beta))b \quad (\alpha,\beta \in \mathfrak{C},\ a,b \in \mathfrak{A})$;

(ii) for each $\alpha \in \mathfrak{C}$, the operators $R(\alpha), L(\alpha) : \mathfrak{A} \to \mathfrak{A}$ defined by $R(\alpha)a = a\alpha$, $L(\alpha)a = \alpha a \quad (a \in \mathfrak{A})$ are continuous.

(3) $\mathfrak{B}$ is a complex Banach algebra.

(4) $\theta : \mathfrak{A} \to \mathfrak{B}$ is a homomorphism.

The conditions for $\mathfrak{A}$ may look a little strange. Let us therefore immediately mention the main example that we have in mind. Suppose that $\mathfrak{C}$ and $\mathfrak{R}$ are Banach algebras, and denote by $\mathfrak{C} \hat{\otimes}_\nu \mathfrak{R}$ the completion of the algebraic tensor product $\mathfrak{C} \otimes \mathfrak{R}$ with respect to a cross norm ν. Then each closed (not necessarily proper) bi-ideal in $\mathfrak{C} \hat{\otimes}_\nu \mathfrak{R}$ satisfies the conditions for $\mathfrak{A}$. Taking $\mathfrak{R} = \mathbb{C}$, we see that this includes the case of closed bi-ideals in $\mathfrak{G}$.

A second example is the case of a Banach algebra $\mathfrak{A}$ containing $\mathfrak{G}$ as a subalgebra.

THEOREM 4.1. Let situation 4.0 occur. Suppose that there are sequences (α_k), (β_k) in $\mathfrak{A}$ such that

$$(5)_{\text{right}} \qquad \alpha_k\alpha_{k-1}\cdots\alpha_1\beta_k = 0 \quad (k \in \mathbb{N}),$$

(respectively, such that

$$(5)_{\text{left}} \qquad \beta_k\alpha_1\cdots\alpha_{k-1}\alpha_k = 0 \quad (k \in \mathbb{N})).$$

For $k > j$, denote by $\mathfrak{J}_{j,k}$ the bi-ideal in $\mathfrak{G}$, algebraically generated by $\alpha_j\alpha_{j-1}\cdots\alpha_1\beta_k$ (respectively, by $\beta_k\alpha_1\cdots\alpha_{j-1}\alpha_j$). Then there exists $j_0 \in \mathbb{N}$ with the following properties.

(a) $\mathfrak{S}(\theta)\theta(\alpha d) = \{0\}$ (respectively, $\theta(d\alpha)\mathfrak{S}(\theta) = \{0\}$) for $k > j_0$, $\alpha \in \mathfrak{J}_{j_0,k}$, and $d \in \mathfrak{A}$.

(b) If $\mathfrak{J}_{j_0,k} = \mathfrak{G}$ for some $k > j_0$, and if $\mathfrak{A}$ has an identity, then θ is continuous.

(c) If $\mathfrak{A}$ is a Banach algebra with a bounded right (respectively, left) approximate identity, and if τ is an idempotent in $\mathfrak{J}_{j_0,k}$ (for some $k > j_0$) such that $a\tau = \tau a$ $(a \in \mathfrak{A})$, then $\theta \mid \tau\mathfrak{A}$ is continuous. In particular, if $\mathfrak{J}_{j_0,k} = \mathfrak{G}$ for some $k > j_0$, then θ is continuous.

<u>Proof</u>. We shall assume that $(5)_{\text{right}}$ holds, and prove the first forms of the result. The other case is similar.

(a) Consider the following sequence of topological vector spaces and linear mappings:

$$\mathfrak{A} \xrightarrow{R(\alpha_j)} \mathfrak{A} \xrightarrow{R(\alpha_{j-1})} \cdots \xrightarrow{R(\alpha_1)} \mathfrak{A} \xrightarrow{\theta} \mathfrak{B} \xrightarrow{\pi_k} \tilde{\mathfrak{B}},$$

where $\tilde{\mathfrak{B}} = \Pi_{a\in\mathfrak{A}}\mathfrak{B}$ with the product topology, and $\pi_k(b) = (b\theta(\beta_k a))_{a\in\mathfrak{A}}$ for $b \in \mathfrak{A}$. Clearly, π_k is continuous for $k \in \mathbb{N}$, and $\pi_k \circ \theta \circ R(\alpha_1) \circ \cdots \circ R(\alpha_j) = \pi_k \circ \theta \circ R(\alpha_j \cdots \alpha_1) = 0$ for $j \geq k$. From [1, Satz 1.4] (see also [2, Theorem 3.1]) we see that there exists $j_0 \in \mathbb{N}$ such that, for all $k > j_0$, the maps

$$\pi_k \circ \theta \circ R(\alpha_{j_0} \cdots \alpha_1) : \mathfrak{A} \to \tilde{\mathfrak{B}}$$

are continuous. It follows that, for each $d \in \mathfrak{A}$ and $k > j_0$, the maps

$$a \mapsto \theta(a\alpha_{j_0} \cdots \alpha_1\beta_k d) : \mathfrak{A} \to \mathfrak{B}$$

are continuous. If $\alpha \in \mathfrak{J}_{j_0,k}$, then $\alpha = \sum_{i=1}^{m} \gamma_i \alpha_{j_0} \cdots \alpha_1 \beta_k \delta_i$ for some $m \in \mathbb{N}, \gamma_1, \ldots, \gamma_m, \delta_1, \ldots, \delta_m \in \mathfrak{a}$. We conclude that the maps

$$a \mapsto \theta(a\alpha d) = \theta(a)\theta(\alpha d) : \mathfrak{A} \to \mathfrak{B}$$

are continuous for $\alpha \in \mathfrak{J}_{j_0,k}$ and $d \in \mathfrak{A}$. Hence, $\mathfrak{S}(\theta)\theta(\alpha d) = \{0\}$ for $k > j_0$, $\alpha \in \mathfrak{J}_{j_0,k}$, and $d \in \mathfrak{A}$.

(b) This is an immediate consequence of (a): take for α the identity of $\mathfrak{a}$ and for d the identity of $\mathfrak{A}$.

(c) Suppose now that $\mathfrak{A}$ is a Banach algebra with a bounded right approximate identity, and let (a_n) be a sequence in $\tau\mathfrak{A}$ (where τ is as in (c)) with $a_n \to 0$. By the Cohen factorization theorem ([6, §11], [11, 17.5]), there exists a sequence $(x_n) \subset \tau\mathfrak{A} = \mathfrak{A}\tau$ with $x_n \to 0$, and $y \in \mathfrak{A}$, such that $a_n = x_n y = (x_n\tau)y = x_n(\tau y)$ for $n \in \mathbb{N}$. By (a), the linear map $a \mapsto \theta(a\tau y) = \theta(a)\theta(\tau y)$, $\mathfrak{A} \to \mathfrak{B}$, is continuous. Hence, $\theta(a_n) = \theta(x_n \tau y) \to 0$, and so $\theta \,|\, \tau\mathfrak{A}$ is continuous.

This concludes the proof of the theorem.

We wish to apply this theorem to algebras $\mathfrak{A}$ related to an AW*-algebra $\mathfrak{a}$. To this end we first recall some of the basic facts from this theory ([21, §3.9], [18], [19]). A projection in a C*-algebra is an element p such that $p = p^2 = p^*$. A C*-algebra $\mathfrak{a}$ is an AW*-algebra if each set of orthogonal projections has a least upper bound and each maximal commutative self-adjoint subalgebra is (topologically) generated by its projections. The class of AW*-algebras includes the class of von Neumann algebras, but there are (even commutative) AW*-algebras which are not von Neumann algebras. By [18, Theorem 4.6], each AW*-algebra $\mathfrak{a}$ has a decomposition

$$\mathfrak{a} = z_I\mathfrak{a} \oplus z_{II}\mathfrak{a} \oplus z_{III}\mathfrak{a},$$

where z_I, z_{II}, z_{III} are central orthogonal projections, and $z_I\mathfrak{a}$, $z_{II}\mathfrak{a}$, $z_{III}\mathfrak{a}$ are AW*-algebras of the corresponding type. By [18, Theorem 4.2], z_I has an orthogonal decomposition $z_I = z_f + z_\infty$ into central projections such that $z_f\mathfrak{a}$ is finite of type I and $z_\infty\mathfrak{a}$ is purely infinite of type I.

THEOREM 4.2. Suppose that in situation 4.0, $\mathfrak{a}$ is an AW*-algebra and $\mathfrak{A}$ is a Banach algebra with a bounded right or left approximate identity. Suppose that $a\tau = \tau a$ for each central projection $\tau \in \mathfrak{a}$ and each $a \in \mathfrak{A}$. Then there exists a central projection $z_n \in z_f\mathfrak{a}$ such that:

(i) $z_n\mathfrak{a} = \oplus_{j=1}^{n} \mathfrak{a}_j$, where $\mathfrak{a}_j = M_j(C_j)$ with $C_j = \{0\}$ or $C_j = C(K_j)$ for a compact (Stonean) Hausdorff space K_j $(j = 1, \ldots, n)$;

(ii) $\theta \,|\, (1 - z_n)\mathfrak{A}$ is continuous.

Proof. It follows from [18, Lemma 4.5 and Lemma 4.12] that there are sequences (α_n) and (β_n) of projections in $\mathfrak{A}$ such that (with z_f as in the preceding remarks)

$$1 - z_f = \alpha_1 + \beta_1 ,$$
$$\alpha_1 = \alpha_2 + \beta_2 ,$$
$$\cdots$$
$$\alpha_k = \alpha_{k+1} + \beta_{k+1} ,$$
$$\alpha_k\beta_k = \beta_k\alpha_k = 0 ,$$
$$\alpha_k \sim \beta_k ,$$

for $k \in \mathbb{N}$. (Two projections, p and q in $\mathfrak{A}$ are equivalent, written $p \sim q$, if there exists $u \in \mathfrak{A}$ with $u^*u = p$, $uu^* = q$.) By induction, we see that, for each $j \in \mathbb{N}$, the bi-ideal in $\mathfrak{A}$ algebraically generated by $\alpha_j \cdots \alpha_1\beta_{j+1} = \beta_{j+1} = \beta_{j+1}\alpha_1 \cdots \alpha_j$ is $(1 - z_f)\mathfrak{A}$. From part (c) of Theorem 4.1, we conclude that $\theta \mid (1 - z_f)\mathfrak{A}$ is continuous.

Let us now investigate $\theta \mid z_f\mathfrak{A}$. Recall that $z_f\mathfrak{A}$ has representation as a C*-product

$$z_f\mathfrak{A} = \prod_{j=1}^{\infty} \mathfrak{A}_j = \{(a_j) : a_j \in \mathfrak{A}_j,\ \sup\|a_j\| < \infty\} ,$$

where $\mathfrak{A}_j = M_j(C_j)$ with $C_j = \{0\}$ or $C_j = C(K_j)$ for a compact (Stonean) Hausdorff space K_j ($j \in \mathbb{N}$): this is an easy consequence of [19, Lemma 18], combined with some elementary facts on finite sets of matrix units (see also [19, p. 465] and [16, p. 52]). Now let p_n be the identity of $\mathfrak{A}_n \cong M_n(C_n)$. Then $p_n = \sum_{j=1}^n e_j^{(n)}$, where $\{e_1^{(n)},\ldots,e_n^{(n)}\}$ is a set of equivalent abelian, orthogonal projections. (A projection p in $\mathfrak{A}$ is abelian if $p\mathfrak{A}p$ is commutative: see [18, p. 241].) For $n,k \in \mathbb{N}$, we put $m(n,k) = [n/2^k]$, where $[\cdot]$ denotes the integral part of a number, and we define

$$\alpha_k = \left(\sum_{j=1}^{m(n,k)} e_j^{(n)} \right)_{n\in\mathbb{N}} , \quad \beta_k = z_f - \alpha_k \quad (k \in \mathbb{N}).$$

Notice that $\alpha_j\alpha_k = \alpha_k$ $(k \geq j)$, so that $(5)_{\text{right}}$ holds. Let $\mathfrak{J}_{j,k}$ be the bi-ideal in $\mathfrak{A}$ algebraically generated by $\alpha_j\alpha_{j-1}\cdots\alpha_1\beta_k$. Then $\mathfrak{J}_{k,k+1}$ is generated by

$$\alpha_k \cdots \alpha_1\beta_{k+1} = \alpha_k\beta_{k+1} = \beta_{k+1}\alpha_k = \beta_{k+1}\alpha_1 \cdots \alpha_k$$
$$= \left(\sum_{j=m(n,k+1)+1}^{m(n,k)} e_j^{(n)} \right)_{n\in\mathbb{N}} .$$

(Here, sums over an empty index set are taken to be zero.) By Theorem 4.1, there exists $k_0 \in \mathbb{N}$ such that $\theta | \tau\mathfrak{A}$ is continuous for each idempotent τ in $\mathfrak{J}_{k_0, k_0+1}$. Let $n = 2^{k_0+1}$ and let $z_n = (p_1, \ldots, p_n, 0, 0, \ldots)$. Then $z_n \mathcal{G} = \oplus_{j=1}^{n} \mathcal{G}_j$, proving (i). A straightforward calculation (compare [24, p. 155/156] in the case of finite type I von Neumann algebras) shows that $z_f - z_n$ can be written as

$$z_f - z_n = \sum_{j=1}^{n} q_j ,$$

where $q_1, \ldots, q_n$ are orthogonal projections with $q_j \lesssim \alpha_{k_0} \beta_{k_0+1}$. Thus, $z_f - z_n$ is a central projection of $\mathcal{G}$ and an element of $\mathfrak{J}_{k_0, k_0+1}$. It follows from Theorem 4.1(c) that $\theta | (z_f - z_n)\mathfrak{A}$ is continuous. Since we already know that $\theta | (1 - z_f)\mathfrak{A}$ is continuous, we see that $\theta | (1 - z_n)\mathfrak{A}$ is continuous, and this completes the proof.

In the special case that $\mathcal{G} = \mathfrak{A}$ is a von Neumann algebra, Theorem 4.2 reduces to a result of J. D. Stein [24].

Let us give an immediate consequence of the theorem.

Let H be an infinite-dimensional Hilbert space, and denote by $\mathcal{K}(H)$ the bi-ideal of compact operators in $\mathcal{L}(H)$. Let $\mathfrak{A}$ be a C*-algebra, and let λ be a C*-norm on $\mathfrak{A} \otimes \mathcal{K}(H)$. By [25, p. 216], λ is a cross-norm, in the sense that $\lambda(a \otimes K) \leq k\|a\| \|K\|$ $(a \in \mathfrak{A}, K \in \mathcal{K}(H))$ for a constant k. Let $\mathfrak{A} \hat{\otimes}_\lambda \mathcal{K}(H)$ be the C*-algebra which is the completion of $\mathfrak{A} \otimes \mathcal{K}(H)$ with respect to λ. Similarly, we can define $\mathfrak{A} \hat{\otimes}_\lambda \mathcal{L}(H)$.

COROLLARY 4.3. Let $\mathfrak{A}$ be a C*-algebra. Then each homomorphism from $\mathfrak{A} \hat{\otimes}_\lambda \mathcal{K}(H)$ or from $\mathfrak{A} \hat{\otimes}_\lambda \mathcal{L}(H)$ into a Banach algebra is automatically continuous.

<u>Proof</u>. The algebra $\mathcal{L}(H)$ is a purely infinite AW*-algebra. Since $\mathfrak{A} \hat{\otimes}_\lambda \mathcal{K}(H)$ has a bounded approximate identity (because it is a C*-algebra), the statement follows from Theorem 4.2.

Of course, we may replace $\mathcal{K}(H)$ in 4.3 by any closed bi-ideal in an AW*-algebra which has no non-trivial finite type I part.

A C*-algebra $\mathfrak{A}$ such that $\mathfrak{A} \cong \mathfrak{A} \hat{\otimes}_\lambda \mathcal{K}(H)$ is sometimes said to be <u>stable</u>: Corollary 4.3 shows that they lie in the class of Question II.

By using the method of proof given in part (a) of Theorem 4.1, one obtains the following fact due to B. E. Johnson [17, Theorem 3.5].

COROLLARY 4.4. Let X be a Banach space such that $X \cong X \oplus X$ (topologically), and let $\mathfrak{A}$ be a closed bi-ideal in $\mathcal{L}(X)$ having a bounded left or right approximate identity. Then each homomorphism from $\mathfrak{A}$ into a Banach algebra is automatically continuous.

The most interesting case is that in which $\mathfrak{A} = \mathfrak{K}(X)$, the bi-ideal of compact operators on X. For a discussion of the existence of bounded approximate identities in $\mathfrak{K}(X)$, see [11, §30], but note that it is now known that there are Banach spaces X such that $\mathfrak{K}(X)$ has neither a right nor a left bounded approximate identity.

We now regret to say that we are going to introduce and name a new class of C*-algebras: for this class, we are able to resolve Questions I, II, and III. A C*-algebra $\mathfrak{A}$ will be called an <u>AW*M-algebra</u> if there is an AW*-algebra $\mathfrak{C}$ such that:

(i) the conditions in 4.0(2) are satisfied;

(ii) $\tau a = a\tau$ for each central projection τ in $\mathfrak{C}$ and each $a \in \mathfrak{A}$;

(iii) if $\mathfrak{C} = z_f\mathfrak{C} \oplus (1-z_f)\mathfrak{C}$, where $z_f\mathfrak{C}$ is the finite type I part of $\mathfrak{C}$ and z_f is the corresponding central projection, then we require that $z_f\mathfrak{C}$ is a closed bi-ideal in a C*-algebra $\mathfrak{B}$ of the form

$$\left.\begin{array}{l} \mathfrak{B} = \prod_{n=1}^{\infty} \mathfrak{B}_n, \text{ where } \mathfrak{B}_n \text{ is *-isomorphic to } M_n(C_n) \\ \text{and } C_n = \{0\} \text{ or } C_n = C(K_n) \text{ for some compact Hausdorff} \\ \text{space } K_n \ (n \in \mathbb{N}). \end{array}\right\} \quad (*)$$

Thus, the class of AW*M-algebras includes the class of closed (not necessarily proper) bi-ideals in AW*-algebras, and it includes the class of closed bi-ideals in algebras of the type $C(K,\mathfrak{C})$, where K is a compact Hausdorff space and $\mathfrak{C}$ is an AW*-algebra. In particular, it includes the commutative C*-algebras $C(K)$.

THEOREM 4.5. Let $\mathfrak{A}$ be an AW*M-algebra.

(a) Let $\theta : \mathfrak{A} \to \mathfrak{B}$ be a homomorphism from $\mathfrak{A}$ into a Banach algebra $\mathfrak{B}$. Then:

(i) $\mathfrak{A} = \mathfrak{A}_1 \oplus \mathfrak{A}_2$, where $\mathfrak{A}_1, \mathfrak{A}_2$ are closed bi-ideals in $\mathfrak{A}$, $\theta \mid \mathfrak{A}_2$ is continuous, and $\mathfrak{A}_1$ is *-isomorphic to a closed bi-ideal in $\prod_{j=1}^{n} M_j(C_j)$ for some $n \in \mathbb{N}$ (C_j as in (*));

(ii) $\mathfrak{S}(\theta) \subset \mathfrak{Q}(\mathfrak{L})$;

(iii) if $\overline{\theta(\mathfrak{A})}$ is semi-simple, θ is continuous.

(b) There exists a discontinuous homomorphism from $\mathfrak{A}$ into some Banach algebra if and only if there is an $n \in \mathbb{N}$ such that $\mathfrak{A}$ has infinitely many non-equivalent irreducible *-representations of dimension n.

Thus, if $\mathfrak{A}$ is an AW*M-algebra, we have a positive answer to Question I, and our conjecture about the C*-algebras which fall into the class of Question II holds for AW*M-algebras.

<u>Proof</u>. (a) Let $\mathfrak{C}$ and z_f be as in the definition of AW*M-algebras.

By Theorem 4.2, $\theta \mid (1 - z_f)\mathfrak{A}$ is continuous. Thus, $\mathfrak{S}(\theta) = \mathfrak{S}(\theta \mid z_f\mathfrak{A})$. Now $z_f\mathfrak{A}$ is a closed bi-ideal in an algebra $\mathfrak{B}$ as in (*). Hence, $z_f\mathfrak{A}$ is in a natural way an $\mathfrak{C}_0$-bimodule. Here, $\mathfrak{C}_0$ is the C*-product $\mathfrak{C}_0 = \Pi_{j=1}^{\infty} M_j$. Notice that $\mathfrak{C}_0$ is a finite AW*-algebra of type I, and that the conditions in Theorem 4.2 are fulfilled for $\mathfrak{C}_0$, $z_f\mathfrak{A}$, and $\theta \mid z_f\mathfrak{A}$. Thus, we obtain from that theorem the existence of $n \in \mathbb{N}$ such that, with $z_n = (id_1, \ldots, id_n, 0, 0, \ldots)$ (id_j being the identity in M_j), $\theta \mid (1 - z_n)z_f\mathfrak{A}$ is continuous, and $z_n z_f\mathfrak{A}$ is a closed bi-ideal in $\Pi_{j=1}^{n} \mathfrak{B}_j$, and hence is *-isomorphic to a closed bi-ideal in $\Pi_{j=1}^{n} M_j(C_j)$. This proves (i) with $\mathfrak{A}_1 = z_n z_f\mathfrak{A}$ and $\mathfrak{A}_2 = (1 - z_n)z_f\mathfrak{A} \oplus (1 - z_f)\mathfrak{A}$.

Corollary 3.7 now implies that $\mathfrak{S}(\theta) = \mathfrak{S}(\theta \mid \mathfrak{A}_1) \subset \mathfrak{Q}(\mathfrak{B})$, and (iii) follows immediately from (ii) and from Lemma 1.1.

(b) If, for each $m \in \mathbb{N}$, $\mathfrak{A}$ has at most a finite number of non-equivalent irreducible *-representations of dimension m, then C_n in (*) can be chosen to be of finite dimension for each $n \in \mathbb{N}$. If θ is a homomorphism from $\mathfrak{A}$ into a Banach algebra, then the ideal $\mathfrak{A}_1$ given in (i) has finite dimension, and so θ must be continuous.

The converse follows directly from Theorem 2.5.

We now solve Question III for AW*M-algebras. We require some preliminary lemmas. If $\mathfrak{B}$ is an algebra, we denote by $\mathrm{lan}(\mathfrak{B})$ the left annihilator of $\mathfrak{B}$, as in [6, 32.1]:

$$\mathrm{lan}(\mathfrak{B}) = \{x \in \mathfrak{B} : xb = 0 \ (b \in \mathfrak{B})\} .$$

LEMMA 4.6. Let $\mathfrak{A}$ be a complex algebra, and let $\theta : \mathfrak{A} \to \mathfrak{B}$ be an epimorphism from $\mathfrak{A}$ onto a Banach algebra $\mathfrak{B}$. Let $\pi : \mathfrak{B} \to \mathfrak{B}/\mathrm{lan}(\mathfrak{B})$ be the canonical epimorphism. If $\mathfrak{A}_1$, $\mathfrak{A}_2$ are bi-ideals in $\mathfrak{A}$ such that $\mathfrak{A} = \mathfrak{A}_1 \oplus \mathfrak{A}_2$, then

$$\mathfrak{B}/\mathrm{lan}(\mathfrak{B}) = \pi(\theta(\mathfrak{A}_1)) \oplus \pi(\theta(\mathfrak{A}_2)) ,$$

and $\pi(\theta(\mathfrak{A}_1))$ and $\pi(\theta(\mathfrak{A}_2))$ are closed in $\mathfrak{B}/\mathrm{lan}(\mathfrak{B})$.

<u>Proof</u>. Since θ and π are epimorphisms, we have

$$\mathfrak{B}/\mathrm{lan}(\mathfrak{B}) = \pi(\theta(\mathfrak{A}_1)) + \pi(\theta(\mathfrak{A}_2)) .$$

If $x \in \pi(\theta(\mathfrak{A}_1)) \cap \pi(\theta(\mathfrak{A}_2))$, then $x = \theta(a_1) + \mathrm{lan}(\mathfrak{B}) = \theta(a_2) + \mathrm{lan}(\mathfrak{B})$ for some $a_1 \in \mathfrak{A}_1$, $a_2 \in \mathfrak{A}_2$. Then $\theta(a_1) = \theta(a_2) + b_0$ with $b_0 \in \mathrm{lan}(\mathfrak{B})$. If $b \in \mathfrak{B}$, then $b = \theta(u_1 + u_2)$ with $u_1 \in \mathfrak{A}_1$, $u_2 \in \mathfrak{A}_2$. We obtain $\theta(a_1)b = \theta(a_1)\theta(u_1) = (\theta(a_2) + b_0)\theta(u_1) = 0$, and thus $\theta(a_1) \in \mathrm{lan}(\mathfrak{B})$. Hence, $x = 0$, and the sum is direct.

We now prove that $\pi(\theta(\mathfrak{A}_1))$ is closed in $\mathfrak{B}$. Take $(a_n) \subset \mathfrak{A}_1$ such that $(\pi(\theta(a_n)))$ converges in $\mathfrak{B}/\mathrm{lan}(\mathfrak{B})$ to $\pi(\theta(x + y))$, say, where $x \in \mathfrak{A}_1$, $y \in \mathfrak{A}_2$. Then there is a sequence $(b_n) \subset \mathrm{lan}(\mathfrak{B})$ such that $\theta(a_n) + b_n \to \theta(x + y)$ in $\mathfrak{B}$.

For arbitrary $u_1 \in \mathfrak{A}_1$, $u_2 \in \mathfrak{A}_2$, we then obtain $\theta(y)\theta(u_1) = 0$ and $0 = (\theta(a_n) + b_n)\theta(u_2) \to \theta(x+y)\theta(u_2) = \theta(y)\theta(u_2)$. Hence, $\theta(y) \in \mathrm{lan}(\mathfrak{B})$, and thus $\lim \pi(\theta(a_n)) \in \pi(\theta(\mathfrak{A}_1))$, showing that $\pi(\theta(\mathfrak{A}_1))$ is closed.

REMARKS 4.7. (a) If, in the situation of the preceding lemma, we have $\mathrm{lan}(\mathfrak{B}) = \{0\}$, then we see that $\mathfrak{B} = \pi(\mathfrak{A}_1) \oplus \pi(\mathfrak{A}_2)$, and $\pi(\mathfrak{A}_1)$ and $\pi(\mathfrak{A}_2)$ are closed in $\mathfrak{B}$.

(b) If $\mathrm{lan}(\mathfrak{B}) \neq \{0\}$, then the statement in (a) is, in general, false. For example, if $\mathfrak{A}$ is an infinite-dimensional Banach space, and if $\mathfrak{A} = \mathfrak{A}_1 \oplus \mathfrak{A}_2$ with linear subspaces $\mathfrak{A}_1, \mathfrak{A}_2$ such that $\mathfrak{A}_1$ is not closed in $\mathfrak{A}$, then $\mathfrak{A}$ is a Banach algebra for the zero multiplication. Set $\mathfrak{B} = \mathfrak{A}$, and let π be the identity map on $\mathfrak{A}$. Then $\theta(\mathfrak{A}_1)$ is not closed in $\mathfrak{B}$.

LEMMA 4.8. Let K be a compact Hausdorff space. Then each epimorphism from $M_n(C(K))$ onto a Banach algebra is continuous.

<u>Proof</u>. Let θ be an epimorphism from $M_n(C(K))$ onto a Banach algebra $\mathfrak{B}$. For $i,j = 1,\dots,n$, set $e_{ij} = (\delta_{ip}\delta_{jq})_{p,q=1,\dots,n}$. Then

$$\{E_{ij} = \theta(e_{ij} \otimes 1) : i,j = 1,\dots,n\}$$

is a set of matrix units in $\mathfrak{B}$. (See [16, p. 52] for the notion of matrix units.) Let $\mathfrak{R} = \{b \in \mathfrak{B} : bE_{ij} = E_{ij}b \ (i,j = 1,\dots,n)\}$. Then $\mathfrak{R}$ is a closed subalgebra of $\mathfrak{B}$, and, by [16, Prop. 6, p. 52], $\mathfrak{B} \cong M_n(\mathfrak{R})$ and $\mathfrak{R} \cong E_{11}\mathfrak{B}E_{11}$. In fact, $E_{11}\mathfrak{B}E_{11} = \{bE_{11} : b \in \mathfrak{R}\}$, and the isomorphism is given by $b \mapsto bE_{11}$. Obviously, all these isomorphisms are topological.

We define $\theta_0 : C(K) \to \mathfrak{R}$ by $\theta_0(f) = \theta(e_{11} \otimes f)$ for $f \in C(K)$. Then we see that $\theta = \mathrm{id}_n \otimes \theta_0$ (with the identification $\mathfrak{B} \cong M_n(\mathfrak{R}) \cong M_n \otimes \mathfrak{R}$), where id_n is the identity of M_n. Since θ is an epimorphism, θ_0 must also be an epimorphism. Now we may (and do) apply the important result of J. Esterle ([15, Theorem 5.3]) that each epimorphism from $C(K)$ onto a Banach algebra is automatically continuous: θ_0, and hence θ, are continuous.

LEMMA 4.9. Let $\mathfrak{C}$ be a unital Banach algebra. Then each closed bi-ideal $\mathfrak{J}$ in $M_n(\mathfrak{C})$ is of the form $\mathfrak{J} = M_n(I)$ for some closed bi-ideal I in $\mathfrak{C}$.

<u>Proof</u>. It is a standard algebraic fact ([16, Prop. 1, p. 40]) that each bi-ideal $\mathfrak{J}$ in $M_n(\mathfrak{C})$ is of the form $\mathfrak{J} = M_n(I)$ for some bi-ideal I in $\mathfrak{C}$. Clearly, if $\mathfrak{J}$ is closed in $M_n(\mathfrak{C})$, then I is closed in $\mathfrak{C}$.

LEMMA 4.10. Let $\theta : \mathfrak{A} \to \mathfrak{B}$ be a homomorphism from a C*-algebra $\mathfrak{A}$ into a Banach algebra $\mathfrak{B}$. If $\mathfrak{S}(\theta)$ consists of nilpotent elements, then $\mathfrak{S}(\theta) = \{0\}$, and θ is continuous.

Proof. Without loss of generality, we may assume that $\mathfrak{A}$ has an identity, 1. By [23, 12.7], θ is continuous if θ is continuous on each C*-subalgebra $\mathfrak{A}_h$ of $\mathfrak{A}$ generated by a single hermitian element h. Now $\mathfrak{A}_h$ is *-isomorphic to $C(K)$, where $K = \sigma(h)$ is a compact subset of $\mathbb{R}$. Since $\mathfrak{S}(\theta \mid \mathfrak{A}_h) \subset \mathfrak{S}(\theta)$, $\mathfrak{S}(\theta \mid \mathfrak{A}_h)$ consists of nilpotent elements. By the proof of [5, Theorem 4.5], we see that $\theta \mid \mathfrak{A}_h$ is continuous, and so θ is continuous.

LEMMA 4.11. Let K be a compact Hausdorff space, and let $\mathfrak{J}$ be a closed bi-ideal in $M_n(C(K))$. Then each epimorphism from $\mathfrak{J}$ onto a Banach algebra is continuous.

Proof. Let θ be an epimorphism from $\mathfrak{J}$ onto a Banach algebra $\mathfrak{B}$.

By Lemma 4.9, $\mathfrak{J} = M_n(I)$, where I is a closed ideal in $C(K)$. Then $I^\# = I \oplus \mathbb{C}1$ is a commuatative C*-algebra which we identify with $C(X)$ for some compact space X. Moreover, $M_n(I^\#) \cong M_n(I) \oplus M_n$, where the multiplication on $M_n(I) \oplus M_n$ is defined by

$$(a_1,m_1)(a_2,m_2) = (a_1a_2 + a_1m_2 + m_1a_2, m_1m_2)$$
$$(a_1, a_2 \in M_n(I),\ m_1,m_2 \in M_n).$$

Set $L = \{b \in \mathfrak{B} : \mathfrak{B} b \mathfrak{B} = \{0\}\}$, so that L is a closed bi-ideal in $\mathfrak{B}$ and $L^3 = \{0\}$. Let $\mathfrak{B}_0$ be the Banach space $(\mathfrak{B}/L) \oplus M_n$, and introduce a multiplication on $\mathfrak{B}_0$ as follows: If $(b_1,m_1), (b_2,m_2) \in \mathfrak{B}_0$, then $b_j = \theta(a_j) + L$ for some $a_j \in \mathfrak{J}$ $(j = 1,2)$, and we set

$$(b_1,m_1)(b_2,m_2) = (\theta(a_1a_2 + a_1m_2 + m_1a_2) + L,\ m_1m_2).$$

Of course, we have to check that this multiplication is well defined. Hence, suppose that also $\theta(\tilde{a}_j) + L = b_j$ for some $\tilde{a}_j \in \mathfrak{J}$. If $c,d \in \mathfrak{B}$ are given, then $c = \theta(u)$, $d = \theta(v)$ for some $u,v \in \mathfrak{J}$, and we obtain

$$c\theta(a_1a_2 - \tilde{a}_1\tilde{a}_2)d = c(\theta(a_1 - \tilde{a}_1)\theta(a_2) - \theta(\tilde{a}_1)\theta(\tilde{a}_2 - a_2))d = 0,$$

$$c\theta(a_1m_2 - \tilde{a}_1m_2)d = c\theta((a_1 - \tilde{a}_1)(m_2v)) = c\theta(a_1 - \tilde{a}_1)\theta(m_2v) = 0,$$

and, similarly, $c\theta(m_1a_2 - m_1\tilde{a}_2)d = 0$. It follows that the product is indeed well defined. With this multiplication, $\mathfrak{B}_0$ is a Banach algebra.

Define $\theta^\# : M_n(I) \oplus M_n \to \mathfrak{B}_0$ by

$$\theta^\#(a,m) = (\theta(a) + L, m) \qquad (a \in M_n(I),\ m \in M_n).$$

Obviously, θ is a surjective linear map, and it is easily checked to be multiplicative. Since $M_n(I) \oplus M_n$ is topologically isomorphic to $M_n(C(X))$, we conclude from Lemma 4.8 that $\theta^{\#}$ is continuous. Hence, $\theta^{\#} | M_n(I) = \pi \circ \theta$ is continuous, where $\pi : \mathfrak{B} \to \mathfrak{B}/L$ is the canonical epimorphism. Therefore, $\mathfrak{S}(\theta) \subset L$. Since $L^3 = \{0\}$, Lemma 4.10 implies that θ is continuous, as required.

We are now ready to prove our main result concerning Question III.

THEOREM 4.12. Each epimorphism from an AW*M-algebra onto a Banach algebra is automatically continuous.

Proof. Let θ be an epimorphism from an AW*M-algebra $\mathfrak{A}$ onto a Banach algebra $\mathfrak{B}$.

By Theorem 4.5, $\mathfrak{A} = \mathfrak{A}_1 \oplus \mathfrak{A}_2$, where $\mathfrak{A}_1, \mathfrak{A}_2$ are closed bi-ideals in $\mathfrak{A}$, $\theta | \mathfrak{A}_2$ is continuous, and $\mathfrak{A}_1$ is *-isomorphic to a closed bi-ideal in $\Pi_{j=1}^{n} M_j(C_j)$. Hence, $\mathfrak{A}_1 = \mathfrak{J}_1 \oplus \cdots \oplus \mathfrak{J}_n$, where $\mathfrak{J}_j$ is a closed bi-ideal in $\mathfrak{A}_1$ which is *-isomorphic to a closed bi-ideal in $M_j(C_j)$ $(j = 1,\ldots,n)$. Let $\mathfrak{R}_j = (\oplus_{\substack{i=1\\ i\neq j}}^{n} \mathfrak{J}_i) \oplus \mathfrak{A}_2$: this is a closed bi-ideal in $\mathfrak{A}$.

By Lemma 4.6, $\mathfrak{B}/\mathrm{lan}(\mathfrak{B}) = \pi(\theta(\mathfrak{J}_j)) \oplus \pi(\theta(\mathfrak{R}_j))$, and $\pi(\theta(\mathfrak{J}_j))$ is closed in the Banach algebra $\mathfrak{B}/\mathrm{lan}(\mathfrak{B})$. Hence, $\pi \circ \theta | \mathfrak{J}_j$ is an epimorphism from $\mathfrak{J}_j$ onto a Banach algebra, and thus it is continuous by Lemma 4.11. Therefore, $\mathfrak{S}(\theta | \mathfrak{J}_j) \subset \ker \pi = \mathrm{lan}(\mathfrak{B})$ $(j = 1,\ldots,n)$. Since $\mathrm{lan}(\mathfrak{B})^2 = \{0\}$, we may apply Lemma 4.10 to deduce that $\theta | \mathfrak{J}_j$ is continuous for $j = 1,\ldots,n$. As $\theta | \mathfrak{A}_2$ is continuous, we obtain the continuity of θ.

Finally, we wish to give an application of Theorem 4.1 in a situation where no idempotents are available. We shall need some notations. Let $(G, +)$ be a compact abelian group. For $f \in C(G)$ we define $\mathfrak{M}_f \in \mathfrak{L}(C(G))$ by $\mathfrak{M}_f(g) = fg$ $(g \in C(G))$. For $s \in G$, let $T_s \in \mathfrak{L}(C(G))$ be the translation operator given by $(T_s g)(t) = g(t+s)$, $(g \in C(G),\ t \in G)$. Let $\mathfrak{a}_G$ be the subalgebra of $\mathfrak{L}(C(G))$ generated (in the algebraic sense) by $\{\mathfrak{M}_f : f \in C(G)\} \cup \{T_s : s \in G\}$. In the case that $(G, +) = (\mathbb{T}, .)$, with $\mathbb{T} = \{z \in \mathbb{C} : |z| = 1\}$, we may also replace $\{\mathfrak{M}_f : f \in C(\mathbb{T})\}$ by the smaller set $\{\mathfrak{M}_f : f \in C^{\infty}(\mathbb{T})\}$.

THEOREM 4.14. Let $\mathfrak{A}_0$ be a complex unital Banach algebra, and suppose that for some infinite compact abelian group G there is a unital homomorphism $\Phi : \mathfrak{a}_G \to \mathfrak{A}_0$. Let $\mathfrak{A}$ be a closed bi-ideal in $\mathfrak{A}_0$ having a bounded left or right approximate identity. Then each homomorphism from $\mathfrak{A}$ into a Banach algebra is automatically continuous.

Proof. By using the mapping Φ in a natural way, we can regard $\mathfrak{A}$ as an $\mathfrak{a}_G$-bimodule satisfying conditions 4.0 (2). Since G is an infinite abelian group, there is a sequence $(t_n) \subset G$ with 0 as its only accumulation point and such that $t_n \neq t_m$ for $n \neq m$. We can then find a sequence (U_n) of compact neighbourhoods of 0 in G such that $(t_n + U_n) \cap (t_m + U_m) = \emptyset$ for $n \neq m$. For each

$n \in \mathbb{N}$, take $f_n \in C(G)$ (respectively, $f_n \in C^\infty(T)$ in the case that $(G, +) = (\mathbb{T}, .)$) such that $f_n(t_n) = 1$, $\operatorname{supp} f_n \subset t_n + U_n$, and $f_n(G) \subset [0,1]$. Set

$$\alpha_n = \sum_{j=n+1}^{\infty} \frac{1}{j} \mathfrak{M}_{f_j}, \quad \beta_n = \sum_{j=1}^{n} \mathfrak{M}_{f_j} \quad (n \in \mathbb{N}).$$

Since $\Sigma_{n+1}^{\infty} j^{-1} f_j$ and $\Sigma_1^n f_j$ belong to $C(G)$ (respectively, $C^\infty(\mathbb{T})$), we see that $\alpha_n, \beta_n \in \mathfrak{a}_G$. Moreover $\alpha_n \ldots \alpha_1 \beta_n = \beta_n \alpha_1 \ldots \alpha_n = 0$ $(n \in \mathbb{N})$.

Let θ be a homomorphism from $\mathfrak{A}$ into a Banach algebra $\mathfrak{B}$. Let j_0 be the integer specified in Theorem 4.1 for our situation. Note that the ideal $\mathfrak{J}_{k,k+1}$ of that theorem is the ideal in $\mathfrak{a}_G$ generated by

$$\alpha_k \ldots \alpha_1 \beta_{k+1} = \beta_{k+1} \alpha_1 \ldots \alpha_k = \mathfrak{M}_g,$$

where $g_k = (k+1)^{-k} f_{k+1}^{k+1}$. Then $g_k(t_{k+1}) = (k+1)^{-k} \neq 0$. Let $U = \operatorname{int}(\operatorname{supp} g_{j_0})$, then $\{U - s_1, \ldots, U - s_n\}$ is an open cover of G for some $s_1, \ldots, s_n \in G$. Let $h = \Sigma_{j=1}^n T_{s_j} g_{j_0}$. Since

$$h(t) = \sum_{j=1}^{n} g(t + s_j) > 0 \quad (t \in G),$$

we see that $(\mathfrak{M}_h)^{-1} = \mathfrak{M}_{h^{-1}}$ exists in $\mathfrak{a}_G$. Moreover,

$$\mathfrak{M}_1 = \mathfrak{M}_{h^{-1}} \sum_{j=1}^{n} T_{s_j} \mathfrak{M}_{g_{j_0}} T_{-s_j} \in \mathfrak{J}_{j_0, j_0+1}.$$

By Theorem 4.1(c), we now obtain the continuity of θ (choosing $\tau = \mathfrak{M}_1$).

COROLLARY 4.14. Let G be an infinite compact abelian group, and let X be a Banach space. Let $\mathfrak{A}$ be a closed subalgebra of $\mathfrak{L}(L^p(G,X))$ (for some p, with $1 \leq p \leq \infty$) containing the canonical image of $\mathfrak{a}_G$. Then each homomorphism from $\mathfrak{A}$ into a Banach algebra is automatically continuous.

Notice that, in the case $G = \mathbb{T}$, we may replace $L^p(G,X)$ by spaces of k-times continuously differentiable functions or by Sobolev spaces.

It is also easy to obtain a corresponding result for $G = \mathbb{R}$. One then has to consider the operators of multiplication by 2π-periodic C^∞-functions. The exact formulation of the analogues of 4.13 and 4.14 is left to the reader.

COROLLARY 4.15. Let G be an infinite compact abelian group, and let H be a Hilbert space. Let $\mathfrak{A}$ be a closed bi-ideal in a C*-subalgebra of $\mathfrak{L}(L^2(G,H))$ containing the canonical image of $\mathfrak{a}_G$. Then each homomorphism from $\mathfrak{A}$ into a Banach algebra is automatically continuous.

Again, a variant with $G = \mathbb{R}$ or even $G = \mathbb{R}^n$ can be proved.

Perhaps we should conclude by recalling, in the context of C*-algebras, the main results which we have proved.

Question I asks whether or not each element of the separating space of a homomorphism from a C*-algebra is necessarily quasi-nilpotent. Question III asks whether or not an epimorphism from a C*-algebra is necessarily continuous, and Question II essentially asks whether or not a homomorphism from a C*-algebra which has only finitely many non-equivalent irreducible *-representations of dimension n for each n is necessarily continuous. We have proved that the answer to each of these questions is positive for AW*M-algebras, a class of C*-algebras which contains all closed bi-ideals in von Neumann algebras and all commutative C*-algebras. No example is known for which the answer to any of these questions is negative.

Obviously, a major class of C*-algebras for which the questions are open is the class of Type 1 C*-algebras; a particularly difficult case seems to be that of simple C*-algebras (without identity), and of these perhaps the most intractable will prove to be the simple C*-algebras which have no non-zero projections. These last algebras are good examples to attack because the resolution of the problem in these cases will surely require some new idea.

References

[1] E. Albrecht and M. Neumann, Automatische Stetigkeitseigenschaften einiger Klassen linearer Operatoren, Math. Ann., 240 (1979), 251-280.

[2] ___________, Automatic continuity for operators of local type, this Volume.

[3] A. S. Amitsur and J. Levitzki, Minimal identities for algebras, Proc. Amer. Math. Soc., 1 (1950), 449-463.

[4] W. G. Bade and P. C. Curtis, Homomorphisms of commutative Banach algebras, Amer. J. Math., 83 (1960), 589-608.

[5] B. A. Barnes, Some theorems concerning the continuity of algebra homomorphisms, Proc. Amer. Math. Soc., 18 (1967), 1035-1037.

[6] F. F. Bonsall and J. Duncan, Complete Normed Algebras, Springer-Verlag, New York, 1973.

[7] L. A. Coburn, The C*-algebra generated by an isometry, II, Trans. Amer. Math. Soc., 137 (1969), 211-217.

[8] H. G. Dales, A discontinuous homomorphism from $C(X)$, Amer. J. Math., 101 (1979), 647-734.

[9] ___________, Automatic continuity: a survey, Bull. London Math. Soc., 10 (1978), 129-183.

[10] J. Dixmier, C*-algebras, North Holland, Amsterdam - New York - Oxford, 1977.

[11] R. S. Doran and J. Wichmann, Approximate identities and factorization in Banach modules, Lecture Notes in Math., 768, Springer-Verlag, 1979.

[12] R. G. Douglas, Banach algebra techniques in the theory of Toeplitz operators, Conference Board of the Mathematical Sciences Regional Conference series in Mathematics, No. 15, Amer. Math. Soc., 1972.

[13] __________, C*-algebra extensions and K-homology, Annals of Mathematics Studies, Vol. 95, Princeton University Press, 1980.

[14] J. Esterle, Injection de semi-groupes divisibles dans des algèbres de convolution et construction d'homomorphismes discontinus de C(K), Proc. London Math. Soc., (3) 36 (1978), 59-85.

[15] __________, Theorems of Gelfand-Mazur type and continuity of epimorphisms from C(K), J. Functional Analysis, 36 (1980), 273-285.

[16] N. Jacobson, Structure of rings, (second edition), Colloquium Publ. Vol. 37, Amer. Math. Soc., Providence, R.I., 1964.

[17] B. E. Johnson, Continuity of homomorphisms of algebras of operators, J. London Math. Soc., 42 (1967), 537-541.

[18] I. Kaplansky, Projections in Banach algebras, Annals of Math., 53 (1951), 235-249.

[19] __________, Algebras of type I, Annals of Math., 56 (1952), 460-472.

[20] J. Levitzki, A theorem on polynomial identities, Proc. Amer. Math. Soc., 1 (1950), 334-341.

[21] G. K. Pederson, C*-algebras and their Automorphism Groups, Academic Press, London 1979.

[22] L. H. Rowen, Polynomial Identities in Ring Theory, Academic Press, New York, 1980.

[23] A. M. Sinclair, Automatic Continuity of Linear Operators, London Math. Soc. Lecture Notes Series Vol. 21, Cambridge University Press, Cambridge, 1976.

[24] J. D. Stein, Jr., Continuity of homomorphisms of von Neumann algebras, Amer. J. Math., 91 (1969), 153-159.

[25] M. Takesaki, Theory of Operator Algebras, I, Springer-Verlag, New York, 1979.

Fachbereich Mathematik der
Universität des Saarlandes
D-6600 Saarbrücken
West Germany

School of Mathematics
University of Leeds
Leeds LS2 9JT
England

COFINITE IDEALS IN BANACH ALGEBRAS, AND FINITE-DIMENSIONAL REPRESENTATIONS OF GROUP ALGEBRAS

H. G. Dales[†] and G. A. Willis

1. Introduction

This note arises from consideration of the question whether or not finite-dimensional representations of a group algebra $L^1(G)$ are automatically continuous. Here, G is a locally compact topological group which is not necessarily abelian. It is perhaps surprising that this question does not seem to have been considered before (see [5], for example). We prove that these representations are continuous when G is an amenable group and in some other cases, but we leave the general question open.

In fact, most of what we have to say can be formulated in more general terms, and we do this in §2. The applications to $L^1(G)$ are given in §3, together with a remark that all finite-dimensional representations of C*-algebras are continuous.

Let A be an algebra, and let I and J be ideals in A. (Throughout, an ideal is a two-sided ideal, unless we say otherwise.) Then

$$IJ = \left\{ \sum_{j=1}^{n} a_j b_j : n \in \mathbb{N},\ a_1,\dots,a_n \in I,\ b_1,\dots,b_n \in J \right\},$$

and $I^n = I(I^{n-1})$ for $n \in \mathbb{N}$. An ideal I is <u>nilpotent</u> if $I^n = \{0\}$ for some n, <u>idempotent</u> if $I = I^2$, <u>cofinite</u> if the quotient space A/I is finite-dimensional, and <u>modular</u> if A/I has an identity.

Let A be a Banach algebra, and let I be a modular ideal with $\dim(A/I) = 1$, so that $I = \ker \varphi$ for a character φ on A. The point derivations on A form a vector space isomorphic to the dual of the vector space I/I^2 (see [2, §6]), and so, if I is not an idempotent ideal, there are non-zero point derivations at φ. Further, if I^2 is either not closed or not cofinite in A, then there are discontinuous point derivations on A at φ. This gives 'the easy way' of constructing discontinuous homorphisms and discontinuous derivations from the Banach algebra A ([2, Proposition 6.6]).

Here, we show in §2 that there is a discontinuous homomorphism with finite-dimensional range from a Banach algebra A if and only if there is a closed, cofinite ideal I in A with I^2 either not closed or not cofinite in A. We also consider whether or not these conditions are equivalent to the further condition that there is a discontinuous derivation from A into a finite-dimensional

[†] The first author was supported by NATO Grant No. RG073.81.

Banach A-bimodule.

2. General results

It is convenient to first give a preliminary remark and to prove two lemmas.

The remark is the following. Let I be a cofinite ideal, not necessarily closed, in a Banach algebra A. Then A/I is a finite dimensional algebra. Adjoin an identity to A/I, if necessary, to obtain X, say, and let X have some vector space norm which makes it into a Banach space. Then A/I acts faithfully as an algebra of bounded linear operators on X, and so A/I is a Banach algebra with respect to the corresponding operator norm. Thus, A/I is always a Banach algebra: if I is in fact closed, then the norm is equivalent to the quotient norm, of course.

In the lemmas, we take the zero algebra to be semi-simple.

LEMMA 2.1. Let I be a cofinite ideal in a Banach algebra A. Then there exists a closed, cofinite ideal J in A such that A/J is a semi-simple algebra and such that $J^n \subset I$ for some $n \in \mathbb{N}$.

<u>Proof</u>. Let R be the (Jacobson) radical of A/I, and let $J = \pi^{-1}(R)$, where $\pi : A \to A/I$ is the quotient map. Then J is a cofinite ideal in A, and, since $A/J \cong (A/I)/R$, A/J is a semi-simple algebra, possibly equal to $\{0\}$. Thus, J is the kernel of an epimorphism from A onto a semi-simple Banach algebra; by [1, 25.10], J is closed in A. Since A/I is finite-dimensional, R is a nilpotent ideal [6, p.202], and so $J^n \subset I$ for some $n \in \mathbb{N}$.

LEMMA 2.2. Let A be a Banach algebra. Suppose that there exists a cofinite ideal I in A such that I^2 is either not closed or not cofinite. Then there exists a closed, cofinite ideal J in A such that A/J is semi-simple and such that <u>either</u>

(α) J^2 is not cofinite in A, <u>or</u>

(β) J^n is cofinite for $n \in \mathbb{N}$, and J^k is not closed in A for some $k \geq 2$.

If A is separable, then alternative (β) cannot occur.

<u>Proof</u>. Let J be the ideal constructed in Lemma 2.1. Then $J^n \subset I^2$ for some $n \in \mathbb{N}$ because eventually $J^m \subset I$.

Suppose first that J^n is not cofinite for some n. Then J^2 is not cofinite in A, and alternative (α) holds. If J^n is cofinite for all $n \in \mathbb{N}$, then I^2 is cofinite, and hence not closed, in A. But this shows that, for some k, J^k is not closed in A. In this case, alternative (β) holds.

Let A be a separable Banach algebra. If J is a closed ideal in A such that J^n is cofinite in A, then, by a theorem of Loy [8], J^n is closed in A. Hence, alternative (β) cannot occur if A is a separable algebra.

An example of Dixon ([4]; cf [3]) shows that alternative (β) may occur for certain non-separable Banach algebras.

THEOREM 2.3. Let A be a Banach algebra. Then the following conditions on A are equivalent:

(i) each homomorphism from A with finite-dimensional range is continuous;

(ii) each cofinite ideal in A is closed;

(iii) I^2 is closed and cofinite in A for each closed, cofinite ideal I of A.

<u>Proof</u>. It is clear that (ii) implies (i).

We show that (i) implies (ii). Suppose that (i) holds, but that (ii) is false. Let I be a cofinite ideal in A which is not closed. By our preliminary remark, A/I is a Banach algebra. It follows that the quotient map from A onto A/I is a discontinuous homomorphism from A onto a finite-dimensional Banach algebra. Thus, we have the required contradiction.

We next show that (iii) implies (ii). Let I be a cofinite ideal in A, and let J be the closed, cofinite ideal constructed in Lemma 2.1. By repeated application of hypothesis (iii), we see that J^{2^k} is closed and cofinite for each $k \in \mathbb{N}$. But, for k sufficiently large, $J^{2^k} \subset I$, and this implies that I is closed in A. Thus, (ii) holds.

Finally, we show that (ii) implies (iii). To obtain a contradiction, assume that (ii) holds, and that there is a closed, cofinite ideal I in A such that I^2 is either not closed or not cofinite in A. By (ii), I^2 is not cofinite, and so it follows from Lemma 2.2 that there is a closed, cofinite ideal J in A such that A/J is semi-simple and such that J^2 is not cofinite in A.

If $J = A$, take K to be a cofinite linear subspace of A which contains A^2 and which is not closed. Then K is an ideal in A, and hence K is a cofinite ideal of A which is not closed, a contradiction of (ii). Thus, we can suppose that $J \neq A$.

The algebra A/J is a non-zero, finite-dimensional, semi-simple algebra: we shall apply the Wedderburn-Artin theorem [6, §4.4] to this algebra. First, A/J has an identity, say e. Let $X = J/J^2$. Then X is an A/J-bimodule, and X may be decomposed as the direct sum of submodules

$$X = (1-e)\cdot X\cdot(1-e) \oplus (1-e)\cdot X\cdot e \oplus e\cdot X\cdot(1-e) \oplus e\cdot X\cdot e,$$

where 1 is the identity operator on X. Since X is infinite-dimensional, at least one of these four direct summands must also be infinite-dimensional.

Suppose that $(1-e)\cdot X\cdot(1-e)$ is infinite-dimensional. Then $Y=\rho^{-1}((1-e)\cdot X\cdot e\oplus e\cdot X\cdot(1-e)\oplus e\cdot X\cdot e)$ has infinite codimension in J, where $\rho: J\to X$ is the projection. Clearly, $(1-e)\cdot\rho(AJ)=\rho(JA)\cdot(1-e)=\{0\}$, and so $AJ\cup JA\subset Y$. Let K be a subspace of J which contains Y, which is cofinite in J, and which is not closed. Then again K is a cofinite ideal in A which is not closed, a contradiction of (ii).

Suppose now that $e\cdot X\cdot(1-e)$ is infinite-dimensional. Let $e=\sum_{j=1}^m e_j$, where the e_j's are orthogonal minimal central idempotents in A/J. Then we may decompose $e\cdot X\cdot(1-e)$ as a direct sum of submodules,

$$e\cdot X\cdot(1-e)=\bigoplus_{j=1}^{m} e_j\cdot X\cdot(1-e),$$

where at least one of the direct summands, say $e_1\cdot X\cdot(1-e)$, is infinite dimensional. Since e_1 is a minimal idempotent, $e_1(A/J)$ is isomorphic to $M_r(\mathbb{C})$, the algebra of $r\times r$-matrices over $\mathbb{C}$, for some $r\in\mathbb{N}$. Let $\{E_{ij}\}$ be a system of matrix units for $e_1(A/J)$, so that $\sum E_{ii}=e_1$ and $E_{ij}E_{pq}=\delta_{jp}E_{iq}$. Then

$$e_1\cdot X\cdot(1-e)=\bigoplus_{i=1}^{r} E_{ii}\cdot X\cdot(1-e),$$

and we may suppose that $E_{11}\cdot X\cdot(1-e)$ is infinite-dimensional.

Let V be a subspace of codimension one in $E_{11}\cdot X\cdot(1-e)$. Since $V\cdot(A/J)=\{0\}$, $(A/J)\cdot V$ is a submodule of $e_1\cdot X\cdot(1-e)$. But $(A/J)\cdot V=\sum E_{i1}\cdot V$, and so $(A/J)\cdot V$ has codimension r in $e_1\cdot X\cdot(1-e)$, noting that E_{i1} is an isomorphism of $E_{11}\cdot X\cdot(1-e)$ onto $E_{ii}\cdot X\cdot(1-e)$. Let

$$W=(A/J)\cdot V\oplus\bigoplus_{j=2}^{m} e_j\cdot X\cdot(1-e))\oplus(1-e)\cdot X\cdot(1-e)\oplus(1-e)\cdot X\cdot e\oplus e\cdot X\cdot e\,.$$

Then W is a submodule with codimension r in X, and so $\rho^{-1}(W)$ is a cofinite ideal in A, where again $\rho: J\to X$ is the projection. Let $K=\rho^{-1}(W)$. Because $E_{11}\cdot X\cdot(1-e)$ is infinite-dimensional, it is clear that we may choose the subspace V so that K is not closed in A. Thus, K is a cofinite ideal in A which is not closed, again a contradiction.

The case when $(1-e)\cdot X\cdot e$ is infinite-dimensional may be treated in a similar way.

Finally, suppose that $e\cdot X\cdot e$ is infinite-dimensional. Since

$$e\cdot X\cdot e=\bigoplus_{p,q=1}^{m} e_p\cdot X\cdot e_q\,,$$

we can choose p, q so that $e_p\cdot X\cdot e_q$ is infinite-dimensional. Let $\{E_{ij}^{(p)}\}$

and $\{E_{ij}^{(q)}\}$ be systems of matrix units for $e_p(A/J)$ and $e_q(A/J)$, respectively, and choose i, j so that $E_{ii}^{(p)} \cdot X \cdot E_{jj}^{(q)}$ is infinite-dimensional. Let V be a subspace of codimension one in $E_{ii}^{(p)} \cdot X \cdot E_{jj}^{(q)}$. Then $(A/J) \cdot V \cdot (A/J) = \sum_{u,v} E_{ui}^{(p)} \cdot V \cdot E_{jv}^{(q)}$ is a submodule of finite codimension in $e_p \cdot X \cdot e_q$. As before, we can use this fact to choose a cofinite ideal K in A which is not closed, a contradiction.

Thus, we have obtained a contradiction in each case, and so we may conclude that (ii) does imply (iii).

It is interesting to consider whether or not the assertion of Theorem 2.3 still holds when conditions (ii) and (iii) are placed only on the modular ideals of A. That is, we consider whether the following statements are equivalent for a Banach algebra A:

(i)' each homomorphism from A onto a finite-dimensional algebra with an identity is continuous;

(ii)' each modular cofinite ideal is closed;

(iii)' I^2 is closed and cofinite for each closed, cofinite, modular ideal I of A.

Once again, it is clear that (i)' and (ii)' are equivalent.

It may also be shown that (iii)' implies (ii)'. The above proof needs to be modified because (in the notation of that proof), although we can apply (iii)' to show that J^2 is a closed and cofinite ideal in A, it may not be a modular ideal, and so (iii)' cannot be applied a second time to show that J^4 is closed and cofinite. However, $\pi^{-1}(R^2) \supset J^2$, and so $\pi^{-1}(R^2)$ is a closed, cofinite, modular ideal in A. By (iii)', $(\pi^{-1}(R^2))^2$ is a closed, cofinite ideal in A. Since $\pi^{-1}(R^4) \supset (\pi^{-1}(R^2))^2$, it follows that $\pi^{-1}(R^4)$ is a closed, cofinite, modular ideal in A. Continuing this argument, we find that $\pi^{-1}(R^{2^k})$ is a closed, cofinite, modular ideal for each k. For k sufficiently large, $\pi^{-1}(R^{2^k}) = I$, and so I is closed, as required.

However, it is not true that (ii)' and (iii)' are equivalent for all Banach algebras A, as the following example shows. Let X be an infinite-dimensional Banach space, and put $A = \mathbb{C} \oplus X$. Define a norm on A by $\|(\lambda,x)\| = |\lambda| + \|x\|$, and define a product on A by setting $(\lambda,x)(\mu,y) = (\lambda\mu,0)$. Then A is a Banach algebra, and X is the only modular ideal of A. Now X is closed in A, and so A satisfies (ii)'. However, $X^2 = \{0\}$, which has infinite codimension in A, and so A does not satisfy (iii)'.

We next consider whether or not the conditions of Theorem 2.3 are also equivalent to the condition that each derivation from the Banach algebra A into a finite-dimensional Banach A-bimodule is continuous.

Let X be a finite-dimensional vector space which is an A-bimodule. A derivation is a linear map $D : A \to X$ such that $D(ab) = Da \cdot b + a \cdot Db$ $(a,b \in A)$. The bimodule X is a Banach bimodule if the bilinear maps $(a,x) \mapsto a \cdot x$ and $(a,x) \mapsto x \cdot a$, $A \times X \to X$, are both continuous. Since X is a finite-dimensional space, to show that these maps are continuous it suffices to show that the maps $a \mapsto a \cdot x$ and $a \mapsto x \cdot a$ are continuous for each $x \in X$.

Let A be a Banach algebra, and consider the condition:

(iv) each derivation from A into a finite-dimensional Banach A-bimodule is continuous.

THEOREM 2.4. Let A be a Banach algebra. Then the conditions (i)-(iii), above, each imply that (iv) holds.

Proof. Let D be a derivation from A into a finite-dimensional Banach A-bimodule X. Let $I = \{a \in A : a \cdot X = X \cdot a = \{0\}\}$. Then I is a closed, cofinite ideal in A. Define a product and a norm on $B = (A/I) \oplus X$ by setting

$$(c_1,x_1)(c_2,x_2) = (c_1c_2, c_1 \cdot x_2 + x_1 \cdot c_2),$$
$$\|(c,x)\| = \|c\| + \|x\| .$$

Then B is a finite-dimensional algebra, and the map $a \mapsto (a+I, D(a))$, $A \to B$, is a homomorphism. Clearly, it is continuous if and only if D is continuous. Thus, if condition (i) holds, then condition (iv) holds.

We do not know whether or not condition (iv) is always equivalent to conditions (i)-(iii), but we can make some progress.

Let A be a Banach algebra which does not satisfy condition (iii). By Lemma 2.2, there is a closed, cofinite ideal J in A such that A/J is semi-simple and such that either (α) J^2 is not cofinite, or (β) J^n is cofinite for $n \in \mathbb{N}$, and J^k is not closed for some $k \geq 2$. In the latter case, we can suppose that k is the least number for which J^k is not closed.

Suppose first that either alternative (α) holds, or that alternative (β) holds with $k=2$. Then we have seen in the course of the proof of Theorem 2.3 that, if alternative (α) holds, then there is a cofinite ideal K in A such that K is not closed and such that $J^2 \subset K \subset J$. If alternative (β) holds with $k=2$, then we obtain an ideal K with the same properties by taking $K = J^2$. The algebra A/K is finite-dimensional: let its radical be X, say. Since the quotient map $A/K \to (A/K)/(J/K) \cong A/J$ is a homomorphism onto a semi-simple algebra, $X \subset J/K$. But $(J/K)^2 = \{0\}$, and X contains each nilpotent ideal of A/K, so that $X = J/K$. Let $\pi : A \to A/K$ be the quotient map.

By Wedderburn's principal theorem [6, p. 374], there is a subalgebra, say B, of A/K such that

$$A/K = B \oplus X, \tag{1}$$

a vector space direct sum. The multiplication in $B \oplus X$ is given by the formula

$$(b_1,x_1)(b_2,x_2) = (b_1b_2, b_1x_2 + x_1b_2), \tag{2}$$

where we recall that $X^2 = \{0\}$.

The finite-dimensional vector space X is an A-bimodule for the operations

$$a \cdot x = \pi(a)x, \; x \cdot a = x\pi(a) \quad (a \in A, \; x \in X).$$

If $x \in X$, then the kernel of the map $a \mapsto a \cdot x$ contains J, and hence is closed. This shows that the map $a \mapsto a \cdot x$, $A \to X$, is continuous. Similarly, the map $a \mapsto x \cdot a$ is continuous, and so X is a Banach A-bimodule. Let $Q : A/K \to X$ be the projection onto the second coordinate of the direct sum, (1). Then it is easily checked from (2) that $D = Q \circ \pi : A \to X$ is a derivation. Also, $(\ker D) \cap J = K$, and so D is a discontinuous map.

The above calculation proves at least the following theorem.

THEOREM 2.5. Let A be a Banach algebra which satisfies at least one of the following: (a) A is separable; (b) A/I is semi-simple whenever I is a cofinite, closed ideal in A. Then conditions (i)-(iv) on A are equivalent.

There remains the case in which we only know that there is a closed, cofinite ideal J in A such that alternative (β) occurs with $k \geq 3$. We can make a little further progress.

Suppose that $k = 3$, and take $K = J^3$. Then A/K is again a finite-dimensional algebra with radical $X = J/K$, and we have the Wedderburn decomposition $A/K = B \oplus X$ for a subalgebra B of A/K, as above. Then X is a B-bimodule. Since B is a semi-simple algebra, X is a semi-simple bimodule, and so the submodule X^2 of X is a direct summand, say $X = Y \oplus X^2$ for some B-bimodule Y [6, p.119]. Thus

$$A/K = B \oplus Y \oplus X^2,$$

as a vector space direct sum. The multiplication in $B \oplus Y \oplus X^2$ is given by

$$(b_1,y_1,x_1)(b_2,y_2,x_2) = (b_1b_2,\; b_1y_2 + y_1b_2,\; b_1x_2 + y_1y_2 + x_1b_2).$$

Again, the finite-dimensional vector space X is an A-bimodule for the above operations. It is a Banach A-bimodule because the kernels of the maps $a \mapsto a \cdot x$ and $a \mapsto x \cdot a$ contains J^2, and hence are closed. Define $Q : A/K \to X = Y \oplus X^2$

by setting

$$Q((b,y,x)) = (y,2x) \qquad (b \in B,\ y \in Y,\ x \in X^2).$$

Then it is again easily checked that $D = Q \circ \pi : A \to X$ is a derivation. Also, $(\ker D) \cap J^2 = K$, and so D is a discontinuous map.

Unfortunately, we cannot see how to continue this argument to the case in which it is only known that alternative (β) occurs with $k \geq 4$: it is no longer true that A necessarily has the structure of a graded algebra, as easy examples show, but maybe some other trick could circumvent the difficulty.

3. Applications

Let G be a locally compact topological group. Then $L^1(G)$ is the space of (equivalence classes of) measurable functions f on G such that

$$\|f\| = \int_G |f(x)|dx < \infty ,$$

where dx denotes the left-invariant Haar measure on G. The Banach space $L^1(G)$ is a Banach algebra with respect to the convolution multiplication given by $(f \star g)(x) = \int_G f(y)g(y^{-1}x)dy$ (see [5]). It is the group algebra of G.

The group G is amenable if there is an invariant mean on G, that is, there is a continuous linear functional m on $L^\infty(G)$ such that (a) $m(\overline{f}) = \overline{m(f)}$ $(f \in L^\infty(G))$, (b) $m(f) \geq 0$ if $f \geq 0$, (c) $m(1) = 1$, (d) $m({}_xf) = m(f)$ $(x \in G,\ f \in L^\infty(G))$. Here, ${}_xf(y) = f(x^{-1}y)$ $(x,y \in G)$. Compact groups and abelian groups are amenable. Some groups which are not amenable are: $SL(n,\mathbb{R})$ when $n \geq 2$, with its usual topology; $SL(n,\mathbb{R})$ and $U(n,\mathbb{C})$ when $n \geq 2$, with the discrete topology; the free group on two or more generators.

The group algebra of a group G always has at least one closed ideal with codimension one. This is the augmentation ideal $I_0(G)$:

$$I_0(G) = \left\{ f \in L^1(G) : \int_G f(x)dx = 0 \right\}.$$

It is shown in [10] that G is amenable if and only if $I_0(G)$ has a (two-sided) bounded approximate identity. This result is extended in one direction in [7], where it is shown that, if G is amenable, then each closed, cofinite ideal in $L^1(G)$ has a bounded approximate identity, and it is extended in another direction in [12], where it is shown that, if G is not amenable, then no proper, closed, cofinite ideal in $L^1(G)$ has a bounded approximate identity.

We now give some applications of our above results to C*-algebras and to group algebras.

LEMMA 3.1. Let A be a C*-algebra or a group algebra. If I is a closed, cofinite ideal in A, then A/I is semi-simple.

Proof. This is a well-known, elementary fact if A is a C*-algebra (e.g., [9, 1.5.5]).

The result may also be well-known in the case that A is a group algebra, but we are unable to give a published reference, and so we sketch a proof.

Let I be a closed, cofinite ideal in a group algebra $L^1(G)$. Since I is translation-invariant, the left regular representation of G on $L^1(G)$ induces a continuous, bounded representation, say ρ, of G on $L^1(G)/I$. Since ρ is bounded and since $L^1(G)/I$ is finite-dimensional, there is a sesquilinear form on $L^1(G)/I$ with respect to which ρ is a unitary representation.

The unitary representation ρ determines a representation $\tilde{\rho}$ of $L^1(G)$ on $L^1(G)/I$ in the usual way, and it may be shown that $\ker \tilde{\rho} = I$: the proof of this uses the fact that $L^1(G)$ always has a bounded approximate identity. Hence, $L^1(G)/I \cong \tilde{\rho}(L^1(G))$. The result now follows because $\tilde{\rho}(L^1(G))$ is generated by the group of unitary operators $\rho(G)$ on $L^1(G)/I$, and so it is isomorphic to a C*-algebra.

Full details of the above result are given in [14].

The following result is now immediate from Theorem 2.5.

THEOREM 3.2. Let A be a C*-algebra or a group algebra. Then conditions (i)-(iv), above, are equivalent for A.

Hence, if A is a C*-algebra or a group algebra, then we may resolve the question of the continuity of finite-dimensional representations of A and of derivations into finite-dimensional Banach A-bimodules by determing I^2 for I a closed, cofinite ideal in A. It is not difficult to determine I^2 when A is a C*-algebra or the group algebra of an amenable group because in these cases each closed, cofinite ideal has a bounded approximate identity, and so it follows from Cohen's factorization theorem [1, 11.11] that $I^2 = I$. Thus, we have the following result.

THEOREM 3.3. Let A be either a C*-algebra or the group algebra of an amenable group. Then each homomorphism with finite-dimensional range from A, and each derivation from A into a finite-dimensional Banach A-bimodule is continuous.

If G is a non-amenable group, then no proper, closed, cofinite ideal in $L^1(G)$ has a bounded approximate identity. However, it is still often possible to show that closed, cofinite ideals in $L^1(G)$ are idempotent. For example, if $I = L^1(G)$, then $I^2 = I$ (because $L^1(G)$ has a bounded approximate identity) and, if I has codimension one in $L^1(G)$, then $I^2 = I$ (see [13] or [14]). Furthermore, certain non-amenable groups, such as $SL(n,\mathbb{R})$ and $GL(n,\mathbb{R})$ with either their usual or discrete topologies, have the property that all of their closed, cofinite ideals are idempotent. This is shown in §3 of [11]. Thus, we have shown that these groups also have the property that all finite-dimensional representations of their group algebras are continuous.

There is no known example of a group algebra which contains a closed, cofinite ideal which is not idempotent. It follows from our results that such an algebra would have discontinuous finite-dimensional representations, and that there would be a discontinuous derivation from the algebra into a finite-dimensional bimodule. The group algebras of the free groups may provide such an example: these algebras are discussed in §4 of the following article, [11].

References

[1] F. F. Bonsall and J. Duncan, Complete Normed Algebras, Springer-Verlag, New York, 1973.

[2] H. G. Dales, Automatic continuity: a survey, Bull. London Math. Soc., 10 (1978), 129-183.

[3] __________, The continuity of traces, this volume.

[4] P. G. Dixon, Non-separable Banach algebras whose squares are pathological, J. Functional Analysis, 26 (1977), 190-200.

[5] E. Hewitt and K. A. Ross, Abstract Harmonic Analysis I, Springer Verlag, Berlin, 1963.

[6] N. Jacobson, Basic Algebra II, W. H. Freeman, San Francisco, 1980.

[7] T.-S. Liu, A. van Rooij and J.-K. Wang, Projections and approximate identities for ideals in group algebras, Trans. Amer. Math. Soc., 175 (1973), 469-482.

[8] R. J. Loy, Multilinear mappings and Banach algebras, J. London Math. Soc., (2), 14 (1976), 423-429.

[9] G. K. Pedersen, C*-algebras and their Automorphism Groups, Academic Press, London, 1979.

[10] H. Reiter, Sur certains idéaux dans $L^1(G)$, C. R. Acad. Sci., Paris, 267 (1968), 882-885.

[11] G. A. Willis, The continuity of derivations from group algebras and factorization in cofinite ideals, this volume.

[12] __________, Approximate units in finite codimensional ideals of group algebras, J. London Math. Soc., (2), 26 (1982), 143-154.

[13] __________, Factorization in codimension one ideals of group algebras, to appear.

[14] G. A. Willis, Thesis, University of Newcastle-upon-Tyne, England, 1980.

School of Mathematics
University of Leeds
Leeds, LS2 9JT, England

Department of Mathematics
University of New South Wales
Kensington, 2033 Australia

THE CONTINUITY OF DERIVATIONS FROM GROUP ALGEBRAS AND FACTORIZATION IN COFINITE IDEALS

G. A. Willis

Question 22 in [3] asks for which locally compact groups G it is true that every derivation from $L^1(G)$ to a Banach $L^1(G)$-bimodule is continuous. This paper summarizes some progress made by the author towards answering this question and raises some further problems which need to be solved before it can be answered completely.

The problem of the continuity of derivations from $L^1(G)$ to a finite dimensional $L^1(G)$-bimodule has already been discussed in this volume in [19]. The group algebra $L^1(G)$ of the locally compact group G is defined in that paper, as is the notion of an amenable group which we will require in §§1, 2.

Throughout, G will be a locally compact group. The augmentation ideal of $L^1(G)$ is denoted by $I_0(G)$ and is defined by

$$I_0(G) = \left\{ f \in L^1(G) \mid \int_G f(x)dx = 0 \right\}.$$

It is a closed bi-ideal with codimension one in $L^1(G)$. If A is a Banach algebra, then A^2 denotes the linear span of all the products in A, $c_0(A)$ is the set of all sequences in A which converge to zero and $A.c_0(A)$ is the linear span of all sequences of the form $(ax_n)_{n=1}^{\infty}$, where $a \in A$ and $(x_n)_{n=1}^{\infty} \in c_0(A)$.

In §1 is stated a theorem of N. P. Jewell which gives conditions on a Banach algebra, A, which guarantee that all derivations from A to a Banach A-bimodule are continuous. The later sections discuss the extent to which group algebras satisfy these conditions.

Most of the results described here are contained in my Ph.D. thesis which was prepared and written under the supervision of Professor B. E. Johnson. I would like to thank him for his supervision and for stimulating my interest in these problems. I am also grateful to John Rose and Oliver King for many discussions about discrete groups and to Christopher Meaney and Tony Dooley for helpful discussions about Lie groups.

1. Question 22 was motivated by a theorem of J. R. Ringrose, in [12], which states that every derivation from a C*-algebra, A, to an A-bimodule is continuous. The same argument as used by Ringrose may be used to show that if G is abelian, then every derivation from $L^1(G)$ to an $L^1(G)$-bimodule is continuous.

A generalization of that argument was given by N. P. Jewell in [7] to prove the following theorem.

THEOREM 1. Let A be a Banach algebra satisfying the following conditions:

(i) if I is a closed bi-ideal with infinite codimension in A, then there are sequences (a_n), (b_n) in A such that $b_n a_1 \dots a_{n-1} \notin I$ but $b_n a_1 \dots a_n \in I$ $(n \geq 2)$;

(ii) each closed, cofinite bi-ideal in A has a bounded left (or right) approximate identity.

Then every derivation from A to a Banach A-bimodule is continuous.

Every C*-algebra satisfies both these conditions. They are also satisfied by abelian group algebras as may be deduced from Theorems 7.2.4 (Wiener's Tauberian theorem) and 2.6.2 of [13]. However, not every group algebra satisfies them because $L^1(G)$ does not satisfy (ii) if G is not amenable. It is not very clear when $L^1(G)$ will satisfy (i) but there are ad hoc methods for showing that it does when something is known about the structure of G.

Let I be a closed, cofinite bi-ideal in a Banach algebra A and consider the following three conditions on I:

(a) I has a left bounded approximate identity;

(b) $I \cdot c_0(I) = c_0(I)$;

(c) $I^2 = I$.

It is clear that (b) implies (c). That (a) implies (b) follows from Cohen's factorization theorem for modules (see Corollary 11.12 of [1]).

Condition (b) has been introduced because it will suffice in the proof of Theorem 1 that the closed, cofinite bi-ideals of A satisfy (b). It is not necessary that they should have a bounded approximate identity. If A is a group algebra, then it is necessary that the closed, cofinite bi-ideals in A satisfy (c). This is because it is shown in [19] that if $L^1(G)$ has a closed, cofinite bi-ideal I with $I^2 \neq I$, then there is a discontinuous derivation from $L^1(G)$ to a finite-dimensional bimodule.

2. Quite a lot can be said about whether $L^1(G)$ satisfies conditions (i) and (ii) of Theorem 1 when G is an amenable locally compact group. For a start, a theorem of Liu, van Rooij and Wang in [8] implies that if G is amenable, then $L^1(G)$ satisfies (ii). They showed that if G is amenable and I is a closed, right ideal such that there is a bounded projection, P, of $L^1(G)$ onto I, then I has a left bounded approximate identity with bound less than or equal to $\|P\|$. Every finite-codimensional subspace of a Banach space has a bounded projection onto it and so, in particular, every cofinite, closed bi-ideal in $L^1(G)$ has a left bounded approximate identity. On the other hand, it was shown by H. Reiter in [10], which preceded the paper of Liu, van Rooij and Wang, that G is amenable if

and only if $I_0(G)$ has a left (or right) bounded approximate identity. Taking this a little further, it is shown in [18] (Theorem 5.2) and in Chapter 2 of [17] that if G is not amenable, then no cofinite, closed right ideal in $L^1(G)$ has a left bounded approximate identity.

In Chapter 4 of [17] it is shown that for many amenable groups, G, $L^1(G)$ also satisfies condition (i). The amenable groups for which this can be shown include soluble, compact, and locally finite groups as well as groups which are a compact extension of an abelian group or an abelian extension of a compact group. The methods used involve the application of some recent theorems from non-commutative harmonic analysis, in particular Proposition 3.1 from [9] and the main theorem of [4]. These theorems play the same role as Wiener's Tauberian theorem plays in the proof of the case when G is abelian.

There are also two combinatorial lemmas in [17]. One of them, Lemma 4.5.3, deals with direct limits of groups and may be used, for example, to show that if G is a locally finite group then $L^1(G)$ satisfies condition (i). The other, Lemma 4.4.4, may be improved to give the following theorem, which we state and prove for discrete groups only. It is possible to make a corresponding assertion about arbitrary locally compact groups. The proof will be the same as that given below with an additional approximate identity argument.

THEOREM 2. Let G be a discrete group and suppose that $G = H_1H_2 \dots H_n$, where $H_1, H_2, \dots, H_n$ are subgroups of G. Let I be a closed bi-ideal in $\ell^1(G)$ such that $I \cap \ell^1(H_k)$ has finite codimension in $\ell^1(H_k)$ for $k = 1,2,\dots,n$. Then I has finite codimension in $\ell^1(G)$.

<u>Proof</u>. For each $k = 1,2,\dots,n$, $H_1H_2 \dots H_k$ is a subset, although not necessarily a subgroup, of G and so $\ell^1(H_1H_2 \dots H_k)$ may be regarded as a subspace, but not necessarily as a subalgebra, of $\ell^1(G)$. We will show, by induction on k, that $I \cap \ell^1(H_1 \dots H_k)$ has finite codimension in $\ell^1(H_1 \dots H_k)$ for each k. The theorem will then follow by taking k equal to n.

The case when k equals 1 is given. Suppose that $I \cap \ell^1(H_1H_2 \dots H_k)$ has finite codimension in $\ell^1(H_1H_2 \dots H_k)$. Let T be an element of $\ell^\infty(H_1 \dots H_kH_{k+1})$ which annihilates $\ell^1(H_1 \dots H_kH_{k+1}) \cap I$, where $\ell^\infty(H_1 \dots H_kH_{k+1})$ is identified with the dual space of $\ell^1(H_1 \dots H_kH_{k+1})$ by putting

$$T(f) = \sum_x T(x)f(x) \quad (f \in \ell^1(H_1 \dots H_kH_{k+1})) .$$

For each x in $H_1 \dots H_k$ define S_x in $\ell^\infty(H_{k+1})$ by

$$S_x(y) = T(xy) \qquad (y \in H_{k+1}) .$$

Then, for each f in $\ell^1(H_{k+1})$,

$$S_x(f) = \sum_{y \in H_{k+1}} S_x(y)f(y) = \sum_{y \in H_{k+1}} T(xy)f(y)$$

$$= \sum_{y \in H_{k+1}} T(xy)(\bar{x} * f)(xy),$$

where $\bar{x}$ denotes the point mass at x,

$$= \sum_{z \in H_1 \dots H_k H_{k+1}} T(z)(\bar{x} * f)(z),$$

because the support of $\bar{x} * f$ is contained in xH_{k+1}. Hence $S_x(f) = T(\bar{x} * f)$ for every f in $\ell^1(H_{k+1})$. Since I is a closed bi-ideal in $\ell^1(G)$ it now follows that S_x annihilates $I \cap \ell^1(H_{k+1})$ for every x in $H_1 \dots H_k$. By hypothesis $I \cap \ell^1(H_{k+1})$ has finite codimension in $\ell^1(H_{k+1})$. Therefore there is a finite set of elements, $y_1, \dots, y_M$ in H_{k+1} such that, for every x in $H_1 \dots H_k$, S_x is uniquely determined by its values at $y_1, \dots, y_M$.

Now, for each y in H_{k+1} define R_y in $\ell^\infty(H_1 \dots H_k)$ by

$$R_y(x) = T(xy) \quad (x \in H_1 \dots H_k).$$

Then, by applying the induction hypothesis, it may be shown in the same way that there is a finite set of elements, $x_1, \dots, x_N$ in $H_1 H_2 \dots H_k$ such that, for every y in H_{k+1}, R_y is uniquely determined by its values at $x_1, \dots, x_N$.

Now let x in $H_1 \dots H_k$ and y in H_{k+1} be arbitrary. Then $T(xy) = S_x(y)$, which is uniquely determined by the values of $S_x(y_1), \dots, S_x(y_M)$. That is $T(xy)$ is uniquely determined by $T(x y_1), \dots, T(xy_M)$. For each j between 1 and M, $T(xy_j) = R_{y_j}(x)$, which is uniquely determined by the values of $R_{y_j}(x_1), \dots, R_{y_j}(x_N)$. Hence, $T(xy)$ is determined by the values of T on $\{x_i y_j \mid i = 1,2,\dots,N, j = 1,2,\dots,M\}$, for every xy in $H_1 \dots H_k H_{k+1}$ and every T which annihilates $I \cap \ell^1(H_1 \dots H_k H_{k+1})$. It follows that the set of all continuous linear functionals on $\ell^1(H_1 \dots H_k H_{k+1})$ which annihilate $I \cap \ell^1(H_1 \dots H_k H_{k+1})$ is finite dimensional. Therefore, by the Hahn-Banach theorem, $I \cap \ell^1(H_1 \dots H_k H_{k+1})$ has finite codimension in $\ell^1(H_1 \dots H_k H_{k+1})$.

These two combinatorial lemmas are used together in Chapter 4 of [17] to show that if G is soluble or is an abelian extension of a compact group then

$L^1(G)$ satisfies condition (i).

The results in [17] suggest that $L^1(G)$ will satisfy condition (i) whenever G is amenable but, unfortunately, the methods used do not suggest how to prove it. All of the methods of [17] exploit some structure of the group under consideration but, as far as I am aware, there are no structure theorems available for an arbitrary amenable group. It is necessary to introduce some notation before making a conjecture which, if true, would provide such a structure theorem.

Let H be a closed subgroup of G. Then $L^1(H)$ may be embedded as a subalgebra of $M(G)$ by defining

$$f(\varphi) = \int_H f(x)\varphi(x)dx \qquad (f \in L^1(H),\ \varphi \in C_0(G)).$$

Now for each closed bi-ideal, I, in $L^1(G)$ we define $I \wedge L^1(H) = \{f \in L^1(H) \mid f * L^1(G) \subseteq I\}$, where the '$*$' means convolution in $M(G)$. Then $I \wedge L^1(H)$ is a closed bi-ideal in $L^1(H)$. If G is a discrete group then $L^1(H)$ will be a subalgebra of $L^1(G)$ and $I \wedge L^1(H)$ will be just $I \cap L^1(H)$. The following problem is suggested by Theorem 2.

<u>Problem 1</u>. Let G be an amenable locally compact group and I be a closed bi-ideal in $L^1(G)$. Suppose that $I \wedge L^1(H)$ has finite codimension in $L^1(H)$ for every closed, abelian subgroup, H, of G. Does it follow that I has finite codimension in $L^1(G)$?

If this question could be answered affirmatively then $L^1(G)$ would satisfy condition (i) whenever G was amenable. It would then follow that for an amenable group G, every derivation from $L^1(G)$ to an $L^1(G)$-bimodule would be continuous.

If H is a finite subgroup of G, then, clearly, $I \wedge L^1(H)$ is cofinite in $L^1(H)$ for every bi-ideal in $L^1(G)$. Hence, if G is an infinite group such that all of its abelian subgroups are finite, then (0) will be a closed bi-ideal with infinite codimension in $L^1(G)$ but $(0) \wedge L^1(H)$ $(= (0))$ will be cofinite in $L^1(H)$ for every abelian H. Thus a second consequence of the above conjecture would be that every infinite amenable group should have an infinite abelian subgroup. This assertion does hold for certain amenable groups. For example, it is not difficult to show that every infinite soluble group has an infinite abelian subgroup and it is Theorem 3.43 in [14] that every infinite locally finite group has an infinite abelian subgroup.

3. We will say that a group G is <u>factorizable</u>, or is an <u>[F]-group</u>, if it has a finite number of abelian subgroups, $H_1, \ldots, H_n$, such that

$$G = H_1 H_2 \ldots H_n .$$

Note that this does not mean just that G is generated by the subgroups $H_1,\dots,H_n$ but that every element x in G has the form $x = y_1y_2\dots y_n$, where y_k is in H_k.

One reason why [F]-groups are interesting in our context is that if G is an [F]-group, then $L^1(G)$ satisfies condition (i). For discrete [F]-groups, this is so because if I is a closed bi-ideal with infinite codimension in $\ell^1(G)$ then, by Theorem 2, $I \cap \ell^1(H_k)$ has infinite codimension in $\ell^1(H_k)$ for at least one of the abelian subgroups, H_k, of G. Hence, there are sequences (a_n) and (b_n) in $\ell^1(H_k)$ such that

$$b_n a_1 \dots a_{n-1} \notin I \cap \ell^1(H_k) \quad \text{but} \quad b_n a_1 \dots a_n \in I \cap \ell^1(H_k).$$

Since $\ell^1(H_k) \subseteq \ell^1(G)$, (a_n) and (b_n) are also the required sequences for I. A similar argument will prove the same result for non-discrete [F]-groups.

It can be shown that every connected semisimple Lie group is an [F]-group. The proof uses only basic facts about Lie groups and is essentially the same as the proof of the special case when the group is $SL(n,\mathbb{R})$ given as Lemma 5.1.2 in [17]. It follows that many non-amenable groups such as $SL(n,\mathbb{R})$ and $SU(n,\mathbb{C})$ (with its discrete topology) are [F]-groups.

The other reason why [F]-groups are interesting is that if G is an [F]-group and I is a closed cofinite bi-ideal in $L^1(G)$ then $I \cdot c_0(I) = c_0(I)$. The following lemma will be required for the proof of this fact.

LEMMA 3. Let G be an [F]-group and C be a countable subset of G. Then there is a countable subgroup, H, of G which contains C and is an [F]-group.

Proof. We have that $G = H_1H_2\dots H_n$, where H_k is an abelian subgroup of G for each k. Hence, for each countable subset, B, of G there is a countable subset, $B^\#$, of G such that every element x in B has the form $x = y_1y_2\dots y_n$, where y_k is in $H_k \cap B^\#$. (To obtain such a set, for each x in B choose particular elements $y_1,\dots,y_n$ such that $x = y_1y_2\dots y_n$ and y_k is in H_k and define $B^\#$ to be the set of all the y_k's chosen. Then $B^\#$ is a countable union of finite sets and so is countable.)

Let B_1 be the subgroup of G generated by C. Then B_1 is countable and so $B_1^\#$ is also countable. Call this set C_1. Now define subsets B_i and C_i of G inductively by putting B_{i+1} equal to the subgroup of G generated by C_i and C_i equal to $B_i^\#$. Then each of the subgroups, B_i, is countable and $B_i \subseteq B_{i+1}$ and $C_i \subseteq B_{i+1}$ for each i.

Hence, if we put $H = \bigcup_i B_i$, then H is a countable subgroup of G. Now each x in H is in B_i for some i and so there are $y_1,\dots,y_n$ in

$C_i \cap H_k$, $k = 1,\dots,n$ such that $x = y_1 y_2 \cdots y_n$. Since $C_i \subseteq B_{i+1} \subseteq H$ it follows that y_k is in $H \cap H_k$ for $k = 1,\dots,n$. Therefore

$$H = (H \cap H_1)(H \cap H_2)\cdots(H \cap H_n),$$

and so H is an [F]-group.

THEOREM 4. Let G be a discrete [F]-group and I be a closed, cofinite bi-ideal in $\ell^1(G)$. Then

$$I \cdot c_0(I) = c_0(I).$$

<u>Proof</u>. We have $G = H_1 H_2 \cdots H_n$, where $H_1, H_2, \dots, H_n$ are abelian subgroups of G. For each k between 1 and n, $\ell^1(H_k) \cap I$ is a closed, cofinite bi-ideal in $\ell^1(H_k)$ and so we may choose elements $z_{k,i}$, $i = 1,2,\dots,N_k$, in $\ell^1(H_k)$ such that

$$\ell^1(H_k) = \ell^1(H_k) \cap I + \operatorname{span}\{z_{k,i} \mid i = 1,2,\dots,N_k\}.$$

These $z_{k,i}$'s which are chosen will remain fixed for the rest of the proof.

We will show to begin with that $I^2 = I$. Choose f in I. For convenience, f will be written in the form

$$f = \sum_{x \in G} f(x)\bar{x},$$

where $\bar{x}$ is the element of $\ell^1(G)$ which takes the value one at x and is zero elsewhere.

Now choose a representative for each of the right cosets in G of H_1 of the form $y = y_2 \cdots y_n$, where $y_k \in H_k$, $k = 2,\dots,n$. This is possible because $G = H_1 H_2 \cdots H_k$. Let Y be the set of all coset representatives chosen. For each such representative y, $\sum_{z \in H_1} f(zy)\bar{z}$ is in $\ell^1(H_1)$ and so there are scalars $\lambda_i(y)$, $i = 1,\dots,N_1$ such that

$$\sum_{z \in H_1} f(zy)\bar{z} - \sum_{i=1}^{N_1} \lambda_i(y) z_{1,i} \in I \cap \ell^1(H_1).$$

Then,

$$f = \sum_{x \in G} f(x)\bar{x} = \sum_{y \in Y} \left(\sum_{z \in H_1} f(zy)\bar{z} - \sum_{i=1}^{N_1} \lambda_i(y)\bar{z}_{1,i} \right) * \bar{y} + \sum_{i=1}^{N_1} \bar{z}_{1,i} * \left(\sum_{y \in Y} \lambda_i(y)\bar{y} \right),$$

where

$$\sum_{y \in Y} \left(\sum_{z \in H_1} f(zy)\bar{z} - \sum_{i=1}^{N_1} \lambda_i(y)\bar{z}_{1,i} \right) * \bar{y}$$

is in $[(I \cap \ell^1(H_1)) * \ell^1(G)]^-$. Since $I \cap \ell^1(H_1)$ is a closed, cofinite bi-ideal in $\ell^1(H_1)$ and since H_1 is abelian, $I \cap \ell^1(H_1)$ has a bounded approximate identity and so, by the module version of Cohen's factorization theorem, there is an element h in $I \cap \ell^1(H_1)$ $(\subseteq I)$ and an element g in $[(I \cap \ell^1(H_1)) * \ell^1(G)]^-$ $(\subseteq I)$ such that

$$\sum_{y \in Y} \left(\sum_{z \in H_1} f(zy)\bar{z} - \sum_{i=1}^{N_1} \lambda_i(y)\bar{z}_{1,i} \right) * \bar{y} = h * g ,$$

which is in I^2.

The coset representatives, y, chosen above belong to certain right cosets in G of H_2 and it is possible to choose representatives of these cosets of H_2 of the form $w = w_3 \cdots w_n$, where $w_k \in H_k$, $k = 3,\ldots,n$. The same argument as above shows that, for each i,

$$\sum_{y \in Y} \lambda_i(y)\bar{y} = \sum_{j=1}^{N_2} \bar{z}_{2,j} * \left(\sum_{w} \lambda_{i,j}(w)\bar{w} \right) + \text{something in } I^2 .$$

Hence,

$$f = \sum_{i=1}^{N_1} \sum_{j=1}^{N_2} \bar{z}_{1,i} * \bar{z}_{2,j} * \left(\sum_{w} \lambda_{i,j}(w)\bar{w} \right) + \text{something in } I^2 .$$

Repeating this argument for $k = 3,\ldots,n$ we eventually find that f is in

$$I^2 + \operatorname{span}\{\bar{z}_{1,i_1} * \bar{z}_{2,i_2} * \cdots * \bar{z}_{n,i_n} \mid 1 \le i_k \le N_k\}$$

and thus that I^2 has finite codimension in I.

Now let f be in the norm closure of I^2, and let $(f_i)_{i=1}^{\infty}$ be a sequence in I^2 which converges to f. For each i choose particular functions $a_{i,j}, b_{i,j}$, $1 \le j \le M_i$, in I such that $f_i = \sum_{j=1}^{M_i} a_{i,j} * b_{i,j}$. Let C be the union of the supports of f and of all the $a_{i,j}$'s and $b_{i,j}$'s. Then C is countable and so by Lemma 3 there is a countable subgroup, H, of G which contains C and is an [F]-group. Since $I \cap \ell^1(H)$ is a closed, cofinite bi-ideal in $\ell^1(H)$ and since H is an [F]-group, we have as above that $(I \cap \ell^1(H))^2$ has finite codimension in $I \cap \ell^1(H)$. Hence, since H is countable, $\ell^1(H)$ is a separable Banach algebra and it follows from the theorem of [2] that $(I \cap \ell^1(H))^2$ is closed. The $a_{i,j}$'s and $b_{i,j}$'s all lie in $I \cap \ell^1(H)$, and so

$$f = \lim_{i \to \infty} f_i \in \{[I \cap \ell^1(H)]^2\}^- = [I \cap \ell^1(H)]^2 \subseteq I^2 .$$

Therefore I^2 is closed

We have that I^2 is a closed, cofinite bi-ideal in $\ell^1(G)$. Hence by [19] Lemma 3.1, $\ell^1(G)/I^2$ is semisimple. Now I/I^2 is a bi-ideal in $\ell^1(G)/I^2$ and $(I/I^2)^2 = (0)$, which implies that $I/I^2 \subseteq \mathrm{Rad}(\ell^1(G)/I^2)$. Therefore, $I^2 = I$.

Now, to show that $I \cdot c_0(I) = c_0(I)$, let $(f_i)_{i=1}^{\infty}$ be a sequence in I which converges to zero. Then the argument at the beginning of this proof when applied to all the f_i's simultaneously shows that

$(f_i)_{i=1}^{\infty} = (g_i)_{i=1}^{\infty}$ + something in $I \cdot c_0(I)$,

where (g_i) converges to zero and g_i is in

$$I \cap \mathrm{span}\{\bar{z}_{1,i_1} * \bar{z}_{2,i_2} * \cdots * \bar{z}_{n,i_n} \mid 1 \le i_k \le N_k\}$$

for every i. Let $h_1, \ldots, h_m$ be a basis for

$$I \cap \mathrm{span}\{\bar{z}_{1,i_1} * \bar{z}_{2,i_2} * \cdots * \bar{z}_{n,i_n} \mid 1 \le i_k \le N_k\}.$$

Then, $g_i = \sum_{j=1}^{m} \lambda_{i,j} h_j$, where, for each j, $(\lambda_{i,j})_{i=1}^{\infty}$ is a sequence of scalars which converges to zero. Since $I^2 = I$, there are, for each j, elements $a_{j,\ell}$, $b_{j,\ell}$, $\ell = 1, \ldots, L_j$, in I such that,

$$h_j = \sum_{\ell=1}^{L_j} a_{j,\ell} * b_{j,\ell}.$$

Hence,

$$g_i = \sum_{j=1}^{m} \sum_{\ell=1}^{L_j} a_{j,\ell} * (\lambda_{i,j}\, b_{j,\ell}),$$

where, for each j and ℓ, $(\lambda_{i,j}\, b_{j,\ell})_{i=1}^{\infty} \in c_0(I)$. Therefore $(g_i)_{i=1}^{\infty}$ is in $I \cdot c_0(I)$ and so $(f_i)_{i=1}^{\infty}$ is in $I \cdot c_0(I)$ as required.

Theorems 1, 2, and 4 together imply the following.

THEOREM 5. Let G be a discrete [F]-group. Then every derivation from $\ell^1(G)$ to a Banach $\ell^1(G)$-bimodule is continuous.

In some particular cases it is possible to prove Theorem 4, and in fact give an explicit factorization of $c_0(I)$, without appealing to the theorem of Christensen in [2], which is not constructive. For example, if G is a discrete [F]-group and I is the augmentation ideal in $\ell^1(G)$, which has codimension one, then the elements $z_{k,j}$, chosen at the beginning of the proof, may all be taken to be the identity element. The first part of the proof then shows that every sequence $(f_i)_{i=1}^{\infty}$ in I which converges to zero has the form

$$f_i = \sum_{i=1}^{n} a_k * b_{k,i},$$

where n is the number of abelian subgroups, $H_1,\dots,H_n$, of G such that $G = H_1H_2\dots H_n$ and $b_{k,i}$ is in $I_0(H_k) * \ell^1(G)$ for each i and a_k is in $I_0(H_k)$ for each k.

If G is a real special linear group, then it is a theorem of Wigner and von Neumann (see [5], §22.22) that $I_0(G)$ is the only closed, cofinite bi-ideal in $\ell^1(G)$. Hence it is possible to give a constructive proof of Theorem 4 for all the closed, cofinite bi-ideals of the group algebras of real special linear groups.

The special unitary groups, $SU(n,\mathbb{C})$, $n \in \mathbb{N}$, are another class of groups for which a constructive proof may be possible. Such a proof may be obtainable by applying a not very well-known automatic continuity theorem. It is shown in [20] that every homomorphism from one compact semisimple Lie group to another is continuous and so, in particular, every finite-dimensional unitary representation of $SU(n,\mathbb{C})$ is continuous. Now every finite-dimensional unitary representation of G determines a finite-dimensional representation of $\ell^1(G)$, the kernel of which is a closed, cofinite bi-ideal in $\ell^1(G)$ and every closed, cofinite bi-ideal of $\ell^1(G)$ arises in this way (see, for example, [17]). Hence the closed, cofinite bi-ideals in $\ell^1(SU(n,\mathbb{C}))$ correspond to the continuous finite-dimensional, unitary representations of $SU(n,\mathbb{C})$ regarded as a compact Lie group. There is a complete classification of the irreducible, continuous, finite-dimensional, unitary representations of $SU(n,\mathbb{C})$ (for the classification for $SU(2,\mathbb{C})$, see [15], Theorem 4.1). It ought to be possible to use this classification to give a constructive proof of Theorem 4 for $\ell^1(SU(n,\mathbb{C}))$.

Similar results to Theorems 2, 4 and 5 may be proved for separable (not necessarily discrete) groups. Together with some structure theorems for connected groups, these results will show that if G is a connected locally compact group, then every derivation from $L^1(G)$ to an $L^1(G)$-bimodule is continuous. The structure theorems referred to are one by Iwasawa, who showed ([6], Lemma 5.2) that if G is a connected locally compact group, then G has an amenable, closed, connected normal subgroup N such that G/N is isomorphic to the direct sum of a finite number of simple Lie groups, and one by Reiter who showed ([11], Ch. 8, §7.1) that if G is a connected, amenable group, then it is the extension of a soluble group by a compact one.

4. In this final section we will be mainly concerned with the question as to whether there is a closed, cofinite bi-ideal in some group algebra with $I^2 \neq I$. As is shown in [19], such an ideal exists if and only if there is a discontinuous derivation from the group algebra to a finite dimensional bimodule.

Let $\mathbb{F}_n$ denote the free group on n generators and $\mathbb{F}_w$ denote the free group on a countably infinite number of generators. It is easily seen that if there is a discrete group whose group algebra contains a closed, cofinite bi-ideal, I, with $I^2 \neq I$, then there is such an ideal in $\ell^1(\mathbb{F}_w)$. In view of this fact, it is of interest to consider the groups $\mathbb{F}_n$, $n = 2,3,\ldots$ and $\mathbb{F}_w$.

It is shown in Chapter 3 of [17] and is the main theorem of [16] that if G is a locally compact group and I is a closed bi-ideal with codimension one in $L^1(G)$, then $I^2 = I$. In fact it is shown that every element of I is a sum of four products of elements of I. A natural question to ask is whether four is the best number possible in this result and even whether every element of I is a product of elements of I. The first case to be answered is the following.

Problem 2. Is every element of $I_0(\mathbb{F}_2)$ a product of two others? If not, is the set of all products of elements of $I_0(\mathbb{F}_2)$ dense in $I_0(\mathbb{F}_2)$?

For each f in $I_0(\mathbb{F}_2)$, $(f * I_0(\mathbb{F}_2))^-$ is contained in the closure of the products in $I_0(\mathbb{F}_2)$. Hence if the products in $I_0(\mathbb{F}_2)$ are not dense, then every singly generated, closed right ideal in $I_0(\mathbb{F}_2)$ would be proper. Now it may be shown that if a_1 and a_2 generate $\mathbb{F}_2$, then $\bar{e} - \bar{a}_1$ and $\bar{e} - \bar{a}_2$ generate $I_0(\mathbb{F}_2)$ as a closed right ideal over itself, where e is the identity element of $\mathbb{F}_2$ and, for each a in $\mathbb{F}_2$, $\bar{a}$ denotes the point mass at a. Therefore, if the products are not dense in $I_0(\mathbb{F}_2)$, then $I_0(\mathbb{F}_2)$ would be generated, as a closed, right ideal, by two elements but would not be singly generated.

It may also be shown that, if $a_1,\ldots,a_n$ generate $\mathbb{F}_n$, then the closed, right ideal in $I_0(\mathbb{F}_n)$ generated by the n elements $\bar{e} - \bar{a}_1,\ldots,\bar{e} - \bar{a}_n$ is the whole of $I_0(\mathbb{F}_n)$. This fact, along with the discussion in the above paragraph, leads us to ask the next question.

Problem 3. Can $I_0(\mathbb{F}_n)$, regarded as a closed, right ideal in itself, be generated by fewer than n elements?

This problem has a connection with a factorization problem for Banach algebras. It has been asked: under what conditions on a Banach algebra A is it possible to show that, if $A^2 = A$, then $A \cdot c_0(A) = c_0(A)$? Now, as has already been mentioned, $I_0(\mathbb{F}_w)^2 = I_0(\mathbb{F}_w)$, and so we might expect that $I_0(\mathbb{F}_w) \cdot c_0(I_0(\mathbb{F}_w)) = c_0(I_0(\mathbb{F}_w))$.

Problem 4. Does $I_0(\mathbb{F}_w) \cdot c_0(I_0(\mathbb{F}_w)) = c_0(I_0(\mathbb{F}_w))$?

Suppose that this is so. Then $(k^{-1}(\bar{e} - \bar{a}_k))_{k=1}^{\infty}$ (where $a_1, a_2, \ldots$ are generators of $\mathbb{F}_w$) is in $c_0(I_0(\mathbb{F}_w))$ and so there are elements $b_1,\ldots,b_m$ in $I_0(\mathbb{F}_w)$ and sequences $(c_{1,k})_{k=1}^{\infty},\ldots,(c_{m,k})_{k=1}^{\infty}$ in $c_0(I_0(\mathbb{F}_w))$ such that

$$k^{-1}(\bar{e} - \bar{a}_k) = \sum_{i=1}^{m} b_i * c_{j,k}, \qquad k = 1,2,3,\ldots .$$

Hence, since $\bar{e} - \bar{a}_1$, $\bar{e} - \bar{a}_2,\ldots$ generate $I_0(\mathbb{F}_w)$, the closed, right ideal in

$I_0(\mathbb{F}_w)$ generated by $b_1, b_2, \ldots, b_m$ is the whole of $I_0(\mathbb{F}_w)$. Each of the algebras $I_0(\mathbb{F}_n)$, $n = 1,2,3,\ldots$, is a quotient of $I_0(\mathbb{F}_w)$ and so, for every n, $I_0(\mathbb{F}_n)$ is generated by m elements. Thus, if the answer to Problem 3 is 'no', then the answer to Problem 4 will also be 'no', even though $I_0(\mathbb{F}_w)^2 = I_0(\mathbb{F}_w)$.

There is another question which is connected with Problem 4. We already know that if I is a closed, cofinite bi-ideal with codimension one in a group algebra, then $I^2 = I$. It may also be shown that if G is a finitely generated discrete group and I is a closed, cofinite bi-ideal with codimension two in $\ell^1(G)$, then $I^2 = I$. However, it is not known whether the same is true for groups which are not finitely generated.

<u>Problem 5</u>. Does $I^2 = I$ whenever I is a closed bi-ideal with codimension two in $\ell^1(\mathbb{F}_w)$?

If there is a closed bi-ideal, I, with codimension two in $\ell^1(\mathbb{F}_w)$ such that $I^2 \neq I$, then the construction given in [19] (this volume) yields a discontinuous derivation, D, from $\ell^1(\mathbb{F}_w)$ into a one-dimensional Banach $\ell^1(\mathbb{F}_w)$-bimodule. The one dimensional bimodule is defined by putting

$$f \cdot z = \varphi_1(f)z \quad \text{and} \quad z \cdot f = \varphi_2(f)z \qquad (z \in \mathbb{C},\ f \in \ell^1(\mathbb{F}_w)),$$

where φ_1 and φ_2 are the uniquely determined multiplicative linear functionals on $\ell^1(\mathbb{F}_w)$ such that $I = \ker \varphi_1 \cap \ker \varphi_2$. Thus D will be a discontinuous linear functional such that

$$D(f * g) = \varphi_1(f)D(g) + \varphi_2(g)D(f) \qquad (f,g \in \ell^1(\mathbb{F}_w)).$$

On the other hand, if $I_0(\mathbb{F}_w) \cdot c_0(I_0(\mathbb{F}_w)) = c_0(I_0(\mathbb{F}_w))$, then $J \cdot c_0(J) = c_0(J)$ for every closed bi-ideal, J, with codimension one in $\ell^1(\mathbb{F}_w)$. (This is so because, if J is a codimension one bi-ideal in $\ell^1(\mathbb{F}_w)$, then there is an automorphism of $\ell^1(\mathbb{F}_w)$ which maps J onto $I_0(\mathbb{F}_w)$, as is shown in Chapter 3 of [17] and in [18]). In particular, $\ker \varphi_1 \cdot c_0(\ker \varphi_1) = c_0(\ker \varphi_1)$. Thus, if $(f_k)_{k=1}^{\infty}$ is in $c_0(\ker \varphi_1)$, then there are elements $b_1, \ldots, b_m$ in $\ker \varphi_1$ and sequences $(c_{1,k})_{k=1}, \ldots, (c_{m,k})_{k=1}$ in $c_0(\ker \varphi_1)$ such that $f_k = \sum_{i=1}^m b_i * c_{i,k}$, $k = 1,2,\ldots$.

Hence

$$\begin{aligned} D(f_k) &= \sum_{i=1}^{m} D(b_i * c_{i,k}) = \sum_{i=1}^{m} \varphi_1(b_i)D(c_{i,k}) + \varphi_2(c_{i,k})D(b_i) \\ &= \sum_{i=1}^{m} \varphi_2(c_{i,k})D(b_i), \quad \text{because } b_i \in \ker \varphi_1, \\ &\xrightarrow{k} 0, \qquad \text{because } \varphi_2 \text{ is continuous} \end{aligned}$$

and $c_{i,k} \xrightarrow{k} 0$, $i = 1,2,\ldots,m$. It follows that the restriction of D to $\ker \varphi_1$ is continuous. Since $\ker \varphi_1$ is closed and has codimension one in $\ell^1(\mathbb{F}_w)$, D is continuous, and so, by the above paragraph, we must have that $I^2 = I$.

Therefore we have shown that, if $I_0(\mathbb{F}_w) \cdot c_0(I_0(\mathbb{F}_w)) = c_0(I_0(\mathbb{F}_w))$, then $I^2 = I$ whenever I is a closed bi-ideal with codimension two in $\ell^1(\mathbb{F}_w)$.

Although we have not shown that Problems 2-5 are equivalent, it seems likely in view of the connections between them which we have demonstrated, that a solution to one of the problems will suggest techniques for solving the others.

So far in this section we have not mentioned bi-ideals with codimension greater than two. We conclude with one problem concerning such ideals. It may be shown, using what we already know about ideals of codimensions one and two, that if G is finitely generated and I is a closed, cofinite bi-ideal in $\ell^1(G)$ such that $\ell^1(G)/I$ is commutative, then $I^2 = I$. Now if I is a closed bi-ideal with codimension three in $\ell^1(G)$, then $\ell^1(G)/I$ is commutative and so it follows immediately that if G is a finitely generated group and I has codimension three, then $I^2 = I$. However, the same argument does not apply to bi-ideals of codimension four or higher.

Problem 6. Is there a closed bi-ideal I of codimension four in $\ell^1(\mathbb{F}_2)$ with $I^2 \neq I$?

It is easy to obtain I, a closed codimension four bi-ideal in $\ell^1(\mathbb{F}_k)$, such that $\ell^1(\mathbb{F}_k)/I$ is not commutative. To obtain such an ideal, let ρ be an irreducible, two-dimensional unitary representation of $\mathbb{F}_2$. Then ρ induces an indecomposable, two-dimensional representation $\tilde{\rho}$ of $\ell^1(\mathbb{F}_2)$, and we take I to be the kernel of $\tilde{\rho}$. Since $\tilde{\rho}$ is indecomposable, $\ell^1(\mathbb{F}_2)/I$ is isomorphic to the algebra of complex 2×2 matrices, which is not commutative.

References

[1] F. F. Bonsall and J. Duncan, Complete Normed Algebras, Erg. der Math. 80, Springer-Verlag (1973).

[2] J. P. R. Christensen, Codimension of some subspaces in a Fréchet algebra, Proc. Amer. Math. Soc., 57 (1976), 276-278.

[3] H. G. Dales, Automatic continuity: a survey, Bull. London Math. Soc., 10 (1978), 129-183.

[4] W. Hauenschild and E. Kaniuth, The generalized Wiener theorem for groups with finite dimensional irreducible representations, J. Functional Analysis, 31 (1979), 13-23.

[5] E. Hewitt and K. Ross, Abstract Harmonic Analysis Vol. I, Springer-Verlag, 1963.

[6] K. Iwasawa, On some types of topological groups, Annals of Math., 50 (1949), 507-558.

[7] N. P. Jewell, Continuity of module and higher derivations, Pacific J. Math., 68 (1977), 91-98.

[8] T.-S. Liu, A. V. Rooij and J.-K. Wang, Projections and approximate identities for ideals in group algebras, Trans. Amer. Math. Soc., 175 (1973), 469-482.

[9] K. Liukkonen and R. Mosak, Harmonic analysis and centres of group algebras, Trans. Amer. Math. Soc., 95 (1974), 147-163.

[10] H. Reiter, Sur certains idéaux dans $L^1(G)$, C. R. Acad. Sci., Paris, 267 (1968), 882-885.

[11] __________, Classical Harmonic Analysis and Locally Compact Groups, Oxford Math. Monographs (1968).

[12] J. R. Ringrose, Automatic continuity of derivations of operator algebras, J. London Math. Soc., (2), 7 (1974), 553-560.

[13] W. Rudin, Fourier analysis on groups, Pure and Applied Mathematics 12, Interscience, (1962).

[14] D. J. S. Robinson, Finiteness Conditions and Generalized Soluble Groups I, Erg. der Math., 62, Springer-Verlag (1972).

[15] M. Sugiura, Unitary representations and harmonic analysis, Kodansha (1975).

[16] G. A. Willis, Factorization in codimension one ideals of group algebras, to appear.

[17] __________, Ph.D. thesis, Newcastle-Upon-Tyne (1980).

[18] __________, Approximate units in finite codimensional ideals of group algebras, J. London Math. Soc., (2), 26 (1982), 143-154.

[19] H. G. Dales and G. A. Willis, Cofinite ideals in Banach algebras, and finite-dimensional representations of group algebras, this Volume.

[20] B. L. van der Waerden, Stetigkeitssätze für halbeinfache Liesche Gruppen, Math. Zeit., 36 (1932), 780-786.

Department of Mathematics
University of New South Wales
Kensington, 2033 Australia

IV. CONTINUITY OF LINEAR FUNCTIONALS

SOME PROBLEMS AND RESULTS ON TRANSLATION - INVARIANT LINEAR FORMS

Gary Hosler Meisters

1. Introduction to the TILF question

Let G denote a locally compact Hausdorff topological group and let $\mathfrak{F}(G)$ denote a locally convex topological vector space (LCTVS) of distributions f on G which is "translation-invariant" in the sense that for each element a in G the translate $f_a = \tau(a)f$ belongs to $\mathfrak{F}(G)$ whenever f does. For functions f on G,

$$f_a(t) = \tau(a)f(t) = f(t-a) \quad \text{for all } t \text{ in } G,$$

while for distributions generally,

$$\langle \tau(a)f, \varphi\rangle = \langle f, \tau(-a)\varphi\rangle \quad \text{for all test functions } \varphi .$$

For general information on distributions see [7], [8], [9], [31], or [35]. For background in analysis on locally compact groups see [7], [9], [12], [27], or [30]. References to specific technical results used in this paper will be cited as they occur.

DEFINITION. A translation-invariant linear form (TILF) on $\mathfrak{F}(G)$ is a (not necessarily continuous) linear functional Φ on $\mathfrak{F}(G)$ such that $\Phi(f_a) = \Phi(f)$ for all a in G and for all f in $\mathfrak{F}(G)$.

Given a translation-invariant "function space" $\mathfrak{F}(G)$ as described above we ask the following question.

Question 1: Does the space $\mathfrak{F}(G)$ have the property that every TILF on $\mathfrak{F}(G)$ is automatically continuous?

This paper is a survey of known results and some open problems connected with this question. The first examples of some spaces that do have the property mentioned in Question 1 were given by myself in [20] and [21]. The central result there was that Question 1 has an affirmative answer for the space $\mathcal{D}(\mathbb{R})$ of all C^∞ functions of compact support on the real line $\mathbb{R}$. An affirmative answer was also established for the Hilbert space $L^2(\mathbb{T})$ on the circle group $\mathbb{T}$ in [26]. But not all spaces $\mathfrak{F}(G)$ have this property: The first examples of discontinuous translation-invariant linear forms were given in [22] with further examples announced in [23] and [36]. The first systematic and fairly exhaustive investigation of discontinuous TILF's was carried out by Gordon S. Woodward for a variety of spaces on noncompact groups in [39]. So far there is no known characterization of spaces $\mathfrak{F}(G)$ which have the property mentioned in Question 1. In fact, the answer to Question 1, and even the reason for that answer, seems to vary greatly from one space to another and is unknown for many spaces $\mathfrak{F}(G)$. Most of the results discussed in this paper (in order to reduce technicalities) are for function spaces on one of the three groups $\mathbb{R}$, $\mathbb{T}$, and $\mathbb{Z}$; the reals, the circle, and the integers, respectively.

In §2 I give a much improved exposition of my 1970 proof for $\mathfrak{D}(\mathbb{R})$. In §3 I present the essentials for $L^2(\mathbb{T})$ and state the generalizations which that method allows. Actually, the only spaces for which Question 1 is known to have an affirmative answer are associated (by corollary or analogous proof) with one or the other of the two spaces $\mathfrak{D}(\mathbb{R})$ and $L^2(\mathbb{T})$. Of particular interest in this respect is the space $C_E(\mathbb{T})$ of all E-spectral continuous functions on the circle group $\mathbb{T}$ where E is a Sidon subset of the integers $\mathbb{Z}$. A function f is called <u>E-spectral</u> (for some subset E of $\mathbb{Z}$) if

$$\hat{f}(k) = \int_0^1 f(t)e^{-2\pi ikt}dt = 0$$

for all integers k not in E. A subset E of $\mathbb{Z}$ is called a <u>Sidon set</u> if every function f in $C_E(\mathbb{T})$ has an absolutely convergent Fourier series

$$f(t) = \sum_{k\in\mathbb{Z}} \hat{f}(k)e^{2\pi ikt} .$$

See Chapter 15 in volume 2 of [9] or §5.7 of [30] for an account of Sidon sets. It was shown in [25] that the answer to Question 1 is affirmative for the space $C_E(\mathbb{T})$ whenever E is a Sidon subset of $\mathbb{Z}$. But if E is not a Sidon set, in particular if $E = \mathbb{Z}$, then Question 1 remains open for $C_E(\mathbb{T})$. (Note that $C_{\mathbb{Z}}(\mathbb{T}) = C(\mathbb{T})$.)

<u>Open Problem</u> 1. For which subsets E of $\mathbb{Z}$ (besides Sidon sets) is it true that every TILF on $C_E(\mathbb{T})$ is automatically continuous? In particular, when $E = \mathbb{Z}$ we have

<u>Open Problem 2</u>. Is every TILF on $C(\mathbb{T})$ automatically continuous?

At present I see no reason why the following statement could not be true: <u>Every</u> TILF <u>on</u> $C_E(\mathbb{T})$ <u>is automatically continuous if and only if</u> E <u>is a Sidon subset of</u> $\mathbb{Z}$. What little is known about the cases $\mathfrak{F}(G) = C_E(\mathbb{T})$ is stated in §4. Some examples of discontinuous TILF's on both compact and noncompact groups are given in §5; and in §6 I mention some connections with automatic continuity of linear operators.

Although, as stated above, the details vary greatly from one space to another, there is a common point of departure based on the following elementary facts from functional analysis. (See [38, p. 122] and [22].) Given a translation-invariant LCTVS $\mathfrak{F} = \mathfrak{F}(G)$ over a group G (here assumed to be commutative), let $\Delta = \Delta[\mathfrak{F}]$

denote the linear subspace of $\mathfrak{F}$ which is generated by all finite differences of the form

$$\Delta_a f = f - f_a = f - \tau(a)f ,$$

as f varies over $\mathfrak{F}$ and as a varies over G. I use the term "hyperplane" to mean a maximal proper linear subspace of $\mathfrak{F}$. The facts we will need are:

(1) Every proper linear subspace of $\mathfrak{F}$ is contained in a hyperplane.
(2) A linear subspace M is a hyperplane iff it has codimension one.
(3) If M is a hyperplane, then M is either closed or dense in $\mathfrak{F}$.
(4) There exists a linear functional $\Phi \not\equiv 0$ with null space $\mathfrak{n}(\Phi) = M$ iff M is a hyperplane.
(5) Φ is continuous iff $\mathfrak{n}(\Phi)$ is closed.
(6) $\Phi \not\equiv 0$ is discontinuous iff $\mathfrak{n}(\Phi)$ is dense.
(7) If $\mathfrak{n}(\Phi_1) = \mathfrak{n}(\Phi_2)$, then there exists a constant c so that $\Phi_1 = c \cdot \Phi_2$.
(8) Φ is translation-invariant iff $\Delta[\mathfrak{F}] \subset \mathfrak{n}(\Phi)$.
(9) If $\Delta[\mathfrak{F}]$ is not closed, then there exist discontinuous translation-invariant linear functionals on $\mathfrak{F}$.

In many cases, one knows (or can easily prove) that $\overline{\Delta(\mathfrak{F})}$ is a (closed) hyperplane M which is in fact the null space in $\mathfrak{F}$ of an invariant (continuous) integral on G. In such cases the automatic continuity of TILF's on $\mathfrak{F}$ reduces to the question: Is $\Delta(\mathfrak{F})$ already closed? Or, in other words, is every element f in M expressible as a <u>finite</u> linear combination of differences

$$f(t) = \sum_{j=1}^{N} [f_j(t) - f_j(t - a_j)]? \tag{1}$$

Thus, in the case that $\mathfrak{F}(G) = C(\mathbb{T})$, the Open Problem 2 above can be rephrased as follows. Can every continuous function $f : \mathbb{R} \to \mathbb{C}$ of period 1 with $\hat{f}(0) = \int_0^1 f(t)dt = 0$ be expressed in the form (1) for some integer $N \geq 1$, real numbers $a_1,\dots,a_N$ and continuous (period 1) functions $f_1,\dots,f_N$, all of which (N, a_j's and f_j's) might depend on the choice of f?

When such a formula as (1) holds true, it is a kind of generalization of the formula $H(x) = G(x+1) - G(x)$ due to Guichard [10]. See also [24] and the references cited there.

2. Automatic continuity of TILF's on $\mathfrak{D}(\mathbb{R})$

Diophantine Approximation seems to play a fundamental role in all positive results on the automatic continuity of TILF's. Therefore I have included references to several valuable sources on this important subject. These are [2], [4], [11, Chapter XI], [33], and [34]. In order to state and prove Theorem 1, which is the key to the automatic continuity of TILF's on $\mathcal{D}(\mathbb{R})$, we will need certain facts concerning "Liouville numbers" which can be found in one or another of these sources. A real number θ is called a Liouville number if it can be rapidly approximated by rationals in the sense that for every positive integer m there exist integers p and q with $q > 1$ such that $|\theta - (p/q)| < q^{-m}$. These numbers θ include all rationals and infinitely many (actually $\mathfrak{c}$) transcendentals, but no algebraic irrationals (Liouville's Theorem [2, p. 1], [4, p. 103], [11, p. 161], [33, p. 193], and [34, p. 114]). Furthermore, the set $\mathbb{L}$ of all Liouville numbers has Lebesgue measure zero. It is easy to show, by use of the elementary inequalities

$$4\|\theta\| \leq |1 - e^{-2\pi i\theta}| \leq 2\pi\|\theta\|$$

(where $\|\theta\|$ denotes the distance from θ to the nearest integer), that an irrational number θ is not a Liouville number if and only if there exist positive constants L and $d \geq 2$ such that the inequality

$$|1 - e^{-2\pi ik\theta}|^{-1} \leq L|k|^{d-1} \tag{2}$$

holds for all nonzero integers k. Thus almost all real numbers θ (in the sense of Lebesgue measure), including all algebraic irrationals of degree $d \geq 2$, satisfy Inequality (2).

Theorem 1 below is a statement about generators for the convolution ring $(\mathcal{E}',*,+)$ of all Schwartz distributions A of compact support on the real line $\mathbb{R}$. The space $\mathcal{E}'$ is the dual of the space $\mathcal{E}$ of all C^∞ functions with arbitrary supports in $\mathbb{R}$. Convolution $*$ and Fourier Transforms $\hat{A}(z)$, $z = x + iy \in \mathbb{C}$, are defined by the formulas $\langle A*B,\varphi\rangle = \langle A_s,\langle B_t,\varphi(s+t)\rangle\rangle$, for all φ in $\mathcal{E}$, and $\hat{A}(z) = \langle A_t, e^{-2\pi izt}\rangle$, for all z in $\mathbb{C}$. We will also need the Paley-Wiener-Schwartz Theorem ([7, p. 189], [8, pp. 211-213], [31, p. 183], or [35, p. 272]) from which it follows that the Fourier transform is an isomorphism of the convolution ring $\mathcal{E}'$ onto the multiplication ring $\mathfrak{M}$ of all entire functions $g(z)$ which satisfy an inequality of the form

$$|g(z)| \leq M(1 + |z|)^m e^{b|\mathrm{Im} z|} \tag{3}$$

for some positive constants M, m, b, depending on g.

Recall also that the Dirac delta "function" δ, defined by $\langle\delta,\varphi\rangle = \varphi(0)$ for all φ in $\mathcal{D}(\mathbb{R})$, is the distributional derivative of the Heaviside function $Y(t)$ [$= 0$ for $t < 0$ and $= 1$ for $t > 0$] and satisfies $\delta^{(p)} * A = A^{(p)}$, $\delta^{(p)\wedge}(z) = (2\pi iz)^p$, and $\delta_a * A = A_a = \tau(a)A$. A linear functional Φ on $\mathcal{D}(\mathbb{R})$ is said to have the period $\alpha \neq 0$ iff $\Phi = \Phi_\alpha$. Such a Φ need not be continuous (i.e., need not be a distribution).

THEOREM 1. (Meisters [20] and [21].) If α and β are nonzero real numbers such that β/α is not a Liouville number, then there exist two (necessarily distinct) distributions A and B, both with compact supports on the real line, such that

$$\delta = A * (Y - Y_\alpha) + B * (Y - Y_\beta) \tag{4a}$$

or equivalently,

$$\begin{aligned} \delta' &= A - A_\alpha + B - B_\beta , \\ &= A * (\delta - \delta_\alpha) + B * (\delta - \delta_\beta) . \end{aligned} \tag{4b}$$

Conversely, if the formula (4a) or (4b) holds for some distributions A and B in $\mathcal{E}'$, then β/α is not a Liouville number.

COROLLARY OF THEOREM 1. (Meisters [20] and [21]). If Φ is a linear functional on $\mathcal{D}(\mathbb{R})$ with two periods α and β whose ratio β/α is not a Liouville number, then there exists a complex constant c such that for every φ in $\mathcal{D}(\mathbb{R})$,

$$\langle \Phi, \varphi \rangle = c \cdot \int_{-\infty}^{+\infty} \varphi(s) ds .$$

In particular, TILF's on $\mathcal{D}(\mathbb{R})$ are automatically continuous.

It follows directly from Theorem 3 (vii) and its proof on pages 183-185 of [21] that there does exist an irrational value of β/α (necessarily a Liouville transcendental) and a corresponding linear functional Φ on $\mathcal{D}(\mathbb{R})$ with periods α and β, but such that Φ is not a constant multiple of integration. Such a Φ has dense periods $m\alpha + n\beta$ and is of course very discontinuous (and certainly not a distribution).

At least three distinct proofs have been given for Theorem 1. These can be found in [21], [25], and [17]. The first one [21] is perhaps the most direct and I give an improved version of it below. The second one [25] uses Fourier series of periodic distributions instead of Fourier transforms of distributions of compact support and thus avoids the Paley-Wiener-Schwartz Theorem. While it is more elementary in its details it is not really shorter because it involves some rather elaborate technical lemmas. The third proof [17] is the shortest of all, but less direct since it depends on a very general theorem of Hörmander [13] on generators for certain rings of analytic functions. Also Jean Dieudonné has outlined a proof of Theorem 1 as Problem 30 on pages 207-208 in Volume VI (Chapter XXII: Harmonic

Analysis) of his "Treatise on Analysis" [7]. Furthermore, Robert D. Richtmyer [28] has investigated the detailed structure of the distributions A and B occurring in Formulas (4a) and (4b). Finally, David Lee Johnson [14] has shown by reduction to our case that TILF's are also automatically continuous on the space $\mathcal{D}(G)$ of Schwartz-Bruhat test functions [3] on an arbitrary Hausdorff Locally compact abelian group G.

Proof of Theorem 1. Formula (4b) follows from (4a) by differentiation (or by convolution with δ') and Formula (4a) follows from (4b) by convolution with the Heaviside function Y.

Since

$$\hat{\delta}_\alpha(z) = \langle \delta_\alpha, e^{-2\pi izt} \rangle = e^{-2\pi iz\alpha},$$

it follows from the Paley-Wiener-Schwartz Theorem [8, pp. 211-213] that there exist distributions A and B in $\mathcal{E}'(\mathbb{R})$ satisfying (4b) if and only if there exist entire functions $\hat{A}(z)$ and $\hat{B}(z)$ in $\mathfrak{M}$, i.e. satisfying an inequality of the form (3), and also satisfying the equation

$$2\pi iz = \hat{A}(z)(1 - e^{-2\pi i\alpha z}) + \hat{B}(z)(1 - e^{-2\pi i\beta z}) . \tag{5}$$

Now let α and β be given positive real numbers such that β/α is not a Liouville number (i.e. suppose that Inequality (2) is satisfied for $\theta = \beta/\alpha$), and define the meromorphic function

$$h(z) = \Sigma' \, a_k/(\alpha z - k) , \tag{6}$$

where Σ' denotes summation over all (positive and negative) nonzero integers k, and where

$$a_k = (\alpha/k)^{d+1}(1 - e^{-2\pi ik\beta/\alpha})^{-1}, \; k \neq 0 .$$

The inequality (2) for $\theta = \beta/\alpha$ implies that

$$A \equiv \Sigma' \, |a_k| \leq L \, \alpha^{d+1} \, \Sigma' \, k^{-2} < +\infty .$$

Consequently, for each $\varepsilon > 0$ and for complex z outside all the disks $|\alpha z - k| < \alpha\varepsilon$, for $k = \pm 1, \pm 2, \ldots$, we have

$$\Sigma' |a_k| \, / \, |\alpha z - k| \leq A/\alpha\varepsilon ,$$

so that the series defining $h(z)$ in (6) converges uniformly outside these disks.

Since ε is arbitrary we see that $h(z)$ is a meromorphic function with simple poles at k/α for $k \neq 0$, and is analytic everywhere else. Furthermore, the principal part of $h(z)$ at k/α (for $k \neq 0$) is $a_k/(\alpha z - k)$. Since $(1 - e^{-2\pi i\alpha z})$ has simple zeros at the points k/α, the function $(1 - e^{-2\pi i\alpha z})$ $h(z)$ is entire and has the value $2\pi i a_k$ at the point k/α for $k \neq 0$. If we now define

$$\begin{aligned} H(z) &\equiv z^{d+2}(1 - e^{-2\pi i\alpha z})h(z) \\ &= 2\pi i z^{d+2}\, e^{-i\pi\alpha z}\, \Sigma'(-1)^k\, a_k\, \frac{\sin\pi(\alpha z - k)}{\pi(\alpha z - k)} \end{aligned}$$

then $H(z)$ is an entire function satisfying

$$H(k/\alpha) = (2\pi i k/\alpha)(1 - e^{-2\pi i k\beta/\alpha})^{-1}, \tag{7}$$

for all integers $k \neq 0$. Now consider the entire function $s(w) \equiv (\sin w)/w$ for $w \neq 0$, and $s(0) \equiv 1$. If $|w| \leq 1$, then there exists a constant $K > 1$ such that $|s(w)| \leq K$. If $|w| \geq 1$, then $|s(w)| \leq |\sin w| \leq e^{|\mathrm{Im}\, w|} \leq Ke^{|\mathrm{Im}\, w|}$. Since $e^{|\mathrm{Im}\, w|} \geq 1$, $|s(w)| \leq Ke^{|\mathrm{Im}\, w|}$ for all complex w. Therefore, for each integer k and with $w = \pi(\alpha z - k)$ we have $\mathrm{Im}\, w = \pi\alpha\, \mathrm{Im}\, z$ so that $|s(w)| \leq Ke^{\pi\alpha|\mathrm{Im}\, z|}$. Consequently

$$|H(z)| \leq (2\pi AK)|z|^{d+2}\, e^{\pi\alpha|\mathrm{Im}\, z|}, \tag{8}$$

where $A = \Sigma'|a_k|$. Since (8) is equivalent to an inequality of the form (3) we see that $H(z)$ is the Fourier transform of some distribution B of compact support on the real line $\mathbb{R}$: $H(z) = \hat{B}(z)$. But then the entire function

$$\hat{G}(z) \equiv 2\pi i z - \hat{B}(z)(1 - e^{-2\pi i\beta z}) \tag{9}$$

obviously belongs to $\mathfrak{M}$ also and so it too is equal to the Fourier transform of some distribution G in $\mathcal{E}'(\mathbb{R})$. Furthermore, because of (7), $\hat{G}(z)$ is zero at $z = k/\alpha$ for all integers k. It follows easily from this fact (see Lemma 2' on p. 177 of [21]) that there exists a distribution A in $\mathcal{E}'(\mathbb{R})$ such that $G = A - A_\alpha$, or

$$\hat{G}(z) = \hat{A}(z)(1 - e^{-2\pi i\alpha z}). \tag{10}$$

Equations (9) and (10) may now be combined to give (5) which, as already noted, is equivalent to (4b).

If A and B were not distinct in (4b), then we would have

$$1 = \hat{A}(z)[(2 - e^{-2\pi i\alpha z} - e^{-2\pi i\beta z})/2\pi iz],$$

which is impossible because the left side has no zeros while the second factor of the right side has infinitely many zeros: For otherwise Hadamard's refinement of the Weierstrass factorization theorem [32, p. 332] would imply that the second factor is identical to $p(z)e^{bz}$ for some polynomial $p(z)$ and complex number b, which is easily seen to be impossible.

Finally, if β/α is a Liouville number and (4b) holds, then for $z = k/\alpha$ (5) yields

$$\hat{B}(k/\alpha) = 2\pi i(k/\alpha)(1 - e^{-2\pi ik\beta/\alpha})^{-1}. \tag{11}$$

But, on the one hand, since $\hat{B}(z) \in \mathfrak{m}$, (3) implies

$$|\hat{B}(k/\alpha)| \leq M(1 + |k/\alpha|)^m \quad \text{for all integers } k, \tag{12}$$

while on the other hand, since β/α is a Liouville number, it is easy to show that for every integer $m > 1$ there are infinitely many integers k satisfying

$$|1 - e^{-2\pi ik\beta/\alpha}|^{-1} > |k|^m. \tag{13}$$

Therefore, (11), (12) and (13) combine to give the contradiction that

$$2\pi|k|^{m+1}/\alpha < |\hat{B}(k/\alpha)| \leq M(1 + |k/\alpha|)^m$$

for infinitely many integers k. This completes the proof of Theorem 1.

I would like to thank Rao Chivukula for pointing out to me that my original (1970) proof of Inequality (8), namely the first paragraph on page 179 of [21], can be replaced by the much simpler argument given here, namely the five sentences preceding (8).

In her Ph.D. thesis [15], Mrs. Edith Kregelius-Petersen has deduced the following generalization of Theorem 1 from Hörmander's theorem in [13].

<u>Generalization of Theorem</u> 1. (Kregelius-Petersen [15] and [16].) If $\alpha_1,\ldots,\alpha_{N+1}$ are $N + 1 \geq 2$ nonzero real numbers, then there are distributions $A_1,\ldots,A_{N+1}$ in $\mathcal{E}'(\mathbb{R})$ such that

$$\delta' = \sum_{j=1}^{N+1} [A_j - \tau(\alpha_j)A_j]$$

if and only if the quotients $\alpha_1/\alpha_{N+1},\ldots,\alpha_N/\alpha_{N+1}$ are not simultaneously rapidly approximable.

Nonzero real numbers $\theta_1,\dots,\theta_N$ are called simultaneously rapidly approximable (s.r.a.) iff for every positive integer m there exist integers $p_1,\dots,p_N$ and $q \geq 2$ such that $|\theta_j - p_j/q| < q^{-m}$ for $j = 1,2,\dots,N$. If $N = 1$, then this definition reduces to that of a Liouville number. If $\theta_1,\dots,\theta_N$ are s.r.a., then each θ_j is a Liouville number; but the converse is false. For if it were true, then the set $\mathbb{L}$ of Liouville numbers would be closed under addition and hence also closed under products

$$\theta_1\theta_2 = \frac{1}{4}\,[(\theta_1 + \theta_2)^2 - (\theta_1 - \theta_2)^2]\,,$$

since $\mathbb{L}$ is closed under integral exponentiation. But $\mathbb{L}$ is not closed under products because every nonzero real number can be written as a product of two Liouville numbers (proved in a letter to me from Wolfgang M. Schmidt).

Upon seeing [16] and [17], Lilian Asam proved, by a different and perhaps simpler method, an n-dimensional version [stated in a letter to me dated November 1979] of the above Generalization of Theorem 1.

Proof of the Corollary to Theorem 1. If Φ is a linear functional on $\mathcal{D}(\mathbb{R})$ with two periods α and β whose ratio β/α is not a Liouville number, then by Theorem 1 there exist two distributions A and B in $\mathcal{E}'(\mathbb{R})$ such that (4b) holds. By convolution with an arbitrary element φ of $\mathcal{D}(\mathbb{R})$ we obtain (for all real t)

$$\varphi'(t) = u(t) - u(t-\alpha) + v(t) - v(t-\beta)$$

where $u = \varphi * A$ and $v = \varphi * B$ belong to $\mathcal{D}(\mathbb{R})$. Therefore, since Φ has periods α and β, $\Phi(\varphi') = 0$ for all φ in $\mathcal{D}(\mathbb{R})$. That is, the null space M of the integration functional on $\mathcal{D}(\mathbb{R})$,

$$\varphi \mapsto \int_{-\infty}^{+\infty} \varphi(t)\,dt\,,$$

is contained in the null space of Φ. But M is a closed hyperplane in $\mathcal{D}$ and consequently (by virtue of the properties of hyperplanes listed in §1) it follows that there is a complex constant c such that

$$\Phi(\varphi) = c\int_{-\infty}^{+\infty} \varphi(t)\,dt \quad \text{for all } \varphi \text{ in } \mathcal{D}(\mathbb{R})\,. \qquad \text{Q.E.D.}$$

3. Automatic continuity of TILF's on $L^2(\mathbb{T})$

For the choice $\mathfrak{F}(G) = L^2(\mathbb{T})$ we also obtain a positive result in the form of Equation (1), but somewhat different from Formula (4b), and the method of proof (see [26]) is completely different from that used to prove Theorem 1 for $\mathcal{D}(\mathbb{R})$.

THEOREM 2. (Meisters & Schmidt [26].) For every f in $L^2(\mathbb{T})$ satisfying $\hat{f}(0) = \int_{\mathbb{T}} f(x)dx = 0$ there is a set N_F consisting of almost all triples $(\alpha_1,\alpha_2,\alpha_3)$ in the unit cube $[0,1]^3$ such that for each triple $(\alpha_1,\alpha_2,\alpha_3)$ in N_f there are functions u_1, u_2, u_3 in $L^2(\mathbb{T})$ satisfying

$$f(t) = \sum_{j=1}^{3} [u_j(t) - u_j(t-\alpha_j)] \tag{14}$$

for almost all t in $\mathbb{T}$. Furthermore, not a single triple $(\alpha_1,\alpha_2,\alpha_3)$ will serve in this manner for all f in $L^2(\mathbb{T})$ satisfying $\hat{f}(0) = 0$, i.e. $\bigcap N_f = \emptyset$, and some such functions f cannot be represented with less than three terms in the sum (14).

The proof of Theorem 2 works just as well when $\mathbb{T}$ is replaced by any compact abelian group G which is either connected or has at most a finite number of connected components. (See [26].)

There is also another way in which Theorm 2 can be generalized; a way which throws some light on the problem of automatic continuity of TILF's on $C(\mathbb{T})$, and leads to the formulation of Open Problem 1 stated in §1. This second generalization also works for connected compact abelian groups G (see [25] and [26]) but will be stated here for the circle group $\mathbb{T}$. If $A(\mathbb{T})$ denotes the space of all absolutely convergent trigonometric series

$$f(t) = \sum_{k\in\mathbb{Z}} a_k e^{2\pi ikt},$$

then $A(\mathbb{T})$ is a Banach space with the norm

$$\|f\|_A = \sum_{k\in\mathbb{Z}} |a_k| = \sum_{k\in\mathbb{Z}} |\hat{f}(k)|,$$

the elements of whose dual space $A'(\mathbb{T})$ are called pseudomeasures on $\mathbb{T}$ (see §12.11 of [9]) and correspond to trigonometric series with bounded coefficients a_k in $\mathbb{C}$. For each real number $p \geq 1$ and for each subset E of $\mathbb{Z}$ (dual group of $\mathbb{T}$) let $\mathfrak{F}_p(\mathbb{T})_E$ denote the translation-invariant vector space of all pseudomeasures S on $\mathbb{T}$ which satisfy $\hat{S} \in \ell^p(\mathbb{Z})$ and supp $\hat{S} \subset E$. Here $\hat{S}(k)$ denotes the kth Fourier coefficient of S which can be defined by $\hat{S}(k) = \langle S, e^{-2\pi ikx}\rangle$ because S is a (continuous) linear functional on $A(\mathbb{T})$. Thus $\mathfrak{F}_p(\mathbb{T})_E$ consists of the Fourier transforms of the elements of $\ell_p(E)$. Elements of $\mathfrak{F}_p(\mathbb{T})_E$ are regarded as objects (functions, psuedomeasures, or distributions) on the circle group $\mathbb{T}$.

Generalization of Theorem 2. (Meisters [24], [25], [26]). Let $p \geq 1$ and $E \subset \mathbb{Z}$ be given. For each S in $\mathfrak{F}_p(\mathbb{T})_E$ satisfying $\hat{S}(0) = 0$ and for each integer $m > p$ there is a set $H \subset \mathbb{T}^m$, consisting of almost all of the m-dimensional torus $\mathbb{T}^m$, such that for each m-tuple $(a_1,\dots,a_m)$ in H there are pseudomeasures $S_1,\dots,S_m$ in $\mathfrak{F}_p(\mathbb{T})_E$ satisfying

$$S = \sum_{j=1}^{m} [S_j - \tau(a_j)S_j] .$$

This is Theorem 6 of [25] except that the statement there is for any connected compact abelian group G. (Unfortunately, in proof-reading that paper I failed to notice that the phrase "...each S in $\mathfrak{F}_p(G)_E$ satisfying $\hat{S}(1) = 0$ and for..." was omitted by the typist from the statement there, between the first and second words of the second sentence, making that statement somewhat mysterious and difficult to understand.)

I am confident that examples can be given to show that the hypothesis "$m > p$" is best possible. It follows from Theorem 2 (in exactly the same way that the Corollary of Theorem 1 follows from Theorem 1) that TILF's are automatically continuous on each of the spaces $\mathfrak{F}_p(G)_E$, $p \geq 1$, $E \subset \hat{G}$ (dual group of G). In particular, TILF's are automatically continuous on $L^2(\mathbb{T})$ and $A(\mathbb{T})$. In [24] I have shown that this same method can be used to establish the automatic continuity of TILF's on each of the spaces $A_\varepsilon(G)$, $0 < \varepsilon < 1$, which consist of all continuous functions f on G with "better than absolutely convergent" Fourier series in the sense that

$$\sum_{\xi \in \hat{G}} |\hat{f}(\xi)|^\varepsilon < \infty .$$

4. Some results for $C(\mathbb{T})$.

As usual, let $C(\mathbb{T})$ denote the Banach space (with supremum norm) of all continuous complex-valued functions on the circle group $\mathbb{T}$. The subspaces $C_E(\mathbb{T})$ of all "E-spectral" continuous functions (for any subset E of $\mathbb{Z}$), and Sidon subsets E of $\mathbb{Z}$ were defined in §1.

THEOREM 3. (Meisters [25].) If E is a Sidon subset of $\mathbb{Z}$, then TILF's are automatically continuous on $C_E(\mathbb{T})$.

Proof. Recall that a subset E of $\mathbb{Z}$ is a Sidon set if and only if $C_E(\mathbb{T}) = A_E(\mathbb{T})$. But $A_E(\mathbb{T}) = \mathfrak{F}_1(\mathbb{T})_E$, so by the Generalization of Theorem 2 we may conclude that for each function f in $C_E(\mathbb{T})$ with $\hat{f}(0) = 0$, there is a pair of numbers α, β in $\mathbb{T}$ and a pair of functions g, h in $C_E(\mathbb{T})$ such that

$$f(t) = g(t) - g(t-\alpha) + h(t) - h(t-\beta)$$

for all t in $\mathbb{T}$. It follows immediately from this (as in the proof of the

Corollary of Theorem 1) that any TILF Φ on $C_E(\mathbb{T})$ has the same (maximal closed) null space as the continuous functional $f \mapsto \hat{f}(0) = \int_{\mathbb{T}} f(x)dx$. Therefore by property (7) of hyperplanes listed in §1, there must be a complex constant c such that $\Phi(f) = c \cdot \hat{f}(0)$ for every f in $C_E(\mathbb{T})$. In particular, Φ is automatically continuous. Q.E.D.

Not a single subset E of $\mathbb{Z}$, other than Sidon sets E, is known to have the property that TILF's on $C_E(\mathbb{T})$ are automatically continuous. We formulate this as

Open Problem 3. True or false? Every TILF on $C_E(\mathbb{T})$ is automatically continuous if and only if E is a Sidon subset of $\mathbb{Z}$.

Obviously, Problems 1, 2, and 3 are closely related. The following partial result was obtained in [25] for Problem 2 (that is, when $E = \mathbb{Z}$).

THEOREM 4. (Meisters [25].) For every N-tuple $a = (a_1, \ldots, a_N)$ in $\mathbb{T}$ there exists a function f in $C(\mathbb{T})$ such that $\hat{f}(0) = 0$ and equation (1) is false, no matter how the functions $f_1, \ldots, f_N$ are chosen in $C(\mathbb{T})$.

Proof. See Theorem 5 and its proof in [25]. In other words, as for $L^2(\mathbb{T})$ in Theorem 2, if Equation (1) holds for $C(\mathbb{T})$, then the numbers $a_1, \ldots, a_N$ must depend on (and change with) the choice of f. Contrast this with Theorem 1 where the numbers α and β can be chosen independently of f.

In her Diplomarbeit [1], Lilian Asam studied TILF's on the space $\mathcal{K}(G)$ of all continuous functions with compact support on a locally compact group G. Her results were extended in [29] to give some nice results on the dimension of the linear space of all TILF's on $\mathcal{K}(G)$. In particular it is proved in [29] that if H is a normal subgroup of G whose relative topology is discrete, then the linear space of TILF's on $\mathcal{K}(G)$ is isomorphic to the linear space of TILF's on $\mathcal{K}(G/H)$. One consequence is that the question of the automatic continuity of TILF's on $\mathcal{K}(\mathbb{R})$ is equivalent to the question (Open Problem 2 stated in §1) of the automatic continuity of TILF's on $\mathcal{K}(\mathbb{T}) = C(\mathbb{T})$.

See also the interesting remarks of R. J. Loy in his paper "On the uniqueness of Riemann integration" in these same Proceedings.

5. Examples of discontinuous TILF's

The known examples of discontinuous TILF's can be divided first of all into those on noncompact groups and those on compact groups. Generally speaking, those on noncompact groups are somewhat easier to obtain.

The first examples of discontinuous TILF's on noncompact groups were given by myself in [22] for the spaces $\ell_1(\mathbb{Z})$, $L^1(\mathbb{R})$, and $L^2(\mathbb{R})$. Then, by using different methods, A. C. Serold [36] obtained examples of discontinuous TILF's on $\ell_p(\mathbb{Z})$ for $1 < p \leq \infty$, as well as on $c(\mathbb{Z})$ and $c_0(\mathbb{Z})$. Independently of Serold, Gordon S. Woodward [39] gave a rather complete analysis of the situation for noncompact groups. I summarize his results as follows.

THEOREM 5. (G. S. Woodward [39].) Let G denote a noncompact (Hausdorff topological) group. Then discontinuous TILF's do exist

(a) on the space L^1 if G is sigma-compact,

(b) on the spaces C_0, C, and L^p for $1 \leq p \leq \infty$ if G is sigma-compact and amenable,

(c) on the spaces C, L^1, and L^∞ if G is abelian, and

(d) on the spaces C_0 and L^p for $1 < p < \infty$ for abelian G if and only if G/H is a torsion group for some open sigma-compact subgroup H of G.

Here C denotes the space of all continuous bounded functions and C_0 the subspace of those functions which "vanish at infinity" (i.e., are less than ε in absolute value outside some compact set). Although Woodward's results do cover many cases, yet they do not cover all cases. Perhaps the most obvious omissions are for spaces on noncompact sigma-compact but non-amenable groups G, such as $SL_n(\mathbb{R})$, where he considered only $L^1(G)$.

The first examples of discontinuous TILF's on compact groups were also given by myself (in [22] and [23]) for the Hilbert spaces $L^2(\mathbb{D})$ and $L^2(G)$ where $\mathbb{D}$ denotes the totally disconnected compact abelian "Cantor discontinuum" group $(\mathbb{Z} \bmod 2)^\omega$, and where (in [23]) more generally G denotes any "non-polythetic" second-countable compact abelian group. I call a compact abelian group polythetic if it contains a finitely generated dense subgroup. This generalizes "monothetic" which means that G contains a dense cyclic subgroup. (See page 85 and Theorem 24.31 on page 389 of [12].) For example, for each integer $n \geq 2$, the (complete) direct product $(\mathbb{Z} \bmod n)^\omega$ is not polythetic. See [24] for a TILF-related proof of the Halmos - Samelson Theorem which states that all second-countable compact and connected abelian groups are "very" monothetic: almost every member generates a dense cyclic subgroup. Since a compact abelian group is monothetic iff its character group is isomorphic with a subgroup of the circle group $\mathbb{T}$ with the discrete topology ([12], Corollary 24.32), there are polythetic CA groups which are not monothetic. However, every connected polythetic CA group G is monothetic.

Proof. Since each continuous character is uniquely determined by its values on any finite set that generates a dense subgroup, G can have at most $\mathfrak{c}^m = \mathfrak{c}$ continuous characters. By Theorem 24.15 of [12], the weight $w(G)$ of G (the least cardinal number of an open basis for G) is equal to the cardinality of $\hat{G}$. But every connected CA group with weight $w(G) \leq \mathfrak{c}$ is monothetic ([12], Theorem

25.14). Q.E.D. Since the connected ν-dimensional torus groups $\mathbb{T}^{\nu}$ are not polythetic and not second countable for $\nu > c$ ([12], Corollary 25.15), the hypothesis "second-countable" in the first sentence of this paragraph cannot be dropped because of our Theorem 2 in §3 which implies the automatic continuity of TILF's on each $L^2(\mathbb{T}^{\nu})$. The results announced in [23] are included in the following theorem ultimately proved jointly by Larry Baggett and the author.

THEOREM 6. (Meisters & Baggett). Let G be a compact abelian group with dual group $\hat{G}$ written multiplicatively with identity 1.

(a) If there exist discontinuous TILF's on $L^2(K)$ where K is a continuous homomorphic image of G, then there exist discontinuous TILF's on $L^2(G)$ also.

(b) If G/C is not second-countable (where C denotes the connected component of the identity in G), then the second-countable non-polythetic group $(\mathbb{Z} \bmod p)^{\omega}$ is a continuous homomorphic image of G for some prime p.

(c) The following two properties are equivalent, are satisfied by every compact abelian torsion group, and imply that there are discontinuous TIlf's on $L^2(G)$:

 (i) There exists a denumberable subset Γ_0 of $\hat{G}\setminus\{1\}$ such that for every finite subset $F = \{a_1,\ldots,a_m\}$ of G there is a continuous character ξ in Γ_0 which annihilates all members of F.

 (ii) There is a closed subgroup H of G such that G/H is second-countable and non-polythetic.

(d) A group G is not polythetic if and only if it satisfies the following property (iii) which is weaker than (i):

 (iii) For every finite subset $F = \{a_1,\ldots,a_m\}$ of G there is a continuous character $\xi \neq 1$ of G such that $\xi(a_1) = \cdots = \xi(a_m) = 1$. (That is, ξ annihilates all members of F.)

(e) The following property (iv) is stronger than property (i) of part (c) and is satisfied by every group G of the form $(\mathbb{Z} \bmod n)^{\alpha}$ where n is a positive integer > 1 and α is any infinite cardinal.

 (iv) There exists a second-countable and non-polythetic CA direct summand G_0 of G.

COROLLARY TO THEOREM 6. (Meisters & Baggett). There are discontinuous TILF's on $L^2(G)$ for all compact abelian groups G for which G/C is infinite (including all infinite torsion G) except possibly when G/C is polythetic (and a fortiori second-countable).

There thus remains the following open problem.

Open Problem 4. Are TILF's automatically continuous on $L^2(G)$ for some totally disconnected, but polythetic, compact abelian group G?

The following important special case was posed to the author by Irving Kaplansky in 1971.

Open Problem 5. Are TILF's automatically continuous on $L^2(\mathbb{Z}_p)$ where $\mathbb{Z}_p$ denotes the totally disconnected, but monothetic, compact abelian group of all p-adic integers? The p-adic integers are described (in a different notation) on pages 107-117 of Hewitt & Ross [12].

For the proof of Theorem 6 the following lemma (interesting in its own right for other cases not yet solved) will be needed.

LEMMA. For any compact abelian group G, the linear subspace $\Delta(L^2(G))$ defined in §1 is dense in the closed hyperplane M which is the null space in $L^2(G)$ of the Haar integral on G, and the following are equivalent.

(1) Discontinuous TILF's exist on $L^2(G)$.

(2) $\Delta(L^2(G)) \neq M$.

(3) There exists a function f in M such that the series

$$\sum_{\xi \in \operatorname{supp} \hat{f} \subset \hat{G}} |\hat{f}(\xi)|^2 / \sum_{j=1}^{m} |1 - \xi(a_j)|^2$$

diverges for every choice of $a_1, \ldots, a_m$ in G and integer $m \geq 1$.

Proof. See Lemma 1 and its proof in [26], and also the argument at the beginning of Theorem 3 in [22] which obviously generalizes to arbitrary compact abelian groups G.

Proof of Theorem 6(a). Recall that K is isomorphic to G/H where H is the closed kernel of a continuous homomorphism h of G onto K; and that if $f : K \to \mathbb{C}$ is a Haar integrable function on K, then

$$\int_G (f \circ h)(x)dx = \int_K f(y)dy .$$

(For a short proof of this latter fact see the proof of Lemma 2 in [26].) Now by the hypothesis of Theorem 6(a) and by part (3) of the above lemma there exists a function $f : K \to \mathbb{C}$ in $M_K = \{g \in L^2(K) : \hat{g}(1) = 0\}$ such that the series

$$\sum_{\chi \in \hat{K}} |\hat{f}(\chi)|^2 / \sum_{j=1}^{m} |1 - \chi(b_j)|^2$$

diverges for all m-tuples $b_1,\dots,b_m$ in K and all integers $m \geq 1$. If we define $F = f \circ h$, then F is a function on G satisfying

$$\hat{F}(1) = \int_G F(x)dx = \int_G (f \circ h)(x)dx = \int_K f(y)dy = \hat{f}(1) = 0 ,$$

and

$$\int_G |F(x)|^2 dx = \int_K |f(y)|^2 dy < \infty ,$$

so that $F \in M_G = \{g \in L^2(G) : \hat{g}(1) = 0\}$. Also if $\chi \in \hat{K}$, then $\chi \circ h \in \hat{G}$ and $\chi \circ h \equiv 1$ on H, the kernel of h.

$$(G/H)\hat{} \cong \hat{K} \cong \mathbb{A}(\hat{G},H) \subset \hat{G} .$$

(See [12], p. 365.) In fact, the mapping $\chi \to \chi \circ h$ is one-to-one from $\hat{K}$ onto $\mathbb{A}(\hat{G},H)$. Furthermore,

$$\hat{f}(\chi) = \int_K f(y)\overline{\chi(y)} = \int_G f(h(x))\overline{\chi(h(x))}dx = \hat{F}(\chi \circ h) .$$

Therefore, if $\chi \in \operatorname{supp} \hat{f}$, then $\chi \circ h \in \operatorname{supp} \hat{F}$, and for any m-tuple $a_1,\dots a_m$ in G the series

$$\begin{aligned} &\sum_{\xi \in \operatorname{supp} \hat{F}} |\hat{F}(\xi)|^2 \Big/ \sum_{j=1}^{m} |1 - \xi(a_j)|^2 \\ &\geq \sum_{\chi \in \operatorname{supp} \hat{f}} |\hat{F}(\chi \circ h)|^2 \Big/ \sum_{j=1}^{m} |1 - \chi \circ h(a_j)|^2 \\ &= \sum_{\chi \in \operatorname{supp} \hat{f}} |\hat{f}(\chi)|^2 \Big/ \sum_{j=1}^{m} |1 - \chi(b_j)|^2 \end{aligned}$$

diverges, where $b_j = h(a_j) \in K$. Thus, by the lemma, part 6(a) is proved.

Proof of Theorem 6(b). Let D denote the totally disconnected factor group G/C and let $\hat{D}$ denote its character group. We assume that D is not second-countable. Since $\hat{D}$ is a torsion group ([12], Theorem 24.26) it follows that $\hat{D}$ is the denumerable union of the subgroups

$$\hat{D}_{(n)} = \{\xi \in \hat{D} : \xi^n = 1\} , \qquad n = 1,2,3,\dots .$$

If all the $\hat{D}_{(n)}$ are finite groups, then $\hat{D}$ is denumerable and consequently D is second-countable. Since by hypothesis this is not the case, it must be true that

$\hat{D}_{(n)}$ is an infinite group for at least one integer $n \geq 2$. For each prime factor p of n, the mapping $\xi \mapsto \xi^{n/p}$ is a homomorphism of $\hat{D}_{(n)}$ into $\hat{D}_{(p)}$. If $\hat{D}_{(p)}$ is finite, then the kernel $K(n/p)$ of this homormorphism is infinite. But $K(n/p) = \hat{D}_{(n/p)}$. It follows that $\hat{D}_{(p)}$ must be infinite for at least one prime factor p of n. Now select any $\xi_1 \not\equiv 1$ in such an infinite $\hat{D}_{(p)}$. Then $\langle \xi_1 \rangle \cong \mathbb{Z} \bmod p$. Next choose any ξ_2 in $\hat{D}_{(p)}$ but not in $\langle \xi_1 \rangle$. Then $\langle \xi_1 \rangle \cap \langle \xi_2 \rangle = \{1\}$; for if $\xi = \xi_1^k = \xi_2^\ell$ belongs to $\langle \xi_1 \rangle \cap \langle \xi_2 \rangle$ and if $\ell \not\equiv 0 \pmod p$ so that $\xi \not\equiv 1$, then there exists an integer j such that $j\ell \equiv 1 \pmod p$ and therefore

$$\xi_2 = (\xi_2^\ell)^j = (\xi_1^k)^j \in \langle \xi_1 \rangle ,$$

contrary to our choice of ξ_2. Therefore

$$\langle \xi_1, \xi_2 \rangle = \langle \xi_1 \rangle + \langle \xi_2 \rangle = \langle \xi_1 \rangle \oplus \langle \xi_2 \rangle .$$

Suppose now that $\xi_1, \ldots, \xi_m$ have been chosen from $\hat{D}_{(p)}$ so that

$$\langle \xi_1, \ldots, \xi_m \rangle = \langle \xi_1 \rangle \oplus \langle \xi_2 \rangle \oplus \cdots \oplus \langle \xi_m \rangle .$$

Choose any ξ_{m+1} in $\hat{D}_{(p)}$ but not in $\langle \xi_1, \ldots, \xi_m \rangle$. This is possible because $\langle \xi_, \ldots, \xi_m \rangle$ is finite. Then suppose that

$$\xi = (\xi_{m+1})^\ell \in \langle \xi_1, \ldots, \xi_m \rangle \quad \text{for some} \quad 2 \leq \ell < p .$$

Then for some j, $j\ell \equiv 1 \pmod p$ so that

$$(\xi_{m+1})^{j\ell} = \xi^j \in \langle \xi_1, \ldots, \xi_m \rangle .$$

That is, $\xi_{m+1} \in \langle \xi_1, \ldots, \xi_m \rangle$ contrary to hypothesis. Thus

$$\langle \xi_{m+1} \rangle \cap \langle \xi_1, \ldots, \xi_m \rangle = \{1\} ,$$

so that

$$\langle \xi_1, \ldots, \xi_m, \xi_{m+1} \rangle = \langle \xi_1 \rangle \oplus \cdots \oplus \langle \xi_m \rangle \oplus \langle \xi_{m+1} \rangle .$$

Hence there exists an infinite sequence $\xi_1, \xi_2, \xi_3, \ldots$ in $\hat{D}_{(p)}$ such that for every $m \geq 1$, $\xi_{m+1} \notin \langle \xi_1, \ldots, \xi_m \rangle$ and

$$\langle \xi_1, \ldots, \xi_m \rangle = \langle \xi_1 \rangle \oplus \cdots \oplus \langle \xi_m \rangle .$$

Thus $H = \langle \xi_1, \xi_2, \xi_3, \ldots \rangle$ is the weak direct product of all the $\langle \xi_j \rangle$ for

$j = 1,2,3,\ldots$. Since each $\langle \xi_j \rangle$ is isomorphic to $\mathbb{Z}$ mod p, it follows that the subgroup H of $\hat{D}_{(p)}$ is isomorphic to the weak direct product $(\mathbb{Z} \bmod p)^{\omega*}$ of denumerably many copies of $\mathbb{Z}$ mod p. (Of course $\hat{D}_{(p)} \cong (\mathbb{Z} \bmod p)^{\alpha*}$ for some cardinal α. What we have just proved for the convenience of the reader can also be deduced from Theorem 8.5 on page 43 of Fuchs "Infinite Abelian Groups".)

Now this group $H = \langle \xi_1, \xi_2, \xi_3, \ldots \rangle$ just constructed is a subgroup of $\hat{D}$ and satisfies ([12], Theorem 23.25)

$$[D/\mathbb{A}(D,H)]^\wedge \cong \mathbb{A}(\hat{D}, \mathbb{A}(D,H)) = H \cong (\mathbb{Z} \bmod p)^{\omega*},$$

from which it follows that

$$D/\mathbb{A}(D,H) \cong (\mathbb{Z} \bmod p)^{\omega}.$$

Since $D = G/C$ it follows that $(\mathbb{Z} \bmod p)^{\omega}$ is a continuous homomorphic image of G.

Finally, $(\mathbb{Z} \bmod p)^{\omega}$ is clearly second-countable and since it is a torsion group it follows from parts (c) and (d) of Theorem 6 that it is not polythetic.

Q.E.D.

Proof of Theorem 6(c). First we prove that (i) implies (ii). Let $\hat{G}_0$ denote the subgroup of $\hat{G}$ generated by Γ_0. Then $\hat{G}_0$ is also denumerable and plays the same role as Γ_0. Now if H denotes $\mathbb{A}(G,\hat{G}_0)$, the annihilator of $\hat{G}_0$ in G [12, p. 365], then $(G/H)^\wedge \cong \mathbb{A}(\hat{G},H) = \mathbb{A}(\hat{G},\mathbb{A}(G,\hat{G}_0)) = \hat{G}_0$. Therefore, G/H is second countable because $w(G/H) = \text{card}(\widehat{G/H}) = \text{card } \hat{G}_0 = \text{card } \Gamma_0$. Also G/H is not polythetic: For suppose that $a_1 + H, \ldots, a_m + H$ are m elements of G/H which generate a dense subgroup. But each continuous character would be uniquely determined by its values on these m elements and yet by Property (i) there is a character $\xi \neq 1$ in $\hat{G}_0 \cong (G/H)^\wedge$ such that $\xi(a_1 + H) = \cdots = \xi(a_m + H) = 1$.

Next we prove that (ii) implies (i). Given that (ii) holds, then $\hat{G}_0 \equiv (G/H)^\wedge$ is a denumerable subgroup of $\hat{G}$, namely $(G/H)^\wedge \cong \mathbb{A}(\hat{G},H)$. (See [12], §23.25.) If $\{a_1,\ldots,a_m\}$ is any finite subset of G, then since G/H is not polythetic there must exist by 6(d) a character ξ in $(G/H)^\wedge$ such that $\xi(a_1+H) = \cdots = \xi(a_m+H) = 1$ so that in particular $\xi(a_1) = \cdots = \xi(a_m) = 1$ and $\xi \neq 1$. Q.E.D.

In order to prove that (i) implies the existence of discountinuous TILF's on $L^2(G)$, it suffices to show (because of the above lemma) that $\Delta \neq M$. Choose any f in $L^2(G)$ such that the support of $\hat{f}$ is exactly the denumerable set Γ_0. Then since $1 \notin \Gamma_0$, $\hat{f}(1) = 0$. That is, $f \in M$. But property (i) shows that every equation of the form

$$\hat{f}(\xi) = \hat{g}_1(\xi)(1 - \overline{\xi(a_1)}) + \cdots + \hat{g}_m(\xi)(1 - \overline{\xi(a_m)})$$

is impossible, because (no matter how the elements $a_1,\dots,a_m$ are chosen in G) there is always a character ξ in $\Gamma_0 = \operatorname{supp} \hat{f}$ such that $\xi(a_1) = \cdots = \xi(a_m) = 1$. Thus, such an f in M is not in Δ.

Finally, if G is an infinite compact abelian torsion group, then G is necessarily totally disconnected (for otherwise each character maps the component of the identity, also torsion, onto the circle group $\mathbb{T}$, which is impossible) and cannot be polythetic (because a finitely generated subgroup of a torsion group is finite). Since subgroups of torsion groups are also torsion groups and since $[G/\mathbb{A}(G,\Gamma)]^\wedge \cong \Gamma$ for any subgroup Γ of $\hat{G}$, including denumerable subgroups, it follows that compact abelian torsion groups must satisfy property (i). Q.E.D.

Proof of Theorem 6(d). Suppose that G is polythetic. Then there is a finite subset $F = \{a_1,\dots a_m\}$ of G such that the closed subgroup $\langle F\rangle$ of G generated by F is equal to G. Therefore, if ξ is any continuous character on G satisfying $\xi(a_1) = \cdots = \xi(a_m) = 1$, then $\xi \equiv 1$. Consequently, G does not satisfy property (iii). On the other hand, suppose that G is not polythetic. Then for each finite subset $F = \{a_1,\dots a_m\}$ of G,

$$\mathbb{A}(\hat{G},F) \cong (G/\langle F\rangle)^\wedge \neq \{1\},$$

so that there must be a continuous character $\xi \not\equiv 1$ which annihilates all members of F. That is, property (iii) is satisfied. Q.E.D.

Proof of Theorem 6(e). Suppose that $G = G_0 \oplus G_1$ where G_0 is second-countable and not polythetic. Then, by 6(d), for every finite set of elements $a_j = b_j + c_j$, $1 \leq j \leq m$, in G with b_j in G_0 and c_j in G_1 there is a character ξ in $\hat{G}_0 \subset \hat{G}$ such that $\xi \not\equiv 1$ and $\xi(b_j) = 1$ for all $1 \leq j \leq m$. But then also $\xi(a_j) = \xi(b_j)\cdot 1 = \xi(b_j) = 1$ for all $1 \leq j \leq m$. Thus G satisfies property (i) of Theorem 6(c) with $\Gamma_0 = \hat{G}_0 \setminus \{1\}$. Q.E.D.

Proof of the Corollary to Theorem 6. We already know from Theorem 2 that there are no discontinuous TILF's on $L^2(G)$ if G/C is finite. If G/C is infinite, then it is either second-countable or it is not second-countable. In the latter case it follows from 6(b) and 6(a) that there are discontinuous TILF's on $L^2(G)$ because there are such TILF's on $L^2((\mathbb{Z} \bmod p)^{\omega})$ by 6(c). In the former case, i.e. when G/C is second-countable, it again follows from 6(c) that there are discontinuous TILF's on $L^2(G)$ if G/C is not polythetic. Thus there remains only the case that G/C is polythetic (and a fortiori second-countable). Q.E.D.

The next examples of discontinuous TILF's on compact groups were given in 1974 by Alain Connes [5] and Peter Ludvik [19]. Connes constructed an example of a function g in $L^1(\mathbb{T})$ with $\hat{g}(0) = \int_{\mathbb{T}} g(x)dx = 0$ but which is not in the linear subspace $\Delta(L^1)$ defined in §1 of this paper. Thus by property (9) of the list of properties of hyperplanes given in §1, there exist discontinuous TILF's on $L^1(\mathbb{T})$. Independently and more generally, Peter Ludvik proved in [19] that there are discontinuous TILF's on $L^1(G)$ for every compactly generated locally compact abelian group G. This complements one of Woodward's results in [39] to show that there are discontinuous TILF's on $L^1(G)$ for every locally compact abelian group G, but leaves the following question open.

Open Problem 6. Are TILF's automatically continuous on $L^1(G)$ for compact nonabelian groups G? The same question is open for $L^p(G)$, $1 < p \leq \infty$ and $C(G)$. For example, there is evidently [37] just one continuous TILF on L^∞ over the n-dimensional rotation group for $n \geq 4$. What about automatic continuity of TILF's on L^∞ over such groups?

6. Automatic continuity of operators

There is a large and growing literature on the automatic continuity of linear operators which commute with translations. See, for example, the excellent survey article by H. G. Dales [6] and the other papers of this Proceedings. I mention here only one such result that connects with TILF's. Starting with the results of Meisters & Schmidt associated with our Theorem 2 above, C. J. Lester proved in [18] that linear operators on $L^2(G)$ which commute with translations (by elements of a connected and compact abelian group G) are automatically continuous. Many analogous theorems can be proved in a similar manner. For example, starting with the results associated with Theorem 1 and its Corollary, it can be proved that every linear mapping S of $\mathcal{D}(\mathbb{R})$ into itself which commutes with translations is automatically continuous. It is probably possible to establish a general result of this type: If TILF's on $\mathfrak{F}(G)$ are automatically continuous, then so are linear operators $S: \mathfrak{F}(G) \to \mathfrak{F}(G)$ which commute with translations.

Rick Loy points out (in a personal communication to me) that some growth condition may be needed on $\mathfrak{F}(G)$ in order to get such a result to work because $C(\mathbb{R})$ has no discontinuous TILF's and yet Barry Johnson [TAMS 128 (1967), 88-102] has shown that $C(\mathbb{R})$ does admit a discontinuous operator commuting with all translations. Furthermore, in the same paper, Barry Johnson has already shown that the converse of such a result is false for $\mathfrak{F}(G) = L^p(\mathbb{R})$.

References

[1] L. Asam, Invariante Linearformen auf Funktionenräumen über lokalkompakten Gruppen, Diplomarbeit, München, 1977.

[2] A. Baker, Transcendental Number Theory, Cambridge Univ. Press, London, 1975.

[3] F. Bruhat, Distributions sur un groupe localement compact et applications à l'étude des représentations des groupes p-adiques, Bull. Soc. Math. France, 89 (1961), 43-75.

[4] J. W. S. Cassels, An Introduction to Diophantine Approximation, Cambridge Univ. Press, London, 1957.

[5] A. Connes, There exist discontinuous translation-invariant linear forms on $L^1(\mathbb{T})$. Personal communication to C. J. Lester and included in Lester's paper in J. London Math. Soc., (2) 11 (1975), 145-146.

[6] H. G. Dales, Automatic Continuity: a survey, Bull. London Math. Soc., 10 (1978), 129-183.

[7] J. Dieudonné, Treatise on Analysis, Volume VI. (Chapter XXII, Harmonic Analysis), Academic Press, New York, 1978. Problem 30, pages 207-208, outlines Meisters' 1971 results on $\mathcal{D}(\mathbb{R})$.

[8] W. F. Donoghue, Distributions and Fourier Transforms, Academic Press, New York, 1969.

[9] R. E. Edwards, Fourier Series, a Modern Introduction, 2 vols. Holt, Rinehart and Winston, New York, 1967.

[10] C. Guichard, Sur la résolution de l'équation aux différences finies $G(x+1) - G(x) = H(x)$, Ann. Sci. Ecole Norm. Sup., 4 (1887), 361-380.

[11] G. H. Hardy and E. M. Wright, An Introduction to the Theory of Numbers, Oxford Univ. Press, London, 1954.

[12] E. Hewitt and K. A. Ross, Abstract Harmonic Analysis, Vol. I, Springer-Verlag, Berlin, 1963.

[13] L. Hörmander, Generators for some rings of analytic functions, Bull. Amer. Math. Soc., 73 (1967), 943-949.

[14] D. L. Johnson, Translation-invariant linear forms on $\mathcal{D}(G)$, Math. Ann., 250 (1980), 109-112.

[15] E. Kregelius-Petersen, On distributions of compact support, Ph.D. thesis, Univ. Nebraska-Lincoln, 1977.

[16] E. Kregelius-Petersen, Simultaneous approximation and a theorem of Hörmander, to appear.

[17] E. Kregelius-Petersen and G. H. Meisters, Non-Liouville numbers and a theorem of Hörmander, J. Functional Analysis, 29 (1978), 142-150.

[18] C. J. Lester, Continuity of operators on $L^2(G)$ and $L^1(G)$ commuting with translations, J. London Math. Soc., (2) 11(1975), 144-146.

[19] P. Ludvik, Discontinuous translation-invariant linear functionals on $L^1(G)$, Studia Mathematica, 56 (1976), 21-30.

[20] G. H. Meisters, Translation-invariant linear forms and a formula for the Dirac measure, Bull. Amer. Math. Soc., 77 (1971), 120-122.

[21] __________, Translation-invariant linear forms and a formula for the Dirac measure, J. Functional Analysis, 8 (1971), 173-188.

[22] __________, Some discontinuous translation-invariant linear forms, J. Functional Analysis, 12 (1973), 199-210.

[23] __________, Discontinuous invariant linear forms on $L^2(G)$, Notices A.M.S., Vol. 19, No. 6, October 1972. Abstract 72T-B247, page A-693.

[24] __________, A Guichard theorem on connected monothetic groups, Studia Mathematica, 43 (1973), 161-163.

[25] __________, Periodic distributions and non-Liouville numbers, J. Functional Analysis, 26 (1977), 68-88.

[26] G. H. Meisters and W. M. Schmidt, Translation-invariant linear forms on $L^2(G)$ for compact abelian groups G, J. Functional Analysis, 11 (1972), 407-424.

[27] H. Reiter, Classical Harmonic Analysis and Locally Compact Groups, Oxford Univ. Press, London, 1968.

[28] R. D. Richtmyer, On the structure of some distributions discovered by Meisters, J. Functional Analysis, 9 (1972), 336-348.

[29] W. Roelcke, L. Asam, S. Dierolf, and P. Dierolf, Discontinuous translation-invariant linear forms on $\mathcal{K}(G)$, Math. Ann., 239 (1979), 219-222.

[30] W. Rudin, Fourier Analysis on Groups, Interscience, New York, 1962.

[31] __________, Functional Analysis, McGraw-Hill, New York, 1973.

[32] S. Saks and A. Zygmund, Analytic Functions, 2nd ed., Warszawa 1965.

[33] W. M. Schmidt, Approximation to algebraic numbers, L'Enseignement Mathématique, 17 (1971), 187-253.

[34] __________, Diophantine Approximation, Lecture Notes in Mathematics No. 785, Springer-Verlag, Berlin, 1980.

[35] L. Schwartz, Théorie des Distributions, Hermann, Paris, 1966.

[36] A. C. Serold, Discontinuous translation-invariant linear forms exist on $\ell_p(\mathbb{Z})$, $1 < p \leq \infty$, $C(\mathbb{Z})$ and $C_0(\mathbb{Z})$. Notices A.M.S., Vol. 19, No. 2, February 1972, Abstract 72 T-B 90, page A-321.

[37] D. Sullivan, For $n > 3$ there is only one finitely additive rotationally invariant measure on the n-sphere defined on all Lebesgue measurable subsets, Bulletin Amer. Math. Soc. (N.S.), 4 (1981), 121-123.

[38] A. E. Taylor and D. C. Lay, Introduction to Functional Analysis, 2nd ed., Wiley, N.Y. 1980.

[39] G. S. Woodward, Translation-invariant linear forms on $C_0(G)$, $C(G)$, $L^p(G)$ for noncompact groups, J. Functional Analysis, 16 (1974), 205-220.

Department of Mathematics & Statistics
University of Nebraska-Lincoln
Lincoln, NE 68588, U.S.A.

ON THE UNIQUENESS OF RIEMANN INTEGRATION

Richard J. Loy

Denote by $\mathcal{K}(\mathbb{R})$ the space of continuous functions of compact support in $\mathbb{R}$ with the inductive limit topology. For $f \in \mathcal{K}(\mathbb{R})$ and $\alpha \in \mathbb{R}$ define the translate $f_\alpha : t \mapsto f(t-\alpha)$. Let φ be a non-zero linear functional on $\mathcal{K}(\mathbb{R})$ which is translation invariant: $\varphi(f_\alpha) = \varphi(f)$ for $f \in \mathcal{K}(\mathbb{R})$, $\alpha \in \mathbb{R}$. The obvious example of such a φ is the Riemann integral of f which we will denote $\int f$. Further, by elementary properties of Riemann integration this is continuous on $\mathcal{K}(\mathbb{R})$. Conversely, if the translation invariant functional φ is supposed to be continuous then $\varphi(\cdot) = \lambda \int(\cdot)$ for some constant λ. Again, the proof is elementary and relies on the simple result that if $f \in \mathcal{K}(\mathbb{R})$ and $\int f = 0$ then $f = g'$ for some $g \in \mathcal{K}(\mathbb{R})$.

The problem of concern to us here is whether the continuity hypothesis on φ is superfluous: if φ is a translation invariant linear functional on $\mathcal{K}(\mathbb{R})$ is φ necessarily continuous? My own interest in the problem dates from around 1968 when I came across it in the notes of Edwards [2], or rather a prepublication version of the same. I am still looking for the answer!

Let us define

$$\Delta(\mathbb{R}) = \operatorname{span}\{f_\alpha - f : f \in \mathcal{K}(\mathbb{R}),\ \alpha \in \mathbb{R}\}.$$

Then $\Delta(\mathbb{R}) \subset \ker\varphi$ is equivalent to the translation invariance of φ. The uniqueness result above shows via the Hahn-Banach theorem that $\overline{\Delta(\mathbb{R})} = \{f \in \mathcal{K}(\mathbb{R}) : \int f = 0\}$. If $\Delta(\mathbb{R})$ is not closed, or has infinite codimension, then there are discontinuous φ with $\Delta(\mathbb{R}) \subset \ker\varphi$. Indeed, we will show below that $\Delta(\mathbb{R})$ must have infinite codimension if such a discontinuous φ exists.

Thus we see that the automatic continuity of translation invariant linear functionals on $\mathcal{K}(\mathbb{R})$ is equivalent to any (and hence all) of the following

$$\operatorname{codim}\Delta(\mathbb{R}) = 1 \tag{i}$$

$$\Delta(\mathbb{R}) = \{f \in \mathcal{K}(\mathbb{R}) : \textstyle\int f = 0\} \tag{ii}$$

$$\Delta(\mathbb{R}) = \{f \in \mathcal{K}(\mathbb{R}) : f = g' \text{ for some } g \in \mathcal{K}(\mathbb{R})\}. \tag{iii}$$

The first significant result was that of Meisters [11] who showed, in particular, that on $\mathcal{D}(\mathbb{R}) = C^\infty(\mathbb{R}) \cap \mathcal{K}(\mathbb{R})$ the analogue of (iii) held. Indeed,

$$\{f \in \mathcal{D}(\mathbb{R}) : f = g' \text{ for some } g \in \mathcal{D}(\mathbb{R})\}$$
$$= \{u_\alpha - u + v_\beta - v : u, v \in \mathcal{D}(\mathbb{R})\} \tag{iv}$$

precisely when the quotient α/β is a non-Liouville number, and so (iv) certainly holds for almost all α/β. Suffice it to say at this point that the proof involves study of the Fourier transforms of elements of $\mathcal{D}(\mathbb{R})$ and this is just not possible for the larger space $\mathcal{K}(\mathbb{R})$. One should compare with the case of the unit circle given below.

Thus if we take $f \in \mathcal{D}(\mathbb{R})$ with $\int f = 0$ and choose $\beta = 1$ then, for suitable α between 0 and 1, there exist $u,v \in \mathcal{D}(\mathbb{R})$ with $f = u_\alpha - u + v_1 - v$. Taking the translates of this equation by each $k \in \mathbb{Z}$ and adding we obtain

$$\Sigma f_k = (\Sigma u_k)_\alpha - \Sigma u_k + \Sigma (v_{k+1} - v_k).$$

Since f,u,v have compact support the sums are well defined, and clearly $\Sigma (v_{k+1} - v_k) = 0$. Further, each of $\Sigma f_k, \Sigma u_k$ is C^∞ and of period 1. Thus if we restrict to the interval $[0,1)$, which we will identify with the unit circle $\mathbb{T}$, we have

$$\Sigma f_k = (\Sigma u_k)_\alpha - \Sigma u_k$$

(where translation by α is now modulo 1) holding between functions in $C^\infty(\mathbb{T})$, and, in particular, Σf_k has zero integral on $\mathbb{T}$. Conversely, any function in $C^\infty(\mathbb{T})$ with zero integral is obtainable in this manner — extend the function to $\mathbb{R}$ by periodicity then multiply by a suitable C^∞ cut-off function to obtain the desired element of $\mathcal{D}(\mathbb{R})$. So we deduce

$$\{f \in C^\infty(\mathbb{T}) : \int f = 0\} = \{g_\alpha - g : g \in C^\infty(\mathbb{T})\} \qquad \text{(v)}$$

for all non-Liouville $\alpha \in \mathbb{T}$. This result was obtained independently by Herman [5] in his study of diffeomorphisms of $\mathbb{T}$. The proof is simple enough to reproduce here; it is dual to the argument of Meisters [12].

Given $f \in C^\infty(\mathbb{T})$ with $\int f = 0$ the only possibility for $g \in C^\infty(\mathbb{T})$ (or even $C(\mathbb{T})$) to satisfy (v) must be determined by its Fourier coefficients

$$\hat{g}(k) = \hat{f}(k)\,(1 - e^{2\pi i k\alpha})^{-1} \qquad (k \in \mathbb{Z}\setminus\{0\})$$

($\hat{f}(0) = \int f = 0$, take $\hat{g}(0) = 0$ without loss of generality). If α is a non-Liouville number (in Herman's terminology α satisfies a diophantine condition) then there is some $\beta > 0$ such that

$$\sum_{k\neq 0} |k|^{-1-\beta}\,|1 - e^{2\pi i k\alpha}|^{-1} < \infty.$$

But since $f \in C^\infty(\mathbb{T})$, $\hat{f}(k) = \underline{O}(|k|^{-r})$ for any $r > 0$ and so

$$g(t) = \Sigma\,\hat{g}(k)e^{2\pi i k t} \qquad (t \in \mathbb{T})$$

defines a C^∞ function on $\mathbb{T}$ as required. Just as Meisters had shown for $\mathbb{R}$ the supposition that α be a non-Liouville is essential: for α Liouville the right side of (v) is meagre in the left side.

As can be judged from this argument things are much simpler on $\mathbb{T}$, essentially because of compactness, and perhaps we should consider our basic problem there first. In fact this is all we need to do! Roelcke et al [14] have shown that codim $\Delta(\mathbb{R}) =$ codim $\Delta(\mathbb{T})$ where, of course,

$$\Delta(\mathbb{T}) = \text{span}\{f_\alpha - f : f \in C(\mathbb{T}),\ \alpha \in \mathbb{T}\}. \tag{vi}$$

We will henceforth consider the more tractable case of $C(\mathbb{T})$ and for convenience introduce the notation $C(\mathbb{T})_0 = \{f \in C(\mathbb{T}) : \int f = 0\}$. Here things look a little brighter, at least for a start. If $\alpha \in \mathbb{T}$ is irrational then a result of Gottschalk and Hedlund [4] (see also Shapiro [15]) gives the following. For $f \in C(\mathbb{T})_0$ there is $g \in C(\mathbb{T})$ with $f = g_\alpha - g$ if and only if $\sup_n |\Sigma_1^n f(r\alpha)| < \infty$. This looks like a direct method for investigating whether the continuous case of (v) holds. It does, however, have some less pleasant companions. Given α, we can choose $f \in C(\mathbb{T})_0$ such that there is no $g \in C(\mathbb{T})$ with $f = g_\alpha - g$. Indeed, there are $f \in C(\mathbb{T})_0$ such that $f = g_\alpha - g$ fails for any Borel g for α in a non-meagre set [5], and for each α the range of $g \mapsto g_\alpha - g$ in $C(\mathbb{T})_0$ is meagre even when g is allowed to be Borel [6]. So clearly (v) fails hopelessly for the continuous case.

There is yet a further result which shows that the desired result, if true, must allow the translates to be dependent on the function concerned, in contrast to the C^∞ case. Let $\alpha_1, \ldots, \alpha_n \in \mathbb{T}$ be fixed and consider the mapping of $C(\mathbb{T})_0^n$ into $C(\mathbb{T})_0$ given by

$$D : (g_1, \ldots, g_n) \mapsto \sum_{i=1}^{n} \{(g_i)_{\alpha_i} - g_i\}.$$

By an open mapping theorem argument Meisters [12] has shown that D has meagre range. It is then immediate that even allowing countably many (but fixed) α we cannot span $C(\mathbb{T})_0$. On the other hand it is worth remarking on a result of Feldman and Moore [3]. If $\alpha, \beta \in \mathbb{T}$ are algebraically independent irrationals then <u>any</u> Borel function $f : \mathbb{T} \to \mathbb{R}$ is of the form $g_\alpha - g + h_\beta - h$ for some Borel g, h. Here of course g, h will be unbounded, indeed non-integrable, in general, even for continuous f. Having two functions here is crucial, if $f = g_\alpha - g$ for continuous f and Borel g then $\int f = 0$ by a result of Anosov [1].

Coming back to the result of Gottschalk and Hedlund, let $f \in C(\mathbb{T})_0$. Then

$$f(t) = \sum_k \hat{f}(k) e^{2\pi i k t}$$

for almost all t. Thus for almost all α

$$\sum_{r=1}^{n} f(r\alpha) = \sum_{k \neq 0} \hat{f}(k) \frac{1 - \exp(2\pi i k(n+1)\alpha)}{1 - \exp(2\pi i k \alpha)}$$

for all n. Unfortunately the obvious estimates of the right side give no more information than the more direct approach of requiring $\sum \hat{f}(k)\,(1-\exp(2\pi i k\alpha))^{-1}$ to be absolutely convergent. And that cannot work in general. We remark that if α is irrational the individual ergodic theorem shows that

$$\frac{1}{n}\sum_{r=1}^{n} f(r\alpha)$$

converges to 0 as $n\to\infty$. Given f, all we want is to be able to choose α such that the convergence is sufficiently rapid. By the law of the iterated logarithm the best one could hope for in general (that is, with sums $\frac{1}{n}\sum f(x_r)$ where $(x_r)\in \mathbb{T}^{\infty}$) is convergence $\underline{O}(\sqrt{\log\log n/n})$. In our case (that is, $x_r = r\alpha$) the best estimate is $\underline{O}(M(\log n/n))$ where M is the modulus of continuity of f. See Kuipers and Niederreiter [7] for details. These estimates are not sufficient for our purpose, but of course they make no allowance for (x_r) to depend on f (this would obviously make a considerable difference in the first case).

All in all the problem for general f in $C(\mathbb{T})_0$ continues to be intractable. But we can definitely improve on (v) which only deals with C^{∞} functions. Indeed for any $\varepsilon>0$, if $f\in C(\mathbb{T})_0$ is $C^{1+\varepsilon}$ then $\hat{f}(k)=\underline{O}(|k|^{-1-\varepsilon})$ and the identical argument to the C^{∞} case shows that if $\alpha\in\mathbb{T}$ is a Roth number (as are almost all α) then $f=g_{\alpha}-g$ for some $g\in C(\mathbb{T})$. In fact even more than this is true [5], but for our purposes this is a dead end because of the category result of Meisters above. There is one positive result, however, which breaks new ground in that the translates are no longer fixed. Meisters and Schmidt [13] have shown that if $f\in A(\mathbb{T})$ (that is, $\hat{f}\in \ell^1$) with $\int f=0$ then for almost all $(\alpha,\beta)\in\mathbb{T}^2$ there are $g,h\in A(\mathbb{T})$ with $f=(g_{\alpha}-g)+(h_{\beta}-h)$. Furthermore the pair (α,β) cannot be chosen independently of f. This does not preclude that g alone suffices or that independence of (α,β) from f may be possible when g,h are allowed to lie in the larger $C(\mathbb{T})$.

On the other hand Connes (see [8]) and Ludvik [9] have shown that the span of differences of translates of functions in $L^1(\mathbb{T})$ is <u>not</u> of codimension one in $L^1(\mathbb{T})$, and noting that the operators of translation and differentiation commute it follows that this is also the case for the space of absolutely continuous functions on $\mathbb{T}$. Again, this does not preclude the possibility that every absolutely continuous function with zero integral lies in $\Delta(\mathbb{T})$. The same argument would give the desired result in $C(\mathbb{T})$ if we had it in $C^1(\mathbb{T})$, but it is not even known whether $C^1(\mathbb{T})_0\subset\Delta(\mathbb{T})$.

After these piecemeal results, let us conclude with some remarks on the overall structure of $\Delta(\mathbb{T})$. For each n define

$$X_n=\{f\in\Delta(\mathbb{T}) : f=\sum_{i=1}^{n}\{(g_i)_{\alpha_i}-g_i\} \text{ for some } \alpha_1,\dots,\alpha_n \text{ with } \|g_i\|\le n\|f\|,\ i=1,2,\dots n\}.$$

It follows that $\Delta(\mathbb{T}) = \cup X_n$ is a dense analytic subspace of $C(\mathbb{T})_0$, and as such either $\Delta(\mathbb{T}) = C(\mathbb{T})_0$ or codim $\Delta(\mathbb{T})$ is uncountable. Hence the analogous result for $\Delta(\mathbb{R})$ follows by the result of [14], or directly by the obvious modification of argument. If we suppose that $\Delta(\mathbb{T}) = C(\mathbb{T})_0$ then some X_n is non-meagre in $C(\mathbb{T})_0$ and so by the Pettis Lemma (see [9] for example) $X_n - X_n$ is a neighbourhood of zero in $C(\mathbb{T})_0$. But $X_n - X_n \subset X_{2n}$ which is closed under scalar multiplication, whence $X_{2n} = C(\mathbb{T})_0$. In particular a fixed number of summands suffices for representing a function in $C(\mathbb{T})_0$ as a sum of differences of translates. And we have norm estimates, so that

$$|f| = \inf\{\Sigma \|g_i\| : f = \Sigma (g_i)_{\alpha_i} - g_i \quad \text{for some} \quad \alpha_i\}$$

is a norm on $C(\mathbb{T})_0$ equivalent to the supremum norm $\|\cdot\|$.

References

[1] D. V. Anosov, On an additive functional homology equation connected with an ergodic rotation of the circle, Translations Math. U.S.S.R. Izvestija, 74 (1973), 1257-1271.

[2] R. E. Edwards, What is the Riemann integral? Notes on Pure Mathematics #1, Australian National University, 1974.

[3] J. Feldman and C. C. Moore, Ergodic equivalence relations, cohomology, and von Neumann algebras I, Trans. Amer. Math. Soc., 234 (1977), 289-324.

[4] W. Gottschalk and G. Hedlund, Topological Dynamics, Amer. Math. Soc. Colloq. Publ. Vol. 36, American Mathematical Society, Providence, 1955.

[5] R. Herman, Sur la conjugaison différentiable des difféomorphismes du cercle a des rotations, Publ. Math. No. 49, I.H.E.S. (1979).

[6] R. Jones and W. Parry, Compact abelian group extensions of dynamical systems II, Comp. Math., 25 (1972), 135-147.

[7] L. Kuipers and H. Niederreiter, Uniform Distribution of Sequences, John Wiley and Sons, New York, 1974.

[8] C. J. Lester, Continuity of operators on $L_2(G)$ and $L_1(G)$ commuting with translations, J. London Math. Soc., (2), 11 (1975), 144-146.

[9] R. J. Loy, Multilinear mappings and Banach algebras, J. London Math. Soc., (2) 14 (1976), 423-429.

[10] P. Ludvik, Discontinuous translation invariant linear functionals on $L^1(G)$, Studia Math., 56 (1976), 21-30.

[11] G. H. Meisters, Translation-invariant linear forms and a formula for the Dirac measure, J. Functional Analysis, 8 (1971), 173-188.

[12] G. H. Meisters, Periodic distributions and non-Liouville numbers, J. Functional Analysis, 26 (1977), 68-76.

[13] G. H. Meisters and W. M. Schmidt, Translation-invariant linear forms on $L^2(G)$ for compact abelian groups G, J. Functional Analysis, 11 (1972), 407-424.

[14] W. Roelcke. L. Asam, S. Dierolf and P. Dierolf, Discontinuous translation-invariant linear forms on $\mathcal{K}(G)$, Math. Ann., 239 (1979), 219-222.

[15] L. Shapiro, Regularities of Distribution, Studies in Proability and Ergodic Theory, Advances in Mathematics Supplementary Studies, Vol. 2 (G.-C. Rota Ed.), Academic Press, New York, 1978.

Department of Mathematics
Faculty of Science
The Australian National University
Canberra ACT 2600
Australia

THE CONTINUITY OF TRACES

H. G. Dales[†]

Let A be a Banach star algebra. We consider conditions on A which imply that each trace on A is necessarily continuous. If each trace on A is continuous, then it is easily seen that A^2 is closed and of finite codimension in A, and it has been guessed that this necessary condition on A for all traces to be continuous might be sufficient: indeed, the possibility that, if A^2 is closed and of finite codimension, then each positive functional on A is continuous was suggested by Varopoulos in 1964, and was raised as Question 25 in [2].

In this note, I shall give a counter-example to the guess: an example of a Banach star algebra A (with isometric involution) for which A^2 is closed and of finite codimension, but on which there is a discontinuous trace. The example is a modification of a well-known example of P. G. Dixon ([3]). However, the example is not very satisfactory, in that neither it nor simple variants seem to resolve either of two related questions: If A^3 is closed and of finite codimension in A, or if $A = A^2$, is every trace on A necessarily continuous?

The example is non-separable. We shall note the easy fact that, if A is a separable Banach star algebra, then each trace on A is necessarily continuous if and only if A^2 is closed and of finite codimension in A. Again, however, this leaves open a more interesting question: If A is a separable Banach star algebra in which A^2 is closed and of finite codimension, is every positive functional on A necessarily continuous?

In §1, I give the basic definitions and describe the modifications of Dixon's example that are required to exhibit a Banach star algebra A in which A^2 is of finite codimension in A but is not closed: this may be of interest in itself. In §2, I give the counter-example to the guess, and I make a few further remarks in §3.

I am grateful to Peter Dixon for valuable comments.

1. Throughout, all linear spaces and algebras are taken over the complex field, $\mathbb{C}$. Let $n \in \mathbb{N}$, and let A be an algebra. Then A^n is the space spanned by products of n elements of A: it is a bi-ideal in A. If Y is a subspace of the linear space X, then Y is <u>cofinite</u> in X if the quotient space X/Y is finite dimensional, and we write $\operatorname{codim} Y = \dim(X/Y)$.

[†] Supported by NATO Grant No. RG 073.81.

DEFINITION. Let X be a linear space, A <u>linear involution</u> on X is a map $x \mapsto x^*$, $X \to X$, such that

(i) $(\alpha x + \beta y)^* = \bar{\alpha}x^* + \bar{\beta}y^*$,

(ii) $(x^*)^* = x$,

for $x,y \in X$ and $\alpha,\beta \in \mathbb{C}$. An element x is <u>self-adjoint</u> if $x = x^*$, and a subspace Y of X is <u>star-closed</u> if $x^* \in Y$ $(x \in Y)$.

LEMMA 1. Let X be a linear space with a linear involution. Then X has a Hamel basis consisting of self-adjoint elements.

<u>Proof</u>. Let $Y = \{x + x^*, i(x - x^*) : x \in X\}$. Then Y spans X and Y consists of self-adjoint elements. Take a maximal linearly independent subset of Y. This is the required Hamel basis.

DEFINITION. Let A be an algebra. An <u>involution</u> on A is a linear involution with the additional property that

(iii) $(ab)^* = b^*a^*$ $(a,b \in A)$.

A <u>Banach star algebra</u> is a Banach algebra with an involution.

It is not required in the definition that the involution on a Banach star algebra be continuous. Basic results on Banach star algebras are given in Bonsall and Duncan ([1, §12 and Chapter V]).

Let X be an infinite-dimensional Banach space, and let X_1 be any subspace of X. In [3], Dixon constructs a Banach algebra A such that

$$A \supset X,\ AX = XA = 0,\ A^2 \cap X = X_1, \text{ and } A^2 + X = A. \qquad (1)$$

Now suppose that X is a Banach space with an isometric linear involution $*$, and suppose that X_1 is any star-closed subspace of X. We shall modify the construction in [3] to prove the following result.

THEOREM 2. There is a Banach star algebra A with a isometric involution such that (1) holds and such that the involution on A agrees with the given linear involution on X.

<u>Proof</u>. Throughout, we shall use the direct sum notation $\oplus$ as in [3] to denote the ℓ^1-direct sum of Banach spaces. A purely algebraic direct sum of linear spaces is denoted by $\odot$.

The construction in [3] gives Banach algebras A_α and E, and Banach spaces B_α, C_α, D_α for $0 \leq \alpha \leq \omega_1$ with

$$A_\alpha = X \oplus E \oplus B_\alpha \oplus C_\alpha \oplus D_\alpha \quad (0 \leq \alpha \leq \omega_1).$$

Here, E is a one-dimensional subspace generated by an element e, and $\|e\| = 1$. The multiplication in A_α is given by the following rules:

$$\left.\begin{aligned} &e^2 = e,\ be = b\ (b \in B_\alpha),\ ec = c\ (c \in C_\alpha), \\ &A_\alpha B_\alpha = C_\alpha A_\alpha = A_\alpha X = XA_\alpha = A_\alpha D_\alpha = D_\alpha A_\alpha = 0\,. \end{aligned}\right\} \tag{2}$$

The product $B_\alpha C_\alpha$ will be described below. If $\beta < \alpha$, then B_β, C_β, and D_β are closed subspaces of B_α, C_α, and D_α, respectively, and A_β is a closed subalgebra of A_α. The construction also involves a subspace D'_α of D_α which is defined for $0 \leq \alpha < \omega_1$.

We shall prove inductively that there is an involution $*$ on each algebra A_α such that $(A_\alpha, *)$ is a Banach star algebra with isometric involution and such that the following properties hold for each α and each $\beta < \alpha$:

$$\left.\begin{aligned} &* \text{ agrees with the given linear involution on } X; \\ &* \text{ extends the involution on } A_\beta; \\ &E^* = E \text{ and } D^* = D; \\ &B^*_\alpha = C_\alpha \text{ and } C^*_\alpha = B_\alpha; \\ &D'_\alpha \text{ is a star-closed subspace of } D_\alpha. \end{aligned}\right\} \tag{3}$$

Since $A = A_{\omega_1}$, this will give an algebra with the properties required in the theorem.

We shall use the following construction. Let S be an index set, let $\{a_s : s \in S\}$ be a subset of A_α, and let $T = \oplus\{\mathbb{C}a_s : s \in S\}$, a certain closed subspace of A_α. We shall specify $a^*_s \in A_\alpha$ so that $\|a^*_s\| = \|a_s\|$ for $s \in S$. Having done this, we shall extend the map $*$ to T by conjugate-linearity and continuity: if $\sum\alpha_s a_s \in T$, then $(\sum\alpha_s a_s)^* = \sum\bar{\alpha}_s a^*_s$. Then $* : T \to A$ will be an isometry satisfying (i), above.

We start by recalling the definition of A_0. In [3], a Hamel basis Y_0 of X_1 was chosen with $\|y\| = 1$ $(y \in Y_0)$. Since X_1 is a star-closed subspace of X, $*$ is a linear involution on X_1, and so, by Lemma 1, we can suppose further that

$$y^* = y \quad (y \in Y_0)\,. \tag{4}$$

Now $A_0 = X \oplus E \oplus B_0 \oplus C_0 \oplus D_0$ where $B_0 = \oplus\{\mathbb{C}b_y : Y_0\}$, $C_0 = \oplus\{\mathbb{C}c_y : y \in Y_0\}$, and $D_0 = D'_0 = \oplus\{\mathbb{C}b_y c_z : y, z \in Y_0,\ y \neq z\}$ for certain elements b_y, c_y, $b_y c_z$ with $\|b_y\| = \|c_y\| = \|b_y c_z\| = 1$. The multiplication on A_0 is defined by (2) and the further condition that $b_y c_y = y$ $(y \in Y_0)$. We define $*$ on A_0. It is already defined on X, and we set $e^* = e$. Set

$$b_y^* = c_y,\ c_y^* = b_y \quad \text{and} \quad (b_y c_z)^* = b_z c_y \tag{5}$$

for $y, z \in Y_0$ with $y \neq z$, and extend $*$ to B_0, C_0, and D_0, respectively. It is clear that $*$ is an isometric linear involution on A_0 which satisfies conditions (3).

It remains to prove that

$$(ab)^* = b^*a^* \quad (a, b \in A_0). \tag{6}$$

Since $A_0X = XA_0 = A_0D_0 = D_0A_0 = 0$, it is sufficient to suppose that $a, b \in E \oplus B_0 \oplus C_0$. Firstly, suppose that $b \in B_0$. Then $(eb)^* = 0 = b^*e^*$ because $b^* \in C_0$ and $A_0B_0 = C_0A_0 = 0$, and $(be)^* = b^* = e^*b^*$ because $be = b$ and $eb^* = b^*$. Similarly, the result holds if $b \in C_0$. Thus, (6) holds if $e \in \{a, b\}$. Secondly, suppose that $a, b \in B_0$ or that $a, b \in C_0$, or that $a \in C_0$, $b \in B_0$. Then $ab = 0 = b^*a^*$ because $A_0B_0 = C_0A_0 = 0$, and so (6) holds. Finally, suppose that $a \in B_0$ and $b \in C_0$. If $y \in Y_0$, then $(b_y c_y)^* = y^*$ and $c_y^* b_y^* = b_y c_y = y$, and so $(b_y c_y)^* = c_y^* b_y^*$ by (4). If $y, z \in Y_0$ with $y \neq z$, then $(b_y c_z)^* = c_z^* b_y^*$ by (5). Thus, (6) holds in this case. We have checked (6) in each case, and so A_0 has the required properties.

Now suppose that the involution has been constructed on A_α to have the required properties.

Equation (5) of [3] defines a linear subspace $F_{\alpha+1}$ of A_α, and we see that $F_{\alpha+1} = D_\alpha \cap (A_\alpha + X)$. Since D_α, A_α, and X are star-closed subspaces of A_α, it follows that $F_{\alpha+1}$ is also star-closed. Now $G_{\alpha+1}$ is a linear subspace of D_α chosen so that $F_{\alpha+1} \odot G_{\alpha+1} = D_\alpha$ and so that $G_{\alpha+1}$ is a subspace of D'_α. Since $F_{\alpha+1}$, D_α, and (by the inductive hypothesis) D'_α are star-closed, it follows that $G_{\alpha+1}$ can be taken to be star-closed. In [3], a Hamel basis, $Y_{\alpha+1}$, of $G_{\alpha+1}$ was chosen with $\|y\| = 1$ $(y \in Y_{\alpha+1})$: by Lemma 1, we can suppose further that

$$y^* = y \quad (y \in Y_{\alpha+1}), \tag{7}$$

a condition analogous to condition (4).

As in [3], $B_{\alpha+1} = B_\alpha \oplus B'_\alpha$ and $C_{\alpha+1} = C_\alpha \oplus C'_{\alpha+1}$, where $B'_{\alpha+1} = \oplus\{\mathbb{C}b_y : y \in Y_{\alpha+1}\}$ and $C'_{\alpha+1} = \oplus\{\mathbb{C}c_y : y \in Y_{\alpha+1}\}$. Also $D_{\alpha+1} = D_\alpha \oplus D'_{\alpha+1}$, where $D'_{\alpha+1} = \oplus\{\mathbb{C}b_y c_z : y \in Y_\beta, z \in Y_\gamma, y \neq z, \max\{\beta, \gamma\} = \alpha + 1\}$. The element $b_y c_z$ is the product of b_y and c_z, in that order, whilst $b_y c_y = y$ $(y \in Y_{\alpha+1})$. We extend $*$ to $B_{\alpha+1}$, $C_{\alpha+1}$, and $D_{\alpha+1}$ by setting $b_y^* = c_y$, $c_y^* = b_y$, and $(b_y c_z)^* = b_z c_y$, respectively, as in (5). The verification that $A_{\alpha+1}$ satisfies the inductive conditions is formally the same as the verification that A_0 satisfies the conditions: we must use (7) in place of (4). We must also specifically note that $D'_{\alpha+1}$ is a star-closed subspace of $D_{\alpha+1}$.

If α is a limit ordinal, the obvious construction is used. Here, B_α is the completion of the normed algebra $\bigcup\{B_\beta : \beta < \alpha\}$, etc., and we have an isometric map $* : \bigcup_{\beta<\alpha} B_\beta \to \bigcup_{\beta<\alpha} C_\beta$ which may be extended to B_α by continuity. Thus, we have defined $*$ on A_α. The space D'_α is chosen to be a subspace of D_α such that $(\bigcup_{\beta<\alpha} D_\beta) \odot D'_\alpha = D_\alpha$: since each D_β is a star-closed subspace, we can suppose that D'_α is also star-closed. Clearly, the inductive hypotheses are satisfied at α.

This completes the inductive construction. We set $A = A_{\omega_1}$, so that A is a Banach star algebra with isometric involution.

It was proved in [3] that A satisfies (1), and so the theorem is proved.

Note that the algebra A is not separable.

2.

DEFINITION. Let A be an algebra with an involution. A positive functional on A is a linear functional φ on A such that $\varphi(a^*a) \geq 0$ $(a \in A)$. A trace on A is a positive functional τ on A with the additional property that $\tau(a^*a) = \tau(aa^*)$ $(a \in A)$.

Note that, if a and b are self-adjoint elements of A, then

$$(a+ib)^*(a+ib) - (a+ib)(a+ib)^* = 2i(ab-ba),$$

and so it follows that, if τ is a trace on A, then

$$\tau(ab) = \tau(ba) \quad (a,b \in A). \tag{8}$$

We are concerned with conditions on a Banach star algebra A which ensure that positive functionals or traces on A are automatically continuous. The following is clear.

LEMMA 3. Let A be a Banach star algebra. If each trace on A is continuous, then A^2 is closed and cofinite in A.

Proof. If either A^2 is not closed or A^2 is not cofinite in A, then there exists a discontinuous linear functional on A which is zero on A^2. Such a linear functional is clearly a trace.

EXAMPLE 4. There is a Banach star algebra B (with an isometric involution) such that B^2 is closed and of codimension one in B, but such that there is a discontinuous trace on B.

Proof. Start with an infinite-dimensional Banach space X with an isometric linear involution $*$, and take a Hamel basis of the form $\{x_2, h_\nu\}$, consisting of self-adjoint elements, such that there is a sequence $(y_n) \subset \mathrm{lin}\{h_\nu\}$ with $y_n \to x_2$. [For example, let $X = C[0,1]$ with $F^*(t) = \overline{f(t)}$ $(f \in X,\ t \in [0,1])$, take x_2 to be the coordinate functional, and let $\{h_\nu\} \supset \{x_2^n : n \geq 2\}$.] Define $\tau(x_2) = 1$, $\tau(h_\nu) = 0$, and extend τ to X linearly. Let $X_1 = \ker \tau$. Then X_1 is a star-closed, dense linear subspace of codimension one in X.

By Theorem 2, there is a Banach star algebra A with an isometric involution such that equations (1) hold, and such that the involution extends the linear involution on X. Extend τ to be a linear functional on A by setting $\tau(a) = 0$ $(a \in A^2)$: equations (1) show that τ is consistently defined on A.

Take an element $x_1 \notin A$, set $\|x_1\| = 1$, and set $B = \mathbb{C}x_1 \oplus A$, so that B is a Banach space. Define a multiplication on B by setting $x_1^2 = x_2$ and $Ax_1 = x_1A = 0$. The multiplication is associative because $x_2A = Ax_2 = 0$, and so B is a Banach algebra. Set $(\alpha x_1 + a)^* = \overline{\alpha}x_1 + a^*$: this is clearly an isometric involution on B.

We have $A = \mathbb{C}x_2 \odot A^2$ from (1), and $\overline{A^2} = A$. Thus,

$$\overline{B^3} = B^2 = A, \quad B^3 = A^2,$$

and so B^2 is closed and of codimension one in B, whereas B^3 is not closed.

Extend the linear functional τ from A to B by choosing $\tau(x_1)$ arbitrarily. If $b_j = \alpha_j x_1 + A$ are elements of B for $j = 1,2$, then $b_1^* b_1 = |\alpha_1|^2 x_2 + A^2$, so that $\tau(b_1^* b_1) = |\alpha_1|^2$, and $b_1 b_2 = \alpha_1\alpha_2 x_2 + A^2$, so that $\tau(b_1 b_2) = \tau(b_2 b_1)$. Thus, τ is a trace on B. Clearly, τ is discontinuous.

Note that the trace in the above example has the additional property that $\tau(abc) = \tau(acb)$ for $a,b,c \in B$: in fact $\tau | B^3 = 0$. The example can be easily modified to give, for each $k \in \mathbb{N}$, an algebra B with B^2 closed and $\operatorname{codim} B^2 = k$, but such that there is a discontinuous trace on B.

3. The original guess can be modified by successively strengthening the hypotheses on A. Let A be a Banach star algebra. Suppose that:

(a) A^3 is closed and of finite codimension in A;

or

(b) $A = A^2$.

Is each trace on A necessarily continuous?

There are some indications that the answer to question (a) may be positive. The first remark is essentially contained in Loy ([4]).

PROPOSITION 5. Let τ be a trace on a Banach star algebra A. Then there exists a constant K such that

$$|\tau(abc)| \leq K\|a\| \, \|b\| \, \|c\| \quad (a,b,c \in A).$$

Proof. For each fixed $a,b \in A$, the map $x \mapsto \tau(axb)$ is continuous on A ([1, 37.8]), and so, by (8), the map $(a,b,c) \mapsto \tau(abc)$ is separately continuous on the Banach space $A \times A \times A$. The result follows from the uniform boundedness theorem.

It is not, of course, true that a trace on a Banach star algebra A necessarily satisfies the additional condition that $\tau(abc) = \tau(acb)$ $(a,b,c \in A)$: the ordinary trace on $M_2(\mathbb{C})$ does not have this property! The next result shows that, for traces with this property, the answer to question (a) is affirmative. Recall that Example 4 shows that there are discontinuous traces on algebras A with A^2 closed and cofinite, which do have this additional property. Thus, we see that, if we assume that A^3 is closed and cofinite, we may obtain different results in this area from those that follow from the assumption that A^2 is closed and cofinite.

PROPOSITION 6. Let A be a Banach star algebra such that A^3 is closed and cofinite in A. Let τ be a trace on A such that $\tau(abc) = \tau(acb)$ for a,b,c in A. Then τ is continuous.

Proof. Let D be the set of non-zero linear functionals λ on A^2 with the properties that: (i) $\lambda(a^*a) \geq 0$ $(a \in A)$; (ii) $\lambda(ab) = \lambda(ba)$ $(a,b \in A)$; (iii) $\lambda(abc) = \lambda(acb)$ $(a,b,c \in A)$. A standard argument ([1, 37.13]) shows that, if for each $\lambda \in D$ there is a continuous $\mu \in D$ with $(\lambda - \mu)(a^*a) \geq 0$ $(a \in A)$, then each $\lambda \in D$ is continuous.

Note that, for each functional on A^2 satisfying (i), $|\lambda(abc)|^2 \leq \lambda(c^*c)\lambda(abb^*a)$ for $a,b,c \in A$ ([1, 37.6(ii)]). Now take $\lambda \in D$. If $\lambda|A^4 = 0$, then the above inequality shows that $\lambda(abc) = 0$ $(a,b,c \in A)$, and so $\lambda|A^3 = 0$, whence λ is continuous on A^2 because A^3 is closed and cofinite in A^2: in this case, we take $\mu = \lambda$. If $\lambda|A^4 \neq 0$, then the standard identity $4axb = \sum_{n=1}^{4} i^n(a + i^n b^*)^* x (a + i^n b^*)$ shows that there exist $a_0, b, c \in A$ with $\lambda(a_0^* bca_0) \neq 0$: we can suppose that $\|a_0^* a_0\| < 1$. Let $\mu(x) = \lambda(a_0^* x a_0)$ $(x \in A^2)$. Then μ is a continuous non-zero linear functional on A^2. If $a \in A$, then $\mu(a^*a) = \lambda((aa_0)^*(aa_0)) \geq 0$, and so μ satisfies (i). If $a,b \in A$, then $\mu(ab) = \lambda(a_0^* aba_0) = \lambda(a_0 a_0^* ab) = \lambda(a_0 a_0^* ba) = \mu(ba)$, and so μ satisfies (ii). Similarly, μ satisfies (iii). Thus, $\mu \in D$. By [1, 12.11], there exists $b_0 \in A$ with $b_0^* = b_0$ and $2b_0 - b_0^2 = a_0^* a_0$. Thus, if $a \in A$, $(\lambda - \mu)(a^*a) = \lambda(a^*a - a_0^* a^* aa_0) = \lambda(a^*a - a^*a a_0^* a_0 a) = \lambda((a - b_0 a)^*(a - b_0 a)) \geq 0$. It follows that each $\lambda \in D$ is continuous.

If $\tau \neq 0$, then $\tau \in D$, and so the result follows.

As we remarked, the algebra B constructed in Example 4 is non-separable. For separable algebras, the original guess is true for traces: this was essentially proved by R. J. Loy in his work on the application of the theory of analytic spaces to separable Banach algebras ([4, 2.2]).

PROPOSITION 7. Let A be a separable Banach star algebra. Then each trace on A is continuous if and only if A^2 is closed and cofinite.

Perhaps the most interesting unresolved point is whether or not a positive functional on a separable Banach star algebra A in which A^2 is cofinite (and hence closed) is necessarily continuous.

References

[1] F. F. Bonsall and J. Duncan, Complete Normed Algebras, Springer-Verlag, New York, 1973.

[2] H. G. Dales, Automatic continuity: a survey, Bull. Lond Math. Soc., 10 (1978), 129-183.

[3] P. G. Dixon, Non-separable Banach algebras whose squares are pathological, J. Functional Analysis, 26 (1977), 190-200.

[4] R. J. Loy, Multilinear mappings and Banach algebras, J. London Math. Soc., (2), 14 (1976), 423-429.

School of Mathematics
University of Leeds
Leeds LS2 9JT
England

V. OPEN QUESTIONS

V. OPEN QUESTIONS

We first give a list of open problems in areas considered in this volume. It is essentially a consolidated version of the union of lists of questions discussed during the Conference. References are given to the articles where the questions and partial results are discussed: it may be assumed that these questions were suggested at least by the author of the articles.

Secondly, I shall report on the fate, so far as I know it, of the questions mentioned in the survey article [3].

Garth Dales

I. GENERAL THEORY OF RADICAL BANACH ALGEBRAS

(1) Does there exist a commutative, topologically simple Banach algebra other than $\mathbb{C}$? Does every commutative Banach algebra with bounded approximate identity have a proper closed ideal?

See [E.1], [E.2], [3, Question 7].

(2) Let R be a commutative, radical Banach algebra with an element of finite closed descent. Does R possess real continuous semigroups? Does $\ell^1(\mathbb{Q}^+, e^{-t^2})$ possess real continuous semigroups? Does $\ell^1(\mathbb{Q}^+, \omega)$ possess real continuous semigroups for any radical weight ω?

See [E.1]. Related questions are raised at the end of the article. [Here and throughout, $\mathbb{Q}^+$ denotes the set of (strictly) positive rational numbers.]

(3) Does $L^1(\mathbb{R})$ possess analytic semigroups (a^t) which are bounded on $\{0 < |t| \leq 1, \operatorname{Re} t > 0\}$? Same questions for $L^1(G)$ when G is a non-discrete locally compact group. If $L^1(G)$ has an analytic semigroup $(a^t)_{\operatorname{Re} t > 0}$ which is bounded on the line $\{\operatorname{Re} t = 1\}$, does it follow that G is compact? Is the converse true?

See [E.1] and [16].

(4) (a) A <u>local pseudo-bounded approximate identity</u> in a commutative Banach algebra A is a sequence $(e_n) \subset A$ such that, for some $a \in A$ with $a \neq 0$,

$$\sup\{\|ae_{n_1}^{k_1} \cdots e_{n_p}^{k_p} : n_1, \ldots, n_p, k_1, \ldots, k_p \in \mathbb{N}\} < \infty$$

and

$$a = \lim_{n \to \infty} ae_n .$$

Do such local identities necessarily exist in each radical algebra A which has elements of finite closed descent?

(b) Describe the regular quasi-multipliers in $\ell^1(\mathbb{Q}^+, e^{-t^2})$.

See [E.2].

(5) Let ν be a continuous homomorphism from $C(X)$ into a Banach algebra. Is there a finite family, say $\{P_1, \ldots, P_k\}$, of prime ideals of $C(X)$ such that $\nu \mid P_1 \cap \cdots \cap P_k$ is continuous?

See [3, Question 15].

II. EXAMPLES OF RADICAL BANACH ALGEBRAS

(6) Is there a radical weight ω on $\mathbb{R}^+$ such that the Banach algebra $L^1(\omega)$ has non-standard closed ideals? Let $\omega = e^{-\eta}$ be a radical weight satisfying the additional conditions: (a) η is continuous and convex on $\mathbb{R}^+$; (b) $(\log n(t) - \log t)/(\log t)^{1/2} \to 0$ as $t \to \infty$. Then Domar ([Do.2]) proves that each closed ideal of $L^1(\omega)$ is standard. Can either or both of these conditions be lifted or significantly weakened?

Related questions, concerning the description of closed translation-invariant subspaces, are given at the end of [Do.2].

(7) Determine which elements of a radical Banach algebra $L^1(\omega)$ are polynomial generators.

See [Da.1, §4].

(8) (a) Let ω be a sufficiently nice radical weight. Characterize those $f, g \in L^1(\omega)$ with $\alpha(f) = \alpha(g) = 0$ for which there is an automorphism T of $L^1(\omega)$ with $Tf = g$. In particular, what is the orbit of u (where $u(t) \equiv 1$) under the automorphism group of $L^1(\omega)$? Can one answer these questions for $L^1[0,1]$?

(b) Which separable Banach spaces can be continuously embedded as a subalgebra of $L^1_{loc}(\mathbb{R}^+)$ with polynomial generator u? Can this be done so that the algebra has only standard closed ideals?

See [Gra].

(9) Let ω be such that $L^1(\omega)$ is a semi-simple algebra, say $\omega(t)^{1/t} \to 1$. Let I be a closed ideal of $L^1(\omega)$ such that $\alpha(I) = 0$ and $Z(I) = \emptyset$. Does $I = L^1(\omega)$?

The case $\omega \equiv 1$ is Nyman's Theorem, given in [Da.1, §6]. By using more sophisticated complex analysis Gurariĭ ([10]) proves the result in the case that ω increases on $[0, \infty)$ and

$$\int_0^\infty \frac{\log \omega(t)}{1 + t^2}\, dt < \infty .$$

(10) (a) Exhibit a radical weight sequence ω on $\mathbb{Z}^+ = \{0,1,2,\ldots\}$ such that the Banach algebra $\ell^1(\omega)$ has non-standard closed ideals. More generally, find a local Banach algebra of power series in which the polynomials are dense and which has non-standard closed ideals.

(b) Find the weakest possible conditions on ω which still imply that all closed ideals of $\ell^1(\omega)$ are standard.

(c) Find a radical weight sequence ω and an element $f = \sum_{n=1}^\infty C_n X^n \in \ell^1(\omega)$ with $C_1 \neq 0$ such that polynomials (with zero constant term) in f fail to be dense in the maximal ideal $M_1(\omega)$ of $\ell^1(\omega)$.

See [Bad], [La], [Th], [Al.2], etc. The best conditions on ω so far known for (b) are given by Thomas in [Th]. Essentially, it is required that ω be star-shaped: can one weaken this to 'ω is regulated'? For recent work on (c), see [17].

(11) Let $\mathfrak{m}(M)$ be the multiplier algebra of the maximal ideal M of a local Banach algebra $\ell^1(\omega)$.

(a) It is shown in [La, Propositions 6 and 8] that, if ω is a Domar weight, then $M(\omega_\infty)$ is radical. Is the converse true?

(b) Is the condition 'ω is a D_1-weight' sufficient for $\mathfrak{m}(M)$ to be local ([La])?

(c) Describe the radical of $\mathfrak{m}(M)$ in general. In particular, are the algebras $\mathfrak{m}(M)$ of [Bad, §2] necessarily local?

(12) (a) Let $A = \ell^1(Q^+)$ or $\ell^1(Q^+, e^{-t^2})$ or $\ell^1(Q^+ \cap (0,1])$. Does A factor? Does $A = A^2$?

(b) Let ω be a radical weight. Are all closed ideals of $\ell^1(Q^+,\omega)$ standard?

(c) Give conditions on semigroups S and/or weights ω such that $\ell^1(S,\omega)$ is an integral domain.

See [Gro]. For more general questions on factorization, see Question 25, below.

(13) Let G be an abelian, non-discrete locally compact group. (a) Let E be a compact subset of G which is a Helson set, but which is not a set of synthesis. Let $B = A(G)/J_E$, so that B has radical $N = I_E/J_E$. Then $B/N =$

$A(G)/I_E \cong C(E)$. Does B have a Wedderburn decomposition, in the sense that $B = M \oplus N$ for some subalgebra M of B?

(b) Let $\mathrm{Rad}\, L^1(G)$ be the kernel of the hull of $L^1(G)$ as an ideal in the measure algebra. Are the nilpotent elements dense in $(\mathrm{Rad}\, L^1(G))/L^1(G)$?

See Questions 1 and 2 in [Bac].

(14) Let A be the Volterra algebra $(L^1[0,1],*)$ (or the measure algebra $(M(0,1),*)$). Characterize the sequences (ω_n) such that there exists $a \in A$ with $\|a^n\| = \omega_n$ $(n \in \mathbb{N})$.

Let (ω_n) be such a sequence. Then certainly (i) $\omega_n \geq 0$, (ii) $\omega_{m+n} \leq \omega_m \omega_n$, and (iii) $\omega_n^{1/n} \to 0$.

It is shown in [Al.1] that, if $a \in A$ is positive, then (iv) $(\omega_n^{1/n})$ is monotone decreasing. Can each sequence (ω_n) satisfying (i)-(iv) be attained?

It is shown in [W.1] that either $\omega(n+1)/\omega(n) \to 0$ as $n \to \infty$ or that $\limsup \omega(n+p)/\omega(n) > 0$ for each $p \in \mathbb{N}$. Can the latter alternative occur for a (necessarily non-positive) element of A?

It is shown in [E.1, 8.2 and 8.3] that, given any sequences (λ_n), (μ_n) of positive reals which tend to 0, there exists $a \in A$ with $\limsup \|a^n\|^{1/n}/\lambda_n = \infty$ and $\liminf \|a^n\|^{1/n}/\mu_n = 0$.

(15) For which radical weight sequences ω is there an isometric isomorphism of $\ell^1(\omega)$ into $M(0,1)$?

This problem is suggested by some theorems from harmonic analysis. By [14, 5.3.3], if G is a non-discrete, locally compact abelian group, there is an isometric isomorphism of $\ell^1(\mathbb{Z}^+)$ into $M(G)$. If E is a perfect, independent set in $[0,1]$, and if μ is a continuous probability measure with support in E, then the measures μ^{*n} are mutually singular. Let $\omega_n = \|\mu^{*n}\|$. Then the map $(x_n) \to \sum x_n \mu^{*n}$ is an isometric embedding of $\ell^1(\omega)$ into $M(0,1)$. (G. A. Willis)

III. AUTOMATIC CONTINUITY FOR HOMOMORPHISMS AND DERIVATIONS

(16) Let $\theta : B \to A$ be a homomorphism between Banach algebras A and B. Suppose that $\overline{\theta(B)}$ is semi-simple. Is θ necessarily continuous?

This is perhaps the most interesting open question which remains in automatic continuity theory for Banach algebras. See [3, Question 5], and see [A-Da] for several reformulations and some partial results. See also [Au].

(17) Characterize the class of Banach algebras such that each homomorphism (respectively, epimorphism) from a member of this class into a Banach algebra is

automatically continuous. In particular, which C*-algebras belong to these classes?

See [A-Da].

(18) Is every epimorphism from a Banach algebra onto a semi-prime (respectively, prime) Banach algebra necessarily continuous? Does a commutative Banach algebra with no non-zero nilpotents necessarily have a unique complete norm topology?

See [3, Question 6].

(19) Is every epimorphism from $C^{\infty}([0,1])$ onto a Banach algebra necessarily continuous?

It can be seen that such an epimorphism is continuous if and only if the range is finite dimensional. One implication follows from known automatic continuity results, the other from the nuclearity of $C^{\infty}([0,1])$. See [11].
(K. B. Laursen, M. Neumann)

(20) Let R and S be decomposable operators on Banach spaces X and Y, and let E_R and E_S denote the corresponding spectral capacities. Let $T : X \to Y$ be a linear operator (not assumed to be continuous) such that $TR = ST$. Does it follow that $T(E_R(F)) \subset E_S(F)$ for each closed subset F of C?

If so, general automatic theory shows that T has nice continuity properties. The answer is "yes" if S is a generalized scalar operator. See [A-N.1].

(21) Let A be a (non-commutative) Banach algebra. Suppose that every homomorphism from a Banach algebra into a commutative subalgebra of A is continuous. Is every homomorphism from a Banach algebra into A necessarily continuous?

See [Tr].

(22) Give necessary and sufficient conditions on a Banach algebra A such that each derivation from A into a Banach A-bimodule is continuous. Is it true that, if A is commutative and separable, and if A has a closed prime ideal of infinite codimension, then there is a discontinuous derivation from A into a bimodule? For which locally compact groups G is a derivation from $L^1(G)$ into a Banach $L^1(G)$-bimodule necessarily continuous? Is there a discontinuous derivation from a radical algebra $L^1(\omega)$ into a bimodule?

See [C], [W.2], [3, Questions 21-24].

(23) A Banach algebra A has property (S) if the natural map $A \otimes A \to A^2$ is open (when $A \otimes A$ has the projective norm). (a) If A has finite dimensional radical R, and if A has property (S), does A/R have property (S)? In the particular case when $\dim R = 1$, does $R \cap A^2 = \{0\}$ imply that $R \cap A^2 = \{0\}$?

(b) What is the relation, if any, between property (S) holding in the maximal ideals M of A_+ and in the corresponding maximal modular ideals $M \cap A$ of A?

(c) If A is separable and $\operatorname{codim} A^2 < \infty$, then A has property (S). For general, non-separable, A, this fails. But what if A is commutative and/or has finite dimensional radical?

See [Lo.1].

(24) Are the following two conditions equivalent for all Banach algebras A: (i) each homomorphism from A into a finite-dimensional algebra is automatically continuous; (ii) each derivation from A into a finite-dimensional Banach A-bimodule is continuous.

It is proved in [Da-W] that (i) $\Rightarrow$ (ii).

(25) Let G be an amenable locally compact group and let I be a closed bi-ideal in $L^1(G)$. Suppose that $I \cap L^1(H)$ has finite codimension in $L^1(H)$ for each closed, abelian subgroup H of G. Does it follow that I has finite codimension in $L^1(G)$? In particular, what if G is discrete?

See [W.2].

(26) Let A be a Banach algebra. Then A <u>factors</u> if, for each $a \in A$, there exist $b,c \in A$ with $a = bc$. A pair (a_1,a_2) <u>factors</u> if there exist $b,c_1,c_2 \in A$ with $a_1 = bc_1$ and $a_2 = bc_2$. Set

$$c_0(A) = \{(a_k) \subset A : a_k \to 0\},$$

$$A\cdot c_0(A) = \left\{ \left(\sum_{j=1}^{\infty} a_j b_{jk} \right)_{k\in\mathbb{N}} : a_j \in A,\ (b_{jk})_{k\in\mathbb{N}} \in c_0(A) \quad \text{for } j = 1,\dots,m \right\}.$$

Consider the following properties of A:

(1) A has a left bounded approximate identity;

(2) there exists a constant K such that, for each $a \in A$ and each $\varepsilon > 0$, there exists $u \in A$ with $\|a - ua\| < \varepsilon$ and $\|u\| \le K$;

(3) for each $(a_n) \in c_0(A)$, there exist $b \in A$ and $(c_n) \in c_0(A)$ with

$a_n = bc_n$ $(n \in \mathbb{N})$;

(4) the map $(a,b) \mapsto ab$, $A \times A \mapsto A$, is an open surjection;

(5) $c_0(A)$ factors;

(6) there exists a constant C such that, for each $a \in A$, there exist $b, c \in A$ with $a = bc$ and $\|b\|\|c\| \leq \|a\|$;

(7) $c_0(A) = A \cdot c_0(A)$;

(8) each pair in A factors;

(9) A factors;

(10) $A = A^2$;

(11) A^2 is closed and has finite codimension in A.

Then the following implications hold:

$$\begin{array}{ccccccccc}
 & & & & (6) & & \\
 & & & & \Updownarrow & & \\
 & & & & (5) & & \\
 & & & & \Updownarrow & & \\
 & & & & (4) & & \\
 & & & \nearrow & & \searrow & \\
(1) \Longleftrightarrow (2) & \Rightarrow & (3) & \Rightarrow & (8) & \Rightarrow & (9) \\
 & & & \searrow & & \nearrow & \\
 & & & & (7) \Rightarrow (10) \Rightarrow (11)\,. & &
\end{array}$$

The equivalence "$(1) \Longleftrightarrow (2)$" is [1, 11.2], [7, 9.3], the implication "$(2) \Rightarrow (3)$" is Cohen's factorization theorem in the form [1, 11.12], and the other implications are straightforward. A standard hypothesis for positive automatic continuity results is that (1) holds, although (7) would usually suffice. On the other hand, if (10) fails, there are usually counter-examples to possible automatic continuity results. The problem is to determine whether or not each of the implications in the above diagram can be reversed, especially in the case that A is commutative and separable. (I know that $(11) \not\Rightarrow (10)$!) If we knew that $(10) \Rightarrow (7)$ (for a reasonable class of Banach algebras), a lot of proofs could be simplified, so this is the important one.

Here are some partial results:

(a) Let M be a maximal ideal of $H^\infty(\Delta)$, the bounded analytic functions on the open unit disc, which corresponds to a one-point Gleason part off the Silov boundary. (See [9].) Then M satisfies (a), but neither (7) nor (8): however, M is not separable.

(b) Let $A = \{f \in C^\infty[0\ 1] : f^{(k)}(0) = 0$ for all $k\}$. Then A satisfies (3) but not (1): however, A is not a Banach algebra.

(c) The algebra $\ell^1(\mathbb{Q}^+)$ does not satisfy (1), and it is not known whether or not it satisfies (2) - (11). See Question (12), above, and [Gro].

(d) Examples of non-commutative, separable Banach algebras and of commutative, non-separable Banach algebras which satisfy (9) but not (1) are given in [7, §22]. Both these examples satisfy (7) and (8).

(e) If A is a separable Banach algebra satisfying (10), then there exists $m \in \mathbb{N}$ and $C > 0$ such that each element $a \in A$ can be expressed in the form $a = \sum_{j=1}^m b_j c_j$ with $b_1, \ldots, b_m, c_1, \ldots, c_m \in A$ and $\sum_{j=1}^m \|b_j\| \|c_j\| \leq C\|a\|$.

([13]). See also [6].

(27) Let F_n be the free group on n generators, and let F_ω be the free group on countably many generators; I_0 denotes the augmentation ideal, as in [W.2].

(a) Let $A = I_0(F_2)$. Does A factor? Is the set of all products of elements of A dense in A?

(b) Take $n \in \mathbb{N}$, and let J be a closed right ideal of $I_0(F_n)$. If J is generated by fewer than n elements, does it follow that $J \neq I_0(F_n)$?

(c) Let $A = I_0(F_\omega)$. Does $A \cdot c_0(A) = c_0(A)$?

(d) Is there a closed bi-ideal I of codimension 2 in $\ell^1(F_\omega)$ with $I^2 \neq I$?

(e) Is there a closed bi-ideal I of codimension 4 in $\ell^1(F_2)$ with $I^2 \neq I$?

See [W.2].

IV. CONTINUITY OF LINEAR FUNCTIONALS

(28) Let TILF = 'translation-invariant linear functional'.

(a) Is every TILF on $C(\mathbb{T})$ automatically continuous? For which subsets E of $\mathbb{Z}$, other than Sidon sets, is it true that each TILF on $C_E(T)$ is automatically continuous?

(b) Are TILFs automatically continuous: on $L^2(G)$ for some totally disconnected but polythetic, compact abelian group G; on $L^2(\mathbb{Z}_p)$, where $\mathbb{Z}_p$ denotes the compact abelian group of p-adic integers; on $L^1(G)$ for compact, non-abelian groups G (same question for $L^p(G)$, $C(G)$, etc); on $L^p(T)$ for $1 < p < 2$. (True for $p = 2$, false for $p = 1$, open for $1 < p < 2$?)

(c) Is it possible to prove a general result of the type: if all TILFs on a space $\mathfrak{F}(G)$ are automatically continuous, then all operators $\mathfrak{F}(G) \to \mathfrak{F}(G)$ which commute with translation are automatically continuous?

See [M], [Lo.2].

(29) Fill in the blanks in the table below. Throughout, A is a Banach star algebra.

	Are all positive functionals continuous?	Are all traces continuous?
1. $A = A^2$		
2. A^3 closed & cofinite		
3. A^2 closed & cofinite	No	No
4. A separable, $A = A^2$		
5. A separable, A^2 cofinite		
6. A commutative, A^3 closed and cofinite	Yes	Yes
7. A commutative, A^2 closed and cofinite		
8. A commutative, A separable, A^2 cofinite	Yes	Yes

See [Da.2].

Note that, if A is separable and if A^2 has finite codimension in A, then A^3 has finite codimension in A, and both A^2 and A^3 are closed in A. (See [12].)

Re 2: It is shown in [15, Theorem 13.7] that, if AZ^2 is closed and cofinite in A (where Z is the centre of A), then each positive functional on A is continuous. This can be improved, in that 'AZ^2' can be replaced by 'A^2Z'.

Re 7: No example is known of a commutative Banach algebra A in which A^2 is closed and cofinite (and hence in which A^3 is cofinite), without A^3 also being closed.

V. OTHER QUESTIONS

(30) Is there a discontinuous character on a Fréchet algebra?

See [3, Question 3]. In [5], P. G. Dixon shows that there is an unbounded character on a certain complete, commutative locally convex algebra. This algebra is neither metrizable nor LMC.

(31) Let A be a semi-simple Fréchet algebra. Does A have a unique topology as a Fréchet algebra?

See [3, Question 9].

I now conclude with the status of some questions discussed in [3]. The list of questions in [3] which have been (even partially) solved is rather modest in length, if strong in quality, and amounts to the following:

QUESTION 3. (i) Is every character on a commutative Fréchet algebra necessarily continuous? (ii) Is every character on a commutative, complete, metrizable locally convex algebra necessarily continuous?

(i) Question 3. See (30), above.

QUESTION 14. If $L^1(\omega)$ is a radical algebra, and if ω is sufficiently nice, is every non-zero closed ideal of $L^1(\omega)$ equal to M_a for some $a \geq 0$?

(ii) Question 14. Y. Domar has shown that, for many radical weights ω, it is true that all closed ideals of $L^1(\omega)$ are standard. See [Do.2].

QUESTION 16. Characterize the radical Banach algebras R such that there is a radical homomorphism from $C(X)$ into R.

(iii) Question 16. J. Esterle has characterized the radical Banach algebras R such that there is a discontinuous homomorphism from $C(X)$ into R. See [E.1, Theorem 5.3].

QUESTION 19. Is an epimorphism from $C(X)$ onto a Banach algebra necessarily continuous?

(iv) Question 19. J. Esterle has proved that every epimorphism from $C(X)$ onto a Banach algebra is necessarily continuous. See [8]. A related result for algebras $C^n[0,1]$ is given in [11].

QUESTION 25. Let $\mathfrak{A}$ be a Banach star algebra, and suppose that $\mathfrak{A}^2$ is closed and of finite codimension in $\mathfrak{A}$. Is each positive linear functional on $\mathfrak{A}$ continuous?

(v) Question 25. The answer to the question as it stands is negative: a Banach star algebra A for which A^2 is closed and cofinite, but which has a discontinuous positive linear functional, is constructed in [Da.2]. However, several related questions remain open.

Several of the remaining questions in [3] are given above. To set against the above gains, I must point out a loss, in that one claimed result must be withdrawn. It is stated on page 145 of [3] that N. K. Nikolskii proves in [13] that there is a radical weight sequence such that the Banach algebra $\ell^1(\omega)$ has non-standard closed ideals. In fact, Nikolskii does prove correctly that there are sequences $\omega = (\omega_n)$ such that $\omega_n^{1/n} \to 0$ and such that the Banach space $\ell^1(\omega)$ has non-standard translation-invariant closed subspaces, but his attempts to construct an algebra with this property are in error. It seems (see [18] and [19]), that the approach of Nikolskii cannot produce such ideals.

REFERENCES

[1] F. F. Bonsall and J. Duncan, Complete Normed Algebras, Springer-Verlag, New York, 1973.

[2] G. Brown and W. Moran, Analytic discs in the maximal ideal space of M(G), Pacific J. Math., 75 (1978), 45-57.

[3] H. G. Dales, Automatic continuity: a survey, Bull. London Math. Soc., 10 (1978), 129-183.

[4] H. G. Dales and J. P. McClure, Completion of normed algebras of polynomials, J. Austral. Math. Soc., 20 (1975), 504-510.

[5] P. G. Dixon, Scalar homomorphisms on algebras of infinitely long polynomials with an application to automatic continuity theory, J. London Math. Soc., (2) 19 (1979), 488-496.

[6] __________, Automatic continuity of positive functionals on topological involution algebras, Bull. Austral. Math. Soc., 23 (1981), 265-281.

[7] R. S. Doran and J. Wichmann, Approximate Identities and Factorization in Banach Modules, Lecture Notes in Mathematics, 768, Springer-Verlag, 1979.

[8] J. Esterle, Theorems of Gelfand-Mazur type and continuity of epimorphisms from C(K), J. Functional Analysis, 36 (1980), 273-286.

[9] J. B. Garnett, Bounded Analytic Functions, Academic Press, New York, 1981.

[10] V. P. Gurariĭ, Harmonic analysis in spaces with a weight, Trans. Moscow Math. Soc., 35 (1979), 21-75.

[11] K. B. Laursen, Prime ideals and automatic continuity in algebras of differentiable functions, J. Functional Analysis, 38 (1980), 16-24.

[12] __________, Multilinear mappings and Banach algebras, J. London Math. Soc., (2) 14 (1976), 423-429.

[13] N. K. Nikolskii, Selected problems of weighted approximation and spectral analysis, Proc. Steklov Inst. Math., 120 (1974), 1-278.

[14] W. Rudin, Fourier Analysis on Groups, J. Wiley, New York, 1962.

[15] A. M. Sinclair, Automatic Continuity of Linear Operators, London Math. Soc. Lecture Note Series, 21, Cambridge Univ. Press, 1976.

[16] __________, Continuous Semigroups in Banach Algebras, London Math. Soc. Lecture Note Series, to appear.

[17] D. Söderberg, Generators in radical weighted ℓ^1, Uppsala University Department of Mathematics Report, 9 (1981).

[18] M. P. Thomas, Closed ideals and biorthogonal systems in radical Banach algebras of power series, Proc. Ed. Math. Soc., to appear.

[19] __________, Closed ideals of $\ell^1(\omega_n)$ when $\{\omega_n\}$ is star-shaped, Pacific J. Math., to appear.

Vol. 873: Constructive Mathematics, Proceedings, 1980. Edited by F. Richman. VII, 347 pages. 1981.

Vol. 874: Abelian Group Theory. Proceedings, 1981. Edited by R. Göbel and E. Walker. XXI, 447 pages. 1981.

Vol. 875: H. Zieschang, Finite Groups of Mapping Classes of Surfaces. VIII, 340 pages. 1981.

Vol. 876: J. P. Bickel, N. El Karoui and M. Yor. Ecole d'Eté de Probabilités de Saint-Flour IX – 1979. Edited by P. L. Hennequin. XI, 280 pages. 1981.

Vol. 877: J. Erven, B.-J. Falkowski, Low Order Cohomology and Applications. VI, 126 pages. 1981.

Vol. 878: Numerical Solution of Nonlinear Equations. Proceedings, 1980. Edited by E. L. Allgower, K. Glashoff, and H.-O. Peitgen. XIV, 440 pages. 1981.

Vol. 879: V. V. Sazonov, Normal Approximation – Some Recent Advances. VII, 105 pages. 1981.

Vol. 880: Non Commutative Harmonic Analysis and Lie Groups. Proceedings, 1980. Edited by J. Carmona and M. Vergne. IV, 553 pages. 1981.

Vol. 881: R. Lutz, M. Goze, Nonstandard Analysis. XIV, 261 pages. 1981.

Vol. 882: Integral Representations and Applications. Proceedings, 1980. Edited by K. Roggenkamp. XII, 479 pages. 1981.

Vol. 883: Cylindric Set Algebras. By L. Henkin, J. D. Monk, A. Tarski, H. Andréka, and I. Németi. VII, 323 pages. 1981.

Vol. 884: Combinatorial Mathematics VIII. Proceedings, 1980. Edited by K. L. McAvaney. XIII, 359 pages. 1981.

Vol. 885: Combinatorics and Graph Theory. Edited by S. B. Rao. Proceedings, 1980. VII, 500 pages. 1981.

Vol. 886: Fixed Point Theory. Proceedings, 1980. Edited by E. Fadell and G. Fournier. XII, 511 pages. 1981.

Vol. 887: F. van Oystaeyen, A. Verschoren, Non-commutative Algebraic Geometry, VI, 404 pages. 1981.

Vol. 888: Padé Approximation and its Applications. Proceedings, 1980. Edited by M. G. de Bruin and H. van Rossum. VI, 383 pages. 1981.

Vol. 889: J. Bourgain, New Classes of $\mathcal{L}^p$-Spaces. V, 143 pages. 1981.

Vol. 890: Model Theory and Arithmetic. Proceedings, 1979/80. Edited by C. Berline, K. McAloon, and J.-P. Ressayre. VI, 306 pages. 1981.

Vol. 891: Logic Symposia, Hakone, 1979, 1980. Proceedings, 1979, 1980. Edited by G. H. Müller, G. Takeuti, and T. Tugué. XI, 394 pages. 1981.

Vol. 892: H. Cajar, Billingsley Dimension in Probability Spaces. III, 106 pages. 1981.

Vol. 893: Geometries and Groups. Proceedings. Edited by M. Aigner and D. Jungnickel. X, 250 pages. 1981.

Vol. 894: Geometry Symposium. Utrecht 1980, Proceedings. Edited by E. Looijenga, D. Siersma, and F. Takens. V, 153 pages. 1981.

Vol. 895: J.A. Hillman, Alexander Ideals of Links. V, 178 pages. 1981.

Vol. 896: B. Angéniol, Familles de Cycles Algébriques – Schéma de Chow. VI, 140 pages. 1981.

Vol. 897: W. Buchholz, S. Feferman, W. Pohlers, W. Sieg, Iterated Inductive Definitions and Subsystems of Analysis: Recent Proof-Theoretical Studies. V, 383 pages. 1981.

Vol. 898: Dynamical Systems and Turbulence, Warwick, 1980. Proceedings. Edited by D. Rand and L.-S. Young. VI, 390 pages. 1981.

Vol. 899: Analytic Number Theory. Proceedings, 1980. Edited by M.I. Knopp. X, 478 pages. 1981.

Vol. 900: P. Deligne, J. S. Milne, A. Ogus, and K.-Y. Shih, Hodg Cycles, Motives, and Shimura Varieties. V, 414 pages. 1982.

Vol. 901: Séminaire Bourbaki vol. 1980/81 Exposés 561–578. II 299 pages. 1981.

Vol. 902: F. Dumortier, P.R. Rodrigues, and R. Roussarie, Gern of Diffeomorphisms in the Plane. IV, 197 pages. 1981.

Vol. 903: Representations of Algebras. Proceedings, 1980. Edit by M. Auslander and E. Lluis. XV, 371 pages. 1981.

Vol. 904: K. Donner, Extension of Positive Operators and Korovk Theorems. XII, 182 pages. 1982.

Vol. 905: Differential Geometric Methods in Mathematical Physic Proceedings, 1980. Edited by H.-D. Doebner, S.J. Andersson, a H.R. Petry. VI, 309 pages. 1982.

Vol. 906: Séminaire de Théorie du Potentiel, Paris, No. 6. Procee ings. Edité par F. Hirsch et G. Mokobodzki. IV, 328 pages. 1982.

Vol. 907: P. Schenzel, Dualisierende Komplexe in der lokal Algebra und Buchsbaum-Ringe. VII, 161 Seiten. 1982.

Vol. 908: Harmonic Analysis. Proceedings, 1981. Edited by F. Ri and G. Weiss. V, 325 pages. 1982.

Vol. 909: Numerical Analysis. Proceedings, 1981. Edited by J Hennart. VII, 247 pages. 1982.

Vol. 910: S.S. Abhyankar, Weighted Expansions for Canonical I singularization. VII, 236 pages. 1982.

Vol. 911: O.G. Jørsboe, L. Mejlbro, The Carleson-Hunt Theorem Fourier Series. IV, 123 pages. 1982.

Vol. 912: Numerical Analysis. Proceedings, 1981. Edited by G. Watson. XIII, 245 pages. 1982.

Vol. 913: O. Tammi, Extremum Problems for Bounded Unival Functions II. VI, 168 pages. 1982.

Vol. 914: M. L. Warshauer, The Witt Group of Degree k Maps a Asymmetric Inner Product Spaces. IV, 269 pages. 1982.

Vol. 915: Categorical Aspects of Topology and Analysis. Proce ings, 1981. Edited by B. Banaschewski. XI, 385 pages. 1982.

Vol. 916: K.-U. Grusa, Zweidimensionale, interpolierende Lg-Spli und ihre Anwendungen. VIII, 238 Seiten. 1982.

Vol. 917: Brauer Groups in Ring Theory and Algebraic Geometry. I ceedings, 1981. Edited by F. van Oystaeyen and A. Verschoren. 300 pages. 1982.

Vol. 918: Z. Semadeni, Schauder Bases in Banach Space Continuous Functions. V, 136 pages. 1982.

Vol. 919: Séminaire Pierre Lelong – Henri Skoda (Analyse) Ann 1980/81 et Colloque de Wimereux, Mai 1981. Proceedings. E par P. Lelong et H. Skoda. VII, 383 pages. 1982.

Vol. 920: Séminaire de Probabilités XVI, 1980/81. Proceedi Edité par J. Azéma et M. Yor. V, 622 pages. 1982.

Vol. 921: Séminaire de Probabilités XVI, 1980/81. Supplément: (métrie Différentielle Stochastique. Proceedings. Edité par J. Az et M. Yor. III, 285 pages. 1982.

Vol. 922: B. Dacorogna, Weak Continuity and Weak Lower S continuity of Non-Linear Functionals. V, 120 pages. 1982.

Vol. 923: Functional Analysis in Markov Processes. Proceed 1981. Edited by M. Fukushima. V, 307 pages. 1982.

Vol. 924: Séminaire d'Algèbre Paul Dubreil et Marie-Paule Mall Proceedings, 1981. Edité par M.-P. Malliavin. V, 461 pages. 1

Vol. 925: The Riemann Problem, Complete Integrability and metic Applications. Proceedings, 1979-1980. Edited by D. Chu sky and G. Chudnovsky. VI, 373 pages. 1982.

Vol. 926: Geometric Techniques in Gauge Theories. Proceec 1981. Edited by R. Martini and E.M.de Jager. IX, 219 pages.